Handbook of Qur'ānic Hermeneutics

De Gruyter Handbook

Handbook of Qurʾānic Hermeneutics

Vol. 5: Contemporary Qurʾānic Hermeneutics

Edited by
Georges Tamer

DE GRUYTER

ISBN 978-3-11-131732-8
e-ISBN (PDF) 978-3-11-132006-9
e-ISBN (EPUB) 978-3-11-132021-2

Library of Congress Control Number: 2024950661

Bibliographic information published by the Deutsche Nationalbibliothek
The Deutsche Nationalbibliothek lists this publication in the Deutsche Nationalbibliografie; detailed bibliographic data are available on the internet at http://dnb.dnb.de.

Cover image: Calligraphy: Mr. Zuheir Elia (Erlangen, Germany). Q 3:7:
"wa-mā yaʿlamu taʾwīlahu illā Llāhu wa-r-rāsikhūna fī-l-ʿilmi"
Typesetting: Integra Software Services Pvt. Ltd.
Printing and binding: CPI books GmbH, Leck

www.degruyter.com
Questions about General Product Safety Regulation:
productsafety@degruyterbrill.com

MIX
Papier | Fördert gute Waldnutzung
FSC® C083411

Contents

Georges Tamer

Introduction

In the fifth volume of the *Handbook of Qur'ānic Hermeneutics*, we encounter contemporary hermeneutical approaches to the Qur'ān. The field of Qur'ānic hermeneutics, which has become much more diverse in contemporary times, has undergone significant changes. Some scholars have challenged traditional approaches and embraced innovative methodologies. The formerly prevalent idea of a center and periphery in the Muslim world has been, to a certain extent, rendered obsolete, with diverse Islamic centers influencing each other in a cycle of cross-fertilization. In the following, I will present some important aspects of contemporary Qur'ānic hermeneutics, a number of which are novel. They include geographic pluralism, the adoption of Western concepts, feminist exegesis, methodological innovations, religious pluralism, the impact of new media, and the role of Qur'ānic interpretation in shaping Islamic identity in the global context. The names of the scholars discussed are given as they are predominantly used by the scholars themselves or in current research.

Contextual pluralism: Mainly flourishing in Arab countries, Iran, and Türkiye until the mid-twentieth century, contemporary Qur'ānic hermeneutics now exhibits geographic pluralism. Muslim scholars are no longer confined to a single geographical center and are pursuing Qur'ānic hermeneutics all over Europe, Australia, and North and Central America, contributing to a more global discourse on the Qur'ān. This geographic diversity reflects a departure from the past, showcasing more interconnected and dynamic approaches to the Qur'ān, which have developed in different cultural and political contexts.

Adoption of Western concepts: A notable shift in contemporary Qur'ānic hermeneutics is the increased acceptance and incorporation of concepts, approaches, and methods developed in the West. The traditional authority of the Qur'ān is no longer viewed as an obstacle to adopting hermeneutical methods. Instead, the divine character of the text is acknowledged within the context of a divine-human discourse, which emphasizes the human role in communication. This marks a shift from rigid traditionalism and a willingness to an engagement with modern intellectual achievements.

– *Feminist exegesis*: One groundbreaking development in contemporary Qur'ānic hermeneutics is the emergence of feminist exegesis. This approach reevaluates traditional interpretations through a gender-sensitive lens, challenging patriarchal readings and offering new insights into the Qur'ān. The feminist perspective introduces a fresh dimension to Qur'ānic scholarship, addressing long-standing

Acknowledgments: I would like to thank Mr. Ramy Abdin, Mr. Andreas Knöll, and Mr. Yaaqub Kutterer for their assistance editing the contributions in this volume. I would also like to thank Ms. Katrin Mittmann and Dr. Torsten Wollina from the publishing house Walter de Gruyter and Dr. Lydia J. White for their kind and cooperative support.

https://doi.org/10.1515/9783111320069-001

gender biases and contributing to a more inclusive understanding of the Islamic scripture.

- *Methodological innovations*: Contemporary scholars such as Muḥammad Aḥmad Khalaf Allāh, Mohammed Arkoun, Naṣr Ḥāmid Abū Zayd, and Muḥammad Mujtahid Shabestarī have introduced new methodological approaches that seek to extract timeless principles from the historical context of the Qur'ānic message. This break with traditional exegesis reflects a desire to reconcile the enduring nature of the Qur'ān with a modern intellectual outlook, challenging the long-standing exegetical paradigm.
- *Challenges and resistance*: Despite these innovative developments, contemporary Qur'ānic hermeneutics faces resistance from the entrenched paradigm of traditional exegesis. Scholars advocating for a more historical understanding of the text encounter opposition as they are seen as a threat to the prevailing interpretative monopoly. Governments in Muslim-majority states, reliant on a unified understanding of Islam for social cohesion, are also resisting the plurality of interpretations that contemporary hermeneutics brings.
- *Religious pluralism, new media, and global identity*: Contemporary Qur'ānic hermeneutics is engaging with religious pluralism, using new media as tools for disseminating diverse interpretations. The intersections between Qur'ānic interpretation and global identity are becoming evident as scholars strive to strengthen Islamic identity in an increasingly interconnected world.

The increasing plurality of hermeneutical approaches to the Qur'ān is leading to general differences among them, particularly in relation to the following topics:

- *The nature of revelation*: While Abdulkarīm Surūsh sees the Qur'ān as a result of human experience inspired by God, Ḥasan Ḥanafī and Mohamed Talbi firmly maintain that the Qur'ān is divine in nature. This difference in the perception of revelation influences their respective hermeneutical approaches.
- *The political dimension of Qur'ānic hermeneutics*: Scholars like Ḥanafī, Mūsā ṣ-Ṣadr, and Muhammad Shahrur have highlighted the potential ideological utilization of Qur'ānic interpretation as a political weapon. This ideological dimension of hermeneutics is especially pronounced in the interpretation of verses with sociopolitical implications, as the discussion of Turkish translations of the Qur'ān in this volume demonstrates.
- *Historical critique and pluralism*: Abū Zayd's and Mahmoud Ayoub's hermeneutical approaches involve historical critiques of traditional sources, promoting religious pluralism through the Qur'ān. This contrasts with, for instance, Ḥanafī's more politically oriented interpretation, where differences in discourse are seen as ideologically driven, or Yaşar Nuri Öztürk's hermeneutics, which is barely interested in the historical circumstances of the emergence of the Qur'ān.
- *End-time prophecies*: Imran Hosein's unique approach focuses on end-time prophecies within the Qur'ān. His hermeneutics intertwines with eschatology, offering

a distinctive perspective on the relevance of the Qur'ān to contemporary world events.
- *Gender as an analytical tool*: Scholars like Amina Wadud and Fatima Mernissi are using gender as a crucial analytical tool in Qur'ānic hermeneutics. Their work challenges traditional interpretations that have perpetuated gender imbalances within Islamic discourse.

Nevertheless, the hermeneutical approaches to the Qur'ān presented in the following chapters clearly share some features. They can be summarized as follows:
- *Open interpretation and accessibility*: Scholars like Abdulkarīm Surūsh, Abū Zayd, and Amina Wadud advocate for an open interpretation of the Qur'ān, challenging the notion that it should be reserved solely for religious scholars. They emphasize accessibility, encouraging broader engagement with the Qur'ān outside of traditional scholarly circles.
- *Considering the historical context*: Many scholars, including Abdulkarīm Surūsh, Abū Zayd, and Talbi, emphasize the importance of considering the historical context in which the Qur'ān was revealed. This approach seeks to contextualize its verses within the sociopolitical and cultural circumstances of the time.
- *Multidimensional approach*: Scholars such as Arkoun and Shabestarī adopt a multidimensional hermeneutical approach to the Qur'ān, combining religious anthropology, philosophical theology, historical analysis, and linguistic methods to provide a comprehensive understanding of their scripture.
- *Progressive and inclusive vision*: Most of the scholars presented in this volume advocate for an inclusive vision of Islam that integrates the various elements of Muslim tradition. They combine theological, historical, anthropological, and philosophical perspectives, reflecting a progressive and holistic approach.
- *Critical engagement with tradition*: Most of the scholars discussed in this volume engage critically and selectively with traditional sources, questioning established interpretations and encouraging the reevaluation of historical narratives. This critical approach contributes to a more nuanced understanding of the Qur'ān.

This brief presentation of various positions in contemporary Qur'ānic hermeneutics thus reveals a rich tapestry of perspectives, reflecting both the diversity of and common themes within Islam. All of the approaches described here underscore the dynamic nature of Qur'ānic hermeneutics, shaped by the world's evolving intellectual, cultural, and political landscapes. The following overview will concisely present the main characteristics of each one of these approaches.

The second half of the twentieth century saw three influential scholars in Salafi and Wahhābī circles rise to prominence: ʿAbd al-ʿAzīz b. Bāz, Nāṣir ad-Dīn al-Albānī, and Muḥammad b. Ṣāliḥ al-ʿUthaymīn. The influence of Muḥammad b. ʿUthaymīn (d. 2001/1421) can be attributed to his engagement with matters of everyday life, the legal rulings he issued, and his participation in interactive radio shows, through which he disseminated his views to a global audience. His alignment with Saudi Arabia's official politics and his discussions of women's roles, pilgrimage rites, and matters of Islamic creed helped him to make a broad impact.

Two key factors contributed to Ibn ʿUthaymīn's significance. First, his extensive scholarly output, spanning Islamic law and creed, and, second, his unique position within the Wahhābī establishment. Unlike Ibn Bāz, who held prominent political positions, Ibn ʿUthaymīn's focus on academic pursuits in ʿUnayza allowed him to maintain his independence from political influences, ensuring his scholarly integrity.

Ibn ʿUthaymīn contributed significantly to the realm of Qurʾānic hermeneutics. While he did not leave a complete commentary, the interpretations he presented in various recorded lectures were compiled posthumously by the Ibn ʿUthaymīn Foundation. His approach followed Ibn Taymiyya's methodology, emphasizing the authority of Muḥammad's companions (*aṣ-ṣaḥāba*) in interpreting the Qurʾān. However, his explanations bear the influence of the cultural, social, and religious dynamics of Saudi Arabia during his time. His *Tafsīr al-Qurʾān*, a publication still in progress, covers around 70 percent of the Qurʾānic text. Transcriptions and compilations of his works have involved former students and close associates of Ibn ʿUthaymīn, maintaining the integrity of his teachings. His *Tafsīr*, presented in individual volumes, reflects the oral nature of Ibn ʿUthaymīn's original presentations.

To comprehend Mohammed Arkoun's (d. 2010/1431) Qurʾānic hermeneutics, it is crucial to grasp the political and intellectual landscape in which he was immersed. Arkoun lived in a time that was marked by significant sociopolitical changes within the Muslim world. His work was a response to the challenges faced by Islamic societies, urging them to reevaluate the Islamic intellectual tradition. Evolving political scenarios and intellectual challenges acted as catalysts for Arkoun's call to rethink Islamic thought.

Arkoun's Qurʾānic hermeneutics is deeply rooted within his philosophical perspective, as highlighted in his *Pour une Critique de la Raison Islamique* (Toward a Critique of Islamic Reason). His emphasis on a comprehensive vision of Islam seeks to reconcile and integrate various elements of Muslim tradition that had been marginalized or neglected. Religious anthropology and philosophical theology were the tools Arkoun used to merge the theological needs of believers, the philological imperatives of historians, the explanatory perspectives of anthropologists, and the critical scrutiny of philosophers.

Arkoun's critique of Qurʾānic exegesis revolves around the notion that revelation is not a normative discourse sent to Earth but a gift that imparts meaning to existence. Accordingly, revelation is intrinsically tied to its historical context and was shaped by the actors of that era. Arkoun distinguishes between the linguistic events that took place during the Prophet's time (Qurʾānic fact) and the established religion encapsulated in the corpus of the holy text (Islamic fact). This distinction provides a foundation for employing various methods to interpret the Qurʾān, creating a progressive, multidimensional approach.

The deployment of linguistic, historical, hermeneutical, and anthropological categories in Arkoun's Qur'ānic hermeneutics provides a superior framework for analysis. He eloquently demonstrates his multidimensional method in *Lectures du Coran* (Readings of the Qur'ān), which presents a nuanced understanding of the Qur'ān on multiple levels. By categorizing the Qur'ān as a discourse, Arkoun provides a comprehensive toolset for interpreting the scripture, making his method both progressive and adaptable to various perspectives. Thus, his Qur'ānic hermeneutics represents a significant departure from conventional approaches. It is a pioneering effort to bridge the gap between tradition and modernity within Islamic thought.

Naṣr Ḥāmid Abū Zayd (d. 2010/1431) was as a pivotal figure in the realm of contemporary Islamic thought, challenging traditional Muslim scholarship through his radical hermeneutical views. Abū Zayd's hermeneutical journey started with his critique of the two opposing currents of political Islam in Egypt: the "official" Islam endorsed by state-affiliated scholars and institutions and the Islamist opposition. He argued that both streams manipulated the Qur'ān to their own advantage, presenting a simplistic, ahistorical, and definitive understanding while disregarding fundamental principles of interpretation.

In his pursuit of a hermeneutical understanding, Abū Zayd drew on various intellectual traditions, incorporating Muʿtazilite thought, Islamic mysticism, and the linguistics of ʿAbd al-Qāhir al-Jurjānī (d. 1078/471). His explorations aimed to uncover similarities between Arab-Islamic ideas and modern hermeneutical thinking. Abū Zayd often drew parallels between the two strands with regard to the concept of "signs" (*āyāt*) and methods for deciphering them, drawing connections between Islamic semiotics and modern semiotic theories.

At the core of Abū Zayd's hermeneutics lies the application of a semiotic framework to the Qur'ān, emphasizing a communicative relationship between God and humanity. Abū Zayd conceptualized the Qur'ān as the outcome of a communicative process between the divine "sender" (*mursil*) and the human "receiver" (*mutalaqqī*), portraying God as the "speaker" (*mutakallim*) and humans as the "addressees" (*mukhāṭab*). The Arabic language, in this context, serves as the "code" (*shifra*) for transmitting the divine message.

Abū Zayd challenges the conventional notion of revelation as dictation by presenting it as a nonverbal form of communication. He views the Qur'ān not as a dictated text but as a collaborative effort between the divine and human, where the receiver must express the ideas conveyed in their own words. While refraining from labeling Muḥammad as the author, Abū Zayd perceives the language used in the Qur'ān as part of the human aspect of the revelatory process – temporal and open to human analysis. His work inspires us to reflect on the dynamic relationship between God, language, and human interpretation. While his ideas remain radical for many traditional Muslim scholars, they have sparked significant discussions and paved the way for a more nuanced exploration of the Qur'ānic text in the contemporary context.

Wahba az-Zuḥaylī (d. 2015/1436) left a significant mark on contemporary Qur'ānic exegesis with his interconnected trilogy of *tafsīr* works: *at-Tafsīr al-munīr* (The Enlightening Commentary), *at-Tafsīr al-wajīz* (The Concise Commentary), and *at-Tafsīr al-wasīṭ* (The Intermediate Commentary). His most extensive work, *at-Tafsīr al-munīr*, provides the foundation for his concise and intermediary commentaries.

At-Tafsīr al-wajīz focuses on Qur'ānic semantics, offering a straightforward interpretation in simple Arabic. Az-Zuḥaylī touches on the circumstances of the revelation (*asbāb an-nuzūl*) based on authentic hadiths, emphasizing the Qur'ān not only as divine revelation but also as a guide for believers to enact its principles. He advocates for measured Qur'ānic recitation (*tartīl*) and underscores the importance of practical knowledge in Qur'ānic elocution (*tajwīd*).

At-Tafsīr al-wasīṭ, intended for a more educated audience, provides a detailed yet compact commentary, delving into Qur'ānic semantics and utilizing Islamic traditions for broader understanding. Az-Zuḥaylī offers background information from prophetic traditions and divides each sura up thematically, enhancing the reader's comprehension.

At-Tafsīr al-munīr caters to experts in Qur'ānic exegesis. It incorporates various disciplines such as theology, Arabic philology, Qur'ānic recitation, and jurisprudence. Az-Zuḥaylī employs a systematic exegetical method. He covers different thematic units, provides an overview of each sura, conducts comparative recitation analysis, contextualizes verses, explains meanings and legal judgments, and examines rhetoric and grammar. His approach combines traditional sources with a rational interpretation (*at-tafsīr bi-l-ma'thūr wa-l-ma'qūl*). While not explicitly defined, reason appears to be intertwined with Islamic tradition, suggesting that the two complement rather than contradict each other. The commentary also introduces scholarly aspects, seemingly aimed at convincing non-Muslims of the Qur'ān's validity. However, while az-Zuḥaylī's inclusion of scientific aspects may be perceived as an attempt to appeal to a broader audience, the primary focus remains religious morality and guidance for leading a devout life.

Fāṭima Mernīssī (d. 2015/1437, referred to in the following as "Mernissi," as she is commonly known) was a pioneering figure in Islamic feminism. Drawing on the views of Abū Ḥāmid al-Ghazālī and Sigmund Freud, Mernissi developed a concept of active female sexuality. Her Qur'ānic hermeneutics revolves around the concept of *fitna*, a term laden with multiple connotations in Islamic culture. While traditionally associated with disorder, chaos, and even civil strife, Mernissi highlights its alternative meaning – a beautiful woman with a *femme fatale* allure, capable of unsettling men and disrupting their self-control. Through this lens, she challenges the prevailing narrative surrounding female sexuality and seeks to empower women by reframing their role in society. However, Mernissi's approach, though lauded for its emphasis on female agency and its reexamination of history from a female perspective, is not without contradictions. Some critics argue that her analysis of the hadiths lacks reflexivity, as she employs a "pick-and-choose" method, disregarding the problematic nature of her source material.

Central to Mernissi's Qur'ānic hermeneutics is her critique of a male elite manipulating sacred texts to perpetuate the exclusion of women. She contends that the marginalization of women is not intrinsic to the Qur'ān, the Prophet, or Islamic tradition. Rather, it is a deliberate fabrication by a male elite to maintain its dominance. Mernissi boldly asserts that the manipulation of holy texts has been a structural feature of power dynamics within Muslim societies. Thus, her hermeneutics offers a distinctive perspective that challenges patriarchal structures, unveiling the potential for active female agency within Islamic discourse. While her work is not without inconsistencies and criticisms, it remains a groundbreaking contribution to the field of Islamic feminism. Mernissi's call for a reexamination of sacred texts and a rejection of manipulated traditions reverberates as an invitation to question power dynamics within Muslim societies and to strive for more equitable and inclusive interpretations of Islamic teachings.

The hermeneutical approach taken by Yaşar Nuri Öztürk (d. 2016/1437) is characterized by its selectivity in interpreting specific Qur'ānic verses that he deems relevant to his understanding. Unlike more comprehensive approaches, he neglects the historical context of the verses and allows the text to explain itself. This results in what can be termed "thematic exegesis," where interpretation is guided by the thematic relevance of the verses rather than by considering the broader historical or contextual background. Öztürk's hermeneutics can therefore be perceived as somewhat arbitrary, as he opts to interpret only certain passages from the Qur'ān, disregarding any systematic or theoretical foundation for his selection. This approach, though it provides thematic insights, lacks the methodological rigor often associated with traditional Qur'ānic exegesis. In this sense, Öztürk's hermeneutics appears willful and driven more by thematic preferences than by a consistent theoretical framework.

One essential characteristic of Öztürk's hermeneutics is its complete disregard for the historical context in which the Qur'ānic verses were revealed. Unlike other scholars, who emphasize how important it is to understand the circumstances, Öztürk believes that the Qur'ān inherently explains itself without the need for historical contextualization. Öztürk's hermeneutics departs from traditional approaches and established methods. Öztürk's hermeneutical stance extends beyond the Qur'ān to criticize other Islamic sources such as hadith and *sīra*. He expresses skepticism about their accuracy and advocates for a return to the Qur'ān as the primary and reliable source of guidance. This critical stance aligns with his overarching emphasis on the Qur'ān as a self-sufficient and self-explanatory scripture.

Central to Öztürk's hermeneutics is his endorsement of *ijtihād*, a concept encouraging independent reasoning and interpretation. This viewpoint asserts that the Qur'ān should not merely be adhered to through traditional interpretation but should be continually reevaluated and adapted to contemporary contexts. Öztürk emphasizes the need for a rational exegesis that allows the Qur'ān to evolve and remain relevant over time.

The scholarly contributions made by Mohamed Talbi (d. 2017/1438) go beyond the recounting of Muslim history during the Aghlabid Dynasty, delving into the realm of modern Islamic thought. His Qur'ānic hermeneutics, grounded in the principles of liberalism and pluralism, reflects his attempt to contemporize the divine revelation. While he diverges from traditional Qur'ānic exegesis, he also pays homage to its merits, demonstrating his nuanced approach to understanding the Qur'ān's relevance in the modern context.

Talbi approaches the Qur'ān as a believer, viewing it as the ultimate source of truth that humanity should seek to comprehend through reason. For him, the Qur'ān is as a wellspring of profound wisdom that guides individuals in their pursuit of truth and understanding. Simultaneously working as a historian, Talbi incorporates critical thinking into his hermeneutics. He perceives history as an ongoing process of creation, a dynamic force shaping the understanding of the Qur'ān. This historical lens allows Talbi to bridge the gap between the ancient text and the contemporary world, fostering a more nuanced interpretation.

Talbi's Qur'ānic understanding is further shaped by the concept of the "right path" (*aṣ-ṣirāṭ al-mustaqīm*). He asserts that this path aligns with the pursuit of science and knowledge. Talbi advocates for *ijtihād* as an everlasting quest for knowledge, emphasizing that this endeavor should not be confined to specific groups, but should be open to all seekers of truth. Accordingly, he emphasizes that the existence of human civilization hinges on values such as love, pluralism, and freedom. Furthermore, he posits that religion cannot thrive without these fundamental values, which represent the essence of humanity. In this sense, he envisions a harmonious coexistence where individuals with diverse minds, cultures, and trajectories embrace liberty, accept pluralism, and find inner peace through their love for God.

Talbi's approach mirrors that of Abū Isḥāq ash-Shāṭibī (d. 1388/790), who articulated the concept of *maqāsid ash-sharīʿa* (the objectives of Islamic law). Talbi seeks to implement this concept in his Qur'ānic hermeneutics, exploring the fundamental intentions behind the Qur'ān. His Qur'ānic hermeneutics therefore exemplifies a harmonious blend of tradition and modernity. By approaching the Qur'ān with faith, historical consciousness, a commitment to knowledge, and an unwavering dedication to universal values, Talbi endeavors to unravel the timeless wisdom embedded within the Qur'ān to meet the challenges of the contemporary era.

Muhammad Shahrur (d. 2019/1441) took a revolutionary path in interpreting the Qur'ān. Departing from centuries-old traditional interpretations, Muhammad Shahrur based his method on the principle that human knowledge originates in the material world, aligning it with contemporary philosophy that emphasizes the progression from empirical to theoretical knowledge. He rejected the Sufi approach of knowledge through illumination and emphasized the human intellect's capacity to comprehend the material world, asserting that there was no contradiction between the Qur'ān and empirical philosophy.

Muhammad Shahrur labels his method a new Qur'ānic reading, emphasizing the linguistic dimension. Based on the concept of remembrance (*dhikr*), Muhammad Shahrur sees the Qur'ān as the transformation of the divine into human language, drawing inspiration from linguistic scholars like Abū l-Fatḥ b. Jinnī and ʿAbd al-Qāhir al-Jurjānī, as well as modern linguistics, which denies synonymy. Rejecting synonymy is a crucial aspect of his hermeneutics, which is also influenced by Ibn Fāris al-Lughawī.

In Muhammad Shahrur view, every individual guided by reason can understand the Qur'ān. For him, Islam (*islām*) is universal submission to God, promoting peaceful coexistence among world religions. Muhammad Shahrur also challenges the traditional understanding of *kitāb* (book) and *qur'ān*, asserting that the Qur'ān is only one part of the *Kitāb*, which was God's revelation to the Prophet. His method extends to contrasting conceptual pairs, such as *ulūhiyya* (Divinity) and *rubūbiyya* (Lordship), which he frames within his dialectical worldview. Muhammad Shahrur introduces mathematical concepts like limits and hyperbolic functions to interpret Qur'ānic principles, especially in relation to inheritance laws and capital punishment. His integration of Marxist language and mathematical theories raises questions but highlights his innovative approach. He confronts political Islam, advocating for freedom of worship, democracy, and *shūrā* (mutual consultation) as core Islamic values. His unconventional interpretations have drawn criticism from the establishment, but Muhammad Shahrur legacy means challenging traditional approaches, encouraging alternative readings while remaining a committed believer.

Mahmoud Ayoub (d. 2021/1443) stood out in the field of contemporary Qur'āṇic exegesis due to his pluralistic interpretation of the Qur'ān. Like many of his contemporaries, Ayoub engaged in a critical examination of traditional Islamic sources, employing a selective hermeneutics that fosters interreligious dialogue. His Qur'ānic hermeneutics represents a nuanced approach to the Qur'ān, navigating a delicate balance between tradition and the evolving context of the contemporary world. His thought and hermeneutical approach developed within the context of Western religious studies, influenced by the development of the Christian theology of religions and religious pluralism. Moreover, Ayoub's hermeneutical perspective emphasizes Qur'ānic exegesis as a crucial medium through which the scripture shapes the Muslim community.

Ayoub approaches the Qur'ān through a pluralistic lens. This involves a critical engagement with traditional Islamic sources and a selective hermeneutics that encourages dialogue between different religious perspectives. His pluralistic approach aims to transcend rigid boundaries and foster understanding among diverse religious traditions. While upholding the traditional understanding of the Qur'ān as the word of God, he accepts diverse interpretations that contextualize divine revelation for all times and places. In his view, the Qur'ān is both transcendent and worldly, guiding the emergence of the Muslim community and offering responses to specific challenges faced by early Muslims. Furthermore, Ayoub argues that for the Qur'ān to maintain its guiding role in human life, it must speak to all epochs and conditions. This necessitates reading the Qur'ān in accordance with contemporary circumstances and challenges. Thus, Ayoub's hermeneutics seeks to ensure that the Qur'ān remains relevant by adapting its guidance to the ever-changing context of human experience.

Ayoub's hermeneutical writings can be categorized into two groups with slightly different objectives. First, he produced objective works on the general topic of *tafsīr*, intending to make this genre more widely known in the West. Second, Ayoub wrote about specific topics within Qur'ānic exegesis, where he not only objectively presented traditional views but also took positions on interpretations that he deemed reasonable. His method combined traditional exegetical tools with critical examinations of previous interpretations from a historical perspective. Through his hermeneutical approach, Ayoub sought in particular to champion the concept of religious pluralism.

Ḥasan Ḥanafī's (d. 2021/1443) approach to interpreting the Qur'ān not only reflects a fusion of Arab-Islamic heritage (*turāth*) and modernity (*ḥadātha*) but also underscores the political dimensions inherent in the interpretative process. Fundamental to his perspective is the belief that God's revelation in the Qur'ān bestows sanctity upon the text while maintaining an enigmatic quality that requires continual reinterpretation.

Ḥanafī's Qur'ānic hermeneutics is deeply rooted in a nuanced understanding of the divine. He emphasizes God's manifestation through practice. For him, the Qur'ān's sanctity derives from its role as a guiding force intrinsically linked to the formation of the Muslim community. The interpretive process is seen as a form of dynamic engagement with the divine, where the Qur'ān's meaning remains perpetually veiled, allowing for ongoing ontological creation.

According to Ḥanafī, the Qur'ānic exegete is not merely a passive recipient of meaning but an active participant in the creation of knowledge. Interpreting the Qur'ān becomes a transformative process, where the givenness of the text metamorphoses into knowledge within the interpreter's spirit. This view underscores the dynamic nature of Qur'ānic interpretation, rejecting static approaches and embracing a continuous evolution of understanding.

Ḥanafī contends that Qur'ānic interpretation is intrinsically tied to the sociopolitical context of the exegete. The act of interpreting becomes an ideological weapon wielded by different sociopolitical powers. Whether maintaining the status quo or initiating revolutionary change, interpretation becomes a means to articulate and preserve specific interests. This perspective presents differences in Qur'ānic discourse as inherently ideological in nature.

Furthermore, Ḥanafī distinguishes between longitudinal and latitudinal Qur'ān commentaries. Longitudinal commentaries follow a systematic temporal evolution, risking a loss of thematic integrality. On the other hand, latitudinal commentaries allow for the simultaneous gathering of Qur'ānic facts, fostering varied interpretations. According to Ḥanafī, the latter approach avoids the pitfalls of the former, ensuring more coherent and comprehensive Qur'ānic interpretation. He proposes a shift toward thematic interpretation, emphasizing how necessary it is to understanding and explaining the Qur'ān. Thematic interpretation allows not only for analysis but also synthesis, enabling interpreters to go beyond dividing the text up into parts and to unite them into a coherent whole. Moreover, thematic interpretation, Ḥanafī says, leads to a deeper level of understanding and consciousness.

For Ḥanafī, the conflict between various Qur'ānic interpretations is essentially a sociopolitical conflict rather than a theoretical one. The theoretical aspect serves as an epistemological cover, with interpretation functioning as an ideological weapon. Whether maintaining or challenging the status quo, the act of interpretation reflects the sociopolitical convictions of the interpreter.

Ḥanafī thus advocates for a Qur'ānic hermeneutics aimed at reconstructing Muslim thought and society. His method involves discernment and intuition, factoring in the needs of human beings and the welfare of the *umma*. When brought into harmony, consciousness and factuality lead to interactions that transform into operative plans, projects, and ideas. The ultimate goals are to revitalize the *umma* and transition into a modern era. Thus, Ḥanafī's Qur'ānic hermeneutics represents a dynamic interplay between tradition, politics, and societal reconstruction.

Mūsā ṣ-Ṣadr (b. 1928/1346, disappeared in 1978/1398) did not write any books on Qur'ānic hermeneutics. However, the vast corpus of his collected writings and speeches contains numerous references to his interpretations of Qur'ānic verses. Aṣ-Ṣadr's hermeneutical approach is shaped by the interplay of exoteric (*ẓāhirī*) and esoteric (*bāṭinī*) orientations within Muslim exegesis. He integrates both approaches and synthesizes his intellectual worldview, drawing influence from Neoplatonic patterns. Aṣ-Ṣadr's hermeneutics takes a problem-based approach, influenced by his academic background in law and economics. Thus, he adapts interpretive approaches related to evolving issues, showcasing a flexibility and responsiveness to changing contexts. His hermeneutics is rooted in humanitarianism, emphasizing human existence as one of religion's core values. Aṣ-Ṣadr is open to diversity and believes in the sanctity of humanity. Rejecting traditional semantic assumptions, he advocates for a pluralistic understanding of Qur'ānic meanings. According to him, Qur'ānic meanings go beyond specific language use or systems of signification, and verses can have allegorical, metaphorical, literal, and scientific meanings all at once. He rejects esoteric and mystical interpretations that contradict customary understanding.

One important feature of aṣ-Ṣadr's Qur'ānic hermeneutics is his emphasis on humanity as the core of shared religious and ethical values. In this sense, he rejects the dichotomy between rational and religious, good and evil. He is not afraid of scientific progress and believes that it is the result of continuous interaction between humans, nature, and the word of God. Emphasizing the importance of historical contexts, aṣ-Ṣadr presents interpretations that prioritize social understanding while also embracing the modern sciences as complementary to religious understanding. Additionally, aṣ-Ṣadr addresses the need for coexistence among different religious communities. With a pluralistic perspective on religious truth, he views all religions as paths toward serving humanity and rejects exclusivism. Interpreting Qur'ānic injunctions as guidelines for balanced growth in human societies, he provides fresh interpretations on polygyny, temporary marriage, and paternal intervention into marriage, distancing himself from traditional approaches.

Süleyman Ateş's (b. 1933/1352) hermeneutical magnum opus, *Yüce Kur'ân'ın Çağdaş Tefsîri* (The Contemporary Commentary on the Noble Quran), not only established his authority as an exegete but also sparked discussions and critiques within religious circles. Ateş, with his religious education rooted in both traditional madrasa-like institutions and state-funded establishments, employs a dual approach in interpreting the Qur'ān. He combines classical sources and methods with modern and scientific considerations, creating a distinctive blend of tradition and modernity. Despite his classical training and experience in the Arab world, Ateş is often perceived by Turkish contemporaries as more modernist than classical in his approach.

Inspired by thinkers like Muḥammad ʿAbduh and Rashīd Riḍā, Ateş places a strong emphasis on the ethical role of the Qur'ān and the spiritual guidance it provides its readers. His commitment to ethical teachings and spiritual insights aligns with broader modernist inclinations toward reforming Islam.

In his commentary, Ateş emphasizes elucidating Qur'ānic verses by making intratextual references to other passages within the Qur'ān while also taking into account the relevant occasions of revelation (*asbāb an-nuzūl*) and extensively relying on hadith. Moreover, he presents contested interpretations and debates among scholars, occasionally deviating from classical opinions. Ateş compares Qur'ānic verses with materials from Jewish and Christian traditions, revealing the comparative dimension of his work. His commentary challenges conventional readings on issues such as abrogation, historical criticism applied to Qur'ānic narratives, and hadith. As a result, he has faced criticism, which often revolves around perceived inconsistencies in his views on the Qur'ānic verses. However, his approach is often characterized as a call to return to the Qur'ān. In this sense, he shares common ground with modernist contemporaries advocating to reconnect with the essence of the scripture.

Muḥammad Mujtahid Shabestarī (b. 1936/1354) presents a distinctive approach to the Qur'ān, aiming to shift the way it is perceived – from heavenly document to worldly phenomenon. Despite not producing a formal *tafsīr* work, Shabestarī advocates for a novel approach to Qur'ānic studies, challenging conventional readings prevalent in traditional educational institutions. He criticizes the efforts of Iranian scholars like Ṭāliqānī and Sharīʿatī, claiming that they are insufficient to gain a profound understanding of the Qur'ān and Islam. Shabestarī perceives the prevailing official reading as overly focused on jurisprudence, leading him to call for a broader interpretive approach to comprehending religion and the Qur'ān.

Shabestarī introduces the concept of *ilāhiyyāt-i tafahhumī*, or hermeneutical theology, emphasizing the role played by interpreters' preconceptions, interests, and expectations in shaping their perspectives on the Qur'ān. His approach aligns with modern hermeneutical principles, acknowledging diverse interpretations and the need for critical reflection on different readings in order to arrive at a nuanced understanding of religions and scriptures.

Central to Shabestarī's philosophy and crucial to his understanding of the Qur'ān is the idea that the encounter between religious conviction and modernity requires scrutiny. He addresses the tension between religious belief and modern life, particularly in relation to scientific and philosophical assumptions. Contrary to traditional scholars, Shabestarī rejects the notion of a fundamental contradiction between rational comprehension and divine decree. He argues that conflicts arise from human assumptions, whether religious or secular, highlighting the need for hermeneutics to reconcile these perspectives.

Furthermore, Shabestarī develops the concept of the "human reading of religion" (*qirā'at-i insānī az dīn*), placing humans at the center of religious understanding. This perspective considers religion to be a personal, contextual choice that encompasses various facets of human existence, including philosophy and law. Shabestarī challenges the traditional separation between God and humanity, asserting that religious texts can be interpreted in diverse ways shaped by historical experiences and contextual conditions.

Crucially, Shabestarī suggests that the Qur'ān, despite its divine origin, is inherently connected to Muḥammad's experiences and perceptions. He contends that the Qur'ān is a dynamic, historical text that engages in an ongoing dialogue with humanity. Shabestarī's approach includes a reevaluation of Islamic norms, which he regards as spatiotemporal constructs that are subject to change based on evolving societal conditions. His perspective on the Qur'ān emphasizes the importance of hermeneutics, individual interpretation, and a dynamic understanding of religious texts.

The Qur'ānic hermeneutics of Imran Hosein (b. 1942/1361) is distinctive, emphasizing the Qur'ān as the absolute truth and foundation of Muslim epistemology. Hosein argues that understanding this truth requires believers to recite and reflect upon the Qur'ān, making them receptive to divine guidance. Hosein criticizes the prevailing Western political ideology of secularism, advocating for a foundation of faith in human civilization. He condemns the global economic system, asserting that it is prohibited in Islamic law, and proposes an alternative currency based on gold and silver.

Hosein's approach to studying the Qur'ān involves holistic examination, drawing parallels with the astronomer studying stars to grasp constellations. He echoes his teacher Fazlur Rahman Ansari's idea of the Qur'ān's semantic unity, emphasizing the interconnectedness of verses and chapters, and denouncing fragmentary exegesis. Practical aspects like living virtuously, proficiency in Arabic, and constant recitation, he says, are integral to successful Qur'ānic study.

In Hosein's methodology, the Qur'ān takes precedence over hadith, with the latter considered a source of relative truth. He advocates for a holistic reading of the sunna. His multidisciplinary approach includes insights from various sciences and the analysis of global changes throughout history. Hosein's hermeneutics, deeply embedded within the end-times framework, emphasizes spirituality, moral conduct, and adherence to an Islamic way of life in Qur'ānic recitation. His approach, referred to as "end-times exegesis," incorporates a mystical aspect influenced by Sufi tradition.

Muhammad Quraish Shihab (b. 1944/1363) places great emphasis on the exceptional literary quality and superiority of the Qur'ān. His approach to Qur'ānic hermeneutics reflects the ideas of reformist Islamic scholars, utilizing Qur'ānic exegesis as an intellectual foundation to reshape religious traditions in a modern and pluralistic Muslim society.

In line with ʿAbduh's perspective, Shihab views the Qur'ān as a primary guide for religious direction, described as guidance for humanity. Consequently, he believes that Qur'ānic interpretation should address real problems within Muslim society. However, he acknowledges the difficulty of understanding the Qur'ān's guidance solely through an exclusively scholarly endeavor focusing on vocabulary, syntax, and interpretative principles.

Shihab therefore employs two methods in his Qur'ānic exegesis. Analytical (*taḥlīlī*) interpretation provides detailed explanations of various aspects relevant to a specific verse, including the context of its revelation, its content, its placement in relation to other verses, and its literary quality. However, Shihab expresses a preference for the second method, thematic (*mawḍūʿī*) interpretation, which aims to provide Muslims with a sense of "living" guidance. In this approach, the exegete examines a real problem in Muslim society by gathering verses related to the topic in order to draw conclusions about what the Qur'ān might have to say about the matter. According to Shihab, thematic interpretation clarifies the Qur'ān's guidance.

Additionally, Shihab incorporates the approach of *uṣūl al-fiqh* (principles of Islamic jurisprudence) to explain ambiguities in the Qur'ānic text and make them accessible. His thematic interpretation of the Qur'ān, inspired by Egyptian scholars like Maḥmūd Shaltūt, has gained widespread popularity. However, he has encountered criticism for his reformist views on issues like the headscarf, which he does not believe is prescribed by the Qur'ān, and his call for moderation in dealing with heterodox groups like the Ahmadiyyah.

Shihab's efforts to write a new Indonesian commentary on the Qur'ān may have been motivated by his dissatisfaction with existing commentaries. His work is more detailed and systematic than previous works, although his source selection, with a focus on modern commentators, has drawn attention. Despite citing new hermeneutical approaches, his commentary seems more interested in linguistic details than understanding the Qur'ān's revelation in its historical context. Shihab's fascination with scientific and mathematical approaches aligns with the scientific interpretation (*at-tafsīr al-ʿilmī*) of the Qur'ān, supporting the idea of its inimitability (*iʿjāz al-Qur'ān*).

ʿAbdulkarīm Surūsh (commonly known as Abdulkarim Soroush) (b. 1945/1365) presents a unique perspective on the Qur'ān, viewing it through the lens of temporal change. His concept of *ijtihād* encapsulates the idea of evolving religious understanding, both in a general sense and, more specifically, in the realm of Qur'ānic hermeneutics. Soroush does not seek to fully comprehend religion or God, but rather aims to approach religious truth. He believes that, even if errors are made in this pursuit, they hold historical significance. This understanding leads him to question all religious authority, including the concept of *valī-yi faqīh* (the Guardian Jurist).

Central to Soroush's hermeneutics is an acknowledgment that the Qur'ān is open to various interpretations. He challenges the notion of unquestionable religious authority and emphasizes the importance of engaging critically with religious texts. This approach resonates with the developing nature of religious understanding, accommodating a diverse range of interpretations over time.

Soroush assigns a crucial role to the Prophet in his hermeneutical framework. He argues that the Prophet not only received the divine message but also played a role in shaping it through his experiences and evolving personality. The Qur'ān, according to him, is not a static revelation but a dynamic process that followed the Prophet's development. This perspective challenges the traditional view of the Prophet as a passive recipient of divine messages, suggesting a more interactive relationship between the Prophet and revelation.

According to Soroush, the idea of sequential revelation in the Qur'ān mirrors the religious transformations that occurred over time. The Qur'ān, being a product of its historical context, reflects the changing nature of religion. Soroush contends that, just as the Qur'ān emerged in response to the changing circumstances of the Prophet's life, religion itself is subject to change over time.

Soroush's questioning of religious authority stems from his belief in the interpretive flexibility of the Qur'ān. He argues that rigid adherence to religious authority stifles the dynamic potential of religious thought. By advocating for continuous critical engagement with religious texts, Soroush encourages a more fluid and adaptable understanding of Islam, emphasizing the dynamic nature of Qur'ān in response to temporal changes.

Jāved Aḥmad Ghāmidī (b. 1951/1370) draws heavily on the works and approaches of his two predecessors, Ḥamīd ad-Dīn Farāhī and Amīn Aḥsan Iṣlāḥī. Farāhī in particular had a significant influence on Ghāmidī's hermeneutics. Ghāmidī firmly believes in the unparalleled eloquence of the Qur'ān, asserting that its content does not deviate from the language spoken by Muḥammad, which is the language of the people of Mecca. Consequently, correctly understanding the Qur'ān is contingent upon proper knowledge and a genuine appreciation of the Arabic language. An excellent command of classical Arabic is therefore a prerequisite for meaningfully reflecting on the Qur'ān. Moreover, Ghāmidī contends that familiarity with classical Arabic literature, especially pre-Islamic poetry, is essential as it exemplifies the eloquence found in the Qur'ān. He argues that the Qur'ān serves as a guide for all moral actions, clearly delineating the distinction between good and evil. Each sura is considered a cohesive and coherent unit, and the arrangement of its verses is not arbitrary.

Moreover, Ghāmidī emphasizes that the Qur'ān's message must be contextualized within the circumstances of its revelation. According to traditional accounts, the Qur'ān was revealed to Muḥammad in seventh-century Arabia. On the basis of this historical fact, it is evident that Muḥammad explained the Qur'ān whenever and wherever the need arose, and his companions and subsequent scholars similarly continued this practice.

Ghāmidī's hermeneutical approach is based on the belief that the Qur'ān's meanings are embedded within the linguistic and historical context of its revelation. Its language, cultural nuances, and historical background are integral to grasping the intended meanings of the Qur'ānic verses. In this regard, Ghāmidī places significant importance on linguistic and historical studies, arguing that they are indispensable in order to comprehensively understand the Qur'ān.

The work of Amina Wadud (b. 1952/1371) is often described as a radical theological engagement akin to liberation theology. Notably, Wadud has been recognized as the first Muslim academic to systematically employ the category of gender as an analytical tool in Qur'ānic hermeneutics. Central to her theological exploration is the Qur'ān which she prioritizes as the foremost example of divine guidance while acknowledging the sunna as a secondary primary source in Islam. According to Wadud, the Qur'ānic signs (*ayāt*) are metaphors pointing to the ultimate reality which transcends all symbols and signs, whether in the language of revelation, creation, or nature.

Wadud's hermeneutics is characterized by its ongoing, dynamic nature. Rooted in discussions and dialogues with activists and fellow scholars, reflective thinking, and the continuous reevaluation of principles, her approach is adaptive and responsive. She adheres to the classical exegetical principle that the best interpreter of the Qur'ān is the Qur'ān itself (*tafsīr al-Qur'ān bi-l-Qur'ān*). However, she criticizes traditional exegetes for reducing this method to a simplistic interpretation, resulting in fragmented and restricted Qur'ānic exegesis.

One fundamental aspect of Wadud's hermeneutics is her rejection of any connection between God and oppression (*ẓulm*). She firmly asserts that God does not oppress, which leads her to a radical theological commitment to justice and mutuality among human beings. Wadud's theological stance rejects all forms of injustice and emphasizes that theological interpretations must be in harmony with principles of justice.

Wadud's hermeneutical lens is particularly focused on challenging the gender disparities within traditional Islamic exegesis. She contends that traditional, male-dominated exegesis has resulted in skewed interpretations. For her, correcting this gender imbalance is a moral imperative. She advocates reclaiming the idea that women's experiences matter and calls for the Qur'ān to be read through the lens of gender. This involves reading "before the text, behind the text, and through the text" to identify hegemonic rhetorical strategies and power relations along gender lines.

There is a widespread belief that female scholars in the Islamic tradition did not author any commentaries on the Qur'ān until the twentieth century. This belief is echoed not only in popular discourse but also within academic circles. However, there is a lack of historical studies examining the genuine contributions of women in the realm of Qur'ānic hermeneutics. Bibliographic and biobibliographic sources mention works by female scholars, but the existence of manuscripts or printed editions remains uncertain. However, in the twentieth and twenty-first centuries, influential exegetical works arguing for gender-inclusive interpretations have emerged.

The tenuous situation of female-authored Qur'ān commentaries before the twentieth century can be attributed to various factors, including societal norms, limited educational opportunities for women, and the pervasive influence of male scholars in the Islamic scholarly tradition. However, contemporary scholars have begun to challenge this gender gap in Qur'ānic hermeneutics. Amina Wadud and Asma Barlas are notable figures who have played pivotal roles in shaping the discourse on Qur'ānic exegesis. They offer feminist perspectives and critique traditional interpretations. Their works are contributing to a growing body of literature seeking a more inclusive and gender-responsive approach.

Contemporary scholars advocating for gender equality in Qur'ānic interpretation base their arguments on a moral framework derived from the Qur'ān's foundational statements about the essence and purpose of humanity. The creation narrative, which emphasizes the shared origin of man and woman from a single soul (*nafs wāḥida*) (Q 4:1), serves as a cornerstone for these interpretations. This statement aligns with the central belief in divine unity (*tawḥīd*), affirming the equality of all creatures before God.

Although the Qur'ān considers women and men to be essentially equal beings, definitions of gender equality and justice remains elusive. The chapter on Qur'ānic hermeneutics by and for women sheds light on the complexities of defining gender equality within the Qur'ānic context. It contributes to a more nuanced understanding of Islamic theology and gender relations, prompting a reevaluation of historical assumptions and encouraging a more inclusive engagement with diverse voices in Islamic scholarship.

The chapter on fundamentalist Qur'ānic hermeneutics addresses the discourse on the multifaceted concept of fundamentalism within the Islamic tradition. The author discusses fundamentalism being perceived as a response to external threats to religious identity and explores its association with the complete implementation of *sharīʿa* law, often seen as a reaction to Western globalization. It critically examines the hermeneutics employed by the Muslim Brotherhood and the Salafis, two prominent groups within Islamic fundamentalism.

The hermeneutics of the Muslim Brotherhood, exemplified by figures like Rashīd Riḍā and Abū l-Aʿlā Mawdūdī, is deeply rooted in a reinterpretation of the Qur'ān that responds to perceived political threats. Riḍā's Qur'ānic exegesis greatly influenced Ḥasan al-Bannā, illustrating the transmission of hermeneutical ideas within the Brotherhood. Additionally, Mawdūdī's commentary further shaped the Brotherhood's ideology, reinforcing the belief that Islam is under political threat, whether from external Western forces or internal challenges within the Muslim world. Premodern exegesis, which largely confined itself to religious matters, underwent a transformation during nineteenth-century Western colonialism, prompting scholars like Riḍā to apply Qur'ānic interpretation to contemporary societal issues and thereby contributing to the politicization of the Brotherhood's Qur'ānic hermeneutics. Thus, in response to Western threats, leaders such as Muṣṭafā s-Sibāʿī, the founder of the Syrian branch of the Brotherhood, focused on issues such as the Western treatment of women, presenting it as a historical narrative of inequality, subjugation, and dependency. This highlights how the political landscape has shaped the interpretation of the Qur'ān, framing it as a source of resistance against perceived external threats.

In contrast to the Muslim Brotherhood, Salafis, in particular those influenced by Ibn Taymiyya, focus their Qur'ānic exegesis on countering dogmatic threats. Ibn Taymiyya, though he did not author a complete commentary, laid the groundwork for a textualist approach. His method involves identifying and rectifying what he considers erroneous readings of the Qur'ān, emphasizing a return to the original meanings of the text and rejecting interpretations that deviate from its intended message. Salafi hermeneutical approaches are rooted in the belief that, over time, misunderstandings and theological errors have emerged in Qur'ānic interpretation. They contend that the Prophet Muḥammad himself clarified and interpreted the Qur'ān for his followers, with significant differences in interpretation arising only after the first three generations of Muslims. Salafis attribute the introduction of faulty readings to influences external to the tradition, asserting that reason, rather than the transmission of tradition, led to those misinterpretations. The outcome of this hermeneutical method is a literal, textual, and doctrinally "pure" *tafsīr* that aims to avoid the perceived errors of other exegetes.

The chapter on the relationship between the transcendent and the contingent in the Qur'ān presents different perspectives on contemporary Qur'ānic hermeneutics that reflect the complex interplay between history, spirituality, and divine transcendence. It examines five modern scholars' theories of revelation, ontology, and Qur'ānic interpretation. All of them examine the relationship between history and the Qur'ān, seeking to uncover new meanings and engaging critically with the Islamic exegetical tradition. However, their approaches to achieving these goals vary. Three of them – Naṣr Ḥāmid Abū Zayd, Abdulkarīm Surūsh, and Wajīh Qānṣū, who are referred to as modernists – advocate history-based hermeneutics as essential to overcoming the perceived stagnation of modern Islamic thought. The other two, Shafīq Jarādī and Nayla Ṭabbāra, seek to incorporate the divine into their concepts of history.

As the chapter shows, Abū Zayd perceives the Qur'ān as a divine message conveyed within the sociocultural context of the Prophet Muḥammad. He advocates for a more open interpretation of the Qur'ān, challenging the supposed infallibility of earlier scholars and the Prophet's companions. Abū Zayd asserts that the meaning of the Qur'ān emerges when the reader's horizon fuses with the text, emphasizing hermeneutic subjectivity. Soroush, on the other hand, views revelation (*waḥy*) as a human religious experience rather than divine revelation. He claims that this experience is not exclusive to prophets, thereby challenging the essential character of Islam. Qānṣū, for his part, considers revelation as the process through which God revealed Himself to the Prophet. Rejecting a clear distinction between divine revelation and the Prophet's experience, Qānṣū argues against drawing a line separating the two.

While critical of historical hermeneutics, Jarādī does not romanticize the past or defend the idea that the Qur'ān's divinity is the result of mystical occurrences. He acknowledges the human language of the Qur'ān, which, he says, reflects its historical context. However, he argues that language is just one expression of divinity, and humans are not entirely detached from God or confined to historical contingency. His perspective challenges the essence of human nature.

Ṭabbāra views history as part of divine action, emphasizing a multidimensional interpretation of the Qur'ān. Her hermeneutical approach focuses less on religious metaphysics and more on the Qur'ān's diverse interpretations. The Qur'ān reflects the sociocultural context of its time, but Ṭabbāra does not believe that this contradicts the conviction that the Qur'ān is God's word. Her hermeneutical starting point is neither historical nor contextual, but rooted in a religious pluralism that accepts religious difference as divinely ordained while affirming the truth of Islam.

The recent positions discussed here offer unique insights into contemporary interpretations of the Qur'ān, ranging from more historical and subjective approaches to the belief that divine intervention has shaped historical events. Some of the analyses in this handbook contribute to a rich discourse on the multifaceted nature of Qur'ānic hermeneutics in the Arab context, prompting ongoing reflections on the role of history and the transcendence in shaping religious interpretations.

When exploring modern Qur'ānic hermeneutics in Türkiye, on the other hand, one encounters a nuanced landscape shaped by various historical, cultural, and intellectual forces. Within the Turkish context, the examination of Qur'ānic interpretation reveals three distinct strands of hermeneutics: explicit, objective, and normative Qur'ānic hermeneutics. These distinctions serve as foundational pillars for understanding the diverse approaches to interpreting the Qur'ān within Turkish scholarly circles.

The development of Qur'ānic hermeneutics in Türkiye is deeply intertwined with the historical trajectory of the region. The fall of the Ottoman Empire and the establishment of the Turkish Republic marked a pivotal juncture that significantly influenced the direction taken by Qur'ānic interpretation. The political movements that emerged during this period subjected Qur'ānic hermeneutics to a process of profound reflection and revision, setting the stage for the integration of both traditional and Western modes of thought.

Central to the development of Qur'ānic hermeneutics in Türkiye have been the academic Qur'ānic schools, which serve as bastions of both traditional and modern intellectual currents. While rooted in Islamic tradition, these institutions have also embraced Western methodologies and perspectives, reflecting the enduring influence of Western academia in shaping Turkish intellectual discourse.

One notable manifestation of modern Qur'ānic hermeneutics in Türkiye is the Ankara School, whose influence reverberates within scholarly circles. This school, characterized by its diverse intellectual landscape, defies easy categorization, with its hermeneutical approach in the 1990s showcasing a plurality of perspectives rather than a singular unified school of thought. Scholars like Mehmet Paçacı, Ömer Özsoy, and Burhanettin Tatar occupy prominent positions within the Ankara School, each of them contributing distinct hermeneutical viewpoints to the broader discourse. The chapter on the Ankara School elucidates the varied hermeneutical perspectives within the Ankara School, offering insights into the interpretative methodologies employed by these scholars and highlighting the rich tapestry of intellectual inquiry within Turkish Qur'ānic hermeneutics. The hermeneutical landscape in Türkiye is marked by the dynamic interplay between tradition and modernity, as scholars grapple with the complexities of interpreting the Qur'ān within contemporary contexts.

The chapter on Turkish translations of the Qur'ān traces shifts in interpretative methodologies and the impact of political ideologies on translations over time. The Young Turk movement marked a significant turning point in the evolution of Turkish Qur'ānic hermeneutics. With the rise of this movement, prayer, worship, and the Islamic call to prayer were all translated into Turkish, a move criticized for deviating from tradition. The early Turkish Republic sought to present the Qur'ān in an understandable language, free from superstition, and to ultimately eliminate the ulema (religious scholars) serving as intermediaries between the Qur'ān and the people.

The demand for a "Turkish Qur'ān" for Turks and a rejection of need to learn Arabic reflected a growing emphasis on Turkish nationalism. The period leading up to the establishment of the Turkish Republic witnessed the emergence of several Turkish translations of the Qur'ān. The government's goal was to prevent potential misinterpretation by removing intermediaries, allowing the Qur'ān to speak directly to the people.

To address the need for a Turkish Qur'ān, Elmalı Muhammed Hamdi Yazır was commissioned to produce a translation and commentary. His work, published between 1935 and 1939, became the most widely used Turkish translation. However, the first Turkish Qur'ān translation, "Tanrı Buyruğu" by Ömer Rıza Doğrul, was published in 1934.

Contemporary Turkish Qur'ān translations typically feature an introduction and an index, presenting the translation as *Muṣḥaf meāl, Nuzul meāl,* or thematic *meāl,* aiming to convey the meaning of the Qur'ān – i.e., the intended goal of revelation. These translations differ in the methods they apply, such as "interlinear"/"literal" and "interpretive" translation. Early translations used interlinear methods with smaller Turkish script placed below the Arabic words, leading to potential misunderstandings due to syntactic differences.

The political landscape of Türkiye has played a crucial role in shaping Qur'ānic translations. The Kemalist era emphasized secularism, while subsequent governments have influenced the prominence of Islam in translations. Various translations have aligned with the ideologies of ruling parties, impacting the interpretation of key verses. Different translators, influenced by individual *ijtihād*, have contributed to the diversity of interpretation. Furthermore, issues related to gender equality in different interpretations, such as the principle of male superiority and female subordination, have been called into question.

Exploring modern Qur'ānic hermeneutics in Indonesia and Malaysia reveals a rich tapestry of exegetical works influenced by historical, cultural, and intellectual contexts. Modern Qur'ānic hermeneutics in Indonesia has been influenced by Mahmud Yunus's influential work. His three-part commentary seeks to provide a concise, rational interpretation free from *isrā'īliyyāt* elements, appealing to the modern sensibilities of educated readers. In the latter half of the twentieth century, Indonesia witnessed the emergence of more extensive exegetical works, exemplified by Hasbi ash-Shiddieqy's *Tafsir an-Nūr* and Buya Hamka's *Tafsir al-Azhar*. These commentaries aimed to provide access to Qur'ānic exegesis for Indonesian Muslims unfamiliar with Arabic, enabling them to engage with the sacred text directly. However, two distinctive approaches emerged in Indonesian Qur'ānic hermeneutics in the late twentieth century. The first focused on commentators responding to modern scientific knowledge, aiming to reconcile religious teachings with contemporary scientific understandings. The second approach is evident in Qur'ān commentaries commissioned by the Indonesian government in the twenty-first century. These works deal with various scientific themes, such as food and drink, time, microorganisms, and the oceans, reflecting concerted efforts to bridge the gap between religious teachings and scientific advancements.

In Malaysia, modern Qur'ānic hermeneutics is closely tied to the reformist movements initiated by Muḥammad ʿAbduh and Rashīd Riḍā. This connection is evident in exegetical works that often embrace reformist ideologies. While evidence of exegetical activity in what is now Malaysia prior to the beginning of the twentieth century is limited, the country produced over fifty Qur'ān commentaries in the twentieth century. Renowned scholars like Muhammad Sa'id b. Umar, Syed Syeikh al-Hadi, and Syeikh Uthman Jalaluddin al-Kelantani made significant contributions to this corpus. Their commentaries serve several purposes, such as acting as bridges between classical Malay exegesis and modern works, and they showcase diverse approaches, with some providing original interpretations in Malay and others translations of Arabic commentaries. Nik Abdul Aziz and Abdul Hadi Awang's exegetical works illustrate the growth of thematic exegesis (*at-tafsīr al-mawḍūʿī*) in Malaysia, focusing on specific themes and suras. Comprehensive thematic exegesis has dominated in the twenty-first century so far.

The study of modern Qur'ānic hermeneutics in Indonesia and Malaysia reveals nuanced approaches shaped by different historical, cultural, and intellectual contexts. While both nations share a connection to reformist movements, the distinct trajectories of their exegetical works demonstrate the diversity of contemporary Islamic thought in Southeast Asia. The interplay between traditional elements and responses to modern challenges has characterized the development of Qur'ānic interpretation in these dynamic societies.

The final chapter in this volume sketches the landscape of progressive Qur'ānic hermeneutics in the twenty-first century, giving examples of how contemporary Muslim scholars in Germany and Australia are approaching the Qur'ānic interpretation. The modern era has witnessed a departure from traditional methods, with a particular emphasis on thematic exegesis and a shift toward a more comprehensive and user-oriented approach. This chapter examines the evolution of Qur'ānic hermeneutics in the twentieth century, its contextual developments, and introduces representatives of progressive Qur'ānic interpretation in the twenty-first century.

In contrast to premodern exegesis that derived general behavioral norms from individual verses, twentieth-century Qur'ānic hermeneutics embarked on a journey of thematically-driven interpretation. The focus has shifted from deriving universal norms from individual verses to establishing overarching principles that guide the interpretation of specific themes within the Qur'ān. This trend, observed not only in the Islamic world but also in the European context, has emphasized a cross-textual approach that groups the Qur'ān into overarching themes rather than dissecting it verse by verse. The roots of this shift can be traced back to the early Islamic reformist movements of the late eighteenth and nineteenth centuries, which continued to gain prominence throughout the twentieth century. Further developments in the twentieth century included a more apologetic orientation in Qur'ānic exegesis. In response to encounters and confrontations with the West, interpreters have increasingly emphasized the compatibility of Islam with scientific knowledge, refuting claims of irrationality. The rise of the mass media has also facilitated the user-oriented popularization of Qur'ānic exegesis, reaching beyond scholarly circles.

The developing landscape of Qur'ānic hermeneutics in the twentieth century gave rise to two primary strands: a political Islamist orientation and a progressive one. Political Islamists emphasized the compatibility of Islam with science and rationality in response to Western encounters. In contrast, the progressive strand sought to present a more inclusive and open interpretation of the Qur'ān, accommodating modern perspectives and intellectual diversity.

One distinctive characteristic of contemporary Qur'ānic hermeneutics is the emergence of non-theologians, including intellectuals from a range of disciplines, as key proponents of progressive interpretation. This shift has broadened the scope of Qur'ānic hermeneutics, introducing innovative and multidisciplinary perspectives. This last chapter ends with a discussion of the hermeneutical approaches taken by Abdel-Hakim Ourghi, Milad Karimi, and Mouhanad Khorshide in Germany, and Abdullah Saeed in Australia. These Muslim scholars are contributing to a more inclusive and dynamic understanding of the Qur'ān, emphasizing its relevance in the modern world. They serve as examples of the ongoing dialogue between tradition and modernity in Qur'ānic hermeneutics in Islam.

The chapters in this volume were written by authors from different regions and academic disciplines, who are at different stages of their academic careers. Each volume of the *Handbook* demonstrates that the history of Qur'ānic hermeneutics is extremely diverse, and its scholarly treatment must be similarly diverse. Massimo Campanini's chapter on Ḥasan Ḥanafī had to be completed posthumously due to the regrettable death of the author in 2020.

Unfortunately, some of the scholars who were supposed to be covered in this volume are missing. They include Muḥammad ʿĀbid al-Jābirī (d. 2010/1431) and Farid Esack (b. 1959/1374). The reason why there are no chapters on them in this volume is that the colleagues who undertook to write these chapters did not perform their duty, despite several reminders. It was no longer possible to wait for them or to find substitute authors.

No authors could be found for studies on other scholars who should have been considered in this volume, including ʿAbd al-ʿAzīz b. ad-Dardīr (*at-Tafsīr al-mawḍūʿī li-āyāt at-tawḥīd fī l-Qur'ān al-karīm*, Cairo 1990), Amīr ʿAbd al-ʿAzīz (*at-Tafsīr ash-shāmil li-l-Qur'ān al-karīm*, 6 volumes, Cairo 2000), and Anwar al-Bāz (*at-Tafsīr at-tarbawī li-l-Qur'ān al-karīm*, 3 volumes, Cairo 2007). Although it is almost impossible to cover all hermeneutical approaches to the Qur'ān today or in any other era of Qur'ānic scholarship exhaustively, the outlined shortcomings will hopefully be corrected in a future, updated version of this volume.

Mohammad Gharaibeh

Ibn ʿUthaymīn

1 Introduction

In the second half of the twentieth century, three scholars dominated the ideological and theological debates in Salafi and Wahhābī circles. Those are: ʿAbd al-ʿAzīz b. Bāz (d. 1999/1420), who was the second state mufti of Saudi Arabia and the first leading Wahhābī scholar who did not belong to the Āl ash-Shaykh family; Nāṣir ad-Dīn al-Albānī (d. 1999/1420), who managed to become the highest authority in hadith studies till his death and who influenced the theological discourse around hadith in Saudi Arabia and beyond till his death; and Muḥammad b. Ṣāliḥ al-ʿUthaymīn (d. 2001/1421, also known as Ibn ʿUthaymīn), who has become the most acknowledged authority in matters of creed and Islamic jurisprudence.

The present article will take a look into his *Tafsīr al-Qurʾān* which was and still is compiled by the Ibn ʿUthaymīn Foundation. The Foundation, which was established by Ibn ʿUthaymīn's sons and students, goes through multiple recordings of lessons that Ibn ʿUthaymīn specifically dedicated to the interpretation of the Qurʾān or lessons that contain among other interpretations of the Qurʾān. Though Ibn ʿUthaymīn's principles and methods in interpreting the Qurʾān generally follow those of Taqī d-Dīn Aḥmad b. Taymiyya (d. 1328/728), his explanation reflects the cultural, social, and religious context of Saudi Arabia in the second half of the twentieth century.

2 Life and Work

About seven years after the death of Muḥammad b. Ṣāliḥ al-ʿUthaymīn, the religious Saudi TV-Channel *al-Majd* broadcasted a documentary program about the life of the Wahhābī scholar including interviews of his students, friends, and family members. The producer of this episode of the series called *Ḥayāt Insān* decided to open the episode with a quotation of a tale that was widely spread in Wahhābī circles:

> A man saw in his dream Ibn Taymiyya passing him a luminary while he was on his way to the mosque. After the man woke up, he visited a dream reader who told him that a child will be born for him who is going to be a leading religious scholar. Hoping that the prophecy becomes true, the man named his first born Ṣāliḥ. Although Ṣāliḥ was a just and pious man he did not become a religious scholar. However, his son Muḥammad [Ibn ʿUthaymīn] happened to be the fulfillment of this prophecy.[1]

1 This episode was broadcasted on 16.11.2008 under the title "Ḥayāt al-faḍīla al-ʿallāma Muḥammad b. Ṣāliḥ b. ʿUthaymīn," islamway.com, accessed March 3, 2021, http://www.islamway.com/?iw_s=Lesso

https://doi.org/10.1515/9783111320069-002

Certainly, this tale has no historical value and does not tell who Ibn ʿUthaymīn really was. However, it reveals how Ibn ʿUthaymīn was perceived in his time and after his death, as a leading Islamic scholar whose righteousness was prophesied. Moreover, the appearance of Ibn Taymiyya, who is one of the most important thinkers for the intellectual culture of the Wahhābī movement, shows Ibn ʿUthaymīn as the intellectual heir of the great medieval scholar Ibn Taymiyya.

The following account gives a short overview of his life and tries to answer the question why Ibn ʿUthaymīn became such an important scholar.

2.1 Biographical Sketch

Born on March 9, 1929/Ramaḍān 9, 1347 in ʿUnayza, Ibn ʿUthaymīn whose full name is Abū ʿAbd Allāh Muḥammad b. Ṣāliḥ b. Muḥammad b. Sulaymān al-Wuhaybī sh-Shamīmī was a descendant of the famous Arab tribe of al-Wuhaba. Ibn ʿUthaymīn began his religious career with memorizing the Qurʾān under the advice of his maternal grandfather and finished memorizing it with the age of thirteen. Meanwhile, he learned writing and reading in a traditional school (*kuttāb*). Ibn ʿUthaymīn received his first religious lessons in the Great Mosque of ʿUnayza. The leader and preacher (*imām wa-khaṭīb*) of the mosque at that time was ʿAbd ar-Raḥmān as-Saʿdī (d. 1956/1376) who was married to a great-aunt of Ibn ʿUthaymīn and who became his principal teacher.[2] But instead of being taught by as-Saʿdī directly, the basic courses were taught by some of as-Saʿdī's elder students.[3] After he reached a certain educational level, as-Saʿdī took over the education of Ibn ʿUthaymīn, who was already recognized due to his outstanding intelligence compared to his peers, as Walīd al-Ḥusayn reported, who is one of the biographers of Ibn ʿUthaymīn.[4] Studying intensively with him, as-Saʿdī had certainly a huge influence on Ibn ʿUthaymīn. Being one of the most famous exegetes of the Najd region, as-Saʿdī was also one of the first scholars who included views of other Islamic legal schools (*madhāhib*) in addition to the dominant Ḥanbalī school into his writings and teachings. Impressed by as-Saʿdī's "flexibility" and "intellectual openness," Ibn ʿUthaymīn learned with his principal teacher in the field of Qurʾānic exegesis (*tafsīr*), science of hadith, the biography of the prophet (*sīra*), Islamic creed (*ʿaqīda*), Islamic law (*fiqh*), the principles of law (*uṣūl al-fiqh*), law

n&iw_a=view&lesson_id=86327. Other episodes of the series *Ḥayāt Insān* treat the life of ʿAbd al-ʿAzīz b. Bāz, ʿAbd ar-Raḥmān as-Saʿdī and Muḥammad al-Amīn ash-Shinqīṭī.

2 Walīd b. Aḥmad al-Ḥusayn, *al-Jāmiʿ li-ḥayāt al-ʿallāma Muḥammad b. Ṣāliḥ al-ʿUthaymīn al-ʿilmiyya wa-l-ʿamaliyya wa-mā qīla fīhi min al-marāthī* (Leeds: Majallat al-Ḥikma, 2002), 71.

3 To this educational practice see Guido Steinberg, *Religion und Staat in Saudi-Arabien. Die wahhabitischen Gelehrten 1902–1953* (Würzburg: Ergon, 2002), 86f.

4 Al-Ḥusayn, *al-Jāmiʿ*, 10.

of succession (*farāʾiḍ*), and grammar (*naḥw*).[5] After the establishment of the first official Saudi Religious Institute (*maʿhad ʿilmī*) in 1952/1371, Ibn ʿUthaymīn studied in Riyadh for two years during 1952–1954/1371–1373. In Riyadh, he met the Mauritanian scholar Muḥammad al-Amīn ash-Shinqīṭī (d. 1974/1393) who influenced Ibn ʿUthaymīn in the science of grammar and rhetoric (*naḥw wa-balāgha*). He also met ʿAbd al-ʿAzīz b. Bāz for the first time who was described by Ibn ʿUthaymīn as his second principal teacher. Although the regular length of study at the institute was four years, Ibn ʿUthaymīn studied the second and the fourth year in the summer break of the first and the third year, so that he finished his study after two years. It can be assumed, that Ibn ʿUthaymīn did not spend the whole two years in Riyadh, since he stated that he also continued his study with ʿAbd ar-Raḥmān as-Saʿdī during that time. After his study in Riyadh however, he never left ʿUnayza for a longer period, except for the month of Ramaḍān and during the pilgrimage that he used to spend in the Hijaz giving public lectures each year from 1982/1402 on.

After the death of as-Saʿdī, Ibn ʿUthaymīn assumed his teaching position in the Great Mosque of ʿUnayza, in addition to the position as imam and preacher.[6] A quarter century later, in 1982/1402, he reached a high rank among the Wahhābī scholars, so that he was appointed the most prestigious teaching position in the Great Mosque of Mecca and Medina during pilgrimage and the month Ramaḍān.[7] Shortly after he accepted these teaching positions, Ibn ʿUthaymīn's reputation in Saudi Arabia constantly increased. In the last years before his death, around five hundred people, among them students, scholars, and interested people, regularly attended his teaching sessions in ʿUnayza.[8] Beside his public lessons, Ibn ʿUthaymīn taught in the religious institute of ʿUnayza since 1975/1395 until he received a teaching position at the University in Burayda at the sharia faculty in 1977/1397 that he held until his death on January 11 2001/Shawwāl 16 1421.[9]

In contrast to other high-ranked Wahhābī scholars, Ibn ʿUthaymīn held just one official position. He only accepted the appointment as a member of the Council of Senior Scholars (*hayʾat kibār al-ʿulamāʾ*) in 1987/1407.[10] It seems that Ibn ʿUthaymīn decided early to put all his efforts into becoming a well-recognized and leading scholar, rather than being involved in politics, a strategy Ibn ʿUthaymīn probably learned from his principal teacher who himself had been perceived as the leading scholar of

---

5 Muḥammad b. ʿUthaymīn, *Sharḥ al-uṣūl* (Dammam: Dār Ibn al-Jawzī, 2010/1431), 6.

6 Al-Ḥusayn, *al-Jāmiʿ*, 67.

7 Ibn ʿUthaymīn, *Sharḥ al-uṣūl*, 8.

8 Al-Ḥusayn, *al-Jāmiʿ*, 50.

9 Ibid., 67; Ibn ʿUthaymīn, *Sharḥ al-uṣūl*, 8.

10 Muhammad al-Atawneh, *Wahhābī Islam Facing the Challenges of Modernity. Dār al-Iftāʾ in the Modern Saudi State* (Leiden: Brill, 2010), 18.

his time.[11] A look at the biography of as-Saʿdī reveals, that he, although he became a well-recognized scholar, in particular kept distance to the political elite in general.[12]

2.2 Ibn ʿUthaymīn's Outstanding Position

Certainly, several reasons could be mentioned why Ibn ʿUthaymīn had become one of the leading Wahhābī scholars of the second half of the twentieth century. In the following, two possible answers will be presented in more detail.

2.2.1 His Scholarly Achievements

As many other scholars, Ibn ʿUthaymīn was commenting on nearly every aspect of daily life and was issuing legal Islamic rulings (*fatāwā*). In his public lessons and even more in the interactive radio shows, that were broadcasted internationally, Ibn ʿUthaymīn had the possibility of spreading his views. Not rarely supporting the official politics in Saudi Arabia, Ibn ʿUthaymīn talked about women's matter and behavior in public, about the pilgrimage rites or matters of the Islamic creed. In general, his lectures were recorded and the Ibn ʿUthaymīn Foundation, that was founded shortly after his death in 2001/1421 by his sons and students, put huge effort in making them available on the homepage of the foundation and printing them on the long-term.[13]

Beside the huge number of recorded lessons, especially in his early years, Ibn ʿUthaymīn wrote books mainly about Islamic law and creed. However, the works that have been written by him are relatively rare comparing them to those that have been compiled by his students under his advice or later after his death. One work in particular should be mentioned here in more detail. In contrast to his other works that are mostly commentaries on works of other mostly pre-modern scholars,[14] *al-Qawāʿid al-muthlā fī ṣifāt Allāh taʿālā wa-asmāʾihi al-ḥusnā* is compiled by Ibn ʿUthaymīn himself. Since there are very few self-written works of other Wahhābī scholars, *al-Qawāʿid* has reached the rank of a standard work by the likes of Ibn Taymiyya, Shams ad-Dīn Muḥammad b. Qayyim al-Jawziyya (d. 1350/751), or Muḥammad b. ʿAbd al-Wahhāb (d. 1792/1203), as will be shown below. The popular and simple style of Ibn ʿUthaymīn

11 Steinberg, *Religion*, 259.

12 Ibid., 260.

13 An overview of the recorded material is provided at the official homepage of the foundation: "Muḥammad b. Ṣāliḥ al-ʿUthaymīn," binothaimeen.net, accessed December 30, 2022, https://binothaimeen.net/.

14 Commenting on works of famous other scholars was and still is a trend in Saudi Arabia. Against this background, writing a book must be seen as a testimony of large professional competence.

that he used in his lessons and writings in general and in *al-Qawāʿid* in particular was also an important factor why the book achieved broad acceptance.[15]

2.2.2 Strategic Behavior?

The second reason why Ibn ʿUthaymīn became so important lies in the circumstances of his education and his distance from politics. During his youth, he still had the opportunity to study the traditional way that enabled him to establish close personal ties to his teachers. At the same time, the official educational system was yet to be established and Ibn ʿUthaymīn benefited from the possibility to study in Riyadh with some foreign scholars from Mauritania and Egypt, such as Muḥammad al-Amīn ash-Shinqīṭī or ʿAbd ar-Razzāq ʿAfīfī (d. 1994/1415). Staying most of his life in ʿUnayza while many other scholars moved to the political center and capital Riyadh where official positions were waiting to be filled, may seem to be a disadvantage at first sight. However, rather than being a small fish in a big sea, after the death of his teacher as-Saʿdī, Ibn ʿUthaymīn ascended in ʿUnayza to become a big fish in a small sea. He assumed all the teaching positions of his mentor and in particular his position as a leading scholar in the city ʿUnayza and in the region of Qaṣīm in general was never challenged.

The fact that his position in the Qaṣīm region was indeed never questioned also shows Ibn ʿUthaymīn's high position within the Wahhābī establishment. In comparison to his teacher and later friend Ibn Bāz, Ibn ʿUthaymīn's reputation as a highly capable scholar becomes clearer. Ibn Bāz succeeded his main teacher Muḥammad b. Ibrāhīm Āl ash-Shaykh as the great mufti of Saudi Arabia in 1993/1414[16] and can, therefore, be perceived as the leading figure withing the Wahhābī establishment in the second half of the twentieth century. The position of the state mufti, that has been introduced with and probably for Muḥammad b. Ibrāhīm Āl as-Shaykh, had not been occupied after his death in 1969/1389 for more than twenty years. The appointment of Ibn Bāz was hence a historical step and put a scholar that did not belong to the Āl ash-Shaykh on the top of the official Wahhābī hierarchy. Ibn Bāz was the head of the Institution of *iftāʾ* and Scholarly Research, and the Council of Senior Scholars (*hayʾat kibār al-ʿulamāʾ*) as well as the director of the Islamic University in Medina. Both positions were previously held by Muḥammad b. Ibrāhīm, too.[17] However, being that close to the political center and occupying all these rather political than religious posi-

15 For an extended discussion of this work and a translation into German see Mohammad Gharaibeh, *Zur Attributenlehre der Wahhābiyya unter besonderer Berücksichtigung der Schriften Ibn ʿUthaimīns (1929–2001)* (Berlin: EB-Verlag, 2011), 84–95, 357–413.

16 Al-Atawneh, *Wahhābī Islam*, 32f.

17 ʿAbd ar-Raḥmān ar-Raḥma, *al-Injāz fī tarjamat al-imām ʿAbd al-ʿAzīz b. Bāz ḥayātuhu wa-juhūduhu al-ʿilmiyya wa-l-ʿamaliyya wa-d-daʿwa wa-āthāruhu al-ḥamīda* (Dammam: Dār Ibn al-Jawzī, 2007) 137–65.

tions let Ibn Bāz be influenced by the political pragmatism of the Āl Saʿūd. The institutionalization of the *ʿulamāʾ* and their integration into the administrative system had put them under the influence of the king, in general.[18] The fact that Ibn ʿUthaymīn did not accept many appointments for official positions and that he stayed in ʿUnayza focusing more on the academic side of his career may have helped him save his independence to a certain level. An example should illustrate the different standings of the two important scholars, Ibn Bāz and Ibn ʿUthaymīn.

It was due to a ruling (*fatwā*) of the Council of Senior Scholars which was under the leading of Ibn Bāz at this time that allowed the Āl Saʿūd to call American troops for assistance during the Iraq invasion of Kuwait 1990–92/1411–12.[19] This decision led to protest among the Saudi population and many other religious scholars criticized the council's decision. Ibn Bāz, in particular, lost some of his reputation among the conservatives of the Saudi society. Although Ibn ʿUthaymīn was a member of the council and supported its decision as well, his reputation did not suffer from this event. In this context, on May 18, 1991/ Dhū l-Qaʿda 4, 1411, four hundred scholars signed a petition and handed it over to King Fahd requesting more political changes and reforms.[20] Apparently, the *ʿulamāʾ* assumed they could increase their influence on society and on political decision making in return for their support of the *fatwā* and during this political crisis. However, King Fahd urged Ibn Bāz as the President of the Council of Senior Scholars to condemn the petition he himself signed earlier. That must have meant a huge loss of face for Ibn Bāz. Ibn ʿUthaymīn, on the contrary, did not lose any of his high reputation during these events.[21]

In the view of many, Ibn ʿUthaymīn remained a great scholar till the present and even though he did not hold much official positions, he seemed to be the most influential scholar of the second half of the twentieth century in Saudi Arabia. His writings are recognized internationally and have been studied by plenty of students in Saudi universities in Master and PhD theses. A huge amount of internet forums refers to his opinions and many of his writings are translated into English, French and German.

---

18 See for this discussion in a broader context Ayman al-Yassini, *Religion and State in the Kingdom of Saudi Arabia* (Boulder: Westview Press, 1985), 59f.; Aharon Layish, "ʿUlamāʾ and politics in Saudi Arabia," in *Islam and politics in the Modern Middle East*, ed. Metin Heper and Raphael Israeli (London: Routledge, 1984), 55–57; Othman Yakub al-Rawwaf, "The concept of five crises in political development: Relevance to the Kingdom of Saudi Arabia" (PhD diss., Duke University, 1981), 527; al-Atawneh, *Wahhābī Islam*, 36.

19 Henner Fürtig, *Demokratie in Saudi-Arabien? Die Āl Saʿūd und die Folgen des zweiten Golfkrieges* (Berlin: Verlag Das Arabische Buch, 1995), 20f.

20 Mansoor Jassem Alshamsi, *Islam and Political Reform in Saudi Arabia. The Quest for Political Change and Reform* (New York: Routledge, 2011), 99f.

21 Fürtig, *Demokratie*, 23.

3 Ibn ʿUthaymīn's Works

As mentioned above, Ibn ʿUthaymīn was primarily acknowledged for his scholarly engagement. Accordingly, his output is immense. At the same time, Ibn ʿUthaymīn did not author many books by himself. Rather, the primary medium of his scholarly engagement were lessons given in different contexts; in form of lectures, fatwas, or sermons in mosques, in radio programs, or TV-shows. Already during the lifetime of Ibn ʿUthaymīn, students of his took it upon themselves to transcribe recordings of these lectures, fatwas and sermons and turn them into books to be printed and published. Ibn ʿUthaymīn used to receive these transcriptions, revised and authorized them before they were printed by a publisher house. After his death, some of his students and his sons founded the Ibn ʿUthaymīn Foundation that was concerned with spreading the intellectual heritage of him. The Ibn ʿUthaymīn Foundation continues to transcribe recordings of Ibn ʿUthaymīn to publish them. They also act as the official direction to authorize transcriptions of projects that were initiated by other individuals.[22] Therefore, the number of his publications is still growing, and the present overview of his writings can only be a momentary overview. Currently, the foundation lists the (written) publication of Ibn ʿUthaymīn in ten disciplines: 1) *at-Tafsīr wa-ʿulūm al-Qurʾān* with thirty-six titles; 2) *al-Ḥadīth wa-muṣṭalaḥuhu* with seventeen titles; 3) *al-ʿAqīda* with twenty-eight titles; 4) *al-Fiqh wa-uṣūluhu wa-qawāʿiduhu* with forty-seven titles; 5) *ʿUlūm al-lugha* with six titles; 6) *al-Khuṭab* with one title; 7) *al-Fatāwā wa-l-liqāʾāt* with eighteen titles; 8) *Mawḍūʿāt ʿāmma* with fourteen titles; 9) *as-Sīra an-nabawiyya* with two titles; 10) *Ādāb ṭālib al-ʿilm* with four titles.[23]

The majority of his scholarly engagement, as can be seen by the number of publications, was focused on the disciplines of Islamic law (*fiqh*) and Islamic creed (*ʿaqīda*) together with the interpretation of the Qurʾān (*tafsīr*) and the hadith. The topics he talked about though are very many-faceted. This is due to the fact that he also frequently answered questions of individuals who either joined his teaching sessions at mosques, or who called during a radio program and TV-show. Those were mostly questions on daily matters that Ibn ʿUthaymīn would tie to topics of either Islamic jurisprudence and Islamic creed and, therefore, contributed to an Islamization of daily life in Saudi Arabia in particular and other countries in general, where people would follow his opinions.

---

22 See for an overview of their activities and the works they have so far published the website of the Ibn ʿUthaymīn foundation: "al-Mawqiʿ ar-rasmī li-Muʾassasa faḍīlat ash-shaykh Muḥammad b. Ṣāliḥ al-ʿUthaymīn al-khayriyya," binothaimeen.net, accessed December 30, 2022, https://binothaimeen.net/foundothemen/.

23 "Al-Mawqiʿ ar-rasmī li-Muʾassasa faḍīlat ash-shaykh Muḥammad b. Ṣāliḥ al-ʿUthaymīn al-khayriyya. Mabīʿāt al-kutub," binothaimeen.net, accessed December 30, 2022, https://binothaimeen.net/foundothemen/content/Mnue/lib.

3.1 Islamic Jurisprudence (*fiqh*)

In the discipline of Islamic jurisprudence (*fiqh*) Ibn ʿUthaymīn was especially known for two works. These are the five-volume commentary on *Zād al-mustaqniʿ fī ikhtiṣār al-Muqniʿ* of the Ḥanbalī scholar Sharaf ad-Dīn Musā l-Ḥajjāwī (d. 1555/962) called *ash-Sharḥ al-mumtiʿ ʿalā Zād al-mustaqniʿ*. Al-Ḥajjāwī's work in return is an abridgement of *al-Muqniʿ* of Muwaffaq ad-Dīn b. Qudāma al-Maqdisī (d. 1223/620), a work that can be seen as a standard work among Ḥanbalī jurists that was also used to train jurists in Saudi Arabia in the twentieth century.[24] The second work, that Ibn ʿUthaymīn was associated with the most, is also a commentary on a popular hadith collection that contains all law-related traditions. The *Bulūgh al-marām fī adillat al-aḥkām* was authored by the Cairene Shāfiʿī scholar Aḥmad b. Ḥajar al-ʿAsqalānī (d. 1449/852) and was frequently read in Saudi Arabia in the second half of the twentieth century. Ibn ʿUthaymīn commented on this work in two series which were both recorded. The first series started in 1981/1401 while the other started in 1996/1428. For the current edition, the second recording served as the base recording while bits and pieces were also taken from the first series.[25]

The special features of those commentaries as well as Ibn ʿUthaymīn's law-related other works is the fact that he did not fixate on the Ḥanbalī school of law in general. Moreover, he took his liberty to find individual solutions for juridical problems (*ijtihād*). In the context of Saudi Arabia of the early second half of the twentieth century this was still not something common to do. Ibn ʿUthaymīn was influenced in this by his principal teacher, ʿAbd ar-Raḥmān as-Saʿdī. He was perceived to be one of the first, if not the first, scholar to include rulings and opinions of the other three Sunni law schools, the Mālikiyya, Ḥanafiyya and Shāfiʿiyya.[26] As-Saʿdī followed this path especially after the study of the writings of Ibn Taymiyya and his student Ibn Qayyim al-Jawziyya.[27] Ibn ʿUthaymīn was greatly influenced by as-Saʿdī's method. In over forty legal questions Ibn ʿUthaymīn had preferred the opinion of Ibn Taymiyya over other established Ḥanbalī opinions. Moreover, Ibn ʿUthaymīn had challenged the dominant Ḥanbalī opinion in over 317 questions.[28] Ibn ʿUthaymīn also combined the method of the jurists with that of the traditionists (*ahl al-ḥadīth*). As a consequence, Ibn ʿUthay-

24 Other works that were used to train jurists especially in the first half of the twentieth century include *al-ʿUmda*, *al-Kāfī*, and *al-Mughnī* all written by Ibn Qudāma as well, in addition to *al-Iqnāʿ* of al-Ḥajjāwī. See Steinberg, *Religion*, 108–14.

25 See the introduction to Muḥammad b. Ṣāliḥ b. ʿUthaymīn, *ash-Sharḥ al-mukhtaṣar ʿalā Bulūgh al-marām*, ed. Muʾassasat ash-Shaykh Muḥammad b. Ṣāliḥ al-ʿUthaymīn al-Khayriyya (Qaṣīm: Muʾassasat ash-Shaykh Muḥammad b. Ṣāliḥ al-ʿUthaymīn al-Khayriyya, 2019²), 1: 5f.

26 ʿAbd Allāh b. Sābiḥ aṭ-Ṭayyār, *ash-Shaykh ʿAbd ar-Raḥmān b. Nāṣir as-Saʿdī mufassiran* (Dammam: Dār Ibn al-Jawzī, 2000), 43f.

27 ʿAbd Allāh b. Sābiḥ aṭ-Ṭayyār, *Athar ʿallāma al-Qaṣīm ash-Shaykh ʿAbd ar-Raḥmān b. Nāṣir as-Saʿdī ʿalā l-ḥaraka al-ʿilmiyya al-muʿāṣira* (Dammam: Dār Ibn al-Jawzī, 1992), 19.

28 Al-Ḥusayn, *al-Jāmiʿ*, 100–04.

mīn often dismissed the opinions of the other law schools as well in favor of a tradition (*ḥadīth*) that he – or others – found to be stronger, both in terms of its authenticity and its statement.[29] Although Ibn ʿUthaymīn had studied hadith during his education, one cannot say that he was a specialist in hadith matters. Therefore, he most often relied on the rulings of the early scholars such as Muḥammad al-Bukhārī (d. 870/256) and Muslim b. al-Ḥajjāj (d. 822/261) in their *Ṣaḥīḥ* collections, as well as the Medieval scholar Ibn Ḥajar al-ʿAsqalānī.[30]

This method of Ibn ʿUthaymīn ultimately led to the fact that he changed his opinion in some questions several times. There were some incidents in which Ibn ʿUthaymīn changed his opinion during a session after a student or another present individual turned his attention to the weakness of a tradition Ibn ʿUthaymīn used to rely on. In other cases, scholars corresponded with Ibn ʿUthaymīn over some other traditions so that Ibn ʿUthaymīn changed his opinion afterwards.[31]

Interestingly, none of these incidents have reduced Ibn ʿUthaymīn's reputation. Quite the opposite. In the public eye, Ibn ʿUthaymīn was attributed with honesty and sincerity since he did not insist on his personal opinion out of stubbornness or arrogance. Instead, he submitted himself to the 'truth,' as al-Ḥusayn puts it. Especially, when he was to contradict Ibn Taymiyya's opinion, he gained even more prestige and authority.[32]

3.2 Islamic Creed (*ʿaqīda*)

As the Wahhābī movement became internationally more recognized during the twentieth century, more and more polemics and refutations were compiled against some of the key beliefs of the movement. One of the central parts were the *tawḥīd* concept of Muḥammad b. ʿAbd al-Wahhāb and his threefold division of *tawḥīd* into *tawḥīd ar-rubūbiyya*, *tawḥīd al-ulūhiyya*, and *tawḥīd al-asmāʾ wa-ṣ-ṣifāt*.[33] This division, though, was not invented by Ibn ʿAbd al-Wahhāb himself. It goes back to the Medieval Ḥanbalī reformist scholar Taqī d-Dīn b. Taymiyya. However, the way Ibn ʿAbd al-Wahhāb used this concept to denounce his opponents and members of other Arab tribes on the Arab peninsular was rather political than purely theological. In the twentieth century, many scholars from the other Arabic countries, especially Egypt, wrote refutations against this concept. Moreover, in the middle of the twentieth century one part of this

29 Ibid., 80.

30 Ibid., 79f., 109f.

31 See for some examples of that ibid., 80–85.

32 Ibid., 97–99.

33 See for the concept of his *tawḥīd* the most cited book of him, Ibn ʿAbd al-Wahhāb, *Kitāb at-Tawḥīd*, ed. Aḥmad b. ʿAlī l-Qufaylī (Cairo: Maktabat al-ʿUlūm wa-l-Ḥikam, 2008).

concept came to the fore, which was previously almost entirely ignored, i.e. the *tawḥīd al-asmā' wa-ṣ-ṣifāt*.

This part of the *tawḥīd* is concerned with the Divine attributes and how they relate to God's essence. Wahhābī scholars surprisingly had only little to say about this for the most part of the first half of the twentieth century.[34] The few statements that could be found state that one must affirm all attributes mentioned in the Qur'ān and the Prophetic tradition literally as they were revealed in the texts. This eventually led, in the eyes of the non-Wahhābī critics, to an anthropomorphic image of God. Only in the second half of the twentieth century did Wahhābī scholars react to those accusations with more complex writings and responses. One of the most elaborated and prolific scholars who specialized on this part of the *tawḥīd* was Ibn ʿUthaymīn. In the discipline of Islamic creed, he authored many treatises – most of which were actually commentaries on past writings – taking up this topic in an attempt to defend the Wahhābī position which was claimed to be the position of the pious ancestors and the Sunni community (*ahl as-sunna wa-l-jamāʿa*). Although he also talked about other topics related to Islamic creed, Ibn ʿUthaymīn was recognized and highly respected among his Wahhābī colleagues for his expertise on the *tawḥīd al-asmā' wa-ṣ-ṣifāt*. The following description of his works in Islamic creed will therefore focus on those writings.

Most writings of Ibn ʿUthaymīn are commentaries on the writings of mainly four authors, Ibn Taymiyya, Ibn Qayyim al-Jawziyya, Ibn Qudāma al-Maqdisī and Muḥammad as-Saffārīnī (d. 1774/1188) who all belonged to the Ḥanbalī school of law. This is for mainly two reasons. First, during the educational years, the writings of these authors were read and taught by almost all scholars during the second half of the twentieth century so that a commentary on them demonstrated the mastery of the content. Second, by commenting on these writings, Wahhābī scholars could demonstrate, or at least imply, that their position is identical to those of accepted past scholars. For example, Ibn ʿUthaymīn commented on the *Nūniyya* of Ibn Qayyim al-Jawziyya which primarily targets practices of Sufis that were controversial. Ibn ʿUthaymīn's commentary on this work is entitled *Tawḍīḥ al-kāfiya ash-shāfiya* and was well received among Wahhābī scholars. The writings that primarily focused on the Divine attributes, though, were the ones of Ibn Taymiyya, Ibn Qudāma and as-Saffārīnī. Especially Ibn Taymiyya provided the Wahhābī scholars with rich content to dwell on the *tawḥīd al-asmā' wa-ṣ-ṣifāt*.

In total, Ibn ʿUthaymīn commented on three works of Ibn Taymiyya. The commentary on *al-Fatāwā al-ḥamawiyya* entitled *Fatḥ rabb al-bariyya bi-talkhīṣ al-ḥamawiyya* was even one of the first written works of Ibn ʿUthaymīn. He began writing this

34 For an overview of what Wahhābī scholars used to study during the first half of the twentieth century see Steinberg, *Religion*, 89–92.

commentary at the age of twenty-three years in 1950/1370[35] and it was published almost ten years later in 1960/1380.[36] Another work of Ibn Taymiyya is his *al-ʿAqīda al-wāsiṭiyya* which also primarily addresses the Divine attributes and their relation to God's essence. Ibn ʿUthaymīn's commentary on this was compiled during the years 1955–56/1375–76. It is entitled simply *Sharḥ al-ʿAqīda al-wāsiṭiyya* and is much more detailed than the rather short *Fatḥ rabb al-bariyya*.[37] The last work of Ibn Taymiyya that was commented on by Ibn ʿUthaymīn is the *Risāla at-tadmuriyya*. Ibn ʿUthaymīn entitled his commentary *Taqrīb at-Tadmuriyya*. It was published relatively late and goes back to a series of lectures that Ibn ʿUthaymīn held. A student of his transcribed the recordings and Ibn ʿUthaymīn authorized it in December 1991/Rajab 1381.[38] The work was published several times after that.

The following two commentaries were compiled on writings of Ibn Qudāma al-Maqdisī and as-Saffārīnī. The *Taʿlīq mukhtaṣar ʿalā Kitāb Lumʿat al-iʿtiqād al-hādī ilā sabīl ar-rashād* was compiled to be a teaching manual for the religious institute (*al-maʿhad al-ʿilmī*) in ʿUnayza, in which Ibn ʿUthaymīn had taught. He finished this commentary in 1972/1392.[39] The other commentary goes back to a number of teaching sessions Ibn ʿUthaymīn conducted to teach this book of the Ḥanbalī scholar as-Saffārīnī. The recording that was used to transcribe the content and to publish it as a book was taken in the years 1989–1990/1409–1410.[40] The base text is a versification that addresses the Islamic creed. The title of Ibn ʿUthaymīn's commentary on it is *Sharḥ al-ʿAqīda as-saffārīniyya ad-Durra al-muḍiyya fī ʿaqd ahl al-firqa al-marḍiyya*.

One additional work that needs to be mentioned here is an original one of Ibn ʿUthaymīn. His *al-Qawāʿid al-muthlā fī ṣifāt Allāh taʿālā wa-asmāʾihi al-ḥusnā* is one of the most known, acknowledged, accepted and received works of Ibn ʿUthaymīn. It has

35 See the Lecture by Sāmī ṣ-Ṣuqayr, "Ibn ʿUthaymīn wa-ṭalabuhu li-l-ʿilm," in the Ibn ʿUthaymīn-Mosque in ʿUnayza on June 30, 2007, min. 20:40–27:00, https://al-soger.com/ar/show/sounds/1c10f490-cf87-4dcc-9218-910b48ad53e0/%D8%A7%D8%A8%D9%86_%D8%B9%D8%AB%D9%8A%D9%85%D9%8A%D9%86_%D9%88_%D8%AD%D9%8A%D8%A7%D8%AA%D8%A9_%D8%A7%D9%84%D8%B9%D9%84%D9%85%D9%8A%D8%A9 [access 11.03.2024].

36 See Muḥammad b. Sāliḥ b. ʿUthaymīn, *Fatḥ rabb al-bariyya bi-talkhīṣ al-ḥamawiyya*, ed. Muʾassasat ash-Shaykh Muḥammad b. Ṣāliḥ al-ʿUthaymīn al-Khayriyya (Dammam: Dār Ibn al-Jawzī, 2003), 6.

37 See the edition in two vols.: Muḥammad b. Sāliḥ b. ʿUthaymīn, *Sharḥ al-ʿAqīda al-wāsiṭiyya*, ed. Saʿd b. Fawwāz aṣ-Ṣumayl, 2 vols. (Dammam: Dār Ibn Jawzī, 2010[6]).

38 Muḥammad b. Sāliḥ b. ʿUthaymīn, *Taqrīb at-Tadmuriyya*, ed. Muʾassasat ash-Shaykh Muḥammad b. Ṣāliḥ al-ʿUthaymīn al-Khayriyya (Riyadh: Madār al-Waṭan li-n-Nashr, 2012), 151.

39 Muḥammad b. Sāliḥ b. ʿUthaymīn, *Taʿlīq mukhtaṣar ʿalā Kitāb Lumʿat al-iʿtiqād al-hādī ilā sabīl ar-rashād*, ed. Muʾassasat ash-Shaykh Muḥammad b. Ṣāliḥ al-ʿUthaymīn al-Khayriyya (Riyadh: Madār al-Waṭan li-n-Nashr, n.d.), 91.

40 Muḥammad b. Sāliḥ b. ʿUthaymīn, *Sharḥ al-ʿAqīda as-saffārīniyya ad-Durra al-muḍiyya fī ʿaqd ahl al-firqa al-marḍiyya*, ed. Muʾassasat ash-Shaykh Muḥammad b. Ṣāliḥ al-ʿUthaymīn al-Khayriyya (Riyadh: Madār al-Waṭan li-n-Nashr, 2005), i.

been the subject of commentaries itself[41] and has become sort of a standard work on the topic of the Divine attributes within Wahhābī and Salafi circles.

The book was written by Ibn ʿUthaymīn to defend himself from accusations that were held against him by Ḥumūd b. ʿAbd Allāh at-Tuwayjirī (d. 1992/1413), ʿAbd Allāh b. Ibrāhīm al-Qarʿāwī (b. 1954/1373), and ʿAlī b. ʿAbd Allāh al-Ḥawwās (d. 1990/1411).[42] In one of his teaching sessions, Ibn ʿUthaymīn spoke about the verse Q 57:3 "And He is with you wherever you are," explaining that God literally is with his creation. In the eyes of these three scholars, Ibn ʿUthaymīn violated the dogma of God being above (*ʿulū*) and over his throne. To believe that he is with his creation, would eventually lead to a pantheism, a position that Wahhābī scholars attribute to Sufism and condemn strongly. At-Tuwayjirī and al-Ḥawās authored a refutation of Ibn ʿUthaymīn's statement, which were entitled *Ithbāt ʿulū Allāh wa-mubāyanatihi li-khalqihi wa-r-radd ʿalā man zaʿama anna maʿiyyat Allāh li-l-khalq dhātiyyatun* (by at-Tuwayjirī) and *an-Nuqūl al-wāḍiḥa al-jaliyya ʿan as-salaf aṣ-ṣāliḥ fī maʿnā l-maʿiyya al-ilāhiyya al-ḥaqīqiyya* (by al-Ḥawwāṣ).[43] Ibn ʿUthaymīn reacted upon these accusations with his *al-Qawāʿid al-muthlā* which should demonstrate that he maintained the right belief, i.e. the belief of the *ahl as-sunna wa-l-jamāʿa* and the pious ancestors (*as-salaf aṣ-ṣāliḥ*). Although his opponents weren't really convinced by *al-Qawāʿid*, Ibn ʿUthaymīn came out stronger and was deemed even more knowledgeable after this incident. This not least because of the support he received by other highly respected scholars such as Ibn Bāz.[44]

4 Ibn ʿUthaymīn's Exegetical Writings and his Qurʾān Commentary

In the discipline of *tafsīr*, Ibn ʿUthaymīn had multiple contributions. Mostly, though, in the form of lectures, either in the religious institutes, where he was teaching, or in public in mosques and other places. Besides his *tafsīr* lectures that were later transcribed of records, he also held lectures on the principles of the interpretation of the Qurʾān. Although Ibn ʿUthaymīn was mostly known for his engagements in Islamic ju-

41 Ash-Shāyiʿ lists four commentaries on al-Qawāʿid: 1) *at-Taʿlīqāt al-fuḍlā min al-Qawāʿid al-muthlā* by Muḥammad aṣ-Ṣughayr b. Qāʾid al-Maqṭarī; 2) *Sharḥ al-Qawāʿid al-muthlā* by Kamāl al-Maṣrī; 3) *Fatḥ al-ʿalī l-aʿlā bi-sharḥ al-Qawāʿid al-muthlā* by ʿUbayd b. ʿAbd Allāh b. Sulaymān al-Jābirī; 4) *al-Mujlā fī sharḥ al-Qawāʿid al-muthlā* by Kāmila al-Kawārī; see Muḥammad b. ʿAbd al-ʿAzīz b. Muḥammad ash-Shāyiʿ, *Dalīl al-maktaba al-ʿaqdiyya. Muʿjam mawḍūʿī li-l-kutub wa-r-rasāʾil wa-l-buḥūth fī l-ʿaqīda* (Riyad: Dār Zidnī], 2007), 38.

42 Al-Ḥusayn, *al-Jāmiʿ*, 79.

43 ʿAbd Allāh b. ʿAbd ar-Raḥmān b. Ṣāliḥ al-Bassām, *ʿUlamāʾ Najd khilāl thamāniya qurūn* (Riyad: Dār al-ʿĀṣima, 1998²), 5: 230.

44 See for a more detailed account on this controversy: Gharaibeh, *Zur Attributenlehre*, 84–95.

risprudence and Islamic creed, his Qur'ān interpretations – which also frequently were included in the discussion of topics in the aforementioned disciplines – received great attention.

4.1 The Principles and Methods of Qur'ān Interpretation

Out of his lectures and writings two works were published that deal with the principles and methods of Qur'ān interpretations. Those are his *Uṣūl fī t-tafsīr* and his *Sharḥ Muqaddimat at-tafsīr*. The latter is a commentary on Ibn Taymiyya's *Muqaddimat at-tafsīr* and was transcribed and edited by a student of Ibn ʿUthaymīn, ʿAbd Allāh aṭ-Ṭayyār. The first, however, was written by Ibn ʿUthaymīn himself upon a request by some of his students. Ibn ʿUthaymīn had taught *tafsīr* at a class in the Islamic al-Imām Muḥammad b. Saʿūd University and prepared some notes on this matter. After a while he was asked to compile a treatise out of these notes. The result is the *Uṣūl fī t-tafsīr*.[45]

The *Uṣūl fī t-tafsīr* is a short treatise on the *ʿulūm al-Qur'ān* and addresses topics related to the Qur'ān more generally. It addresses for example how the Qur'ān was revealed, what verse was reveled first, what verse was reveled with and without any reason, the relation between the meaning of the verse and the reason of its revelation, and what verses were revealed in Mecca and Medina. Moreover, the treatise includes topics related to the interpretation of the Qur'ān. Ibn ʿUthaymīn explains what the sources of the interpretation should be, and who the most famous interpreters among the Companions (*ṣaḥāba*) and the Followers (*tābiʿūn*) were. Moreover, he speaks about verses that on the first glimpse contradict each other, the oaths in the Qur'ān, and how to deal with *isrāʾīliyyāt*.[46]

As for the methodology and the principles of the interpretation of the Qur'ān, Ibn ʿUthaymīn generally seems to follow Ibn Taymiyya. This means that he believes that the Prophet is the first interpreter of the Qur'ān and that he had explained both, the words and the meaning of the Qur'ān to the Companions. From this belief, Ibn ʿUthaymīn draws the conclusion that the Companions also only memorized what they have understood, and since they memorized the entire Qur'ān – even though some of the Companions had memorized only parts of it –, this means that the meaning of the entire Qur'ān was clear to the Companions. Moreover, Ibn ʿUthaymīn supports Ibn Taymiyya's argument that the Companions need to be seen as authorities in interpreting the Qur'ān as well, since the Qur'ān commands all who read the text to think about it and to understand it. Therefore, the Companions would not have only transmitted the text without discussing and understanding the meaning.[47]

---

45 Muḥammad b. Sāliḥ b. ʿUthaymīn, *Uṣūl fī t-tafsīr* (N.p: al-Maktaba al-Islāmiyya, n.d.), 3.
46 See Ibn ʿUtahymīn, *Uṣūl.*
47 Muḥammad b. Sāliḥ b. ʿUthaymīn, *Sharḥ Muqadimmat at-tafsīr* (Riyad: Dār al-Waṭn, 2005), 21–27.

From this presumption, Ibn ʿUthaymīn concludes that the interpretation of the Qurʾān should be first and foremost with intertextual references to the Qurʾān itself. Many verses of the Qurʾān add some meaning or clarify the meaning of other verses. Some of them also specify what has been expressed elsewhere more generally, or even abrogate previously revealed statements. Since the Prophet is the first interpreter and the one who explained the meaning to his Companions, the Prophetic tradition needs to be consulted as well. Moreover, for Ibn ʿUthaymīn, the Prophetic tradition can only support the understanding of the Qurʾān. It cannot abrogate verses of the Qurʾān.[48] From what was mentioned earlier, Ibn ʿUthaymīn concludes that if no Prophetic tradition can be found to understand a certain verse, it is obligatory to look into the statements and opinions of the Companions. Since the Prophet had explained to them the meaning of the Qurʾān and since they had understood its meaning, the Prophetic explanations are contained within the sayings and interpretations of the Companions.[49] A consequence of the same argument, is that the Companions had transmitted the Qurʾānic text to the next generation, i.e. the Followers (*tabiʿūn*), together with all the explanations of the Prophet. Therefore, it is necessary to take into consideration the opinions of this generation as well, when facing difficulties to understand the Qurʾān.[50]

This method on the one hand, shall give scholars a path to follow when engaging with the Qurʾān. On the other hand, it also and probably more importantly is meant to remove the basis of the methodology of those who interpret the Qurʾān with reason (*tafsīr al-Qurʾān bi-r-raʾy*) without following the above-mentioned steps.[51]

4.2 Ibn ʿUthaymīn's *Tafsīr al-Qurʾān*

Up to this time,[52] the Ibn ʿUthaymīn Foundation has provided Ibn ʿUthaymīn's interpretation of the Qurʾān for almost 70% of the entire text. Only the following suras are still missing: Q 7–17, 19–23, 44–48, and 58–77.[53] They are still working on publishing the rest, which is actually not an easy task. Ibn ʿUthaymīn did not compile a Qurʾān commentary as his principal teacher ʿAbd ar-Raḥmān as-Saʿdī had done. And one can even doubt, that Ibn ʿUthaymīn has ever had the intention to do so. Rather, he explained the Qurʾān during sessions and classes on different occasions. In these ses-

48 Ibid., 127–29.
49 Ibid., 129–37.
50 Ibid., 138–40.
51 Ibid., 140–42.
52 This article was written in 2022.
53 See "Al-Mawqiʿ ar-rasmī li-Muʾassasa faḍīlat ash-shaykh Muḥammad b. Ṣāliḥ al-ʿUthaymīn al-khayriyya. At-Tafsīr wa-ʿulūm al-Qurʾān," binothaimeen.net, accessed December 30, 2022, https://binothaimeen.net/foundothemen/content/Mnue/lib?tid=1030.

sions he did not proceed from the beginning to the end, rather he chose parts of the Qur'ān depending on the occasion. For some suras there are no interpretation sessions. It is also difficult to assess how Ibn ʿUthaymīn proceeded his sessions, since the later compilation of the material brings material together that sometimes were distributed in various lessons. In the introduction to the interpretation of some suras, the editors mentioned that Ibn ʿUthaymīn used the *Tafsīr al-Jalālayn*[54] as basis of his lessons.[55] This, however, does not apply to the interpretation of all suras. Other suras go back to a series of lectures on the interpretation of the Qur'ān (*tafsīr*) that Ibn ʿUthaymīn gave in the Great Mosque of ʿUnayza beginning in 1989/1409 till 1998/1419.[56] Some of them were interpreted in a series of 'open door meetings' (*liqā'āt al-bāb al-maftūḥ*).[57] For some suras there are no information about when and how Ibn ʿUthaymīn interpreted them or from where the Foundation took the material.[58] Interestingly, one can notice some similarities in the style of how Ibn ʿUthaymīn proceeded the interpretation that seem to go back to the occasion of his lectures. Those parts of his *Tafsīr* that were based on the *Tafsīr al-Jalālayn* for example follow mostly the same structure. Ibn ʿUthaymīn begins with a linguistic analysis of the verse, then engages in explaining the meaning and mentions differences of opinions. He concludes the interpretation of a verse with enumerating and summarizing some benefits that can be deduced from it.[59] The other parts seem to be more intuitive and less struc-

54 A popular *tafsīr* work that contains the interpretations of two scholars, Jalāl ad-Dīn Muḥammad b. Aḥmad b. Muḥammad al-Maḥallī (d. 1460/864) and Jalāl ad-Dīn ʿAbd ar-Raḥmān b. abī Bakr b. Muḥammad as-Suyūṭī (d. 1505/911). The title *at-Tafsīr al-Jalālayn* refers to their names Jalāl ad-Dīn.

55 This applies to the following suras: 24, *an-Nūr* (Light) – 35, *Fāṭir* (The Originator); 39, *az-Zumar* (The Throngs) – 43, *az-Zukhruf* (Gold Ornaments). See Muḥammad b. Sāliḥ b. ʿUthaymīn, *Tafsīr al-Qur'ān al-karīm*, ed. Mu'assasat ash-Shaykh Muḥammad b. Ṣāliḥ al-ʿUthaymīn al-Khayriyya. Dammam: Dār Ibn al-Jawzī, 2005-.

56 This applies to the following *suras*: 3, *Āl al-ʿImrān* (The House of ʿImrān) from 1989–1991; 4, *an-Nisā'* (Women) from 1992–1996; 18, *al-Kahf* (The Cave) in 1998. Ṣāliḥ aṣ-Ṣāliḥ, a student of Ibn ʿUthaymīn, transcribed the recordings of this course and Ibn ʿUthaymīn authorized the transcription. Since Ibn ʿUthaymīn did not complete the entire sura during this course, aṣ-Ṣāliḥ added material of other recordings of Ibn ʿUthaymīn to complete the *tafsīr*. See Muḥammad b. Sāliḥ b. ʿUthaymīn, *Tafsīr al-Qur'ān al-karīm. Sūrat al-Kahf*, ed. Mu'assasat ash-Shaykh Muḥammad b. Ṣāliḥ al-ʿUthaymīn al-Khayriyya (Dammam: Dār Ibn al-Jawzī, 2022), 5–6.; Muḥammad b. Sāliḥ b. ʿUthaymīn, *Tafsīr al-Qur'ān al-karīm. Sūrat Yāsīn*, ed. Mu'assasat ash-Shaykh Muḥammad b. Ṣāliḥ al-ʿUthaymīn al-Khayriyya (Riyadh: Dār ath-Thurayā li-n-Nashr wa-t-Tawzīʿ, n.d.); Muḥammad b. Sāliḥ b. ʿUthaymīn, *Tafsīr al-Qur'ān al-karīm. Sūrat Ṣād*, ed. Mu'assasat ash-Shaykh Muḥammad b. Ṣāliḥ al-ʿUthaymīn al-Khayriyya (Riyadh: Dār ath-Thurayā li-n-Nashr wa-t-Tawzīʿ, n.d.)

57 This applies to the suras: 49, *al-Ḥujurāt* (The Private Apartments) – 57, *al-Ḥadīd* (Iron), see Muḥammad b. Sāliḥ b. ʿUthaymīn. *Tafsīr al-Qur'ān al-karīm. Al-Ḥujurāt, Qāf, adh-Dhāriyāt, aṭ-Ṭūr, an-Najm, al-Qamar, ar-Raḥmān, al-Wāqiʿa, al-Ḥadīd*, ed. Mu'assasat ash-Shaykh Muḥammad b. Ṣāliḥ al-ʿUthaymīn al-Khayriyya (Riyadh: Dār ath-Thurayā li-n-Nashr wa-t-Tawzīʿ, 2002), 5–6; as well as to the so-called *Juz' ʿAmma*, see Ibn ʿUthaymīn, *Tafsīr al-Qur'ān al-karīm. Juz' ʿAmma*, 6.

58 This is true for sura 5, *al-Mā'ida* (The Table Spread) and sura 6, *al-Anʿām* (The Cattle).

59 See the first example below.

tured. Ibn ʿUthaymīn explains the meaning of the verse as he understands it, apparently in the moment in which he speaks to his audience.

Usually, the Foundation commissions different individuals to transcribe the recordings and to compile the material. For the volume sura 1, *al-Fātiḥa* (The Opening) – sura 2, *al-Baqara* (The Cow) for example they commissioned Khālid b. Ḥāmid b. Khalīl, while for the volume sura 4, *an-Nisāʾ* (Women) they commissioned Ibrāhīm b. Muḥammad Qāsim al-Mayman with the transcription and edition. The Foundation unfortunately did not mention every individual they commissioned, but one can assume that the individuals were among the closer students of Ibn ʿUthaymīn himself.[60]

This all means that the *Tafsīr al-Qurʾān* is published in different individual volumes and not as one multi-volume work. This also means that the end product remains an enterprise of the Ibn ʿUthaymīn Foundation and is not something Ibn ʿUthaymīn has aimed for.

4.3 Examples of Ibn Uthaymīn's Interpretation of the Qurʾān

The above-mentioned principles and methods seem plausible and logical from the point of view of a Sunni (Salafi) scholar. However, in practice it is not that easy to implement them in every case since the Prophetic traditions as well as the opinions of the Companions and Followers are not that numerous to solely build on them the interpretation of the Qurʾān. Moreover, they can only little help to understand the text in the light of the present-day challenges. The examples of Ibn ʿUthaymīn's *Tafsīr al-Qurʾān* shall demonstrate how the Wahhābī scholar approached the interpretation of the Qurʾān in practice.

4.3.1 How Men and Women Should Behave

This example shows how norms and values that exist in the Saudi society, especially the specific clothing rules for women, are read into the Qurʾān. The interpretation of Ibn ʿUthaymīn includes verses 30–31 of sura 24, *an-Nūr* (Light). The English translation goes as follows[61]:

> (30) [O Prophet!] Tell the believing men to lower their gaze and guard their chastity. That is purer for them. Surely Allah is All-Aware of what they do. (31) And tell the believing women to

60 See Muḥammad b. Sāliḥ b. ʿUthaymīn, *Tafsīr al-Qurʾān al-karīm. Al-Fātiḥa – al-Baqara*, ed. Muʾassasat ash-Shaykh Muḥammad b. Ṣāliḥ al-ʿUthaymīn al-Khayriyya (Dammam: Dār Ibn al-Jawzī, n.d.) i–ii; and Muḥammad b. Sāliḥ b. ʿUthaymīn, *Tafsīr al-Qurʾān al-karīm. Sūrat an-Nisāʾ*, ed. Muʾassasat ash-Shaykh Muḥammad b. Ṣāliḥ al-ʿUthaymīn al-Khayriyya (Dammam: Dār Ibn al-Jawzī, 2009), 6.

61 The translation of the following passages is done by the author of this contribution.

> lower their gaze and guard their chastity, and not to reveal their adornments except what normally appears. Let them draw their veils over their chests, and not reveal their adornments except to their husbands, their fathers, their fathers-in-law, their sons, their stepsons, their brothers, their brothers' sons or sisters' sons, their fellow women, those in their possession, male attendants with no desire, or children who are still unaware of women's nakedness. Let them not stomp their feet, drawing attention to their hidden adornments. Turn to Allah in repentance all together, O believers, so that you may be successful.

Ibn ʿUthaymīn begins his interpretation with a linguistic analysis which means that he explains the grammar of the sentences and adds semantic clarification for words that weren't clear. Here, Ibn ʿUthaymīn often refers to other verses of the Qurʾān, to explain the meaning intertextually. In the next step, he begins to explain the meaning of certain passages. For example, regarding the first imperative "say," Ibn ʿUthaymīn states that this generally implies that the Prophet is commanded to speak to all humankind. However, since the verse continues by addressing the believers (*al-muʾminūn* and *al-muʾmināt*), the Prophet is only instructed to speak to Muslims with regard to the following content.[62]

Regarding the part "Tell the believing men to lower their gaze and guard their chastity," Ibn ʿUthaymīn explains that men have to lower the gaze. However, this is not and cannot always be an obligation (*wājib*). Rather, it is generally permissible (*jāʾiz*) to gaze. But since lowering the gaze is a means to prevent worse things (*sadd adh-dharāʾiʿ*), it can also be understood as an obligation with exceptions when it is necessary to gaze. In contrast, guarding the chastity is always an obligation.[63]

This is also true for women, who are advised to lower their gaze as well. In addition, however, they are not allowed "to reveal their adornments except what normally appears." Here Ibn ʿUthaymīn cites two opinions of what the exception should include. Some scholars allow for a woman to show her face and her hands, while others interpret the phrase "what normally appears" as the clothes that women wear to cover themselves. This implies that neither the face nor the hands are to be shown. Moreover, Ibn ʿUthaymīn explains that the "adornment" (*zīna*) refers to jewelry and other adornments that women put on, not to their outer appearance. He cites a number of verses in which the Qurʾān uses the word *zīna* in almost all cases with regard to additional adornment, whether it is money, offspring or jewelry etc. Therefore, only the women's clothes can be meant in this verse.[64] In addition, the women are obliged to cover their face and chest. This is because the word used for veil is the plural of *khimār* which covers the head of a women. One can either understand this verse as a command to cover the chest beginning from the neck of a women or to

---

62 See Muḥammad b. Sāliḥ b. ʿUthaymīn, *Tafsīr al-Qurʾān al-karīm. Sūrat an-Nūr*, ed. Muʾassasat ash-Shaykh Muḥammad b. Ṣāliḥ al-ʿUthaymīn al-Khayriyya (N.p.: Dār ad-Durra li-n-Nashr wa-t-Tawzīʿ, 2015), 164f.

63 See ibid., 165f.

64 See ibid., 166f.

cover the chest beginning from the head including the face. The latter, according to Ibn ʿUthaymīn is the right understanding, since the *khimār* begins from the head, so, how should a women cover her chest without also covering her head, he asks. Moreover, for Ibn ʿUthaymīn, a woman's face is actually more tempting than her chest, so that it should be covered all the more.[65]

After having gone through the entire verse in this manner, Ibn ʿUthaymīn adds a section on the benefits of these two verses. He speaks primarily about the fear of God (*taqwā*) that these verses imply and that should always be considered when engaging with the other sex. In addition, those who are not fearful do not possess a complete faith (*nāqiṣ al-īmān*). Eventually, the mentioned success at the end of the verse 31 can only be accomplished with the fear of God (*taqwā*).[66]

4.3.2 "Then Established Himself on the Throne"

This example demonstrates, how Ibn ʿUthaymīn speaks out of his Wahhābī identity and not seriously considers other opinions to be worth thinking about. In many verses, the Qur'ān mentions that God establishes Himself on or rises over the throne. In sura 57, *al-Ḥadīd* (Iron), Ibn ʿUthaymīn comments on verse 4:

> He is the One Who created the heavens and the earth in six Days, then established Himself on the throne. He knows whatever goes into the earth and whatever comes out of it, and whatever descends from the sky and whatever ascends into it. And He is with you wherever you are. For Allah is All-Seeing of what you do.

This commentary was part of the "open door meeting" and, therefore, is not as structured as the first example, which followed the outline of the *Tafsīr al-Jalālayn*. Ibn ʿUthaymīn begins by explaining why God created the heavens and the earth in six days instead of creating them in one moment.[67] Then he turns to the rising on the throne, where he states that this happened in a way that befits God. The throne is a created thing the material of which is unknown. So is the size of the throne, which only God knows. A Prophetic tradition states that the ratio of the heavens and the earth in comparison to the throne is like a ring that is thrown into a desert.[68] Ibn ʿUthaymīn moves on to state that the verb *istawā* indicates that God is above the throne since in all cases in the Qur'ān this verb takes this meaning. So, whoever interprets this expression in the sense that God has claimed the throne (*istawlā*) is therefore not correct. The Qur'ān is revealed in the Arabic language, Ibn ʿUthaymīn continues, and in the Arabic language "rising above" is always the meaning of *istawā*. Any

65 See ibid., 167f.
66 See ibid., 196–99.
67 See Ibn ʿUthaymīn, *Tafsīr al-Qur'ān al-karīm. Al-Hujurāt*, 363–65.
68 See ibid., 365f.

other interpretation is a lie and a violation of the apparent meaning (*ẓāhir*) of the word, in the eyes of Ibn ʿUthaymīn. According to him, mostly the Ashʿariyya establishes a meaning of the expressions that contradicts the apparent meaning (*ẓāhir*). And they, in the eyes of Ibn ʿUthaymīn, have gone astray. Ibn ʿUthaymīn poses a rhetorical question to demonstrate the falseness of their position.

> Who claimed the throne after the heavens and the earth had been created? If it is God, then this means that He hadn't claimed it before the creation of the heavens and the earth. Who had claimed the throne then before? The Ashʿariyya now must answer this question with someone else instead of God, otherwise the falseness of their own opinion would become obvious.[69]

After this remark Ibn ʿUthaymīn moves on to explain the next verse.

5 Concluding Remarks

Ibn ʿUthaymīn is undeniably one of the most influential and acknowledged Wahhābī scholars of the second half of the twentieth century in Saudi Arabia, and in Salafi circles outside the kingdom. He was particularly known for his accomplishments in the disciplines of Islamic jurisprudence and Islamic creed. His accomplishments in the field of *tafsīr* are not less crucial. However, Ibn ʿUthaymīn did not leave a complete *tafsīr* work. Rather, he left his contributions in many recordings of numerous lessons and lectures that were dedicated on the interpretation of certain suras or that contain the interpretation of the Qurʾān. After his death, the Ibn ʿUthaymīn Foundation began to put the material together and published the interpretation of suras in books. For this, they commissioned mostly former students of Ibn ʿUthaymīn to transcribe the recordings and to bring together material from other recordings to fill potential gaps. The result is a publication of about 70% of the Qurʾān so far.

As for Ibn ʿUthaymīn's methodology and principles in interpreting the Qurʾān, he follows generally Ibn Taymiyya. Moreover, Ibn ʿUthaymīn had prepared a commentary on Ibn Taymiyya's *Muqaddimat at-tafsīr*, and also wrote a treatise on the *ʿulūm al-Qurʾān* entitled *Uṣūl fī t-tafsīr*. Ibn ʿUthaymīn's style in interpreting the Qurʾān, though, depended strongly on the original format in which he produced his *tafsīr*, i.e., the oral presentation of it. Some of his lessons were based on the *Tafsīr al-Jalālayn*, and therefore seem to follow its outline. Other lessons were given more intuitive and are less structured. While Ibn ʿUthaymīn claims to follow the method established by Ibn Taymiyya and others, which is supposed to be the method of the *ahl as-sunna wa-l-jamāʿa*, the examples of his *Tafsīr* demonstrate how he brings his Wahhābī identity into his *Tafsīr al-Qurʾān*.

69 See ibid., 366f.

6 Bibliography

6.1 Sources

Ibn ʿAbd al-Wahhāb, Muḥammad. *Kitāb at-Tawḥīd*. Ed. Aḥmad b. ʿAlī l-Qufaylī. Cairo: Maktabat al-ʿUlūm wa-l-Ḥikam, 2008.

Ibn ʿUthaymīn, Muḥammad b. Sāliḥ. *Tafsīr al-Qur'ān al-karīm. Al-Hujurāt, Qāf, adh-Dhāriyāt, aṭ-Ṭūr, an-Najm, al-Qamar, ar-Raḥmān, al-Wāqiʿa, al-Ḥadīd*. Ed. Mu'assasat ash-Shaykh Muḥammad b. Ṣāliḥ al-ʿUthaymīn al-Khayriyya. Riyadh: Dār ath-Thurayā li-n-Nashr wa-t-Tawzīʿ, 2002.

Ibn ʿUthaymīn, Muḥammad b. Sāliḥ. *Tafsīr al-Qur'ān al-karīm. Juz' ʿAmma*. Ed. Fahd b. Nāṣir as-Sulaymān. Riyadh: Dār ath-Thurayā li-n-Nashr wa-t-Tawzīʿ, 2002[2].

Ibn ʿUthaymīn, Muḥammad b. Sāliḥ. *Fatḥ rabb al-bariyya bi-talkhīṣ al-ḥamawiyya*. Ed. Mu'assasat ash-Shaykh Muḥammad b. Ṣāliḥ al-ʿUthaymīn al-Khayriyya. Dammam: Dār Ibn al-Jawzī, 2003.

Ibn ʿUthaymīn, Muḥammad b. Sāliḥ. *Sharḥ al-ʿAqīda as-saffārīniyya ad-Durra al-muḍiyya fī ʿaqd ahl al-firqa al-marḍiyya*. Ed. Mu'assasat ash-Shaykh Muḥammad b. Ṣāliḥ al-ʿUthaymīn al-Khayriyya. Riyadh: Madār al-Waṭan li-n-Nashr, 2005.

Ibn ʿUthaymīn, Muḥammad b. Sāliḥ. *Tafsīr al-Qur'ān al-karīm. Sūrat an-Nisāʾ*. Ed. Mu'assasat ash-Shaykh Muḥammad b. Ṣāliḥ al-ʿUthaymīn al-Khayriyya. Dammam: Dār Ibn al-Jawzī, 2009.

Ibn ʿUthaymīn, Muḥammad b. Sāliḥ. *Sharḥ al-ʿAqīda al-wāsiṭiyya*. Ed. Saʿd b. Fawwāz aṣ-Ṣumayl. 2 vols. Dammam: Dār Ibn Jawzī, 2010[6].

Ibn ʿUthaymīn, Muḥammad b. Sāliḥ. *Sharḥ al-uṣūl*. Dammam: Dār Ibn al-Jawzī, 2010/1431.

Ibn ʿUthaymīn, Muḥammad b. Sāliḥ, *Sharḥ Muqadimmat at-tafsīr*, Riyad: Dār al-Waṭn, 2005

Ibn ʿUthaymīn, Muḥammad b. Sāliḥ. *Taqrīb at-Tadmuriyya*. Ed. Mu'assasat ash-Shaykh Muḥammad b. Ṣāliḥ al-ʿUthaymīn al-Khayriyya. Riyadh: Madār al-Waṭan li-n-Nashr, 2012.

Ibn ʿUthaymīn, Muḥammad b. Sāliḥ. *Tafsīr al-Qur'ān al-karīm. Sūrat al-Aḥzāb*. Ed. Mu'assasat ash-Shaykh Muḥammad b. Ṣāliḥ al-ʿUthaymīn al-Khayriyya. Dammam: Dār Ibn al-Jawzī, 2015.

Ibn ʿUthaymīn, Muḥammad b. Sāliḥ. *Tafsīr al-Qur'ān al-karīm. Sūrat an-Nūr*. Ed. Mu'assasat ash-Shaykh Muḥammad b. Ṣāliḥ al-ʿUthaymīn al-Khayriyya. N.p.: Dār ad-Durra li-n-Nashr wa-t-Tawzīʿ, 2015.

Ibn ʿUthaymīn, Muḥammad b. Ṣāliḥ. *Ash-Sharḥ al-mukhtaṣar ʿalā Bulūgh al-marām*. Ed. Mu'assasat ash-Shaykh Muḥammad b. Ṣāliḥ al-ʿUthaymīn al-Khayriyya. 3 vols. Qaṣīm: Mu'assasat ash-Shaykh Muḥammad b. Ṣāliḥ al-ʿUthaymīn al-Khayriyya, 2019[2].

Ibn ʿUthaymīn, Muḥammad b. Sāliḥ. *Tafsīr al-Qur'ān al-karīm. Sūrat al-Kahf*. Ed. Mu'assasat ash-Shaykh Muḥammad b. Ṣāliḥ al-ʿUthaymīn al-Khayriyya. Dammam: Dār Ibn al-Jawzī, 2022.

Ibn ʿUthaymīn, Muḥammad b. Sāliḥ. *Tafsīr al-Qur'ān al-karīm. Al-Fātiḥa – al-Baqara*. Ed. Mu'assasat ash-Shaykh Muḥammad b. Ṣāliḥ al-ʿUthaymīn al-Khayriyya. Dammam: Dār Ibn al-Jawzī, n.d.

Ibn ʿUthaymīn, Muḥammad b. Sāliḥ. *Taʿlīq mukhtaṣar ʿalā Kitāb Lumʿat al-iʿtiqād al-hādī ilā sabīl ar-rashād*. Ed. Mu'assasat ash-Shaykh Muḥammad b. Ṣāliḥ al-ʿUthaymīn al-Khayriyya. Riyadh: Madār al-Waṭan li-n-Nashr, n.d.

Ibn ʿUthaymīn, Muḥammad b. Sāliḥ. *Uṣūl fī t-tafsīr*. N.p: al-Maktaba al-Islāmiyya, n.d.

6.2 Secondary Literature

Alshamsi, Mansoor Jassem. *Islam and Political Reform in Saudi Arabia. The Quest for Political Change and Reform*. New York: Routledge, 2011.

Al-Atawneh, Muhammad. *Wahhābī Islam Facing the Challenges of Modernity. Dār al-iftā' in the Modern Saudi State*. Leiden: Brill, 2010.

Al-Bassām, ʿAbd Allāh b. ʿAbd ar-Raḥmān b. Ṣāliḥ. *ʿUlamāʾ Najd khilāl thamāniya qurūn*. 6 vols. Riyadh: Dār al-ʿĀṣima, 1998^{2}.

Fürtig, Henner. *Demokratie in Saudi-Arabien? Die Āl Saʿūd und die Folgen des zweiten Golfkrieges*. Berlin: Verlag Das Arabische Buch, 1995.

Gharaibeh, Mohammad. *Zur Attributenlehre der Wahhābiyya unter besonderer Berücksichtigung der Schriften Ibn ʿUthaimīns (1929–2001)*. Berlin: EB-Verlag, 2011.

Al-Ḥusayn, Walīd b. Aḥmad. *Al-Jāmiʿ li-ḥayāt al-ʿallāma Muḥammad b. Ṣāliḥ al-ʿUthaymīn al-ʿilmiyya wa-l-ʿamaliyya wa-mā qīla fīhi min al-marāthī*. Leeds: Majallat al-Ḥikma, 2002.

Layish, Aharon. "ʿUlamāʾ and politics in Saudi Arabia." In *Islam and politics in the Modern Middle East*, edited by Metin Heper and Raphael Israeli. London: Routledge, 1984: 29–63.

Ar-Raḥma, ʿAbd ar-Raḥmān. *Al-Injāz fī tarjamat al-imām ʿAbd al-ʿAzīz b. Bāz ḥayātuhu wa-juhūduhu al-ʿilmiyya wa-l-ʿamaliyya wa-d-daʿwa wa-āthāruhu al-ḥamīda*. Dammam: Dār Ibn al-Jawzī, 2007.

Al-Rawwaf, Othman Yakub. "The concept of five crises in political development: Relevance to the Kingdom of Saudi Arabia." PhD diss., Duke University, 1981.

Ash-Shāyiʿ, Muḥammad b. ʿAbd al-ʿAzīz b. Muḥammad. *Dalīl al-maktaba al-ʿaqdiyya. Muʿjam mawḍūʿī li-l-kutub wa-r-rasāʾil wa-l-buḥūth fī l-ʿaqīda*. Riyad: Dār Zidnī, 2007.

Steinberg, Guido. *Religion und Staat in Saudi-Arabien. Die wahhabitischen Gelehrten 1902–1953*. Würzburg: Ergon, 2002.

Aṭ-Ṭayyār, ʿAbd Allāh b. Sābiḥ. *Athar ʿallāma al-Qaṣīm ash-Shaykh ʿAbd ar-Raḥmān b. Nāṣir as-Saʿdī ʿalā l-ḥaraka al-ʿilmiyya al-muʿāṣira*. Dammam: Dār Ibn al-Jawzī, 1992.

Aṭ-Ṭayyār, ʿAbd Allāh b. Sābiḥ. *Ash-Shaykh ʿAbd ar-Raḥmān b. Nāṣir as-Saʿdī mufassiran*. Dammam: Dār Ibn al-Jawzī, 2000.

Al-Yassini, Ayman. *Religion and State in the Kingdom of Saudi Arabia*. Boulder: Westview Press, 1985.

6.3 Internet Sources

Binothaimeen.net. "Muḥammad b. Ṣāliḥ al-ʿUthaymīn." Accessed December 30, 2022. https://binothaimeen.net/.

Binothaimeen.net. "Al-Mawqiʿ ar-rasmī li-Muʾassasa faḍīlat ash-shaykh Muḥammad b. Ṣāliḥ al-ʿUthaymīn al-khayriyya." Accessed December 30, 2022. https://binothaimeen.net/foundothemen/.

Binothaimeen.net. "Al-Mawqiʿ ar-rasmī li-Muʾassasa faḍīlat ash-shaykh Muḥammad b. Ṣāliḥ al-ʿUthaymīn al-khayriyya. Mabīʿāt al-kutub." Accessed December 30, 2022. https://binothaimeen.net/foundothemen/content/Mnue/lib.

Binothaimeen.net. "Al-Mawqiʿ ar-rasmī li-Muʾassasa faḍīlat ash-shaykh Muḥammad b. Ṣāliḥ al-ʿUthaymīn al-khayriyya. At-Tafsīr wa-ʿulūm al-Qurʾān." Accessed December 30, 2022. https://binothaimeen.net/foundothemen/content/Mnue/lib?tid=1030.

Islamway.com. "Ḥayāt al-faḍīla al-ʿallāma Muḥammad b. Ṣāliḥ b. ʿUthaymīn." Accessed March 11, 2024. http://www.islamway.com/?iw_s=Lesson&iw_a=view&lesson_id=86327.

Aṣ-Ṣuqayr, Sāmī. "Ibn ʿUthaymīn wa-ṭalabuhu li-l-ʿilm." Lecture in the Ibn ʿUthaymīn-Mosque in ʿUnayza on June 30, 2007. https://al-soger.com/ar/show/sounds/1c10f490-cf87-4dcc-9218-910b48ad53e0/%D8%A7%D8%A8%D9%86_%D8%B9%D8%AB%D9%8A%D9%85%D9%8A%D9%86_%D9%88_%D8%AD%D9%8A%D8%A7%D8%AA%D8%A9_%D8%A7%D9%84%D8%B9%D9%84%D9%85%D9%8A%D8%A9 [Accessed on March 11, 2024]

Jaʿfar Faḍlallāh

Muḥammad Ḥussein Faḍlallāh

1 Introduction: Life, Works, and Political and Intellectual Context

Sayyid Muḥammad Ḥussein Faḍlallāh was born in the city of Najaf, Iraq, on November 16, 1935/Shaʿban 19, 1354.[1] He received his primary education at an academic school until the fourth grade, after which he decided to devote himself entirely to the religious sciences, which he both studied and taught, at a time when he was committed to diversifying his cultural interests. Hence, he did not restrict himself to the official seminary program of study, namely jurisprudence (*fiqh*) and legal theory (*uṣūl al-fiqh*), but also attended various lectures on subjects such as exegesis and philosophy.

He also opened himself up to contemporary Arab thought and literature through the Egyptian *al-Muṣawwar* magazine, *ar-Risāla* by Ḥasan az-Zayyāt, the magazines published by Egyptian author Taha Hussein, and some Lebanese magazines that would arrive in Najaf from time to time, as well as Iraqi newspapers. He also exposed himself to Western thought through the translations of writings by Western thinkers of various ideologies that were available at the time. This is what provided Faḍlallāh with a broad outlook that went beyond the role and interests of a traditional cleric.

From a very young age, Faḍlallāh was obsessed with endorsing Islamic civilization and culture, and this is reflected in his very first writings,[2] now considered to be Qurʾānic studies of a substantive nature. The first of his books to be published was *Uslūb ad-daʿwa fī-l-Qurʾān* (The Method of Propagation in the Qurʾān). He collated his writings for publication in *al-Aḍwāʾ* magazine, which was issued in Iraq, which were later compiled in the book *Qaḍāyānā ʿalā ḍawʾ al-Islām* (Our Issues in Light of Islam). He then wrote *al-Ḥiwār fī-l-Qurʾān* (Dialogue in the Qurʾān), *al-Islam wa-manṭiq al-quwwa* (Islam and the Concept of Power), and *Khuṭuwāt ʿalā ṭarīq al-Islam* (Steps on the Path of Islam). He also composed numerous articles theorizing the nascent Islamic movement in Lebanon for *al-Munṭalaq* magazine, which were later compiled in the book *al-Ḥaraka al-islāmiyya: humūm wa-qaḍāyā* (The Islamic Movement: Concerns

1 For further information, see: "Sīrat as-sayd," *bayynat.org*, accessed May 6, 2024, https://www.bayynat.org.lb/article/category/1974/%D8%B3%D9%8A%D8%B1%D8%A9-%D8%A7%D9%84%D8%B3%D9%8A%D8%AF/ar%D8%9B; See also: *Muʾassasat al-marjaʿiyya injāzāt wa-āmāl* (Beirut: 2012[4]), 5ff.

2 Ibid., 58–60; For a bibliography of Faḍlallāh's writings, see also "al-Maktaba," *bayynat.org*, accessed May 6, 2024, https://www.bayynat.org.lb/library/category/25922/%D8%A7%D9%84%D9%85%D9%83%D8%AA%D8%A8%D8%A9/ar.

Translated into English by Nazmina Dhanji

https://doi.org/10.1515/9783111320069-003

and Issues). His writings and statements on Islamic unity were also collected in the book *Aḥādīth fī qaḍāyā al-ikhtilāf wa-l-waḥda* (Discourses on Issues of Difference and Unity), and his work *Fī āfāq al-ḥiwār al-islāmī l-masīḥī* (On the Horizons of Islamic-Christian Dialogue) was published as well.

During the second period of his life, after he permanently resettled from Najaf to Lebanon, he gave lectures in jurisprudence and legal theory (*uṣūl al-fiqh*) in Beirut and Damascus. A large number of these lectures and discussions were subsequently published, covering topics such as the judiciary, pilgrimage, marriage, laws of employment, divorce, and other jurisprudential topics that reflected his innovative approach, which distinguished him from the traditional approach. He also published a book about important theories of *ijtihād* titled *al-Ijtihād bayna asr al-māḍī wa-āfāq al-mustaqbal* (Independent Legal Reasoning between the Shackles of the Past and the Horizons of the Future).

Faḍlallāh stood out due to his activism, and he participated in many conferences outside Lebanon too. His door was open to people from all levels of society. The political challenges that Lebanon and the region went through, and his involvement in them in the position of an Islamic theorist rather than a politician, made him a sought-after Islamic authority for many political parties and media outlets, and his contributions were later published in books.

He also had four anthologies of poems published, titled *Ya ẓilāl al-Islām* (O Shadows of Islam), *Qaṣā'id al-Islam wa-l-ḥayāt* (Poems of Islam and Life), *'Alā shāṭi' al-wijdān* (On the Shore of Conscience), and *Fī durūb as-sab'īn* (On the Paths of the Seventy). These anthologies included poems that he wrote in Najaf as well as in Lebanon, and alternated between metered poetry and free verse.

Sayyid Faḍlallāh passed away in Beirut on July 4, 2010/Rajab 22, 1431, after suffering an illness.

2 The Personality of the Exegete

There is no doubt that the personality of a writer influences their intellectual output, whatever field they may belong to. There are several features that distinguish the personality of Faḍlallāh as the author of the exegetical work *Min waḥy al-Qur'ān* (From the Inspiration of the Qur'ān):

1. His jurisprudential legal reasoning (*al-ijtihād al-fiqhī*), meaning his ability to derive the Islamic vision, idea, and theory from their primary sources, specifically the Qur'ān and the hadith. In short, the diligent mind that is dissatisfied with superficial meaning, and which loves to delve deep into the folds of meaning and the depths of content, making connections and analyzing. It is able to distance itself from the conclusions of its precursors and present its own conclusions independently, which may either disagree or correlate with theirs.

2. His intellectual openness and awareness of diverse sciences, intellectual views, and readings from multiple viewpoints, which were well ahead of their time.
3. The outreach aspect, by which I mean his desire to generally educate the masses, which is what motivated Faḍlallāh to disseminate knowledge to different classes of people and diverse segments of society. This is something that the author committed himself to very early on, especially in Najaf, at the beginning of his studies in the religious sciences, where it was the norm for students to teach others what they had just learned and to convey knowledge to whoever needed it.
4. The fact that he had been a writer and a poet since childhood, having clung to reading and poetry as a form of emotional expression during the various stages of his life. This paved the way for his approach to the Arabic text as a direct literary and sensory experience rather than a mere simulation and regurgitation of the meanings provided by dictionaries and lexicons. It also gave Faḍlallāh the advantage of not needing to restrict his exegetical approach to a rigid intellectual treatment separate from the text, instead allowing his literary prowess to shine through in the beauty of his expression, his literary spirit, the power of his eloquence, and his sense of poetry, allowing the reader to interact with the interpretive text in multiple dimensions, and not just with an abstract idea that revolves around an isolated Qur'ānic word or expression.
5. His conduct as an Islamic activist: Faḍlallāh believed that Islam is a way of life and must take its rightful place in the ranks of politics, society, economics, security, culture, etc. His distinguishing characteristic here was that he was actively involved in the workings of daily life and in activism, and did not merely practice his faith from a position far removed from reality. As an exegete, this provided him with a deep, lived experience on the frontlines of reality and in direct contact with the issues of modern society, its current problems, its mentality, and rapid developments. It also equipped him with an awareness of the Islamic movement and its need for authentic and contemporary Islamic intellectual theorizing, as well as an awareness of the conflicts that the Islamic community were facing at that particular point in time, with its ongoing civilizational, economic, political, and cultural changes.

3 The Main Feature of Faḍlallāh's Commentary *Min waḥy al-Qur'ān*

In his introduction to *Min waḥy al-Qur'ān*, Muḥammad Hādī Ma'rifat writes,

> It is a comprehensive educational and social exegesis, and is considered one of the most wonderful and most comprehensive exegetical works, emanating from a dynamic spirit vibrant with ancient Islamic vitality, in which the author, Sayyid Muhammad Hussein Faḍlallāh, one of the most

brilliant Islamic scholars in Lebanon, is working on reviving the Qur'ānic atmosphere in all areas of material and spiritual life.[3]

Min waḥy al-Qur'ān can be classified as a holistic exegesis that aims to interpret each verse individually, though it is written in the style of a thematic exegesis, focusing on treating specific subject matters in light of other verses connected and relevant to the original verse in question. Hence, whenever the need arises, Faḍlallāh can launch into a thematic exegesis as required. This is essentially the approach that he adopted in his treatment of several subjects: doctrinal, legal, and intellectual matters of contemporary importance, such as polygamy, universal authority (*al-wilāya at-takwīniyya*), the death penalty, usury in banks, the female form, etc.

The composition of *Min waḥy al-Qur'ān* went through three stages: During the first stage, it consisted of exegetical lectures, successively edited for publication in twenty-four small volumes, the first of which was published in 1979/1399. The second stage was one of revision and expansion, in which the author added hundreds of pages to the first edition. The exegesis contained the same number of parts, but in larger volumes. It was published in 1998/1419. The third stage saw new additions made to the second edition written by the author. Although this version included the addition of approximately two thousand pages, the number of volumes was reduced to twenty instead of twenty-four, after reformatting, reducing white space, and making the book more reader friendly. In order to provide the reader with broader jurisprudential understanding of the verses, the author added content drawn from his own jurisprudential research when interpreting certain verses.[4]

4 Faḍlallāh's Hermeneutical Method

Faḍlallāh's personality is by no means absent from his intellectual masterpieces, and it is what gives *Min waḥy al-Qur'ān* its flavor. It is therefore only natural that we find the various dimensions of his personality, the myriad intellectual interests that he had, and the position that he occupied within his historical milieu reflected in his exegetical approach. In order to examine these aspects of his life and work, I will focus on several sub-topics, which I will present below:

3 Muḥammad Hādī Ma'rifat, *at-Tafsīr wa-l-mufassirūn fī thawbihi al-qashīb* (Mashhad: Mu'assasat aṭ-Ṭab' wa-n-Nashr fī l-Āstāna ar-Raḍawiyya al-Muqaddasa, 2004), 2: 474.

4 Muḥammad Ḥussein Faḍlallāh, *Tafsīr min waḥy al-Qur'ān* (Beirut: Dār al-Malāk li-ṭ-Ṭibā'a wa-n-Nashr wa-t-Tawzī', 2017), 1: 9.

4.1 The Sanctity of the Qur'ānic Text and Human Understanding

Faḍlallāh believed that the belief in the sacredness of the Qur'ān did not prevent it from being studied for all its intrinsic characteristics and artistic merits, and that understanding the Qur'ān was subject to the rules of the Arabic language, with all its implications and inferences. For, as Faḍlallāh writes, God sent it down in Arabic, with its language, style, and discursive approach directed at the people, that they might ponder it with the intellectual, mental, and cultural tools that they have at their disposal. This is what makes human understanding vulnerable to misunderstanding the Qur'ānic text's face-value meanings. Such interpretations may be supported by apparent evidence, or they may remain fleeting guesses in the reader's mind. This is what the *uṣūlī* scholars (experts in legal theory) convey in their description of the Qur'ān as "having a definitive chain of transmission yet being speculative significance," which is why they consider differences in Qur'ānic interpretation to be legitimate and do not consider fallible human understanding to be sacred or incontestable.[5]

It is self-evident that the issue of understanding the Qur'ān requires as much impartiality and detachment from prior convictions as possible, especially since many of the exegete's previous beliefs may have been formed from multiple sources and included as fundamental components of his or her intellectual makeup. This is especially important in doctrinal, legal, and ethical fields related to religious knowledge, where the apparent meaning of the Qur'ān may differ from the exegete's own beliefs, and subjecting the Qur'ān to them may therefore end up becoming an exercise in subjecting the Qur'ān to philosophy, mysticism, common jurisprudence, etc.

But is it possible for human beings to rid themselves of all subjectivity? Faḍlallāh asserts that this is something that requires great effort, and that it is very difficult for someone to completely detach themselves from their own self and subjectivities. Moreover, he says, it is natural for there to be a deficiency in people's cultural and educational backgrounds when they attempt to understand things. That is when subjective elements come into play, even for most intellectuals, for whom there is a clear intermingling between objective truth and subjective elements.[6]

5 Muḥammad Ḥussein Faḍlallāh, *al-Ijtihād bayna asr al-māḍī wa-āfāq al-mustaqbal* (Beirut: al-Markaz ath-Thaqāfī l-ʿArabī, 2009), 176.
6 Ibid., 235.

4.2 The Limited Text of the Qurʾān and the Breadth of Understanding

The Qurʾān cannot be described in terms used to refer to God, such as "Unlimited," "Infinite," or "Absolute," because it is limited in its words, styles, and rules, just as humans, animals, and all natural phenomena in the sky and on Earth are limited. However, its limited language does not prevent us from understanding its conventional content in a scholarly way, in the sense of resorting to the interpretation of its apparent meaning wherever the apparent meaning indicates something rationally unacceptable or impossible. This approach assumes that, if it did present something impossible, it would mean that something else had been intended, as what God says aligns with what the mind deems possible.

This is what opened the door to interpretation (*taʾwīl*) by way of metaphor in cases that Muslim scholars deemed impossible, i.e., certain things that could not be attributed to God, such as the issue of predestination, anthropomorphism, and the like. They therefore decided to interpret anthropomorphic words such as "face," "hand," "being seated on the Throne," and "looking at God" as meaning essence, power, giving, dominating, and looking at God's signs and the manifestations of His grandeur respectively, in accordance with the rhetorical devices of metaphor and allegory.[7]

At the same time, Faḍlallāh does not negate the reader's role in the text of grasping meanings beyond

> the literal linguistic interpretation of the word, which does not roam in the broad horizons of how meaning naturally moves through the mind or out in reality, but rather remains frozen within the linguistic meaning. In actual fact, we find that meaning, by nature, has horizons that a person cannot help but imagine with the entirety of his cultural and educational background in all its dimensions.[8]

He gives the example of the word "forest," which does not conjure up the "mere image of a forest limited to its physical meaning, but rather the conceptual imagination roams to all the areas in which forests exist and to all the animals that they are home to, both prey and predator, and to all the negative and positive feelings that humans associate with it."[9] Furthermore, Faḍlallāh notes that

> the scope of imagination in its depth and breadth expands and develops according to the information that the speaker or listener stores within the dimensions and inferences of the meaning. Indeed, we may find that the listener understands the word through his broad cultural perceptions more than the speaker may understand it, since he may be limited in his cultural experience of the earth as regards the meaning indicated by the word.[10]

---

7 Ibid., 176.
8 Ibid., 173.
9 Ibid.
10 Ibid., 174.

On this basis, Faḍlallāh does not deny that the meaning may have "several depths according to the awareness of the recipient, for example, the concept of monotheism (there is no god but God). This statement indicates monotheism, and a person may understand it in a simplistic way, or he may understand it in its deep meaning, with all the details outlined by the theologians."[11]

Additionally, Faḍlallāh believes that awareness of the process of life is part and parcel of the process of understanding the Qur'ān because the Qur'ān is the book of the Islamic mission, which was not revealed in the form of a complete book. Rather, it kept pace with the movement of the mission of Islam, addressing every problem that emerged as the historical occurrences in the text were taking place. The mission – being in itself a life experience – can be extended to all other experiences in the future, which requires a thorough understanding of it, "not only by way of the dictionary, which merely provides the meanings of words, but through reality itself."[12] This means that – according to Faḍlallāh – the activists "who live Islam as the very foundation of thought, emotion, and life" are the most capable of understanding it since they "delve into it on the basis of activism in real life."[13]

4.3 The Pivotal Role of the Arabic Language in Exegesis

The fact that the Qur'ānic text is in the Arabic language is a pivotal focus in Faḍlallāh's own exegetical methodology as well as in his discussions of the exegetical approaches and methodologies presented by other exegetes, and of his own position on interpretation (*ta'wīl*). This is to the point that he comes across as zealous regarding anything that strikes at the centrality of the Arabic language and renders it marginal to the meaning offered by the exegete or interpreter. According to Faḍlallāh, we find several parameters in this field:

1. The Arabic language, he says, goes beyond the direct, literal, and linguistic meaning that focuses on the abstract word and instead stands on a foundation that moves through diversity in its styles, "in [its] expression, understanding, and atmosphere, in terms of artistic characteristics that may carry within it suggestion, connotation, inference, and indication"[14] that surpass the literal meaning of the individual words. The basis for this is that the historical aspect of their usage and their peculiarities can also add many more shades of meaning to them, giving them a new flavor. This is what has been termed "customary understanding" or

---

11 Ibid., 233.

12 Muḥammad Ḥussein Faḍlallāh, *an-Nadwa* (Beirut: Dār al-Malāk li-ṭ-Ṭibā'a wa-n-Nashr wa-t-Tawzī', 2000), 4: 403.

13 Ibid.

14 Faḍlallāh, *Min waḥy al-Qur'ān*, 1: 12.

"customary flavor,"[15] referring to the understanding that is based on the societal milieu of the time and the way that meanings are expressed and understood by its members.

2. Faḍlallāh considers the law of mutual understanding and comprehension between the speaker and the listener to be a foundation of interpretation; consequently, his exegesis deals with the text on the principle of

> clarity of significance, whether it is actual or metaphorical, whereby the Qur'ānic speech acts as a proof in having conveyed its message and legislation to the people. There is no room for verbal or semantic complexity, even in stylistic features of metaphor, metonymy, or structural composition, such that the distance between the obliger and the obligated, or between the literal content of the word and the goal intended by the speaker, is so vast that it requires great mental effort to make connections between them, because this would depart from the explanatory approach that the very concept of understanding presumes, and on which the concept of language rests due to the fluidity of its nature.[16]

Thus, the role of the interpreter or researcher is

> to determine the actual lines along which the meaning of the verse proceeds, through the various moods of Arabic, bearing in mind the rule of mutual understanding and comprehension, since the Qur'ānic verses do not proceed along the lines of verbal or moral complexity, or merely symbolic indication that does not clearly impart understanding to the people.[17]

3. There is no objection to resorting to a meaning that is not immediately apparent from the Qur'ān's speech, but it is only acceptable through "presumptive evidence within the speech itself or surrounding it."[18] This is what is known as communicative and current presumptive evidence, meaning that these "presumptions" must have a clear relationship to the text:

> It is not enough to ascribe meanings to the text based solely on personal impressions . . . because the text represents an existing state that expresses the awareness of a certain concept with all its related internal or external elements. The tools of expression must therefore be available to represent that whole concept. That is why the process of communication and mutual understanding must be such that the significance is both linked to the text and its surrounding circumstances as well as to the human being, whereby, when the listener hears it and contemplates it all, he understands the concept that is being expressed in all the components of its significance.[19]

15 Ibid.
16 Ibid., 13.
17 Ibid., 14.
18 Faḍlallāh, *al-Ijtihād*, 281.
19 Ibid.

It is perhaps for this reason that Faḍlallāh has strong reservations

> about much of what is raised in Qur'ānic exegesis in accordance with scientific theories, whose validity within their respective scientific fields is still subject to controversy, which forces us to pause carefully before interpreting the Qur'ān according to this or that theory, since it may contribute to making exegesis more of an exercise in projecting meaning onto the Qur'ānic text than one that extracts meaning from it.[20]

4. Faḍlallāh does not accept the idea that propounds[21] the necessity of esoteric exegesis, i.e., that God's intention cannot be comprehended by words or language in general, and calls for discerning meanings to be derived outside the framework of what is apparent from the text. He focuses on the fact that the Qur'ān was revealed "'in clear Arabic language' (Q 26:195), i.e., that God wanted to clarify to us what He wants through language and whatever meanings language can carry. As for the meanings that language cannot express, there is no room to communicate them by way of it, and the idea that God would have intended those specific meanings is untenable." The fact that God's knowledge is infinite simply means that "God has not chosen to reveal to us all of this knowledge in the Qur'ān,"[22] in order to justify the need for what is broader than the scope of language in mutual understanding and comprehension.
5. Faḍlallāh does not object to pursuing "every new insight that can make us aware of the rules of the Arabic language in understanding the Qur'ān, from a linguistic standpoint, the standpoint of rhetorical rules, and so on." However, as mentioned above, he rejects what is not compatible with that which the rules of understanding the Qur'ānic text naturally necessitate, as contravening them would make it "cumbersome for us to understand the Qur'ān, and we would not be open to its apparent meaning."[23]
6. The Arabic language is distinguished, like other languages, by the fact that its individual and syntactic connotations are subject to special rules of linguistic vocabulary, grammar, and rhetoric, which makes it possible to exert effort in mastering it, according to various scholarly theories that have consolidated its rules in various locations. This is because the beginnings of the study of the Arabic language did not originate in a sacred place, but rather arose through various diligent efforts in observation, analysis and human understanding, which allows it to be subject to critique when it comes to determining correct and incorrect usage.[24]

---

20 Faḍlallāh, *Min waḥy al-Qur'ān*, 1: 166.
21 This issue appears in a question that was posed to the author; see Faḍlallāh, *al-Ijtihād*, 233.
22 Ibid., 233.
23 Faḍlallāh, *an-Nadwa*, 1: 249.
24 Faḍlallāh, *al-Ijtihād*, 171 (with a minor amendment).

Based on all this, we can conclude that Faḍlallāh formulated his exegetical approach and his position toward interpretation (*ta'wīl*) and the relationship between the Qur'ānic text and the sources of Islamic or human knowledge in general based on the authenticity of the Qur'ānic text and its Arabic-ness in the sense mentioned above, and their pivotal role in all of this. The following points all take this position into account.

4.4 Faḍlallāh's Position on Interpretation (*ta'wīl*)

Based on the fact that the Qur'ān is in the Arabic language, as outlined above, Faḍlallāh rejects any understanding of the Qur'ān as a book of symbols and terms that are stylistically far removed from the general context of Arabic language styles, because the possibility that the intended meaning of the verse is something other than its apparent meaning might cause the interpreter to lose confidence in the text at face value. In doing so, he distinguishes himself from many Shiite and Sufi exegetes, especially the proponents of esoteric interpretation.

It is true that words are symbols of meaning that the human mind evokes according to the relationship it forms between a specific word and its meaning, but the symbolism that Faḍlallāh rejects is that where this relationship between word and meaning is absent and where the emerging relationship between the reader and the text becomes distanced from the objective linguistic regulations that facilitate understanding.

However, Faḍlallāh does point out the excessive sensitivity that exists in adopting interpretation (*ta'wīl*) as a subjective process that disregards the standards of the Arabic language concerning the relationship between words and their meanings. He believes that even if the Arabic language does not prohibit the inflection of a word into something other than its apparent meaning, "the commitment to interpreting the apparent meaning depends first on there being mental or verbal evidence to support the interpretation of a word contrary to its apparent meaning, such that it would be impossible for its apparent meaning to have been inferred from a rational or semantic standpoint at face value."[25] Second, he believes that the new meaning must be compatible with the original meaning and with the general context of the speech, as in the case of metaphor or allegory. In this field, there must be a specific or absolute relationship between the real and figurative meanings. Or, if it is a metonymy, then there must be a customary correlation between the original meaning and the meaning that is intended to be conveyed by the word by way of metonymy. Adhering to

25 Ibid., 281.

these two conditions, then, leaves no room for interpretation (*ta'wīl*) because its use would end up contravening the rules of rhetoric."[26]

Here, Faḍlallāh discusses what is common among exegetes, especially Sufis and mystics, in regard to what is known as the "inner meanings of the Qur'ān" and the multiple layers to understanding it, where transcending the text to those symbolic meanings is portrayed as a necessary task in the process of understanding. This assumes that the speaker intended multiple meanings within a single expression, which is not the usual way in which people communicate. It also presumes that there is an inner meaning intended by the speaker that is not at all indicated in the text, which implies that said meaning only exists in the speaker's perception and would need further evidence in order to be considered.

If what is meant by the notion of the inner meanings of the Qur'ān, however, is the diversity of understandings according to the different eras, inputs, and cultures of the audience or researcher, then this meaning is plausible. However, it does not subsume a multiplicity of meanings, but rather a multiplicity of possibilities "that – when they differ – cannot be applied to the expression altogether at once but are subject to the yardstick of correct or incorrect,"[27] just as any possibility that is inconsistent with the rules of the semantics of the text is inacceptable.

By the same token, if what is meant is that

> the content of the Qur'ān manifests in real life through its broader meanings, subtle implications, movement, and its methodological variance, then this too is plausible. It simply means that the Qur'ān internalizes all the elements of meaning in its kinetic role in the life of the human being, and in its sensory and spiritual inferences. It does not, however, embody any kind of pluralism; instead, it represents a state of semantic expansiveness in its comprehensiveness, its extensiveness, its vastness, and in the fluidity of its inferences."[28]

Nonetheless, the idea of interpretation (*ta'wīl*) in this sense, or what Faḍlallāh calls "symbolic" exegesis, raises several problems. The first is that it "negates the great goal of the Qur'ān, which operates on the basis that the verses revealed to people are a means of opening them up to working their minds and their thinking in order to guide them, and to activating mindfulness and awareness of God through them."[29] Second, it creates connotations that correspond with the mentality of the exegete as an individual, but which may be far from the connotations of the word itself according to what the rules of the language and custom permit. Moreover, the meanings superimposed onto it go beyond the word to new, broader meanings that may have cultural dimensions that are foreign to the text itself in general, thus rendering the word a symbol that the exegete seeks to interpret within the cultural contexts that

---

26 Ibid., 172.

27 Ibid., 287.

28 Ibid., 287 (with a minor amendment).

29 Faḍlallāh, *Min waḥy al-Qur'ān*, 1: 13.

feed into his own way of thinking. This eliminates the intrinsic relationship between the word and the meaning, so that there is no longer any need to discuss the contingent significance of the word and its relationship to the context. Thus, the text loses its actual value because it turns the word into a mere symbol that cancels out its relationship with reality, bringing it closer to the realm of imagination. Third, it subjects the relationship between language and the world to the meanings derived from cultural and mental perception, where language no longer expresses the external, objective, existing real world, because it is the product of diverse perceptions and concepts, and language becomes the expression of all subjective, cultural manifestations. Fourth, this approach has resulted in many negative intellectual and jurisprudential conclusions in Qur'ānic hermeneutics and has resulted in great complications in dealing with the text at the level of understanding.

It remains to be said that the notion of "ambiguous" verses (*mutashābih*) mentioned in the Qur'ān[30] does not entail "the symbolic" or obscure meaning that is separate from speech, but rather refers to an expression that may have more than one aspect to its meaning, or one in which the overtones may differ. Sometimes the issue may be directed toward the practical aspect of these "ambiguous" verses and how they are to be applied in reality, and not to their semantic aspect, where those who seek to distort the words from their proper meaning and exploit them, desiring to sow discord and interpret them by making them align with ideas that they wish to stir up in people in order to keep them from the straight path.[31]

4.5 Inference

Faḍlallāh starts with the same dilemma faced by anyone dealing with a limited text at a time when that text is supposed to address issues that are broader than the scope of meanings that it directly conveys. If one of the meanings of interpretation (*ta'wīl*) tends to go beyond the text and to meanings that are generated in the mind of the *reader*, even if they do not yield to the direct meaning of the word or come close to it, Faḍlallāh in his exegesis proposes the method of inference.

By inference, Faḍlallāh does not mean "the desire for anything other than the apparent meaning of the word, but rather inferring meaning through the intended meaning of the word itself, according to the situation that has emerged through its usage."[32] This means that the reader maintains the relationship of the word to its original meaning, but "infers" new meanings from that intended meaning, drawing on various justifications, some of which will be mentioned below.

---

30 Q 3:7.

31 Faḍlallāh, *Min waḥy al-Qur'ān*, 1: 14.

32 Faḍlallāh, *al-Ijtihād*, 1148.

Faḍlallāh proves the validity of the inference method by making two points: first, a theoretical point, based on the authority of Imam Muḥammad al-Bāqir – the Fifth Imam of the Twelver Shīʿa – who said,

> If a verse was revealed to a community and upon the death of that community, the verse died too, then nothing of the Qurʾān would remain; but the Qurʾān will continue from its beginning until its end as long as the heavens and the earth endure. And every community has a verse that they recite, [and] that applies to them, whether good or evil.[33]

Faḍlallāh, commenting on this, says,

> And what we can infer from this is that the Qurʾān, in all that it mentions about the lives of previous nations in all their successful and failed experiences, their points of weakness and strength, and the experiences of the prophets with their communities, applies to us too. We must take away and draw inspiration from it and infer whatever new issues crop up from generation to generation, from issues related to war and peace, to corruption and social reform, to God's laws in the universe.
>
> The Qurʾān, then, was not revealed to merely be the book for the era in which the Prophet emerged; rather it was revealed to be the book of life, given that the Prophet is the seal of all the prophets with no prophet after him. That is why his book must necessarily be the book for all times to follow him. Every generation must then draw inspiration from the Qurʾān for the issues that they encounter within it. When we study the Battle of Badr, we draw inspiration from the fundamental convictions that the Muslims had at Badr, and we draw inspiration from their victories. When we study the Battle of Uhud, in which the Muslims were victorious initially and subsequently defeated, we infer from it the reasons for their victory and the reasons behind their defeat – why were they victorious and why were they defeated? This is what the Qurʾān addresses in Sura Āl ʿImrān and so on.[34]

The second point is practical, as it has been narrated on the authority of Imam Jaʿfar aṣ-Ṣādiq – the Sixth Imam of the Twelver Shīʿa. When asked about the Qurʾānic passage, ". . . if anyone kills a person – unless in retribution for murder or spreading corruption in the land – it is as if he kills all mankind, while if anyone saves a life, it is as if he is saving the lives of all mankind,"[35] the Imam replied, "Whoever brings someone out from error into guidance, it is as if he has saved his life, and whoever brings him from guidance into error has indeed killed him."[36] As Faḍlallāh comments on this response: "The apparent meaning of saving a life in the verse is material revival, but the Imam (peace be upon him) inferred spiritual revival from material revival, which is guidance. When someone transfers a person from a state of misguidance to guid-

33 Muḥammad b. Masʿūd al-ʿAyyāshī, *Tafsīr al-ʿAyyāshī*, ed. Rasūlī Maḥallātī (Tehran: al-Maktaba al-ʿIlmiyya al-Islāmiyya, 1960), 1: 10.
34 Faḍlallāh, *an-Nadwa*, 10: 501.
35 Q 5: 32.
36 Faḍlallāh, *an-Nadwa*, 10: 501.

ance, he is saving their life, for misguidance is spiritual death and guidance is spiritual life."[37]

This constitutes the general approach that Faḍlallāh takes in his exegesis, and in fact the title of his exegetical work *Min waḥy al-Qur'ān* (From the Inspiration of the Qur'ān) was based on this very concept. This is manifest in the chapters based on the verses about pedagogy, teaching, society, and politics, where the verse is designated as a source of guidance on how to deal with phenomena that the exegete and his era consider to be the fundamental, practical issues facing contemporary society. This is the reason that Faḍlallāh considered the Qur'ān to be "the book of the Islamic movement and Islamic propagation":[38] God centered the Islamic movement through the Qur'ān, and when Muslims experienced any problems during times of war or peace, verses of the Qur'ān would be revealed, commenting on the experience.[39]

Thus, we find that the Qur'ān does not stop at the historical particulars that provided the impetus for its revelation, but rather extends to all living examples in all times, just as it is fluid in its general concepts, pointing to how reality proceeds through issues of right and wrong, and legality and illegality, so that it may:

- act as proof along the lines of rectitude and deviation in actual lived Islam, which has come a long way since the time of its revelation;
- address every new phase by talking about a previous similar phase; and
- direct people to symbols of truth in the future, and distance them from symbols of falsehood by directing their attention to similar symbols in the past

because the Qur'ān represents the vast reality that encompasses all of time and rises above it.

Faḍlallāh concludes here that this inferring Qur'ānic method of exegesis is one that can transport someone from the verse into other worlds by nature of the goals to which it leads, broadening their perspective within the scope of the general principles of Islam and the Arabic language. This is what shifts believers from a purely material understanding to a spiritual one, and from that of the historical experience of the society in which the Qur'ān was revealed – addressing its specific problems, challenges, and issues – to one that encompasses new experiences, where believers face the challenges and problems of their current reality. This is what gives the Qur'ān its "kinetic" quality in addition to its legislative, directive, and advisory qualities, among others.[40]

According to Faḍlallāh, this method of inference does not fall outside the constraints of language, the correlations of speech, and the internal and external presumptions that come with speech. It does, however, require a comprehensive knowl-

37 Ibid.

38 Ibid., 1: 235.

39 Ibid., 1: 235.

40 Faḍlallāh, *Min waḥy al-Qur'ān*, 1: 21.

edge of the Qur'ān, Islam, the laws of God, the laws of life, and so on. Faḍlallāh, referring to one such term justifying inference, says,

> If we know that the Qur'ān's treatment of past situations springs from God's natural systems (*sunan*) regulating the universe and the human being, and not through any specificity in time, place, or people, then we can adopt them in similar conditions, allowing us to act in accordance with them in the present and future, since God's natural system (*sunna*) neither changes nor deviates: "You will never find any change in God's practice; you will never find any deviation there."[41]
>
> And just as the spiritual aspects converge with the material aspects in more than one sense, allowing for the inference of the spiritual from the material, it is similarly possible to infer the general from the specific since specificity does not represent something vital in the concept or ruling being presented in reality.'[42]

Faḍlallāh affirms his commitment to the pivotal role played by language in the method of inference that he employs, saying,

> In light of this, we do not need to deviate from the usual rules of the Arabic language in its exegesis, nor to distance it from general issues in order to focus on a particular fact or a particular legal symbol to arrive at the truth, because the general lines scattered within it, and the living and active models within it are perfectly capable of inspiring us with what we want in the world of evidence and proof.[43]

4.6 The Relationship of the Qur'ānic Text to the Sources of Islamic Knowledge

Faḍlallāh believes that the Qur'ānic text acts as a normative authority for any knowledge derived from other sources, be they Islamic sources such as the hadith or other sources related to Islamic fields, such as philosophy, mysticism, and so on. The normative authority for the Qur'ānic text, in turn, is based on what Faḍlallāh lays down as the normative authority of the Arabic language, with all the issues discussed previously. Additionally, the fact that it is a book of guidance means that any content interpreting the Qur'ān must also be consistent with the nature of the Qur'ān's purpose. This point is evident in several practical applications of Faḍlallāh's exegesis:

1. Rejecting any hadith whose content contradicts the content of the Holy Qur'ān or the concepts that it highlights.
2. Liberating exegesis from its restriction to what is found in the narratives about the causes of revelation (*asbāb an-nuzūl*), many of which have not been authenticated or serve more as applications of the verse than definitions of the interpre-

---

41 Q 35: 42; Faḍlallāh, *Min waḥy al-Qur'ān*, 1: 23.
42 Faḍlallāh, *Min waḥy al-Qur'ān*, 1: 23.
43 Ibid., 22.

tive meaning. Additionally, a fair number of them have resulted from the efforts of exegetes or narrators engaged in attempts to apply the verses to events that occurred or were being narrated, and thus stem from their differing goals and mentalities. The approach to exegesis in *Min waḥy al-Qur'ān* constitutes a distinct step in terms of critiquing the *asbāb an-nuzūl* literature based on the extent to which it is logically or circumstantially consistent with Qur'ānic meaning.

3. Ridding the Islamic faith of all that has attached itself to it by way of philosophical concepts that are either immersed in abstraction or do not come from the Qur'ān or from dialectic theology (*kalām*), turning them into established and non-negotiable truths when they actually contradict a number of clear Qur'ānic concepts and teachings, and where their adherents have felt compelled to interpret the Qur'ān in favor of those ideas, instead of holding the Qur'ān up as the yardstick for their correctness and their truth. Hopefully, those who examine the exegesis in *Min waḥy al-Qur'ān* will clearly find that Faḍlallāh's discussion of many doctrinal ideas, even his rejection and denial of them, is based on his exegetical approach, which serves as a benchmark.
4. Taking leave of philosophical, Sufi, or linguistic preconceptions because they only lead to the apparent meaning of the Qur'ān being dragged toward the convictions held by the exegete, who then seeks to apply them to the Qur'ānic verses. What is required of exegetes, however, is to acknowledge the cultural biases that they carry within themselves and not to claim absolute truth when carrying out their exegesis.
5. Holding reservations about the

> *uṣūlī* discussion that claims that the [jurisdiction in a verse of the] Qur'ān can be made specific by single-transmission narration (*khabar al-wāḥid*). Since the Qur'ān was sent down for all people, so it must be that even the specificities of the Qur'ān were meant for all people; otherwise, it would mean that the Qur'ān can exist only as a generality, and a narrative can come whispering into the ear of a man from the desert that what is meant by a Qur'ānic verse is such and such, without it being made clear and told to all people. This cannot possibly be specification![44]

5 Conclusion

What we can conclude from all this is that Faḍlallāh's hermeneutic approach comprises several binaries. According to him, a close relationship between the text and real life is paramount, as real, lived experiences provide the exegete with comprehensive information and enable Faḍlallāh to infer meaning from indirect indications that go beyond the linguistic boundaries of the text without having to delve into interpre-

44 Faḍlallāh, *al-Ijtihād*, 234.

tation (*taʾwīl*) outside the rules of language or the customary dimensions of mutual understanding and comprehension. These experiences also allow readers to become aware of the direction of the verse so that they can apply it to their own current experiences, whether in the fields of outreach, education, culture, politics, society, etc. All of this is done without the need to distance reality from the text by way of *taʾwīl*, which places the exegete's mentality in charge of the text instead of the other way around.

Additionally, Faḍlallāh demonstrates his strong emphasis on the authority of the Qurʾān and its normative power in Islamic knowledge without eliminating the noble hadith, instead placing it in its definitive position in relation to the Qurʾān as an explanatory or functional contributor to the text that facilitates inference, but not where the apparent text of the Qurʾān or its meaning have been sidelined in favor of the hadith.

Finally, Faḍlallāh appeals to the mechanisms of independent juristic reasoning (*al-ijtihād al-fiqhī*) centered on the apparent meaning of the text and the rules of understanding it to achieve a more profound and comprehensive approach, while at the same time raising probabilities that broaden the horizons of the researcher or jurist in their research.

6 Bibliography

6.1 Sources

Abdel Haleem, M. A. *The Qurʾān: A New Translation.* New York: Oxford University Press, 2005.

Al-ʿAyyāshī, Muḥammad b. Masʿūd. *Tafsīr al-ʿAyyāshī*. Ed. Rasūlī Maḥallātī. 2 vols. Tehran: al-Maktaba al-ʿIlmiyya al-Islāmiyya, 1960.

Faḍlallāh, Muḥammad Ḥussein. *Al-Ijtihād bayna usar al-māḍī wa-āfāq al-mustaqbal*. Beirut: al-Markaz ath-Thaqāfī al-ʿArabī, 2009.

Faḍlallāh, Muḥammad Ḥussein. *Tafsīr min waḥy al-Qurʾān*. 25 vols. Beirut: Dār al-Malāk li-ṭ-Ṭabāʿa wa-n-Nashr wa-t-Tawzīʿ, 2017.

Faḍlallāh, Muḥammad Ḥussein. *An-Nadwa*. 20 vols. Beirut: Dār al-Malāk li-ṭ-Ṭabāʿa wa-n-Nashr wa-t-Tawzīʿ, 2000.

Maʿrifat, Muḥammad Hādī. *At-Tafsīr wa l-mufassirūn fī thawbihi al-qashīb*. 2 vols. Mashhad: Muʾassasat aṭ-Ṭabʿ wa-n-Nashr fī l-Āstānat ar-Raḍawiyya al-Muqaddasa, 2004.

6.2 Online Sources

"Al-Maktaba." *bayynat.org*. Accessed May 6, 2024. https://www.bayynat.org.lb/library/category/25922/%D8%A7%D9%84%D9%85%D9%83%D8%AA%D8%A8%D8%A9/ar.

"Sīrat as-sayd." *bayynat.org*. Accessed May 6, 2024. https://www.bayynat.org.lb/article/category/1974/%D8%B3%D9%8A%D8%B1%D8%A9-%D8%A7%D9%84%D8%B3%D9%8A%D8%AF/ar%D8%9B.

Ursula Günther

Mohammed Arkoun

1 Introduction

Though it is impossible to capture Mohammed Arkoun's life's work in a single sentence, his *critique of Islamic reason* can be identified as its central project. This critique and its attendant demands to rethink and reread Islam as a cultural and religious system is the leitmotif of his writings, and this, of course, includes interpretations of the Qur'ān. Arkoun's approach represents a radical paradigm shift in approaches to Islam. It is based on the innovative conceptual pillars of *Qur'ānic* and *Islamic fact/event*,[1] closely connected to the notion of orthodoxy, societies of the Book/book,[2] the *unthought* and the *unthinkable*, and the *imaginaire*. Apart from the latter, whose origins trace back to the French Annales school of historians, all concepts bear his mark. He applied them to Islamic studies while simultaneously introducing methodological innovations from discourse analysis, semiotics, structuralism, structural anthropology, and the Annales school. These concepts are central to understanding his hermeneutical approach and outlining the structure of his thought. Underlying his critique of Islamic reason is Arkoun's desire to contribute to an archaeology of the hidden, repressed, and marginalized elements of Islam in order to uncover and set free what he called the "exhaustive tradition."[3] This article brings together interconnected concepts from Arkoun's work with a particular focus on his approach to the Qur'ān. An overview of his life and career, and the political and intellectual context within which he created and developed his project to its final complexity, gives key insights into the direction of his thinking. This is followed by a short introduction to his concepts, as they describe the structure of his thought and provide his hermeneutical methods. A subsequent section is then dedicated to the hermeneutical approaches in Arkoun's main works,[4] while the final

1 Since the French terms *fait coranique* and *fait islamique* are not consistently translated in English texts, I use both variations.

2 Arkoun uses an orthographic trick to illustrate the ambiguity of "book": Holy Book vs. book in the sense of a cultural product. This perspective takes into account the transcendent character of the Qur'ān as a revelation (the Book) as well as its character as a historical and literary document with its own literary, social, and political meanings. This poses a challenge for modern scholarship, all the more so as this perspective is meant to be taken both with regard to Islamic contexts and to other religious traditions in order to fully develop its comparative potential. See, e.g., Mohammed Arkoun, *The Unthought in Contemporary Islamic Thought* (London: Saqi, 2002), 37f.

3 Mohammed Arkoun, "L'Islam actuel devant sa tradition," in *Christianisme, Judaisme et Islam: Fidélité et ouverture*, ed. Joseph Doré (Paris: Editions du Cerf, 1999), 103–51.

4 A detailed presentation of his main works would exceed the limits of this article. Such a presentation can be found in Ursula Günther, *Mohammed Arkoun: Ein moderner Kritiker der islamischen Vernunft* (Würzburg: Ergon, 2004).

https://doi.org/10.1515/9783111320069-004

section appraises the impact of his thought, providing a conclusion and reflecting on future developments.

2 Biographical Considerations

2.1 Life

Mohammed Arkoun was born on February 1, 1928/Shaʿbān 9, 1346, in Taourirt-Mimoun, Greater Kabylia, in French-occupied Algeria.[5] He grew up in modest circumstances as the eldest son of a large Sunni family. His father ran a grocery store in ʿAyn al-Arbaʿā', a wealthy village of French settlers east of Oran, and took him there at age nine to train him for the business. Arkoun owed the opportunity to continue his schooling beyond primary level to his uncle's influence. He described this period, which he initially spent with the White Fathers and later as a boarding student at the *lycée,* the French equivalent to high school in Oran, as inspiring because it exposed him not just to French literature and history but also to Latin literature and culture, the writings of the African Church Fathers, and Christianity, which nourished his confidence in the possibility of finding "a continuous historical memory encompassing not only the Maghreb, but the entire Mediterranean."[6] From 1950/1369 to 1954/1373, he studied Arabic literature at the University of Algiers, where he also attended courses on Arabic philosophy. There, he was confronted with the "cognitive practices of fervent Orientalism"[7] and formulated early impulses to develop countervailing concepts.

Shortly before the Algerian War of Independence, Arkoun left Algeria to continue his studies at the Sorbonne's Institut des Études Islamiques under Régis Blachère, Robert Brunschvig, Henri Laoust, Evariste Lévi-Provençal, and Charles Pellat. He graduated in 1956/1375 with an *Agrégation* in Arabic language and literature. Between 1956/1375 and 1959/1378, he taught in Strasbourg, both at a *lycée* and the university, while simultaneously studying Classical Arabic philosophy. Historian Claude Cahen, then a professor in Strasbourg, introduced him to the new approaches being taken by the Annales school to the history of the Near and Middle East. When his original dissertation project on religious practices in Greater Kabylia, supervised by Jacques Berque and Régis Blachère, had to be abandoned due to the war, Arkoun followed Brunsch-

5 Many details of this paragraph refer to my personal communication with Mohammed Arkoun.

6 Hassan Arfaoui, "Entretien avec Mohammed Arkoun," *Le Monde Arabe dans la Recherche Scientifique/The Arab World in Scientific Research/al-ʿĀlam al-ʿarabī fī-l-baḥth al-ʿilmī* 5 (1995): 13f. In French: "orientalisme flamboyant."

7 Mohammed Arkoun, *Penser l'Islam aujourd'hui: La communication impossible.* Unpublished manuscript, author's private collection. Paris, 1997, 1.

vig's advice to write a thesis on philosopher Abū ʿAlī Miskawayh.[8] His intense engagement with the social and intellectual climate of Buyid society in the tenth century gave him the tools he needed to articulate an intellectual critique of religious thought. This was due in no small part to the innovative methods he had applied in his dissertation, where he had transcended the traditional philological practices of the field by embracing nascent postmodernist ideas and their attendant methodological innovations, such as discourse analysis, semiotics, structuralism, and the approaches of the Annales school. This shift in perspective formed the basis for his academic career and provided important impulses for his critique of Islamic thought.

Following a period of lecturing (as *maître de conférence*) at the Sorbonne between 1961/1380 and 1969/1389, Arkoun was offered the Chair of the Islamic History of Ideas at the University of Lyon (Lyon II), before moving on to become Chair of the Islamic History of Ideas at Vincennes (Paris VIII) in 1971/1391. From 1980/1400 to his retirement in 1993/1413, he taught at the Sorbonne Nouvelle (Paris III), where he simultaneously headed the departments of Arabic Studies and the Islamic History of Ideas. His tenure as editor of *Arabica* saw a considerable conceptual expansion of this traditional journal.

2.2 Political and Intellectual Context

Arkoun's career was shaped by biographical and historical influences, both from his childhood and youth in Algeria and his studies in Paris.[9] He experienced the intellectual climate in mid-1950s and 1960s France as exceptional and incomparable, marked by an overwhelming sense of change and upheaval.[10] He joined the social circles, critical movements, and schools that were emerging and coalescing at that time, such as the Annales, post-structuralism, structural anthropology, historical discourse criticism, and deconstruction. His own academic rise from lecturer to full professor took place amid a generational and administrative shift that saw the establishment of new departments of, for example, Arabic and Islamic Studies at French universities. This afforded his concepts, influenced by the contemporary humanities and social sciences, greater reach. Arkoun combined active research with his public engagement as a critical intellectual in many publications, numerous international lectures (for example the Gifford Lectures), and a strong media presence, and with his membership in a number of civic bodies. He sat on the jury of the Aga Khan Prize for architecture and

8 His thesis was published in Paris by Vrin in 1970 under the title *L'Humanisme Arabe au IVe/Xe siècle: Miskawayh, Philosophe et Historien.*

9 Mohammed Arkoun, *Rethinking Islam Today.* Occasional Papers Series (Washington: Center for Contemporary Arab Studies, Georgetown University, July 1987), 1.

10 Mohammed Arkoun, *La Construction Humaine de l'islam: Entretiens avec Rachid Benzine et Jean-Louis Schlegel* (Paris: Albin Michel, 2012), 47.

the French commission on *laicism* (Commission Stasi) in 2003/1424 as well as on the Board of Governors of the Institute of Ismaili Studies in London, where he also regularly taught. Numerous awards and prizes recognized his achievements: Arkoun was made an Officer of the Legion of Honor and the Palmes Académiques. In 2002/1423, he was awarded the Lévi Della Vida Prize for his life's work, and one year later the Ibn Rushd Fund for Freedom of Thought awarded him a similar honor. In 2012/1433, he posthumously received the al-Idrissi Prize.

Mohammed Arkoun died in Paris on September 14, 2010/Shawwāl 5, 1431.

3 Theological and Intellectual Background and Analysis of Major Hermeneutical Concepts, Methods, and Works

Aside from his dissertation, we mainly encounter Arkoun's extensive body of work in the form of essay collections, whose constituent parts were often first published in hard-to-find journals. These collections usually share a common methodology or unifying theme, but figuring out the intertextual relationships between the works in his oeuvre presents the reader with a challenge. Arkoun himself always emphasized the importance of reading his writings in their entirety and not restricting his work to a single book.[11] Representing this breadth is difficult, as navigating his densely interwoven concepts requires both specific knowledge of other texts and a broad understanding of his oeuvre.[12] Following a brief introduction to the theological and intellectual background to his approach, an introductory outline of his methods will take us through central concepts not so closely associated with one single collection or essay.[13] Only then will we look at the two works that are central to Arkoun's hermeneutical approach, namely *Lectures du Coran* and *Pour une Critique de la Raison Islamique*.[14] The chapter will end with some brief notes on the reception of his work.

---

11 Mohammed Arkoun, *Lectures du Coran* (Paris: Maisonneuve et Larose, 1982; second augmented edition Tunis: Alif – Éditions de la Méditerrannée, 1991; third edition published posthumously with the subtitle *Edition définitive*, Paris: Albin Michel, 2016). I will refer both to the second and third edition with double page references, separated by slashes. Here, see second edition, 32; there is no equivalent in the revised chapter of the third edition.

12 For a detailed introduction to Arkoun's work, see Günther, *Ein moderner Kritiker*, 161–215.

13 The dynamic of the rhizome-like contiguities and mutual causality of Arkoun's concepts cannot be discussed exhaustively; for further details, see Günther, *Ein moderner Kritiker*, 53–106; and Günther, "Mohammed Arkoun: Towards a Radical Rethinking of Islamic Thought," in *Modern Muslim Intellectuals and the Qur'an*, ed. Suha Taji-Farouki (London: Oxford University Press, 2004), 125–67.

14 Though the essay collection *The Unthought in Contemporary Islamic Thought* (London: Saqi, 2002) deserves greater attention – it is after all the first summarization of his thought and concepts for an English-language readership – this would go beyond the scope of this article.

3.1 Theological and Intellectual Background

The concepts Arkoun developed to enable his project of a *critique of Islamic reason* and *rethinking Islamic thought* are central to his hermeneutical approach. They stand at odds with both the Eurocentric reading of Islam and the reading proposed by the guardians of orthodoxy. Ultimately, Arkoun opposes all hegemonic reason. Furthermore, he attempts to establish a counterbalance to the dominant philological method he regards as too one-sided. Though acknowledging its academic merits, he accused it of leaving behind "ruins, as the philologist *will never accept responsibility for the cultural and intellectual consequences* of what he declares to be apocryphal, false, legendary, or imaginary."[15] He intends to face this responsibility by advocating an inclusive vision of Islam that brings back to light all of the elements of Muslim tradition that have been separated, repressed, and discredited, and integrating them into a holistic vision. Using religious anthropology and philosophical theology, he aims to combine "the theological requirement of believers, the philological imperative of the positive (but not positivistic) historian, the explanatory perspective of the anthropologist, and the critical control of the philosopher."[16] He places particular emphasis on the philosophical perspective and position as it offers access to the meta-level required to comprehend the concept of *religious fact* or *religious phenomenon* as a category of Muslim consciousness, without making recourse to theological *a prioris*.[17] It also allows him to unmask the cognitive barriers and boundaries that coalesce in "dogmatic enclosure"[18] and thus to finally overcome them. Arkoun understands Islamic reason as Islamic orthodoxy's claim to have founded a form of reason whose transcendent character makes it superior to previous forms of reason. It corresponds to the normative reason of jurisconsults and religious scholars, and is subject to the influence of the *Islamic fact/event*.[19] Thus, Islamic reason is identical to the orthodox variety of reason that overcame competing forms by claiming divine origin as its premise. Meanwhile, Arkoun develops the postulate of postmodern reason in ever-increasing detail, before summarizing all of it within the term "emerging reason," a form of reason with a critical function:[20] "It is concerned with the philosophical subversion of the use of reason itself and all forms of rationality produced so far as well as those that will be produced in the future so as not to repeat the ideological compromises and derivations of

---

15 Arfaoui, *Entretien avec Mohammed Arkoun*, 16, original emphasis.

16 Arkoun, *Lectures du Coran*, 20/59.

17 Ibid., 38f./66.

18 Mohammed Arkoun, *Humanisme et Islam: Combats et Propositions* (Paris: Vrin, 2005; 2010^2), 156–60; see also Arkoun, *The Unthought*, 37–65, especially 61f.

19 Mohammed Arkoun, *Pour une Critique de la Raison Islamique* (Paris: Maisonneuve et Larose, 1984), 311f.

20 See, e.g., Arkoun, *The Unthought*, 22–24.

the precedent postures and performances of reason."[21] It is from this perspective that he undertakes to deconstruct both Islamic and Western reason.[22] *Emerging reason* can be regarded as a form of meta-reason that allows a comparative approach to cultural systems and historical experience to be taken because it makes no implicit claim to truth and is beholden to no doctrine. Further, it does not impose polar structures of thought, but is context-sensitive and flexible, uniting multiple perspectives. It seeks to rehabilitate, reactivate, and reintegrate all significant contributions from every tradition of thought. Arkoun seeks nothing less than to develop a reason "capable of encompassing the human condition."[23]

Arkoun diagnoses an epistemological breach between theological and academic research, asserting that the former seeks to maintain and affirm the community of believers while the latter frequently neglects the question of meaning that all religions have in common. He looks to religious anthropology to provide the overarching perspective and frame of reference to break open apparent incompatibilities and polarities, and to take account of the spiritual dimension of religion without becoming enmeshed in theological aporia.[24]

3.2 Analysis of the Major Hermeneutical Concepts

3.2.1 Logocentric and Dogmatic Enclosure

Arkoun's thought focuses on the "logocentric" and "dogmatic enclosure" of Islamic reason and discourse dating to the tenth or eleventh century. This rupture in Islamic intellectual history created a closed realm of the *thinkable*, and this had a lasting impact on Islamic thought, as "the great commentaries which made themselves authoritative over the historical development of the living tradition are used as orthodox instances of authority in order to reproduce the fixed inherited interpretations."[25] This concept guarantees the consistency and coherence of the corresponding tradition or orthodoxy since the members of the respective group refer to the same fixed framework of perception and representation as a given. It can be applied to

21 Ibid., 23f.

22 Mohammed Arkoun, *La Question Ethique et Juridique dans la Pensée Islamique* (Paris: Vrin, 2010), 169–73.

23 Arkoun, *The Unthought*, 22.

24 See, e.g., Arkoun, *Lectures du Coran*, 54f/85–88; Günther, *Ein moderner Kritiker*, 62.

25 Arkoun, *The Unthought*, 61. Needless to say that, beyond the limits of the *thinkable* is the *unthinkable*, i.e., the realm of taboos at a given time. The notion of the *unthought* describes the *unthinkable* in its entirety as well as all that is outside the limits of knowledge at a particular time. Arkoun emphasizes that these concepts are historical rather than philosophical categories as they are subject to change depending on the context. See, e.g., Arkoun, *Rethinking Islam Today*, 13.

> the totality of the articles of faith, representations, tenets and themes which allow a system of belief and unbelief (q.v.) to operate freely without any competing action from inside or outside. A strategy of refusal, consisting of an arsenal of discursive constraints and procedures, permits the protection and, if necessary, the mobilization of what is presumptuously called faith (q.v.). [. . .] No green light has ever been given to a deconstruction of the axioms, tenets and themes that hold together and establish the adventurous cohesion of every faith.[26]

For Arkoun, it is obvious that Islamic reason and Islamic thought operate within the confines of *dogmatic enclosure.*

3.2.2 The Notion of Orthodoxy

Orthodoxy is "one of the keys to rethink the whole theology of Islam."[27] It goes without saying that *dogmatic enclosure* and orthodoxy are interconnected and interdependent. The logical consequence of this approach is a reconsideration of revelation. Thus, the notion of orthodoxy is always used in a theological sense, yet has never been *thought* in a historical sense; however, it is daring to point out that it is the result of a slow historical process of selection, elimination, and the diffusion of names, works, schools, and ideas according to the objectives that the group, community, or power in place in each case aims to achieve. This is how tradition is formed, which works like a security system for the religious or national community.[28]

The following diagram attempts to illustrate the implications of orthodoxy:

3.2.3 Qur'ānic and Islamic Fact/Event

Arkoun's understanding of revelation, according to which it "is not a normative speech (*parole*) that came down from heaven [. . .,] it is an offer of meaning for existence and can be revised (see the abrogation and abrogated verses in the Qur'ān),"[29] is closely tied to these interwoven concepts as they describe, first, the historical process of a new religion coming into being, effected and supported by social, political, and cultural actors.[30] Second, they allow for differentiation between an oral dis-

---

26 Mohammed Arkoun, "Contemporary Critical Practices and the Qur'ān," in *EQ*, ed. Jane Dammen McAuliffe (Leiden: Brill, 2001), 1: 427.
27 Mohammed Arkoun, "Algeria," in *The Politics of Islamic Revivalism: Diversity and Unity*, ed. Shireen T. Hunter (Bloomington etc.: Indiana University Press, 1988), 171–86.
28 Mohammed Arkoun, *Penser l'Islam Aujourd'hui* (Algiers: Editions Laphomic/Enal, 1993), 290, footnote 1.
29 Mohammed Arkoun, "Le concept de société du Livre-livre," in *Interpréter: Hommage amical à Claude Geffré*, ed. Jean-Pierre Josua and Nicholas-Jean Séd (Paris: Edition du Cerf, 1992), 217.
30 Mohammed Arkoun, *La Pensée Arabe, Que Sais-Je?* no. 915 (Paris: Presses Universitaires de France, 1975), 7–11.

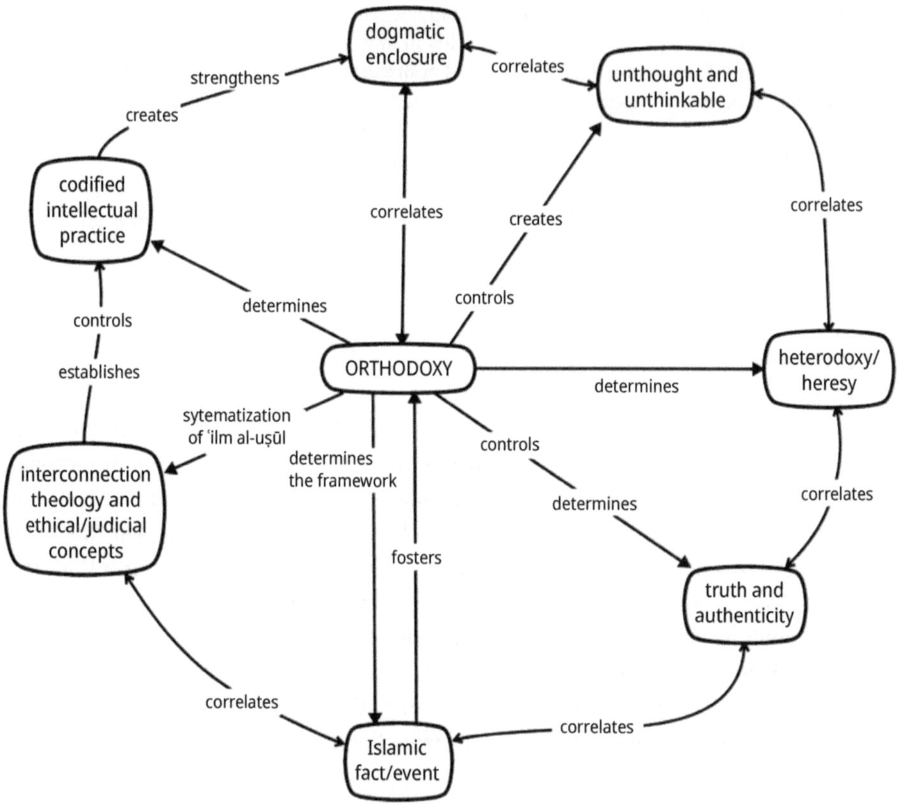

Figure 1: Implications of orthodoxy. Graphic created by the author, see also Günther, *Ein moderner Kritiker*, 61.

course, i.e., a linguistic event during the lifetime of the Prophet Muḥammad – in other words, the *Qur'ānic fact/event* – and the consolidation of the new religion with the closed corpus of the sacred text, i.e., the *Islamic fact/event*. This represents a historical, cultural, and linguistic transformation of major import.[31] Deployed as analytical categories, these concepts provide an overarching framework for historical, linguistic, hermeneutical, and anthropological analysis. This reading is impressively demonstrated in *Lectures du Coran*.[32] In particular, Arkoun emphasizes that this neither implies any criticism of religion as such nor adopts any theological or dogmatic postulates. His vision of a philosophy of theology or religious anthropology makes the

31 Mohammed Arkoun, "Du Dialogue Inter-religieux à la Reconnaissance du Fait Religieux." *Diogène* 182 (1998): 103–26, here 104.

32 Arkoun, *Lectures du Coran*, 77–101/113–42 and 103–25/143–70, and only in the definitive edition: 429–67.

argument that any analysis of religion or the religious question needs to be embedded within the overarching context of the question of meaning.[33]

The linguistic, semiotic, and historical conditions of the genesis of the *Qur'ānic fact/event* explain the difference between the *Qur'ānic* and the *Islamic fact/event* and allow the Qur'ān to be read on different levels. At the same time, they strip the cognitive limits of *dogmatic enclosure* of their efficacy. The Qur'ān can be read both as a document of revelation and as a historical or literary document that reflects specific cultural and historical phenomena, the correlation of sociopolitical forces, currents in the history of thought, and new or changing epistemes of *Qur'ānic* and *Islamic discourse* and mental structures.[34]

The following diagram attempts to compare the characteristic features of each *fact/event*:

3.3 Methods of Arkoun's Hermeneutical Approach

It should be amply clear by now that Arkoun rejects any narrow methodological constraints and pursues a multidisciplinary approach. Instead of bowing to the traditional logic of his discipline, he draws on a broad set of methods, which he adapts to his subject, usually in creative combinations. This can be summarized as follows:

Discourse analysis allows the analytical category of the *episteme* developed by Michel Foucault[35] to be applied to the history of Islamic thought and the revelation of the underlying historical dialectic without the need to pass judgment on its contents.[36] The hidden cognitive schema underlying everyday knowledge, science, and philosophy allows for comprehensive engagement with the historical and sociocultural conditions of "Islamic thought" by focusing on discursive structures and the implicit postulates they are founded on. This gives us insights into the basic discourse of a given era.[37]

Applying semiotics, Arkoun analyzes structures and forces at the sociohistorical level. He can thus maintain a methodological distance from the sacred texts and

33 Arkoun, "Du Dialogue Inter-religieux," 104.

34 See Arkoun, *Lectures du Coran*, 12/48f.

35 Arkoun, *Pour une Critique*, 319–22.

36 See, e.g., Mohammed Arkoun and Louis Gardet, *L'Islam, Hier – Demain* (Paris: Buchet-Castel, 1978; 1982²), 136; Arkoun, *Pour une Critique*, 66, 305f. All references are examples of a greater presence throughout the work.

37 Arkoun, *Rethinking Islam*, 5. Arkoun's definition of this concept is similar to Foucault's, but his division of the history of knowledge into epistemes of classical vs. modern thought differs from Foucault's. This is not explicitly acknowledged. See also Arkoun, *Pour une Critique*, 305–07; Arkoun, *Rethinking Islam*, 11.

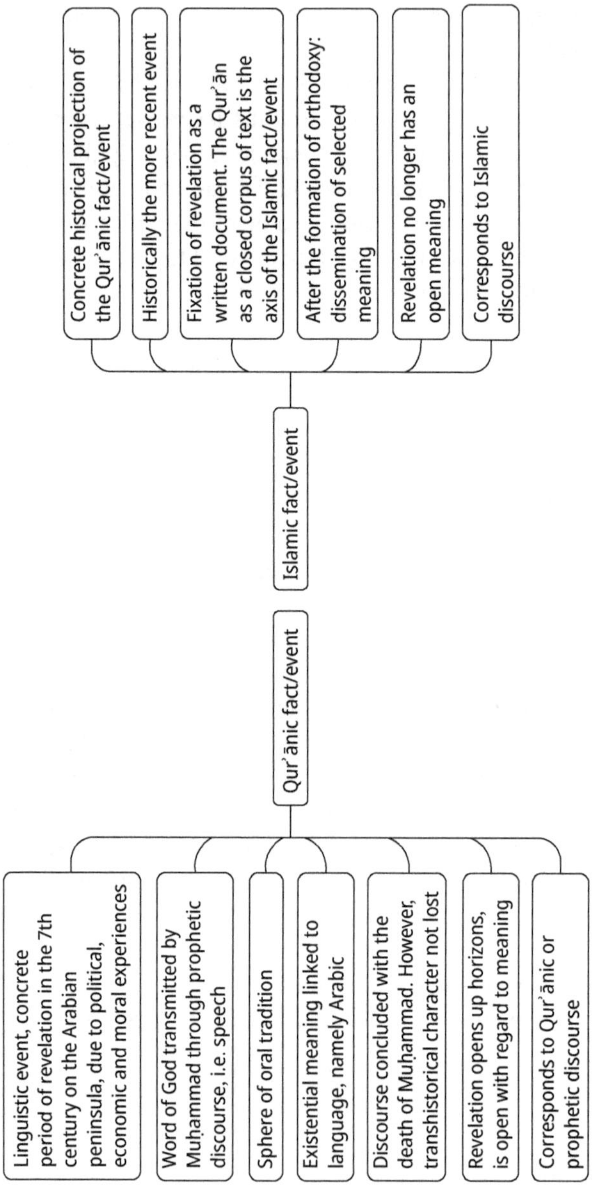

Figure 2: Qur'ānic and Islamic fact/event. Tabular comparison created by the author. See also Günther, *Ein moderner Kritiker*, 87.

avoids adopting theological or historical judgments.[38] He semiologically draws out the linguistic, narrative, and discursive mechanisms that are decisive in fixing meaning.[39] Texts and their meanings can be located within their respective sociopolitical contexts by means of structural linguistics. This has the twofold advantage of avoiding a linear historical perspective[40] and the use of theological *a priori* assumptions or dogmatic patterns in their interpretation. Structural linguistics reverses the order of traditional exegesis by first producing a linguistic definition of a corpus and then developing theological questions.[41] This allows a distinction to be made between *Qur'ānic* and *Islamic discourse.*[42]

The Blochian progressive-regressive method serves to uncover mechanisms and manipulations that go hand in hand with changing historical and sociopolitical contexts while also making space once more for that which has been repressed, marginalized, and forgotten in historiography, and reactivating it in the current context.[43] Finally, Arkoun argues for inquiries from historical sociology – particularly those concerning values and ethics – to be adopted into both Islamic studies and Qur'ānic-hermeneutical research approaches in order to understand which social groups, classes, or communities champion which values and how the visions proclaimed in the sources are presented, as well as to determine whether any traces of morals or values other than those reflected in the literature can be found.[44]

3.4 Analysis of Major Hermeneutical Works

3.4.1 *Lectures du Coran*

Lectures du Coran can be regarded as the core of Arkoun's work.[45] In the introduction, he defines the premises underlying his approach to Qur'ānic studies. He is critical of both traditional Western Orientalist and Islamic research, and advocates a pluridisci-

38 See, e.g., Mohammed Arkoun, *L'Islam: morale et politique* (Paris: Desclée de Brouwer, 1986; 1990²), 55; Arkoun, *Lectures du Coran*, 16f.; Arkoun, *Penser l'Islam Aujourd'hui*, 253, 288.

39 See, e.g., Arkoun, *Lectures du Coran*, 12/48f.

40 See, e.g., Mohammed Arkoun, *Essais sur la Pensée Islamique* (Paris: Maisonneuve-Larose, 1984³), 170.

41 See, e.g., Mohammed Arkoun and Maurice Borrmans, *L'Islam, Religion et Société*, interviews by Mario Arioso (Paris: Editions du Cerf, 1982), 3; Arkoun, *Penser l'Islam Aujourd'hui*, 287f.

42 See, e.g., Arkoun, *Penser l'Islam Aujourd'hui*, 287f.

43 See, e.g., Arkoun, *L'Islam: morale et politique*: 16, 22; and *Essays sur la Pensée*, 7–12.

44 See, e.g., Arkoun, *L'Islam: morale et politique*, 113.

45 This essay collection will be presented in greater detail due to its immediate focus on the Qur'ān. It was published posthumously in an *Édition définitive* in 2016, in which Arkoun's widow made available eight of nine re-edited chapters as well as three previously unpublished manuscripts and one edited essay from *Pour une Critique de la raison Islamique*. The only unchanged chapter is *Lecture de la Fā-*

plinary approach to the Qur'ān. Arkoun further develops his project of a plural reading by expanding his methodology with borrowings from linguistics, anthropology, and contemporary history while adopting a philosophical perspective. Addressing the question of the Qur'ān's divine origin, Arkoun tackles its strategic importance with regard to new readings and its significance in changing epistemes through the emergence of the *Islamic fact/event*. He also proposes guidelines and concepts for new readings and a thorough examination of the Arabic language system between 550 and 632 CE as well as the myths, rites, and religions of the Ancient Orient. Moreover, he develops the concept of societies of the Book/book (*sociétés du Livre/livre*), a conscious expansion of the term *ahl al-kitāb*. His close linguistic and semiotic analysis of the opening sura *al-Fātiḥa* (The Opening) is complemented by his historical and anthropological perspective. Arkoun includes an analysis of the work of Qur'ānic exegete Fakhr ad-Dīn ar-Rāzī in order to demonstrate how religious tenets and dogmas determine the direction of thought and thus of discourses, expounding on how to potentially develop a theory of symbolic language. A similar semiotic and linguistic analysis of the textually heterogeneous sura 18, *al-Kahf* (The Cave), demonstrates Arkoun's profound knowledge of classical exegesis. He discusses the approaches and interpretations of ar-Rāzī and aṭ-Ṭabarī, showing how they shaped the Muslim *imaginaire* and produced a homogenous space of conceptions and projections that must be studied and understood in connection with the concept of the *Qur'ānic fact/event*.

An analysis of two Qur'ānic passages that address rules of inheritance (Q 4:12 as well as Q 2:180–82 and 240), drawing out connections between faith and the social *imaginaire*. This shows how the needs of a society as well as prevailing customs and power strategies are reflected in Qur'ānic exegesis. *Ijtihād* as a decision-making process is identified as holding the key to a critique of *Islamic reason* that could be accepted by traditional religious scholars and jurisconsults.

In another contribution to a semiotics of religious language in *Qur'ānic discourse*, Arkoun analyzes the lexical and marvelous aspects of the Qur'ān linguistically and semiotically, and relates them to the consciousness produced by *Qur'ānic discourse*. Further, he analyzes characteristics of this discourse in the vocabulary related to authority and power used in reference to the actions of the Prophet Muḥammad, tracing the emergence of *Islamic reason*. His analysis of the dialectics of pilgrimage as a central symbol of Islam with pre-Islamic history and significance also provides clues as to the emergence of religious symbolism, the *unthinkable* and the *unthought*, and the transition from the *Qur'ānic* to the *Islamic fact/event*. An essay on the concept of revelation in sura 9, *at-Tawba* (Repentance), explains the historical dialectic between reve-

tiḥa. Titles that have been changed in the *Édition définitive* are indicated separately from the original title by slashes. Double page references refer to the respective editions.

lation and truth. Thus, Arkoun succeeds in establishing the historicity of the *religious discourse.*[46]

Arkoun revisits the issue of revelation in a second essay, addressing it as a complex cultural and historical phenomenon within the context of a religious phenomenon. Throughout the collection, he repeatedly attempts to locate the significance and function of Islam in society, contrasting the traditional Muslim position with that of Islamic studies. The normative frames on both sides, he says, could be transcended if those positions were contrasted with real Muslim societies in order to draw attention to their respective different historical and structural realities. At the same time, it is important to distinguish between the values proclaimed in a society, the norms that it sets itself, and the values that it lives by. These are not always identical, as Arkoun lays out in the context of the emergence of the *Qur'ānic* and the *Islamic fact/event*, and the incursion of modernity.

The posthumous edition includes a further essay analyzing sura 9, which can be understood as an introductory text opening up an exegetical horizon, trying to dissolve the confrontational tension between historicity and transcendence. Arkoun demonstrates how productive the freedom to undertake an exegesis beyond orthodox boundaries can be. His systematic historical and anthropological reading of sura 9 illuminates the metaphorical dimension of the *Qur'ānic discourse* without dogmatic, political, or national limitations. This dimension addresses a topic that Arkoun still situates within the field of the *unthought* and *unthinkable.*

3.4.2 *Pour une Critique de la Raison Islamique*

In the essays collected in *Pour une Critique de la Raison Islamique*, Arkoun explores the following question: "Under which discernible conditions does the idea of truth take shape to the point that it determines an individual fate or produces a collective history?"[47] This collection is closely related to *Lectures du Coran*, as both volumes can be read as variations on the theme of a critique of Islamic reason, a project that pervades Arkoun's entire oeuvre. Following an extensive introduction, the collection consists of two parts, titled "Approches de la Pensée Islamique" and "Lecture de l'Espace Maghrébin," containing nine and three essays respectively.[48]

In this collection, Arkoun outlines the project of applied Islamic studies following Roger Bastides's *applied anthropology*. What distinguishes it from traditional Islamic Studies is its methodology and its clearly defined hermeneutical intent. The concept is

46 Arkoun, *Lectures du Coran*, 257–81/379–427.

47 Arkoun, *Pour une Critique*, 33.

48 I will only outline the articles relevant to Arkoun's hermeneutic concepts, not those addressing the specific situation in the Maghreb. More detail on the latter can be found in Günther, *Ein moderner Kritiker*, 196.

not, however, intended as a competing model, but as a necessary complement to philological methods. It is a reflective, critical form of research that addresses its own limitations and disconnections, and the forgotten or repressed, in order to create space and produce the preconditions for free Islamic thought. As a set of guidelines on the aims of applied Islamic studies, Arkoun presents six axioms.[49] The concept of *Islamic reason* that is so central to Arkoun's thinking emerges as a manifestation of religious reason. The adjective *Islamic* refers both to the sociocultural place of this type of reason, located within a set of unquestioningly accepted religious postulates, and to the claim passed down by Muslims that they have created a form of reason superior to others that function outside Qurʾānic revelation.[50] This concept must also be studied against the backdrop of Islamic consciousness, its historical development, and its present crisis.[51] Arkoun develops ten hypotheses that can be applied to his entire oeuvre and may be regarded as a distillate of the findings of the modern social sciences and humanities. They point to his desire for the reintegration (*remembrement*) of areas of Islamic thought that have been separated, repressed, or unthought in order to fully understand all aspects of its meaning and to restore its original plurality.[52] Arkoun's vision of the future lies beyond the self-imposed limitations of thought.

3.5 Reception and Influence in the Past and Present

Until the 1990s, Arkoun was barely acknowledged as a critical thinker. It was only with the rise of Islamism and the new enmity to Islam that his thought began to draw wider interest. He was regarded as an innovative, original intellectual and influential scholar applying postmodern theory to classical Islamic texts with the intention of helping to liberate Islam from both Western and Muslim misconceptions and to deconstruct monopolies on the interpretation of Islamic history.[53]

Since the beginning of the twenty-first century, interest in his work has continued to grow in both the Global North and the Global South. Ursula Günther's dissertation was the first introduction to and systematic presentation of his thinking and the concepts he had developed.[54] Other studies have compared Arkoun to thinkers such as Fa-

---

49 Arkoun, *Pour une Critique*, 43–63.

50 Ibid., 65–99.

51 Ibid., 101–27.

52 Ibid., 129–53.

53 See, e.g., Ron von Haleber, *Islam en Humanisme: De Wereld van Mohammed Arkoun* (Amsterdam: VU Uitgeverij, 1991); Anke von Kügelgen, "Mohammed Arkoun und die islamische Philosophie" (MA thesis, Freie Universität Berlin, Berlin, 1990); Nāʾila Abī Nādir, "Madkhal ilā dirāsat fikr Muḥammad Arkūn" (MA thesis, Université Libanaise, Beirut, 1994). A detailed presentation of the reception of his work until the end of the twentieth century can be found in Günther, *Ein moderner Kritiker*, 143–59, 247–49.

54 Günther, *Ein moderner Kritiker*.

zlur Raḥmān, Naṣr Ḥāmid Abū Zaid,[55] Ḥasan Ḥanafī, Madjid Nurcholish,[56] and Muḥammad ʿĀbid al-Jābirī.[57] The emphasis in the latest publications has been on his approach to the Qurʾān and theological issues. Arkoun's scholarly renown and standing is further borne out by the number of full and partial translations of his work into numerous European languages as well as Indonesian, Turkish, and Arabic. The proportion of his work that has been translated into Arabic is remarkable, with some of his essays available exclusively in that language.[58] Many of these translations are the work of his former student Hāshim Ṣāliḥ, whose extensive explanations of Arkoun's terminology have contributed greatly to the dissemination of his thought in the Arabic-speaking world.[59]

Since Arkoun's death, his work has been studied by an increasing number of mainly Muslim students in master's and doctoral theses. This is most likely due to a combination of factors: First, the annual awarding of the Prix Mohammed Arkoun by the Mohammed Arkoun Foundation in Casablanca provides a certain incentive. Second, many of his former students, now themselves teaching in universities, are disseminating his thought and fostering engagement with it. This is especially the case in Arabic-speaking countries. For example, Professor Nāʾila Abī Nādir at the Lebanese University in Beirut has established Arkoun's reevaluation of Islamic thought as a permanent part of the philosophy curriculum. Her frequent lecturing in the Arabic-speaking world is spreading awareness of his critique of Islamic reason and his accompanying reading of the Qurʾān. In the Maghreb, Professor of Philosophy Fāriḥ Masraḥī, who teaches at the University of Batna in Algeria, is also continuing Arkoun's intellectual legacy by supervising theses that engage with his thought. Whether and how the increasing reach of Arkoun's approaches will influence coming hermeneutical debates and whether this will also translate into changing practices for religious scholars and imams remains to be seen.

4 Conclusion

When we stringently follow Arkoun's project to its logical conclusion, it becomes evident that it can only be carried out in an intellectual and hermeneutical context that views the diversity of readings and the attendant plurality and complexity of meaning

55 Katarina Völker, *Quran and Reform: Rahman, Arkoun, Abu Zayd* (Frankfurt a.M.: Peter Lang, 2017).
56 Carool Kersten, *Cosmopolitans and Heretics: New Muslim Intellectuals and the Study of Islam* (London/New York: Hurst Publishers, 2011).
57 Nāʾila Abī Nādir, *at-Turāth wa-l-manhaj bayna Arkūn wa-l-Jābirī* (Beirut: ash-Shabaka al-ʿArabiyya li-l-Abḥāth wa-n-Nashr, 2008).
58 Such as, e.g., Muḥammad Arkūn, *Min al-Ijtihād ilā naqd al-ʿaql al-islāmī* (Beirut: Dār as-Sāqī, 1991; 1993^2).
59 See also Olivier Dubois, "Contribution à une Bibliographie de Mohammed Arkoun," *AR* 60 (2013): 473–515, here 477–79. Unfortunately, this bibliography is incomplete and not always correct.

not as a threat, but as an enrichment, as an environment where gradual changes to the conception of normality are accepted as fact. This type of diversity provides access to what Arkoun calls the *exhaustive tradition*. Most of his own work is situated within the conceptual realm, and he offers little guidance for translating theory into practice. However, processes of pluralization have become increasingly characteristic of recent decades, and they do not spare religious or political monopolies on interpretation. This calls the continued efficacy of any form of orthodoxy into question.

Whether and how the impact of Arkoun's occasionally subversive rethinking of Islam will develop in the future and whether it will extend its reach beyond a narrow elite is likely to depend on whether the coming generation of Muslim scholars and intellectuals are motivated to further explore and elaborate on, or possibly modify, the space beyond dogmatic boundaries that Arkoun has opened up, and thus shape present and future discourses. This could be accompanied by a new engagement with forms of Muslim identity, and it could productively revisit the renewed debate on intellectual belonging, external attributions, and old, new, or modified allegiances in both the Global North and Global South.

5 Bibliography

5.1 Sources

Arkoun, Mohammed. *L'Humanisme Arabe au IVᵉ/Xᵉ siècle: Miskawayh, Philosophe et Historien*. Paris: Vrin, 1970 (2nd revised edition 1982, 3rd edition 2005).

Arkoun, Mohammed. *Essais sur la Pensée Islamique*. Paris: Maisonneuve et Larose, 1973; 1977[2]; 1984[3].

Arkoun, Mohammed. *La Pensée Arabe*. Que Sais-Je? No. 915. Paris: Presses universitaires de France, 1975 (at least eight more editions were published until 2010 and one new edition in the series *Quadrige*, Paris: Presses universitaires de France, 2012; 2014[2]).

Arkoun, Mohammed. *Lectures du Coran*. Paris: Maisonneuve et Larose, 1982 (2nd augmented edition Tunis: Alif – Editions de la Méditerrannée, 1991; 3rd edition published posthumously with the subtitle *Edition définitive*. Paris: Albin Michel, 2016).

Arkoun, Mohammed. *Pour une Critique de la Raison Islamique*. Paris: Maisonneuve et Larose, 1984.

Arkoun, Mohammed. *L'Islam: morale et politique*. Paris: Desclée de Brouwer, 1986; 1990[2].

Arkoun, Mohammed. *Rethinking Islam Today*. Occasional Papers Series. Washington: Center for Contemporary Arab Studies, Georgetown University, July 1987 (later republished in *Liberal Islam: A Sourcebook*, edited by Charles Kurzman. New York/Oxford: Oxford University Press, 1988: 205–21).

Arkoun, Mohammed. "Algeria." In *The Politics of Islamic Revivalism: Diversity and Unity*, edited by Shireen T. Hunter. Bloomington: Indiana University Press, 1988: 171–86.

Arkoun, Mohammed. "Le concept de société du Livre-livre." In *Interpréter: Hommage amical à Claude Geffré*, edited by Jean-Pierre Jossua and Nicholas-Jean Séd. Paris: Éditions du Cerf, 1992: 211–23.

Arkoun, Mohammed. *Min al-Ijtihād ilā naqd al-ʿaql al-islāmī*. Beirut: Dār as-Sāqī, 1991; 1993[2].

Arkoun, Mohammed. *Penser l'Islam Aujourd'hui*. Algiers: Laphomic/Enal, 1993.

Arkoun, Mohammed. *Penser l'Islam aujourd'hui: La communication impossible*. Unpublished manuscript, author's private collection. Paris, 1997.

Arkoun, Mohammed. “Du Dialogue Inter-religieux à la Reconnaissance du Fait Religieux.” *Diogène* 182 (1998): 103–26.
Arkoun, Mohammed. “L’Islam Actuel devant sa Tradition.” In *Christianisme, Judaisme et Islam: Fidélité et ouverture*, edited by Joseph Doré. Paris: Éditions du Cerf, 1999: 103–51.
Arkoun, Mohammed. “Contemporary Critical Practices and the Qurʾān.” In *EQ*, edited by Jane Dammen McAuliffe. 6 vols. Leiden: Brill, vol. 1, 2001: 412–31.
Arkoun, Mohammed. *The Unthought in Contemporary Islamic Thought*. London: Saqi, 2002 (revised as *Islam: To Reform or to Subvert?* London: Saqi, 2006).
Arkoun, Mohammed. *Humanisme et Islam: Combats et Propositions*. Paris: Vrin, 2005; 2010[2].
Arkoun, Mohammed. *La Question Ethique et Juridique dans la Pensée Islamique*. Paris: Vrin, 2010.
Arkoun, Mohammed. *La Construction Humaine de l’islam: Entretiens avec Rachid Benzine et Jean-Louis Schlegel*. Paris: Albin Michel, 2012.
Arkoun, Mohammed and Louis Gardet. *L’Islam, Hier – Demain*. Paris: Buchet-Castel, 1978; 1982[2].
Arkoun, Mohammed and Maurice Borrmans. *L’Islam, Religion et Société*. Interviews by Mario Arioso. Paris: Editions du Cerf, 1982.

5.2 Secondary Literature

Abī Nādir, Nāʾila. *Madkhal ilā dirāsat fikr Muḥammad Arkūn*. MA thesis, Université Libanaise/Lebanese University, Faculté des Lettres et des Sciences Humaines. Beirut,1994.
Abī Nādir, Nāʾila. *At-Turāth wa-l-manhaj bayna Arkūn wa-l-Jābirī*. Beirut: ash-Shabaka al-ʿArabiyya li-l-Abḥāth wa-n-Nashr, 2008.
Arfaoui, Hassan. “Entretien avec Mohammed Arkoun.” *Le Monde Arabe dans la Recherche Scientifique/The Arab World in Scientific Research/al-ʿĀlam al-ʿarabī fī-l-baḥth al-ʿilmī* 5 (1995): 7–32.
Benzine, Rachid. *Les nouveaux Penseurs de l’Islam*. Paris: Albin Michel, 2004.
Brisson, Thomas. “Leading Arab Intellectuals in the West: The Cases of Mohamed Arkoun and Edward Said.” In *Intellectuals and civil society in the Middle East: Liberalism, Modernity and Political Discourse*, edited by Mohamed Bamyeh. London: I.B. Tauris & Co LTD, 2012: 159–80.
Dubois, Olivier. “Contribution à une Bibliographie de Mohammed Arkoun.” *AR* 60 (2013): 473–515.
El-Ayadi, Mohammed. “Mohammed Arkoun ou l’Ambition d’une Modernité Intellectuelle.” In *Penseurs maghrébins contemporains*, edited by Collectif. Casablanca: Cérès-Editions, 1993: 43–71.
El Makrini, Naïma. *Regards Croisés sur les Conditions d’une Modernité Arabo-Musulmane: Mohammed Arkoun et Mohammed al-Jabri*. Louvain-La-Neuve: L’Harmattan, 2015.
Günther, Ursula. *Mohammed Arkoun: Ein moderner Kritiker der islamischen Vernunft*. Würzburg: Ergon, 2004.
Günther, Ursula. “Mohammed Arkoun: Towards a Radical Rethinking of Islamic Thought.” In *Modern Muslim Intellectuals and the Qur’an*, edited by Suha Taji-Farouki. London: Oxford University Press, 2004: 125–67.
Haleber, Ron, met medewerking van P.S. van Koningsveld. *Islam en Humanisme: De Wereld van Mohammed Arkoun*. Amsterdam: VU Uitgeverij,1991.
Kassab, Elizabeth Suzanne. “From the Unthought and the Unthinkable to the Thinkable: Mohammed Arkoun.” In *Contemporary Arab Thought: Cultural Critique in Comparative Perspective*, edited by Elizabeth Suzanne Kassab. New York: Columbia University Press, 2010: 174–83.
Kersten, Carool. *Cosmopolitans and Heretics: New Muslim Intellectuals and the Study of Islam*. London/ New York: Hurst Publishers, 2011.
Lacoste, Yves. “L’Islam et les islams: Entretien avec Mohammed Arkoun.” *Hérodote* 35 (1984): 19–34.
Lee, Robert D. “Arkoun and Authenticity.” *Peuples Méditerranéens* 50 (January–March 1994): 75–106.

Malti-Douglas, Fedwa. “Arkoun, Mohammed.” In *The Oxford Encyclopedia of the Modern Islamic World*, edited by John L. Esposito. 4 Vols. New York/Oxford: Oxford University Press, vol. 1, 1995: 139–40.

Masraḥī, Fāriḥ. *Al-Hadātha fī fikr Muḥammad Arkūn*. Beirut/ Algier: Dār al-ʿArabiyya li-l-ʿUlūm – Nāshirūn, 2006.

Mincheva, Dilyana. “Quranic Hermeneutics: The Work of Muhammad Arkoun and Nasr Abu Zayd.” In *The Politics of Muslim Intellectual Discourse in the West: The Emergence of a Western-Islamic Public Sphere*, edited by Dilyana Mincheva. Brighton/Chicago/Toronto: Sussex Academic Press, 2016: 33–41.

Saafi Hamda, Kalthoum. “Vers une Plateforme de Modernisation de la Pensée Islamique: Une nouvelle Alternative à travers la Pensée de Abdelmagid Charfi, Mohammed Arkoun et Nasr Hamid Abou Zeid.” PhD dissertation, Universite Sorbonne Nouvelle – Paris 3, 2011. www.theses.fr/2011PA030197.pdf, accessed August 28, 2024, 69–107.

Von Kügelgen, Anke. *Mohammed Arkoun und die islamische Philosophie*. MA thesis, Freie Universität Berlin, Berlin 1990.

Völker, Katarina. *Quran and Reform: Rahman, Arkoun, Abu Zayd*. Frankfurt/Main: Peter Lang, 2017.

Stefan Wild and Thomas Hildebrandt

Naṣr Ḥāmid Abū Zayd

1 Introduction

The Egyptian scholar Naṣr Ḥāmid Abū Zayd was one of the leading Muslim thinkers of the twentieth and early twenty-first century on Qurʾānic hermeneutics. He combined ideas from areas such as literary studies, linguistics, Western semiotics and hermeneutics, Islamic mysticism and theology and employed them in order to expand the debate about the Qurʾān and develop new models for its understanding. He also was a dedicated citizen, contributing to the religious and political discourse in his country, and he had to pay dearly for this when his views caused one of the major uproars in modern Egyptian intellectual life, forcing him into exile.

Abū Zayd took the Arabic word *taʾwīl*, usually rendered as "allegorical interpretation," as an approximation of the word "hermeneutics" and challenged traditional Muslim scholarship with a fusion of classical Islamic and modern hermeneutical views. His ideal of a hermeneutic *taʾwīl* was radical, but also deeply humanistic and influenced by his sympathies for the "Islamic Left." While inspiring a large number of liberal thinkers, his visions remain far too radical for many, if not most traditional Muslim scholars in Egypt and elsewhere.

2 Life and Career

2.1 From Egypt to Japan and back

Abū Zayd was a man of humble origins.[1] He was born in 1943/1362 in the village of Quḥāfa, located in the Nile Delta, and went through a traditional upbringing. He was only fourteen when his father died in 1957/1376. At this age, Naṣr had not only memorized the Qurʾān, but was also trained as a *qāriʾ*, being able to recite all of its surahs in one of the official melodic expressions. At the age of twelve, Naṣr had also been imprisoned due to alleged sympathies for the Muslim Brotherhood. In the 1960s he worked as a telephonist in Cairo, supporting his family, but in 1968/1388 he enrolled as a student at the Department of Arabic Language and Literature at Cairo University, from where he obtained a B.A. and later an M.A., taking a linguistically informed look

1 For autobiographical accounts, see his *Ein Leben mit dem Islam.* Narrated by Navid Kermani (Freiburg: Herder, 1999); *Voice of an Exile: Reflections on Islam.* With Esther R. Nelson (Westport, Conn./London: Praeger, 2004); and *Anā Naṣr Abū Zayd.* Written by Jamāl ʿAmr (Alexandria: Dār al-ʿAyn, 2013).

https://doi.org/10.1515/9783111320069-005

at questions of Qur'ānic exegesis. He managed to receive a fellowship for the American University in Cairo (1975/1395) and another one for the University of Pennsylvania in the United States (1978–1980/1398–1400). In 1981/1401, he received his Ph.D. in Cairo and became assistant professor of Islamic and rhetorical studies at his department. An exceptional scholarship allowed him to spend four years at the Osaka University of Foreign Studies in Japan (1985–1989/1405–1409).

In Osaka, Abū Zayd read the semiotic works of the Soviet structuralist Jurij M. Lotman (1922–1993/1340–1413) and was strongly influenced by this non-Muslim's reading of Islam. His most influential source, however, was the Japanese scholar of Islamic studies and comparative religion, Toshihiko Izutsu (1914–1993/1332–1413), who had translated the Qur'ān into Japanese and with whom Abū Zayd cooperated closely during his stay. Abū Zayd avidly read Izutsu's works, including the article "Revelation as a Linguistic Concept in Islam" and the book *God and Man in the Koran: Semantics of the Koranic Weltanschauung*.

After his return to Egypt in 1989/1409, Abū Zayd resumed work at his department and published a number of studies on Qur'ānic hermeneutics, the history of Islamic theology, Arabic linguistics, women's rights and the contemporary religious discourse in his country.

2.2 The Expulsion of a Critic

During his early years as a student, Abū Zayd's main teacher had been Ḥasan Ḥanafī (1935–2021/1354–1442), a professor of philosophy and an important authority in modern Islamic thought. Ḥanafī had returned from years at the Sorbonne in 1966/1386. He encouraged criticism and debated freely with his students, and many loved him for that. Ḥanafī was also a major advocate of the "Islamic Left." He problematized the outworn parole "Islam is the solution" and stressed the necessity of "enlightenment." He also put a great emphasis on hermeneutics. Ḥanafī's encyclopedic project *at-Turāth wa-t-tajdīd* (Heritage and Renewal) was an important milestone. During the 1990s, Ḥanafī was publicly accused of being an apostate, but he and his entourage managed to downplay this criticism and he remained largely undisturbed. Abū Zayd, however, did not have this recourse or this kind of supporters: when faced with similar accusitions, his life was in real danger.

Abū Zayd's problems began in May 1992/Dhū l-Qaʿda 1412, after he had submitted his application for promotion as full professor at his department.[2] At first, everything seemed in order: the relevant committee presented three reports, two in favour of the

2 For an overview of the case, with additional literature, see Kilian Bälz, "Submitting Faith to Judicial Scrutiny Through the Family Trial: The Abû Zayd Case," *Die Welt des Islams* 37, no. 2 (1997): 135–55; Jörn Thielmann, *Naṣr Ḥāmid Abū Zaid und die wiedererfundene ḥisba: Šarīʿa und Qānūn im heutigen Ägypten* (Würzburg: Ergon, 2003).

promotion, one against. But another committee soon denounced him as an apostate and unfit for a professorship. Even worse, Islamist groups were alerted and started to call for Abū Zayd's execution. These threats had to be taken seriously, as in the same year Islamists had killed the well-known Egyptian secularist Farag Foda in broad daylight in Cairo after similar accusations.

One of the main agitators against Abū Zayd was ʿAbd aṣ-Ṣabūr Shāhīn (1929–2010/1347–1431), an influential preacher in Cairo and a professor of Arabic linguistics at *Dār al-ʿUlūm*. He was the one person in the mentioned committee who had voted against the promotion. Shāhīn also happened to be a high-ranking member of the Islamic Investment Foundation of the El Rayan Group, i.e. a proponent of "Petro-Islam." The company had appeared in the late 1980s in the Egyptian-Saudi financial market and had played a vicious role by running a pyramid scheme that deprived thousands of pious Egyptians of their life-savings.

In 1993/1413, Shāhīn and a group of Islamist allies took the case to the Giza Family Court, demanding the nullification of Abū Zayd's marriage to his wife Ibtihāl Yūnis, a professor of French literature, also at Cairo University, arguing that *sharīʿa* law forbids a marriage between a Muslim woman and a non-Muslim man. The court rejected the demand, arguing that the plaintiffs had no personal interest in the matter, but on June 14, 1995/Muḥarram 16, 1415, Cairo's Court of Appeals reversed that ruling and found Abū Zayd to be an apostate, dissolving his marriage. On August 5, 1995/Rabīʿ al-Awwal 9, 1416, the Egyptian Supreme Court upheld that decision. The reactionary forces had won the day by using the classical principle of *ḥisba* – an activity to uphold community morals – as a loophole. Their goal was not the forced divorce, but the official verdict that Abū Zayd was an "apostate of Islam." It was not long before calls for Abū Zayd's execution increased, now also outside of Egypt.

It had become clear that some of Abū Zayd's stances were unacceptable to conservative Muslim scholars in Egypt and elsewhere in the Muslim world. Certainly, there were many intellectuals in favour of him: they were eager to discuss his views and wanted him to stay in Egypt – but they were not strong and determined enough to shield him. Fearing for their lives, Abū Zayd and his wife left Egypt on October 25, 1995/Jumādā l-Ākhira 1, 1416, and asked for asylum in Europe.

2.3 Exile, Return and Death

The couple first went to Spain, then to the Netherlands, where Abū Zayd found shelter in a teaching position at the University of Leiden which later appointed him to the Cleveringa Chair during the academic year of 2000–2001/1421–1422, named after an opponent of Nazi rule in the 1940s. In 2004/1425, he was granted the Ibn Rushd Chair of Humanism at the University of Humanistic Studies at Utrecht. For a while, he also worked at the Wissenschaftskolleg zu Berlin and took part in a number of projects on intercultural hermeneutics. He was invited to speak in many countries and also travelled to the United States where he was awarded the Franklin Delano Roosevelt Freedom of Worship Medal in 2002/1423. In 2005/1426, he received the Ibn Rushd Prize for Freedom of Thought in Berlin. During these years, Abū Zayd remained an active critic of the political and social situation in Egypt and elsewhere in the Islamic world.

Abū Zayd did not return to Egypt for a long time, but his wife came and went often. However, during the turn of the year 2002–2003/1424–1425, he spent two short weeks there and later started to make preparations for what he called a "decent return,"[3] but he tried in vain to reclaim his role at his department, his pension and his full rights as a citizen. Abū Zayd's last trip was an invitation to the Indonesian University in Jakarta in 2010/1431. During this visit he fell ill, infected by an unknown virus. He was sent to Cairo, where he was taken to hospital, and died there on July 5, 2010/Rajab 23, 1431, at the age of 66. He was buried on the same day in Quḥāfa.

3 Hermeneutical Thinking

3.1 From the Literary Exegesis of the Qur'ān to the Muʿtazilite Concept of *majāz*

The "case of Abū Zayd" has often overshadowed the more intricate details of his hermeneutical thinking. The campaign against him was mainly a reaction to his book *Naqd al-khiṭāb ad-dīnī* (Critique of the Religious Discourse) of 1992/1413.[4] In it, Abū Zayd describes the two conflicting trends of political Islam in Egypt – the "official" Islam of those scholars and institutions close to the state and the Islamist opposition – as two sides of the same coin. According to him, both of them used the Qur'ān for their own advantages, offering a simplistic, ahistorical and definite understanding of the divine book, while ignoring some of the most basic principles of interpretation. This assessment, including an open critique of the mentioned investment company, was Abū Zayd's real transgression, not his hermeneutical ideas underlying it.[5]

These ideas, however, cannot be fully separated from his analysis, and they were of such a delicate nature, that it was easy to use them against him. Early on, Abū Zayd had ventured into a heavily contested field which related back to another major controversy over religious issues in Egyptian academic life: back then, the accused were Amīn al-Khūlī (1895–1966/1313–1386), a professor of Arabic literature at Cairo University, and one of his students. During the 1930s and 1940s, al-Khūlī was the main pioneer of the so-called "literary exegesis" (*at-tafsīr al-adabī*), calling for an application

3 Cf. Abu Zaid, *Voice of an Exile*, 153–63.

4 A summary can be found, i.a., in Elizabeth Suzanne Kassab, *Contemporary Arab Thought: Cultural Critique in Comparative Perspective* (New York: Columbia University Press, 2009), 184–88.

5 Cf. Abu Zaid, *Voice of an Exile*, 5.

of the methods of literary studies to the analysis of the Qur'ān. His argument was that the Qur'ān was the greatest literary work in the history of the Arabic language and, as such, should be interpreted by using the same methods as with regard to any other literary work. In this context, al-Khūlī also reinterpreted the classical idea of the "inimitability" or "miraculous nature" (*iʿjāz*) of the Qur'ān. Originally, this concept refers to the superior nature of the Qur'ānic language, with all its spiritual depth, exclusive wisdom and inner beauty, and it was the prime argument for the Qur'ān's divine origin and the prophethood of Muḥammad, claiming that no human being alone could have come up with words of that magnitude. For al-Khūlī, though, the *iʿjāz* worked on a different level: the true miracle of the Qur'ān, for him, was its ability to speak to the hearts of the people in Muḥammad's surrounding and affect them emotionally in such a way that they turned into dedicated believers and did what was necessary for the establishment of Islam. With this aim in mind, dubbed the "psychological *iʿjāz*" by al-Khūlī, the Qur'ān did not offer precise descriptions of past events or timeless rules of human conduct, but resorted to artistic means. Instead of focusing on the Qur'ān's literal statements, any modern exegesis should therefore strive for a deeper understanding by uncovering the true mechanisms and spiritual goals of this literary kind of speech.

When al-Khūlī's student Muḥammad Aḥmad Khalafallāh (1916–1997/1334–1418) submitted his Ph.D. thesis about "The Art of Storytelling in the Noble Qur'ān" in 1947/1366, putting these ideas into praxis, a scandal erupted. Khalafallāh spoke of things such as God's artistic freedom and his creative imagination in telling stories and parables made for a didactical purpose. He also claimed that God drew on the existing knowledge and the methods of communication of that time – in other words: the cultural imaginary prevalent on the Arabian Peninsula and the vocabulary and grammatical rules of the Arabic language. For this, Khalafallāh described the Qur'ān as human in expression and style.

Both men suffered heavily, as they were accused of being apostates and had to endure personal threats and major setbacks in their careers. As a consequence, the literary exegesis of the Qur'ān was set aside at their department for about three decades for fear of further drawback. This changed in 1976/1396 when Abū Zayd, backed by his teachers, put forward his M.A. thesis about "The Rational Trend in Qur'ānic Interpretation: A Study on the Issue of Metaphorical Expression (*majāz*) as Viewed by the Muʿtazila." In it, he takes a close look at that famous school of theologians, whom he had come to be interested in through his teacher Ḥanafī, and lauds their theoretical ideas about language as a means to a metaphorical exegesis of the Qur'ān, based on human reason. Historically, the Muʿtazilites were at their peak between the ninth and eleventh centuries, until they became increasingly outlawed by the conservative Sunni mainstream. The renewed interest in this school during the twentieth century by many progressive Muslim thinkers was connected with aspirations for rational reform. For Abū Zayd, it was his entrance into the field of classical Muslim views about language, revelation, and exegesis.[6]

A proper knowledge of the Arabic language was always central to the Muslim sciences of the Qur'ān, but the classical theological schools, on a more theoretical level, also professed different views about the origins and inner workings of language in general. In fact, theology and linguistics were closely intertwined, with highly contested issues relating to the Qur'ān looming largely in the background. The famous controversy about the question of the "createdness" of the Qur'ān and the related question of God's proper attributes were linked to various theories about language. In his thesis, Abū Zayd examines the Muʿtazilite idea that language is based on human "convention" (*iṣṭilāḥ*), not on divine "instruction" (*tawqīf*), as claimed by their rivals, the Ashʿarites, and he explains how these views relate to their respective approaches to the Qur'ān: while both sides agreed that certain forms of metaphorical expression could be found throughout the Qur'ān, the Muʿtazila assigned a significantly greater role to human reason in deciding where this is the case and how it should be interpreted. For Abū Zayd, the Muʿtazilites also displayed a deeper view of the relationship between literal statements and their "semantic meaning" (*dalāla*) in the mind of a reader or hearer. They knew that, in order to comprehend any "speech" (*kalām*), including the speech of God, it was essential to have a certain insight into the underlying "intention" (*qaṣd*) and "state" (*ḥāl*) of that speaker, which, for them, could only be achieved by means of human reason. Abū Zayd also emphasizes the psychological role of *majāz* in captivating an audience, a clear link to the views of al-Khūlī.

3.2 From Muʿtazilite Theology to Islamic Mysticism and Ashʿarite Linguistics

Abū Zayd's view of the Muʿtazila is full of linguistic considerations in support of that school, but it also contains some criticism: he observes that the Muʿtazilites did not shy away from using the exegetical concept of *majāz* in order to justify some of their dogmatic and political positions when they were in contradiction to the actual wording of the Qur'ān. In his Ph.D. thesis of 1983/1403, Abū Zayd found a remedy against this self-serving approach in the mystical pantheism of the Andalusian poet and author Muḥyī d-Dīn b. ʿArabī (d. 1240/638). Titled "The Philosophy of Exegesis (*falsafat at-ta'wīl*): Studies on the Interpretation of the Qur'ān by Muḥyī d-Dīn Ibn ʿArabī," the study deals with one of the most comprehensive and controversial works of Muslim mysticism. Abū Zayd describes the approach of Ibn ʿArabī through the lens of modern

6 For Abū Zayd's view of the Muʿtazila, see Thomas Hildebrandt, "Between Muʿtazila and Mysticism: How much of a Muʿtazilite is Naṣr Ḥāmid Abū Zayd?" in *A Common Rationality: Muʿtazilism in Islam and Judaism*, ed. Camilla Adang, Sabine Schmidtke and David Sklare (Würzburg: Ergon, 2007), 495–512.

semiotic and hermeneutical thought, as offered by authors like Hans-Georg Gadamer (1900–2002/1318–1423), highlighting a number of similarities between them. According to him, *both* of these sides followed a dynamic, relativistic and individualistic approach to the problems of understanding and truth. In this sense, Abū Zayd admires the exegetical depth of Ibn ʿArabī's famous work *al-Futūḥāt al-makkiyya* (Meccan Revelations), as it refrains from giving definite explanations for specific verses, but regards the Qurʾān through a variety of lenses, offering a multitude of highly complex layers to its understanding.

Results like these inspired Abū Zayd to further searches in the Islamic heritage for similarities between Arab-Islamic ideas and modern hermeneutical thought. He often described these parallels with regard to the concept of "signs" and the mechanisms through which they have to be deciphered. Semiotics, of course, is the "science of signs" – in Arabic: *ʿilm al-ʿalāmāt* – and, according to Islamic theology, the world is also full of "signs" (*āyāt*) for the existence of its Creator, with every single verse of the Qurʾān also called *āya*, "sign." Abū Zayd now declared that a common feature between classical Muslim linguists was their profoundly semiotic world-view: they all agreed that man existed within a divinely created epistemological order, in which he was tasked to decode the surrounding signs through a process of "reading" (*qirāʾa*), in order to be able to live up to the principles laid down in the revelation.

Among the different linguistic traditions, Abū Zayd soon found his new hero in the Ashʿarite philologist ʿAbd al-Qāhir al-Jurjānī (d. 1078/471) whom he lauded for an approach that was firmly rooted in aesthetics. In contrast to the Muʿtazilites, who were willing to accept only those types of Qurʾānic *majāz* which suited their dogmatical interests and, for this, attempted to define language as a precise tool of the rational mind, al-Jurjānī focused on the infinite possibilities of language. He introduced theories about the artistic "composition" (*naẓm*) of speech, as professed in the realm of Arabic poetry, to the study of the Qurʾān and, through this, managed to offer an explanation of its *iʿjāz* that gave full recognition to often overlooked features, such as beauty and creativity, in contrast to the dogmatic and legalist approach of the Sunni mainstream. According to Abū Zayd, al-Jurjānī also displayed a unique sense for the hermeneutical intricacies of the process of "understanding" (*fahm*), when dealing with texts, which prompted him to liken the eleventh-century Persian scholar to the Swiss linguist Ferdinand de Saussure (1857–1913/1273–1331), one of the founding fathers of modern semiotics.

Al-Jurjānī's achievements, though, in the view of Abū Zayd, were still exceeded by Ibn ʿArabī: In the exegetical philosophy of this mystical giant the whole world was a kaleidoscope of God-given signs, directed to man by his Creator and resulting in a process of never-ending communication. Concepts like "text" (*naṣṣ*), "reading" (*qirāʾa*), "understanding" (*fahm*) and "interpretation" (*taʾwīl*) were key to this world-view, in which only the "knowing mystic" (*ʿārif*) – i.e. the person with the ability to see through the surface of things – was able to differentiate between the "outer" (*ẓāhir*) and "inner" (*bāṭin*) meanings of all phenomena.[7]

3.3 The Concept of *naṣṣ* as Focal Point of Islam

In al-Jurjānī, Abū Zayd had found a scholar from to the Ashʿarite school of theology who could be seen as a predecessor to some of his own key hermeneutical ideas. This strategy of reimagining and embracing some of the most respected authorities in the history of Sunni Islam was also employed in what is widely regarded as Abū Zayd's most important work, "The Concept of Text (*mafhūm an-naṣṣ*): A Study on the Sciences of the Qurʾān," of 1990/1410.[8] The two main Sunni authorities, built upon here, are az-Zarkashī (d. 1392/794) and as-Suyūṭī (d. 1505/911), whose respective books on the said sciences are among the most influential works in this field. Abū Zayd describes their approach not only in terms of its literary nature, but he also analyzes their contributions to a historical contextualization of the Qurʾān, as to be seen in concepts such as the "causes of revelation" (*asbāb an-nuzūl*), "abrogation" (*naskh*) and others.

While many of the previously mentioned ideas, e.g. on the origins of language and the phenomenon of *majāz*, reemerge in *Mafhūm an-naṣṣ*, the cornerstone of Abū Zayd's argument here lies in his application of the semiotic concept of communication to revelation and to the relationship between God and man. He describes the Qurʾān as resulting from a communicative relationship between its divine "sender" (*mursil*) and its human "receiver" (*mutalaqqī*), between God as the "speaker" (*mutakallim*) and man as the "addressed" (*mukhāṭab*), with the Arabic language being used as the "code" (*shifra*) for transmitting the message. In this model, clearly influenced by theorists such as Lotman and Izutsu, Abū Zayd also argues that the process of human *taʾwīl* must be seen as "the other side of the text." Muslim culture, in his depiction, always revolved around the concept of *naṣṣ*, with methodological reflections about the legitimate mechanisms of interpretation being part and parcel of Islamic civilization. Qurʾānic and literary studies could be proven to be closely intertwined, and a return to this hermeneutical line of thought constituted no obstacle whatsoever.

7 For his views on al-Jurjānī and Ibn ʿArabī, see Abū Zayd's article "al-ʿAlāmāt fī t-turāth: Dirāsa istikshāfiyya," in *Ishkāliyyāt al-qirāʾa wa-āliyyāt at-taʾwīl* (Beirut: al-Markaz ath-Thaqāfī l-ʿArabī, 1992^{2}), 51–118.

8 For relevant studies, see Stefan Wild, "Die andere Seite des Textes: Naṣr Ḥāmid Abū Zaid und der Koran," *Die Welt des Islams* 33, no. 2 (1993): 256–61; Navid Kermani, *Offenbarung als Kommunikation: Das Konzept* waḥy *in Naṣr Ḥāmid Abū Zayds* Mafhūm an-naṣṣ (Frankfurt am Main: Peter Lang, 1996).

3.4 Human and Historical Aspects of the Qur'ān

In the course of *Mafhūm an-naṣṣ*, and in a number of other works from that period, Abū Zayd often reflects on questions regarding the historicity of the revelation and its human dimensions. Traditional Sunni theologians had elevated the Qur'ān as "preexistent" (*qadīm*) and "uncreated" (*ghayr makhlūq*), existing long before history and being nothing but the "speech of God" (*kalām Allāh*), revealed to Muḥammad over a period of roughly twenty-three years, with the angel Gabriel acting as a transmitter and mediator. Revelation, here, was imagined as a form of dictation in which the prophet's role was that of a recipient, grasping the words, which were transmitted to him, in his heart and passing them on, in a literal manner, to his followers who, then, collected them and turned them into a book in exactly the way they had existed, long before time, in God's so-called "preserved tablet" (*al-lawḥ al-maḥfūẓ*).

Abū Zayd challenges these ideas by taking a closer look at the concept of "revelation" (*waḥy*) in Islamic thought and in the history of the Arabic language. Drawing on examples from pre-Islamic poetry, he argues that the term *waḥy* alluded not to dictation, but rather to a non-verbal form of communication in which the recipient had to express the transmitted ideas with his own words. Abū Zayd does not describe Muḥammad as the author of the Qur'ān, but he explains that the actual speech he came up with must be seen as part of the human side of the revelatory process and, as such, as temporal and open to human analysis.

Another major theme of Abū Zayd is the idea of a back-and-forth between God and the people in Muḥammad's surroundings during the years of his mission. With the Qur'ān being revealed over a long period of time, it provoked a variety of human reactions, which the Qur'ān, in turn, took into account and replied to – as can be observed in all verses containing expressions such as "they say" or, addressing Muḥammad, "tell them." Such evidence of an exchange, happening in time, for Abū Zayd is proof for what he describes as the historical and human dimensions of the revelation: God did not simply transmit certain prefabricated commands, but he rather listened to people's reactions and sent further ideas accordingly.

In the course of this reasoning, Abū Zayd often referred to the controversy about the "createdness" of the Qur'ān. The main proponents of this idea, the Muʿtazilites, had crafted their theories in the context of their discussion of the divine attributes. They insisted that the Qur'ān had to be seen as created, as otherwise Islam's fundamental concept of God's unity and uniqueness (*tawḥīd*), i.e. his role of being the sole preexistent being in the universe, would be compromised. Abū Zayd took these ideas a step further, discussing them under the title "The Defamed Concept of Historicity," where he argues that the Qur'ān, indeed, resulted from the mentioned back-and-forth, but it was also, in a more general sense, a product of the culture on the Arabian Peninsula during the seventh century AD. Abū Zayd, here, seizes upon the Muʿtazilite defense of monotheism and turns it into an argument for the historical nature of the revelation.[9]

3.5 From the Textuality and Historicity of the Qur'ān to Its Discoursivity

Abū Zayd's forced exile, including all the hassle and media attention accompanying it, took a heavy toll on his abilities to focus on his academic work. He often had to repeat things which he had already explained before. At the same time, his audience changed, now including many non-Arabic speakers from Western countries with considerably less knowledge about the complex details of classical Islamic thought. Against this background, Abū Zayd sometimes had to resort to more superficial and generalizing types of argumentation, also putting on a good face to his being passed around by some as the new "Luther of Islam," while remaining steadfast in his spirit of exploration and intellectual curiosity. But the new environment also provided him with opportunities to express his views with less strings attached, and it advanced his creativity as a theological thinker.

One of Abū Zayd's late topics brought back memories from life in an Egyptian village: the role of the Qur'ān in everyday Muslim life. He wrote an extended article about this subject for the *Encyclopaedia of the Qur'ān* and gave a theological summary of his findings in his inaugural lecture for the Cleveringa Chair in Leiden. Under the title "God and Man in Communication" Abū Zayd expanded his argument about the communicative nature of the Qur'ān. This communication, he says, did not only occur during the original revelation at the time of the prophet, but it was rather permanently revived and reenacted by the believers, individually and collectively, during the central Muslim rituals, such as recitation, prayer, fasting and pilgrimage. Abū Zayd draws the picture of an aesthetically and sensually guided form of living *with* the Qur'ān, long before and in contrast to the rational and doctrinal efforts by the scholars. In reality, the Qur'ān was not simply taken in by means of rational analysis, it was rather something to be experienced through listening and artistic recitation, and in doing so Muslims found themselves in a constant and multi-faceted exchange with their Creator, often even switching roles between speaker and listener. Mystical influences are abundant in this model, as there remain no clear boundaries between the human and the divine, the temporal and the transcendent, just as semiotic influences revolving around the concept of signs and the necessity of their decoding. Highlighting the impact of the Qur'ān on everyday Muslim life and stressing its foundational character for Islamic civilization, Abū Zayd here also refines his argument about

9 Cf. "Mafhūm 'at-tārīkhiyya' al-muftarā 'alayh," in Naṣr Ḥāmid Abū Zayd, *at-Tafkīr fī zaman at-takfīr: Ḍidd al-jahl wa-z-zayf wa-l-khurāfa* (Cairo: Sīnā li-n-Nashr, 1995), 197–230.

the Qur'ān being a product of its historical context: it is, he says, as much a product of its culture and time as a starting-point and producer of an entirely new culture.

While the historicity of the Qur'ān loses some of its importance here, the same later happened with regard to its textuality. Abū Zayd had worked hard to establish the idea of a textual character of the Qur'ān in order to make it accessible for textual analysis and the methodological tools of literary studies. But the Arabic equivalent for the term "text," the concept of *naṣṣ*, is a double-edged sword: in Islamic theology, just as in the Arabic language, it not only expresses the idea of something, usually words, that can be "read," but it is also – and prominently – used in a legal sense, as a *naṣṣ* is also understood to be a binding provision or a definite stipulation. Thus, the central norms of Islamic law are derived from what God has laid down in the revelation as "clear *naṣṣ*." Taking a step back from this concept, Abū Zayd now increasingly spoke of the Qur'ān as a "discourse" (*khiṭāb*), resorting to a term which is – due to the influences of (mainly French) discourse analysis – omnipresent in Arabic intellectual matters.

Abū Zayd explained the dimensions of the concept of *khiṭāb* in another inaugural lecture, this time for the Ibn Rushd Chair of Humanism in Utrecht, under the title "Rethinking the Qur'an: Towards a Humanistic Hermeneutics." The lecture starts off with a discussion of the Algerian-French thinker Mohammed Arkoun (1928–2010/1347–1431) who, since the 1980s, differentiated between the original "discourse," which God had directed to Muḥammad in the process of revelation, and the official "codex" (*muṣḥaf*), which was produced by his companions through systematic collection and redaction. For Arkoun, this difference was of crucial importance, as in the moment of codification Islam had gained its own foundational document behind which there was no going back. Abū Zayd criticizes this view, arguing that the "closed codex," which Arkoun talks about, was the Qur'ān of the social and intellectual elite, while for the ordinary believers the Qur'ān always remained an open source for their personal and direct communications with God. And even on its literary level, the Qur'ān contained many elements of a "living discourse" which should be rediscovered and revived by means of modern discourse analysis in order to reach a more democratic and humanistic understanding. The Qur'ān was a polyphonic text and a living phenomenon, containing a plethora of different voices, levels and arguments, and any scholarly attempt to narrow it down to a "clear *naṣṣ*" or a "closed *muṣḥaf*" made it prone to bias and manipulation.

4 Conclusion

Abū Zayd's ideas remain controversial in the Muslim world, as his statements are often hard to accept for conservative minds, especially when taken out of context or summarized for Western audiences, like in: "*Shari'a* law is human law. There is nothing divine about it." And: "There is no way to test or prove that the Qur'an is the Word of God."[10]

His efforts to revive the literary exegesis of the Qur'ān, in the line of al-Khūlī, and to infuse it with elements from modern semiotics and philosophical hermeneutics are Abū Zayd's central contribution to modern Islamic thought. Since he never wrote a systematical summary of his hermeneutical thinking, his ideas on this topic must be inferred from his numerous studies on the history of Qur'ānic exegesis. The underlying factor, in all of these studies, is his wish to steer away from one-sided dogmatic, legal and political interpretations of the Qur'ān towards a liberal, modern and humanistic approach. For Abū Zayd, the Qur'ān is not just a historical account or a legal text, dictated by God in order to be applied in a literal manner, and it does not provide clear answers to all human questions. The Qur'ān rather stems from a divine source, while using human language as its means of communication, and it is historically and culturally determined, while also kicking off an entirely new culture. Scholarly and individual approaches often stand in contradiction to each other, and too often the Qur'ān is being instrumentalized in order to shut down honest debate, rather than to encourage it.

With this last aim in mind, the late Abū Zayd spoke of establishing an institution in the Muslim world solely dealing with questions concerning the Qur'ān. He wanted to take over its scientific direction in cooperation with his friend ʿAlī Mabrūk (1961–2016/1381–1437), another student of Ḥanafī, who later became professor of Islamic and Arabic thought.[11] This wish was never fulfilled, as were so many others, but it showed Abū Zayd's unwavering interest in the Qur'ān and his hope to continue to explore the never-ending forms of communication it evokes.

5 Bibliography

5.1 Sources

Abū Zayd, Naṣr Ḥāmid. "Al-Ittijāh al-ʿaqlī fī t-tafsīr: Dirāsa fī qaḍiyyat al-majāz ʿinda l-muʿtazila." M.A. thesis of 1976. Beirut: al-Markaz ath-Thaqāfī l-ʿArabī, 1981.

Abū Zayd, Naṣr Ḥāmid. *Falsafat at-taʾwīl: Dirāsa fī taʾwīl al-Qurʾān ʿinda Muḥyī d-Dīn b. ʿArabī*. Beirut: Dār al-Waḥda, 1983.

Abū Zayd, Naṣr Ḥāmid. *Mafhūm an-naṣṣ: Dirāsa fī ʿulūm al-Qurʾān*. Cairo: al-Hayʾa al-Miṣriyya al-ʿĀmma li-l-Kitāb, 1990.

Abū Zayd, Naṣr Ḥāmid. *At-Tafkīr fī zaman at-takfīr: Ḍidd al-jahl wa-z-zayf wa-l-khurāfa*. Cairo: Sīnā li-n-Nashr, 1995.

Abū Zayd, Naṣr Ḥāmid. *Naqd al-khiṭāb ad-dīnī*. Cairo: Sīnā li-n-Nashr, 1992.

Abū Zayd, Naṣr Ḥāmid. *Ishkāliyyāt al-qirāʾa wa-āliyyāt at-taʾwīl*. Beirut: al-Markaz ath-Thaqāfī l-ʿArabī, 1992².

10 Abu Zaid, *Voice of an Exile*, 89, 96.

11 Cf. Abū Zayd, *Anā Naṣr*, 219.

Abu Zaid, Nasr Hamid. *Ein Leben mit dem Islam.* Narrated by Navid Kermani. Freiburg: Herder, 1999.
Abu Zayd, Nasr Hamid. "The Qur'ān: God and Man in Communication." Inaugural Lecture. Leiden University, 2000. Accessed January 06, 2023. https://hdl.handle.net/1887/5337.
Abû Zayd, Nasr. *Rethinking the Qur'ân: Towards a Humanistic Hermeneutics.* Utrecht: Humanistics University Press, 2004.
Abu Zaid, Nasr, with Esther R. Nelson. *Voice of an Exile: Reflections on Islam.* Westport, Conn./London: Praeger, 2004.
Abū Zayd, Naṣr Ḥāmid. *Anā Naṣr Abū Zayd.* Written by Jamāl ʿAmr. Alexandria: Dār al-ʿAyn, 2013.

5.2 Secondary Literature

Bälz, Kilian. "Submitting Faith to Judicial Scrutiny Through the Family Trial: The Abû Zayd Case." *Die Welt des Islams* 37, no. 2 (1997): 135–55.
Hildebrandt, Thomas. "Between Muʿtazila and Mysticism: How much of a Muʿtazilite is Naṣr Ḥāmid Abū Zayd?" In *A Common Rationality: Muʿtazilism in Islam and Judaism*, edited by Camilla Adang, Sabine Schmidtke and David Sklare. Würzburg: Ergon, 2007: 495–512.
Izutsu, Toshihiko. "Revelation as a Linguistic Concept in Islam." In *The Journal of the Japanese Society of Medieval Philosophy* 5 (1962): 122–67.
Izutsu, Toshihiko. *God and Man in the Koran: Semantics of the Koranic Weltanschauung.* Tokio: Keio University, 1964.
Kassab, Elizabeth Suzanne. *Contemporary Arab Thought: Cultural Critique in Comparative Perspective.* New York: Columbia University Press, 2009: 183–94.
Khalafallāh, Muḥammad Aḥmad. *Al-Fann al-qaṣaṣī fī l-Qur'ān al-karīm.* Cairo: Maktabat an-Nahḍa al-Miṣriyya, 1950/51.
Kermani, Navid. *Offenbarung als Kommunikation: Das Konzept* waḥy *in Naṣr Ḥāmid Abū Zayds* Mafhūm an-naṣṣ. Frankfurt am Main: Peter Lang, 1996.
Thielmann, Jörn. *Naṣr Ḥāmid Abū Zaid und die wiedererfundene ḥisba: Šarīʿa und Qānūn im heutigen Ägypten.* Würzburg: Ergon, 2003.
Wild, Stefan. "Die andere Seite des Textes: Naṣr Ḥāmid Abū Zaid und der Koran." *Die Welt des Islams* 33, no. 2 (1993): 256–61.

Ramy Abdin

Wahba az-Zuḥaylī

1 Short Introduction

Wahba b. Muṣṭafā z-Zuḥaylī (d. 2015/1436), a Damascene Islamic scholar of the twentieth and twenty-first centuries, can be described in a simple phrase: a scholar of modernized Islamic tradition. Although he is best known as an expert on sharia and secular law in the Arab World, he also left behind an impressive work of Qurʾānic exegesis, *at-Tafsīr al-munīr*, which will be the main focus of this chapter. In addition to a discussion of his exegetical work, the chapter will include a short biography and an overview of his work in order to shed some light on his scholarly career and thus his exegetical thought.

2 Life, Context, and Achievements

Unlike many past Islamic scholars who grew up in an environment of knowledge, paving the way for their careers, az-Zuḥaylī came from a simple family from the Syrian town of Dayr ʿAṭiyya, located north of Damascus.[1]

Born in 1932/1351, az-Zuḥaylī spent his early childhood studying the Qurʾān in a *kuttāb* (Qurʾān school) under the guidance of a *ḥāfiẓa*[2] until he was proficient in the discipline of Qurʾānic recitation (*tajwīd*). He then started primary school in his local region, after which he moved to Damascus, where he spent six years studying at the madrasa Kulliyat ash-sharīʿa, marking the beginning of his education in the Islamic disciplines.[3] In 1953/1372, az-Zuḥaylī began studying at the prestigious University of al-Azhar in Egypt, graduating five years later with highest marks in the field of Arabic language and sharia law. In addition to studying religious studies, he earned a degree in secular law from Ain Shams University in Cairo. As an undergraduate, he continued his studies at Cairo University's Institute of Sharia (*Maʿhad ash-sharīʿa al-islāmiyya*) in the Faculty of Law, enrolling in a master's program that he completed two years later.[4] He followed up his master's degree with a PhD thesis titled "*Āthār al-ḥarb fī l-fiqh al-islāmī – dirāsa muqārana*" (The Effects of War in Islamic Jurispru-

1 Badīʿ as-Sayyid al-Laḥḥām, *Wahba az-Zuḥaylī: al-ʿĀlim al-faqīh al-mufassir* (Damascus: Dār al-Qalam, 2001), 7.

2 A female Qurʾān teacher who memorized the Qurʾān.

3 Al-Laḥḥām, *Wahba az-Zuḥaylī*, 9.

4 Ibid., 9f.

https://doi.org/10.1515/9783111320069-006

dence – a Comparative Study),[5] in which he examined the legality of peace and war from an Islamic and secular perspective, presenting a synthetic Islamic view on international relations in the context of war in the modern age. It was indeed a pioneering work, as Thomas Pierret has pointed out, as it shed light on a very controversial issue that scholars had never looked at in so much detail before. Thus, it is no accident that az-Zuḥaylī's thesis is still an important source on the subject of war in Islam in relation to modern international law, which is why it has been translated and reprinted numerous times.[6] As Pierret writes, "The most original aspects of al-Zuhayli's work [were] his attempts at making *fiqh* relevant to legal realms in which it remained underdeveloped."[7]

Az-Zuḥaylī's interest in modern secular law was likely influenced by the political situation of his time. The Ottoman Empire, which existed only formally at the beginning of the twentieth century, completely collapsed in 1922 and, with it, the only remaining Islamic Sunni caliphate.[8] Thus, az-Zuḥaylī lived in the *post-Caliphate era*, politically characterized by the Western idea of the nation-state. He also witnessed the establishment of the French mandate for Syria and Lebanon, which took effect in 1923/1341. Although there were attempts at achieving Syrian independence in the 1930s, it was not declared de facto independent until 1946/1365.[9] In Egypt, az-Zuḥaylī must have been affected by the consequences of the Egyptian Revolution and the subsequent Arab Nationalist Movement led by Jamāl ʿAbd an-Nāṣir (d. 1970/1390), who initially overthrew the monarchy of the Muḥammad ʿAlī dynasty. The following years were shaped by the political ideology of Pan-Arabism, whose decline was reinforced by Israel's victory over Egypt and Syria in the Six-Day War in 1967/1387.[10] Hence, it can be argued that az-Zuḥaylī's efforts to gain a profound understanding of both systems of law, sharia and secular law, were an attempt to harmonize sharia and modern law, or even to make it possible to implement the rules of sharia within the legal system of a nation-state. This aspect can be demonstrated by examining his works in the field of sharia, which will follow in the next section.

After receiving his PhD in 1963/1383, az-Zuḥaylī was appointed an instructor at the Sharia Faculty of the University of Damascus until he was finally promoted to the position of Professor in 1975/1395. He then served as a visiting professor at many universities around the Arab world, such as Khartoum, Benghazi, and Riyadh, mainly

---

5 Ibid., 10f.

6 Thomas Pierret, *Religion and State in Syria: The Sunni Ulama from Coup to Revolution* (New York: Cambridge University Press, 2013), 113f.

7 Ibid., 114.

8 Stanford Jay Shaw, "Ottoman Empire: Historical Empire, Eurasia and Africa," in *Encyclopedia Britannica*, accessed August 30, 2021, https://www.britannica.com/place/Ottoman-Empire.

9 Howard Hayes Scullard et al., "Syria – The French Mandate," in *Encyclopedia Britannica*, accessed September 2, 2021, https://www.britannica.com/place/Syria/The-French-mandate.

10 Robert St. John, "Gamal Abdel Nasser: President of Egypt," in *Encyclopedia Britannica*, accessed September 2, 2021, https://www.britannica.com/biography/Gamal-Abdel-Nasser.

teaching in the faculties of law and sharia.[11] Alongside his teaching activities, he also took on a number of academic administrative duties over the years. From 1967/1387 to 1969/1389, he was the Dean of the Sharia Faculty of Damascus. Later he was appointed the Principal of Sharia and Law at the United Arab Emirates University, and, until his retirement, az-Zuḥaylī acted as Principal of *Fiqh* at Damascus University.[12]

In August 2015/Shawwāl 1436, he passed away in Damascus at the age of eighty-three.[13] He left behind a legacy of innovative perspectives in the field of sharia, which can be found in his numerous works, as well as a vast exegetical work on the Qurʾān.

Overall az-Zuḥaylī had a successful career as an Islamic academic. However, his scholarly activities were not limited to teaching at universities, as he held numerous memberships in international Islamic organizations, such as the International Islamic Fiqh Academy (Saudi Arabia) and the Royal Aal al-Bayt Institute for Islamic Thought (Jordan).[14] Furthermore he was a signatory to the Amman Message, which was issued in 2004/1425,[15] as well as to an open letter entitled "A Common Word between Us and You," signed by prominent figures of Islam and Christianity in 2007/1428.[16]

Az-Zuḥaylī can generally be described as a legal scholar who advocated for a modern view of sharia with the intention of paving the way for an Islamic renaissance in his time. He once stated, "I used to think that I would become an Islamic reformer [. . .], although I faced many obstacles [in] that direction. So, I took the path of writing in order [to] deliver what I [really] want to deliver by means of written script, not rhetoric [. . .]."[17] He did not see modernity and tradition as contradictory – on the contrary, he believed that tradition must be preserved, although it must change with time. This very idea was expressly emphasized in his works of *fiqh*.

---

11 Al-Laḥḥām, *Wahba az-Zuḥaylī*, 17f.

12 Ibid., 18f.

13 "Wafāt al-ʿallāma 'Wahba az-Zuḥaylī' al-mufassir al-faqīh al-uṣūlī as-sūrī," *Sabq Online Newspaper* (August 9, 2015), accessed January 24, 2024, https://sabq.org/saudia/kndgde-2.

14 Al-Laḥḥām, *Wahba az-Zuḥaylī*, 19–21.

15 "List of the Signatories at the International Islamic Fiqh Academy Seventeenth Session," *The Amman Message*, accessed August 29, 2021, https://ammanmessage.com/list-of-the-signatories-at-the-international-islamic-fiqh-academy-seventeenth-session/.

16 "A Common Word Between Us & You: Signatories (October 13, 2007)," *A Common Word*, accessed August 29, 2021, https://www.acommonword.com/signatories/.

17 "Ḥadīth adh-dhikrayāt maʿ D. Wahba az-Zuḥaylī," YouTube video, 52:50, posted by "drjasemtv," July 1, 2011, https://www.youtube.com/watch?v=JiypwMdXEuY=14m18s, my translation. Here, az-Zuḥaylī was a guest on a program called "*ʿulamāʾ mubdiʿūn*" (Creative Scholars) in which he described his career and achievements as an Islamic scholar.

3 Key Aspects of Az-Zuḥaylī's Work in Sharia Law

Az-Zuḥaylī authored more than thirty books, in particular in the field of sharia.[18] It would thus take more than one article to adequately analyze his extensive oeuvre. However, I would like to highlight some aspects of his legal work in order to provide some insights into his ideas about Islamic law.

In many of his works, he emphasizes the importance of sharia in a secular legal system and explores how Islamic law can provide an answer to unprecedented questions in the modern era,[19] such as the banking system and the problem of usury. For example, he states that the banking system is in itself *ḥarām* (prohibited) due to its contamination with usury, which, as a consequence, essentially prohibits Muslims from depositing funds into bank accounts. So how does a Muslim store his or her earnings without exposing it to risks like theft? Az-Zuḥaylī views the use of bank accounts as an unavoidable necessity, which is why he assesses storing money in a bank account as only *makrūh* (detestable). Nevertheless, he says, Muslims must use Arabic or Islamic banks to store their wealth in order to benefit the economies of Islamic countries.[20]

Az-Zuḥaylī calls for the transformation and renewal of *itjtihād*, arguing that the companions of the Prophet Muḥammad, including later prestigious scholars such as Muḥammad b. Idrīs ash-Shāfiʿī (d. 820/204), often revised their legal judgments.[21] He states, "It is possible for the new *mujtahid*s to change the judgments pursued by the former Imams, who were sincere in the highest degree in demanding the truth [. . .]."[22] This very transformation should provide an Islamic answer to recent developments such as modern technology, bureaucracy, and global finance, which past Islamic scholars had been unable to reflect upon.[23]

Furthermore, he distinguishes between *fixed* and *changing* sources of sharia, which he calls *thawābit* and *mutaghayyirāt*. *Fixed* legal sources in Islam are unchangeable and therefore must be implicitly preserved. This includes unambiguous texts (*nuṣūṣ*) from the Qurʾān and the hadith, whose legal validity is not limited to certain periods of time or particular causes, nor are they dependent on specific social customs.[24] However, the normativity of other texts falls under the category of these limitations. One considerable factor that potentially causes a *fiqh* rule to change is the

18 Al-Laḥḥām, *Wahba az-Zuḥaylī*, 22–42.

19 Wahba az-Zuḥaylī, *Taghayyur al-ijtihād* (Damascus: Dār al-Maktabay, 2000), 7–10.

20 Wahba az-Zuḥaylī, *al-Muʿāmalāt al-māliyya al-muʿāṣira: Buḥūth wa-fatāwā l-ḥulūl* (Damascus: Dār al-Fikr, 2002), 126f.

21 Wahba az-Zuḥaylī, *Subul al-istifāda min an-nawāzil wa-l-fatāwā wa-l-ʿamal al-fiqhī fī t-taṭbīqāt al-muʿāṣira* (Damascus: Dār al-Maktabay, 2001), 19.

22 Az-Zuḥaylī, *Taghayyur al-ijtihād*, 11.

23 Ibid., 14.

24 Az-Zuḥaylī, *Subul al-istifāda*, 19–21.

circumstance of *fasād* (corruption) as well as the reality of *taṭawwurāt* (developments). The living conditions of a society can be altered to such an extent that a legal rule that is already established becomes inutile or even harmful. This requires the *mujtahid* to rescind the rule and to potentially replace it with another judgment in order to reestablish society's well-being.[25]

In his works, az-Zuḥaylī also addresses the subject of comparative *fiqh* (*al-fiqh al-muqāran*). Publications like *al-Fiqh al-islāmī wa-adillatuhu* (Islamic Jurisprudence and its Proofs),[26] *Mawsūʿat al-fiqh al-islāmī wa-l-qaḍāyā l-muʿāṣira* (The Encyclopedia of Islamic Jurisprudence and Modern Issues),[27] and *Uṣūl al-fiqh al-islāmī* (The Principles of Islamic Jurisprudence)[28] mark his interest in comparative legal studies in the field of sharia. Nonetheless, az-Zuḥaylī dedicated a book series to the Sunni Schools of Islamic Jurisprudence, in which he composed one volume for each *madhhab*,[29] thereby providing a general overview of their systems of rule. Each volume bears the title *al-Fiqh al-muyassar* (Simplified Islamic Jurisprudence) in addition to the particular name of the Sunni School (*al-Fiqh al-ḥanafī l-muyassar*,[30] *al-Fiqh al-mālikī l-muyassar*,[31] *al-Fiqh ash-shāfiʿī l-muyassar*,[32] and *al-Fiqh al-ḥanbalī l-muyassar*[33]).

Az-Zuḥaylī's legal oeuvre is manifold, comprehensive, and reformist all at once. The variety of his legal work demonstrates his efforts to build a bridge between tradition and modern issues. Although he adheres to the teachings of the traditional Sunni schools, he aims to take them further by producing new answers to unprecedented issues while maintaining their legal methodology and, on top of that, rigorously reflecting on current global developments. This aspect of his innovative method appears especially clearly in works like *Subul al-istifāda min an-nawāzil wa-l-fatāwā wa-l-ʿamal al-fiqhī fī t-taṭbīqāt al-muʿāṣira* (Ways to Benefit from [new Legal] Issues, Fatwas, and the Implementation of Juristic Practice in Modernity) and *al-Muʿāmalāt al-māliyya al-ḥadītha wa-l-fatāwā l-muʿāṣira* (New Financial Transactions and Modern

25 Ibid., 23–28.

26 Wahba az-Zuḥaylī, *al-Fiqh al-islāmī wa-adillatuhu: ash-Shāmil li-l-adilla ash-sharʿiyya wa-l-ārāʾ al-madhhabiyya wa-ahamm an-naẓariyyāt al-fiqhiyya wa-taḥqīq al-aḥādīth an-nabawiyya wa-takhrījuhā*, 6 vols. (Damascus: Dār al-Fikr, 1985²).

27 Wahba az-Zuḥaylī, *Mawsūʿat al-fiqh al-Islāmī wa-l-qaḍāyā l-muʿāṣira*, 14 vols. (Damascus: Dār al-Fikr, 2010).

28 Wahba az-Zuḥaylī, *Uṣūl al-fiqh al-islāmī* (Damascus: Dār al-Fikr, 1986).

29 The traditional Sunni school of Islamic jurisprudence.

30 Wahba az-Zuḥaylī, *al-Fiqh ḥanafī l-muyassar*, 2 vols. (Damascus: Dār al-Fikr, 2010).

31 Wahba az-Zuḥaylī, *al-Fiqh mālikī l-muyassar*, 4 vols. (Damascus: Dār al-Kalim aṭ-Ṭayyib, 2010).

32 Wahba az-Zuḥaylī, *al-Fiqh ash-shāfiʿī l-muyassar*, 2 vols. (Damascus: Dār al-Fikr, 2008).

33 Wahba az-Zuḥaylī, *al-Fiqh al-ḥanbalī l-muyassar*, 4 vols. (Damascus: Dār al-Qalam, 1997).

Fatwas).[34] Outside his interest in sharia and Islamic legal theory, he also authored several works on hadith[35] and on the history of religion.[36]

4 Az-Zuḥaylī's Exegetical Approaches According to his *Tafsīr* Works

4.1 Exegetical Works

After a brief overview of az-Zuḥaylī's legal thought, we have finally arrived at the main subject. As in his aforementioned works, he was also productive in the field of Qur'ānic exegesis. Az-Zuḥaylī authored three *tafsīr* works, which are interconnected. His most extensive exegetical work is the *Tafsīr al-munīr* (Enlightening Commentary), on which the other two Qur'ān commentaries are based. The *Tafsīr al-wajīz* (Concise Commentary) and the *Tafsīr al-wasīṭ* (Intermediate Commentary) are mere summaries of *at-Tafsīr al-munīr*, though the *Tafsīr al-wasīṭ* is more comprehensive than the *Tafsīr al-wajīz*. The *Tafsīr al-wasīṭ* consists of three volumes, unlike the *Tafsīr al-wajīz*, which only comprises a single volume. However, the *Tafsīr al-munīr* is a sixteen-volume commentary and is therefore an extensive work of Qur'ānic exegesis, which is why the emphasis of this chapter is on the latter commentary. Nonetheless, I will briefly provide an overview of the other two exegetical works in order to give general insights into his *tafsīr* approach.

4.2 *At-Tafsīr al-wajīz* (The Concise Commentary)

The *Tafsīr al-wajīz* is a commentary with an emphasis on the semantics of the Qur'ān. It clarifies the textual meaning of each verse in simple Arabic language without delving into ambiguities, different recitation variants (*qirā'āt*), or even the normative issues that are occasionally addressed in the Qur'ān. Az-Zuḥaylī sometimes mentions the circumstances of the revelation (*asbāb an-nuzūl*), but only if they have been transmitted in an authentic hadith.[37] In the introduction he writes,

34 Al-Laḥḥām, *Wahba az-Zuḥaylī*, 35.
35 Ibid., 33.
36 Ibid., 36.
37 Wahba az-Zuḥaylī, *at-Tafsīr al-wajīz: 'alā hāmish al-Qur'ān al-'aẓīm wa-ma'ahu asbāb an-nuzūl wa-qawā'id at-tartīl* (Damascus: Dār al-Fikr li-ṭ-Ṭibā'a wa-t-Tawzī' wa-n-Nashr, 1996[2]), e.g., 5, 53, 274, 348, and 538.

> The Concise Commentary is confined to a summarized clarification of the verse, as well as to the authentic [transmitted] occasion of the revelation if it is available, and at the same time avoids the *isrāʾīliyyāt* [external sources from the Jewish and Christian traditions]. It also commits precisely to the accomplishments of the old and new *tafsīr* works and what they include from traditional and rational Islamic commentary (*at-tafsīr bi-l-maʾthūr wa-l-maʿqūl*), which is [supposed to be] in harmony with the original indications of the Arabic language (*uṣūl al-bayān*) as well as with the objectives of sharia (*maqāṣid ash-sharīʿa*) and the way of the pious predecessors (*manhaj as-salaf aṣ-ṣāliḥ*) in terms of belief (*iʿtiqād*).[38]

In the introduction to the commentary, he also discloses his view on the intentions of the Qurʾānic revelations, which gives some indication of his exegetical thought:

> The main objective of the revelation of the noble Qurʾān is to implement its verses [signs] (*āyāt*), bring about moral behavior, and engender a commitment to its judgments and ways of life, so that those who act accordingly will become the best nation brought forth to mankind.[39]

Moreover, az-Zuḥaylī regards the Qurʾān as a divine revelation and also understands it as a practical scripture that should lead believers to manifest the Qurʾān's meanings through their actions. Furthermore, he advocates for adherence to the Islamic obligation of measured Qurʾānic recitation (*tartīl*), which requires both theoretical and practical knowledge of the basic rules of Qurʾānic elocution (*tajwīd*).[40]

Notable in this *tafsīr* is the regular mentions of a virtue (*faḍl*) at the beginning of numerous Qurʾānic chapters (sura). These virtues sometimes illustrate a certain benefit of the sura, which is elucidated in a particular hadith, proving and emphasizing the importance of the chapter. At times, az-Zuḥaylī uses "virtue" to refer simply to the specific occasion of revelation or to a prophetic practice in which the sura was involved. For example, the Prophet Muḥammad used to recite sura 56, *al-Wāqiʿa* (The Event), while conducting the dawn prayer (*ṣalāt al-fajr*);[41] the significance of sura 18, *al-Kahf* (The Cave), is that it protects against the *Dajjāl* (the great Deceiver) when it is recited by the believer;[42] and, according to a hadith of the Prophet Muḥammad, sura 36, *Yā Sīn*, is supposed to be recited for the dead since it addresses eschatological as well as theological subjects like resurrection on the Day of Judgment and the Oneness of God.[43]

---

38 Ibid., "Introduction" (no page number is given; however, this is the first page of text in this book that begins with the Basmala formula, just before az-Zuḥaylī's Qurʾān commentary begins).
39 Ibid.
40 Ibid.
41 Ibid., 535.
42 Ibid., 294.
43 Ibid., 441.

4.3 *At-Tafsīr al-wasīṭ* (The Intermediate Commentary)

With his *at-Tafsīr al-wasīṭ*, az-Zuḥaylī presents a more detailed but nevertheless concise Qur'ān commentary. Compared to the *Tafsīr al-wajīz*, which is supposed to be a commentary for the common people, the *Tafsīr al-wasīṭ* is intended to be an exegetical work for the more educated.[44] It also emphasizes the field of Qur'ānic semantics, though with more extensive explanations of the verses. Az-Zuḥaylī addresses in more detail many of the more uncommon or ambiguous Arabic terms that occasionally occur in the Qur'ān, but does not restrict Islamic traditions concerning the circumstances of revelatory descent (*asbāb an-nuzūl*) portrayed in many suras to just a small number of prophetic traditions (*aḥādīth*), as he does in the *Tafsīr al-wajīz*. Hence, the *Tafsīr al-wasīṭ* provides more background information to the suras, obtained from traditional Islamic sources, namely the hadith. Sometimes the *tafsīr* provides a very brief linguistic analysis of a certain verse without delving into morphological or rhetorical examinations.[45]

Notable is the thematic division of each sura. To be more specific, the Qur'ānic chapters have themselves been divided into different parts, depending on which subject the sura addresses. For example, at the beginning of verse 67 in sura 2, *al-Baqara* (The Cow), the Qur'ān recounts the story of the Jews and the cow that is supposed to be sacrificed on the order of God. Az-Zuḥaylī dedicates a whole chapter to this short narrative in his interpretation of the sura, presenting a lengthy explication supplemented with additional information.[46] Even a much shorter sura like sura 67, *al-Mulk* (Sovereignty), is divided into four thematic parts: 1) "From the proofs of divine power" (verses 1–11), 2) "The promise for the believers and the threat against the disbelievers" (verses 12–19), 3) "The challenge against the idol worshipers and the appearance of absolute divine power" (verses 20–24), and 4) "The one who knows about the resurrection and the threat against its deniers" (verses 25–30).[47]

I classify this exegetical work as a thematic commentary providing extended textual explanations in each chapter. Like the *Tafsīr al-wajīz*, it has been written in simplified Arabic, a feature which characterizes az-Zuḥaylī's exegetical approach, namely that of verbal simplicity.

Although az-Zuḥaylī's commentary generally relies on highly recognized Sunni *tafsīr* works (*ummahāt at-tafsīr*),[48] which he seems to use for his own elucidation, he hardly refers to any other exegetes or works in *Tafsīr al-wajīz* or *Tafsīr al-wasīṭ*. However, in the introduction to the *Tafsīr al-munīr* he gives some indication of his references.

---

44 Wahba az-Zuḥaylī, *at-Tafsīr al-wasīṭ* (Damascus: Dār al-Fikr, 2001), 1: 6.

45 Ibid. For example: 1: 12–15, 174f.; 2: 1509–15; and 3: 2570f.

46 Ibid., 1: 31–33.

47 Ibid., 3: 2696–707.

48 Ibid., 1: 6f.

4.4 *At-Tafsīr al-munīr* (The Enlightening Commentary)

Az-Zuḥaylī's most comprehensive *tafsīr* work is intended for the experts of Qur'ānic exegesis. It is a commentary written from the perspective of various Islamic disciplines. Thus, the *tafsīr* is a kind of universal explanation of the Qur'ān, providing different levels of sura analysis. Although there is no perceptible difference between this and the other two exegetical works when it comes to methodology, this *tafsīr* was evidently intended as a source for not just Qur'ānic exegetes but also scholars in the fields of theology, Arabic philology, Qur'ānic recitation, and sharia as well.[49] In the introduction, az-Zuḥaylī provides an indication of his exegetical orientation by shedding light on the sources that he used for his *tafsīr*. Regarding the theological and normative aspect of his work, az-Zuḥaylī remarks,

> [. . .] I distinguished [this exegetical work] by emphasizing aspects of creed and prophethood (*al-'aqā'id wa-n-nubuwwiyyāt*), ethics and morals (*al-akhlāq wa-l-mawā'iẓ*), as well as the clarification of God's signs in the universe (*tawḍīḥ āyāt Allāh fī l-kawn*), like ar-Rāzī in the *Tafsīr al-kabīr*, Abū Ḥayyān al-Andalusī in the *Baḥr al-muḥīṭ*, al-Alūsī in the *Rūḥ al-ma'ānī*, and [also like in] *al-Kashshāf* by az-Zamakhsharī.[50]

In terms of the narrative aspects of traditional Islamic sources, he writes, "[I also distinguished the *tafsīr*] by the clarification of the Qur'ānic narratives [about the prophets] (*qiṣaṣ*) and historic accounts (*akhbār at-tārīkh*) like the commentary of al-Khāzin or al-Baghawī."[51] And on the aspect of Islamic law, he lists the commentaries of al-Qurṭubī (d. 1273/671), Ibn Kathīr (d. 1373/774), al-Jaṣṣāṣ (d. 981/370), and Qāḍī Abū Bakr Ibn al-'Arabī (d. 1148/543) as his sources.[52] For his philological analysis, he refers to the exegetical work of Abū l-Qāsim az-Zamakhsharī (d. 1144/538) and Abū Ḥayyān al-Andalusī (d. 1344/745), while in terms of the discipline of Qur'ānic recitation (*al-qirā'āt*), he names exegetes like 'Abd Allāh b. Aḥmad an-Nasafī (d. 1310/710) and Abū Bakr Muḥammad b. al-Anbārī (d. 940/328).[53]

The last thematic aspect covered by the *Tafsīr al-munīr* is nevertheless exceptional. In the words of az-Zuḥaylī: "[I also distinguished this commentary] by [considering] [modern] sciences (*'ulūm*) and scientific theories regarding the universe (*naẓariyyāt 'ilmiyya kawniyya*) like Ṭanṭāwī Jawharī in his book *al-Jawāhir fī tafsīr al-Qur'ān al-karīm*."[54]

---

49 Wahba az-Zuḥaylī, *at-Tafsīr al-munīr fī l-'aqīda wa-sh-sharī'a wa-l-manhaj* (Damascus: Dār al-Fikr, 2001), e.g., 1: 55–60; 3: 99–105; 8: 521–29; and 12: 210–14.

50 Ibid., 1: 13.

51 Ibid.

52 Ibid.

53 Ibid., 13f.

54 Ibid., 14.

As previously mentioned, *At-Tafsīr al-munīr* contains various levels of analysis, which az-Zuḥaylī structures as an exegetical method (*manhaj*) in the following way:

1. Separating the verses into thematic units (*waḥdāt mawḍūʿiyya)* with clarifying titles
2. Providing an initial, brief overview of the content of each sura
3. Conducting a comparative analysis of the recitation method
4. Contextualizing the verses (*asbāb an-nuzūl*) using the most authentic traditional sources and biographies about the Prophet (*sīra*)
5. Explaining the meaning (*at-tafsīr wa-l-bayān*)
6. Emphasizing the legal judgments that can be derived from the Qurʾān (*al-aḥkām al-mustanbaṭa min al-Qurʾān*)
7. Examining the rhetorical and grammatical aspects of numerous verses (*al-balāgha wa-l-iʿrāb*) as a means of clarification[55]

Furthermore, he provides an historical overview of how the Qur'ān was compiled and then eventually evolved into a canonized book.[56]

5 Az-Zuḥaylī's Exegetical Method

Az-Zuḥaylī describes his exegetical method as *at-tafsīr bi-l-maʾthūr wa-l-maʿqūl* (exegesis based on tradition and reason). Traditional *tafsīr*, namely *at-tafsīr bi-l-maʾthūr*, only accepts Islamic traditions such as the hadith and the statements of the companions (*ṣaḥāba*) as a source for explaining the Qurʾān. Moreover, it uses other Qurʾānic references to explain the Qurʾān itself, indicating that the verses are interconnected, though az-Zuḥaylī does not seem to be content with the kind of solely traditional commentary that has already been written repeatedly in the past; he thus also describes his *tafsīr* as *maʿqūl* (rational). However, he does not reveal what "reason" is supposed to mean in his commentary, nor does az-Zuḥaylī present a definition of reason contrary to the idea of Islamic tradition, which he discusses by mentioning several Qurʾān exegetes. Nevertheless, I presume that he views the Qurʾān's correspondence with scientific discoveries as a manifestation of reason, as he believes the meaning of the Qurʾān to be in accordance with science and what he calls "solid theories" (*naẓariyyāt thābita*).[57] This gives an indication of his understanding of reason, which consequently must be reason based on empirical evidence and analytics.

55 Ibid., 11f.
56 Ibid., 22–27.
57 Ibid., 36.

The contrast between reason and tradition is frequently reflected upon in many Islamic disciplines,[58] although this contrast does not seem to appear within az-Zuḥaylī's exegetical work. I therefore propose that he did not see reason and tradition as two separate sources of Islam through which the Qur'ān is to be read and understood. Rather, he saw them as two interrelated factors that are mutually compatible. In other words, tradition resembles reason and vice versa. For him, it was unthinkable that tradition, in particular the Qur'ān, could contradict rational thinking. Errors in reasoning would therefore be a result of misunderstanding tradition.

He views the Qur'ān as the word of God (*kalām Allāh*). Thus, he holds firm to the concept of *al-iʿjāz* (inimitability), meaning that the Qur'ān cannot be imitated by God's creation, namely human beings or demons (*jinn*). Az-Zuḥaylī locates inimitability specifically within the Arabic language itself. He emphasizes the divine book's philological uniqueness on the basis of its creative metrics (*an-naẓm al-badīʿ*), mode of language (*al-uslūb*), and divine eloquence (*al-jazāla*), but the content of the suras likewise supposedly reveals the divine origin of the Qur'ān. He claims that the Prophet Muḥammad received detailed information about former Prophets and their related nations (*umam*), which he could not have obtained except through divine revelation, as he and his nation were a community of a people who did not read or write. Further, the Qur'ān is unique due to its *universality*, for it addresses numerous aspects of human life. Its semantics are always in harmony with reason (*ʿaql*), logic (*manṭiq*), science (*ʿilm*), and emotion (*ʿāṭifa*), and, moreover, touches the human soul due to its power of persuasion.[59]

Az-Zuḥaylī's *tafsīr* represents an updated and compendious version of traditional Islamic Qur'ānic exegesis. It is neither allegoric interpretation nor commentary, which, according to him, follows a certain creed or ideology. Instead, he seeks to bind Muslims and non-Muslims with what he refers to as the only divine revelation. Although his commentary is not intended for beginner students, he aims to render it accessible to a broad audience of scholars by including various insights from important Islamic disciplines in his *tafsīr al-munīr*. How exactly non-Muslim scholars are to be addressed remains unknown. Muslims on the other hand are induced to practice the Qur'ān by living by sharia rules and Islamic values. However, it is indeed a rewarding piece of work for the reader of traditional *tafsīr* literature with a reformist touch.

58 Seyyed Hossein Nasr and Oliver Leaman, eds., *History of Islamic Philosophy* (London/New York: Routledge, 2008), 75, 81–84, 132, 518, 537, 636, 1079.
59 Az-Zuḥaylī, *at-Tafsīr al-munīr*, 1: 32–36.

6 Conclusion

As previously indicated, az-Zuḥaylī was a traditional Sunni Muslim who was keen to preserve Islamic tradition. This aspect also appears in his biographical analysis of prominent early Muslim personalities.[60] However, I have argued that he adopted a very critical position toward the West by advocating for an Islamic revival, especially through the field of sharia, which explains his admiration for Jamāl ad-Dīn al-Afghānī (d. 1897/1314). His work on sharia reinforces his inclination toward the old Islamic tradition, though coupled with his interest in modern, reformist Islamic thought. It reflects his idea of continuing tradition in order to establish a modern Islamic collective that is not only able to keep pace with Western civilization but also to surpass it. This objective would have been almost in line with Nāṣir's political ideology and dream of a united Arab nation, which az-Zuḥaylī was probably influenced by – hence his additional occupation with secular law. Nevertheless, he believed that reflecting on Islamic heritage was the main factor in reforming Islam within its own tradition, paving the way for an Islam that has the capacity to find its place in the modern world.

His exegetical approach was undoubtedly influenced by his activity as a legal scholar, as it is an approach aimed at the unambiguity of the Qur'ān. This aspect is especially evident in his explanation of the *ḥurūf al-qaṭ'iyya*, the enigmatic letters, with which several suras begin, such as *Alif Lām Mīm* – as in sura 2, *al-Baqara* (The Cow) and sura 3, *Āl 'Imrān* (The House of 'Imrān) – or *Ḥā' Mīm* – as in sura 40, *Ghāfir* (The Forgiver) and sura 43, *az-Zukhruf* (Gold Ornaments). Although he admits that the letters are ambiguous, he nevertheless favors the idea that they signify a particular meaning.[61] He refers to the Qur'ān as a clear message expressed in a pure Arabic,[62] yet he does not deny the Qur'ānic usage of rhetorical devices such as metaphor or metonymy.[63] However, the Qur'ān describes itself as a scripture that also contains ambiguous verses (*mutashābihāt*).[64] Az-Zuḥaylī addresses this very ambiguity, but he never dares to offer his own interpretation of ambiguous passages in the Qur'ān. His *tafsīr* works are merely explanatory commentaries mirroring a very long tradition of Qur'ānic exegesis. Thus, readers seeking an innovative exegetical work will be disappointed, as az- Zuḥaylī does not reinvent the wheel in the discipline of *tafsīr*, but rather reproduces a traditional exegetical work by providing an admittedly well-summarized and encompassing reconstruction of traditional exegesis.

60 See works such as Wahba az-Zuḥaylī, *Usāma b. Zayd: Ḥibb rasūl Allāh wa-bnu ḥibbih* (Damascus: Dār al-Qalam, 1996[3]); Wahba az-Zuḥaylī, *'Ubāda b. Ṣāmit: Ṣāḥibī kabīr wa-fātiḥ mujāhid* (Damascus: Dār al-Qalam, 1988[3]); Wahba az-Zuḥaylī, *Sa'īd b. Musayyib: Sayyid at-tābi'īn* (Damascus: Dār al-Qalam, 1992[5]).

61 Az-Zuḥaylī, *at-Tafsīr al-munīr*, 1: 40f.

62 Ibid., 36–38.

63 Ibid., 42–45.

64 Q 3:7.

His description of his *tafsīr* as *maʿqūl* (rational) might be understood as an attempt to produce an innovative commentary by bringing up the aspects of science and reason, as if the Qurʾān were not only God's revelation but also a book of scientific proofs.[65] But why is the aspect of science so relevant? Because it gives the impression that az-Zuḥaylī is trying to convince non-Muslims of the scientificity of the Qurʾān, especially as a Muslim who already believes in its divine origin. Nonetheless, the Qurʾān is not a scientific work, but a religious scripture that addresses aspects of the afterlife as well as normative issues related to society. Instead of bringing up the scientific aspect in the Qurʾān, it is instead expected that an exegete will deeply reflect upon theological issues, educating believers on how to lead a religious life. Since az-Zuḥaylī indeed discusses religious morality, his efforts to build a bridge between a religious scripture and modern science appear to have been unnecessary.

7 Bibliography

7.1 Sources

Az-Zuḥaylī, Wahba. *Al-Fiqh al-islāmī wa-adillatuhu: ash-Shāmil li-l-adilla ash-sharʿiyya wa-l-ārāʾ al-madhhabiyya wa-ahamm an-naẓariyyāt al-fiqhiyya wa-taḥqīq al-aḥādīth an-nabawwiyya wa-takhrījuhā*. 6 vols. Damascus: Dār al-Fikr, 1985[2].

Az-Zuḥaylī, Wahba. *Uṣūl al-fiqh al-Islāmī*. Damascus: Dār al-Fikr, 1986.

Az-Zuḥaylī, Wahba. *ʿUbāda b. Ṣāmit: Ṣāḥibī kabīr wa-fātiḥ mujāhid*. Damascus: Dār al-Qalam, 1988[3].

Az-Zuḥaylī, Wahba. *Saʿīd b. Musayyib: Sayyid at-tābiʿīn*. Damascus: Dār al-Qalam, 1992[5].

Az-Zuḥaylī, Wahba. *At-Tafsīr al-wajīz: ʿalā hāmish al-Qurʾān al-ʿaẓīm wa-maʿahu asbāb an-nuzūl wa-qawāʿid at-tartīl*. Damascus: Dār al-Fikr li-ṭ-Ṭibāʿa wa-t-Tawzīʿ wa-n-Nashr, 1996[2].

Az-Zuḥaylī, Wahba. *Usāma b. Zayd: Ḥibb rasūl Allāh wa-bnu ḥibbih*. Damascus: Dār al-Qalam, 1996[3].

Az-Zuḥaylī, Wahba. *Al-Fiqh al-ḥanbalī al-muyassar*. 4 vols. Damascus: Dār al-Qalam, 1997.

Az-Zuḥaylī, Wahba. *Taghayyur al-ijtihād*. Damascus: Dār al-Maktabay, 2000.

Az-Zuḥaylī, Wahba. *Subul al-istifāda min an-nawāzil wa-l-fatāwā wa-l-ʿamal al-fiqhī fī t-taṭbīqāt al-muʿāṣara*. Damascus: Dār al-Maktabay, 2001.

Az-Zuḥaylī, Wahba. *At-Tafsīr al-munīr fī l-ʿaqīda wa-sh-sharīʿa wa-l-manhaj*. 17 vols. Damascus: Dār al-Fikr, 2001.

Az-Zuḥaylī, Wahba. *At-Tafsīr al-wasīṭ*. 3 vols. Damascus: Dār al-Fikr, 2001.

Az-Zuḥaylī, Wahba. *Al-Muʿāmalāt al-māliyya al-muʿāṣira: Buḥūth wa-fatāwā l-ḥulūl*. Damascus: Dār al-Fikr, 2002.

Az-Zuḥaylī, Wahba. *Al-Fiqh al-ḥanbalī l-muyassar*. 4 vols. Damascus: Dār al-Qalam, 1997.

Az-Zuḥaylī, Wahba. *Al-Fiqh ash-shāfiʿī l-muyassar*. 2 vols. Damascus: Dār al-Fikr, 2008.

Az-Zuḥaylī, Wahba. *Al-Fiqh ḥanafī l-muyassar*. 2 vols. Damascus: Dār al-Fikr, 2010.

Az-Zuḥaylī, Wahba. *Al-Fiqh mālikī l-muyassar*. 4 vols. Damascus: Dār al-Kalim aṭ-Ṭayyib, 2010.

Az-Zuḥaylī, Wahba. *Mawsūʿat al-fiqh al-islāmī wa-l-qaḍāyā l-muʿāṣira*. 14 vols. Damascus: Dār al-Fikr, 2010.

65 Az-Zuḥaylī, *at-Tafsīr al-munīr*, 1: 36.

7.2 Secondary Literature

Armstrong, Karen. *Islam: A Short History*. New York: The Modern Library, 2002.

Geoffroy, Éric. *Introduction to Sufism: The Inner Path of Islam*. Indiana: World Wisdom, Inc., 2010.

Harris, William. "The View from Zahle: Security and Economic Conditions in the Central Bekaa 1980–1985." *MEJ* 39, no. 3 (1985): 270–86.

Hossein Nasr, Seyyed, and Oliver Leaman, eds. *History of Islamic Philosophy*. London/New York: Routledge, 2008.

Keddie, N.R. "AFĠĀNĪ, JAMĀL-AL-DĪN." In *EIr*, edited by Ehsan Yarshater. *Iranica online*, 1996–. Last updated July 22, 2011. https://www.iranicaonline.org/articles/afgani-jamal-al-din.

Al-Laḥḥām, Badīʿ as-Sayyid. *Wahba az-Zuḥaylī: al-ʿĀlim al-faqīh al-mufassir*. Damascus: Dār al-Qalam, 2001.

Pierret, Thomas. *Religion and State in Syria: The Sunni Ulama from Coup to Revolution*. New York: Cambridge University Press, 2013.

Scullard, Howard Hayes, et al. "Syria – The French Mandate." In *Encyclopedia Britannica*. Accessed September 2, 2021. https://www.britannica.com/place/Syria/The-French-mandate.

Shaw, Stanford Jay. "Ottoman Empire: Historical Empire, Eurasia and Africa." In *Encyclopedia Britannica*. Accessed August 8, 2021. https://www.britannica.com/place/Ottoman-Empire.

St. John, Robert. "Gamal Abdel Nasser: president of Egypt." In *Encyclopedia Britannica*. Accessed on September 2, 2021. https://www.britannica.com/biography/Gamal-Abdel-Nasser.

7.3 Online Sources

"A Common Word Between Us & You: Signatories (October 13, 2007)." *A Common Word*. Accessed August 29, 2021. https://www.acommonword.com/signatories/.

"Ḥadīth adh-dhikrayāt maʿ D. Wahba az-Zuḥaylī." YouTube video, 52:50. Posted by "drjasemtv," July 1, 2011. https://www.youtube.com/watch?v=JiypwMdXEuY=14m18s.

"List of the Signatories at the International Islamic Fiqh Academy Seventeenth Session." *The Amman Message*. Accessed August 29, 2021. https://ammanmessage.com/list-of-the-signatories-at-the-international-islamic-fiqh-academy-seventeenth-session/.

"Wafāt al-ʿallāma 'Wahba az-Zuḥaylī' al-mufassir al-faqīh al-uṣūlī as-sūrī." *Sabq Online Newspaper* (August 9, 2015). Accessed January 24, 2024. https://sabq.org/saudia/kndgde-2.

Irene Schneider

Fatima Mernissi

1 Introduction

Fatima Mernissi (1940–2015/1359–1437) was born in Fez, Morocco, and spent the majority of her life there. She was an influential scholar and feminist activist, who strove for gender equality and gender justice not only in her home country but in the whole Middle East and North Africa (MENA) region. She was perhaps *the* celebrity in the area of feminism, well known and widely read, not only in the Muslim world but also in Europe and the United States, where she graduated and lived for several years. As one of the first scholars in the area of gender equality and female sexuality (together, for example, with Nawal El-Saadawi in Egypt) she has been called an "icon" of feminism,[1] a "powerful voice" and an "inspirational figure for generations of women," and has encouraged many women as well as men to follow her path and fight against the inequality and oppression of women.[2] The attributions given to her vary between being an "Arab,"[3] a "secular" as well as an "Islamic" feminist, which describe not only her broad and differentiated approach to the topic of gender equality, but also reflects the development in her scholarly life.[4] As a trained scholar of sociology she focused from the beginning of her scientific career on social developments and social questions in her country, listening to the voices of not only middle-class women but also peasant women and factory workers in Morocco. At the same time, she always understood Islam as an important and indispensable part of the heritage of the MENA region, and it served as a framework for her scholarly work. Her recourse to Islam as a religion and her approach to interpreting Islamic sources, especially Qur'ān and hadith, in a gender-equal way intensified during her lifetime. With respect to this, the Gulf War in the 1990s and the rise of political Islam and Salafism as a turning point of the relations between the United States and Europe and the Arab world may have played a role in this.

In her publications, Mernissi argued that to understand the roots of the current situation of women's oppression in the Muslim world one has to go back and research and understand the Islamic past. Taking the veil as a symbol – it appears in several

1 Suleman Taufiq, "Obituary: Fatima Mernissi. Icon of Arab feminism," trans. Nina Coon, Qantara.de, last modified December 2, 2015, https://en.qantara.de/content/obituary-fatima-mernissi-icon-of-arab-feminism.

2 Zakia Salime, "Remembering Fatema Mernissi," Jadaliyya, last modified December 18, 2015, https://www.jadaliyya.com/Details/32791.

3 Taufiq, "Obituary".

4 Raja Rhouni, *Secular and Islamic Feminist Critiques in the Work of Fatima Mernissi* (Leiden/Boston: Brill, 2010), see for example, 3.

https://doi.org/10.1515/9783111320069-007

titles of her publications – she had a lifelong struggle with the question of why equal rights for women were so difficult to gain. According to her, the veil was originally a means of restricting the powerful female sexuality feared by the male elite. Her interpretation that men are weak and cannot resist women's charm, was not new – she admits this herself – but it was provocative. It turned the allegedly strong patriarchal men into fragile tyrants fighting for their position in society. At the same time, she fought furiously to deconstruct the cliché of "the Muslim woman" as a generally oppressed and passive human being. Stressing women's agency, she dealt in her research with early Islamic history and the Prophet's wives as role models, excavated *The Forgotten Queens of Islam* (1994/1415), and was finally able to empower many women and men in her lifetime to fight for gender equality.

Besides her academic publications, Mernissi also wrote fiction. Well worth mentioning – and recommending to the observant reader – is her book *Dreams of Trespass* (1994/1414), in which she describes growing up in an upper-class harem in the conservative Moroccan city of Fez. In relating the daily lives of women in this harem, she wanted again to deconstruct the predominant Western (male) perception of harems as exotic places of unrestricted female sexuality – always available to men. Sexuality, the cultural rootedness of female and male sexuality, and the historical difference of Christian and Islamic culture in dealing with female sexuality were some of her main interests.

Mernissi's scientific work, and especially her hermeneutical approach to the Qurʾān and sunna, focused not only on gender relations in the past and present but also on the question of power. She has been described as a "pioneering voice in redefining sharia and the Islamic textual tradition of prophetic rhetoric—hadith—as interpretive, contingent, and subject to human scrutiny."[5] Her approach has, in turn, been praised and disputed, as will be shown below, but she must surely be regarded as one of the first female scholars dealing in detail with female sexuality and the unequal position of women in the Arab and Muslim world, as well as recognizing the position of female subordination without falling prey to Orientalism and adopting the Western path to modernity as the only possible solution. The solution to the problem lay, as she saw it, in a new interpretation of the classical sources. This is why she has been described as "a marvelous philosopher, a creative spirit, and one of our era's most important feminists and Muslim intellectuals."[6]

5 Salime, "Remembering Fatema Mernissi".

6 Sarah Eltantawi, "A Woman for All Seasons. The World has lost a renowned Muslim Feminist with the passing of Fatima Mernissi," *Newsweek Middle East*, December 9, 2015, 56.

2 Life, Works, Political and Intellectual Context

Fatima (sometimes Fatema) Mernissi (Fāṭima Marnīsī) was born on September 27, 1940/Shaʿbān 25, 1359 in Fez, Morocco[7] into an upper-class house organized as a harem with special rooms and rules for the women in the house. It was a space whose "productive limitations" she beautifully documented in her 1994/1414 book, *Dreams of Trespass: Tales of a Harem Girlhood* (in the United States originally *The Harem Within: Tales of a Moroccan Girlhood,* also known under this title in the UK). Of the walls that circumscribed her movements as a child, she observed: "But women dreamed of trespassing all the time. The world beyond the gate was their obsession. They fantasized all day long about parading in unfamiliar streets."[8] Both her mother and her grandmother were illiterate. She mentions them often, whereas the father is not much present in her writings. Her primary education took place in a mixed school established by the nationalist movement. She started to study law at university but gave her studies up and left. In 1965–66/1385–86 she went to London, afterwards to Paris, and in 1970/1390 to the United States where she gained her doctorate at Brandeis University in Massachusetts. She later turned her thesis into a book with the title *Beyond the Veil: Male–Female Dynamics in Muslim Society* (published in 1975/1395, revised in Britain 1985/1405 and the United States 1987/1407). When she returned to Morocco in 1973/1393, she worked as a lecturer and research scholar at the University Institute for Scientific Research, Rabat. Her thesis became a classic in the field of anthropology and sociology on women in the Arab world. However, she published her most famous book in 1987/1407 under the title *Le Harem Politique*, published in English under the title *The Veil and the Male Elite* in 1991/1411. The book was banned in several Muslim countries, among them her home country Morocco. As a sociologist she did fieldwork for the United Nations, UNESCO and the International Labour Organization, focusing on women and work: in *Doing Daily Battles: Interviews with Moroccan Women* (1991/1411) she interviewed peasant women, women laborers, clairvoyants and maidservants.

Mernissi wrote all her books in English or French, never in Arabic, stating that she could not express herself freely in Arabic. The rough education in the Qurʾān school and the restrictions she experienced as a girl were the reasons for this, and she described that she felt as if a "policeman is standing beside me" when talking in Arabic.[9]

Fascinated by the character Scheherazade in the *Arabian Nights*, the powerful female narrator who manages to save her own life by captivating the king's attentive-

7 For her biography see Ursula Günther, *Die Frau in der Revolte: Fatima Mernissis feministische Gesellschaftskritik* (Hamburg: Deutsches Orient-Institut, 1993), 22–29.

8 Fatima Mernissi, *Dreams of Trespass: Tales of a Harem Girlhood* (Reading: Addison Wesley Publishing, 1994), 1–2.

9 Günther, *Frau in der Revolte*, 27.

ness with her fairy tales, Mernissi published in 2001/1422 a book titled *Scheherazade Goes West: Different Cultures, Different Harems,* the title of which already points to her aim: again, to deconstruct Western clichés of Muslim weak and passive women as well as Muslim institutions, such as the harem as an exotic place of unrestricted male sexual pleasure.

In the 2000s, Mernissi began to deal with the question of "digital Islam" and published an article titled "The Satellite, the Prince, and Scheherazade: The Rise of Women as Communicators in Digital Islam" (2003/1424) – again connecting to Scheherazade. Another influential book, offering a new perspective on Muslim history, was titled *Hidden from History: The Forgotten Queens of Islam* (1994/1414). Once more, Mernissi proved herself a pioneer in women's studies, this time in political history, by excavating women in power from Muslim history who had been overlooked in the male-dominated chronicles and history books, setting "her-story" against the classical version of history. In her book *Islam and Democracy: Fear of the Modern World* (1992/1412), Mernissi studies the socio-political context of the Arab-Islamic world shortly after the Gulf war. She deals with the question of whether the newly emerging Islamism dominating the Middle East could ever be compatible with the democratic processes used in Western societies. She analyzes Islamist approaches to rule and discusses "fears" of modernity in the Muslim world, i.e. democracy, the UN, freedom of thought and individualism. She pleads that there should be freedom of thought and the chance to disagree with the state. She also makes suggestions of how Muslims and especially feminists could use the Islamic past to prove that Islam generally is not against women's rights.[10]

3 Analysis of the Major Hermeneutical Works

Two books are important with respect to Mernissi's approach to Qurʾān interpretation and hermeneutics as well as the reading and interpreting of hadith:

3.1 Beyond the Veil

In *Beyond the Veil,* her PhD thesis, Mernissi discusses extensively what she calls in her introduction, added in 2011/1432, the "fixation of our 21st-century, secular, Western

10 For a critical review see, for example, Shanon Shah, review of *Islam and Democracy: Fear of the Modern World,* by Fatima Mernissi, trans. Mary Jo Lakeland, *Kyoto Review of Southeast Asia* 5 (March 2004), https://kyotoreview.org/issue-5/book-review-islam-and-democracy-fear-of-the-modern-world/.

man": the veil.[11] She points to the necessity to go back to focusing on why the Muslim veil was introduced in the seventh-century and to inquire how, and with what consequences, the emerging desire for sexual equality will be met in modern Arab-Muslim societies.[12] She approaches her topic in part one of her book through looking at the traditional Muslim view of women and their place in the social order, and in part two in looking at what she calls anomic effects of modernization on male–female dynamics. In the first part, she digs into the Islamic past and refers to rulings in Qur'ān and hadith. In the second part she relies on sociological data, on interviews she had with fourteen women and on four hundred letters sent to a government broadcast counselling service. Pointing to the fact that people tend to perceive women's liberation as a spiritual and not a material problem, and that changes in conditions for women were perceived by Muslim male literature as involving solely religious problems, she argues that women's liberation is an outspoken economic issue, too – as changes in the twentieth century, especially in socialist countries, have shown. "Liberation is a costly affair for any society, and women's liberation is primarily a question of the allocation of resources,"[13] she writes. On the other hand, describing Islam's basically positive attitude toward sexuality – and the basically negative attitude toward sexuality in the Western or the Christian world – she allows herself to speculate that women's liberation in an Arab context is likely to take a faster and more radical path than in Western countries.[14] Why is this the case? To understand her argument, it is necessary to look at her interpretation of the role of the veil within "active female sexuality" in the Islamic culture.

Relying on the theologian and jurist Abū Ḥāmid al-Ghazālī (d.1111/505) for the Islamic perspective and Sigmund Freud (d. 1939/1358) for the Western "Christian" view, Mernissi develops the concept of an active female sexuality and argues that what is feared in the Islamic culture is *fitna*. *Fitna* can mean disorder, chaos, even civil war but it also has the meaning of "a beautiful woman, the connotation of a femme fatale who makes men lose their self-control."[15] The woman's power and agency (*kayd*: lit. intrigue, ruse, her transliteration *qayd* is misleading, I. Schneider), she argues, is according to al-Ghazālī the most destructive element in Muslim social order and the social order "then appears as an attempt to subjugate her power and neutralize its disruptive effects."[16] This male fear of strong female sexuality is reflected in the hadith: "When a man and a woman are isolated in the presence of each other, Satan is bound to be their third companion." So Muslim scholars and men had to solve the problem of how to regulate and control female sexuality in the social system, through introduc-

11 Mernissi, *Beyond the Veil: Male-Female Dynamics in Muslim Society* (London: Saqibooks, 2011), 7.
12 Ibid., 8.
13 Ibid., 181.
14 Ibid., 193.
15 Ibid., 41.
16 Ibid., 43.

ing the veil, sex-segregation and other measures. Islam has, according to Mernissi, a progressive vision of history and she describes the traditional view of the pre-Islamic time as "barbarism" (*jāhiliyya*). Sexuality in the *jāhiliyya* was "promiscuous, lax, and uncontrolled" but became under Islam regulated and now obeyed rules.[17] It developed into a "civilized sexuality." At the same time, Mernissi is well aware that only woman's sexuality was civilized by Islam, male sexuality remained promiscuous – as an example she mentions polygamy – and lax – here her example is the male's prerogative to repudiate his wife at will without being obliged to give any reason. This contradiction she sees not only in seventh -century "family legislation" but also in the modern Moroccan Code, comparing both classical Islamic legal rulings, especially the ones anchored in Qur'ān, and the modern family law code in Morocco. Qur'ānic rulings with regard to polygamy and repudiation are, she concludes, still in force in the legislation in Morocco. With regard to the pre-Islamic ruling, she states: "Islam banished all practices in which the sexual self-determination of women was asserted."[18]

Mernissi's approaches are not always consistent, and they contain certain contradictions. They have been praised on the one hand for their focus on female agency and for re-reading history from a different – the women's – point of view, and criticized on the other. With regard to her analysis of hadith-literature it is striking that Mernissi, perhaps because she is a sociologist by training and not a historian, takes the hadith-material at face value without reflecting the problematic nature of her source material. She uses the approach of "pick-and-choose" in relation to the Qur'ānic verses as well as hadiths she considers appropriate and ignores others. Whereas modern mainstream research and the classical view of the emergence of the Qur'ān agree to a general extent, as the Qur'ān was revealed in the lifetime of the Prophet and put together in the decades after his death, hadith-collections needed, according to modern research, much longer to come into existence. The hadith-collections, especially the canonical works of al-Bukhārī and Muslim, have a much later origin. Although, according to classical Muslim belief, they contain "authentic" word and deeds of the Prophet, they must, according to modern research, be analyzed cautiously and with special sensitivity, taking into consideration contradictions and inconsistencies. Although having knowledge of and quoting the research of Joseph Schacht, for example, Mernissi ignores his contribution to hadith-research in which he warns that one should not take the canonical hadith-literature at face value and uses a thoroughly source-critical approach. As other independent sources for the early history of Islam are scarce, the social, legal and political history in seventh-century Arabia cannot easily be reconstructed.[19] With regard to the position of women in pre-Islamic and early Islamic time – an important topic for many Muslim

17 Ibid., 57.

18 Ibid., 79.

19 Irene Schneider, *Kinderverkauf und Schuldknechtschaft. Untersuchungen zur Frühen Phase des Islamischen Rechts* (Stuttgart: Franz Steiner, 1999), 62–74.

feminists today – not much can be said for sure and only very cautious conclusions can be drawn from this problematic source material.[20] Her reconstruction of the early Islamic (as well as the pre-Islamic) history is therefore anecdotal instead of historical-critical, and largely unsubstantiated.

Reviewers doubted that the combination of analyzing early Islamic sources on the one hand and looking at letters sent to a government broadcast counselling service in the twentieth century on the other presents an adequate way of dealing with the issue at hand,[21] questioning whether the author's conclusions are correct. Anthropologists have criticized Mernissi's discussion of matrilineality in the pre-Islamic social order and the notion that "ancestral fears can now be awakened by the module of sexuality purportedly prevalent in pre-Islamic times."[22] The lack of an exposition of family structure in Morocco and the fact that Mernissi's source material for the present time, the letters and interviews, are not incorporated into a fuller discussion of the sociology of the family are also seen as critical points.[23] The methodological approach has been considered "unconvincing."[24] It has even been concluded that *Beyond the Veil* remains largely unsubstantiated[25] and the book is explicitly not recommended for use in introductory classes, as it can be misleading to the uninformed.[26]

3.2 The Veil and the Male Elite

"Is Islam opposed to women's rights?" Mernissi asks in the preface to the English edition of *The Veil and the Male Elite*,[27] the book that is often considered her most important and which surely is the central one with regard to hermeneutics of the sacred texts. Her answer, which she develops throughout the book, is a clear "no." It is not Islam that is opposed, but a certain male (mis)interpretation of "true" Islam. Muslim men saying that Muslim women cannot be granted full human rights are "simply betting on our ignorance of the past,"[28] she states provocatively. Instead: "We Muslim

20 Irene Schneider, *Women in the Islamic World. From Earliest Times to the Arab Spring* (Princeton: Wiener, 2014), 35–36.

21 Carroll McC. Pastner, review of *Beyond the Veil: Male-Female Dynamics in a Modern Muslim Society*, by Fatima Mernissi, *Anthropological Quarterly* 51, no. 1 (January 1978): 87. Hildred Geertz, review of *Beyond the Veil: Male-Female Dynamics in a Modern Muslim Society*, by Fatima Mernissi, *Middle East Studies Association Bulletin* 11, no. 1 (February 1977): 50.

22 McC. Pastner, review of *Beyond the Veil*, 87.

23 Ibid., 88.

24 Geertz, review of *Beyond the Veil*, 50.

25 McC. Pastner, review of *Beyond the Veil*, 88; see also Geertz, review of *Beyond the Veil*, 49.

26 Geertz, revie of *Beyond the Veil*, 50.

27 Fatima Mernissi, *The Veil and the Male Elite: A Feminist Interpretation of Women's Rights in Islam*, trans. Mary Jo Lakeland (New York, NY: Basic Books, 1991), vi.

28 Ibid., vii.

women can walk into the modern world with pride, knowing that the quest for dignity, democracy, and human rights, for full participation in the political and social affairs of our country, stems from no imported Western values, but is a true part of the Muslim tradition."[29] These convictions guide Mernissi's approach to the Muslim past, which she develops in this book, looking at the veil and the seclusion of women in the Islamic past. It is "neither because of the Koran, nor the Prophet, nor the Islamic tradition, but simply because those rights conflict with the interests of a male elite"[30] that these customs are being revived. This male elite has manipulated the "sacred texts" to fabricate "false traditions," particularly whenever there has been a need to "legitimate" certain actions or attitudes. "Not only have the sacred texts always been manipulated, but manipulation of them is a structural characteristic of the practice of power in Muslim societies."[31]

Admitting herself that her book is not a work of history but what she calls a "narrative of recollection,"[32] Mernissi still claims to offer a new interpretation of the sources, Qur'ān and hadith. As a sociologist she informs the reader about her indebtedness to two of her colleagues of Université Mohammad V: Alem Moulay Ahmed al-Khamlichi, university professor and at the same time a traditional Muslim scholar, and the philosopher Ali Oumli. Both supported her in what can be called a very traditional approach to the sources, only with a reversed direction: arguing that women's rights as discussed in the twentieth century were already anchored in early Islamic times, that the Prophet himself defended gender equality and condemned violence against women, that Islam brought many rights for women in comparison with the pre-Islamic time, the *jāhiliyya*, that women in the Prophet's time and especially the Prophet's wives were what one could call feminist activists, fighting for more freedom and rights for women. She starts with an analysis of the hadith, which runs as follows: "Those who entrust their affairs to a woman will never know prosperity." This hadith is to be found again in the canonical hadith literature and therefore has throughout the centuries been considered as "sound" in the sense of "authentic" in the Muslim community. It is still today on everyone's lips when there is talk about the role of women in politics. Mernissi of course rejects this position but rejects it with the methodological instruments of the classical science of hadith. Her reinterpretation is based on a description of the circumstances under which Muḥammad allegedly uttered this hadith as well as on a depreciation of the transmitter, Nufayʿ b. Masrūḥ Abū Bakra. Approaching Qur'ānic verses, she describes her own spiritual experience with the reading of the text and her shock when confronted with misogynist traditions concerning women and the reading of the Qur'ān. She focuses on hijab and interprets it "before the background of the '*asbāb an-nuzūl*,'" the "circumstances of the revelation"

29 Ibid., viii.
30 Ibid., ix.
31 Ibid., 8–9.
32 Ibid., 10.

interpreting the hijab not as a "scarf" but as a barrier. In the Prophet's wedding night with Zaynab bint Jaḥsh Q 53:33 was revealed, according to which Mernissi understands hijab not as a barrier between a man and a woman but between two men: there was a group of people who did not leave the Prophet's place after the wedding ceremony had ended but stayed and annoyed the Prophet.[33] The hijab descended in the bedroom of the wedded pair, Mernissi explains, to protect the newly wedded pair's intimacy against this small group of tactless guests. The verse revealed in this special situation, Q 53:33, urges men to talk to the Prophet's wives only from behind a curtain. There are three dimensions to it, she argues: the visual, the special and the ethical.[34] She considers the veil as a key concept of the Muslim tradition just like "sin" in the Christian context and argues that it simply cannot be reduced to a "scrap of cloth."[35] A negative connotation in the sense of an impediment or something that prevents one from seeing God is rooted in e.g. Q 56:13 or in Q 5:41 – here she refers to aṭ-Ṭabarī's *Tafsīr* – and reflects the difficulties the Quraysh had in understanding the monotheistic message of Muḥammad.[36] In the twentieth century, she argues, Muslims in search of identity put the accent on the "confinement of women as a solution for a pressing crisis." A woman's body became what she calls a "symbolic representation of community,"[37] the veil a measure to restrict women's movement and prevent them from having equal rights to men.

Mernissi's descriptions of the people in the early Islamic time, including the Prophet and his wives, are lively and intensive, her judgments of their actions and characters are clear and decisive. As in *Beyond the Veil*, she ignores the problem of source critique and the possible ascriptions by later historians. The Prophet himself is depicted as a gentle, loving husband, sometimes even as a shy or timid man, who, for example, did not want to ask the tactless people still present in his house after the wedding party to leave. He never reacted violently against women but avoided situations of tension with his wives, she argues. How can this be brought into line with Q 4:34 and the husband's permission in this Qurʾānic verse to hit his wife? Mernissi refers here to the literature of "circumstances of revelation," according to which a woman who had been hit by her husband approached the Prophet and asked him for permission to hit back. He at first granted permission to her to hit back – and take vengeance – but shortly after this Q 4:34 was revealed. The Prophet now explained that he wanted something – but God wanted something different and of course God's will has to be obeyed. According to the verse, the husband should first admonish his wife, then avoid her in bed and as a last measure hit her. The interpretation of this verse has been challenging ever since, not only, but especially for feminist Qurʾānic

33 Ibid., 85–86.
34 Ibid., 93.
35 Ibid., 95.
36 Ibid., 97.
37 Ibid., 99.

exegetes.[38] Furthermore, neither the Prophet's advice to hit back nor God's order to allow only the male to hit the female can, strictly speaking, be seen as being in accordance with non-violence in relations between men and women. It is at this point, when describing the situation when the woman who had been slapped by her husband approached the Prophet, that Mernissi admits frankly that a "thorough, methodological study of the religious literature" still has to be done.[39]

Struggling with the different verses about gender relations in the Qur'ān, she points to Q 33:35 as establishing gender equality but has to admit that in Q 4:34 a hierarchy is established.[40] How to solve this contradiction? It is in this context that the person and personality of ʿUmar b. al-Khaṭṭāb, the second caliph, as described by Mernissi, is developed as the Prophet's counterpart. She describes ʿUmar as a misogynist who forced the Prophet into decisions against women's equal rights (see especially chapter 8). To him a saying is attributed according to which the Quraysh, Muḥammad's tribe, dominated their women, but when they arrived in Medina, they discovered that the Anṣār, the helpers among the inhabitants of Medina, let themselves be dominated by their wives – and now the Qurashī women started to copy these habits.[41] ʿUmar had reason to be alarmed, Mernissi argues, because his wife supported such a habit. She concludes that Anṣārī women split the community between women supported by the Prophet and the male population under the guidance of ʿUmar, who exerted pressure to keep their dominant position in the early Islamic society. Her conclusion might be considered surprising: "And this time God decided against his own Prophet, for the very survival of monotheism was threatened."[42] The Prophet was forced to accept this and give in to the openly misogynist behavior of ʿUmar and his peers.

Against the backdrop of the scarcity and contradictory nature of the source material this result can hardly be considered historically proven but it makes up a nice and coherent narrative. And there is more to it: in this narrative the Prophet as a peace-loving and non-violent person is defending women's rights as they are conceptualized in the twentieth century, but is prevented by his contemporary from implementing a gender-equal version of Islam. Mernissi describes it as the "Prophet's egalitarian project."[43] Men were prepared to accept Islam as a "revolution in relations in public life," but they did not want it to change anything in the personal relations with their wives and daughters. Men continued to try to suppress the egalitarian dimension of Islam.[44]

---

38 Schneider, *Women*, 36–57.
39 Mernissi, *The Veil and the Male Elite*, 155.
40 Ibid., 156.
41 Ibid., 143.
42 Ibid., 145.
43 Ibid., 142.
44 Ibid., 126.

Mernissi obviously sees neither any problems in projecting back ideas and concepts of the twentieth century into a seventh-century Arabian cultural context nor any need to try to reconstruct the social and legal environment of this epoch based on a historically thorough analysis of the problematic source material. She develops, then, a very "modern" picture of the behavior and character of the Prophet, his wives and his Companions, with the Prophet on the good side – supporting women and being explicitly against violence – and ʿUmar on the bad side, who is described as an aggressive and misogynist person without scruples.

She is well aware that her arguments constitute a narrative aiming to anchor gender equality as well as democracy in early Islamic time, to implement the "egalitarian project" of the twentieth century in early Islamic time, which became ruined by the male elite under the guidance of the misogynist ʿUmar. Her interpretation might be an easy-to-read narrative with many anecdotes aiming to draw a lively picture of the prophet, his environment, his behavior and his – allegedly non-violent, woman-friendly – disposition, and many character studies of his wives – young and unexperienced but strong ʿĀʾisha bint Abī Bakr, and Umm Salama Hind bint Abī Umayya who discussed political and gender questions with the Prophet and many others – but it is not convincing. The reason is that her interpretation is based on the traditional methodology of *isnād*-criticism, which Mernissi does not do in the necessarily thorough way, and on what can be called a re-reading of the texts in the opposite direction to what classical and traditionally trained scholars had done for the last 1400 years. Thus, her approach is open to attack from these traditional trained scholars. She openly admits her aim to see classical Islam as a source of modern ideals like human rights and democracy. The idea that she reads very modern concepts into ancient texts, does not bother her. She is – and this is what she has been criticized for – on a mission to establish a "true" reading of Islam which, however, has nothing to set against the same "true" reading of conservative, gender-hierarchical approaches.[45] She has been accused, therefore, of wanting to take what is best in the West and inject it into Islam.[46] It has also been argued that, by presenting "her reinvention of early Muslim society and the ideal of gender equality in Muḥammad's time as the truth that has been hidden or 'veiled' by the Muslim male elite," Mernissi reinforces the power of tradition.[47] Despite her "innovative contextual approach," her Islamic feminist methodology is criticized and her work falls prey to essentialism and the search for truth in the texts.[48]

---

45 Lamia Ben Youssef Zayzafoon, *The Production of the Muslim Woman. Negotiating Text, History, and Ideology* (Lanham etc.: Lexington Books, 2005), 22, quoted from: Rhouni, *Secular and Islamic*, 2.
46 Marlene Kanawati, review of *The Veil and the Male Elite: A Feminist Interpretation of Women's Rights in Islam*, by Fatima Mernissi, trans. Mary Jo Lakeland, *IJMES* 25, no. 3 (1993): 501.
47 Zayzafoon, *Production*, 22.
48 Rhouni, *Secular and Islamic*, 16.

4 Conclusion

Mernissi has provoked different reactions – praise and critique. However, it is undeniable that she was and still is the Grand Dame of Arab and Islamic feminism, a hero and idol for many young women (and men) in the Arab countries who are in search of their believed-to-be-lost Islamic identity. Her influence on her generation and the young people following her is rooted not so much in her sometimes disputable reconstruction of Islamic history, but rather in the overall approach that there is a modern heritage with Islamic roots which does not oblige Muslims to turn to Western values – as they have been there before in Islam.

5 Bibliography

5.1 Sources

Bibliography of Fatima Mernissi, see: Archive.org. "Fatima Mernissi." https://web.archive.org/web/20160504231954/http://www.mernissi.net/.

Mernissi, Fatima. *Doing Daily Battle: Interviews with Moroccan Women*, trans. Mary Jo Lakeland. New Brunswick: Rutgers University Press, 1989.

Mernissi, Fatima. *The Veil and the Male Elite: A Feminist Interpretation of Women's Rights in Islam*, trans. Mary Jo Lakeland. New York, NY: Basic Books, 1991.

Mernissi, Fatima. *Islam and Democracy: Fear of the Modern World*, trans. Mary Jo Lakeland. Reading, Mass. etc.: Addison-Wesley, 1992.

Mernissi, Fatima. *Dreams of Trespass: Tales of a Harem Girlhood*. Reading, Mass.: Addison-Wesley Publishing, 1994.

Mernissi, Fatima. *The Forgotten Queens of Islam*, trans. Mary Jo Lakeland. Minneapolis: University Minnesota Press, 2006[4].

Mernissi, Fatima. *Beyond the Veil: Male-Female Dynamics in Muslim Society*. London, Saqibooks, 2011.

5.2 Secondary Literature

Eltantawi, Sarah. "A Woman for All Seasons. The World has lost a renowned Muslim Feminist with the passing of Fatima Mernissi." *Newsweek Middle East*, December 9, 2015.

Fernea, Elizabeth. Review of *Beyond the Veil: Male-Female Dynamics in a Modern Muslim Society*, by Fatima Mernissi. *Social Science Quarterly* 57, no. 4 (March 1977): 925–26.

Geertz, Hildred. Review of *Beyond the Veil: Male-Female Dynamics in a Modern Muslim Society*, by Fatima Mernissi. *Middle East Studies Association Bulletin* 11, no. 1 (February 1977): 49–50.

Günther, Ursula. *Die Frau in der Revolte: Fatima Mernissis feministische Gesellschaftskritik*. Hamburg: Deutsches Orient-Institut, 1993.

Kanawati, Marlene. Review of *The Veil and the Male Elite: A Feminist Interpretation of Women's Rights in Islam*, by Fatima Mernissi, trans Mary Jo Lakeland. *IJMES* 25, no. 3 (1993): 501–03.

McC. Pastner, Carroll. Review of *Beyond the Veil: Male-Female Dynamics in a Modern Muslim Society*, by Fatima Mernissi. *Anthropological Quarterly* 51, no. 1 (January 1978): 87–88.

Mir-Hosseini, Ziba. "Honouring Fatima Mernissi." *Musawa* (February 2016): 1–5.

Rhouni, Raja. *Secular and Islamic Feminist Critiques in the Work of Fatima Mernissi*. Leiden/Boston: Brill, 2010.

Salime, Zakia. "Remembering Fatema Mernissi." Jadaliyya. Last modified December 18. 2015. https://www.jadaliyya.com/Details/32791.

Schneider, Irene. *Kinderverkauf und Schuldknechtschaft. Untersuchungen zur Frühen Phase des Islamischen Rechts*. Stuttgart: Franz Steiner, 1999.

Schneider, Irene. *Women in the Islamic World. From Earliest Times to the Arab Spring*. Princeton: Wiener, 2014.

Shah, Shanon. Review of *Islam and Democracy: Fear of the Modern World*, by Fatima Mernissi, trans. Mary Jo Lakeland. *Kyoto Review of Southeast Asia* 5 (March 2004). https://kyotoreview.org/issue-5/book-review-islam-and-democracy-fear-of-the-modern-world/.

Taufiq, Suleman. "Obituary: Fatima Mernissi. Icon of Arab feminism," trans. Nina Coon. Qantara.de. Last modified December 2, 2015. https://en.qantara.de/content/obituary-fatima-mernissi-icon-of-arab-feminism.

Zayzafoon, Lamia Ben Youssef. *The Production of the Muslim Woman. Negotiating Text, History, and Ideology*. Lanham etc.: Lexington Books, 2005.

Kathrin Eith

Yaşar Nuri Öztürk

1 Introduction

Yaşar Nuri Öztürk (1945–2016/1364–1437) was a Turkish theologian, politician and author. Especially in the 1990s and 2000s, he worked his way into Turkish society as a representative of modern Islam. He is considered the pioneer of a movement back to the Qur'ān (*Kur'an'a dönüş hareketi*) in Turkey. Starting from a contemporary interpretation of the Qur'ān, he set himself the goal of making a professing Muslim's way of life compatible with the laic understanding of politics and the state. In doing so, he criticized the state dominance of religious life and institutions just as clearly as he took issue with the emerging forces of political Islam in the party landscape and government of Turkey. His public presence as a frequent talk-show guest and columnist for major daily newspapers made him at times the most prominent representative of both Qur'ānism and the criticism of Turkey's ossified religious establishment.

2 Life, Work and Intellectual Background

2.1 Life

After his death on June 22, 2016/Ramaḍān 17, 1437, it was often claimed that Yaşar Nuri Öztürk[1] was born on June 22, 1951/Ramaḍān 17, 1370.[2] In fact, Öztürk was born on February 5, 1945/Ṣafar 22, 1364, in the village of Tahsinli in Bayburt Province, his mother's home.[3] On his father's side, he comes from the family of the Niyazoğlu, whose roots are said to reach back to Baghdad. His father originated from the coastal town Sürmene east of Trabzon.[4]

Instead of attending primary school, Öztürk was taught at home by his father, and at the age of nine knew the Qur'ān by heart.[5] He then underwent extracurricular

1 On Öztürk's life, see Kathrin Eith, *Koranexegese bei Yaşar Nuri Öztürk: Ein traditionskritischer Entwurf des Islams in der Türkei* (Würzburg: Ergon, 2013), ch. 3.1, "Life," 69–92.

2 Reports on Öztürk's death and funeral in various Turkish daily newspapers, e.g.: "Yaşar Nuri Öztürk vefat etti!," last modified June 22, 2016, https://www.sabah.com.tr/yasam/2016/06/22/yasar-nuri-ozturk-vefat-etti.

3 Yaşar Nuri Öztürk, *Kur'an'ı Anlamaya Doğru* (Istanbul: Yeni Boyut, 1990), ch. "Tereddütler ve Sorular," 193–95. The book contains 68 of Öztürk's weekly columns published in the newspaper *Hürriyet* since about 1988.

4 Ibid.

5 Ibid.

https://doi.org/10.1515/9783111320069-008

training in Islamic sciences and Arabic literature with Islamic scholars along the lines of an Ottoman madrasa's traditional curriculum.[6] It was not until he was seventeen years old that he completed his secondary schooling and attended an İmam Hatip School, a state school for the training of religious personnel, in Trabzon.[7] After a further year of extracurricular training with local sheikhs of no official authorization to teach, he finally graduated from the İmam Hatip School. Already in his Trabzon years, Öztürk was able to gain experience as an assistant preacher (*hatip*) due to a lack of personnel in local mosques.[8]

In 1968/1388, Öztürk moved to Istanbul to continue his religious studies at the Yüksek İslam Enstitüsü (High Institute of Islam), which had only recently been established.[9] At the same time he completed his law studies at Istanbul University. After his graduation he worked for a short period as a lawyer but decided to return to the Yüksek İslam Enstitütsü in 1976/1396, where he was employed first as university lecturer (*eğitim görevlisi*) and later as docent (*doçent*).[10] He received his doctorate in 1980 with a thesis that was published in 1982/1402 under the title *Kutsal Gönüllü Veli: Kuşadalı İbrahim Halveti (Hayatı, Tasavvufî Düşünceleri, Mektupları)*.[11] Öztürk left Turkey for research stays in Grenoble and New York (1985/1405) and then continued to teach at the Faculty of Theology of Istanbul's Marmara University, into which the High Institute of Islam had been transformed in 1982/1402. When a second Faculty of Theology in Istanbul was established at Istanbul University in 1992/1412 in the course of the nationwide expansion of academic theology, Öztürk was appointed professor.[12] After nine years as professor of theology and also dean of the faculty (1993–2002/1413–1423), he successfully ran for the Cumhuriyet Halk Partisi (CHP, Republican People's Party)

6 Obituary in the daily *Sabah*, "Yaşar Nuri Öztürk vefat etti!": "on yıllık klâsik medrese eğitiminden sonra . . .".

7 Under the Demokrat Parti (DP) government, seven İmam Hatip schools had been opened across the country in 1950/51 with the aim of providing state-directed training for religious personnel. By 1958, there were reportedly a total of eighteen such schools. See Bahaettin Akşit, "Islamic Education in Turkey: Medrese Reform in Late Ottoman Times and Imam-Hatip Schools in the Republic," in *Islam in Modern Turkey: Religion, Politics and Literature in a Secular State*, ed. Richard Tapper (London: Tauris, 1994), 145–70.

8 Margot Badran, "The Religious Face of Secularism," *Al-Ahram Weekly online*, no. 519, February 1–7, 2001, https://web.archive.org/web/20080412113732/http://weekly.ahram.org.eg/2001/519/intrvw.htm.

9 Ibid. Until 1982, the Faculty of Theology of Ankara University was the only institution in Turkey to offer theological studies at university level, the founding idea being comparative religious studies, which apparently was not in line with Öztürk's objectives. See Mehmet Paçacı and Yasin Aktay: "75 Years of Higher Religious Education in Turkey," *MW* 89, no. 3–4 (1999): 389–413.

10 Öztürk, *Kur'an'ı Anlamaya Doğru*, 195.

11 Yaşar Nuri Öztürk, *Kutsal Gönüllü Veli: Kuşadalı İbrahim Halveti (Hayatı, Tasavvufi Düşünceleri, Mektupları)* ["The Saint with the Blessed Heart: Kuşadalı İbrahim Halveti [Life, Sufic Thought, Letters]"] (Istanbul: Fatih, 1982).

12 "Tarihçe," İstanbul Üniversitesi İlahiyat Fakültesi, accessed October 18, 2020, https://ilahiyat.istanbul.edu.tr/tr/content/fakultemiz/tarihce.

in the 2002/1423 parliamentary elections. He suspended his professorship for entering the parliament in Ankara as a CHP representative, which then was the only party in parliament other than, and opposed to, the Adalet ve Kalkınma Partisi (AKP, Justice and Progress Party).[13] Two years later he resigned from the CHP and founded a party of his own, Halkın Yükselişi Partisi (HYP, Party of the People's Upswin). Until 2007/1428, he was the party's only deputy and until 2009/1430 its leader.[14] When he was diagnosed with stomach cancer in 2011/1432, Öztürk largely withdrew from professional and public life. Remaining in Istanbul, he however continued to publish. At his death in 2016/1437, aged seventy-one, he left behind a daughter and two sons from two divorced marriages.

2.2 Work

A most productive writer, Yaşar Nuri Öztürk authored an extraordinarily extensive and diverse range of publications.[15] In particular, it comprises over fifty monographs published between 1989/1409 and 2015/1436 by the Yeni Boyut ("New Dimension") publishing house he founded, among which several bestselling titles. Many of his works have been published in numerous editions within a short period of time, with some of them being heavily revised and updated. *Kur'an'daki İslam* alone saw thirty-nine editions between 1992/1412 and 2001/1422, and his translation of the Qurʾān was reprinted some hundred times within the first ten years.[16] His books target a Turkish-speaking audience interested in religious themes, although translations of individual works into English, German and Persian have also become available.[17]

Öztürk's early publications, starting in the 1970s, concentrate primarily on Sufism. In addition to his dissertation on Kuşadalı İbrahim Halvetî (1982/1402) he wrote on the ninth-century Persian-Iraqi Sufi and poet Ibn Manṣūr al-Ḥallāj (1976/1396) and the nineteenth-century Ottoman mystic Muhammed Tevfik Bosnevî (1981/1401) as well as

13 AKP had won 66% of the mandates with 34% of the electoral votes due to the 10% hurdle. Haberler.com, "2002 Genel Seçim Sonuçları," accessed October 18, 2020, https://haberler.com/secim/2002/.

14 HYP's subsequent chairman was Ragıp Önder Günay (2009–2018). HYP finally dissolved in 2018.

15 Detailed information in Eith, *Koranexegese*, ch. 3.2, "Works," 92–100. For a list of Öztürk's publications, see ibid, 377–82.

16 *Kur'an'ı Kerim Meâli (Türkçe Çeviri)*, trans. Yaşar Nuri Öztürk (Istanbul: Yeni Boyut, 1993), 104 editions between 1993 and 1999; *Kur'an'ı Kerim ve Türkçe Meâli (Metinli)*, trans. Yaşar Nuri Öztürk (Istanbul: Yeni Boyut, 1994), 102 editions between 1994 and 2002; Yaşar Nuri Öztürk, *Kur'an'ın Temel Buyrukları (Emirler ve Yasaklar)* (Istanbul: Yeni Boyut, 1997), fourteen editions between 1997 and 2006, and three editions of an expanded two-volume edition until 2012; Yaşar Nuri Öztürk, *İslam Nasıl Yozlaştırıldı (Vahyin Dininden Sapmalar, Hurafeler, Bid'atlar)* (Istanbul: Yeni Boyut, 2000), thirteen editions within one year of publication (2000–2001).

17 E.g., Yaşar Nuri Öztürk, *Shahīd-i rāh-i ḥaqīqat u ʿishq: Ḥallāj-i Manṣūr*, trans. Tawfīq Subḥānī (Tehran: Rūzana, 2003). For English and German translations, see bibliography.

on the Bektashi Order (1990/1410) and on the thirteenth-century Persian poet, scholar and mystic Jalāl ad-Dīn Rūmī (1992/1412).[18] The volume *The Eye of the Heart: An Introduction to Sufism and The Major Tariqats of Anatolia and the Balkans* evolved from his stay in the USA (English and Turkish eds. 1988/1408).[19] In the course of time, Öztürk also placed articles in scholarly journals, primarily in that of his own university institute.[20] However, he mainly addressed both in his writings and numerous television appearances the general public.[21] For years he travelled Turkey and Europe for his lecturing activities.[22]

The social debate about the significance of religion in the secular state has accompanied the Turkish Republic ever since its foundation, and it flared up again in the course of the economic and media liberalization since the second half of the 1980s. Öztürk then demonstrated a keen sense of current issues and new approaches to their possible solution, which included, by virtue of a certain charisma of his, many much-noted interventions. Critics soon called the omnipresent theologian, always clean-shaven and wearing a suit and tie, a *sosyete müftüsü*, "Islamic scholar of high society."[23] Supporters of his were particularly enthusiastic about his "style" (*üslup*), namely the eloquence and freshness of his oral and literary expression.[24]

In Öztürk's work, three phases of development can be observed parallel to his professional activities. As a young academic, he devoted himself to mysticism and the intellectual life of the early Islamic period. The small booklets of some sixty pages that he pub-

---

18 Öztürk, *Kutsal Gönüllü Veli* (rev. ed. Istanbul: Yeni Boyut, 1994); Yaşar Nuri Öztürk, *Büyük Türk Mutasavvıfı Muhammed Tevfik Bosnevî (Hayatı, Mektupları, Halifeleri)* (Istanbul: Fatih, 1981; rev. ed. Istanbul: Yeni Boyut, 1991); Yaşar Nuri Öztürk, *Hak ve Aşk Şehidi Hallâc-ı Mansûr ve Eseri* (Istanbul: Fatih, 1976; new ed. Istanbul: Yeni Boyut, 1997; two-volume ed., 2011); Yaşar Nuri Öztürk, *Tarihi Boyunca Bektaşilik* (Istanbul: Yeni Boyut, 1990); Yaşar Nuri Öztürk, *Mevlâna Celâleddin Rûmî ve İnsan* (Istanbul: Yeni Boyut, 1992).

19 Öztürk taught at the Theological Seminary of Barrytown, New York. – Yaşar Nuri Öztürk, *The Eye of the Heart: An Introduction to Sufism and the Major Tariqats of Anatolia and the Balkans*, trans. Richard Blakney (Istanbul: Redhouse, 1988); Turkish ed.: *Tasavvufun Rûhu ve Tarîkatlar* (Istanbul: Sidre, 1988; rev. ed. Yeni Boyut, 1990).

20 E.g., Yaşar Nuri Öztürk, "Kur'an'a göre hukuk devleti ve insan hakları," *İstanbul Üniversitesi İlahiyat Fakültesi Dergisi* 1 (1999): 1–38.

21 E.g., *İnanç Dünyası* (World of Faith), weekly broadcast on TRT, 1987 and on; *Yaşar Nuri ile Sohbet* (Talk with Yaşar Nuri), series of daily fifteen-minute Ramadan broadcasts, 1990s (cf. Neşe Ihtiyar, "Televangelism and the Case of Yaşar Nuri Öztürk," *International Institute for the Study of Islam in the Modern World (ISIM) Newsletter* 9 [2002]: 22); appearances in the talk shows *Ayşe Özgün*, 1996–2001 on Fridays, in Hulki Cevizoğlu's *Cevizkabuğu*, 1998, *Saba Tümer ile Bugün*, ca. 2012, and in Gülgün Feyman's *Söz ve Işık*, 2015. For details, see Eith, *Koranexegese*, 79–82.

22 Öztürk presented his lectures and newspaper columns in book form in several volumes, including Öztürk, *Kur'an'ı Anlamaya Doğru* and Yaşar Nuri Öztürk, *Konferanslarım* (Istanbul: Yeni Boyut, 1996).

23 Eith, *Koranexegese*, 83; Buket Türkmen, "Le retour au Coran de Yaşar Nuri Öztürk: vers un Islam national?," in *Formes nouvelles de l'Islam en Turquie*, ed. Gérard Groc (Paris: ERISM, 1999), 81–90, here 87; Hulki Cevizoğlu, *Yaşar Nuri Öztürk: O, Dini Yeniden Yapılandırmak İsteyen bir Reformist mi, Özgün bir Yorumcu mu?* (Istanbul: Beyaz, 1998), 15.

24 Canan Öztürk et al., eds., *Halkın Diliyle Yaşar Nuri: Neler Söylediler* (Istanbul: Yeni Boyut, 1997).

lished during the 1980s as supplements to the high-circulation daily newspaper *Hürriyet* later became monographs with similar titles.[25] As a publicist and established professor, in the 1990s, Öztürk focused on the significance of the Qur'ān for the contemporary religious life of Muslims in Turkey, achieving thereby great recognition and reach.[26] Drawing on the prominence he thus had gained and, moreover, his position as Member of Parliament, Öztürk in the 2000s increasingly devoted himself to political issues and became an important and highly regarded voice in public discourse.[27] With all that, he never lost sight of his early religious concerns, namely the renewal of Islam (*tecdit*, from Arabic *tajdīd*) and the unmasking of false ideas (*hurafe*, from Arabic *ḫurāfa*]).[28]

3 Analysis of Öztürk's Qur'ānic Exegesis

3.1 Main Writings

In Turkey, and also in Öztürk's work, the 1990s are the decade of intensified intellectual engagement with the Qur'ān.[29] His turn to the central source of the Islamic belief first shows in his book *400 Soruda İslam*.[30] This is also the first title that was pub-

25 E.g., Yaşar Nuri Öztürk, *Ehlibeyt'in Annesi Hazreti Fatima*, supplement to *Hürriyet*, 1985 (31 pp.); enhanced monograph edition (Istanbul: Yeni Boyut, 1999); Yaşar Nuri Öztürk, *Asrısaadet'in Büyük Kadınları*, supplement to *Hürriyet*, 1987 (64 pp.); enhanced monograph edition (Istanbul: Yeni Boyut, 1990).

26 For examples of religious titles of the 1990s, see below, chapter 3.1.

27 Examples of political titles of the 2000s: Yaşar Nuri Öztürk, *Kur'an Verileri Açısından Laiklik* (Secularism from the Perspective of the Prescriptions of the Koran) (Istanbul: Yeni Boyut, 2003); Yaşar Nuri Öztürk, *Batı Sömürgeciliği ve İslam Dünyası (Huntington ve Steinbach'ın Eseri)* (Western Imperialism and the Islamic World [The Works of Huntington and Steinbach]) (Istanbul: Yeni Boyut, 2003); Yaşar Nuri Öztürk, *Atatürk'ten Sonra CHP* (The CHP after Atatürk) (Istanbul: Yeni Boyut, 2004); Yaşar Nuri Öztürk, *Kur'an Açısından Küresel Afetler* (Global Disasters from the Perspective of the Qur'ān) (Istanbul: Yeni Boyut, 2008); Yaşar Nuri Öztürk, *Kur'an Penceresinden Kurtuluş Savaşına Bir Bakış* (A Look at the War of Liberation from the Perspective of the Qur'ān) (Istanbul: Yeni Boyut, 2012).

28 Yaşar Nuri Öztürk, *Arapçılığa Karşı Akılcılığın Öncüsü İmamı Azam Ebu Hanife: Esas Fikirleri Gölgelenen Önder* (Abu Hanifa as a Leading Figure of Intellectualism and Critic of Arabism: A Leader Whose Basic Ideas are Overshadowed) (Istanbul: Yeni Boyut, 2011); Yaşar Nuri Öztürk, *İnsanlığı Kemiren İhânet Dincilik: Zulümleriyle Dini Kirletenlerin Tarihi* (Religionism [i.e. Fundamentalism], a Scourge of Mankind: A History of Those Who Taint Religion through Oppression) (Istanbul: Yeni Boyut, 2010); Yaşar Nuri Öztürk, *Maûn Suresi Böyle Buyurdu: Din Maskeli Zulme Tanrının Vuruşu* (What the Sura *al-Māʿūn* Commands: God's Blow against Oppression in the Guise of Religion) (Istanbul: Yeni Boyut, 2012); Yaşar Nuri Öztürk, *Din Maskeli Allah Düşmanlığı Şirk ve Şirke Tepkinin Felsefeleşmesi: Deizm* (Deism: *Shirk* as Religiously Disguised Hostility to God, and the Philosophy of Criticizing *Shirk*) (Istanbul: Yeni Boyut, 2013).

29 Ömer Özsoy, "Koranhermeneutik als Diskussionsthema in der Türkei," online resource 2004, accessed October 13, 2020, http://akademie-rs.de/fileadmin/user_upload/pdf_archive/schmid/TFCI/Fachgespraech_Oezsoy/Artikel_zsoy_7-05.pdf; Eith, *Koranexegese*, 66, 389.

30 Yaşar Nuri Öztürk, *400 Soruda İslam* (400 Questions about Islam) (Istanbul: Yeni Boyut, 1989).

lished by Öztürk's specially founded publishing house Yeni Boyut in the series *Bütün Eserleri* (Complete Works). A quarter of the book is dedicated to the Qurʾān.[31] The title "400 Questions about Islam" refers to the dialogue form that Öztürk chose here. An imaginary interviewer asks an expert (Öztürk) questions about Islam. This staged talk bears witness to Öztürk's goal of creating a deeper understanding of Islam, on the actual basis of the Qurʾān, among a secularized and Kemalist-influenced Turkish reading public that had begun to emancipate itself from traditional and fit into modern life.[32] Öztürk's goal is the self-empowerment of the religious individual as opposed to *devşirme-şirket kurumu*, that is a "religious company" he conceives as a patronizing monopoly of religious knowledge and a stronghold of false and superstitious ideas.[33] In doing so, he takes up an old Sufi concern to oppose a sort of Islam that values the fulfilment of provisions and the outward ostentation of godliness more highly than inner spirituality. In addition, the question of the relationship between secularity and religiosity has been one of the long-running debates in the Turkish republic.

With the same intention, *Kuran'ı Anlamaya Doğru* (Towards Understanding the Qurʾān, 1990/1410) was published shortly after, as well as the first edition of *Kur'an'ın Temel Kavramları* (Basic Concepts of the Qurʾān, 1991/1411), a lexicon-style handbook with initially 170 entries.[34] In the latter, Öztürk cites ar-Rāghib al-Iṣfahānī's (d. 502/1108–9/502) *al-Mufradāt li-l-gharīb al-Qurʾān* and Fazlur Rahman's (1919–1988/1337–1408) *Major Themes of the Qurʾān* as models for his "Basic Concepts."[35] By 2008/1429, the book had undergone twenty-two new editions, some of which were expanded, and became Öztürk's second best-selling work.

Öztürk's major work of exegesis is his *Kur'an'daki İslam* (1992/1412). The provocative title "Islam in the Qurʾān" implies the existence of two "Islams" – i.e., a true, Qurʾān-based one, and a false, "non-Qurʾānic" one. On 750 pages, Öztürk comments on selected passages or even individual terms of the script. For this purpose, he again chooses the question/answer form, with each of the 114 chapters dedicated to one sura. For many of the shorter suras only one question is asked, for example concerning the meaning of the sura's name. Mostly, however, several thematically different questions per sura are treated with reference to different verses or keywords. In *Kur'an'daki İslam*, Öztürk presents the chapters – each one referring to a particular sura – not according to the traditional order of their length, but in the "order of the

31 Ibid., ch. 2, "Kur'an," questions nos. 101–200, 85–148.

32 This title was the first of several of Öztürk's works to be translated into German: *400 Fragen zum Islam – 400 Antworten: Ein Handbuch*, trans. Nevfel Cumart (Düsseldorf: Grupello, 2000). It was followed by *Rumi und die Islamische Mystik: Über das Menschenbild im Islam*, trans. Nevfel Cumart (Düsseldorf: Grupello, 2002) and *Der Verfälschte Islam*, trans. Nevfel Cumart (Düsseldorf: Grupello, 2007).

33 Yaşar Nuri Öztürk, *Kur'an'daki İslam* (Istanbul: Yeni Boyut, 1992), preface, 9.

34 Nuri Öztürk, *Kur'an'ı Anlamaya Doğru* (see note 3 above); Yaşar Nuri Öztürk, *Kur'an'ın Temel Kavramları* (Istanbul: Yeni Boyut, 1991).

35 Öztürk, *Kur'an'ın Temel Kavamları* (2001[21]), preface, 13.

revelation" (*iniş sıralı*). The author thus makes use of two parallel counting systems and, for better orientation, provides a chronologically ordered table of contents and a name index of the suras. It appears that the unspoken yardstick for that specific order of revelation is the chronology suggested by Muḥammad ʿIzzat Darwaza in his "Modern Qurʾān Commentary" (*at-Tafsīr al-ḥadīth: as-suwar murattaba ḥasaba an-nuzūl*, 1962–1964/1382–1384), with which it largely coincides.[36]

Kur'an'daki İslam thus comes as a one-volume, partial commentary of the Qurʾān that through thematic interpretation and dialogue form is far from a traditional *tafsīr* format. The greatest formal peculiarity, however, is the absence of the Qurʾān's text itself, which conventionally is reproduced alongside the commentary. Instead, Öztürk has included only individual terms and short passages from the Qurʾān in Turkish translation.

As a supplement to *Kur'an'daki İslam*, Öztürk presented his own complete Turkish translation of the Qurʾān in 1993/1413, entitled *Kur'an'ı Kerim Meâli (Türkçe Çeviri)* (Approximate Meaning of the Holy Qurʾān [Turkish Translation]) and initially containing the Turkish version only.[37] One year later, a synopsis of this translation and the Arabic text appeared both as paperback and hardcover editions.[38] Öztürk's first foray into translating Arabic texts into Turkish was devoted to a work by the Egyptian Islamic theorist Sayyid Quṭb (1972/1392).[39] After that, he used to give translations of Qurʾānic texts in his works.[40] However, the complete *Kur'an'ı Kerim Meâli* differs significantly from those Qurʾān passages he already gave in his *Kur'an'daki İslam*. Only the

36 Muḥammad ʿIzzat Darwaza, *at-Tafsīr al-ḥadīth: as-suwar murattaba ḥasaba an-nuzūl* (Cairo: ʿĪsā al-Bābī al-Ḥalabī, 1962–1964). Darwaza's "Contemporary Commentary" was the first to arrange the Qurʾān's text in a presumed chronological order of revelation – the Medinan parts of which deviate significantly from the sequence of suras in the standard Cairo edition – but had received little response. The Turkish translation of Darwaza's *Tafsīr* has been available since 1997 and saw new editions in 2014 (cf. Johanna Pink, *Muslim Qurʾānic Interpretation Today: Media, Genealogies and Interpretive Communities* [Bristol: Equinox, 2019], 134f.). On Öztürk's models for his version of the chronological order of suras see Eith, *Koranexegese*, 261–68.

37 *Kur'an'ı Kerîm Meâli (Türkçe Çeviri)*, trans. Öztürk (see note 16 above).

38 *Kur'an'ı Kerîm ve Türkçe Meâli (Metinli)*, trans. Öztürk (see note 16 above).

39 Seyyid Kutub, *İslam Kapitalizm Uyuşmazlığı* (Istanbul: Hareket, 1972), translated by Öztürk under the pseudonym Abdurrahman Niyazoğlu and republished under his actual name as *İslam Kapitalizm Çatışması* (Istanbul: Düşünce, 1980), Arabic original: Sayyid Quṭb, *Maʿrakat al-Islām wa-r-raʾsmāliyya* (The Battle between Islam and Capitalism) (Cairo: Dār al-Kitāb al-ʿArabī, 1951). He also translated under the name Niyazoğlu the book *Ishtirākiyyat al-Islām* (as Islamic Socialism) (Damascus: Jāmiʿat Dimashq, 1959) by the Syrian jurist and Muslim Brother Muṣṭafā s-Sibāʿī (1915–1964), Turkish title: *İslâm Sosyalizmi* (Istanbul: Hareket, 1974), likewise republished under Öztürk's actual name (Istanbul: Yeni Boyut, 2010^2).

40 Öztürk undoubtedly had a knowledge of Arabic through his extracurricular education, although Arabic was not regularly taught in the Turkish school system at his time.

third edition of Öztürk's translation, published in 1994/1414, supplied *Kur'an'daki İslam* with the Qur'ānic text in the said presumed order of revelation.[41]

The translation of the Qur'ān and the partial commentary, together with Öztürk's increased media presence, have stimulated a public debate in Turkey. The press began to speak of him as the "head of the movement back to the Qur'ān."[42] In this phase, Öztürk, as a programmatic supplement to *Kur'an'daki İslam*, presented the essay volume *Dinde Yeniden Yapılanmak (Kur'an'a Dönüş)* as a programmatic supplement to Kur'an'daki İslam.[43] It was translated into English and, in parts, into German, too.[44] In the volume's thirteen essays Öztürk, reflecting his intentions most clearly, unfolds core themes signaled by chapter headings such as *Kur'an'a Dönüşün Anlamı* (The Meaning of the Return to the Qur'ān) and *Yeniden Yapılanmanın Karakteri* (The Character of Renewal) as well as the relationship between Atatürk and Islam (*Atatürk ve Din*).[45]

In 1996/1416, Öztürk published the booklet *Fatiha Suresi Tefsiri*, which contains his translation of the Qur'ān's initial sura into Turkish, a general introduction and a verse-by-verse commentary.[46] While quoting numerous other Qur'ānic passages and also a variety of traditional *tafsīr* works pertaining to the topic of divine *tevhit* ("oneness"), he indulges in long-winded digressions on his central concern, the unmasking of *şirk*.[47] Here, *şirk* is understood, as is usual in modernist Qur'ānic exegesis, as the worshipping of "idols," whether in the concrete or figurative sense, instead of the One God. In support of his own ideas, Öztürk allows the physicist and author Ahmet Yüksel Özemre to have his say on today's "hidden şirk" in lengthy quotations,[48] like he

41 *Surelerin İniş Sırasına göre Kur'an'ı Kerîm Meâli (İniş Sıralı)*, trans. Yaşar Nuri Öztürk (Istanbul: Yeni Boyut, 1994). On the history of Qur'ān translations into Turkish, see, among others, Hidayet Aydar, "Kur'an IX, Tercümesi," in *TDVIA*, ed. İslâm Araştırmaları Merkezi (Ankara: TDV, 2002), 26: 404–9.

42 Blurb to several of Öztürk's books.

43 Yaşar Nuri Öztürk, *Dinde Yeniden Yapılanmak (Kur'an'a Dönüş)* (Istanbul: Yeni Boyut, 1996), quoted from the 13th ed., 1998.

44 Yaşar Nuri Öztürk, *Reconstruction of Religious Life in Islam: Returning to the Koran*, trans. Ali Hayrani Öz (Istanbul: Yeni Boyut, 1999). According to the preface to the English edition, the Turkish version is one of the best-selling books in the history of Turkey, with 1.5 million copies sold in twelve editions to date; Yaşar Nuri Öztürk, "Was Rückkehr zum Koran bedeutet," in *Alter Text – neuer Kontext: Koranhermeneutik in der Türkei heute: Ausgewählte Texte, übersetzt und kommentiert*, ed. and trans. Felix Körner (Freiburg im Breisgau: Herder, 2006), 210–36; see also Körner's commentary therein, "Reform durch Koran – ein Nachspann," 237–43.

45 Translations of some of the chapter headings.

46 Yaşar Nuri Öztürk, *Fatiha Suresi Tefsiri* (Commentary on Sura al-Fatiha) (Istanbul: Yeni Boyut, 1996), 80 pp., quoted from the fifth ed., 1997, preface, 5.

47 For more on *Fatiha Suresi Tefsiri*, see Eith, *Koranexegese*, 112–14; Kathrin Eith, "Yaşar Nuri Öztürk: A Contemporary Turkish Tafsir Theorist," in *Tafsir and Islamic Intellectual History: Exploring the Boundaries of a Genre*, ed. Andreas Görke and Johanna Pink (Oxford: Oxford University Press, 2014), 443–65.

48 Öztürk, *Fatiha Suresi Tefsiri*, 65–68, 70 –72, quoting Ahmet Yüksel Özemre, *İslam'da Aklın Önemi ve Sınırı* (Meaning and Limits of Reason in Islam) (Istanbul: Denge, 1996). Özemre (1935–2008) was a

frequently uses to seek support from the religious-philosophical thinking of contemporary authors. Although *Fatiha Suresi Tefsiri* was to be the prelude to a commentary of the entire Qur'ān and saw several reprints, Öztürk apparently discontinued work on a comprehensive *tafsīr* afterwards.[49]

In 1997/1417, the first edition of *Kur'an'ın Temel Buyrukları* (Elementary Commandments of the Qur'ān) was published.[50] It is structured similarly to *Kur'an'ın Temel Kavramları* in that the Qur'ānic commandments, and after that the interdictions, are treated likewise according to the assumed chronology of revelation, which implies that the earlier a sura had been revealed, the more important its commandments and interdictions. Here, as in general, Öztürk holds the view that the Qur'ān in its entirety is to be understood as an ethical framework for all times and questions of human life. It is thus more important to grasp the spirit of the Qur'ān as a source of guidance than to obey the few specific rules and prohibitions contained in the text. Correspondingly, to follow an inner attitude is deemed more important than outwardly visible piety.

Another criticism of Öztürk's of present-day Islam, reading that in Turkey Islamic belief is all too easily and often exploited by political parties and state institutions for their self-interested purposes, is made the key message of his book *İslam Nasıl Yozlaştırıldı* (How Islam has Been Distorted) of 2000/1420.[51] It is claimed therein that there is a trans-historical continuity between power-political calculations of the Umayyad dynasty and contemporary politics.[52] Furthermore, four basic evils of societal degeneration are identified that consist in "superstition" (*hurafe* [*ḫurāfa*]), "unacceptable innovations" (*bid'at* [*bid'a*]), "political instrumentalization" (*siyaset* [*siyāsa*]) and "deification of man" (*rableştirme*).[53] These general considerations are followed by an alphabetically arranged survey of what Öztürk deems the most important objects of such distortion throughout the history of Islam. Each of the sixty-four chapters consists of a short introduction and a major part, entitled "Bid'atlar, Hurafeler," that exposes issues affecting

---

Turkish nuclear physicist and university professor of theoretical physics; he wrote several books on religious topics. Accessed October 13, 2020, https://www.ozemre.com.

49 In the short preface to *Fatiha Suresi Tefsiri* (p. 5), Öztürk states his intention to gradually complete the *tafsīr*, to be supplemented by an introduction including methodological explanations.

50 Öztürk, *Kur'an'ın Temel Buyrukları (Emirler ve Yasaklar)* (see note 16 above; 2006[14]), reprinted in two vols. under the title *Surelerin İniş Sırasına Göre Kur'anın Temel Buyrukları: Emirler ve Yasaklar* (Istanbul: Yeni Boyut, 2012[17]). On Öztürk's motivations to refer to the chronological order of the suras, see Eith, *Koranexegese*, 268–72.

51 Öztürk, *İslam Nasıl Yozlaştırıldı (Vahyin Dininden Sapmalar, Hurafeler, Bid'atlar)* (see note 16 above), quoted from the 13th ed., 2001. This served as the basis for the abridged German version, *Der Verfälschte Islam*, trans. Cumart.

52 See, among others, the description of the role of the Umayyads in the chapters "Cihad" (2001[13], 136f.), "Hakimiyet" (247–63) and "Hilafet ve Halife" (264–76).

53 These are the titles of the four introductory chapters in *İslam Nasıl Yozlaştırıldı*, table of contents, 5.

the object under discussion.[54] At first sight, this seems to resemble a backward-looking Salafi pamphlet. However, Öztürk once again wants to encourage contemporary Turkish Muslims to open themselves to a correct, because Qur'ān-compliant, and at the same time modern understanding of Islam.

3.2 Hermeneutical Methods

The main exegetical works of Öztürk from the 1990s show various approaches.

Characteristic for Öztürk's methodology is above all a thematic and selective way of exegesis. He picks up exactly those passages, phrases or keywords of the Qur'ānic text that he wishes to have significance. This procedure is quite natural for him and apparently does not require theoretical justification. As a rule, it is not his aim to comment on the whole of suras. For instance, he cites Fazlur Rahman, a thinker to be regarded as a mastermind of thematic exegesis and one of great influence on contemporary Turkish theology, as a model for his work *Kur'an'ın Temel Kavramları* (Basic Concepts of the Qur'ān).[55]

With his selective approach, Öztürk practices the interpretation of the Qur'ān through the Qur'ān (*tafsīr al-Qur'ān bi-l-Qur'ān*).[56] For him, cross-linking different passages because of similar or identical wording or thematic context is a central means of interpretation since Öztürk considers the Qur'ān a coherent text.[57] This understanding appears to be heavily influenced by Muhammed Elmalılı's *tafsīr* work *Hak Dini Kur'an Dili* (1935–1939/1354–1358), who in turn based his understanding of the composition and coherence of the Qur'ānic text on Fakhr ad-Dīn ar-Rāzī's (d. 1210/606) commentary *Mafāṭīḥ al-ghayb*.[58] The idea handed down this way is that unconditional

54 E.g., chapter "Kadın ve Kadın Hakları" (332–81), treating twenty-one issues.

55 On the impact of Fazlur Rahman on modernist exegetes, see Pink, *Muslim Qur'ānic Interpretation Today*, 129–30. On Fazlur Rahman's "double movement" method, see Rotraud Wielandt, "Exegesis of the Qur'ān: Early Modern and Contemporary," in *EQ*, ed. Jane Dammen McAuliffe (Leiden: Brill, 2002), 2: 124–42, also available online, accessed December 18, 2020, http://dx.doi.org/10.1163/1875-3922_q3_EQCOM_00059.

56 The method of *tafsīr al-Qur'ān bi-l-Qur'ān* was not new anymore when Muḥammad Abū Zayd ad-Damanhūrī published in 1930 a Qur'ān commentary entitled *al-Hidāya wa-l-'irfān fī tafsīr al-Qur'ān bi-l-Qur'ān*. See Wielandt, "Exegesis: Early Modern and Contemporary." For a short description of the said method, see Nimet Seker, *Der Koran als Rede und Text: Hermeneutik Sunnitischer Koranexegeten zwischen Textkohärenz und Offenbarungskontext* (Berlin: EB, 2019), 65.

57 On the *munāsabāt* exegesis considering the canonical form of the *muṣḥaf* as a proof of its divinely inspired linear composition, see Seker, *Rede und Text*, 79f. See also p. 82 on the literary terms "coherence" and "composition."

58 Muhammed Hamdi Yazır Elmalılı, *Hak Dini Kur'an Dili: Yeni Meâlli Türkçe Tefsir* (Ankara: Diyanet Reisliği, 1935–1939).

precedence is due to the Qur'ān over all sorts of learning since the Qur'ān's form is of divine origin, too; no secondary aids are necessary for grasping its meaning.

Öztürk expressly wants to encourage laypersons to directly confront the allegedly difficult-to-understand Qur'ānic text. As for him the Qur'ān alone is the source of religious knowledge, both laymen and scholars can dispense with the *Tafsīr* tradition, the Hadith and other accounts of the life of Muḥammad.[59] Öztürk, on the other hand, demonstrates by numerous quotations that he himself is perfectly familiar with the most important works of the historical *tafsīr*.[60] He often cites this kind of literature as a starting point to then arrive at alternative results. This is what he puts forward as the necessary and legitimate "renewal" of Islam (*tecdit [tajdīd], yeniden yapılanma*).

Öztürk's critical attitude towards Hadith is based on Muslim writings past and present.[61] He sees the Hadith as a gateway for the infiltration of Islam with ideas that he denounces as un-Islamic. While he does not go as far as rejecting the corpus of tradition in general, he pleads for the authenticity of traditions to be measured by their compatibility with the message of the Qur'ān. He so refuses the conventional assessment of traditions and chains of transmitters in terms of how strong or weak they are, as is practiced in the literary genre of *cerh ve ta'dîl* (*'ilm ma'rifat ar-rijāl*).[62]

For Öztürk, the original order of revelation plays a crucial role in understanding the text, because he argues, the earliest commandments of God to mankind are therefore the most important ones. "Recite!" thus comes before "Prostrate yourself!" and "Recite the Qur'ān!" is prior to "Pray!"[63] For this argument to be effective, the suras need to be understood as units of revelation. Neither is Muḥammad's life journey relevant for understanding the text properly, nor are the traditional occasions of revelation of individual verses or commandments (*sebeb-i nüzul* [*asbāb an-nuzūl*]). In this logic, it is only consequent that Öztürk rejects the possibility of abrogation (*nesih* [*naskh*]), i.e., the annulment of previously revealed verses by later ones.[64]

The unadulterated tradition and guaranteed authenticity of the Qur'ānic text, in turn, appears to be guaranteed by Muḥammad's editorial supervision of the creation

59 See Eith, *Koranexegese*, ch. "Schriftfassung des Korans," 244–50.

60 Among the medieval authors he quotes most frequently are ar-Rāghib al-Iṣfahānī (d. 1188/502), Abū l-Qāsim ash-Shāṭibī (d. 1194/590), Abū l-Faraj b. al-Jawzī (d. 1201/597), Ibn Qayyim al-Jawziyya (d. 1350/751) and Jalāl ad-Dīn as-Suyūṭī (d. 1505/1911).

61 Öztürk, *Kur'an'daki İslam*, 131–54. Among others, Öztürk cites Abū Sa'īd al-Khudrī (d. 74/693), Ibn 'Abd al-Barr (d. 1070–71/463), Shams ad-Dīn adh-Dhahabī (d. 1347–48/748), Ibn Hajar al-'Asqalānī (d. 1449/852), as-Suyūṭī (d. 1505/911), Rashīd Riḍā (d. 1935/1354), Aḥmad Amīn (d. 1954/1373) and Maḥmūd Abū Rayya (d. 1970/1390).

62 Eith, *Koranexegese*, 274–306, esp. 286, based on Öztürk's *Kur'an'daki İslam*, *İslam Nasıl Yozlaştırıldı*, and *400 Soruda İslam*.

63 Eith, *Koranexegese*, 208–12, 258–63; Öztürk, *Kur'an'daki İslam*, 11 (Q 96:1–5), 351–52 (Q 29:45); Öztürk, *Kur'an'ın Temel Buyrukları*, 8–9, 17–22; Öztürk, *İslam Nasıl Yozlaştırıldı*, 390–91 (Q 73:4 and 20).

64 Eith, *Koranexegese*, 210f.; Öztürk, *Kur'an'daki İslam*, 396–98 (Q 2:106), including an extensive quotation from Hüseyin Atay.

of the first-ever Qur'ān codices Sing. (*mushaf* [*muṣḥaf*]). The order of revelation of the suras (implicitly: in the sense of Darwaza) goes thereby undisputed.[65] So after all no effort is put into any attempt to "historicize" the Qur'ānic text, be it through text-critical research or a hermeneutical interpretation of the revelation within the historical horizon of Muḥammad's audience.

Öztürk is an advocate of *ictihat* (*ijtihād*) or individual rational exegesis instead of any mere handing-down of tradition (*taklit* [*taqlīd*]). Through idiosyncratic interpretation of Quranic passages, he promotes the necessity to continue *ictihat*.[66] Any competent interpretation would draw on the state-of-the-art knowledge of all mankind. He thereby reiterates Sufism's ideas of perfection (*tekâmül* [*takāmul*]), according to which the knowledge of mankind is continually spiraling to ever higher levels.[67] Therefore, he claims, those who were genuinely searching for the meaning of Scripture were at any point in time at the peak of knowledge. For Öztürk, this cumulative human knowledge also includes ideas such as human rights and democracy, which he deems universally valid achievements that are in no way ideological impositions by Western colonial powers but perfectly comply with the Qur'ān.[68]

In Öztürk's view, there is no danger of subjectivity, arbitrariness and "wrong" interpretation, because the general objectives of revelation – notably the *makasit* (*maqāṣid*, "inherent aims")/*amaç hükümler* – and the concrete regulations it includes – *vesâil* (*wasā'il*, "means")/*araç hükümler* – provide clear guidelines to exegesis.[69] Above all, he speaks of the basic principles (*prensip*) or "spirit" (*ruh*) of revelation, to which it is important to do justice with every interpretation. Occasionally Öztürk names one such principle, e.g., God's "facilitation" (*kolaylaştırma, hafifletme*) of faith (instead of any ostensible impediment to exercise Islam).[70] In the exegesis of any particular sura or verse, the fulfillment of the principles that underlie a Qur'ānic ruling in a timely manner weighs more than unconditional obedience to its literal claim.[71]

---

65 Eith, *Koranexegese*, 183f.; Öztürk, *400 Soruda İslam*, ch. 2, 130f. (question 64), 137f. (questions 78–80).

66 Eith, *Koranexegese*, 306–17. Öztürk relies mostly on Q 2:104, translating *lā taqūlū rā'inā* as "'Bizi davar gibi güt' demeyin" and "Davar sürüsüne dönüşmeyin!" ("Do not say 'Herd us like animals/Do not turn (yourself) into a herd of cattle"); see Öztürk, *Kur'an'daki İslam*, 399f.; Öztürk, *İslam Nasıl Yozlaştırıldı*, 134f.; Öztürk, *400 Soruda İslam*, ch. 1, 63f. (question 67), ch. 3, 207f. (questions 98–100).

67 Eith, *Koranexegese*, 196–98; Öztürk, *Kur'an'ın Temel Kavramları*, 371–73; Öztürk, *İslam Nasıl Yozlaştırıldı*, 450f.

68 Eith, *Koranexegese*, 197; Öztürk, *Kur'an'daki İslam*, 486 (Q 39:23); Öztürk, *Kur'an'a Dönüş*, 46–50; Öztürk, *Kur'an'ın Temel Kavramları*, 390f.

69 E.g., Öztürk, *Kur'an'daki İslam*, 452, 555.

70 Ibid., 16 (Q 73:20), 41f. (Q 94:5–6), 284 (Q 44:58), 561 (Q 4:27).

71 E.g., ibid., 554–56 (Q 4:11), regarding the share of inheritance of wives and daughters. The division of inheritance is an implementation (*vâsil/araç*) of the Qur'ānic principle (*maksat/amaç*) of justice (*adalet*) between husband and wife. The concrete form of the principle can nowadays also, deviating from the regulation in the Qur'ān, take place to equal shares of inheritance, the main thing being that it does not come to injustice.

So, while these principles provide unchanging guideposts for interpretation, definite interpretations can, and indeed must, vary. Öztürk argues that the higher proportion of verses in the Qur'ān in need of interpretation (*müteşabih* [*mutashābiḥ*]) than those that are clear and unambiguous (*muhkem* [*muḥkam*]) is not a flaw, but a token of the real miracle (*icaz* [*i'jāz*]) of revelation.[72] Through this ambiguity, God has ensured the everlasting topicality and validity of his message. Each epoch thus faces the task and the possibility to work out the current meaning of God's word.[73]

Despite his intense concern with Qur'ān interpretation, Öztürk abstained from calling himself a traditional *tafsīr* author (*müfessir* [*mufassir*]). Rather he saw himself as a renewer (*müceddit* [*mujaddid*]), messenger, or admonisher (*uyarıcı*) such as appears anew in every epoch of Islamic history in order to restate the relevance of the divine revelation.[74] Öztürk's interpretations of certain passages of the Qur'ān have therefore, in many a case, already been advocated by previous modernist exegetes. He agrees with them, provided they offer a timely understanding of the Qur'ān. Of these, the abovementioned Elmalılı Muhammad Hamdi Yazır (*Hak Dini Kur'an Dili*, 1935–1939/1354–1358) as well as Süleyman Ateş (*Yüce Kur'an'ın Çağdaş Tefsiri*, 1988–1991/1408–1411) have heavily influenced Öztürk's positions.[75] He also frequently received the ideas of Arab modernists like Muḥammad 'Abduh and Rashīd Riḍā.[76] In addition, it is one of Öztürk's peculiarities that he was pleased to deal with the latest publications of contemporary Turkish authors and to quote them extensively wherever he felt it suited his intentions.[77]

With regard to women's rights from the Qur'ān's perspective, Öztürk starts from the principle (*maksat* [*maqṣad*]) of justice (*adalet* ['*adāla*]). Today's regulations in the fields of inheritance law, divorce law or testimony would thus aim at establishing

72 Eith, *Koranexegese*, 192–206. He quantifies the proportion between *müteşabih* and *muhkem* by Q 80:20 or even Q 90:10, meaning that most of Scripture requires incessant interpretation (while statements on *tevhit*, the uniqueness of God, are particularly lucid).

73 Ibid., 187f.; Öztürk, *İslam Nasıl Yozlaştırıldı*, 397, 406.

74 Öztürk's book *Çıplak Uyarı* (Stark Warning) (Istanbul: Yeni Boyut, 1992) as well as newspaper comments, e.g., "Yine Uyarıyoruz" (We Warn You Again) and "Reform ve Tecdit" (Reform and Renewal), reprinted in his *İslam'ı Anlamaya Doğru* (Istanbul: Yeni Boyut, 1996), 85–87 and 164–66, refer to this role of his as a "warner."

75 Elmalılı, *Hak Dini Kur'an Dili* (see note 59 above); on Elmalılı see Benjamin Flöhr, *Ein traditionalistischer Korandeuter im Dienste des Kemalismus: Elmalılı Muhammed Hamdi Yazır (1878–1942)* (Berlin: Klaus Schwarz Verlag, 2015); Süleyman Ateş, *Yüce Kur'an'ın Çağdaş Tefsiri* (Istanbul: Yeni Ufuklar, 1988–1991); on Ateş see Abdullah Takım, *Koranexegese im 20. Jahrhundert: Islamische Tradition und neue Ansätze in Süleyman Ateş's "Zeitgenössischem Korankommentar"* (Istanbul: Yeni Ufuklar, 2007).

76 Muḥammad Rashīd Riḍā, *Tafsīr al-Qur'ān al-ḥakīm* [*aw*] *Tafsīr al-manār* [. . .] *allatī jarā 'alayhā* [. . .] *Muḥammad 'Abduh* (Cairo: Dār al-Manār, 1927–1934).

77 This is the case with Özemre, mentioned above, and also with Edip Yüksel (b. 1957), through whose writings Öztürk adopted ideas of Rashād Khalīfa (1935–1990) suggesting a hidden structuring of the Qur'ān based on the number 19.

equality (*eşitlik*) between the sexes.[78] His thesis that the number of daily ritual prayers is not clearly fixed at five has caused much controversy. Only three daily prayer times, he holds, are specified in the Qur'ān, meaning that three daily prayers were sufficient.[79] Öztürk also debated the possibility of reincarnation, which according to him cannot be completely ruled out as an interpretation of *tanâsuh* (*tanāsuḥ*).[80] He even did not consider the existence of extraterrestrial life forms *dâbbe* (*dābba*) to be fundamentally impossible.[81]

Neither was Öztürk afraid to take up original and controversial interpretations of his contemporaries. For example, he argued that the Qur'ānic word *saqar* (Q 54:48) stands for electricity, that *lawḥ* (Q 85:22) can also imply a computer screen, and that *siccîl* (*sijjīl*; Q 105:4) equals pathogens (*mikrop*).[82] Finally, he aroused offence with the idea that some non-Muslims, who unknowingly lead a perfectly Qur'ān-compliant life, might be better believers than many a professing Muslim, and can even enter paradise.[83] Thus, Öztürk does not see his main task in developing an original, new understanding of the Qur'ān. Rather is he concerned with fighting what he deems the fundamental evil within Islam, i.e., the violation of the "oneness of God" (*tevhit*). To this end, he emphasizes the exclusiveness of the Qur'ān as divine revelation.

4 Conclusion

The death and burial of Yaşar Nuri Öztürk in June 2016/Ramaḍān 1437 has caused a great deal of response in Turkish media. In a volume published by his son Mustafa Tahir Öztürk in 2017/1438, over sixty public figures, among them writers, journalists and artists, academic colleagues, government officials, politicians and the entrepreneurs, bid farewell in personally written obituaries to the "Hodja of Turkey."[84] Öztürk was lauded both as secularly learned man (*bilim adamı*) and traditional religious scholar (*âlim*).

On the one hand, Öztürk indeed claimed and possessed the professional authority of an academic theologian (*ilahiyatçı*). Despite his belonging to a generation that had

---

78 Öztürk, *İslam Nasıl Yozlaştırıldı*, ch. "Kadın ve Kadın Hakları," 332–81.

79 Ibid., 472–74; Öztürk, *Kur'an'daki İslam*, 586–91 (Q 4:103 and Q 2:238), 199–200 (Q 17:79), "teheccüd." This view is based on an unpublished "report" (*rapor*) by Hüseyin Atay, which Öztürk quotes here.

80 Öztürk, *Kur'an'daki İslam*, 170f. (Q 56:60–62), 276–77 (Q 42:30). Öztürk concurs with Süleyman Ateş, who considered the meaning "reincarnation" for *tenasüh* and dealt with it in detail in his *tafsīr*. However, Öztürk also points out that ar-Rāzī or Elmalılı do not mention this interpretive possibility.

81 Ibid., 198 (Q 17:54, 70), 276 (Q 42:29).

82 Ibid., 55, citing ʿIkrimā (d. 725/197) and Muḥammad b. Isḥāq (d. 768/151) according to the *tafsīr* of Abū Jaʿfar aṭ-Ṭabarī (d. 923/310) and Muḥammad ʿAbduh (d. 1905).

83 Öztürk, *İslam Nasıl Yozlaştırıldı*, ch. "Cennet," 128–32.

84 Mustafa Tahir Öztürk, ed., *Türkiye'nin Hocası Prof. Dr. Yaşar Nuri Öztürk Anısına* (Istanbul: Yeni Boyut, 2017).

little or no opportunity to receive theological training during their school and university education in Turkey, Öztürk managed to gain a – partly non-formal – "hybrid education"[85] as lawyer and theologian, obtained his doctorate and a professorship and even served as dean of the Istanbul University's Theological Faculty for almost a decade. It is against this background of Öztürk, that critics like Felix Körner and Ömer Özsoy, taking the angle of the "Ankara School" of theology, judge his work by the academic standards of historical criticism and therefore express basic reservations: "Öztürk's method is that of self-positioning through illustrative Qur'ān quotations, not that of argumentation from within the Qur'ān. [. . .] Instead of convincing arguments, he resorts to snappy formulations."[86]

Yet, classing Öztürk solely as an academic does not entirely do justice to his self-paced mission and work. As the traditionally trained *âlim* that he was, too, Öztürk sought more inclusive ways of spreading his message. Finally, as a public intellectual, he strove to pass on his findings to the widest possible audience through the print media and television. Also, he later undertook a "march through the democratic institutions" in order to really initiate societal transformation. With all that, he did not make efforts to found a "school," company or network, nor did he preach and gather followers in the manner of a religious leader. While many expressed that Öztürk speaks from the bottom of their heart, he did not let his audience pay homage to him or follow him unconditionally.[87] At political level, Öztürk responded to the ongoing debate about the relationship between state and religion, one of the unsolved fundamental issues of the Turkish Republic since its foundation, seeing it as his task to "reconcile Islam and Kemalism" (*Muhammed ile Mustafa'yı bağdaşmak*).[88] However, the political course that Turkey took during the final years of Öztürk's life, and especially so after his death in 2016/1437, has left the country's society more deeply divided than before, and it appears that the authoritarian currents of Islam in Turkey that Öztürk ever was fighting have gained the upper hand for the time being.

According to the criteria of twenty-first century research on modern Muslim's *tafsīr* activities, which does not necessarily place flawless scientific work in the foreground, Öztürk was a modernist in that he referred to the principles of revelation, drew on the chronological sequence of the suras without questioning their inner com-

85 Pink, *Muslim Qur'ānic Interpretation Today*, 126.

86 Körner, "Reform durch Koran – ein Nachspann," 237–43, here 240. See Eith, *Koranexegese*, 365.

87 Canan Öztürk, *Neler Söylediler*, see letters of 1993, 357, 382, 395f.; letters of 1995, 498, 500–2, 517f., 528f., etc.

88 Öztürk frequently used this slogan in the media after leaving the CHP in 2004 while forming the HYP. "Atatürk ve Din" is the headline of a repeatedly given lecture of his as well as the issue of books like *Kur'an Verileri Açısından Laiklik*, 2003, or *Kur'an Penceresinden Kurtuluş Savaşına bir Bakış*, 2012.

position, and rejected the possibility of abrogation.[89] Implicitly, however, approaches of "postmodern"[90] exegesis can be distinguished in Öztürk's work, too: To him, any understanding that emerges from encountering the Qur'ān is "subjective" or "time-bound;" it so is legitimate and therefore correct, but never irrefutable. Every believer can, and should, establish a personal connection to the Qur'ān and strive for a subjective understanding. The whole of historic interpretations may provide orientation with that, yet without dogmatic guidelines and intellectual prohibitions.[91] Whereas, before Öztürk, translating the Qur'ān into Turkish was the almost exclusive privilege of presidents or commissions of the Directorate of Religious Affairs, his example paved the way for a large number of individuals to undertake Turkish Qur'ān translations of their own as well.[92] This is possibly the most visible fruit of the return to the Qur'ān in Turkey, which Öztürk so vehemently advocated.

5 Bibliography

5.1 Sources: Works by Yaşar Nuri Öztürk

5.1.1 Monographs and Selected Articles

Öztürk, Yaşar Nuri. *Hak ve Aşk Şehidi Hallâc-ı Mansûr ve Eseri*. Istanbul: Fatih, 1976. (New ed. Istanbul: Yeni Boyut, 1997; two-volume ed., 2011).

Öztürk, Yaşar Nuri. *Büyük Türk Mutasavvıfı Muhammed Tevfik Bosnevî (Hayatı, Mektupları, Halifeleri)*. Istanbul: Fatih, 1981. (rev. ed. Istanbul: Yeni Boyut, 1991).

Öztürk, Yaşar Nuri. *Kutsal Gönüllü Veli: Kuşadalı İbrahim Halveti (Hayatı, Tasavvufî Düşünceleri, Mektupları*. Istanbul: Fatih, 1982.

Öztürk, Yaşar Nuri. *400 Soruda İslam*. Istanbul: Yeni Boyut, 1989. (Referenced: 2001[14]).

Öztürk, Yaşar Nuri. *Kur'an'ı Anlamaya Doğru*. Istanbul: Yeni Boyut, 1990, (1999[10]).

Öztürk, Yaşar Nuri. *Tarihi Boyunca Bektaşilik*. Istanbul: Yeni Boyut, 1990.

Öztürk, Yaşar Nuri. *Kur'an'ın Temel Kavramları*. Istanbul: Yeni Boyut, 1991. (2001[21]).

Öztürk, Yaşar Nuri. *Çıplak Uyarı*. Istanbul: Yeni Boyut, 1992.

Öztürk, Yaşar Nuri. *Kur'an'daki İslam*. Istanbul: Yeni Boyut, 1992. (2001[39]).

Öztürk, Yaşar Nuri. *Mevlâna Celâleddin Rûmî ve İnsan*. Istanbul: Yeni Boyut, 1992. (2011[9]).

Öztürk, Yaşar Nuri. *Dinde Yeniden Yapılanmak (Kur'an'a Dönüş)*. Istanbul: Yeni Boyut, 1996. (1998[13]).

Öztürk, Yaşar Nuri. *Fatiha Suresi Tefsiri*. Istanbul: Yeni Boyut, 1996. (1997[5]).

89 Pink, *Muslim Qur'ānic Interpretation Today*, ch. 4, "Modernism and its Paradigms," 124–71.

90 Ibid., 266.

91 Ibid., section "Postmodern Uncertainties and Subjective Approaches," 265–69.

92 Moreover, even the latest translation by the state authority Diyanet İşleri Başkanlığı (*Kur'an-ı Kerim Meâli*, trans. Halil Altuntaş and Muzaffer Şahin [Ankara: DİB, 2001]) appears to be a response to the popularity of current non-official translations, and Öztürk's in particular. "Qur'an translation of the week #31: Kur'an-ı Kerim Meali by Halil Altuntaş and Muzaffer Şahin," GloQur – The Global Qur'an, December 17, 2020, https://gloqur.de/quran-translation-of-the-week-31-kuran-i-kerim-meali-by-halil-altuntas-and-muzaffer-sahin/.

Öztürk, Yaşar Nuri. *İslam'ı Anlamaya Doğru*. Istanbul: Yeni Boyut, 1996. (1998[6]).
Öztürk, Yaşar Nuri. *Konferanslarım*. Istanbul: Yeni Boyut, 1996.
Öztürk, Yaşar Nuri. *Kur'an'ın Temel Buyrukları (Emirler ve Yasaklar)*. Istanbul: Yeni Boyut, 1997. (2006[14]).
Öztürk, Yaşar Nuri. *Ehlibeyt'in Annesi Hazreti Fatima*. Istanbul: Yeni Boyut, 1999.
Öztürk, Yaşar Nuri. "Kur'an'a göre hukuk devleti ve insan hakları." *İstanbul Üniversitesi İlahiyat Fakültesi Dergisi* 1 (1999): 1–38.
Öztürk, Yaşar Nuri. *İslam Nasıl Yozlaştırıldı (Vahyin Dininden Sapmalar, Hurafeler, Bid'atlar)*. Istanbul: Yeni Boyut, 2000. (2001[13]).
Öztürk, Yaşar Nuri. *Batı Sömürgeciliği ve İslam Dünyası (Huntington ve Steinbach'ın Eseri)*. Istanbul: Yeni Boyut, 2003.
Öztürk, Yaşar Nuri. *Kur'an Verileri Açısından Laiklik*. Istanbul: Yeni Boyut, 2003.
Öztürk, Yaşar Nuri. *Atatürk'ten Sonra CHP*. Istanbul: Yeni Boyut, 2004.
Öztürk, Yaşar Nuri. *Kur'an Açısından Küresel Âfetler*. Istanbul: Yeni Boyut, 2008.
Öztürk, Yaşar Nuri. *İnsanlığı Kemiren İhânet Dincilik: Zulümleriyle Dini Kirletenlerin Tarihi*. Istanbul: Yeni Boyut, 2010.
Öztürk, Yaşar Nuri. *Arapçılığa Karşı Akılcılığın Öncüsü İmamı Azam Ebu Hanife: Esas Fikirleri Gölgelenen Önder*. Istanbul: Yeni Boyut, 2011.
Öztürk, Yaşar Nuri. *Kur'an Penceresinden Kurtuluş Savaşına Bir Bakış*. Istanbul: Yeni Boyut, 2012.
Öztürk, Yaşar Nuri. *Maûn Suresi Böyle Buyurdu: Din Maskeli Zulme Tanrının Vuruşu*. Istanbul: Yeni Boyut, 2012.
Öztürk, Yaşar Nuri. *Din Maskeli Allah Düşmanlığı Şirk ve Şirke Tepkinin Felsefeleşmesi: Deizm*. Istanbul: Yeni Boyut, 2013.

5.1.2 Translations of the Qur'ān

Kur'an-ı Kerim Meâli, trans. Halil Altuntaş and Muzaffer Şahin. Ankara: DİB, 2001.
Kur'an'ı Kerim Meâli (Türkçe Çeviri), trans. Öztürk Yaşar Nuri. Istanbul: Yeni Boyut, 1993. (Referenced: 1999[104]).
Kur'an'ı Kerim ve Türkçe Meâli (Metinli), trans. Yaşar Nuri Öztürk. Istanbul: Yeni Boyut, 1994. (2002[102]).
Surelerin İniş Sırasına göre Kur'an'ı Kerim Meâli (İniş sıralı), trans. Yaşar Nuri Öztürk. Istanbul: Yeni Boyut, 1994. (2008[74]).

5.1.3 Translations of Other Arabic Writings

Kutub, Seyyid: *İslam Kapitalizm Çatışması*, trans. under the pseudonym Abdurrahman Niyazoğlu. Istanbul: Düşünce, 1980. (1981[2]; under Öztürk's own name: Istanbul: Bir, 1985[3]; Istanbul: Birleşik, 1995[6]; originally published by Quṭb, Sayyid. *Ma'rakat al-Islām wa-r-ra'smāliyya*. Cairo: Dār al-Kitāb al-'Arabī, 1951).
Sibaî, Mustafa: *İslâm Sosyalizmi*, trans. under the pseudonym Abdurrahman Niyazoğlu. Istanbul: Hareket, 1974. (Under Öztürk's own name: Istanbul: Yeni Boyut, 2010[2]; originally published by as-Sibā'ī, Muṣṭafā. *Ishtirākiyyat al-Islām*. Damascus: Jāmi'at Dimashq, 1959).

5.1.4 Translations of Öztürk's Works into English

Öztürk, Yaşar Nuri. *The Eye of the Heart: An Introduction to Sufism and the Major Tariqats of Anatolia and the Balkans*, trans. Richard Blakney. Istanbul: Redhouse, 1988. (First published as *Tasavvufun Rûhu ve Tarîkatlar*. Istanbul: Sidre, 1988; Yeni Boyut, 1990[2]; 1999[7]).
Öztürk, Yaşar Nuri. *Reconstruction of Religious Life in Islam: Returning to the Koran*, trans. Ali Hayrani Öz. Istanbul: Yeni Boyut, 1999. (First published as *Dinde Yeniden Yapılanmak: Kur'an'a Dönüş*. Istanbul: Yeni Boyut, 1996; 1998[13]).

5.1.5 Translations of Öztürk's Works into German

Öztürk, Yaşar Nuri. *400 Fragen zum Islam, 400 Antworten: Ein Handbuch*, trans. Nevfel Cumart. Düsseldorf: Grupello, 2000. (2019[4]; first published as *400 Soruda İslam*, see 5.1.1).
Öztürk, Yaşar Nuri. *Rumi und die islamische Mystik: Über das Menschenbild im Islam*, trans. Nevfel Cumart. Düsseldorf: Grupello, 2002. (2014[2]; first published as *Mevlâna Celâleddin Rûmî ve İnsan*, see 5.1.1).
Öztürk, Yaşar Nuri. "Was Rückkehr zum Koran bedeutet." In *Alter Text – neuer Kontext: Koranhermeneutik in der Türkei heute: Ausgewählte Texte, übersetzt und kommentiert*, edited and translated by Felix Körner. Freiburg im Breisgau: Herder, 2006: 209–36. (First published in *Dinde Yeniden Yapılanmak [Kur'an'a Dönüş]*, see 5.5.1, 1997[2]: 17–40).
Öztürk, Yaşar Nuri. *Der verfälschte Islam*, trans. Nevfel Cumart. Düsseldorf: Grupello, 2007. (2014[2]; first published as *İslam Nasıl Yozlaştırıldı*, see 5.1.1).
[Not achieved: Öztürk, Yaşar Nuri. *Islam und Ökologie: Die globalen Herausforderungen unserer Zeit aus der Sicht des Korans*, trans. Nevfel Cumart. Düsseldorf: Grupello, 2015. (First published as *Kur'an Açısından Küresel Âfetler*, see 5.1.1).]

5.1.6 Translations of Öztürk's Works into Persian

Öztürk, Yaşar Nuri. *Shahīd-i rāh-i ḥaqīqat u ʿishq: Ḥallāj-i Manṣūr*, trans. Tawfīq Subḥānī. Tehran: Rūzana, 2003.

5.2 Secondary Literature

Akşit, Bahaettin. "Islamic Education in Turkey: Medrese Reform in Late Ottoman Times and Imam-Hatip-Schools in the Republic." In *Islam in Modern Turkey: Religion, Politics and Literature in a Secular State*, edited by Richard Tapper. London: Tauris, 1994: 145–70.
Ateş, Süleyman. *Yüce Kur'an'ın Çağdaş Tefsiri*. 12 vols. Istanbul: Yeni Ufuklar, 1988–1991.
Aydar, Hidayet. "Kur'an IX, Tercümesi." In *TDVIA*, edited by İslâm Araştırmaları Merkezi, 44 vols. Ankara: TDV, vol. 26. 2002: 404–9.
Badran, Margot. "The Religious Face of Secularism." *Al-Ahram Weekly Online*, no. 519, February 1–7, 2001. https://web.archive.org/web/20080412113732/http://weekly.ahram.org.eg/2001/519/intrvw.htm.
Cevizoğlu, Hulki. *Yaşar Nuri Öztürk: O, Dini Yeniden Yapılandırmak İsteyen bir Reformist mi, Özgün bir Yorumcu mu?* Istanbul: Beyaz, 1998.

Cevizoğlu, Hulki. *Yaşar Nuri Öztürk'e Soruyorum*. Istanbul: Ceviz Kabuğu, 2001.

Darwaza, Muḥammad ʿIzzat. *At-Tafsīr al-ḥadīth: as-suwar murattaba ḥasaba an-nuzūl*. Cairo: ʿĪsa al-Bābī l-Ḥalabī, 1962–1964.

Eith, Kathrin. *Koranexegese bei Yaşar Nuri Öztürk: Ein traditionskritischer Entwurf des Islams in der Türkei*. Würzburg: Ergon, 2013.

Eith, Kathrin. "Yaşar Nuri Öztürk: A Contemporary Turkish Tafsir Theorist." In *Tafsir and Islamic Intellectual History: Exploring the Boundaries of a Genre*, edited by Andreas Görke and Johanna Pink. Oxford: Oxford University Press, 2014: 443–65.

Elmalılı, Muhammed Hamdi Yazır. *Hak Dini Kur'an Dili: Yeni Meâlli Türkçe Tefsir*. 9 vols. Ankara: Diyanet Reisliği, 1935–1939.

Flöhr, Benjamin. *Ein traditionalistischer Korandeuter im Dienste des Kemalismus: Elmalılı Muhammed Hamdi Yazır (1878–1942)*. Berlin: Klaus Schwarz, 2015.

GloQur – The Global Qur'an. "Qur'an translation of the week #31: Kur'an-ı Kerim Meali by Halil Altuntaş and Muzaffer Şahin." December 17, 2020. https://gloqur.de/quran-translation-of-the-week-31-kuran-i-kerim-meali-by-halil-altuntas-and-muzaffer-sahin/.

Haberler.com. "2002 Genel Seçim Sonuçları." Accessed October 18, 2020. https://haberler.com/secim/2002/.

İhtiyar, Neşe. "Televangelism and the Case of Yaşar Nuri Öztürk." [Summary of unpublished master thesis, University of Bochum.] *International Institute for the Study of Islam in the Modern World (ISIM) Newsletter* 9 (2002): 22 pages.

İstanbul Üniversitesi İlahiyat Fakültesi. "Tarihçe." Accessed October 18, 2020, https://ilahiyat.istanbul.edu.tr/tr/content/fakultemiz/tarihce.

Körner, Felix. *Revisionist Koran Hermeneutics in Contemporary Turkish University Theology: Rethinking Islam*. Würzburg: Ergon, 2005.

Körner, Felix, ed. *Alter Text – neuer Kontext: Koranhermeneutik in der Türkei heute. Ausgewählte Texte, übersetzt und kommentiert*. Freiburg im Breisgau: Herder, 2006.

Körner, Felix. "Reform durch Koran – ein Nachspann." In *Alter Text – neuer Kontext: Koranhermeneutik in der Türkei heute. Ausgewählte Texte, übersetzt und kommentiert*, edited and translated by Felix Körner. Freiburg im Breisgau: Herder, 2006: 237–43.

Özemre, Ahmet Yüksel *İslam'da Aklın Önemi ve Sınırı*. Istanbul: Denge, 1996.

Özsoy, Ömer. "Koranhermeneutik als Diskussionsthema in der Türkei." Online resource, 2004. Accessed October 13, 2020, http://akademie-rs.de/fileadmin/user_upload/pdf_archive/schmid/TFCI/Fachgespraech_Oezsoy/Artikel__zsoy_7-05.pdf.

Öztürk, Canan et al., eds. *Halkın Diliyle Yaşar Nuri: Neler Söylediler*. Istanbul: Yeni Boyut, 1997.

Öztürk, Mustafa Tahir, ed. *Türkiye'nin Hocası Prof. Dr. Yaşar Nuri Öztürk Anısına*. Istanbul: Yeni Boyut, 2017.

Paçacı, Mehmet, and Yasin Aktay. "75 Years of Higher Religious Education in Turkey." *MW* 89, no. 3–4 (1999): 389–413.

Pink, Johanna. *Muslim Qur'ānic Interpretation Today: Media, Genealogies and Interpretive Communities*. Bristol: Equinox, 2019.

Riḍā, Muḥammad Rashīd. *Tafsīr al-Qur'ān al-ḥakīm* [*aw*] *Tafsīr al-manār* [. . .] *allatī jarā ʿalayhā* [. . .] *Muḥammad ʿAbduh*. 12 vols. Cairo: Dār al-Manār, 1927–1934.

Sabah.com. "Yaşar Nuri Öztürk vefat etti!." June 22, 2016. https://www.sabah.com.tr/yasam/2016/06/22/yasar-nuri-ozturk-vefat-etti.

Seker, Nimet. *Der Koran als Rede und Text: Hermeneutik sunnitischer Koranexegeten zwischen Textkohärenz und Offenbarungskontext*. Berlin: EB, 2019.

Takım, Abdullah. *Koranexegese im 20. Jahrhundert: Islamische Tradition und neue Ansätze in Süleyman Ateş's "Zeitgenössischem Korankommentar."* Istanbul: Yeni Ufuklar, 2007.

Türkmen, Buket. "Le retour au Coran de Yaşar Nuri Öztürk: Vers un Islam national?" In *Formes nouvelles de l'Islam en Turquie*, edited by Gérard Groc. Paris: ERISM, 1999: 81–90.

Wielandt, Rotraud. "Exegesis of the Qurʾān: Early Modern and Contemporary." In *EQ*, edited by Jane Dammen McAuliffe. 6 vols. Leiden: Brill, vol. 2, 2002: 124–42; also available online: accessed December 18, 2020, http://dx.doi.org/10.1163/1875-3922_q3_EQCOM_00059.

Moncef Ben Abdeljelil

Mohamed Talbi

1 Introduction

Mohamed Talbi (d. 2017/1438) was a Tunisian professor of medieval Islamic history by training, a researcher in the field of Islamic studies, and an emblematic intellectual who developed a few challenging perspectives in the study of the Qur'ān. Living for almost a century (ninety-six years), he witnessed the unfolding of four periods that were of particular significance in contemporary history of Tunisia: the colonial period, the emergence of the independent state under Habib Bourguiba, the subsequent period under Z. Ben Ali, and the "Arab Spring," an exceptional revolutionary event that was sparked by a massive Tunisian uprising against despotism. A historian by training, Talbi engaged significantly with each of these periods. The critical issue is understanding how he went from being a professional historian to becoming in a later phase of his career a faithful Muslim intellectual on a mission to formulate new readings of the Qur'ān.

A specialist in classical Arabic letters, with a focus on medieval sources pertaining to Aghlabid Africa (800–909/184–296), Talbi began exploring the makings of Islam and Muslim life in the western region of the Islamic world. He gradually came to concentrate on the complex field of political and intellectual history in *Ifrīqiyā*, where theology, *fiqh*, schism(s), ethics, political governance, and everyday life are but a few study subjects of study that provide valuable insights into local responses to the advent of Islam. Talbi focused on the Qur'ān, a sacred, foundational book for Muslims, with a threefold aim: a) to study the reception of the Qur'ān in the region through commentaries and beyond; b) to explore "cultural Muslimness" in the Aghlabid period, and, in a later phase of his scholarship, c) to contribute to the renewal of Muslim thinking by reinterpreting the Qur'ān. Most of his extensive works, which were informed by his meticulous exploration of the genuine sources, involved combining classical periods of Muslim history with contemporary matters, which Talbi addressed as a faithful Muslim. In doing so, the approach of interpretive history proved to be a reliable intellectual tool that Talbi used in the process of developing his own approach to the Qur'ān in contemporary contexts.

At the outset, it can be said that Talbi behaved in two ways throughout his scholarship: a) as a researcher who spent considerable time telling the story of Islam and Muslims in the Aghlabid *Ifrīqiyā*, and b) as a Muslim thinker who aimed to innovate contemporary Muslim thinking in response to new challenges. In both ways, he took his own hermeneutical stance in relation to the Qur'ān. He related the two sides of this long endeavor in a lengthy, detailed interview presented by three academics and published as *'Iyāl Allāh*.

https://doi.org/10.1515/9783111320069-009

In the following, I will introduce, first, Talbi's career and works and, second, his hermeneutical approach to the Qur'ān. The concluding remarks constitute a kind of a brief assessment of his works on the Qur'ān from a "Qur'ānic Studies" perspective, as Talbi has never claimed to be a specialist in this field.

2 Life and Works

2.1 Life and Career

Talbi was born on September 16, 1921/Muḥarram 13, 1340 in Tunis. His father was of Algerian origin. Interestingly, he grew up in a pious family, where he learnt Arabic, became acquainted with religious *adhkār* (contemplative exercises practiced by Sufis to remember God by repeating His name), theological discussions, and the *Qādiriyya* Sufi order, to which his family belongs. He took his first steps in education at a French school, where he learnt Arabic as a second language. As was common at the time, after taking classes at the French school, he attended a *kuttāb*, where he memorized large parts of the Qur'ān. He later enrolled at al-Madrasa al-ʿAlawiyya (Alawi College) in the city of Tunis, entered the High Institute for Arabic Studies, founded by the French, and received his diploma in Arabic Language. The first job he took was as a primary school teacher, a role he performed passionately, before spending a few years teaching at aṣ-Ṣādiqiyya (the Sadiki College) and the Institut des Hautes Études de Tunis, both located in Tunis. Immediately after the Second World War, in 1947/1366, he went to France and enrolled at Sorbonne University, where he successfully passed the difficult *Agrégation* exam in Arabic in 1952/1371. In 1966/1386, he defended a remarkable thesis on "The Aghlabid Emirate: A Political History," which has remained an outstanding work of reference in the field. From 1962–65/1382–85, Talbi was appointed Director of La Maison de la Tunisie in Paris. In 1966/1386, he became the Director of the Faculty of Letters and Human Sciences in Tunis and was Professor-Director at the same faculty from 1968–70/1388–90. From 1969–89/1389–1410, he took the lead on a prestigious academic publication, *Les Cahiers de Tunisie*. In 1974/1394, he established the STHU (La Société Tunisienne des Historiens Universitaires) and was its first President. The STHU ran until 1994/1415 and was then banned by the regime for unknown reasons. Talbi also acted as the Head of the History Department at the CRES (Centre de Recherches Economiques et Sociales) in Tunis. He was one of the founding members of the Tunisian Academy of Sciences, Letters, and Arts, Beit al-Hikma, in Carthage, Tunisia. As a permanent member of the Christian-Muslim Dialogue Group (GRIC), he facilitated honest interreligious exchange as much as possible, which had some impact on his works and fine-tuned his approach to the Qur'ān. From 1983–91/1403–12, Talbi was appointed Director of the National Cultural Committee. In 1995/1416, he joined the Council for National Liberties in Tunisia. For just a few months in 2011/1432, he was also appointed President of the Tunisian Academy, Beit al-Hikma.

His extensive work is held in high esteem by the international academic community and covers Classical Arabic studies, with a particular focus on literature; the early history of Islam; the intellectual, political, and social history of the Aghlabids; the monumental legacy of ʿAbd ar-Raḥmān b. Khaldūn (d. 1406/808); and contemporary Islamic thought. The honors he received included: Grand Officer of the Order of the Republic, Tunisia; Commander of the Order of the Merit, Tunisia; Officer of the Order of the Independence, Tunisia; Officer of the Legion of Honor, France; and Commander of the Order of the Civil Merit, Spain. In addition, Talbi was awarded the Premio Letterario Internazionale del Mediterraneo, Italy (1979/1399); the Léopold Lucas Prize, Germany (1985/1406); the Cultural National Prize, Tunisia (1987/1408); the Literature National Prize, Tunisia (1991/1412); the Hiroshima Prize for Peace and Culture, Sweden (1994/1415); the Senator Giovanni Agnelli International Prize, Italy (1997/1418); and the Ibn Khaldun Award in Humanities, Qatar (2017/1438).

2.2 Works

Talbi's extensive works are difficult to index. In the following, I have relied on a special document produced by the Tunisian Academy of Sciences, Letters, and Arts, Beit al-Hikma, to mark the fortieth day following Talbi's death.[1] In addition to Talbi's biography and works, the document includes contributions from colleagues, friends, scholars, and officials in the form of *témoignages*, testimonial addresses.

Talbi's works can be divided into five categories:

a) He authored and edited thirty-eight books, starting with Abū l-Ḥasan ʿAlī b. Sīda's *al-Mukhaṣṣas* in 1956 and ending with his own *aṣ-Ṣirāʾ al-lāhūtī fi-l-Qayrawān ayyām al-Aghāliba* (The Theological Conflict in Qayrawān in Aghlabid Times) in 2017.[2] Seminal works that demonstrate his hermeneutical endeavor include *Un respect têtu* (A Stubborn Respect) (1982);[3] *Réflexions sur le Coran* (Reflections on the Qurʾān) (1989);[4] *ʿIyāl Allāh: Afkār jadīda fī ʿalāqāt al-Muslim bi-nafsihi wa-bi-l-ākharīn* (The Children of God: New Ideas about a Muslim's Relationships with Himself and with Others) (1992),[5] re-

1 "Mohamed Talbi, (Septembre 16, 1921–Mai 1, 2017)," special document prepared by the Académie Tunisienne des Sciences, des Lettres et des Arts Beit al-Hikma, Carthage, 2017.

2 Abū l-Ḥasan ʿAlī b. Sīda, *al-Mukhaṣṣas* éd. Mohamed Talbi (Tunis: al-Maṭbaʿa al-ʿAṣriyya, 1956); Mohamed Talbi: *aṣ-Ṣirāʾ al-lāhūtī fi-l-Qayrawān ayyām al-Aghāliba* (Tunis: Sotumedia Editions, 2017).

3 Mohamed Talbi and Olivier Clément, *Un respect têtu* (Paris: Nouvelle Cité, 1982).

4 Mohamed Talbi and Maurice Bucaille, *Réflexions sur le Coran* (Paris: Éditions Seghers, 1989).

5 Mohamed Talbi, *ʿIyāl Allāh: Afkār jadīda fī ʿalāqāt al-Muslim bi-nafsihi wa-bi-l-ākharīn* (Tunis: Dār Sirās li-n-Nashr, 1992).

edited in French as *Plaidoyer pour un islam modern* (Call for a Modern Islam) (1998);[6] *Études sur la Tolérance* (Studies on Tolerance) (1998);[7] *Ummat al-wasaṭ wa-taḥaddiyāt al-muʿāṣira* (The Community of the Middle Way and the Challenges of Contemporary Times) (1996);[8] *Al-Islām: Ḥurriyya wa-Ḥiwār* (Islam: Liberty and Dialogue) (1999);[9] *Universalité du Coran* (The Universality of the Qurʾān) (2002);[10] *Réflexions d'un Musulman Contemporain* (Reflections of a Contemporary Muslim) (2005);[11] *Li-Yaṭmaʾinna qalbī: Qaḍiyyat al-īmān* (So that My Heart Becomes Serene: The Question of Faith) (2010);[12] *Ma religion c'est la liberté: L'Islam et les défis de la contemporanéité* (My Religion is Liberty: Islam Facing the Challenges of Contemporary Times) (2011);[13] *L'Islam n'est pas voile, il est culte: Rénovation de la pensée musulmane* (Islam is Not a Veil, it is Worship: The Renewal of Muslim Thought) (2014);[14] *Qaḍiyyat al-ḥaqīqa* (The Question of Truth) (2015);[15] *Médiations (tadabbur) sur le Coran: Vérité – Rationalité – I'jâz Scientifique* (Meditations on the Qurʾān: Truth – Rationality – Scientific Excellence) (2016);[16] and *Dalīl al-Muslim al-Qurʾānī* (The Guide of the Qurʾānic Muslim) (2016).[17]

b) The second category consists of Talbi's contributions to edited works. He authored about nine chapters in different books, most of which address the social history of the Maghreb region as well as everyday life in and the economic history of Aghlabid *Ifrīqiyā*. His chapter on *ʿīd al-aḍḥā* (Eid), published as "Aïd al-Adhā ou Aïd El-Kébir" (1980), may be of relevance to the topic of this article.[18]

c) The third category includes forty-eight articles published in academic journals or conference proceedings in the three languages Arabic, French, and English. Some of these are discussions of Talbi's approach to the Qurʾān, such as: "La vie en communauté (le

6 Mohamed Talbi, *Plaidoyer pour un islam modern* (Rabat: Desclée de Brouwe, 1998).

7 Abdelfattah Amor, Néji Baccouche, and Mohamed Talbi, *Études sur la Tolérance* (Carthage: Beit al-Ḥikma, 1995).

8 Mohamed Talbi, *Ummat al-wasaṭ wa-taḥaddiyāt al-muʿāṣira* (Tunis: Dār Sirās li-n-Nashr, 1996).

9 Mohamed Talbi, *Al-Islām: Ḥurriyya wa-Ḥiwār* (Beirut: Dār an-Nahār, 1999).

10 Mohamed Talbi, *Universalité du Coran* (Arles: Actes Sud, 2002).

11 Mohamed Talbi, *Réflexions d'un Musulman Contemporain* (Casablanca: Les éditions Le Fennec, 2005).

12 Mohamed Talbi, *Li-Yaṭmaʾinna qalbī: Qaḍiyyat al-īmān* (Tunis: CERES éditions, 2010).

13 Mohamed Talbi, *Ma religion c'est la liberté: L'Islam et les défis de la contemporanéité* (Tunis: Éditions Nirvana, 2011).

14 Mohamed Talbi, *L'Islam n'est pas voile, il est culte: Rénovation de la pensée musulmane* (Carthage: Éditions Carthaginoiseries, 2014).

15 Mohamed Talbi, *Qaḍiyyat al-ḥaqīqa* (Tunis: Editions Finzi, 2015).

16 Mohamed Talbi, *Médiations (tadabbur) sur le Coran: Vérité – Rationalité – I'jâz Scientifique* (Tunis: Éditions Mohamed Talbi, 2016).

17 Mohamed Talbi, *Dalīl al-Muslim al-Qurʾānī* (Tunis: Masānīʾ Fīnzī li-t-Ṭibāʿa, 2016).

18 Mohamed Talbi, "Aïd al-Adhā ou Aïd El-Kébir," in *Célébrer Dieu* (Éditions Univers-Médina, Paris, 1980), 117–20.

point de vue musulman)" (Living in Community [a Muslim Point of View]) (1984);[19] "Religious Liberty: A Muslim Perspective" (1985);[20] "Le Coran dans la vie quotidienne des Musulmans" (The Qur'ān in the Daily Life of the Muslim) (1985);[21] "Liberté religieuse et Transmission de la Foi" (Religious Liberty and the Transmission of Faith), published in German as "Religiöse Freiheit und Vermittlung des Glaubens" (1987);[22] "Pour un futur solidaire: la foi facteur de paix" (Toward a Common Future: Faith as a Factor in Peace) (1986);[23] "La liberté religieuse: droit de l'homme ou vocation de l'homme?" (Religious Liberty: Man's Right or his Vocation?) (1988);[24] "Qaḍiyyat ta'dīb al-mar'a bi-ḍ-ḍarb: Qirā'a tārikhiyya li-l-āyatayni 34 wa-35 min Sūrat an-Nisā'" (The Question of Disciplining a Woman by Beating: A Historical Reading of Verses 34 and 35 of sura 4, an-Nisā' [Women]);[25] "Liberté et Dignité de l'homme: Le Coran parole structurante de la Ummah" (Liberty and Dignity of Man: The Qur'ān as a Structuring Speech of the Umma) (1992);[26] and "Unavoidable Dialogue in a Pluralist World: A Personal Account" (1995).[27]

d) The fourth category comprises ninety-seven book reviews on a variety of historical and intellectual topics.

e) The last category relates to Talbi's articles in daily and weekly newspapers and magazines, forty-two in Arabic, and eleven in French, as compiled by the Tunisian Academy of Sciences, Letters, and Arts, Beit al-Ḥikma, Carthage, Tunisia.[28] The last on this list of articles seems to be relevant to our interest here, with the suggestive title: "Ud'ū ilā sabīli rabbika" (Call Others to God's Path) (1992/1412).[29]

19 Mohamed Talbi, "La vie en communauté (le point de vue musulman)," *Islamochristiana* 10 (1984): 85–90.

20 Mohamed Talbi, "Religious Liberty: A Muslim Perspective," in *Proceedings of the Second World Congress on Religious Liberty* (Berrien Springs, MI: University Printers, 1985), 156–68.

21 Mohamed Talbi, "Le Coran dans la vie quotidienne des Musulmans," *Supplément au Bulletin de Littérature Ecclésiastique* 2 (1985): 40–49.

22 Mohamed Talbi, "Religiöse Freiheit und Vermittlung des Glaubens (Libertè religieuse et transmission de la foi)," in *Dialog mit Afrika und dem Islam: Zwei Vorträge von Léopold Sédar Senghor und Mohamed Talbi*, ed. Hans Peter Rüger (Tübingen: J.C.B. Mohr, 1987), 27–69.

23 Mohamed Talbi, "Pour un futur solidaire: la foi facteur de paix," *Roma-Incontro, Rivista mensile di informazione e dibattio* 10, no. 8–9 (August–September 1986), 35.

24 Mohamed Talbi, "La liberté religieuse: droit de l'homme ou vocation de l'homme?" *Conscience et Liberté* 36, no. 2 (1988), 21–34.

25 Mohamed Talbi, "Qaḍiyyat ta'dīb al-mar'a bi-ḍ-ḍarb: Qirā'a tārikhiyya li-l-āyatayni 34 wa-35 min Sūrat an-Nisā'," *Le Maghreb* 182, December 22, 1989, 19–23; and *Le Maghreb* 183, December 29, 1989, 19–23.

26 Mohamed Talbi, "Liberté et Dignité de l'homme: Le Coran parole structurante de la Ummah," in *Würde und Recht des Menschen: Festschrift für Johannes Schwartländer zum 70. Geburtstag*, ed. Heiner Bielefeldt, Winfried Brugger, and Klaus Dicke (Würzburg: Königshausen & Neumann, 1992), 347–61.

27 Mohamed Talbi, "Unavoidable Dialogue in a Pluralist World: A Personal Account," *Encounters: Journal of Inter-Cultural Perspectives* 1 (1995), 56–69.

28 "Mohamed Talbi, (Septembre 16, 1921–Mai 1, 2017)," special document prepared by the Académie Tunisienne des Sciences, des Lettres et des Arts Beit al-Hikma, Carthage, 2017.

29 Mohamed Talbi, "Ud'ū ilā sabīli rabbika," *al-Jazīra*, January 30, 1992, 7, 18.

Such extensive intellectual and academic production over the many years of Talbi's career tells us about the evolution of his interest in the Qur'ān and the genesis of his approach, and may also explain the views he later held while interpreting various verses in a challenging contemporary context.

3 Hermeneutical Enterprise

In this section, I will summarize Talbi's approach to the Qur'ān in two ways. The first is an attempt to identify the foundational postulates he stated in many of his early writings; the second includes his own understanding of key verses of the Qur'ān. Such an approach reveals both his global perspective as a Muslim and the trajectory of his thinking. Interestingly, Talbi insists on two key values, liberty and pluralism. While deconstructing his thought, it is unnecessary to discuss whether he rejected traditional approaches to the Qur'ān, for neither in Talbi's intentions nor in his attitudes toward classical exegesis did he clearly state any such rejection. On the contrary, he had much appreciation for commentaries by prominent classical figures, including Abū Jaʿfar aṭ-Ṭabarī (d. 922/310), Muqātil b. Sulaymān (d. 767/150), Nāṣir ad-Dīn al-Bayḍāwī (d. 1286/685), Fakhr ad-Dīn ar-Rāzī (d. 1209/606), Abū l-Qāsim az-Zamakhsharī (d. 1143/538), and even Yaḥyā b. Sallām (d. 815/200). His deviation from classical approaches confirms Talbi's quest for what the Qur'ān can tell Muslims *today* and what they can take from it. It is an attempt to reframe the Qur'ān as a divine revelation in a modern context.

3.1 Foundational Postulates

The first postulate is Talbi's statement that he thinks as a believer for such is his mindset.[30] He goes on to say that no neutrality of any kind is possible in approaching the Qur'ān.[31] To any Muslim, this book signifies the word of God, which was fully revealed to the Prophet Muḥammad, who shared it with his community as a message free of any alteration. In this respect, Talbi considers and comments on the Qur'ān "from within." The Qur'ān, in his opinion, remains the source of truth, which must be further revealed and unpacked by human reason.[32]

---

30 See Talbi, *Li-Yaṭma'inna qalbī*; Talbi, *Réflexions d'un Musulman Contemporain*; Talbi and Clément, *Un respect têtu*.

31 Talbi, *Médiations (tadabbur) sur le Coran*.

32 Ibid.

The second postulate is his assertion that he is a historian whose perspective cannot be anything other than critical. Being critical means being aware of the context, the construction(s) of reality, and man's mission to develop a purpose for his own sake. A historian is tasked with unpacking past events, rooting the word of God in emerging realities, and further exploring new possibilities of what God means to say to His followers. According to Talbi, "history does not change by means of discovered documents; rather, it changes through the new questions we pose."[33] History is therefore a continuous movement toward evolution, creation that does not end.[34] "History," he goes on to explain, "is the most global of the human sciences; it fills memory and nurtures imagination, and it is through history that we discover how man has behaved while enduring suffering; it is through history, indeed, that we progress toward the ideal state of humanity. History is thus ascendance and permanent creation."[35]

The third postulate concerns Talbi's understanding of the "right path," which has a systematic connection to his approach to the Qur'ān. The "right path," he states, is the path of science and knowledge.[36] *Ijtihād* in this respect remains a permanent quest for knowledge – an enterprise that should not, Talbi says, be restricted to only a few groups of people. It is man's mission in this world.[37]

The last postulate relates to his belief in the essence of liberty, pluralism, and love for God. Religion, according to Talbi, cannot exist without these values, which represent humans' formative essence. As humans with different minds, belonging to various cultures, and following different trajectories, we come to be, by essence, creatures of liberty who have to accept pluralism and may achieve inner peace through love for God. "Without God, I am nothing,"[38] Talbi writes.

3.2 Approaching the Qur'ān: A Few Hermeneutical Examples

Throughout his works, Talbi repeatedly expresses that the approach he takes to the Qur'ān is a *maqāṣid* (purposive) one.[39] This kind of approach was initiated by Abū Isḥāq ash-Shāṭibī (d. 1388/790), an Andalusian Sunni legal scholar, and then taken up by Muḥammad ash-Shawkānī (d. 1834/1250), a prominent exegete and legal scholar from Yemen, before later being re-elaborated by two outstanding Maghribis, Shaykh Muḥammad aṭ-Ṭāhir b. ʿĀshūr (d. 1973/1393) from Tunisia and ʿAllāl al-Fāsī (d. 1974/

33 Talbi, *ʿIyāl Allāh*, 51.
34 Ibid., 36.
35 Ibid., 61.
36 Talbi, *Médiations (tadabbur) sur le Coran*; Talbi, *Dalīl al-Muslim al-Qur'ān*.
37 Ibid., 46.
38 Mohamed Talbi, "Sans Dieu, je ne suis rien," *France Catholique-Ecclesia*, January 28, 1985, 10.
39 Talbi, *ʿIyāl Allāh*, 70.

1394) from Morocco. It consists in exploring what are referred to as the "higher objectives and intents of Islamic law"[40] while raising the problem of *maṣlaḥa* on the one hand and, on the other, addressing the question of *taʿlīl* (justification) in the sharia, or the reasons that might be behind the divine rules of the Qurʾān. In response to both the esoteric Shiite reading of the Qurʾān and the Ẓāhirite legal school founded by Abū Sulaymān Dāwūd b. ʿAlī ẓ-Ẓāhirī (d. 884/270) from Iraq and Abū Muḥammad b. Ḥazm (d. 1064/456) from Andalusia, *maqāṣid* theory is an attempt to think and derive rules according to what the divine message conveys as the final aim of the Qurʾānic message. Muslim scholars who have launched this new legal thinking have gone beyond the formal literality of scripture to explore the *raison d'être* of the sharia. Ash-Shāṭibī has found three categories: *aḍ-ḍarūriyyāt* (the essentials), *al-ḥājiyyāt* (the compulsories), and *at-taḥsīniyyāt* (the optionals). He goes on to detail the first category by stating five aims: *ḥifẓ ad-dīn* (preserving religion, i.e., Islam); *ḥifẓ an-nafs* (preserving human life); *ḥifẓ an-nasl* (preserving mankind), *ḥifẓ al-māl* (preserving one's property); and *ḥifẓ al-ʿaql* (preserving intellect).[41] The same five aims were taken on by later Muslim scholars with minor changes.

Interestingly, Talbi – as one of many other contemporary Muslim scholars, including Fazlur Rahman, Nurcholish Majid, and Abdelmajid Charfi – adopted *maqāṣid* theory, which he sometimes called the "vectorial" reading of the Qurʾān. By this he meant a progressive reading that could meet the new challenges of the time. He believed that the Qurʾān should speak to living generations, *not* to past ones.[42] Speaking to his contemporaries, Talbi initiates an original reflection on the Qurʾān regarding Muslimness and individual rights, the nature of the community, and the contribution he expects the community to make to humanity. The transversal tie that binds the three levels remains, in Talbi's thinking, liberty, faith, and pluralism/alterity.[43]

What does it mean to be Muslim in a certain context? Talbi has repeatedly insisted on four aspects of what makes somebody a genuine Muslim. Beyond the beliefs that a Muslim must hold, he/she: a) enjoys liberty, b) is conscious of his/her mission in this world, c) thinks within the lines of the divine wisdom contained in the Qurʾān, and d) internalizes humility and a sense of alterity. In sum, a Muslim is a creature of God and therefore embodies the image of God to become His *khalīfa*, vicegerent.

In his last book, *Dalīl al-Muslim al-Qurʾānī* (The Guide of the Qurʾānic Muslim), which he published shortly before he passed away, Talbi introduced his vision by stating, "God is the womb of the universe. Nothing could come out of the womb but love.

40 Mohamed Talbi, "Qaḍiyyat taʾdīb al-marʾa bi-ḍ-ḍarb."

41 Abū Isḥāq ash-Shāṭibī, *al-Muwāfiqāt fī uṣūl ash-sharīʿa*, ed. ʿAbd Allāh Darrāz and ʿAbd as-Salām ʿAbd ash-Shāfī (Riyadh: Wizārat ash-Shuʾūn al-Islāmiyya wa-l-Awqāf wa-d-Daʿwa wa-l-Irshād, n.d.), 2: 8f.

42 Talbi, *Ummat al-wasaṭ wa-taḥaddiyāt al-muʿāṣira*, 7.

43 See Talbi *Plaidoyer pour un islam modern*; Talbi, *Réflexions d'un Musulman Contemporain*; Talbi, *Li-Yaṭmaʾinna qalbī*; Talbi, *Ma religion c'est la liberté*; Talbi, *Ummat al-wasaṭ*.

God has created the universe in the nature of love, for He does not need it. Therefore, man's submission to Him is an act of appreciation and love."[44] Further, a Muslim is, according to Talbi, a person as described in Q 2:1–5. Interestingly, he goes on to distinguish infidelity (*kufr*) from sinfulness (*maʿṣiya*). An infidel is, Talbi states, the one who rejects the Book; a sinner is one who commits any violation of God's rules, such as the act of not performing prayer. Significantly, Talbi confirms people's freedom to reject the Book. He recalls in this respect verse Q 2:256: "There is no compulsion in religion" (*lā ikrāha fī-d-dīn*).[45] When it comes to sin, he insists on an ethical attitude by recalling Q 49:12: "O ye who believe! Shun much suspicion; for lo! some suspicion is a crime. And spy not, neither backbite one another. Would one of you love to eat the flesh of his dead brother? Ye abhor that (so abhor the other)! And keep your duty (to Allah). Lo! Allah is Relenting, Merciful." Talbi goes on to state that "the one whom God has covered (*satara ʿalayhi*) in this life would be covered in the Hereafter as well. Interpreting Q 2:284; 5:18 and 40; 48:14, God is the most forgiving and the most merciful."[46] The point that Talbi is making here concerns a moral and social attitude toward both "infidels" and "sinners." He recognizes infidels' right to be as they chose. Liberty is a right fully respected by the Qurʾān, according to Talbi; sinners, on the other hand, are judged by God, for it is He who will pass judgment on their deeds; no human should try to claim that responsibility.

The important conclusion that Talbi tries to draw from his explorations is his rejection of the sharia. The entire first chapter on the aim of his final book is an explicit statement of this claim.[47] However, this does not mean abolishing the Qurʾānic rituals, nor does it mean a vision similar to that of the Sufis and/or extremist Shiites. Rather, it is a call for a new hermeneutical approach to the Qurʾānic text that does not rely extensively on legal corpuses.

Neither belief nor related rituals are held or performed out of a fear of God. Without liberty, those acts are meaningless. Talbi states that fearing God means fearing His majesty and greatness.[48] Adherence to Islam therefore requires both liberty and faith, which mean assuming full responsibility for one's actions. "As a Muslim," Talbi goes on to say, "I recall God's sayings 'There is no compulsion in religion' (Q 2:256) and 'Each soul earneth only on its own account, nor doth any laden bear another's load. Then unto your Lord is your return and He will tell you that wherein ye differed' (Q 6:164)."[49]

---

44 Talbi, *Dalīl al-Muslim al-Qurʾānī*, 4.

45 Qurʿānic translations follow Muhammad Marmaduke Pickthall, *The Meaning of the Glorious Koran: An Explanatory Translation* (New York: A.A. Knopf, 1930).

46 Talbi, *Dalīl al-Muslim al-Qurʾānī*, 17.

47 Ibid., 8–10.

48 Ibid., 4.

49 Talbi, *ʿIyāl Allāh*, 84.

He dedicates a full chapter of his *Ummat al-wasaṭ* (The Community of the Middle Way) to the question of individual liberty, titled "Is Religious Freedom a Right Among Human Rights or Man's Destiny?" His starting point is Q 10:99, which reads: "And if thy Lord willed, all who are in the earth would have believed together. Wouldst thou (Muhammad) compel men until they are believers?" He uses it to discuss his belief that religious freedom is neither tolerance nor complaisance; it is also far from being a kind of relativism. Nor should freedom be confused with mutual respect, which is a highly appreciated principle that implies fairness and submission to God, for the Almighty stated that He respects mankind. Talbi furthers his analysis on the basis of Q 17:15 and Q 88:21–22, concluding that "God has avoided requiring man to believe in His religion; He even forbade the Prophet from doing so. In so doing, God teaches us respect for human dignity."[50] The main conclusion that Talbi draws is that liberty is a constitutive element of man's existence. Liberty is an integral part of man's "primitiveness" (*fiṭra*).[51]

Religious freedom, according to Talbi, is also the root of all liberties, for it allows man to be God's vicegerent. He returns to Q 33:72 "We offered the Trust to the heavens, and the earth, and the mountains; but they refused to bear it, and were apprehensive of it; but the human being accepted it. He was unfair and ignorant." Talbi considers that such trust (*amāna*) is God's master plan (*al-mukhaṭṭaṭ al-ilāhī*) for the universe and man's mission, a mission that might be regarded as an "impossible one"; however, man has accepted it. Man, Talbi remarks, has made this choice because of the liberty he holds. What, then, does "the right path to follow" (*aṣ-ṣirāṭ al-mustaqīm*) mean? Talbi sees it as an open way for a quest for truth, knowledge, and *ijtihād*, in harmony with God's guidance. It is a continuous endeavor aimed at achieving God's plans.[52] The right path, Talbi continues, is constant movement, allowing us to accept new eras and face emerging challenges with innovative minds. This is how one belongs to the contemporary times (*al-muʿāṣara*).[53]

In the current context, where modernity prevails, it is no longer possible, according to Talbi, to consider religious freedom as any form of "tolerance," for tolerance is not a right, while due respect is, indeed, a right. Religious freedom is, again, a right that everybody deserves. In practice, religious freedom does find its way into the secular sphere, where the state expresses its neutrality and its incompetence in the domain of religion. Yet, in several Muslim countries, such a move toward what Ibn Khaldūn described as the mutation from *khilāfa* to *mulk*, a kind of a lived secularity, is still a slow process; such secularity does not seem to have been accepted by many Muslim societies, with the exception of Türkiye.[54] In order to solve the question of religious

50 Ibid., 45.

51 Talbi, *Ummat al-wasaṭ*, 32–44.

52 Talbi, *ʿIyāl Allāh*, 45.

53 Talbi, *Ummat al-wasaṭ*, 9.

54 Ibid., 42.

freedom in Muslim societies, Talbi concludes by alluding to what he calls "ethical responsibility," which means that everyone should benefit from being free when deciding what he/she wants to be. It is not up to the law or the state to decide on this matter; rather, it is up to the individual's consciousness to follow a path where integrity, harmony, and the real life actually make sense. We should adopt this perspective as we move forward, he says.[55] What does a Muslim community look like today?

The concept of *umma*, as Talbi discusses, is both problematic and difficult to capture using terms such as "nation," "community," "people," and "population" in Arabic and Muslim contexts. The reason for this, he says, is the concept's lack of clarity from Arab standpoints. He goes on to note that *umma* is almost impossible to translate into either French or English because of its specificities.[56] To elaborate further on the concept, he proposes going back to Q 21:92 and Q 23:52, asserting that the concept of *umma* is a Qur'ānic one. It means one body of people that is *not* divided and that adheres to the central idea of divinity (*annahā iltifāf ḥawla ar-rubūbiyya*) by means of worship and piety. Therefore, the tie that connects such people to each other is not ethnic, national, cultural, civilizational, patriotic, political, or geographic, etc., but allegiance to God through worship and piety.[57]

Obviously, Talbi thinks that the members of the *umma* should be free and equal, in all respects. Piety is the unique criteria that distinguishes Muslims. The closer a Muslim gets to God, the higher his ranking. Like liberty, equity seems to be Muslims' objective as well. Remarkably, this affects the issue of gender and constitutes an unconventional attitude toward the conditions and rights of women. Throughout his chapter on "The Question of Beating Women," Talbi explores the context of Q 4:34–35. Interestingly, he comes to two conclusions: a feminist movement has existed since the time of the Prophet, emerging in Medina, and the verse rejects such harsh acts against women. He draws his interpretation from Q 30:21. Talbi states his support for gender equality.[58] Further, the *umma*, he writes, is the middle nation (*ummat al-wasaṭ*), tasked with a mission, according to the Qur'ān. Talbi calls such an *umma* the "*ummat al-umam*," or the "community of communities." The two categories need some explanation in relation to his Qur'ānic hermeneutics.

Ummat al-wasaṭ, which is also the title of one of Talbi's books, is borrowed from Q 2:143, which reads:

> Thus, We have appointed you a middle nation, that ye may be witnesses against mankind, and that the messenger may be a witness against you. And We appointed the *qiblah* which ye formerly observed only that We might know him who followeth the messenger, from him who turneth on his heels. In truth it was a hard (test) save for those whom Allah guided. But it was not

---

55 Ibid., 43.

56 See Talbi, *'Iyāl Allāh*, 36–42.

57 Ibid, 38.

58 See Talbi, *Ummat al-wasaṭ*, 137 ("*Ta'dīb al-mar'a bi-ḍ-ḍarb*").

Allah's purpose that your faith should be in vain, for Allah is full of pity, Merciful toward mankind.

Talbi proceeds by first setting out the different readings and the commentaries given by Fakhr ad-Dīn ar-Razī (d. 1209/605) and Abū l-Fidā' b. Kathīr (d. 1373/774), and by Muḥammad ʿAbduh and Rashīd Riḍā in in the *Tafsīr al-manār* in relation to the "causes of revelation" (*asbāb an-nuzūl*) and commentaries on the change in the *qibla's* direction. He rejects almost all exegetical opinions, concluding:

With Muḥammad, a charter (*mīthāq*) with a specific community had come to an end, and a new one started without any limitation in time, except that of the end of worldly life (*qiyām as-sāʿa*). The characteristic of the new charter (*mīthāq*) is a new discourse open in time to all humans, while the message in the Old and New Testament was specific to the Banī Isrāʾīl. Therefore, Muḥammad is "the illiterate Prophet" who witnessed this critical change (of the *qibla*), and was sent to all mankind, none of which was explicitly announced in the previous sacred Books. [. . .] God never does speak to the Arabs in the Qurʾān.[59]

Such a change in discourse meant, according to Talbi, a decisive turn in the divine "plan" at a time when humanity showed a level of maturity and development, making it ready to receive a unique divine message, which it could communicate and disseminate around the world as intended. God therefore tasked the "Last Prophet" with conveying such a message, ending one era and commencing a new one.[60] This new *umma, ummat al-wasaṭ*, was initially open to the community of the Book; the divine message then asked other ethnicities to join in addition to the Arabs, such as the Persians and the Africans. It is, according to Talbi, a community without borders or limitations. Its unity derives from its strong ties with God, *al-ʿurwa al-wuthqā*)the most trustworthy handhold), as Q 3:103 says: "And hold fast, all of you together, to the cable of Allah, and do not separate"; or, as in Q 2:256: "he who rejecteth false deities and believeth in Allah hath grasped a firm handhold which will never break."[61] Therefore, according to Talbi, the new umma deserves to be tasked by God with calling people to Him (*ummat ad-daʿwa*), which started with Ibrāhīm; its mission is an intermediary one. It connects with Ibrāhīm but also engages with a new, open era; in doing so, it is an *umma* of the "middle way."

Talbi states that the same *umma* is also *ummat ash-shahāda*, the *umma* of testimony, following the same verse, Q 2:143: "ye may be witnesses against mankind." Such a new mission relates to Muḥammad's testimony. For Talbi, this means that the act of witnessing relates to the task of conveying God's message, without interruption. The responsibility of the *umma* in this respect is based on its duty to ensure human dignity and to follow the right path (*al-istiqāma*),[62] as previously explained. *Ummat al-*

59 See ibid., 24.
60 Ibid.
61 Ibid., 26.
62 Ibid., 28f.

wasaṭ engages in an ongoing process aimed at attaining the human virtues (*al-faḍīla*), continuously striving toward a state of excellence and perfection. Being of this nature, *ummat al-wasaṭ* is for Talbi the community of communities, which opens up new horizons for both humanism and alterity.

3.2.1 What Does *Ummat al-Umam* Mean?

Talbi discusses Q 49:13, which reads: "O mankind! Lo! We have created you male and female, and have made you nations and tribes that ye may know one another. Lo! the noblest of you, in the sight of Allah, is the best in conduct. Lo! Allah is Knower, Aware." The main idea to be drawn from the verse is a threefold shift that Talbi insists upon, which concerns three areas: a) the global condition that humanity is currently experiencing, which makes the concept of *umma* more complex; b) the distinction between *ummat ad-daʿwa* (the community of the call) and *ummat al-istijāba* (the community of followers); and c) the challenges of plurality and otherness, which is to be internalized by the *umma*, if it seeks harmony and interaction with the whole body of mankind.

Talbi starts by pointing to the contemporary mapping of human civilizations, stating that historians have permanently moved beyond the era of civilizations confined to specific regions. Geography does not mean much today. The point he is making concerns the dynamic process of inter-national communication by all conceivable means. *Umma* has become a part of this new state of humankind's development. Beyond the boundaries that have been set for political and economic purposes, all "nations" are but one, because they are all addressed by the Qur'ānic call. Islam is indeed open to all of them. The concept of the "community of the call" (*ummat ad-daʿwa*) means that every human is the intended recipient of the call. In response, the whole of humanity is *ummat al-istijāba*, "the community of followers," when people decide to accept the Qur'ānic call.[63] In fact, the *umma* goes beyond geographic borders, for it relies on a spiritual tie, which is faith. Across different societies and cultures, a larger *umma* starts to take shape; it heeds the same message, enjoys the same rights, and holds the same ambitions. This global dimension raises two major issues: diversity within the new *umma* and the absolute need for a pluralist attitude.

Being *dīn wa-dunyā* (a religion and a lived world), Islam, according to Talbi, brings together faith and deeds in harmony. However, Islam is not religion or politics. In a living world, Muslims must share both space and the task of improving human living conditions with other nations and people. The diversity of humankind is not only cited in the Qur'ān but is also a concrete reality that a Muslim, wherever he/she happens to live, must deal with. Exclusion would lead nowhere. Talbi continues: "It

63 Ibid., 63–67.

has become impossible for any nation, unless it marginalizes itself and moves against the sense of history, to isolate itself in its own system. Any isolation today is but an act of suicide, in a world which has changed much within our borders, and even inside our homes and minds, without any hope of return."[64] Common, shared existence and all it implies becomes a must for all of humankind, including Muslims. The *umma* must therefore accept its diversity, which is a concrete fact within its body and the tangible way that all nations are configured among non-Muslims. His explicit statement in the form of a strong claim underlines the issue of "Islamic" plurality. Talbi says: "Living in a community of communities, expressing respect for the variety of differences among peoples, and achieving harmony among them, are not matters of fact in reality but attempts to renew the strong ties with the spirit of Islam in its pure meaning."[65]

3.2.2 How Does Islam Stand for All Religions?

Diversity and pluralism stem from the fact that Muslims live with other nations and religious groups, which represents a real and pressing challenge. Tensions, Talbi notes, are inevitable. But they do not affect Muslims' essentials. Among these groups, he goes on to claim, there is a group that requires further attention, the People of the Book (*ahl al-kitāb*). Those who are truly informed of the Judeo-Christian legacy will recognize that Islam is but a continuation of that legacy. According to Muslims, Islam is the seal of this tradition and its final development. To argue this point, Talbi calls attention to Q 2:258:

> The messenger believeth in that which hath been revealed unto him from his Lord and (so do) believers. Each one believeth in God and His angels and His scriptures and His messengers – We make no distinction between any of His messengers – and they say: We hear, and we obey. (Grant us) Thy forgiveness, our Lord. Unto Thee is the journeying.[66]

However, Talbi knew very well that the reality was more complex. Being himself a member of the GRIC (Groupe de Recherches Islamo-chrétien), and leading many seminars on Muslim-Christian dialogue, he encountered both tensions and irresolvable situations. Still, he believed in what he called an "honest and genuine dialogue." His book *Un respect têtu,*[67] written in collaboration with Olivier Clément, represents a significant contribution in this regard.

What about other religious bodies, communities, and groups? Talbi's answer is clear and is also based on the Qur'ān. He mentions that some of these groups have

---

64 See ibid., 71.
65 See ibid., 72.
66 See ibid., 78f.
67 Talbi and Clément, *Un respect têtu.*

been reconnected with the *ahl al-kitāb* (People of the Book), so the space of Muslimness has been enlarged. On the other hand, Talbi says, we are not aware of all of the messages that God would have sent to successive generations of people. Q 20:78 offers proof of this. Further, he remarks that, according to Q 29:19–20, God requests that we explore the secrets of the world. The conclusion, in his opinion, is that we must continue to explore religions in Africa, Asia, and the United States in order to comprehend the ways in which God is worshipped. In doing so, we recognize the existence of a third space larger than the previous ones, where the word of God is also venerated. The third space, Talbi says, is the space of the "community that is in God's service" (*dā'irat ummat khidmat Allāh*).[68]

Interestingly, Talbi does not ignore the remaining members of humankind who do not believe at all. What does he think should be Muslims' attitude toward this category? He adopts an ethical attitude, according to which all humans should share key virtues for the welfare of humankind. According to Q 28:77, he argues, we are required to act to improve the world. Many non-believers are righteous. As we are all sitting in the same boat of life, Muslims should engage with them for peace and mutual betterment.[69] At this point, Talbi addresses the largest dimension of the Muslim community, being as humanist as the context allows. The four circles/spaces that Talbi has defined illustrate his willingness to move away from very "restrictive Muslimness" to a much broader one through the process of *inclusion*. The philosophy of human rights seems to be the basis for this intellectual orientation.

4 Conclusion

Three concluding perspectives are of relevance to Talbi's hermeneutical approach to the Qur'ān: First, his approach to the Qur'ān, based on the above-mentioned postulates, remains *spiritual*. The historicity throughout his works supported his faith as a Muslim, which was deeply rooted in a Sunni theological school and colored by African Malikism. Second, the essence of his hermeneutics is profoundly concerned with the challenge posed by modernity, which Muslims since the medieval times, he says, have not been able to meet due to their extensive sharia heritage. The Muslim Brotherhood Organization in modern times has even worsened the situation. Talbi's rejection of sharia derives from his desire to find an innovative interpretation of the Qur'ān that is in harmony with the new modern era and its complex philosophies. Third, though far from offering a systematic exegesis of the Qur'ān, Talbi selected from the Qur'ān the verses that he believed were relevant to the debate he was engaging with. His selective method reveals much about his orientation and demonstrates his historical-

68 See Talbi, *Ummat al-wasaṭ*, 82f.
69 See ibid., 84f.

critical method for exploring original meanings, before extending his reflections to constitute that which one might call a "spiritual hermeneutical approach." Interestingly, such an approach emerged from within his Muslim consciousness.

5 Bibliography

5.1 Sources

Amor, Abdelfattah, Néji Baccouche, and Mohamed Talbi. *Études sur la Tolérance*. Carthage: Beit al-Ḥikma, 1995.

Pickthall, Muhammad Marmaduke. *The Meaning of the Glorious Koran: An Explanatory Translation*. New York: A.A. Knopf, 1930.

Ash-Shāṭibī, Abū Isḥāq. *Al-Muwāfiqāt fī uṣūl ash-sharīʿa*. Ed. ʿAbd Allāh Darrāz and ʿAbd as-Salām ʿAbd ash-Shāfī, 4 vols. Riyadh: Wizārat ash-Shuʾūn al-Islāmiyya wa-l-Awqāf wa-d-Daʿwa wa-l-Irshād, n.d.

Talbi, Mohamed. "Aïd al-Adhā ou Aïd El-Kébir." In *Célébrer Dieu*. Éditions Univers-Médina, Paris, 1980: 117–20.

Talbi, Mohamed. "La vie en communauté (le point de vue musulman)." *Islamochristiana* 10 (1984): 85–90.

Talbi, Mohamed. "Le Coran dans la vie quotidienne des Musulmans." *Supplément au Bulletin de Littérature Ecclésiastique* 2 (1985): 40–49.

Talbi, Mohamed. "Religious Liberty: A Muslim Perspective." In *Proceedings of the Second World Congress on Religious Liberty*. Berrien Springs, MI: University Printers, 1985: 156–68.

Talbi, Mohamed. "Sans Dieu, je ne suis rien." *France Catholique-Ecclesia*, January 28, 1985.

Talbi, Mohamed. "Pour un futur solidaire: la foi facteur de paix." *Roma-Incontro, Rivista mensile di informazione e dibattio* 10, no. 8–9 (August–September 1986): 35.

Talbi, Mohamed. "Religiöse Freiheit und Vermittlung des Glaubens (Libertè religieuse et transmission de la foi)." In *Dialog mit Afrika und dem Islam: Zwei Vorträge von Léopold Sédar Senghor und Mohamed Talbi*, edited by Hans Peter Rüger. Tübingen: J.C.B. Mohr, 1987: 27–69.

Talbi, Mohamed. "La liberté religieuse: droit de l'homme ou vocation de l'homme?" *Conscience et Liberté* 36, no. 2 (1988): 21–34.

Talbi, Mohamed. *ʿIyāl Allāh: Afkār jadīda fī ʿalāqāt al-Muslim bi-nafsihi wa-bi-l-ākharīn*. Tunis: Dār Sirās li-n-Nashr, 1992.

Talbi, Mohamed. "Liberté et Dignité de l'homme: Le Coran parole structurante de la Ummah." In *Würde und Recht des Menschen: Festschrift für Johannes Schwartländer zum 70. Geburtstag*, ed. Heiner Bielefeldt, Winfried Brugger, and Klaus Dicke. Würzburg: Königshausen & Neumann, 1992: 347–61.

Talbi, Mohamed. "Unavoidable Dialogue in a Pluralist World: A Personal Account." *Encounters: Journal of Inter-Cultural Perspectives* 1 (1995): 56–69.

Talbi, Mohamed. *Ummat al-wasaṭ wa-taḥaddiyāt al-muʿāṣira*. Tunis: Dār Sirās li-n-Nashr, 1996.

Talbi, Mohamed. *Plaidoyer pour un islam modern* (Rabat: Desclée de Brouwe, 1998).

Talbi, Mohamed. *Al-Islām: Ḥurriyya wa-Ḥiwār*. Beirut: Dār an-Nahār, 1999.

Talbi, Mohamed. *Universalité du Coran*. Arles: Actes Sud, 2002.

Talbi, Mohamed. *Réflexions d'un Musulman Contemporain*. Casablanca: Les éditions Le Fennec, 2005.

Talbi, Mohamed. *Li-Yaṭmaʾinna qalbī: Qaḍiyyat al-Īmān*. Tunis: CERES éditions, 2010.

Talbi, Mohamed. *Ma religion c'est la liberté: L'Islam et les défis de la contemporanéité*. Tunis: Éditions Nirvana, 2011.

Talbi, Mohamed. *L'Islam n'est pas voile, il est culte: Rénovation de la pensée musulmane*. Carthage: Éditions Carthaginoiseries, 2014.

Mohamed Talbi, *Qaḍiyyat al-ḥaqīqa*. Tunis: Editions Finzi 2015.
Talbi, Mohamed. *Dalīl al-Muslim al-Qur'ānī*. Tunis: Manshūrāt al-Mu'allif, 2016.
Mohamed Talbi, *Médiations (tadabbur) sur le Coran: Vérité – Rationalité – I'jâz Scientifique* (Tunis: Éditions Mohamed Talbi, 2016).
Talbi, Mohamed, and Maurice Bucaille, *Réflexions sur le Coran*. Paris: Éditions Seghers, 1989.
Talbi, Mohamed, and Olivier Clément. *Un respect têtu*. Paris: Nouvelle Cité, 1995.

5.2 Secondary Literature

Académie Tunisienne des Sciences, des Lettres et des Arts Beit al-Hikma. "Mohamed Talbi, (Septembre 16, 1921–Mai 1, 2017)." Special document produced to mark the fortieth day following Talbi's death, Carthage, 2017.

Josep Puig Montada

Muhammad Shahrur

1 An Unusual Biography for a Qur'ānic Scholar

The Syrian thinker Muhammad Shahrur (Muḥammad Shaḥrūr) was interviewed by Dale F. Eickelman in March 1996/1417 in Damascus. The interview gives us many details about his life and writings, he talks about his first book, printed in 1990/1410, and the reactions against it among the preachers in Damascus, and he adds: "To date, seventeen books, from Syria, Jordan, Lebanon, and Saudi Arabia, have been written against my first book."[1]

That one book caused such an uproar that seventeen books were written to refute it in six years following the publication, is quite remarkable, but before going into the apparently heretic contents of the book, let us outline the biography of its author.

Muhammad Shahrur was born in 1938/1357 in Damascus, he was the fifth child of a dyer who despite being a religious man sent his son to the primary state school, and not to the Qur'ānic school.[2] After graduating from high school in the scientific section, he applied for a government scholarship to study abroad so that he was sent to Saratow, near Moscow, to study Civil Engineering (1957–1964/1376–1383) in 1957/1376. He married a Russian student in 1963/1382 and they had a son in 1964/1383, the year he graduated. At that time, he was emotionally involved with the idea of *ʿurūba*, Panarabism, but after the failure of the union between Egypt and Syria in 1961/1380, he realized the need for a rational theory of *ʿurūba* and became more interested in Islam. He became convinced that the religious bond is much stronger than the linguistic one.

After returning to Syria in 1964/1383, he was appointed assistant professor at the School of Civil Engineering at the University of Damascus. He and his Russian wife divorced and he married a Muslim woman. The war of 1967/1387 deeply affected him and made him think that something was wrong with the way Arabs understood reality. In 1968/1388, he was admitted to the University College in Dublin and he received the MSc (1969/1389) and PhD degrees (1972/1392) in soil mechanics and foundation engineering. Living in Dublin he read many English texts in logic and philosophy of science; he mentions *Science and the Modern World* by A. N. Whitehead as one of the books that influenced him.

Muhammad Shahrur came back to Syria to become a Professor of Civil Engineering also at the University of Damascus (1972–1999/1392–1420). At the same time, he be-

1 Muḥammad Shaḥrūr, *The Qur'an, Morality and Critical Reason: The Essential Muhammad Shahrur*, ed. and trans. Andreas Christmann (Leiden: Brill, 2009), 516.

2 For his biography, I rely on his CV as published on https://shahrour.org/, (accessed January 30, 2024) and on Christmann's Introduction to Shaḥrūr, *The Qur'an, Morality and Critical Reason*, xix–xxii.

https://doi.org/10.1515/9783111320069-010

came even more engaged in understanding Islam and interpreting the Qur'ān. He consulted the traditional Qur'ānic commentaries and was disappointed so that he decided to go an independent and personal way; following words might help us to understand his method: "In 1980/1400, during Ramadan, I began to define the difference between a historically specific act (*al-fiʿl*) and the notion of action in general (*al-ʿamal*). This was the first distinction which I recognized. It took me all of Ramadan to sort out the Qur'anic verses related to these two terms."[3]

Shahrur's education was scientific, in mathematics and natural sciences, and he openly disagreed with the methods of the established scholars at Islamic institutions. He recognized that he had not been trained in the traditional Islamic disciplines, but he sought help from linguists, and he points out Jaʿfar Dakk al-Bāb,[4] who was a fellow student in Moscow. After nine years of work on the manuscript, he considered it ready for publication, and the Damascene house Dār al-Ahālī li-n-Nashr wa-t-Tawzīʿ published *al-Kitāb wa-l-Qur'ān. Qirā'a muʿāṣira* (The Book and the Koran. A Contemporary Reading) at Shahrur's expenses in 1990/1410. The book was a bestseller but at the same time, it unleashed strong reactions from conservative religious bodies. However, Shahrur did not back down and continued writing and publishing new monographs:

Ad-Dawla wa-l-mujtamaʿ (The State and the Society). Damascus: [Dār] al-Ahālī li-n-Nashr wa-t-Tawzīʿ, 1994.

Al-Islām wa-l-īmān: Manẓūmat al-qiyam (Islam and Faith: The Versification of Values). Damascus: [Dār] al-Ahālī li-n-Nashr wa-t-Tawzīʿ, 1996.

Mashrūʿ mīthāq al-ʿamal al-Islāmī (Proposal for a Covenant of Islamic Action). Damascus: [Dār] al-Ahālī li-n-Nashr wa-t-Tawzīʿ, 1999.

Naḥwa uṣūl jadīda li-l-fiqh al-islāmī: Fiqh al-mar'a (Towards New Principles of Islamic Jurisprudence with Regards to Women). Damascus: [Dār] al-Ahālī li-n-Nashr wa-t-Tawzīʿ, 2000.

Tajfīf manābiʿ al-irhāb (Draining the Sources of Terrorism). Damascus: [Dār] al-Ahālī li-n-Nashr wa-t-Tawzīʿ, 2008.

The Qur'an, Morality and Critical Reason. The Essential Muhammad Shahrur, ed. and trans. Andreas Christmann, Leiden: Brill, 2009.

3 Ibid., 512.

4 *Al-Kitāb wa-l-Qur'ān* (Damascus: al-Ahālī li-n-Nashr wa-t-Tawzīʿ, 1990), 47. He is the author of the essay *The Secrets of the Arabic Language*, in Arabic, printed on pp. 735–810 of this book, as well as an introduction to it.

Al-Qaṣaṣ al-Qurʾānī: Qirāʾa muʿāṣira, vol 1, *al-Madkhal ilā l-qiṣaṣ wa-qiṣṣat Ādam* (Qurʾānic Narrative: A Contemporary Reading. Volume 1: Introduction to the Narrative and to the Story of Adam). Beirut: Dār as-Sāqī, 2010.[5]

Al-Qaṣaṣ al-Qurʾānī: Qirāʾa muʿāṣira, vol. 2, *Min Nūḥ ilā Yūsuf* (Qurʾānic Narrative: A Contemporary Reading. Volume 2: From Noah to Josef). Beirut: Dār as-Sāqī, 2011.

As-Sunna ar-rasūliyya wa-s-sunna an-nabawiyya: Ruʾya jadīda (The Messenger's and the Prophet's Tradition: New Vision). Beirut: Dār as-Sāqī, 2012.[6]

Shahrur founded a short-lived Center for Contemporary Intellectual Studies, sponsored by the United Arab Emirates. The UAE named Shahrur among the winners of the 2017/1438 Sheikh Zayed Book Award and their TV channels gave him the possibility to spread his ideas.

The torture of children in a prison of the city of Daraa, near the Jordan border, in March 2011/Rabīʿ al-Awwal 1432 sparkled the civil uprising from March through July 2011/1432, which has been followed by the ongoing Syrian civil war, which made Shahrur move to Abu Dhabi, where he passed away on December 21, 2019/Rabīʿ ath-Thānī 24, 1441 at the age of eighty-one years.

Shahrur approached the Qurʾān in a way that Andreas Christmann has described as 'defamiliarization,' borrowing the concept from the Russian Formalists and the Prague School.[7] Indeed, Shahrur wants us to shake our heads with a reading completely different from the traditional interpretation made by the *ʿulamāʾ* in their commentaries for many centuries, no doubt a risky endeavor.

2 Cosmology and Linguistics

Shahrur wanted to go a new way in reading the sacred text. His method is based on the principle that human knowledge has its source in the material world, which he aims to proof with a Qurʾānic verse.[8] The quotation reads: "And it is God who brought you forth from your mothers' wombs, and He appointed for you hearing, and sight, and hearts, that haply so you will be thankful" (Q 16:78).[9] Shahrur links his method to

5 Shahrur comments that he wrote the book reacting to television programs that limited the Qurʾānic narrative to the emotional dimension, and that did not go beyond the legendary aspect (Shaḥrūr, *al-Qaṣaṣ al-Qurʾānī*, 1: 13).

6 Many of his publications are available online from his own website, https://shahrour.org/.

7 Andreas Christmann, "'The Form is Permanent, but the Content Moves': The Qurʾanic Text and its Interpretation(s) in Mohamad Shahrour's 'Al-Kitāb wa 'l Qurʾān'," *WI* 43 (2003): 143–72.

8 Muḥammad Shaḥrūr, *al-Kitāb wa-l-Qurʾān*, 42.

9 I follow A.J. Arberry's translation according to http://corpus.quran.com/ (accessed January 30, 2024). *The Koran Interpreted*, trans. Arthur J. Arberry (London: Allen & Unwin/New York: Macmillan, [1955]).

contemporary Islamic philosophy that achieves theoretical, abstract knowledge starting from the sense data, and attacks the Sufi way of knowledge by illumination or intuitive experience. He underscores the capacity of the human reason to apprehend the material world so that the "hidden world" is just the dimension of the material world which the human being has not apprehended yet. For Shahrur there is no contradiction between the Qur'ān and this kind of philosophy, as he goes so far as to allow only scientists to carry out any interpretation of the Qur'ān. Shahrur's principles are general, since he adds to them his personal commitment to the contemporary theory that the universe was generated by a "frightful explosion," referring to the big-bang theory.[10]

Shahrur clearly states that his method consists of a new reading of the Qur'ān. Moreover, he labels the Qur'ān as "Remembrance[11] to designate the "transmutation of the Qur'ān into a human linguistic form uttered in Arabic language,"[12] insisting on the linguistic dimension of his method. His method is linguistic insofar as terms and meanings are contrasted, since Shahrur intends to rely on two sources. One is the traditional linguistic method of Abū l-Fatḥ b. Jinnī (d. 1002/392) and ʿAbd al-Qāhir al-Jurjānī (d. 1078/471) although his reliance on them is in the form of a "general overview." The other is the "outcomes of modern linguistics" that should prove the inexistence of synonymy, as he subscribes: "No human language contains synonymy (*tarāduf*) and the contrary is correct, namely, in the course of historical development the same word either vanishes or acquires a new meaning, in addition to the first meaning."[13]

Shahrur declares that Ibn Fāris al-Lughawī (d. 1004/395), who denied synonymy, and his lexicon *Maqāyīs al-lugha*[14] became the main support for his research. The negation of synonymy is essential in his method.[15] Shahrur is rather expeditious with his statement of denial and, since there is no synonymy, he has a free hand to introduce particular meanings for his interpretation:

10 Shaḥrūr, *al-Kitāb wa-l-Qur'ān*, 43. His forced combination of modern concepts and Qur'ānic statements produces interpretations difficult to admit. For instance, he explains the sentence "His Throne was upon the waters" (Q 11:7) by the theory that after the big bang, the elements were generated, and hydrogen was one of them. Since hydrogen means "the generator of water", Shahrur assents to the Qur'ānic statement: *al-Kitāb wa-l-Qur'ān*, 165.

11 He quotes "It is We who have sent down the Remembrance, and We watch over it" (Q 15:9), and "and We have sent down to thee the Remembrance that thou mayest make clear to mankind what was sent down to them; and so haply they will reflect" (Q 16:44).

12 Shaḥrūr, *al-Kitāb wa-l-Qur'ān*, 61.

13 Ibid., 44.

14 Aḥmad b. Fāris al-Lughawī al-Qazwīnī, *Muʿjam maqāyīs al-lugha*, ed. ʿAbd as-Salām Muḥammad Hārūn (Cairo: Muṣṭafā l-Bābī l-Ḥalabī, 1969–72²). His lexicon is based on five books: *Kitāb al-ʿAin* of al-Khalīl b. Aḥmad, *Gharīb al-Ḥadīth* and *al-Gharīb al-Muṣannaf* of Abū ʿUbayd al-Qāsim b. Salām al-Harawī, *Kitāb al-Manṭiq* of Ibn al-Sikkīt and *Jamharat al-Lugha* of Ibn Durayd, as he indicates it on pp. 3f. This edition is used by Shahrur.

15 Shaḥrūr, *The Qur'an*, 9.

> If Islam is valid for every time and place, the starting point must be the thesis (*faraḍiyya*) that the Book has descended to us, and it has come to our generation in the midst of the twentieth century as if the Prophet (ṣ) had just died and had brought us the Book. The reader will clearly notice that for our understanding of the Book we stand on the grounds of the twentieth century without ignoring the historical development.[16]

Any human being is capable of understanding the Book using his reason and any verse of the Book can be understood according to reason because its wording (*ṣīgha*) is in "plain Arabic language"[17] as the Qur'ān itself defines it (Q 26:195). We should recall that the Qur'ān mentions 119 times the active participle *mubīn,* translated here as "plain," and generally as "clear," insisting that any of the contemporary Arabs could understand it.

It is noteworthy that the negation that *inzāl* and *tanzīl* are synonyms in regards to the revelation of the Qur'ān develops into the third chapter of *al-Kitāb wa-l-Qur'ān.*[18] Accordingly, the term *inzāl* is the subjective 'comprehension' of what has been objectively 'conveyed' by God, the *tanzīl* is this side. Shahrur claims: "Explaining the difference between *inzāl* and *tanzīl* is considered one of the main keys to understand the Book on its two sides: Prophethood and the Message."[19]

This distinction allows him flexibility in his interpretation. Shahrur examines "The *inzāl* and the *tanzīl* of the manna and the quails,"[20] and quotes Q 2:57 and Q 7:160 for *inzāl,* Q 20:80 and 81 for *tanzīl.* He considers the fact that the contents of Q 2:57 in one verse "We sent down (*anzalnā*) manna and quails upon you: 'Eat of the good things wherewith We have provided you'" is similar to the content of Q 20:80 and 81 uttered in two verses "Children of Israel, We . . . sent down (*nazzalnā*) on you manna and quails:" and "Eat of the good things wherewith We have provided you . . ." and draws the conclusion that the difference between them lies in the opposition *inzāl* and *tanzīl*:

In Q 2:57 *inzāl* expresses that the Jews *understood* that the manna and the quail were intended for their nourishing.

In Q 20:80 and 81 *tanzīl* expresses that "an *objective* transfer has taken place to the manna and the quails outside the conscience of the Israelites."[21] It is up to the reader to accept this and other subtle interpretations by means of which Shahrur puts forward his tenets.

---

16 Shaḥrūr, *al-Kitāb wa-l-Qur'ān,* 44. My translation.

17 Ibid., 45.

18 Ibid., 145–76.

19 Ibid., 147.

20 Ibid., 168.

21 Ibid.

3 Contrastive Semantics and Qur'ānic Authority

His method usually contrasts two terms, as *fiʿl* against *ʿamal* or *inzāl* against *tanzīl* that have been mentioned above. There is a very old discussion on the difference between *islām* and *īmān* (faith), since the terms *islām* and *īmān* are likely synonymous in the Qur'ān.[22] Islam is declared to consist of a combination of faith and works in early traditions so that the difference between them became subject matter of later enquiry.

In the pioneering work of 1990/1410, Shahrur did not oppose *islām-īmān*, however he defined Islam as "the religion of objective existence, outside the human essence."[23] In 1996/1416 Shahrur published *Islam and Faith: The Versification of Values*, where he used a consistent method of collecting Qur'ānic verses related to some concepts and presenting them as evidence for his arguments. Thus, in the beginning of the chapter "Islam and the Muslims," he looks for Qur'ānic passages where both terms, *islām-īmān*, and the participles *mu'minūn/mu'mināt* and *muslimūn/muslimāt* occur. He quotes four verses in which the Qur'ān is very explicit in the distinction between *islām* and *īmān*, then he enumerates nine passages in which the concepts *mu'minūn* and *muslimūn* are taken as equivalent. He comes to the conclusion that "The geniuses, Abraham, Jacob, the Israelite tribes, Joseph, Pharaoh's magicians, the apostles, Noah and Lot were Muslims."[24]

There is an open contradiction in this, and Shahrur aims at solving it. He knows that the "books of foundations" (*uṣūl*) have defined the pillars on which Islam is built. Five pillars are undisputed: faith, ritual prayer (*ṣalāt*), alms-tax (*zakāt*), the fast of Ramadan, the pilgrimage. Shahrur expresses the view that the latter four are holy rites (*shaʿāʾir*) to the performing of which the believers, not only the Muslims, are commanded. He underpins his view with Qur'ānic verses (Q 4:103; 2:110; 2:183; 2:185).[25] However, it remains difficult for Shahrur to justify the exclusion of *jihād* from these rites, since many verses prescribe it, and he mentions them.[26]

Let us keep in mind Shahrur's assertion that Islam is based on the natural, innate reason (*fiṭra*), and he quotes for the purpose: "So set thy face to the religion, a man of pure faith (*ḥanīf*)" (Q 30:30),[27] something valid for most of his arguments, for instance,

---

22 A.J. Wensinck, *The Muslim Creed: Its Genesis and Historical Development* (London: Routledge, 1965), 22f.

23 Shaḥrūr, *al-Kitāb wa-l-Qur'ān*, 716–18.

24 Muḥammad Shaḥrūr, *al-Islām wa-l-Īmān: Manẓūmat al-qiyam* (Damascus: al-Ahālī li-n-Nashr wa-t-Tawzīʿ, 1996), 33.

25 Ibid., 34.

26 The issue of *jihād* is dealt with more extensively on pages 394–466 of *The Qur'an, Morality and Reason*. There, Shahrur seriously restricts the scope of application of this concept.

27 Ibid., 35. "So set thy face to the religion, a man of pure faith – God's original upon which He originated mankind. There is no changing God's creation. That is the right religion; but most men know it not."

for the main pillar. Shahrur produces the following Qur'ānic verses of sura 2, *al-Baqara* (The Cow):[28] "Surely they that believe, and those of Jewry, and the Christians, and those Sabaeans, whoso believes (*āmana*) in God and the Last Day, and works righteousness (*ṣāliḥan*) – their wage awaits them with their Lord, and no fear shall be on them; neither shall they sorrow" (Q 2:62).

In this connection, he further adduces the verses Q 2:112, 2:128, and 4:125, 5:44, 10:90, 21:108, 41:33. The noun 'Islam' is never used in the mentioned verses but the verb *aslama*[29] or its participle. Shahrur applies a method of argumentation based on authority (*argumentum ab auctoritate*), on the indisputable authority of the Qur'ān; he infers:

> From these and many other verses we understand that Islam is the unquestioning acceptance (*taslīm*) of God's existence and of the Afterlife. If we tie this acceptance to doing right (*iḥsān*) and good work, the holder is a Muslim, regardless whether he is a follower of Muḥammad (those who are believers), a follower of Moses (those who are Jews), or a follower of Jesus (the Christians), or a follower of a religion other than these three, like the followers of Mazdeism, Shaivism and Buddhism (Sabeans).[30]

Therefore 'Islām' is universal "assent to God" accepting the Afterlife as true and working righteousness,[31] so that everyone who believes in God and follows the straight path is a 'Muslim.' On the contrary, *īmān* is not faith but Islam in the conventional understanding as it comprises the mentioned set of rituals and specific ceremonies of worship: Muhammadan faith, ritual prayer (*ṣalāt*), alms-tax (*zakāt*), the fast of Ramadan, the pilgrimage. It means that *īmān* is particular but superior, as it completes *islām*. Shahrur has moved away from the conventional interpretation to a reconciliatory one that allows him to defend a peaceful coexistence of world religions.

Shahrur's anti-conventional interpretation affects the very foundations of Islam. The term *Kitāb* is contraposed to the term *Qur'ān*. According to Shahrur, the distinction is made clear in the verse "We have given thee seven of the oft-repeated (*al-mathānī*), and the mighty Koran (Q 15:87)."[32] The *Kitāb*, the Book, contains God's revelation to the Prophet, "This Book is the totality of the passages revealed by God to the Prophet, in form and content,"[33] while *al-Qur'ān* is some part of it, it is attached to the *Kitāb*, in Shahrur's words. *Al-Qur'ān* is *al-Ḥaqq*, the truth, the objective reality and it is only some part of the *Kitāb*, not the whole *Kitāb*.[34]

28 Ibid., 37.

29 "Nay, but whosoever submits his will (*aslama wajhahu*) to God, being a good-doer, his wage is with his Lord, and no fear shall be on them, neither shall they sorrow" (Q 2:112).

30 Shaḥrūr, *al-Islām wa-l-īmān*, 38. My translation.

31 Shaḥrūr, *The Qur'an*, 29–31.

32 Shaḥrūr, *al-Kitāb wa-l-Qur'ān*, 57.

33 Ibid., 53.

34 Ibid., 58.

Kitāb is again divided by Shahrur into two “main books,” the *Kitāb an-nubuwwa* is opposed to *Kitāb ar-risāla.*[35] He justifies his distinction reading following verse:

> It is He who sent down upon thee the Book, wherein are verses clear that are the Essence of the Book (*Umm al-kitāb*) and others ambiguous. As for those in whose hearts is swerving, they follow the ambiguous part, desiring dissension, and desiring its interpretation; and none knows its interpretation, save only God. And those firmly rooted in knowledge say, ‘We believe in it; all is from our Lord’; yet none remembers, but men possessed of minds (Q 3:7).

The verse, unusually long, makes a distinction between definite (*muḥkama*) and ambiguous (*mutashābiha*) verses of the *Kitāb*. While the definite verses are identified, “and they are the mother of the Book,” the ambiguous verses are just the rest, i.e., what is not clear.[36] But we read in the Qur’ān that people who are evil, make the wrong interpretation of what is ambiguous, and Shahrur makes an effort to avoid this. The conventional interpretation is that definite verses refer to universal and permanent commands, and that the ambiguous ones contain the commands related to historical circumstances.[37] As A. Christmann observes,[38] Shahrur’s reading is contrary to the conventional interpretation, as he links the definite verses to the *risāla*, and the ambiguous ones to universal laws described in Muḥammad’s *nubuwwa*.[39]

Nubuwwa (“prophecy”) and (*risāla*) “message” are explained in connection with two other terms, *al-Qur’ān* and *Umm al-kitāb;* the latter “the mother of the Book” is traditionally understood as the eternal, original Qur’ān – Arberry translated it as “the Essence of the Book” – although Shahrur gives it a different meaning: it is that part of the Book which contains the rules of human behavior. *Umm al-kitāb* equals “message,” and *al-Qur’ān*, “prophecy.”

According to him, *al-Qur’ān* is “objective reality in its absolute existence outside the human consciousness.”[40] This objective Qur’ān contains *ambiguous* verses and is clarified by means of the positive sciences, i.e. natural sciences. Shahrur contends that the heirs of the prophets, since the Qur’ān is presented here as prophecy, are not

---

35 Ibid., 54f.

36 Ibid., 55. Shahrur adds a third category, ibid., see footnote 39, or pages 113–22 of his *al-Kitāb wa-l-Qur’ān*.

37 For a detailed information on the issue see: Leah Kinberg, “Muḥkamāt and Mutashābihāt (Koran 3/7): Implication of a Koranic Pair of Terms in Medieval Exegesis,” *AR* 35, no. 2 (1988): 143–72.

38 Shaḥrūr, *The Qur’an*, 26 (footnote 8).

39 Shahrur introduces even a third category “neither definite nor ambiguous” by interpreting the fragment of the verse “and others are ambiguous” (*wa-ukharu mutashābihātun*). He says that the Qur’ān does not read exactly “and the others are ambiguous” so that he concludes that the verses that are not definite are of two kinds, ambiguous and belonging to a third category, neither definite nor ambiguous. Next, he resorts to the expression *wa-tafṣīl al-kitābi* extracted from Q 10:37 and identifies the third category with a fuller explanation of the *Kitāb*, see Shaḥrūr, *al-Kitāb wa-l-Qur’ān*, 56. Shahrur’s methods are creative but leave many readers unsatisfied.

40 Shaḥrūr, *al-Kitāb wa-l-Qur’ān*, 103.

only the Sharia scholars, but the philosophers, the scientists and the experts in philosophy of history, in the origin of the species, in cosmology or electronics; several Qur'ānic verses are adduced to sustain his claim.[41] One might raise the objection that the latter again follow different methods from that of the linguist, so that four methods coexist: the traditional one of the "professional" *'ulamā'*, the linguistic, scientific, and social-historical.

Umm al-kitāb contains definite verses and it is Muḥammad's message; Shahrur qualifies this message as subjective, as he supports his qualification by means of a Qur'ānic quotation: "And We have enjoined upon man goodness to parents" (Q 29:8, and other verses). Goodness to parents, he says, only exists in the human consciousness. If man does not want to respect his parents, he has the capacity not to respect them. All the commandments of the *Umm al-kitāb* are addressed to human beings and do not have any effect on the laws of the universe.[42]

Shahrur's method of contrasting pairs of terms extends to the concepts of *ulūhiyya*, the doctrine of the Divine, and *rubūbiyya*, the doctrine of Lordship. The doctrine of God's unicity *(tawḥīd)* traditionally divides into the unicity of the Divine and that of Lordship. For Shahrur, *Umm al-kitāb*, i.e., the message is *ulūhiyya*, the doctrine of the Divine, and *al-Qur'ān* and the seven-oft-recited verses, or the prophecy is *rubūbiyya*, the doctrine of Lordship. He defines *rubūbiyya* as "the firm relationship between God and all His creatures and equally given to many," and *ulūhiyya*, as "the recognition of God only by someone who has reason as well of His unity and commandments."[43] Thus we see that Shahrur has linked or opposed following concepts:

Kitāb an-nubuwwa:	prophecy	*al-Qur'ān*	objective	reality	Lordship
Kitāb ar-risāla:	message	*Umm al-kitāb*	subjective	man	Divinity

All this linguistic work is possible under one key assumption, namely, that the underlying text is not conditioned by historical time and is valid for every generation. From early days, the Qur'ānic scholars were concerned about the time when the suras and their verses were revealed and the discipline of the conditions of revelation (*asbāb an-nuzūl*) arose. Time played a role in the doctrine of the *nāsikh*, abrogating verse, and *mansūkh*, abrogated verse. As strongly as he denies synonymity, Shahrur rejects taking into consideration the time circumstances, because the Qur'ān contains "the general laws ruling the existence of the universe from the first explosion until the resurrection, the paradise and the hell, and the Day of Reckoning."[44]

41 Ibid., 104.

42 Ibid., 105.

43 Ibid., 122. Incidentally, we observe that, though Shahrur openly rejects Sufism, he forgets that Sufis had already used this distinction but in a different sense.

44 Ibid., 92.

4 Dialectical Materialism and Calculus

Since Shahrur's education was scientific, he would resort to mathematical functions such as limits and hyperbolic functions for his interpretation. On the other side, he had spent seven years in the former Soviet Union when Dialectical Materialism was the official philosophy and information was easily available: Since 1926/1344 *The Great Soviet Encyclopedia* was a widespread publication and it contained an entry on "Unity and Struggle of Opposites," presenting it as the universal law of natural reality and of social history. The reader will likely be amazed by Shahrur's finding: "The Qur'ān has clearly expressed the law of the internal struggle of the opposites in a direct usage with its words: 'It is God who splits the grain and the date-stone, brings forth the living from the dead; He brings forth the dead too from the living. So that then is God; then how are you perverted?' (Q 6:95)."[45]

Shahrur explains the dialectical struggle in nature and history by means of mathematical "terminology." In classical mechanics a vector is an object that has both a magnitude and a direction; the position and displacement of a vector can be represented in two- and three-dimensional Cartesian coordinates. Therefrom Shahrur got his inspiration for his theory that he first expounded in *Towards New Principles of Islamic Jurisprudence with Regards to Women* of the year 2000/1420.[46]

He distinguishes three forms of existence that he assimilates to coordinates: 'being' (*al-kaynūna*), 'progressing' (*as-sayrūra*), and 'becoming' (*aṣ-ṣayrūra*). *Al-kaynūna* is a kind of being that "filled the realm of the empty space of the cosmos even before it turned into the material world we now know." *As-sayrūra* "refers to time. It denotes movement, process, and motion." The last, *aṣ-ṣayrūra* "denotes change, alteration, and transformation" and gives sense to the two others. The concept equivalent to the vector is called 'development' (*taṭawwur*). Life in this universe is necessarily three-dimensional although Shahrur imagines that 'being' (existence) and 'becoming' (transformation) without 'progressing' (time) is possible "at the time of the Last Hour."[47]

5 Mathematical and Qur'ānic Limits

In Islamic law, *ḥudūd*, literally "borders," "limits" refers to a punishment fixed in the Qur'ān and hadith for crimes considered to transgress the limits of God. In his interview by Dale F. Eickelman, Shahrur told him that he conceived the new concept of

45 Shaḥrūr, The Qur'an, 223.
46 Muḥammad Shaḥrūr, *Naḥwa uṣūl jadīda li-l-fiqh al-islāmī: Fiqh al-mar'a* (Damascus: Al-Ahālī li-n-Nashr wa-t-Tawzī', 2000), 27–58.
47 Shaḥrūr, *The Qur'an*, 11f.

limit in the Qur'ān, i.e., *ḥudūd*, when he was teaching the Proctor compaction test at the Faculty of Civil Engineering in Damascus. The test is represented by a hyperbola and on the top of it, Shahrur draw a line that is the mathematical limit. Limits describe how a function behaves near a point, instead of at that point, as they are essential to calculus and mathematical analysis. There are different ways to represent them, analytical, graphic, tabular. The leading concept is clear: we admit the existence of a limit but at the same time we admit that we always will be approaching it and never reach it. The discovery that Shahrur made is that "Just as in mathematics where we have five ways of representing limits, I found five cases in which the notion of God's limits occurred."[48]

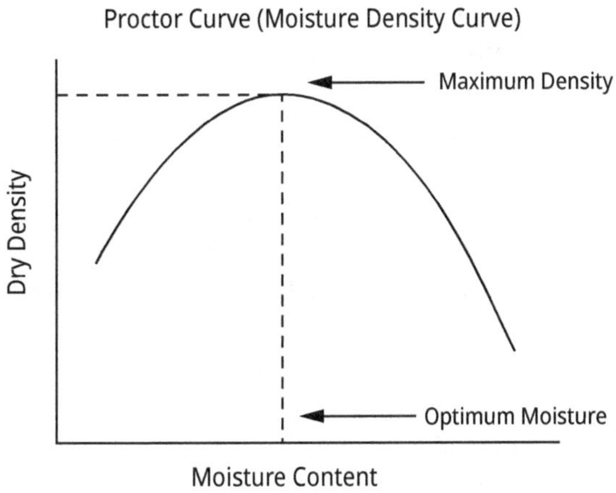

Proctor compaction test.

In mathematics, real sequences, real functions, and sets have upper and lower limits. Shahrur extends the distinction between lower and upper limit to the Qur'ānic limits (*ḥudūd*), and then arranges these in two categories. To the category of lower limit belong the *ḥudūd* concerning a) marriage, b) food, c) women's dress[49] and to that of upper limit, the *ḥudūd* on d) theft, and e) homicide.[50]

However he adds the simultaneous occurrence of both, the upper and the lower limit, in the *ḥudūd* concerning inheritance and polygyny,[51] as well as their occurrence "at one point" for illegal sex (*zinā*).[52] What Shahrur means here is elucidated by the Qur'ānic verse he refers to: "The fornicatress and the fornicator – scourge each one of

48 Ibid., 515.
49 Shaḥrūr, *al-Kitāb wa-l-Qur'ān*, 453–55.
50 Ibid., 455–57.
51 Ibid., 457–62.
52 Ibid., 462f.

them a hundred stripes, and in the matter of God's religion let no tenderness for them seize you if you believe in God and the Last Day; and let a party of the believers witness their chastisement (Q 24:2)."

Further, two particular cases of upper limit are introduced. The one is the case of "a line near to a rectilinear one, i.e., it is close and it does not touch it," and Shahrur means that the man does not touch the woman in a sexual relation. The other has an upper limit, "which cannot be traversed," and a lower limit which can be crossed.[53] Shahrur deals here with the issue of the interest, *ribā*, and he considers that the issue cannot be set apart from that of the legal alms-tax and the free willing charity.

Shahrur has placed the Qur'ānic norms for inheritance in the category comprising an upper and a lower limit: "33.3% is the lower limit for the woman, and 66.6%[54] is the upper limit for the man, and curvilinear movement takes place between both limits, i.e., approaching to them, and not retreating" and he produces as a proof the verse: "[So set thy face to the religion, a man of pure faith] – God's original upon which He originated mankind. There is no changing God's creation. That is the right religion; but most men know it not – (Q 30:30)."[55]

"A man of pure faith" or "inclining to truth" translate *ḥanīfan*, but Shahrur gives the term another meaning, 'curvilinear.' "As for the term 'curvature', this occurs in the Book primarily in the form of the noun Ḥanīf (sing.) or Ḥunafā'"(pl.)."[56] Shahrur arrives at this inference through a deliberation on the dialectical movement that characterizes "the internal oppositions of human life in the field of knowledge, social and economic sciences,"[57] and he distinguishes two opposites: *al-istiqāma wa-l-ḥanīfiyya*, the condition of straight line and the condition of curb line. No need to say that Shahrur produces numerous Qur'ānic quotations to support his view.

The use of Marxist language intermixed with Qur'ānic verses does not look convincing, and the application of the mathematical theory of limits remains rather questionable too in spite of the many geometrical drawings with which the author illustrates his arguments. On the contrary, his solution to the inheritance laws is noteworthy, he applies infinitesimal calculus to find it. "Islamic inheritance law became unfortunately finalized before the introduction of analytical geometry, and the concept of mathematical 'derivatives', which was brought to perfection by the genius of Isaac Newton (1642–1726)."[58]

In the Qur'ān, the inheritance laws are formulated in digital quantities, namely, in fractions. This method cannot avoid cases of surplus or shortage. Shahrur proceeds

53 Ibid., 463–66.

54 Shahrur does not use any special notation for recurrent decimals.

55 Shaḥrūr, *al-Kitāb wa-l-Qur'ān*, 459.

56 Shaḥrūr, *The Qur'an*, 181.

57 Ibid., 447.

58 Ibid., 226.

by using non-digital quantities, and he solves the inheritance issues by means of hyperbolic functions, under the condition that we admit his three laws of limits.[59]

6 Linguistics and Numerology

Another case where two methods are indistinctly applied has to do with the seven-oft-recited verses, that have been mentioned above. In the Qur'ān we read "We have given thee seven of the oft-repeated, and the mighty Qur'ān" (Q 15:87). Most of the commentators identify the seven-oft-recited with sura 1, *al-Fātiḥa* (The Opening), and some with the Qur'ān itself. Although some of the commentators are of the view that this term refers to the seven suras which contain two hundred or more verses, that is, suras 2, 3, 4, 5, 6, 7 and (8–9) or 10, the majority of the early commentators opine that it refers to the *Fātiḥa* the seven verses of which are recited over and over again in every prescribed prayer.[60] In the expression *sab'an min al-mathānī*, the term *al-mathānī* appears as a noun participle, and for a second time as a predicative participle, *mathāniya*: "God has sent down the fairest discourse as a Book, consimilar in its oft-repeated" (Q 39:23).

Shahrur begins with a radical statement, "the Qur'ān is one thing, and the seven-oft-recited is another."[61] He equates their import in spite of the fact that *al-mathānī* is only once mentioned in the whole Qur'ān as he interprets *mathāniya* in Q 39:23 differently: "God has revealed [from time to time] the most beautiful [event] (*aḥsana l-ḥadīthi*) in the form of a[n ambiguous] book (*kitāban* [. . .] *mutashābihan*), [yet] repeating [its teaching in various aspects] (*mathāniya*)."[62]

His argument is therefore that Qur'ān and the seven-oft-recited are both ambiguous, but that the latter are also "repeated." He insists on the alleged original meaning of the lexeme *th-n-y*, "two, twice," and understands the meaning of the expression as "seven times twice-repeated," at the beginning and at the end of something. Moreover, he dismisses the view that the *Fātiḥa* is referred to as *al-mathānī* since he claims that

---

59 Ibid., 258–9. Sharour's three limit laws of inheritance are:
1) The amount of male-female shares is in inverse proportion to the number of male-female heirs: $Y = \frac{1}{X}$. The peak of x is 1.
2) The amount of male-female share is in inverse proportion to the number of male-female heirs: $Y = \frac{2}{X}$. The peak of x is 2)
3) "If only one, her share is a half". The equation is as follows: $Y = \frac{1}{X}$. The peak of x is 1.

60 On the state of the art, see Uri Rubin, "Exegesis and Ḥadīth: The case of the Seven Mathānī," in *Approaches to the Qur'an*, ed. Gerald R. Hawting and Abdul Kader Shareef (London: Routledge, 1993),141–56.

61 Shaḥrūr, *al-Kitāb wa-l-Qur'ān*, 96.

62 Translation by Shaḥrūr, *The Qur'an*, 128.

the "seven twice-repeated are the seven opening sentences (*fawātiḥ*) of the suras."[63] He identifies these *fawātiḥ* with the 'disconnected' letters (*al-muqaṭṭaʿāt*) that open twenty-nine of the Qurʾān's 114 suras. He argues that *al-muqaṭṭaʿāt* close them too because of the meaning 'two' inherent to *al-mathānī*. Then he finds out that *al-muqaṭṭaʿāt* occur in seven different combinations: "Alif-Lām-Mīm-Rāʾ(which also occurs as Alif-Lām-Mīm and Alif-Lām-Rāʾ), 2) Alif-Lām-Mīm-Ṣād, 3) Kāf-Hāʾ-Yāʾ-ʿAin-Ṣād, 4) Yāʾ-Sīn, 5) Ṭāʾ-Hāʾ, 6) Ṭāʾ-Sīn-Mīm (which also occurs as Ṭāʾ-Sīn), 7) Ḥāʾ-Mīm."[64]

Shahrur affirms that *al-muqaṭṭaʿāt* express "the chapter's message, in a different linguistic format."[65] Next he digresses on the history of revelation, and defends the existence of three categories of verses: definite, ambiguous and explanatory. He combines these categories with the above mentioned division of the *Kitāb* into *al-Qurʾān* and *Umm al-kitāb* and concludes: "That is why the two books and the three categories of verses are so harmoniously intertwined in the text, and why they are so precisely, mathematically arranged in a manner that has, until now, escaped the penetrating view of so many commentators on the divine text."[66]

Unsurprisingly his interpretation has been met with widespread incredulity but Shahrur is fiercely convinced of his discovery. As A. Christmann has shown, the many reactions in form of publications against his interpretation originate from very different standpoints and serve very different interests. Very sadly, the vast majority of the attacks did not focus on methodological flaws but were ideologically motivated and politically conditioned. A. Christmann rightly remarks that the refuters were "providers of both normative and deviant images."[67] We should ask ourselves why did Shahrur engage in such a strenuous work that nevertheless left even sympathetic readers unsatisfied, and I dare to give just an intimation.

7 Liberalism vs. Political Islam

Shahrur obviously belonged to those Muslim liberals who aspired to bring the Arab countries under the rule of law and to make values as democracy and human rights their own, vindicating the observance of the human rights as proclaimed by the UN.[68]

63 Shaḥrūr, *al-Kitāb wa-l-Qurʾān*, 97.

64 Shaḥrūr, *The Qur'an*, 129. The same content is found in Shaḥrūr, *al-Kitāb wa-l-Qurʾān*, 97, but there he adds that *al-muqaṭṭaʿāt* comprise all the Arabic letters except Qāf – Rāʾ – Nūn, claiming that these three letters are the foundations (*uṣūl*) of the Qurʾān.

65 Shaḥrūr, *The Qur'an*, 130.

66 Ibid., 132.

67 Andreas Christmann, "'73 Proofs of Dilettantism': The Construction of Norm and Deviancy in the Responses to Mohamad Shahrour's Book al-Kitāb wa'l-Qurʾān: Qirāʾa Muʿāṣira," *WI* 45 (2005): 73.

68 Muḥammad Shaḥrūr, *Tajfīf manābiʿ al-irhāb* (Damascus: al-Ahālī li-n-Nashr wa-t-Tawzīʿ, 2008), 301.

However, they still have to face the so-called Political Islam, besides the servitude of the *'ulamā'* to interpret the Qur'ān following the directions of the state, but also the concessions made by the state to appease them.

To counterbalance them, he proclaimed freedom of worship, *ḥurriyyat al-'aqīda*, as the core value of Islam,[69] and produced favorable verses or fragments to support his view (the best-known is Q 2:256). He longed for a state in which legislation ignores the name sharia (*sharī'a*) but is issued in compliance with "God's limits that are found in the *Umm al-kitāb*."[70]

The term '*shūrā*' comes out only once in the Qur'ān: "and those who answer their Lord, and perform the prayer, their affair being counsel (*shūrā*) between them . . ." (Q 42:38). However, Shahrur established the woolly term as one pillar of *īmān* (that for him is the particular form of *islām*, as indicated above) and defined it as 'mutual consultation', therefrom he linked the concept to democracy.[71]

He extensively dealt with the connection between freedom, democracy and *shūrā* in chapter 6 of his book *The State and the Society*.[72] There, Shahrur gives his own interpretation of the verses Q 42:38 (Meccan), and 3:159 (Medinan); in the latter God tells the Messenger to seek advice "in the affair."[73] The commentators explain that both verses inform us that the Messenger sought advice from his Companions and used to consult with them concerning wars and other matters, and Shahrur expressed this view on the verse Q 42:38:

> Hence, we observe that the *shūrā* is a fundamental part of faith (*īmān*) and of the response to the Lord of the worlds, together with the ritual prayer and the expenses. In the verse Q 42:38, *shūrā* is a general principle within the structure of faith and Islam, [coming] before any other practice at the political and economic level, it is only a dogmatic practice (negative defense).[74]

Shahrur argues that *shūrā* belongs to the fundamentals of the Islamic creed and that the now living Muslims must take the place of the past Companions. In my view, it does not matter so much that inferring democracy from the Qur'ānic concept *shūrā* is questionable as Shahrur's determination to defend liberal values. Since he rejected the separation between religion and state, he had no choice but to approach the Qur'ānic text with hermeneutical instruments. Avoiding traditional exegesis, Shahrur could enjoy some freedom in his explanation.

---

69 Shaḥrūr, *The Qur'an*, 335.

70 Shaḥrūr, *al-Kitāb wa-l-Qur'ān*, 724f.

71 Shaḥrūr, *The Qur'an*, 464.

72 Muḥammad Shaḥrūr, *ad-Dawla wa-l-mujtama'* (Damascus: al-Ahālī li-n-Nashr wa-t-Tawzī', 1994), 141–78.

73 "It was by some mercy of God that thou wast gentle to them; hadst thou been harsh and hard of heart, they would have scattered from about thee. So pardon them, and pray forgiveness for them, and take counsel with them in the affair; and when thou art resolved, put thy trust in God; surely God loves those who put their trust" (Q 3:159).

74 Shaḥrūr, *ad-Dawla*, 151.

Twenty years after the publication of *al-Kitāb wa-l-Qur'ān* the revolts of the Arab Spring of 2011/1432 broke out but the protesting youth had not read him nor his refuters and an obviously gap separates the Arab youth from Shahrur as well as his refuters inside the establishment. As in the case of Hasan Hanafi, Shahrur was overtaken by the events but Hanafi remained in his homeland speaking out his sympathy with the youth through his editorial columns, while Shahrur eventually had to emigrate leaving behind a country torn apart.

Still residing in Syria, he wrote *The Messenger's and the Prophet's Tradition: A New Vision* where he faced the challenges arising from September 11, 2001 throughout the Arab Spring and expressed his concern. One could hear the slogan "Islam is the solution" on the streets and on the mass media, and Shahrur rhetorically asked "the solution to what?" In spite of the vagueness of the slogan, or maybe because of the vagueness, the Islamic forces emerged triumphant from the Arab Spring and he made this reflection:

> We do not find it odd that one of the results of the Arab Spring was that the Islamic movements pressed forward through the ballot boxes. This success places them in front of huge responsibilities and they must know that any failure will hurt these movements, and that the failure will be considered Islam's failure. In the same way that they ascribe their success to Islam, their opponents will ascribe their failure to Islam, and this subject is very dangerous.[75]

Shahrur wrote these lines in November 2011/Dhū l-Ḥijja 1432; he addressed the book to the "Islamic movements" and warned them of their "responsibilities." The book deals with the sunna, namely, the sayings traditionally attributed to Muḥammad and known as hadith, in a new way. The Muslim tradition gathered an increasing quantity of sayings that was never contained in spite of their screening by Muslim b. al-Ḥajjāj (d. 875/261) and Muḥammad al-Bukhārī (d. 870/256) in the ninth century. Many of us will agree that Islam cannot be understood without integrating these traditions.

In the new book Shahrur distinguishes between the *sunna rasūliyya* and the *sunna nabawiyya* in parallel to his former distinction between *risāla* (message) and *nubuwwa* (prophecy). He describes the *sunna rasūliyya* as the Message revealed to Muḥammad as found in the *Umm al-kitāb* and breaks up the *sunna nabawiyya* into two sections, one is those elements of the Prophet's biography urging belief and acceptance "because it is part of the Qur'ān," and the other is the efforts (*ijtihādāt*) made by the Prophet in leading the Muslim community in his times.[76]

Shahrur warned against the Islamic movements aiming to install the Islamic caliphate (*al-khilāfa al-islāmiyya*) after succeeding in democratic elections. He argued that there is no trace of such a project neither in the Qur'ān nor in the sunna. Obedience is due to the rule of law (*dawlat al-qānūn*) and not to someone who has the

---

75 Muḥammad Shaḥrūr, *as-Sunna ar-rasūliyya wa-s-sunna an-nabawiyya: Ru'ya jadīda* (Beirut: Dār as-Sāqī, 2012), 9–10. My translation.
76 Ibid., 99f.

power of restraint, and this truth – Shahrur sustains – was proclaimed "by the prophetic mission in the seventh century C.E. and it spread over to most of the countries in the nineteenth and twentieth centuries. The Arabs are the last ones to grasp this concept."[77]

8 Conclusion

In the precedent volumes of the *Handbook of Qur'ānic Hermeneutics*, it has become clear that the tradition of commenting on the Qur'ānic text from the point of view of language, history, or theology, remaining faithful to the canonical order prevails. However, in recent times, analytical and thematical commentaries, better, encyclopedias have appeared but always keeping faithful to the traditional interpretation.

Shahrur has gone into a disruptive direction, as Hasan Hanafi has done with his thematic commentary of 2018, and the posthumous affective (*shuʿūrī*) commentary of 2021. No wonder that both authors have faced a strong reaction from the establishment insofar as they both do not take into consideration the centuries long tradition that has cemented Islam in the impressive theological and juridical construction with which, we are familiar.

Nobody should deny Muhammad Shahrur his right to do so. He approached the text with a set of tools somehow unconventional, as he applied the mathematical and geometrical concept of limits to the doctrine of capital punishments. He often resorted to the opposition of terms to push forward his liberal interpretation, but his arguments were not flawless. To take the text as it is, carries its own risks. Nevertheless, Shahrur has shown that the Muslim can read the Qur'ān in a way differing from the established one and remain a true believer.

9 Bibliography

9.1 Sources

Shaḥrūr, Muḥammad. *Al-Kitāb wa-l-Qur'ān*. Damascus: al-Ahālī li-n-Nashr wa-t-Tawziʿ, 1990.
Shaḥrūr, Muḥammad. *Ad-Dawla wa-l-mujtamaʿ*. Damascus: al-Ahālī li-n-Nashr wa-t-Tawzīʿ, 1994.
Shaḥrūr, Muḥammad. *Al-Islām wa-l-īmān: Manẓūmat al-qiyam*. Damascus: al-Ahālī li-n-Nashr wa-t-Tawzīʿ, 1996.
Shaḥrūr, Muḥammad. *Mashrūʿ mīthāq al-ʿamal al-Islāmī*. Damascus: al-Ahālī li-n-Nashr wa-t-Tawzīʿ, 1999.
Shaḥrūr, Muḥammad. *Naḥwa uṣūl jadīda li-l-fiqh al-islāmī: Fiqh al-mar'a*. Damascus: al-Ahālī li-n-Nashr wa-t-Tawzīʿ, 2000.
Shaḥrūr, Muḥammad. *Tajfīf manābiʿ al-irhāb*. Damascus: al-Ahālī li-n-Nashr wa-t-Tawzī, 2008.

77 Ibid., 211.

Shaḥrūr, Muḥammad. *The Qur'an, Morality and Critical Reason. The Essential Muhammad Shahrur*, ed. and trans. Andreas Christmann. Leiden: Brill, 2009.
Shaḥrūr, Muḥammad. *Al-Madkhal ilā l-qiṣaṣ wa-qiṣṣat Ādam*. Vol. 1, *al-Qaṣaṣ al-Qur'ānī: Qirā'a mu'āṣira*, 2 vols. Beirut: Dār as-Sāqī, 2010.
Shaḥrūr, Muḥammad. *Min Nuaḥ ilā Yūsuf*. Vol. 2, *al-Qaṣaṣ al-Qur'ānī: Qirā'a mu'āṣira*, 2 vols. Beirut: Dār as-Sāqī, 2011.
Shaḥrūr, Muḥammad. *As-Sunna ar-rasūliyya wa-s-sunna an-nabawiyya: Ru'ya jadīda*. Beirut: Dār as-Sāqī, 2012.
The Koran Interpreted, trans. Arthur J. Arberry. London: Allen & Unwin/New York: Macmillan, [1955]. http://corpus.quran.com/.

9.2 Secondary Literature

Campanini, Massimo. "Towards a Philosophical Qur'anology: Structure and Meaning." *JQS* 20, no. 2 (2018): 1–18.
Christmann, Andreas. "'The Form Is Permanent, but the Content Moves': The Qur'anic Text and Its Interpretation(s) in Mohamad Shahrour's 'Al-Kitāb wa 'l Qur'ān'." *WI* 43 (2003): 143–72.
Christmann, Andreas. "'73 Proofs of Dilettantism': The Construction of Norm and Deviancy in the Responses to Mohamad Shahrour's Book al-Kitāb wa'l-Qur'ān: Qirā'a Mu' āṣira." *WI* 45 (2005): 20–73.
Ibn Fāris, Aḥmad al-Lughawī al-Qazwīnī. *Mu'jam maqāyīs al-lugha*. Ed. 'Abd as-Salām Muḥammad Hārūn, 6 vols. Cairo: Muṣṭafā l-Bābī l-Ḥalabī, 1969–72[2]. Reprinted Cairo: Dār al-Fikr li-ṭ-Ṭibā'a wa-n-Nashr, 1979.
Khusen, Moh. "From Sharī'a 'Ayniyya to Sharī'a Ḥudūdiyya. Shahrour's Interpretation on Qur'anic Legal Verses." *Al-Jami'ah* 50, no. 2 (2012): 431–60.
Kinberg, Leah. "Muḥkamāt and Mutashābihāt (Koran 3/7): Implication of a Koranic Pair of Terms in Medieval Exegesis." *AR* 35, no. 2 (1988): 143–72
Marcotte, Roxanne D. "Shahrur, the Status of Women, and Polygamy in Islam." *OM* 20 (2001): 313–28.
Rubin, Uri. "Exegesis and Ḥadīth: The case of the Seven Mathānī." In *Approaches to the Qur'an*, edited by Gerald R. Hawting and Abdul Kader Shareef. London: Routledge, 1993: 141–56.
Wensinck, A.J. *The Muslim Creed: Its Genesis and Historical Development*. London: Routledge, 1965 [1932].

Heidi Hirvonen

Mahmoud Ayoub

1 Introduction

Mahmoud Ayoub (d. 2021/1443) was a prominent scholar of Lebanese descent who specialized in Islam and the history of religion. His areas of expertise included Qurʾānic exegesis, Shiite Islam, and Muslim-Christian relations, all of which influenced his hermeneutical approach to the Qurʾān. This approach was characterized by historical criticism of the sources used in traditional exegesis, a pluralistic reading of the Islamic scripture, and acknowledged selectivity in the interest of promoting interreligious dialogue.

This contribution begins with a concise introduction to Ayoub's life and career, and then moves on to discuss his hermeneutical works and method. His writings related to Qurʾānic exegesis are basically of two types. First, there are academic works written objectively and with the pedagogical aim of making Muslims' own hermeneutical views better known to English-speaking readers. Second, there are writings that combine academic study with taking a clear stand on the correct interpretation of the Qurʾān. The latter typically discuss religious diversity and Muslim-Christian-Jewish relations. The chapter ends with a brief assessment of Ayoub's contribution to contemporary hermeneutics of the Qurʾān and his position in relation to other reformist intellectuals.

2 Life, Works, Political and Intellectual Context

Maḥmūd Muṣṭafā Ayyūb (better known as Mahmoud Ayoub)[1] was born in 1935/1354 in ʿAyn Qānā, South Lebanon. He was educated at the American University of Beirut (BA in philosophy, 1964/1384), the University of Pennsylvania (MA in religious thought, 1966/1386), and Harvard University (PhD in history of religion, 1975/1395). His first teaching positions were at San Diego State University, the University of Toronto, and McGill University. From 1988/1408 to 2008/1429, he was a professor of Islamic studies in the Department of Religion at Temple University, Philadelphia. In addition, he was

1 For this section, see Hartford Seminary, "Mahmoud Ayoub. Honorary Faculty Associate in Islam and Christian-Muslim Relations," accessed July 7, 2020, https://www.hartsem.edu/faculty/mahmoud-ayoub/; Irfan A. Omar, "Pathways to Dialogue: the Journey of Mahmoud Ayoub," *Islam and Christian-Muslim Relations. Special Issue. Islam in Dialogue. Essays in Honour of Professor Mahmoud Ayoub* 15, no. 1 (2004): 9–11; Mahmoud M. Ayoub, *Islam: Faith and History* (London: Oneworld, 2004; reprint 2018), 3–6; Ataullah Siddiqui, *Christian-Muslim Dialogue in the Twentieth Century* (London: Macmillan; New York: St. Martin's Press, 1997), 97f.

https://doi.org/10.1515/9783111320069-011

an adjunct professor at the Duncan Black Macdonald Center, Hartford Seminary, Connecticut; a research fellow at the Middle East Center, University of Pennsylvania; and the Tolson Visiting Professor at the Pacific School of Religion, Berkeley, California. Beginning in 2008/1429, he was a faculty associate in Islam and Christian-Muslim Relations at the Duncan Black Macdonald Center, Hartford Seminary, Connecticut. He retired from teaching in 2016/1437.

Ayoub received many awards and scholarships and was involved in several local and overseas scholarly engagements. For instance, he received the Kent Doctoral Fellowship and the Canada Council Fellowship. He was awarded Fulbright Fellowships in 1994–95/1414–15, for lecturing in Malaysia, and in 2000/1420, for undertaking a research project on Christian-Muslim relations in Egypt and Lebanon. In 1998/1418, Ayoub helped to launch a graduate-level program in Christian-Muslim relations and comparative religion at the Center for Christian-Muslim Studies, University of Balamand, al-Kūra, Lebanon. In 1999/1419, he was involved in the establishment of a comparative religion graduate program in Yogjakarta, Indonesia, as a joint initiative of Temple and Gajah Mahda Universities. Ayoub was a visiting professor at the Center in Balamand and a regular lecturer in Gajah Mahda. He was also the driving force behind establishing the Imam Ali Chair for Shi'i Studies and Dialogue among Islamic Legal Schools at Hartford Seminary in 2015/1436.[2]

Ayoub authored several books, including *Redemptive Suffering in Islam: A Study of the Devotional Aspects of 'Āshūrā' in Twelver Shīʿism* (1978/1398)[3]; *The Qur'ān and Its Interpreters*, 2 vols. (1984/1992, 1404/1412)[4]; *Dirāsāt fī l-ʿalāqāt al-masīḥiyya al-islāmiyya* (*Studies in Christian-Muslim Relations*, 2 vols., in Arabic, 2000/2001, 1420/1421)[5]; *The Crisis of Muslim History: Religion and Politics in Early Islam* (2003/1423)[6]; *Islam: Faith and History* (2004/1424)[7]; and *A Muslim View of Christianity: Essays on Dialogue by Mahmoud Ayoub* (2007/1428, ed. Irfan A. Omar)[8]. His publications also include over fifty scholarly articles and book chapters on various themes, notably on the interpre-

2 Hartford Seminary, "Imam Ali Chair for Shi'i Studies and Dialogue among Islamic Legal Schools," accessed August 3, 2020, https://www.hartsem.edu/macdonald-center/imam-ali-chair-for-shii-studies-and-dialogue-among-islamic-legal-schools/.

3 Mahmoud M. Ayoub, *Redemptive Suffering in Islam: A Study of the Devotional Aspects of 'Āshūrā' in Twelver Shīʿism* (The Hague: Mouton, 1978).

4 Mahmoud M. Ayoub, *The Qur'ān and Its Interpreters*, 2 vols. (Albany: State University of New York Press, 1984–1992).

5 Mahmoud Ayoub, *Dirāsāt fī l-ʿalāqāt al-masīḥiyya al-islāmiyya* (*Studies in Christian-Muslim Relations*), 2 vols. (Al-Kūra, Lebanon: Publications of the Centre for Christian-Muslim Studies in the University of Balamand, 2000–2001).

6 Mahmoud M. Ayoub, *The Crisis of Muslim History: Religion and Politics in Early Islam* (Oxford: Oneworld, 2003).

7 See footnote 1.

8 Mahmoud Ayoub, *A Muslim View of Christianity: Essays on Dialogue by Mahmoud Ayoub*, ed. Irfan A. Omar (Maryknoll, New York: Orbis Books, 2007).

tation of the Qur'ān and Christian-Muslim relations. Ayoub served on several advisory and editorial boards, for instance as an editorial consultant for the *Oxford Dictionary of Islam*.

Ayoub's personal experiences of religious diversity had an indelible effect on his academic interests, as well as his active involvement in Muslim-Christian dialogue. He was born into a devout Shiite Muslim family in a small town characterized by good social relations between its Muslim and Christian populations. As a child, he was educated in a British Presbyterian missionary school for the blind. While in the school, he became a Christian, which led to tensions between him and his family. Later he started to question his faith, began academic study of religions, and finally, during his doctoral studies, returned to his Islamic heritage. Ayoub's upbringing in a religiously mixed community and his inside experience of both Islam and Christianity gave him an enduring interest in the dialogue between the two traditions, as well as a rare ability to understand diverse points of view on sensitive theological issues. On the other hand, they also gave rise to doubts about his Islamic integrity among some of his fellow Muslims.

Since his birth in French Mandate Lebanon shortly before the beginning of the Second World War, Ayoub witnessed all the subsequent political developments of the twentieth and early twenty-first centuries, including the end of colonialism, globalization, the end of the Cold War, and the straining of relations between the West and Muslim countries since the early 1990s. In the context of a growing anti-Islamic trend in the West, Ayoub sought to promote better communication and peaceful co-existence. He was active in numerous dialogue projects and served as one of the US Department of State's ambassadors to the Middle East and Southeast Asia.[9]

Ayoub's thinking and hermeneutical approach evolved in the context of Western academic discourse on religions, including the development of Christian theology of religions and the pluralistic ideas of his former teacher Wilfred Cantwell Smith (1916–2000/1334–1421). In particular, Smith's distinction between personal piety and cumulative traditions and his prioritizing of faith over propositional beliefs[10] influenced the way Ayoub interpreted the Qur'ānic passages on Muslims' relationship to other religious communities. The Christian debate on exclusivism, inclusivism, and pluralism in relation to other religious traditions was reflected in Ayoub's thinking, which criticized the first two and defended pluralism, which he outlined in a way reminiscent of John Hick's initial, theocentric formulation.[11]

---

9 Mahmoud Ayoub, "9/11 and the Need for a New Paradigm for Interfaith Dialogue," *MW* 101, no. 3 (2011): 562–64.

10 See, e.g., Wilfred Cantwell Smith, *Faith and Belief* (Princeton, NJ: Princeton University Press, 1979).

11 For the theocentric phase of Hick's thinking, see, e.g., John Hick, *God and the Universe of Faiths: Essays in the Philosophy of Religion* (London: Macmillan, 1973). For a more detailed discussion of Ayoub's theology of religions and his criticism of exclusivism and inclusivism, see, e.g., Heidi Hirvonen, *Christian-Muslim Dialogue: Perspectives of Four Lebanese Thinkers* (Leiden: Brill, 2013), 116–20.

3 Hermeneutical Works and Method

3.1 The Major Hermeneutical Works

In Ayoub's hermeneutical vision, Qur'ānic exegesis played an essential role as the means by which the Islamic scripture shapes Muslim community. While he represented a traditional notion of the revelation and nature of the Qur'ān as the Word of God, he accepted diversity in its interpretation, which contextualizes the divine revelation to all times and places. Ayoub stressed that the Qur'ān is both transcendent and earthly, a book of guidance that followed intimately the formation of the Muslim community and many of whose verses were revealed in response to specific problems of the first Muslims. To continue to fulfil its guiding role in human life, it must speak to all eras and conditions, which it does in large measure through Qur'ānic exegesis. For this reason, he argued, the Qur'ān has to be read according to contemporary circumstances and challenges.[12] Ayoub's hermeneutical works should be understood in the light of this role he gave to the *tafsīr* genre in the contextualization of the divine Word.

Ayoub's writings on hermeneutics can be roughly divided into two categories with slightly different purposes. Firstly, he produced mainly objective works on the general theme of Qur'ānic *tafsīr* with a clear pedagogical aim of making this genre better known in the West. Secondly, he wrote on particular themes related to Qur'ānic exegesis, in which he not only presented traditional views objectively but also took stand on which interpretations he considered judicious.

The pedagogical approach is represented by Ayoub's articles on Shiite *tafsīr*. His Shiite background is more evident in the subjects he studied than in the views he presented. He considered it important that more attention be paid to the Shiite interpretation of the Qur'ān in Western research, which had previously focused on the Sunni tradition.[13] He contributed towards this end in several articles that study Qur'ānic exegesis from the Shiite point of view, such as "The Speaking Qur'ān and the Silent Qur'ān: A Study of the Principles and Development of Imāmī Shīʿī *tafsīr*" (1988/1408)[14]

12 Mahmoud Ayoub, "Revelation and Salvation: Towards an Islamic View of History," *Al-Serat* (1987): 10–25, available online: https://www.al-islam.org/al-serat/general-al-serat/revelation-and-salvation-towards-islamic-view-history-mahmoud-ayoub-0, accessed August 2, 2020; Mahmoud Ayoub and Georges Khodr, *Naḥwa al-Jidāl al-aḥsan: muḥāwarāt islāmiyya masīḥiyya* (*Towards Better Dialogue: Muslim-Christian Dialogues*) (The Center for Christian-Muslim Studies in the University of Balamand, 1997), 31; Mahmoud M. Ayoub, "Towards an Islamic Christology, II: The Death of Jesus, Reality or Delusion (A Study of the Death of Jesus in Tafsīr Literature)," *MW* 70, no. 2 (1980): 91–121, at 91.

13 Mahmoud Ayoub, "The Speaking Qur'ān and the Silent Qur'ān: A Study of the Principles and Development of Imāmī Shīʿī *tafsīr*," in *Approaches to the History of the Interpretation of the Qur'ān*, ed. Andrew Rippin (Oxford: Clarendon Press, 1988), 177–98, at 177.

14 Ibid.

and "Literary Exegesis of the Qur'ān: The Case of al-Sharīf al-Raḍī" (2000/1420).[15] In these articles, he analysed Shiite exegetical tradition objectively, without taking positions for or against the views expressed therein. In other contexts, he occasionally presented views that run counter to the traditional Shiite understanding of history.[16] However, this did not diminish his appreciation for the Shiite tradition, because he prioritized piety and existential faith over particular doctrines.

Ayoub's major work on Qur'ānic hermeneutics is *The Qur'ān and Its Interpreters* (2 vols., 1984/1992, 1404/1412),[17] which also falls into the category of mostly pedagogical writings. In this work, Ayoub presented the diversity of the *tafsīr* genre without taking sides between the different interpretations. The work introduces English readers to how the major Muslim schools of thought have understood the Qur'ān. The two volumes cover the first three suras of the Islamic scripture. Ayoub acknowledged that completing this project for the entire text of the Qur'ān is not to be expected within any one individual's lifetime.[18] *The Qur'ān and Its Interpreters* acquaints the reader with the thinking of thirteen prominent exegetes writing in Arabic. Traditional exegetical thought is represented by Abū Ja'far Muḥammad b. Jarīr aṭ-Ṭabarī (d. 923/310), 'Alī b. Aḥmad al-Wāḥidī (d.1076/468), Abū l-Qāsim az-Zamakhsharī (d. 1144/538), Fakhr ad-Dīn ar-Rāzī (d.1209/606), Abū 'Abd Allāh Muḥammad b. Aḥmad al-Qurṭubī (d.1273/671), Abū l-Fidā' b. Kathīr (d.1373/774), and Muḥammad b. 'Alī ash-Shawkānī (d.1834/1250) for the Sunnis; by Muḥyī d-Dīn b. 'Arabī (d.1240/638) and Niẓām ad-Dīn an-Nīsābūrī (d.1328/9/728) for the Sufis; and by Ibn Bābawayh al-Qummī (d.939/328) and Abū 'Alī Faḍl b. Ḥasan aṭ-Ṭabarsī (d.1153/548) for the Shiites. Modern exegetical trends are exemplified by Sayyid Quṭb (d.1966/1386) and Sayyid Muḥammad Ḥusayn Ṭabāṭabā'ī (d.1981/1402), a Sunni and a Shiite, respectively. This comprehensive approach, including exegetes of the major Islamic trends both classical and modern, has contributed to the generally positive reception of the work by both students and experts in the field.

Notwithstanding Ayoub's neutral approach, his modern educational background and personal emphases are reflected in *The Qur'ān and Its Interpreters*. His choice of the exegetes and verses discussed follows, at least to a certain extent, his personal

15 Mahmoud Ayoub, "Literary Exegesis of the Qur'ān: The Case of al-Sharīf al-Raḍī," in *Literary Structures of Religious Meaning in the Qur'ān*, ed. Issa J. Boullata (Abingdon, Oxon: Routledge, 2000), 292–309.

16 Concerning the early history of Islam, Ayoub states that the Prophet Muḥammad probably died without clearly appointing a successor. Ayoub, *Islam: Faith and History*, 72. The Shiite tradition, for its part, firmly adheres to the view that the Prophet Muḥammad appointed his cousin and son-in-law 'Alī b. abī Ṭālib as his successor and the first Shiite Imam. In general, however, Ayoub does not take a position on the origins of Shiite Islam. For example, his work *Redemptive Suffering* (1978) focuses on the meaning of Imam Ḥusayn b. 'Alī's suffering for the Shiite community without commenting on the historicity of the stories told about his life. See Ayoub, *Redemptive Suffering*, 16.

17 See the footnote 4.

18 Ayoub, *The Qur'ān and Its Interpreters*, 2: vii.

preferences. In addition, he uses his own translation of the Qur'ān, and the translations of some concepts reflect his own views. For example, he renders the verb *yu'minu* "to have faith" rather than "believe," a choice that was influenced by his former teacher Wilfred Cantwell Smith.[19] Ayoub's background in religious history and his modern conception of history are evident in the way he treats several narratives of the classical *tafsīr* works that attempt to provide background information on Qur'ānic verses. This especially concerns their elaborate stories about the people of Israel and other Biblical characters. Noting discrepancies between these narratives and historical research and commenting on their various Biblical and folklore sources, he frequently describes them as "hagiographical accounts" or "fantastic legends."[20] While Ayoub rarely takes a position on the issues addressed, he occasionally hints at his personal views, especially concerning other religious traditions such as Judaism and Christianity. For instance, when he discusses verses that speak positively of some group of the People of the Book (*ahl al-kitāb*), he points out that most commentators have avoided their literal interpretation.[21] He also indicates that he finds problematic the traditional interpretation that God gave Jesus's appearance to another person, and that person died on behalf of Jesus on the cross.[22] Ayoub elaborated on these themes in his other, shorter writings, which will be discussed next.

3.2 Hermeneutical Method

Ayoub's shorter writings provide a clearer picture of his hermeneutical method, which combines traditional exegetical tools with a critical examination of previous interpretations from a historical perspective. This approach is especially evident in his numerous articles and book chapters on religious diversity and Muslims' relationship to the People of the Book. These include, just to name a few examples, "Religious Pluralism and the Qur'ān" (2012/1433)[23], "Nearest in Amity: Christians in the Qur'ān and contemporary exegetical tradition" (1997/1418)[24], and "Revelation and Salvation: Towards an Islamic View of History" (1987/1407)[25]. As can be inferred from the titles, his approach was thematic; he examined what the Qur'ān and *tafsīr* tradition say on a particular subject. In these and other writings on related topics, Ayoub discussed Qu-

19 Ayoub, *The Qur'ān and Its Interpreters*, 1: 39–40.

20 See, e.g., Ayoub, *The Qur'ān and Its Interpreters*, 1: 89, 232; 2: 130, 141.

21 Ayoub, *The Qur'ān and Its Interpreters*, 2: 298–99.

22 See ibid., 175–78.

23 Mahmoud Ayoub, "Religious Pluralism and the Qur'ān," in *Contemporary Approaches to the Qur'ān and Sunnah*, ed. Mahmoud Ayoub (London: The Institute of Islamic Thought, 2012), 39–56.

24 Mahmoud Ayoub, "Nearest in Amity: Christians in the Qur'ān and contemporary exegetical tradition," *Islam and Christian-Muslim Relations* 8, no. 2 (1997): 145–64.

25 See the footnote 12.

r'ānic verses in terms of traditional exegetical concepts, including explaining problematic passages through other Qur'ānic passages, prophetic hadiths, and occasions of revelation (*asbāb an-nuzūl*). He also commented on the time of the revelation of certain verses and the possibility of their abrogation. In addition to these traditional instruments, however, he relied on modern historical research in order to clarify the context of Qur'ānic revelations and to criticize many traditional interpretations of reliance on hagiographical stories. It should be noted that Ayoub's historical criticism concerned the context of the Qur'ān and the *tafsīr* literature, not the text of the Qur'ān itself, which he traditionally considered to be the literal Word of God. However, unlike many other commentators, he did not adhere to the view that all the statements of the Qur'ān are historically true. According to him, for example, some of the Qur'ān's characterizations of Biblical persons go beyond mere historical facts, such as the claim that Jesus's mother Mary was Aaron's sister or the denial of Jesus's crucifixion.[26]

Ayoub's arguments for religious pluralism provide a representative example of his hermeneutical method. He gave a historical account of the diversity of religious traditions throughout history, referred to Qur'ānic verses that describe human diversity as God's sign and a reason for people to get to know one another (Q 30:22; Q 49:13), and claimed that religious plurality is willed by God. On the basis of another key verse, Q 2:62, he regarded all who have faith and do good deeds as worthy of salvation, irrespective of their religious affiliation: "Surely those who have faith [that is the Muslims], the Jews, the Christians and the Sabaeans; whoever accepts faith in God and the last day and performs good deeds, those shall have their reward with their Lord; no fear shall come upon them, nor will they grieve."

Ayoub countered the common claim of Muslim commentators that the verse in question is abrogated by later revelation. This is not possible, he argued, for two reasons. First, abrogation applies only to legal verses, not narrative ones. Second, since Q 5:48 has almost the same wording, the verse was actually revealed twice, soon after the Prophet Muḥammad came to Medina and almost at the end of his prophetic mission. Ayoub was well aware that many commentators, both classical and modern, have interpreted the verse in question in the framework of Islamic supersessionism over all previous religions and laws. According to this view, the verse can only apply to those Jews, Christians, and others who either lived before Muḥammad or accepted his prophethood. However, Ayoub denied that this more inclusive understanding of abrogation has any clear foundation in the Qur'ān. Rather, he saw that most commentators have gone out of their way to negate the obvious meaning of the passage. Other verses he discussed to defend religious plurality concern God's primordial covenant with humankind before the beginning of time (Q 7:172), the prophets sent to all na-

26 Ayoub, "Towards an Islamic Christology, II," 116–17. See also Mahmoud Ayoub, "Muslim Views of Christianity: Some Modern Examples," *Islamochristiana* 10 (1984): 49–70, at 50.

tions (Q 35:24), and humans' inner, created capacity to have faith in God (*fitra*, Q 30:30).[27] Wilfred Cantwell Smith's influence is evident in Ayoub's distinction between institutionalized religions and the inner dimension of faith. Ayoub regarded institutionalized religions as necessary and as divinely legislated instruments for the realization of the inner dimension of faith and righteous life and argued that faith unites people across the boundaries of institutional religions. Accordingly, he saw interreligious dialogue in terms of sharing in the commonality of faith, living righteously, and believing in all of God's revelations.[28] According to Ayoub's interpretation of the Qur'ān, Christians and Jews do not need to convert to Islam, but they should acknowledge that Muḥammad was a prophet sent by God and that Islam is an authentic religion.[29]

In order to promote dialogue and peaceful relations, Ayoub consciously adopted a selective approach to the Qur'ānic passages on the People of the Book. He tended to limit the applicability of the Qur'ānic verses that criticize Christians and Jews, relating them to certain theological issues (Jesus's divinity, the idea of Israel as the chosen people of God) or a certain group of the People of the Book, such as the Jews of Medina during the time of the Prophet Muḥammad.[30] As for more positive verses, such as those that portray Christians as pious and humble (e.g., Q 5:82–83, 57:27), he regarded them as applicable to all Christians at all times as long as the conditions described in the verses are met. He argued that the positive verses are more numerous and emphatic than the negative ones, that they cannot be abrogated because of their narrative nature, and that the hagiographical tales commentators use to explain them are not historically reliable. In addition, he defended conscious selectiveness to encourage harmonious relations between different religious communities:[31]

> Without denying the need for academic integrity, it must be recognized that we live in a religiously, culturally and ideologically pluralistic world which we can either share or destroy. We must therefore be selective in our choices of sacred texts and concepts, and should give preference to those which encourage greater understanding and co-operation among the people of different faiths and ideologies.[32]

---

27 Ayoub, "Religious Pluralism and the Qur'ān," 42–51; Ayoub, "Nearest in Amity," 148, 150.

28 Ayoub, "Religious Pluralism and the Qur'ān," 49–51; Ayoub, "Muslim Views of Christianity," 70.

29 Mahmoud M. Ayoub, "The Need for Harmony and Collaboration between Muslims and Christians," in *Information and Formation*, ed. O.C. Lim and Theresa Ee-Chooi (Kuala Lumpur: Catholic Research Centre, 1994), 109–19, at 116; Ayoub and Khodr, *Naḥwa al-Jidāl al-aḥsan*, 23.

30 Ayoub, "Nearest in Amity," 162; Ayoub, "Revelation and Salvation;" Mahmoud Ayoub. "The Islamic Context of Christian-Muslim Relations," in *A Muslim View of Christianity: Essays on Dialogue by Mahmoud Ayoub*, ed. Irfan A. Omar (Maryknoll, New York: Orbis Books, 2007), 17–31, at 19.

31 Ayoub, "Nearest in Amity," 157–59, 162. For how Christians have generally been seen in the *tafsīr* genre, see Jane Dammen McAuliffe, *Qur'ānic Christians: An analysis of classical and modern exegesis* (Cambridge: Cambridge University Press, 1991).

32 Ayoub, "Nearest in Amity," 162.

This selective attitude to the Islamic scripture is related to Ayoub's acknowledgement that sacred books are, in any case, open to many kinds of interpretation according to historical circumstances. In order for the Qur'ān to remain a book of guidance, it has to be read according to contemporary circumstances.[33]

Apart from his pluralistic reading of the Islamic scripture, Ayoub's most distinctive hermeneutical features concern interfaces between Islam and Christianity, and his Islamic Christology. He made insightful comparisons between the two traditions' concepts of revelation, reading the Prologue to the Gospel of John in the light of the Qur'ān. He affirmed the often-made analogy that Christ in Christianity is equivalent to the Qur'ān in Islam, although he also saw clear differences between them. He pointed out that although in Christianity, the Word became flesh and in Islam, a book, in both cases the Word descended from the realm of absoluteness to the human realm. This means that the relationship between the temporality and the eternality of the divine Word presents itself in both traditions. Other related parallels he perceived concern the analogical positions of the Virgin Mary and the illiterate Prophet Muḥammad as pure vessels to receive divine revelation, and participation in the Eucharist and recitation of the Qur'ān as means to internalize God's Word.[34] As for Ayoub's Islamic Christology, it denies Jesus's divinity, but not his special humanity, and denies his sacrificial death on the cross, but not his actual death or his general redemptive role as a model of righteous life. While this is not out of the usual, Ayoub's strong opposition to the traditional substitution theories is distinctive. Substitution theories explain the Qur'ān's denial of Jesus's crucifixion (Q 4:157) by the idea that God gave Jesus's appearance to someone else who died in his stead. This was not acceptable to Ayoub, because he thought that this kind of divine deception makes a mockery of God's justice. Ayoub's own view was that the Qur'ān is not making a historical statement at all, but rather denies humans' ability to kill the divine Word:[35] "The Qur'ān is not speaking here about a man, righteous and wronged though he may be, but about the Word of God who was sent to earth and who returned to God. Thus the denial of the killing of Jesus is a denial of the power of men to vanquish and destroy the divine Word, which is forever victorious."[36]

Ayoub's hermeneutical method and insights have become well known both in academia and among those engaged in interreligious dialogue, especially in the West but also in Muslim countries. In addition to his published works, he has had a broad impact through his students and personal contacts. While his pluralistic views may

33 Ayoub, "Revelation and Salvation;" Ayoub and Khodr, *Naḥwa al-Jidāl al-aḥsan*, 31.

34 Ayoub, "The Need for Harmony," 112f.; Mahmoud Mustafa Ayoub, "The Word of God in Islam." *Greek Orthodox Theological Review* 31, no. 1–3 (1986): 69–78, at 73; Ayoub and Khodr, *Naḥwa al-Jidāl al-aḥsan*, 62, 64–65; Ayoub, "Revelation and Salvation".

35 Ayoub, "Towards an Islamic Christology, II," 94, 104f., 116f.

36 Ibid., 117.

not be easily accepted by the majority of either Muslims or Christians, his extensive expertise is widely acknowledged.[37]

4 Conclusion

Ayoub contributed to contemporary hermeneutics of the Qur'ān especially by making the *tafsīr* genre more widely known and by presenting academic tools to evaluate the sources used in classical exegesis. At a time when the level and prestige of traditional scholarship has declined and many Muslims interpret Islam one-sidedly from a legal point of view, he focused on *tafsīr* as the dynamic discourse which makes the Qur'ān relevant for all times and places. On the other hand, he drew attention to nonhistorical sources of *tafsīr*, still at least partly affirmed by modern exegetes, and presented historical methods to evaluate them critically and consistently. In this manner, he sought to combine constructive criticism and serious engagement with traditional Islamic sciences and methodologies.

In comparison to some contemporary Muslim intellectuals who engage with the Islamic scripture by stressing its human dimension, often inspired by Western literary theories and methods, Ayoub's approach was moderate. He did not challenge the traditional understanding of the Qur'ān as the Word of God, nor was he interested in applying Western literary methods to the Islamic scripture. Rather, he tended to locate the contextual human element in the *tafsīr* tradition, which is obviously more acceptable to the majority of Muslims than allowing for a human dimension in the Qur'ān itself.

As for contemporary reformist and progressive intellectuals in general, Ayoub's hermeneutical views shared common features with them, but his theological priorities set him apart. Many of the ethical emphases that he referred to in his writings on Qur'ānic hermeneutics – such as religious freedom, human equality, social justice, and ethical understanding of the legal verses of the Qur'ān – link him to reformist and progressive tendencies. However, Ayoub's main hermeneutical priority remained religious pluralism and interreligious dialogue based on commonality of faith. His clearly formulated pluralistic theology of religions is rather uncommon compared to most contemporary Muslim thinkers. Reformist and progressive intellectuals in general tend to focus on ethical issues, and even though they often point out that the Qur'ān invites Muslims to good relations and dialogue with other religious communities, they tend to leave open such theological questions as the truth-value of other religions and the fate of their adherents in the afterlife. Ayoub was a distinctive thinker who raised challenging questions. While some of his suggestions remain controver-

37 See Omar, "Pathways to Dialogue".

sial, even discussing them publicly will make the intellectual debate of contemporary Muslims richer.

5 Bibliography

5.1 Sources

Ayoub, Mahmoud. *Redemptive Suffering in Islam: A Study of the Devotional Aspects of 'Āshūrā' in Twelver Shī'ism*. Religion and Society 10. The Hague: Mouton, 1978.

Ayoub, Mahmoud. "Towards an Islamic Christology, II: The Death of Jesus, Reality or Delusion (A Study of the Death of Jesus in Tafsīr Literature)." *MW* 70, no. 2 (1980): 91–121.

Ayoub, Mahmoud. "Muslim Views of Christianity: Some Modern Examples." *Islamochristiana* 10 (1984): 49–70.

Ayoub, Mahmoud. *The Qur'ān and Its Interpreters*. 2 Vols. Albany: State University of New York Press, 1984–1992.

Ayoub, Mahmoud. "The Word of God in Islam." *Greek Orthodox Theological Review* 31, no. 1–3 (1986): 69–78.

Ayoub, Mahmoud. "Revelation and Salvation: Towards an Islamic View of History." *Al-Serat* (1987): 10–25. Available online: https://www.al-islam.org/al-serat/general-al-serat/revelation-and-salvation-towards-islamic-view-history-mahmoud-ayoub-0. Accessed August 2, 2020.

Ayoub, Mahmoud. "The Speaking Qur'ān and the Silent Qur'ān: A Study of the Principles and Development of Imāmī Shī'ī *tafsīr*." In *Approaches to the History of the Interpretation of the Qur'ān*, edited by Andrew Rippin. Oxford: Clarendon Press, 1988: 177–98.

Ayoub, Mahmoud. "The Need for Harmony and Collaboration between Muslims and Christians." In *Information and Formation*, edited by O.C. Lim and Theresa Ee-Chooi. Kuala Lumpur: Catholic Research Centre, 1994: 109–19.

Ayoub, Mahmoud. "Nearest in Amity: Christians in the Qur'ān and contemporary exegetical tradition." *Islam and Christian-Muslim Relations* 8, no. 2 (1997): 145–64.

Ayoub, Mahmoud. "Literary Exegesis of the Qur'ān: The Case of al-Sharīf al-Raḍī." In *Literary Structures of Religious Meaning in the Qur'ān*, edited by Issa J. Boullata. Abingdon, Oxon: Routledge, 2000: 292–309.

Ayoub, Mahmoud. *Dirāsāt fī l-'alāqāt al-masīḥiyya al-islāmiyya* (*Studies in Christian-Muslim Relations*). 2 Vols. Al-Kūra, Lebanon: Publications of the Center for Christian-Muslim Studies at the University of Balamand, 2000–2001.

Ayoub, Mahmoud. *The Crisis of Muslim History: Religion and Politics in Early Islam*. Oxford: Oneworld, 2003.

Ayoub, Mahmoud. *Islam: Faith and History*. London: Oneworld, 2004; Reprint 2018.

Ayoub, Mahmoud. "The Islamic Context of Christian-Muslim Relations." In *A Muslim View of Christianity: Essays on Dialogue by Mahmoud Ayoub*, edited by Irfan A. Omar. Maryknoll, New York: Orbis Books, 2007: 17–31.

Ayoub, Mahmoud. "9/11 and the Need for a New Paradigm for Interfaith Dialogue." *MW* 101, no. 3 (2011): 562–64.

Ayoub, Mahmoud. "Religious Pluralism and the Qur'ān." In *Contemporary Approaches to the Qur'ān and Sunnah*, edited by Mahmoud Ayoub. London: The Institute of Islamic Thought, 2012: 39–56.

Ayoub, Mahmoud and Georges Khodr. *Naḥwa al-Jidāl al-aḥsan: muḥāwarāt islāmiyya masīḥiyya* (*Towards Better Dialogue: Muslim-Christian Dialogues*). The Center for Christian-Muslim Studies at the University of Balamand, 1997.

Omar, Irfan A., ed. *A Muslim View of Christianity: Essays on Dialogue by Mahmoud Ayoub*. Maryknoll, New York: Orbis Books, 2007.

5.2 Secondary Literature

Hartford Seminary. "Imam Ali Chair for Shi'i Studies and Dialogue among Islamic Legal Schools." Accessed August 3, 2020. https://www.hartsem.edu/macdonald-center/imam-ali-chair-for-shii-studies-and-dialogue-among-islamic-legal-schools/.

Hartford Seminary. "Mahmoud Ayoub. Honorary Faculty Associate in Islam and Christian-Muslim Relations." Accessed July 7, 2020. https://www.hartsem.edu/faculty/mahmoud-ayoub/.

Hick, John. *God and the Universe of Faiths: Essays in the Philosophy of Religion*. London: Macmillan, 1973.

Hirvonen, Heidi. *Christian-Muslim Dialogue: Perspectives of Four Lebanese Thinkers.* History of Christian-Muslim Relations 18. Leiden: Brill, 2013.

McAuliffe, Jane Dammen. *Qur'ānic Christians: An analysis of classical and modern exegesis*. Cambridge: Cambridge University Press, 1991.

Omar, Irfan A. "Pathways to Dialogue: the Journey of Mahmoud Ayoub." *Islam and Christian-Muslim Relations. Special Issue. Islam in Dialogue. Essays in Honour of Professor Mahmoud Ayoub* 15, no 1 (2004): 9–11.

Siddiqui, Ataullah. *Christian-Muslim Dialogue in the Twentieth Century*. London: Macmillan; New York: St. Martin's Press, 1997.

Smith, Wilfred Cantwell. *Faith and Belief*. Princeton, NJ: Princeton University Press, 1979.

Massimo Campanini (d. 2020)

Ḥasan Ḥanafī

1 A Controversial Thinker

Ḥasan Ḥanafī (d. 2020/1443) was one of the contemporary Arab-Islamic world's outstanding philosophers.[1] His contribution to scholarship was huge and will be lasting. Like most innovative thinkers, he was controversial: conservative Muslim scholars accused him of being too secular, while secularists accused him of remaining too faithful to traditional religion. Both charges were, in a sense, true. Undoubtedly, some aspects of Ḥanafī's thought seem to exclude any mediation by either narrow-minded men of religion who refuse to deal with modernity or extreme modernizers seeking to cast off past heritage (*turāth*). Indeed, his uncompromising attitude is both challenging and open to potential criticism.

2 The Two Pillars of Ḥanafī's Thought

Ḥasan Ḥanafī was born in Cairo in 1935/1354. After obtaining his first academic degrees in Egypt, he moved to Paris and received his PhD from the Sorbonne in 1965/1385, with a thesis entitled "Les Méthodes de l'exégèse: Essai sur les fondements de la compréhension, ʿIlm uṣūl al-fiqh." His supervisor was Robert Brunschvig. A short time later, in 1967/1387, he was appointed Professor of Philosophy at Cairo University, where he taught until his retirement. Following one of his teachers, Catholic philosopher Jean Guitton, Ḥanafī was involved in inter-faith religious dialogue in the aftermath of Vatican Council II. Since 1969/1389, when Gustavo Gutierrez published the famous *Hacia una teología de la liberación,* liberation theology became one of the main paradigms of Ḥanafī's thought and work.[2] He was an engaged and "politicized" intellectual.[3] Trying to put his theories into practice, Ḥanafī promoted the "Islamic Left" (*al-yasār al-islāmī*), both a political movement and an intellectual enterprise. The Is-

1 For an introduction to contemporary Arab and Islamic thought, see Suzanne Kassab, *Contemporary Arab Thought* (New York: Columbia University Press, 2010); and Carool Kersten, *Contemporary Thought in the Muslim World* (Abingdon, NY: Routledge, 2019). There is a general presentation of Ḥanafī's thought in John L. Esposito and John O. Voll, "Hasan Hanafi: The Classical Intellectual," in *Makers of Contemporary Islam*, ed. J. Esposito and J. Voll (New York/Oxford: Oxford University Press, 2001).

2 With an important difference: while Catholic liberation theology was influenced by Marxism, Ḥanafī's Islamic theology of liberation is not. See his essay *"Ad-Dīn wa-th-thawra fī Amrīkā l-Lātīniyya"* (Religion and Revolution in Latin America), in Ḥasan Ḥanafī, *Fī fikrinā l-muʿāṣir* (Beirut: Dār at-Tanwīr, 1983).

3 See his *maius opus* Ḥasan Ḥanafī, *Min al-ʿaqīda ilā th-thawra* (Cairo: Maktabat Madbūlī, 1989).

https://doi.org/10.1515/9783111320069-012

lamic Left envisaged by Ḥanafī shared a number of Jamāl ʿAbd an-Nāṣir's (Nasser) concepts of Arab but indeed also Islamic socialism. The failure to realize an alliance between the Muslim Brotherhood and Nasser after the Free Officers coup of 1952/1371 in Egypt led, according to Fawaz Gerges, to a confrontation between the military and the Islamists that has been perhaps the most important factor in the Middle East predicament past and present.[4] Interviewed by Gerges about Sayyid Quṭb, Ḥanafī declared that "[Quṭb's] work lays the foundation of the Islamic Left, an egalitarian spirit that shows the fundamental role that justice plays in Islam. Qutb reminded Muslims of the most important moral message of Islam – social justice and freedom."[5] In this sense, he could perhaps be labeled as a "Quṭbian Nasserist."[6] No particularly dramatic events took place in Ḥanafī 's life – his years were all spent teaching, writing, and attending conferences and seminars worldwide.

Ḥanafī's intellectual and practical efforts were aimed at restoring to the Muslim people, and to Afro-Asiatic people in general, their place in world history. This, he said, would come about in two ways. First, by emphasizing that the members of the "Third World," specifically Muslim cultures, have an important role to play in building the historical future. This goal requires them to be able to criticize the dominant culture of the West (which developed the discipline of "Orientalism" to subdue the "Orientals") through a new discipline of "Occidentalism."[7] Ḥanafī argued that, in order to recover their place in history, Arab-Islamic peoples would have to develop Occidentalism – that is, a cultural, political, and philosophical consciousness – that would allow "Orientals" to confront the West from a position of creativity. At the end of the process, Orientals would become active participants interacting with Westerners as *subjects*, and no longer as *objects* of exploitation and oppression. From Ḥanafī's point of view, Occidentalism was not a mere cultural project; rather, it was based on his idea that there would be a *bouléversement* (upheaval) of power relations and consequently a correction of the historical vector. The paradigm of Occidentalism has been fiercely criticized even by Arab intellectuals,[8] including Tunisian intellectual aṭ-Ṭāhir Labīb. Labīb accused Ḥanafī of theorizing a newly disguised Orientalism, with

4 Fawaz Gerges, *Making the Arab World: Nasser, Qutb, and the Clash that Shaped the Middle East* (Princeton: Princeton University Press, 2018).

5 Quoted in Gerges, *Making the Arab World*, 221.

6 The definition is mine. Ḥanafī devoted a huge eight volumes of research to *ad-Dīn wa-th-thawra fī Miṣr* (Cairo: Maktabat Madbūlī, 1987). For a negative evaluation of Ḥanafī's political thought, see Martin Riexinger, "Nasserism Revitalized: A Critical Reading of Ḥasan Ḥanafī's Projects 'The Islamic Left' and 'Occidentalism' (And Their Uncritical Reading)," *WI* 47, no. 1 (2007): 63–118. On the relation between Hanafī and Quṭb, see also Shahrough Akhavi, "The Dialectic in Contemporary Egyptian Social Thought: The Scripturalist and Modernist Discourses of Sayyid Qutb and Hasan Hanafī," *IJMES* 19 (1997): 377–401.

7 Ḥasan Ḥanafī, *Muqaddima fī ʿilm al-istighrāb* (Cairo: Dār al-Fanniyya, 1991).

8 See Yudian Wahyudi, "Arab Responses to Hasan Hanafī's 'Muqaddima fī ʿilm al-istighrāb,'" *MW* 93, no. 2 (2003): 233–48.

the same faults and the same implicit "racism": positing one identity *against* another identity. "As the West invented its own Orient, the East invented its own Occident," Labīb said.[9] This risk is in fact real, but Ḥanafī's positive idea was to create a new subjectivity able to engage intersubjectively in dialogue with other subjectivities.

On the other hand, Ḥanafī emphasized God's centrality to revolutionary human praxis: for him, God was the guarantor of a theology of liberation, where the protagonists are human beings seeking justice. Ḥanafī advocated an *anthropologization of theology*, that is, for the language of religion to be translated into the practical human language of liberation. The passage from theology to anthropology would imply "putting God in brackets," which means a phenomenological reduction effected by means of Edmund Husserl's *epoché*.[10] God's essence would be parenthesized, that is "neutralized," and the concept of God situated in a transcendental dimension. God's essence would remain unknowable, escaping humankind's limited understanding. We must suspend judgment on essence, wrote Ḥanafī, even if God remains real and existent. God is, above all, the ultimate goal (*telos*). He is the source of justice, of equality between human beings, of liberation from all chains, and the answer to the demands of the oppressed. God is the source of liberation of Earth in that He is the incentive or motive for political and social action. "God is people" – this is another of Ḥanafī's characteristic expressions, meaning that people's struggle for liberation finds its incentive in God. Finally, God is revolution, because insofar as He is the objective and *telos* of human political and social action, He is the guarantor and prime mover of social development. Ḥanafī formulates a passage of synthesis as follows:

> God continues to be a value in our consciousness. But it is no less true that humankind too is a value. God has been acquired forever, but humankind, land, freedom, democracy, liberation, development, and progress are aims, goals to be reached, needs to be satisfied. It is not a question of abolishing theology in favor of anthropology, but rather of transforming theology into anthropology, changing the axis of culture's direction and ultimate aims. Let all the latent energy that goes by the name of God be poured into the channels of modern times to realize these aims. We must carry out the same mission as that realized by Descartes for European culture and transform a theocentric culture into an anthropocentric one. What is needed is a *cogito*. But this time it will be an active, not a reflective *cogito*.[11]

9 See the debate in aṭ-Ṭāhir Labīb, Ḥilmī Shaʿrāwī and Ḥasan Ḥanafī, *L'altro nella cultura araba*, ed. F. Corrao and S. Pagani (Messina: Mesogea, 2006).

10 Husserl's *epoché* means that, in order to grasp the essence of a thing, the observer must "suspend judgment," that is, abstract from the physical characters of the thing.

11 Ḥasan Ḥanafī, "Théologie ou Anthropologie?" in *Renaissance du Monde Arabe*, ed. Anouar Abdel-Malek, Abdel-Aziz Belal, and Hasan Hanafi (Gembloux: Duculot, 1972), 118. My translation from the original French.

3 Phenomenology and Hermeneutics

The science of Occidentalism and the translation of theology into anthropology are the two cornerstones of Ḥanafī's reflection within his broad framework of the harmonization of heritage and renewal.[12] We must now consider Ḥanafī's exegesis of the Qur'ān, which runs throughout his works, sometimes openly and sometimes in the background. Three moments are worth emphasizing – each of them marking a turning point in his reflection: his phenomenological approach to the *uṣūl ad-dīn* (foundations of religion) in the 1960s, his thematic interpretation of the 1990s, and his translation of revelation into praxis (*min al-waḥy ilā l-wāqiʿ*) in the more recent phase.

The *French Trilogy* is Ḥanafī's famous three-volume enterprise, published in the 1960s. It was intended to apply phenomenology to the study of *uṣūl al-fiqh* (the roots of jurisprudence) and Qur'ānic hermeneutics. It comprised his PhD thesis "Les méthodes d'exégèse: Essai sur la science des fondements de la comprehension" (The Methods of Exegesis: Essay About the Science of the Foundations of Understanding), published in 1965/1385, and *L'exégèse de la phénomenologie: L'état actuel de la méthode phénomenologique et son application au phénomène religieux* (The Exegesis of Phenomenology: The Present Situation of the Phenomenological Method and its Application to Religious Phenomena), published in 1966/1386, along with *La phénomenologie de l'exégèse: Essai d'une herméneutique existentielle à partir du Nouveau Testament* (The Phenomenology of Exegesis: Essay About an Existential Hermeneutics Starting with the New Testament). Ḥanafī himself described the *Trilogy* as an effort to offer an Islamic reading of phenomenology and a phenomenological reading of Islam.

Besides borrowing Husserl's strategy of *epoché* – aware that Martin Heidegger's phenomenology is very different from his master Husserl's because it moves away from the latter, strictly transcendental approach – Ḥanafī proposed that the text of the Qur'ān runs through the existential consciousness of being-in-the-world (*Dasein*) – a Heideggerian concept. Being-in-the-world must be coupled with the intentionality of Truth with a capital T. In this way, phenomenology is not a simple philosophical method that helps us to understand religious phenomena but also, and in particular, the transcendentalization of the text toward its ultimate goal (*telos*). The ultimate goal of phenomenology is the supreme Truth. The supreme Truth is God, and God revealed Himself in the Qur'ān. As Ḥanafī put it, "Returning to the Qur'ān, the source of all religious thought, both theological and otherwise, God does not present Himself on a theoretical level but on a practical level. God is not Logos, but Praxis. God is not rational object, concept, category or idea, but behavior or conduct."[13] Ḥanafī pointed out that the hidden meaning of the text must be fully "disclosed" so that an "ontology of the

12 Suzanne Olsson, *Renewal and Heritage: The Quest for Authenticity in Hasan Hanafi* (Uppsala: Uppsala University Press, 2004).

13 Ḥanafī, "Théologie ou Anthropologie?" 241.

text" can be gradually recreated through interpretation. Interpretation stands for "the way that any givenness in the form of a text turns itself into knowledge inside the spirit of the interpreter."[14] Central is the Heideggerian concept of "disclosure," which allows the interpreter to immediately grasp the truth-reality (*ḥaqq*) of Being/God (God is *ḥaqq*, as the Qur'ān states many times). God as Truth and Reality is grasped in His immediateness as the result of a pure act of comprehension: He is existent, but, put into brackets, He becomes the infinite *telos* of the human struggle for welfare and liberty.[15]

4 Thematic Interpretation of the Qur'ān

Phenomenology remained the ongoing framework for Ḥanafī's later speculation. However, he went further, putting forward the main coordinates for the thematic interpretation of the Qur'ān in an article he published twice, once in 1995/1416 and again in 1996/1417.[16] Thematic interpretation has been proposed by at least two others alongside Ḥanafī, namely by Fazlur Rahman and Muḥammad al-Ghazālī.

Fazlur Rahman (1919–88/1337–1408) – an eminent Pakistani philosopher – stressed the necessity of treating the text endogenously rather than exogenously, arguing that this was required to not only explain the Qur'ān with the Qur'ān but also and above all in order to study it as an interconnected texture. While an "atomistic" approach fails "to understand the underlying unity of the Qur'an, coupled with a practical insistence upon fixing on the words of various verses in isolation,"[17] the thematic approach allows the Qur'ān to be considered as a whole. Rahman's very intention was to make use of the Qur'ān, or more precisely of Qur'ānic hermeneutics, to rebuild Islamic *intellectualism*. Rahman emphasized the centrality of interpretation for practical and political aims. According to Rahman, given that Muslims have remained backward in the fields of pure thinking and philosophical intellectualism, it is imperative to discover a new basis for an Islamic way of thinking, starting with the Qur'ān.

14 Sylvain Camilleri, "The Eccentric Reception of Heidegger in Hanafi's *French Trilogy*," in *Heidegger in the Islamicate World*, ed. K. Moser, U. Gösken, and J. Hayes (London: Rowman & Littlefield, 2019), 89.

15 For a more detailed discussion of these themes, see Massimo Campanini, "Hasan Hanafī e la fenomenologia: per una nuova politica dell'Islam," *OM* 74, no. 1–6 (August 1994): 103–20; Massimo Campanini, "Dall'Unicità di Dio alla rivoluzione: Un percorso fenomenologico di Hasan Hanafi," in *Teologie Politiche: Modelli a confronto*, ed. G. Filoramo (Brescia: Morcelliana, 2005), 215–40; and Massimo Campanini, *Philosophical Perspectives in Modern Qur'anic Exegesis* (Sheffield UK/Bristol CT: Equinox Publishing, 2016).

16 Ḥanafī, *Islam in the Modern World* (Cairo: Anglo-Egyptian Bookshop, 1995), 1: 407–28; and Ḥanafī, "Method of Thematic Interpretation of the Qur'an," in *The Qur'an as Text*, ed. S. Wild (Leiden: Brill, 1996), 195–212.

17 Fazlur Rahman, *Islam and Modernity* (Chicago/London: The University of Chicago Press, 1984), 2f.

Muḥammad al-Ghazālī (d. 1996/1416), one of the main representatives of *wasaṭiyya*,[18] attempted a thematic reading of the Qurʾān, but, in my view, he failed to achieve his purpose.[19] On the one hand, he was not able to effectively demonstrate the internal coherence of the suras, while, on the other, his merely descriptive approach to the suras impeded the apprehension of their interconnected texture, even when effective. To give but one example: in al-Ghazālī's analysis, the multifarious contents of a complex sura like sura 2, *al-Baqara* (The Cow), remain separate from the others without a general framework capturing their (supposed) linkage.

Ḥanafī's thematic proposal starts with the observation that any conflict of interpretations is essentially a sociopolitical conflict, not a theoretical one.[20] Indeed, theory is only an epistemological cover. Each interpretation expresses the sociopolitical convictions of the interpreter. Interpretation is an ideological weapon wielded by different sociopolitical powers to either uphold or alter the status quo – conservatives strive to maintain the status quo, while revolutionaries seek to change it. Here, we see the same idea that supported the previously discussed idea of the anthropologization of theology. Starting from these premises, Ḥanafī first draws a distinction between longitudinal and latitudinal commentaries on the Qurʾān. Longitudinal commentaries present the thematic development of the Qurʾān as a succession that follows a systematic temporal evolution. Latitudinal commentaries, on the other hand, allow the "facts" in the Qurʾān to be gathered together in their simultaneousness and coexistence, at the same time varying their interpretation. Longitudinal commentaries entail numerous disadvantages:

1) When the same theme in several suras is disrupted by other themes, it leads to a loss in the integrality of the first theme, meaning that the theme becomes lost in different parts.
2) Repeating the same theme several times without accumulating any meaning to build the global concept disperses and reduces the impact of the argument.
3) The absence of a structure for the theme, rational, real, or both – a structure that allows the theme to stand on its own, validating and verifying itself from within, rather than relying on external sources from reason and in nature – clearly weakens the argument.

Operating on these premises, it is impossible to formulate a coherent ideology. Turning the perspective upside down through thematic interpretation, we achieve a number of important results:

---

18 *Wasaṭiyya* is an intellectual trend born in Egypt in the final decades of twentieth century, claiming to apply a "middle way" approach to religion and to Qurʾānic hermeneutics, far from exaggerate secularism and aggressive fundamentalism.

19 Muḥammad al-Ghazālī, *at-Tafsīr al-mawḍūʿī li-l-Qurʾān al-karīm* (Cairo: Dār ash-Shurūq, 2008[10]).

20 What follows is a paraphrase of the author's text: Ḥanafī, "Method of Thematic Interpretation."

1) First of all, we not only explain the text (*Auslegung*) but also understand it (*Verstehen*).
2) This means that the interpreter is not only a receiver of meaning but also a giver. He or she receives meaning and puts it in a rational and real structure, realizing the original identity between revelation, reason, and nature.
3) Interpretation is not only analytical but also synthetic, dividing the whole into its parts as well as bringing the parts together as a whole.
4) Interpreting is discovering something new between the lines, adding the unknown to common knowledge, going to a deeper level of the text, and thereby connecting with another, deeper level of consciousness.

Methodologically speaking, according to Ḥanafī, revelation is once again put in brackets (again through Husserl's *epoché*), being neither affirmed nor denied. The interpreter does not ask the question widely debated in nineteenth-century Orientalism of the Qur'ān's divine origin, whether it came from God or from Muḥammad through oral transmission. Like any other text subject to interpretation, the Qur'ān is considered a legal code, a literary work, a philosophical text, and a historical document. It does not have any special sacred character. In this way, the road is open to the full application of the modern discipline of hermeneutics.

From the point of view of praxis, Ḥanafī argues that the conflict of interpretation is a conflict of interests, even if the linguistic interpretation of the text changes. It means that the issue of interpretation has a political dimension. There is not just one interpretation of a text, but many interpretations, given the differences in understanding between different interpreters, conditioned by their prejudices. An interpretation of a text is, then, essentially pluralistic. The text is a vehicle for human interests and even passions. The text is merely the form they take. The interpreter fills it with content from his space and time.

Though they shared the same approach, Ḥanafī did not apply his method of thematic interpretation systematically, as Rahman did in *Major Themes of the Qur'ān*.[21] However, it could still be methodologically fruitful, like Rahman's, because thematic interpretation avoids the pitfalls of abrogating some verses with verses that were revealed later in the chronology (*naskh*), which longitudinal interpretation is obliged to assess. For example, such abrogation implies that the warlike Medinan verses abrogate the peaceful Meccan verses, legitimating aggressive jihad through the forceful manipulation of the text.

21 Fazlur Rahman, *Major Themes of the Qur'an* (Chicago: Chicago University Press, 2001³).

5 From the Text to the Fact

In the last phase of his creative reflection, Ḥanafī stressed in particular the concrete necessity of a Qurʾānic hermeneutics aimed at reconstructing Muslim thought and society. In doing so, he was reproposing the paradigm of theological translation in anthropology, this time focused on the Qurʾān. The idea of the Islamic Left (*al-yasār al-islāmī*) is again at the center of the author's reflections when he emphasizes the necessity of moving from prophetic inspiration to factual reality (*min al-waḥy ilā l-wāqiʿ*).[22] Ḥanafī reiterates:

> The Islamic Left is not only a speculative, Islamic, social, political, purposeful trend that joins together past and present, authenticity and contemporaneity, heritage and renewal, like all reformist movements; rather, it is also a rational attitude and a method in Qurʾānic commentary that joins together the text and human welfare, the aims of religious law – firmly rooted in the fundamental principles of jurisprudence – and their implementation in the factual reality of everyday life, both in time and place.

This clear definition is elaborated by Ḥanafī in different ways. First of all, he argues that, though the Islamic Left's method is not new insofar as it encompasses the most authoritative trends of the previous traditional Qurʾānic commentary, it is fully conscious of the necessity of applying critical historical method. However, history is employed exclusively in the study of the causes of revelation (*asbāb an-nuzūl*), the chronological order of the verses, the concept of abrogation (both abrogating and abrogated verses), and the development of the sharia rules over time to align with changing circumstances and situations, thereby ensuring the text's relevance to factual reality. It also involves applying normative rules in relation to their effectiveness. Ḥanafī did not accept the complete historicization of the text claimed by Naṣr Abū Zayd (1943–2010/1362–1431) – one of his students – who argued that the Qurʾān, being the word of God descended in history, is a *fully historical* book. This sensitive point was a matter of controversy between the master (Ḥanafī) and his equally famous pupil (Abū Zayd).[23]

As usual – and as I have stressed since the beginning – Ḥanafī was keen to emphasize the political character of commentary.[24] He directly quoted Abū l-Aʿlā Mawdūdī's *Tarjumān al-Qurʾān*, social commentaries like Muḥammad ʿAbduh's and Rashīd Riḍā's *al-Manār*, and literary commentaries like "martyr" Sayyid Quṭb's *Fī ẓilāl al-Qurʾān*. These titles are significant in Ḥanafī's conception of the intellectual development of Muslim modernism. Here, too, there is a thread connecting the first Salafiyya

---

22 Ḥasan Ḥanafī, *Min al-waḥy ilā l-wāqiʿ* (Cairo: al-Maktab al-Miṣrī li-l-Maṭbūʿāt, 2012). What follows is a paraphrase of the author's text (Chapter 1). I have arranged the text or translated it directly from Arabic.

23 In his widely known *Naqd al-khiṭāb ad-dīnī*, Abū Zayd criticized Ḥanafī's Islamic Left project at length and very harshly and, indeed, Ḥanafī's approach to religion as a whole. Cf. Naṣr Abū Zayd, *Critique of Religious Discourse*, ed. C. Kersten (New Haven/London: Yale University Press, 2018), 148–225.

24 This is a paraphrase of Ḥanafī, *Min al-waḥy ilā l-wāqiʿ*, Chapter 1.

to the Muslim Brothers,[25] going far beyond later radicalization in the wake of the 1960s.

According to Ḥanafī, the method suitable for a renewed *tafsīr* must be grounded upon four points. I quote them here at length[26]:

1) Beginning with discernment and intuition, or vision and [the perception] of time in accordance with the mystical path, or the [perception] of contemporaneity in accordance with the reformist spirit. [. . .]
2) Providing for the innumerable needs of human beings and the welfare of the *umma*. For who has [the right] to possess what? And who has [the right] to waste what? Who can claim power over what? Who has [the right] to subjugate whom? Why did religion become so remote from life? [. . .] Factuality is the core of the text and its first wellspring. Factuality is the matter, and the text is the form. The duty of the commentator is to put the matter into the form in order to shape and realize it.
3) Given that consciousness is the first fundamental, while factuality is the second insofar as it discloses what the needs of human beings are, the essential relationship between consciousness and factuality, from the point of view of the themes and interactions involved, constitutes the third fundamental. It means transforming interactions into operative plans, projects, and ideas. Clearly, the constituents of these projects are consistent with the "*fiqh al-awwaliyyāt*," that is, the jurisprudence of priorities: the land in consideration of [foreign] occupation; development; [struggles against] desertification; the defense of the *umma*'s unity against fragmentation, factionalism, ancestry, tribalism, Bedouinism; dignity in the face of strong social inequalities between rich and poor; the necessity of fair income distribution; a new policy intended not only to promote the value of work, but to confront subjugation, the culture of oppressive power and "pharaonic" authority, aiming to speak words of Truth in the face of any despotic power; consciousness and democracy in the face of authoritarian political regimes; progress in the face of backwardness; reformation in the face of corruption; fostering liberty and utopian politics in the face of stagnation, weakness, and subordination to sectarian politics.
4) The reciter of the Qur'ān share in some way the experience of the commentary by becoming aware of its necessities, complying with its implications, being able to realize its demands, observing its prohibitions, and correcting its negativities.

Ḥanafī's wishful perspective is that, as a result of the conceptualization we have just seen, the *umma* will be set in motion once more, marching on the path of history and bringing new life into its civilizing role, putting an end to its "Middle Ages" and entering its modern era.

---

25 See also Tariq Ramadan, *Le Rénoveau musulman: Un siècle de réformisme islamique* (Paris: Bayard, 1998).

26 Editor's Note: The author did not use quotation marks or provide further bibliographical information for the following passages that we could not find in Ḥanafī's work.

Moving from the revealed text to the fact (*Min al-waḥy ilā l-wāqiʿ*) means that divine revelation is not purely language, discourse, message, or eloquent transmission; rather, it is descent (*tanzīl*) and the connection between heaven and Earth, between God and the world. Ḥanafī established a Qurʾānic basis for this paradigm by putting forward an interpretation of some Qurʾānic *loci* that were quite at odds with common interpretations. According to Ḥanafī, religion is "fact," as the Qurʾān says, "Indeed religion is fact" (*wa-inna ad-dīn la-wāqiʿ*) (Q 51:6). It is interesting to note that Muhammad Abdel Haleem's authoritative rendering in English[27] of the same verse is "The Judgment will come," *dīn* meaning both religion and the Day of Judgment. The difference in readings is telling of the two thinkers' very different outlooks. For Ḥanafī, an engaged thinker, religion is not simply the soul's inspiration, nor does it have "states" (*aḥwāl*) tantamount to the states of religious psychology. Religion, he says, is not merely the fulfilment of desires, of yearnings and aspirations for a better world – perspicuous, essential, unadulterated – nor is it merely an obscure, vague transformation of psychological states. That would render religion a dream impossible to realize, a hope difficult to be made true. Some people are excessive and make religion the mere laziness of the soul, the materialization of illusions, the mythological saga of human civilization. According to Ḥanafī, religion is "fact," it actually exists in the lives of individuals and collectivities. It is not an illusion or myth. It is grounded in individual accountability and the primacy of the search for truth, which was referred to by the ancients as *kasb* or "gain" (in classical theological terminology), which is the goal that human beings are able to achieve on their own. Ḥanafī supports his argument with another Qurʾānic quotation: "[They are] fearful because of what they have gained, and this is a fact for them" (Q 42:22) (here, Abdel Haleem's rendering is: "fearful because of what they have done: punishment is bound to fall on them"). Every human being cannot but suffer the effects of his/her actions. This is the "fact" overseen by legal norms. The human being is free, intelligent, accountable, in this world and in the afterlife.

As Ḥanafī points out literally, nothing is so factually intractable as mountains. Here, he is referring to the Qurʾān, which says: "[When We made the mountain loom high above them like a shadow], they thought it was tangibly towering (*wāqiʿ*) above them" (Q 7:171) (Here, Abdel Haleem's rendering is: "[. . .] they thought it would fall on them"). Factuality is tantamount to the law of gravitation, Ḥanafī argues: everything is attracted to the Earth – water and birds, winds and heat and coolness. When religious discourse, which is grounded in admonition and wise guidance, is subjected to human excitability and sensory deviations, to the human obsession with repentance, its search for forgiveness, its crying and sighing – which are common to all assiduous worshippers – it is no longer factual (Ḥanafī's literal words). "Fact" means

27 Muhammad Abdel Haleem, *The Qurʾān: A New Translation* (Oxford: Oxford University Press, 2015^3).

the life of humans and all that occurs therein – poverty and pain, thirst and need. Therefore, when the Qur'ān speaks of the poor and the miserable, the wayfarers and the travelers – those are the people whom the contemporary exegetes call "the sufferers on Earth." The *mustaḍʿafūn* (the oppressed/weakened) are at the center of Islamic liberation theology in Hanafī, but also in ʿAlī Sharīʿatī (1933–77/1352–97), the celebrated Shiite intellectual father of the Iranian revolution of 1979, and in Quṭb too.

6 Conclusion

My brief study of Ḥasan Ḥanafī's Qur'ānic hermeneutics has led me to wonder, with Terry Pinkard's study of classical German philosophy,[28] whether *modern* religion is possible. The names of philosophers like Fichte and Schelling are found not infrequently on Ḥanafī's pages; it is therefore not out of place to mention German Idealism here. In German Idealism, religion is part of the Absolute Spirit. It has a binding connection with what Schelling or even Hegel would have called the "Infinite." Friedrich Schleiermacher (1768–1834/1182–1250) explained that religion is not grounded merely in morals, or even in metaphysics – it is a "breath" vivifying liberty when it has become nature. The infinite nature of the One is perceived by humans in dumb devotion according to Schleiermacher – who, it should be mentioned, can be considered the founder of modern hermeneutics. In the framework of classical German Idealism, what Novalis, for example, called "Christian" (Hegel held the same view), Ḥanafī would have called "Islamic." The idea is that of religion as the spirit of history, a force stirring up consciences and souls, urging them toward the path of liberation (something tantamount to Fichte's *Streben*). This does not mean that history *must* draw to a positive or teleological conclusion; rather, it means that humans, men and women, *must* devote themselves to positive participation in society.

7 Bibliography

7.1 Sources

Ḥanafī, Ḥasan. *Les méthodes d'exégèse: Essai sur la science des fondements de la compréhension*. Cairo: Publications du Conseil des Arts, des Lettres et des Sciences Humaines, 1965.
Ḥanafī, Ḥasan. *La phénomenologie de l'exégèse: Essai d'une herméneutique existentielle à partir du Nouveau Testament*. Cairo: Anglo-Egyptian Bookshop, 1966.
Ḥanafī, Ḥasan. *L'exégèse de la phénomenologie: L'état actuel de la méthode phénomenologique et son application au phénomène religieux*. Cairo: Dār al-Fikr al-ʿArabī, 1966.

28 Terry Pinkard, *German Philosophy 1760– 1860: The Legacy of Idealism* (Cambridge: Cambridge University Press, 2002).

Ḥanafī, Ḥasan. “Théologie ou Anthropologie?” In *Renaissance du Monde Arabe*, edited by Anouar Abdel-Malek, Abdel-Aziz Belal, and Hasan Hanafi. Gembloux: Duculot, 1972: 233–64.
Ḥanafī, Ḥasan. *Fī fikrinā l-muʿāṣir*. Beirut: Dār at-Tanwīr, 1983.
Ḥanafī, Ḥasan. *Ad-Dīn wa-th-thawra fī Miṣr*. 8 vols. Cairo: Maktabat Madbūlī, 1987.
Ḥanafī, Ḥasan. *Min al-ʿaqīda ilā th-thawra*. Cairo: Maktabat Madbūlī, 1989.
Ḥanafī, Ḥasan. *Muqaddima fī ʿilm al-istighrāb*. Cairo: Dār al-Fanniyya, 1991.
Ḥanafī, Ḥasan. *Islam in the Modern World*. 2 vols. Cairo: Anglo-Egyptian Bookshop, 1995.
Ḥanafī, Ḥasan. “Method of Thematic Interpretation of the Qur’an.” In *The Qur’an as Text*, edited by S. Wild. Leiden: Brill, 1996: 195–212.
Ḥanafī, Ḥasan. *Min al-waḥy ilā l-wāqiʿ*. Cairo: al-Maktab al-Miṣrī li-l-Maṭbūʿāt, 2012.

7.2 Secondary Literature

Abdel Haleem, Muhammad. *The Qur’ān: A New Translation*. Oxford: Oxford University Press, 2015[3].
Abū Zayd, Naṣr. *Critique of Religious Discourse*, edited by C. Kersten. New Haven/London: Yale University Press, 2018.
Akhavi, Shahrough. “The Dialectic in Contemporary Egyptian Social Thought: The Scripturalist and Modernist Discourses of Sayyid Qutb and Hasan Hanafī.” *IJMES* 19 (1997): 377–401.
Camilleri, Sylvain. “The Eccentric Reception of Heidegger in Hanafī’s *French Trilogy*.” In *Heidegger in the Islamicate World*, edited by K. Moser, U. Gösken and J. Hayes. London: Rowman & Littlefield, 2019: 85–97.
Campanini, Massimo. “Hasan Hanafī e la fenomenologia: per una nuova politica dell’Islam.” *OM* 74, no. 1–6 (August 1994): 103–20.
Campanini Massimo. “Dall’Unicità di Dio alla rivoluzione: Un percorso fenomenologico di Hasan Hanafi.” In *Teologie Politiche: Modelli a confronto*, edited by G. Filoramo. Brescia: Morcelliana, 2005: 215–30.
Campanini, Massimo. *Philosophical Perspectives in Modern Qur’anic Exegesis*. Sheffield UK/Bristol CT: Equinox Publishing, 2016.
Esposito, John L., and John O. Voll. “Hasan Hanafi: The Classical Intellectual.” In *Makers of Contemporary Islam*, edited by J. Esposito and J. Voll. New York/Oxford: Oxford University Press, 2001.
Gerges, Fawaz. *Making the Arab World: Nasser, Qutb, and the Clash that Shaped the Middle East*. Princeton: Princeton University Press, 2018.
Al-Ghazālī, Muḥammad. *At-Tafsīr al-mawḍūʿī li-l-Qur’ān al-karīm*. Cairo: Dār ash-Shurūq, 2008[10].
Kassab, Suzanne. *Contemporary Arab Thought*. New York: Columbia University Press, 2010.
Kersten, Carool. *Contemporary Thought in the Muslim World*. Abingdon/New York: Routledge, 2019.
Labīb, aṭ-Ṭāhir, Ḥilmī Shaʿrāwī, and Ḥasan Ḥanafī. *L’altro nella cultura araba*, edited by F. Corrao and S. Pagani. Messina: Mesogea, 2006.
Olsson, Suzanne. *Renewal and Heritage: The Quest for Authenticity in Hasan Hanafi*. Uppsala: Uppsala University Press, 2004.
Pinkard, Terry. *German Philosophy 1760–1860: The Legacy of Idealism*. Cambridge: Cambridge University Press, 2002.
Rahman, Fazlur. *Islam and Modernity*. Chicago/London: The University of Chicago Press, 1984.
Rahman, Fazlur. *Major Themes of the Qur’an*. Chicago: Chicago University Press, 2001[3].
Ramadan, Tariq. *Le Rénoveau musulman: Un siècle de réformisme islamique*. Paris: Bayard, 1998.
Riexinger, Martin. “Nasserism Revitalized: A Critical Reading of Ḥasan Ḥanafī’s Projects ‘The Islamic Left’ and ‘Occidentalism’ (And Their Uncritical Reading).” *WI* 47, no. 1 (2007): 63–118.
Wahyudi, Yudian. “Arab Responses to Hasan Hanafī’s ‘Muqaddima fī ʿilm al-istighrāb’.” *MW* 93, no. 2 (2003): 233–48.

Seyed Hossein Morakabi and Mohammad Mahdi Mojahedi*
Mūsā ṣ-Ṣadr

The true completion of the Qur'ān is equivalent to putting an end to human suffering. — Mūsā ṣ-Ṣadr

1 Introduction

After an introduction to his biography and an analytical literature review, this article delves into the hermeneutics of the Imām Mūsā ṣ-Ṣadr and the intellectual and socio-political background of his interpretation of the Qur'ān. Aṣ-Ṣadr, originally an Iranian cleric, a social activist, leader of Lebanese Shia Muslims, and founder of several religious, political, social, educational, and charitable institutions, had experiences spanning three phases of his life in Iran, Iraq, and Lebanon. Aṣ-Ṣadr emerged from Qom, one of the two major Shia seminary cities, and one of the prominent cultural and social Shia families. After migrating to Lebanon in 1959/1378, he had an extensive range of activities. The irreducible diversity of his audience, and his numerous references to the Qur'ān prompted him to develop a novel and pluralistic hermeneutic of the Qur'ān. Aṣ-Ṣadr's interpretations commence with a humanistic, social, dialogical, pluralistic, and anti-discriminatory stance. However, his problem-oriented approach renders him a transnational and transreligious social leader with patriotic concerns, in his later years. This is also reflected in his fresh approach to understanding tradition.

2 Life and Politico-Intellectual Environment

2.1 Education

Sayyīd Mūsā ṣ-Ṣadr was born in1928/1346 in Qom, Iran. His family had a diverse background, with his paternal grandfather, Sayyid Ismāʿīl aṣ-Ṣadr, hailing from Isfahan, Iran, and his paternal great-grandfather, aṣ-Sayyid Ṣadr ad-Dīn al-Mūsavī al-ʿĀmilī, being born in southern Lebanon. Additionally, his father, aṣ-Sayyid Ṣadr ad-Dīn aṣ-Ṣadr, was born in Kāẓimiyya, Iraq. This large scholarly family was fluent in both Arabic and Persian due to their geographical range from Iran to the Mediterranean coast.[1]

1 Moḥsen Kamāliyān, *Mashāhir-i Khāndān-i Ṣadr* (Qom: Rāze Neī, 2007), 18: 110–12.

*Corresponding author: Dr. Mohammad Mahdi Mojahedi

https://doi.org/10.1515/9783111320069-013

Aṣ-Ṣadr's father was one of the three religious authorities who led the Qom seminary 1937–1945/1356–1364, preventing its demise during the Pahlavi era, which knew the first modern government with secular tendencies in Iran. During the famine of the final years of World War II, aṣ-Ṣadr's father rescued tens of thousands of individuals from hunger in Qom.[2] He also authored a book pertaining to the Prophet Muḥammad that drawing on both Shia and Sunni traditions.[3]

From 1944/1363 to 1959/1378, aṣ-Ṣadr pursued studies in jurisprudence, philosophy, and Qur'ānic exegesis at Qom and Najaf. He was particularly influenced by two renowned Shia exegetes of the twentieth century, namely Sayyid Muḥammad Ḥusayn Ṭabāṭabā'ī (1904–1981/1321–1402), the author of *al-Mīzān* (1955/1374), and as-Sayyid Abū l-Qāsim al-Khū'ī (1899–1992/1317–1413), the author of *al-Bayān*.[4] During the years of 1947–1949, aṣ-Ṣadr was responsible for overseeing several intellectual publications in Qom and Najaf, which is indicative of his profound familiarity with the broad realms of popular and local culture.[5]

One year after the coup d'état of August 15, 1953/Dhū l-Ḥijja 1372, aṣ-Ṣadr earned a Bachelor's degree in "Law in Economics" from the University of Tehran. During this period, he also learned French at the University.[6]

2.2 Facing Diversity

In late 1959, the young aṣ-Ṣadr migrated to Lebanon as the successor of the Shiite leader, as-Sayyid ʿAbd al-Ḥusayn Sharaf ad-Dīn (1872–1957/1290–1377). At that particular moment, Lebanon had just obtained its independence from French colonization sixteen years prior and was still at the inception of its journey towards nation/state-building. During this period, aṣ-Ṣadr established favorable relationships with the leaders of anti-colonial struggles in Lebanon and was invited by them to participate in the prestigious national intellectual forum *an-Nadwa al-lubnānīyya* (The Lebanese Forum) which is widely recognized as one of the most significant identity-building forums in Lebanon.[7] During this period, aṣ-Ṣadr's audience diversified in terms of cultural, religious, and linguistic diversity, resulting in his necessity to seek assistance from all traditional sources to establish a common language for dialogue with them. Simultaneously, aṣ-Ṣadr sought to foster convergence and integration of the marginalized Shia community with Lebanese society.

2 As-Sayyid Mūsa aṣ-Ṣadr, "Az Pedaram Beguyam," *Ruznāme-ye Etelāʿāt* 24654 (2010): 6.

3 As-Sayyid Mūsā ṣ-Ṣadr, *Masīrat al-Imām as-Sayyid Mūsā ṣ-Ṣadr*, ed. Yaʿqūb Ḥasan Ḍāhir (Beirut: Dār Bilāl, 2014) 2: 190.

4 ʿAlī Ganjī, *Ayāme Taḥṣīl: Rawāyatī 'az Dwrāne Ṭalabegīe Sayyid Mūsā Ṣadr* (Tehran: Imam Moussa Sadr Cultural & Research Institute, 2023), 282.

5 Ibid., 311.

6 Ibid., 258.

7 Aṣ-Ṣadr, *Masīra*, 1: 131–54.

2.3 Life as a Permanent Migration

One of the non-epistemological aspects that distinguishes aṣ-Ṣadr's hermeneutical approach from other interpretive approaches is his lifestyle as a constant traveler. This lifestyle is closely linked to his familial tradition. Four generations of his ancestors were not born in their father's homeland, and, like aṣ-Ṣadr, their lives were divided into several periods of migration.[8] Aṣ-Ṣadr's annual travel of nearly 100,000 kilometers within a country with a total land border of less than five hundred kilometers is likely to be an unprecedented feat among Shiite intellectuals and leaders to this day. In his works, aṣ-Ṣadr repeatedly refers to a hadith that introduces creatures as God's dependents and His family (*ʿiyāl*), and that the way to gain divine satisfaction is by benefiting them.[9] Aṣ-Ṣadr contends that his elevation to the position of social leadership and the extent of his popularity can be largely attributed to his extensive travels and his unwavering dedication to visiting every household, church, mosque, university, and so forth, in addition to delivering more than seventy lectures annually.[10] The concept of travel embodies aṣ-Ṣadr's acceptance of a diverse humanity and his refinement of his dialogical and anti-discrimination approach.

2.4 The Leadership

Aṣ-Ṣadr, furthermore, constantly traveled to Muslim majority countries in Asia and Africa, as well as to Christian majority countries in Europe.[11] During the latter half of his tenure in Lebanon, he gradually became a social leader with concern for safeguarding national interests and ensuring harmonious coexistence within the homeland. Thus, through his extensive movements in Lebanon and the world, and by utilizing Lebanon's multicultural capacities, aṣ-Ṣadr was able to present a humanitarian, dialogical, and anti-discriminatory interpretation of Islam and Shiite tradition and put it to the test in practice.

In 1964/1386, aṣ-Ṣadr established the *Jabal ʿĀmil* vo-tech School and the charitable institution *Bayt al-fatāt* (Girl's House).[12] In 1966/1388, he founded the Islamic Supreme Council for Shiites,[13] followed by the Committee for Aid to the South in 1970/1390,[14]

8 Farīdeh Saʿīdī, "Ṣadr, Khāndān," in *Dāneshnāmeh-ye Jahān-e Eslām*, ed. Gholām-ʿAlī Haddād-ʿĀdel (Tehran: Dā ʾerat al Maʿāref-i Bozorg-i Eslāmī, 2021), 29: 419–68.

9 Muḥammad b. Yaʿqūb al-Kūlaynī, *al-ʿŪṣūl mīn al-Kāfī* (Qom: Jamāʿat al-Mudarrisīn fī-l-Hawza al-ʿIlmiyya fī-Qum al-Muqaddasa, 1986), 2: 164.

10 Aṣ-Ṣadr, *Masīra*, 3: 270.

11 Ibid., 1: 81.

12 Ibid., 100.

13 Ibid., 236.

14 Ibid., 2: 278.

which brought together leaders from various Islamic and Christian sects. In 1974/1394, he established the *Ḥarakat al-maḥrūmīn* (Movement of the Deprived),[15] and a year later, its military branch, known as the *Afwāj al-muqāwama al-lubnāniyya* (The Lebanese Resistance Regiments, Abbr. Amal),[16] to resist Israeli incursions into southern Lebanon. In 1977/1397, he founded the Islamic Risāla Scout.[17]

In 1978/1398, three years prior to his enforced disappearance and just a few months after the start of the fifteen-year-long civil war, aṣ-Ṣadr delivered a historic speech at a ceremony marking the beginning of Fasting, Ash Wednesday, at a Franciscan church in Beirut, in the presence of prominent Christian figures.[18] Later, he expressed his desire to invite a cardinal to deliver a sermon during one of the Muslim Friday prayers if the flames of civil war did not ignite.

Upon the onset of the Lebanese Civil War in mid-1975/1395, aṣ-Ṣadr, who had by then become a transreligious social leader, found his achievements to be threatened. He intensified his efforts to quell the flames of this war. Among these efforts were hunger strikes at *aṣ-Ṣafa* Mosque in Beirut[19] and subsequent trips aimed at meeting with interested foreign parties involved in the conflict. In August 1978/1398, while on a trip to Libya with two companions for the same purpose as before, aṣ-Ṣadr was abducted while he was a guest of the country's president, and to this day, no credible news has been received about his fate.

3 An Analytical Literature Review

No independent book by aṣ-Ṣadr was published prior to his forced disappearance. However, his works have been gradually collected and published by the Ṣadr Institute in Beirut over the past few decades. Besides seven articles in the *Darshāyī az Maktab-e islam* (Lessons from the School of Islam) journal, which are related to his last two years in Iran, no other documents have been published from his time in Iran or his studies in Iraq.[20]

The most important collection of his works was first published in 2000/1421. This chapter refers to the second edition of this collection, which is referred to as *Masīra al-Imām Mūsā ṣ-Ṣadr* (The Journey of Imam Mūsā ṣ-Ṣadr). This collection, consisting of twelve volumes, encompasses over 2,600 documents from the period of aṣ-Ṣadr's presence in Lebanon from 1960/1379 to 1978/1398, with 2,300 of them having precise

15 Ibid., 4: 115.
16 Ibid, 6: 29.
17 Ibid., 8: 51.
18 Ibid., 5: 193–201.
19 Ibid., 6: 99.
20 Ibid., 12: 309–43.

dates and eight hundred having complete texts. The rest are mostly excerpts from newspapers reporting on his activities. The vast majority of these eight hundred complete texts are speeches or drafts of speeches addressed to Arab dignitaries, or written introductions to their books.

Although no independent works of Qur'ān interpretation by aṣ-Ṣadr have been published, there are over 2,900 references to Qur'ān verses within the documents. Several of aṣ-Ṣadr's speeches are entirely dedicated to Qur'ān interpretation, with his audience mainly consisting of the general public from diverse religions and nationalities, rather than scholars from theological or academic backgrounds.

The tenth volume of *Masīra* is dedicated to his radio talks on the Qur'ān during Ramadan and his speeches on the topic to members of the *Amal* Movement. Unfortunately, the exact dates of these speeches in the second half of the 1970s are unknown. Although certain books have been published separately as exegetical works, they do not extend beyond the material included in *Masīra*.[21]

It is noteworthy that the speeches contained in volume ten of *Masīra* solely reflect a small portion of aṣ-Ṣadr's hermeneutical approach and cannot be relied upon solely for understanding his method. It is essential to study aṣ-Ṣadr's background and his rich experiences in social and political activism to comprehend his hermeneutical approach fully. Furthermore, any document of Qur'ān interpretation by aṣ-Ṣadr that is found outside of the twelve-volume collection has been used in this article.

4 Hermeneutical Assumptions and Background: A Typology

In order to understand aṣ-Ṣadr's hermeneutics, it is pertinent to take a step towards understanding the sources of his inspiration by reviewing the interpretive methods familiar to him in Iran and the Arab world. This typology aids us in assessing aṣ-Ṣadr's relation to these discourses, methods, and objectives. While this typology follows a historical sequence, it is important to note that these interpretive methodologies did all coexist in aṣ-Ṣadr's intellectual world simultaneously.

Since the earliest times, two rival approaches have taken shape within the tradition of Muslim exegesis: exoteric-oriented (*ẓahirī*) and esoteric-oriented (*baṭinī*). The two approaches have continued to be promoted within Shiite seminaries to this day. These two poles can be deemed as the most significant preconceptions of aṣ-Ṣadr's hermeneutical understanding.

21 E.g., as-Sayyid Mūsā ṣ-Ṣadr, *Dirāsāt li-l-ḥayāt* (Beirut: Markaz al-Imām Mūsā ṣ-Ṣadr li-l-Abḥāth wa-d-Dirāsāt, 1999); as-Sayyid Mūsā ṣ-Ṣadr, *Aḥādīth as-saḥar* (Beirut: Markaz al-Imām Mūsā ṣ-Ṣadr li-l-Abḥāth wa-d-Dirāsāt, 1999).

1. The exoteric-oriented approach: The main premise of this approach is that the meaning and intention of God are manifest in the outward aspects of the Qur'ānic verses. The primary goal of this approach, alongside understanding the world and the principles of religion, has been to discover the sharia based on the apparent aspects of the verses. The general method of this approach involves rhetorical analysis, lexical denotations, and occasionally referring to hadiths and traditions. Since the number of verses containing legal injunctions is estimated to be close to eight percent of the Qur'ān,[22] and these verses are controversially interpretable, practically no precise, and certain legal rulings can be extracted from it. Therefore, this interpretive discourse is to a large extent reliant on tradition and syllogism. The method employed in Ṣadr's articles in the journal *Maktab-i islam*, which is the only documentation related to aṣ-Ṣadr's early period in Iran and focuses on Islamic economics, is of a similar nature. Indeed, about two years later, as-Sayyid Muḥammad Bāqir aṣ-Ṣadr, aṣ-Ṣadr's cousin, as a traditional jurist, reflects Mūsā ṣ-Ṣadr's ideas in his famous book *Iqtiṣādunā* (Our Economy).[23]

2. The esoteric-oriented approach: The primary premise of this approach is that the true meanings of the Qur'ān lie beyond the outward appearance of its verses. The ultimate goal of this approach is to understand the world and the deepest layers of the principles of Islam and legal injunctions. This approach has historically expanded alongside the inclination towards Neoplatonic intellectual patterns in Islamic history.[24] During a part of his Islamic studies at the Qom, aṣ-Ṣadr, in addition to pursuing literal interpretations, became familiar with this method by studying the philosophical and mystical works of Ṭabāṭabā'ī. Although Ṭabāṭabā'ī, in his work *al-Mīzān*, less frequently adopted an esoteric approach, in his teachings on philosophy and mysticism, he instructed mystical interpretive views of figures such as Muḥyī d-Dīn b. ʿArabī (1165–1240/558–638). While his mystical inclination is not hidden, Ṭabāṭabā'ī has used it less in his exegesis and has never turned his back on the Shiite tradition. Ṭabāṭabā'ī's method of interpretation involves "interpreting the Qur'ān by Qur'ān" and a *sīyaqī* (contextualized) interpretation in the lengthy chapters, which had less precedent in the Shiite tradition until that time. A recognized and powerful trend in the Shiite tradition was the atomistic interpretation of Qur'ānic verses and their application in many cases to political conflicts between Shiite Imams and their contemporary political powers. It is this point that Ṭabāṭabā'ī distances himself from the mainstream approach and employs the methods of the "Qur'ān alone" approach, which

22 Ahmad Pakatchi, "Āyāt al-aḥkām," in *Encyclopaedia Islamica*, ed. Farhad Daftary, accessed on February 19, https://referenceworks.brillonline.com/entries/encyclopaedia-islamica/ayat-al-ahkam-COM_0319.
23 Chibli Mallat, *The Renewal of Islamic Law: Muhammad Baqer as-Sadr, Najaf and the Shi'i International* (Cambridge: Cambridge University Press, 1993), 142.
24 e.g., Milad Milani, *The Nature of Sufism: An Ontological Reading of the Mystical in Islam* (New York: Routledge, 2022), 1–22.

will be introduced in the next section. He occasionally discusses hadiths, but he separates these discussions structurally from interpretive discussions.

Ṭabāṭabāʾī considered some Qurʾānic verses to have allegorical significance. For example, he interpreted the story of Adam's creation as allegorical and indicative of the inherent dignity of humankind.[25] As will be discussed further, Ṭabāṭabāʾī's inclination towards allegorical interpretation has been influential in aṣ-Ṣadr's hermeneutics.

On the other hand, aṣ-Ṣadr's familiarity with Ibn ʿArabī's works and approach gradually becomes evident in his understanding of the Qurʾān. While aṣ-Ṣadr's hermeneutics is not extensively reliant on Ibn ʿArabī's method, understanding it is vital for comprehending the developments of his interpretation of the Qurʾān. Most probably, aṣ-Ṣadr has drawn his understanding of the relationship between God, the world, and humanity from the works of Ibn ʿArabī. From Ibn ʿArabī's perspective, the world is the manifestation of the beautiful names of God.[26] Corresponding to this, Ibn ʿArabī considers the path to human proximity to God as "assuming the traits of God's names / the character traits of God."[27] This overarching framework constitutes the philosophical theology and anthropology of aṣ-Ṣadr. John Culp regards Ibn ʿArabī as one of the examples of panentheists among Muslim theologians,[28] but scholars such as William Chittick[29] and Muhammad Rustom[30] strongly doubt whether Ibn ʿArabī can be labeled as a panentheist or a pantheist. During the early years of aṣ-Ṣadr's presence in Lebanon, when invited to speak about the relationship between Christianity and Islam in Lebanon at an-Nadwa al-lubnāniyya, aṣ-Ṣadr introduced Ibn ʿArabī as the pinnacle of Islamic Sufism in the presence of leaders from diverse religious sects in Lebanon. He considered the Sufism as the key to the openness of religions, including Eastern religions, towards one another.[31] This statement is reminiscent of his father's writing, where he says, "I do not consider any book of a higher stature than the *Futuḥāt al-makkiyya* (1231/628)."[32] It is noteworthy that the veneration of Ibn ʿArabī in the academic circles of Qom has always

25 As-Sayyid Muḥammad Ḥusayn Ṭabāṭabāʾī, *al-Mīzān fī tafsīr al-Qurʾān* (Qom: Jamāʿat al-Mudarrisīn fī l-Ḥawza al-ʿIlmiyya fī Qum al-Muqaddasa, n.d), 1: 132.

26 William C. Chittick, *The Sufi Path of Knowledge: Ibn al-ʿArabi's Metaphysics of Imagination* (New York: State University of New York Press, 1989), 5.

27 Ibid., 22.

28 John Culp, "Panentheism," in *The Stanford Encyclopedia of Philosophy* (Fall 2023 Edition), ed. Edward N. Zalta and Uri Nodelman, accessed February 19, 2024, https://plato.stanford.edu/archives/fall2023/entries/panentheism/.

29 William C. Chittick, "Waḥdat al-Wujūd in India," *Ishraq: Islamic Philosophy Yearbook* 3 (2012): 29–40.

30 Mohammed Rustom, "Is Ibn al-ʿArabī's Ontology Pantheistic?," *Journal of Islamic Philosophy* 2 (2006): 53–67.

31 Aṣ-Ṣadr, *Masīra*, 1: 147.

32 As-Sayyid Ṣadr ad-Dīn aṣ-Ṣadr, *al-Mahdī* (Beirut: Dār az-Zahrāʾ, 1978), 245.

been a contentious issue, characterized by a tendency to reject traditional jurisprudential practices and the interpretations of orthodox Shiite scholars of the Qur'ān.

3. The scientific approach: This approach is characterized by the assumption that religious and scientific languages are translatable to one another because they are talking about the same sets of truth. Accordingly, scholars of this approach strive to demonstrate that Qur'ānic statements have been able to predict complex scientific realities much earlier than modern sciences. The purpose of such citations, which emerged with the advent of modernity in the Islamic world, is to prove the scientific miracles of the Qur'ān and thereby establish the scientific validity of the religion, ultimately defending the religious worldview and legal provisions. Yadollah Saḥābī (1906–2002/1322–1423) and Mahdī Bāzargān (1907–1995/1325–1415) were the leading and influential representatives of this group in Iran, alongside Sayyid Maḥmūd Tāleqānī (1911–1979/1329–1399), a prominent religious and political figure in the 1979/1399 revolution, they were the co-founders of the Iranian Freedom Movement.[33] Aṣ-Ṣadr had friendly relations with them. Moreover, traces of this approach can also be found among Arab exegetes such as Muḥammad ʿAbduh (1849–1905/1266–1323) and Mālik b. Nabī (1905–1973/1323–1393), whom aṣ-Ṣadr often quotes. He always accepts the views of Ibn Nabī, but does not always attribute the same acceptance to the views of ʿAbduh.[34]

4. The "Qur'ān Alone" approach: This discourse also emerged with the advent of modernity in Iran and the incapacity of traditional thoughts to respond to the new questions. "Qur'ān Alone" scholars aimed to achieve an updated interpretation of religion by eliminating intermediaries i.e., tradition as text and clergy as an institution, in understanding the Qur'ān. The recent approach aligned the endeavors of "Qur'ān Alone" scholars with the secularist inclinations of the Pahlavi regime.[35] In his book on *Sharīʿat Sanglajī* (1891–1944/1308–1363), a representative of this approach during aṣ-Ṣadr's era, Ali Rahnema frequently compared him to Luther and also to the Protestant principle of "Sola Scriptura".[36]

In this context, Tāleqānī can be considered a bridge between three approaches: "Qur'ān Alone", scientific, and ideological (introduced below). He maintained close and scholarly relationships with the most prominent representatives of these three approaches, and the influence of these relationships is evident in his interpretative works.

---

33 Katajun Amirpur, "The Changing Approach to the Text: Iranian Scholars and the Qur'ān," *Middle Eastern Studies* 41, no. 3 (2005): 337–50.

34 E.g., aṣ-Ṣadr, *Masīra*, 10: 63f.

35 Rotraud Wielandt, "Main Trends of Islamic Theological Thought from the Late Nineteenth Century to Present Times," in *The Oxford Handbook of Islamic Theology*, ed. Sabine Schmidtke (Oxford: Oxford University Press, 2016), 734.

36 Ali Rahnema, *Shiʿi Reformation in Iran: The Life and Theology of Shariʿat Sangelaji* (New York: Routledge, 2015), 4.

5. The Ideological Approach: Following the First World War and the dissolution of the Ottoman Empire, efforts to establish a new Muslim identity began in the Islamic world. Some activists and political and social leaders, particularly inspired by leftist movements or in response to them or in competition with them, sought to use the Qur'ān and the legacy of Muslims as an ideological tool for social and political uprisings and anti-colonial identity formation. The most prominent representative of this approach in Iran was ʿAlī Sharīʿatī (1933–1977/1352–1395). Until the passing of Sharīʿatī, though mentioning his ideas, aṣ-Ṣadr did not mention his name in his works. However, subsequent to his demise, he exhibited great respect towards him. Aṣ-Ṣadr, who had been in contact with Sharīʿatī's father, a prominent cleric, for decades, praised Sharīʿatī and his thoughts in a lengthy speech.[37]

Another notable representative of this approach, during the aṣ-Ṣadr's period, was Sayyid Quṭb (1906–1966/1324–1386). Aṣ-Ṣadr paid attention to the intellectual and political movements in Egypt and had numerous trips to the country, meeting with secular, Islamic, political, and religious leaders. Although aṣ-Ṣadr only mentioned Quṭb's social interpretive views once in his works and praised their content,[38] aṣ-Ṣadr's library[39] contained two series of Quṭb's multi-volume interpretation *Fī ẓilāl al-Qur'ān.*

The large span and transformations of aṣ-Ṣadr's Qur'ān interpretations, have made it hard to categorize aṣ-Ṣadr's synthetic interpretive method into one of the aforementioned categories. What aṣ-Ṣadr has adopted from two first approaches, is a close familiarity with the traditional methodology. However, his interpretive method is not exclusively centered on the sharia, nor is it solely mystical. If we consider the ultimate goal of the scientific approach to interpretation as proving the scientific validity of religion, this is also not aṣ-Ṣadr's ultimate concern. On the other hand, if we interpret the concern of the third and fourth categories as an attempt to create a methodological approach for religiosity in the modern world, this is closer to aṣ-Ṣadr's concern, even though ultimately, he does not exclusively adhere to any of these interpretive methods. Although he was influenced by an ideological approach, aṣ-Ṣadr never moved towards seizing power or establishing a religious government, and this was not merely a political decision due to the diverse environment of Lebanon, but rather based on his own way of understanding religion. This fact becomes more apparent when compared to the interpretive efforts of scholars such as as-Sayyid Muḥammad Ḥusayn Faḍlallah (1935–2010/1354–1431) who were openly justifying, and even promoting, the establishment of an Islamic government in Lebanon during the same pe-

37 Aṣ-Ṣadr, *Masīra*, 8: 334–46.

38 Ibid., 10: 295.

39 During the Lebanon Civil War, his library was partially destroyed, but its remains are available at aṣ-Ṣadr's institute in Beirut.

riod.[40] The ultimate goal of aṣ-Ṣadr's synthetic interpretation method was a conscientious endeavor to bring about relief for human sufferings in both spiritual and material dimensions, which will be elaborated in the following section. The next section will encompass the introduction to the foundations, characteristics, and achievements of aṣ-Ṣadr's hermeneutical method.

4.1 The Principles of aṣ-Ṣadr's Hermeneutics: Problem-Solving and Relief

From 1949/1368 to 1954/1373, aṣ-*Ṣadr* studied "Law in Economics" at the University of Tehran. His academic studies, as one of the first clerics at the University of Tehran, which was founded nearly fifteen years earlier, brought him a methodological gift, i.e., problem-oriented methodology in modern humanities. This subject that accompanied him until the end of his experience in Lebanon and distinguished him from his fellow interpreters, even in the apparent discourse. This is because the religious laws that are the focus of the apparent interpretive approach are not primarily concerned with the mundane affairs, while the concern of law is to organize the affairs of governance, society, and its members. In various periods of his activity and according to changing issues that he faced; he used different interpretive approaches. Furthermore, his interpretation of a single sura or verse changed over time, and in this change, he shifted from one approach to another. In subsequent sections, examples will be provided for both of the latter variants.

4.2 Humanitarianism

Aṣ-Ṣadr's problem-oriented approach to interpretation was based on humanitarianism, and his humanitarianism was rooted in a pluralistic understanding of humanity. His emphasis on the core value of human existence in Lebanon was to recognize the diversity of the Lebanese people. Aṣ-Ṣadr's openness to diversity and his efforts to alleviate the suffering of marginalized communities and establish social institutions were influenced by the three foundational elements of his upbringing family.[41] In addition to his tireless efforts against social, religious, economic, and political discrimination, aṣ-Ṣadr repeatedly emphasized the dignity and divine nature of mankind in

---

40 Martin Kramer, "The Oracle of Hizbullah: Sayyid Muhammad Husayn Fadlallah," in *Spokesmen for the Despised: Fundamentalist Leaders in the Middle East*, ed. R. Scott Appleby (Chicago: University of Chicago Press, 1997), 83–181.

41 Saʿīdī, "Ṣadr, Khāndān," 419–68.

his works.[42] The position of humanity in his understanding is so prominent that he often substitutes the word mankind for God in his interpretation of the Qurʾān. He states that the path of God is the same as the path of humanity,[43] that serving God is serving humanity,[44] that remembering God is remembering humanity,[45] and that faith in God is infinite faith in humanity.[46] In his famous sermon at the Christian fasting ceremony, aṣ-Ṣadr regarded the ultimate goal of religions as the worship of God and service to humanity, considering them as two sides of a single reality.[47] In another instance, he stated that the Islam of a person is in their humanity,[48] where humanity is considered the highest religious value. We will revisit this later. Following this path, aṣ-Ṣadr believed that all religious rituals should serve humanity. His interpretation of Q 2:191 prioritizes the sanctity of human beings and their institutions over all other divine sanctities.[49] While such a belief about believers has roots in Shia tradition,[50] aṣ-Ṣadr's innovation lies in his generalization of this sanctity for all humans regardless of their religion or belief. He emphasized that all human talents and abilities must be respected and acknowledged.[51]

Aṣ-Ṣadr himself recounts a personal experience in 1976/1396 during the month of Ramaḍan, where it is customary for Muslims to complete the recitation of the entire Qurʾān. Amidst the civil war in Lebanon (1975–1990/1395–1410), he went to meet with Anwar as-Sādāt (1918–1981/1337–1401), the President of Egypt, in an effort to seek his assistance in ending the internal conflict. Aṣ-Ṣadr saw as-Sādāt in the midst of his third completion of the Qurʾān and he told Sādāt that a true completion of the Qurʾān is equivalent to putting an end to human suffering.[52]

In this regard, it is impossible to overlook the significance of sura 107, *al-Māʿūn* (Small Kindnesses), as its verses are the most frequently cited in aṣ-Ṣadr's works. Aṣ-Ṣadr interprets the theme of this chapter as a condemnation of the religious people who are indifferent to human suffering and deprivation. He emphasizes the use of the Arabic preposition *ʿan* in the fifth verse of this sura, asserting that does not refer to being careless *in* prayer, but rather refers to being careless *about* prayer i.e., neglecting humanity. Aṣ-Ṣadr contends that neglecting human beings is equivalent to ne-

42 Aṣ-Ṣadr, *Masīra*, 10: 393.
43 Ibid., 6: 155.
44 Ibid., 156.
45 Ibid., 10: 46.
46 Ibid., 5: 117.
47 Ibid., 194.
48 Ibid., 3: 50.
49 Ibid., 10: 106.
50 Cf. Muḥammad b. ʿAlī ṣ-Ṣadūq, *Kitāb al-Khiṣāl* (Qom: Jamāʿat al-Mudarrisīn fī l-Ḥawza al-ʿIlmiyya fī Qum al-Muqaddasa, 1983), 27.
51 Aṣ-Ṣadr, *Masīra*, 5: 196.
52 Ibid., 7: 354.

glecting the remembrance of God, which he views as the remembrance of humanity.[53] Under the interpretation of this same chapter, aṣ-Ṣadr shares a hadith from early Shia history that deeply affected him,[54] stating that if one person in a city goes hungry and sleeps, God will not look at any of the people in that city on the day of justice.[55]

Aṣ-Ṣadr places the human being at the axis of his thought and action; a being who is not only created from the names of God but ultimately also connects to the infinite God. Aṣ-Ṣadr finds this in Q 2:156. When a person dies, he is transferred to a higher level of life and is transformed into eternal ideas that embody his actions in this life.[56] Therefore, aṣ-Ṣadr believes that death is one of the stages of human progress and regrets why Q 2:156, despite being full of enthusiasm, life, and an endless path ahead of humans, reminds us, as Muslims, of burial ceremonies.[57] Consequently, in aṣ-Ṣadr's systematization, even death is considered in the service of humanity. In his view, God is the same as infinity towards which humans and other beings are heading, and the prerequisite for taking steps on this path of infinite progress is to accept that nothing in the world is more noble than a human created from the names of God. Therefore, aṣ-Ṣadr says *Lā ilāha illā llāh* (There is no deity but God) means the infinite potential of human beings.[58]

4.3 Semantic Pluralism

Aṣ-Ṣadr believes that understanding the meaning of the Qur'ān is not limited to the circumstances of revelation or the interpretive methods of the predecessors. Thus, he

53 Ibid., 10: 46.

54 Al-Kulaynī, *al-Kāfī*, 2: 668.

55 Aṣ-Ṣadr, *Masīra*, 10: 274–79.

56 Ibid., 1: 181.

About this subject, aṣ-*Ṣadr* refers to a poem from the *Mathnavī* of Jalāl ad-Dīn Rūmī. Rumi cites a portion of verse 88:28 in the concluding line:

> On death, I left being mineral then grew
> And changed from plant to animal form too,
> Then died to that, to be a human here—
> When did death make me less? What should I fear?
> I'll die to humanness at the next battle,
> Then spread my wings and soar above each angel:
> I must transcend the angels' status too—
> All perishes except God's face proves true.

cf. Jalāl ad-Dīn Rūmī, *The Masnavi*, trans. Mojaddedi Jawid (Oxford: Oxford University Press, 2013).

57 Aṣ-Ṣadr, *Masīra*, 5: 188.

58 As-Sayyid Mūsā ṣ-Ṣadr, *Qulū lā ilāha illā llāh tufliḥu* (Beirut: Markaz al-Imām Mūsā ṣ-Ṣadr li-l-Abḥāth w-d-Dirāsāt, 2016), 50.

frees himself from the constraints of most traditional semantic assumptions and creates a foundation for innovative hermeneutics. According to aṣ-Ṣadr, the reason for this pluralistic semantic in the Qur'ān is different from the reasons already presented in this context. He nurtures a theological argument for his pluralistic semantic, which is likely one of his own hermeneutical initiatives. He introduces the widely held belief among Muslims that the Qur'ān is the direct speech of God as the introduction to his argument. Then he explains that since God's knowledge is infinite and eternal, the meanings of His words must be understood according to the scope of his knowledge. Therefore, we can always extract new meanings from the Qur'ān that are not limited by cultural assumptions or the traditions of previous eras but arise from the semantic implications of Qur'ānic words.[59] Based on this premise, he does not confine customary understanding to Muḥammad's contemporary Arabian Peninsula. Nonetheless, aṣ-Ṣadr assumes this linguistic premise that the Qur'ān was revealed in the Arabic language and culture that evolved over different historical periods, rather than in the language and culture of the Muḥammad era. Aṣ-Ṣadr's theological argument has presuppositions about the relationship between text, mind, language, time, and history that are not accepted in contemporary hermeneutical theories. Aṣ-Ṣadr's pluralistic semantic is in complete harmony with his two previous hermeneutical pillars: relief and humanism.

4.4 Semiotic Pluralism

Aṣ-Ṣadr also believes that the meanings of the Qur'ān, due to the God's omniscient nature, are not confined to a particular type of linguistic usage or limited to a certain system of signification. Based on verse Q 10:61 regarding the knowledge of God, he argues that the words of a verse can simultaneously have allegorical, metaphorical, literal, and scientific meanings, implications, denotations, and significations.[60]

Thus, without seeing any conflict, aṣ-Ṣadr has at times interpreted a verse in several layers of language. He sometimes brings scientific interpretations of Qur'ānic verses, for instance, considering Q 51:47 as a scientific miracle of the Qur'ān predicting the expansion of the material heavens.[61] At the same time, elsewhere he understands the heavens in the language of the Qur'ān not as the physical space above us, but as a collection of values and spiritualities,[62] with one of its manifestations being the heart of the Prophet Muḥammad,[63] and the earth in the language of the Qur'ān as encompassing all space including all galaxies, which we refer to as heaven in literal linguis-

59 Aṣ-Ṣadr, *Masīra*, 10: 5f.
60 Ibid., 10: 6; 11: 18.
61 Ibid., 10: 7.
62 Ibid., 2: 320.
63 Ibid., 10: 356f.

tic usage. It is noteworthy that aṣ-Ṣadr thus associates the heavens with the earth to emphasize that the divine vicegerency of humans includes the material heavens and opens up avenues for human scientific explorations.[64]

Aṣ-Ṣadr's multifaceted interpretations of the Qur'ān extend beyond a mere acknowledgment of scientific progress. He simultaneously holds a broad range of allegorical interpretations of the Qur'ānic text and introduces a pervasive form of fictionalism. Consequently, aṣ-Ṣadr perceives the Qur'ānic narratives not only as historical events but also as allegories that are repeated in the life of every individual.[65] Ṭabāṭabā'ī also asserts the existence of allegory in the language of the Qur'ān, specifically considering the story of Adam and Eve as an allegory and an expression of the fate of every human being on earth.[66] Aṣ-Ṣadr echoes this foundation explicitly.[67] This very foundation has led aṣ-Ṣadr, like his mentor, to believe in the inherent dignity of the mankind and to employ it as a hermeneutical principle.

However, aṣ-Ṣadr's range of affirming the multiple implications of the Qur'ānic verses is not unbounded. For example, he disagrees with the interpretation of ʿAbduh, who interprets the word *abābīl* in 105:3 as microbes, considering it beyond the scope of the customary implications of the words of the verse. Aṣ-Ṣadr argues that if the verse had such an implication, the contemporaries of Muḥammad would have opposed him because they would have inevitably misunderstood the verse.[68] The foundation of aṣ-Ṣadr's stance against esoteric interpretations of the disjoined letters[69] or mystical interpretations, including Henry Corbin's (1903–1978/1321–1398) interpretation of Shiism, as outlined in aṣ-Ṣadr's introduction to the Arabic translation of Corbin's book, contradicts these interpretations with the customary language implications.[70] Therefore, aṣ-Ṣadr only permits new interpretations and multiple implications of the text to the extent that they do not contradict the customary understanding of the text.[71]

64 Ibid., 101.

65 As-Sayyid Mūsā ṣ-Ṣadr, *Nāy wa Ney: Justārhāyī darbāre-yi eslām, ensān, lubnān wa-muqāwamat.* Ed. ʿAlī Ḥujjatī Kermanī. (Tehran: Imam Moussa Sadr Cultural & Research Institute, 2004), 46.

66 Ṭabāṭabā'ī, *al-Mīzān*, 14: 228.

67 Aṣ-Ṣadr, *Masīra*, 10: 101.

68 Ibid., 10: 283.

69 Ibid., 11: 23.

70 Ibid., 1: 182–98.

71 Ibid.

5 Aṣ-Ṣadr's Hermeneutical Features

5.1 Axiological Humanism

From aṣ-Ṣadr's perspective, the Islam of a person is in their humanity.[72] By placing humanity at the core of shared religious and ethical values, he dissolves the intellectual dichotomy of rational good and evil vs. religious good and evil. In other words, according to his foundation, that which is not humane is neither religious nor ethical. Although aṣ-Ṣadr's initiative recalls the fundamental principle of the "concomitance of reason and revelation" and also evokes the interpretation of some Shiite jurists that "religious obligations are graces in rational obligations," a notion to which aṣ-Ṣadr also refers,[73] it is clearly an independent initiative from these statements and opinions. Aṣ-Ṣadr believes in the ethical virtue and holds that the virtues that must be elevated in humans are the beautiful names of God. This will distance humans from their inhumane qualities and make them more humane.[74] This is likely a vital appropriation from Ibn ʿArabī.[75] On the other hand, aṣ-Ṣadr believes that the *ḥanīf* religiosity in Q 30:30, the second most cited verse in his works, is the same as the *fiṭra* i.e., creation of humans with divine attributes.[76] In other words, the recognition of human virtues is the only way to recognize divine virtues. The human-centered feature of aṣ-Ṣadr's hermeneutics is also linked to the divine teaching of the names of God to humans in Q 2:31, which has a precedent in mystical interpretations.

5.2 Openness to Modern Sciences

Aṣ-Ṣadr believes in the possibility of the simultaneous truthfulness of multiple interpretations of the Qurʾān. Thus, he views the Qurʾān as a dialogue between God and humanity. This foundation allows him to prioritize his own social understanding over the contemporary societal context of Muḥammad. He states that God was not and is not exclusive to us or any other culture, just as there was no exclusivity in the person of Muḥammad.[77] This foundation also leads aṣ-Ṣadr to have no fear of the advancement of science. He sees progress as the product of continuous interaction between

---

72 Ibid., 3: 50.
73 Ibid., 11: 81.
74 As-Sayyid Mūsā ṣ-Ṣadr, "Falsafat al-Liqāʾ," *imamsadr.net*, accessed on February 20, 2024, https://imamsadr.net/Home/news-details.php/8167.
75 For further reading on this topic, e.g.: Stephen Hirtenstein, "Human Dignity and Divine Chivalry: Rights, Respect, and Toleration According to Ibn 'Arabi," in *Secularization, Desecularization, and Toleration: Cross-Disciplinary Challenges to a Modern Myth*, ed. Vyacheslav Karpov, Manfred Svensson (Cham (Switzerland): Palgrave Macmillan, 2020): 59–79.
76 Aṣ-Ṣadr, *Masīra*, 1: 135.
77 Ibid., 12: 134f.

human beings, nature, and the word of God, and does not limit the meanings of the Qur'ān to any constraint other than the fluid and expansive understanding of the Arabic language (which he considers as an expansive link between its own period and the period of revelation).[78]

5.3 Reinterpreting Worship: Servitude vs. Adoration

Aṣ-Ṣadr understands worship not in the sense of servitude and subjugation, but instead as synonymous with adoration, the center of existence through the manifestation of the names of God.[79] He views worship as a means of communing with God, and acquiring a portion of His beautiful names, which then permeate and enrich human life, resulting in a more humane existence. Consequently, worship encompasses all aspects of social life, while dismantling the superficial and ritualistic frameworks of worship. Aṣ-Ṣadr allegorically interprets the term *falaq* in sura 113, *al-Falaq* (The Daybreak) and *fāliq* in Q 6:96, believing that individuals and society must seek refuge from the darkness of ignorance, death, prejudice, poverty, evil, callousness, and division into the light of knowledge, life, openness, prosperity, goodness, kindness, and unity to remain safe from the dangers created by these. He refers to this process as worship.[80]

In line with this understanding, freedom also finds fresh meaning for aṣ-Ṣadr. He believes that the more a person embodies divine ethics, which are synonymous with human virtues, the closer they come to infinite freedom. Therefore, he interprets worship, in the aforementioned sense, as both freedom and responsibility towards others.[81]

5.4 Religious Pluralism

Aṣ-Ṣadr, initially as a perpetual immigrant and traveler, and eventually as a social leader, found himself compelled not only to assess Islam in relation to humanism and ethical frameworks but also to present an interpretation suitable for coexistence, addressing not only other religions and sects but also Shiism to his audience, particularly those who looked to him for intellectual guidance. Lebanon's urgent need for coexistence drove him toward new interpretations of the Qur'ān, interpretations that

78 Ibid., 1: 295f.
79 Ibid., 11: 30.
80 Ibid., 10: 20; 11: 32.
81 Ibid., 11: 33.

bore no resemblance to any of the traditional exegetical methods known to aṣ-Ṣadr, except for the method of mystical interpretation.[82]

Aṣ-Ṣadr interprets Q 112:3 in a manner that emphasizes the idea that God has no partiality towards any particular people or individuals and no one, whether it be humans, times, places, religions, cultures, etc., is closer to Him than another. The only means of drawing close to God is through the manifestation of His divine names, the same names that coexist with human virtues.[83] For instance, if someone strives to acquire knowledge or, in other words, to draw near to the name of the Omniscient and Wise God, they can achieve closeness to God. It is not the case that followers of a particular religion are metaphorically the children of God as expressed in Q 5:18, while others are not. Aṣ-Ṣadr holds a pluralistic (neither inclusivist nor exclusivist) perspective on religious truth and concludes from Q 2:62 and Q 2:285 that all religions are paths for serving humanity, likened to clear paths and tributaries in Q 5:48.[84] Aṣ-Ṣadr's harmonious discourse with the exegetes of the Middle and Late Shiite periods, including his mentor Ṭabāṭabā'ī, considers Islam in the Qur'ān as a title for all celestial religions, yet aṣ-Ṣadr also draws on a subtle humanistic approach here. When Shiite exegetes speak of Islam as the religion of Abraham, they often conclude by introducing Shiism as the *exclusive* heir of this entire tradition.[85] However, aṣ-Ṣadr's interpretation lacks this orientation and considers each tradition as a valid tributary to truth. He explicitly states that no Christian or Muslim should abandon their heritage, which can be seen as opposition to proselytizing. Interestingly, he raised this point in the early years of his arrival in Lebanon.[86] Additionally, several months before his enforced disappearance, he stated at the opening of the first Amal Congress,

82 The pluralistic approach to religious understanding in aṣ-Ṣadr's work is reflected in such a way that one is reminded of Ibn ʿArabī's poem:

> My heart can take on
> any form
> For gazelles a meadow
> A cloister for monks
> A temple for idols,
> pilgrim's Ka`ba,
> tablets of Torah,
> scrolls of the Qur'ân

cf. Muḥyī d-Dīn b. ʿArabī, *The Translator of Desires*, trans. Michael Sells (Princeton: Princeton University Press, 2021): 39.

83 Aṣ-Ṣadr, *Masīra*, 12: 58; 10: 39.

84 Ibid., 11: 57.

85 Al-Ḥasan b. Sulaymān al-Ḥillī, *Mukhtaṣar Baṣā'ir ad-darajāt* (Najaf: Manshūrāt al-Maṭbaʿa al-Ḥaydariyya fī n-Najaf, 1991), 180.

86 Aṣ-Ṣadr, *Masīra*, 1: 319.

"When I speak of Muslims, I mean both Muslims and Christians, and Islam is not exclusive to us, just as was not exclusive to the Prophet Muhammad."[87]

6 Hermeneutical Achievements of aṣ-Ṣadr

6.1 Human-Centered and Dialogical Theology: A Pluralistic Approach to the Elimination Deprivation

Aṣ-Ṣadr's human-centered hermeneutics did not merely have ethical implications, but rather, his interpretive horizon was oriented towards recognizing the actual diversities in society and engaging in dialogue with them from a perspective of humanistic discourse within theology.[88] Aṣ-Ṣadr equally values unity, dialogue, and diversity, considering their simultaneous existence necessary.[89] He believes that if diversities are eliminated, dialogue will be lost,[90] and if dialogue is eliminated, society will be deprived of the interests and functions of individuals, identities, religions and cultures.[91] He contends that if absolute faith and pursuit of unity are eliminated, selfishness, arrogance, and widespread conflict will ensue.[92]

Here, aṣ-Ṣadr's definition of human beings as creatures fashioned from the names of God comes into play. He defines the effort to elevate deprived and suffering individuals as the subject of unity[93] and considers the true perfection of humanity to be familiarity with others, and the difference between the believing and non-believing community in terms of existing interactions in society to elevate mankind.[94] Aṣ-Ṣadr considers humanity as a unified entity and interprets the word *li-taʿārafū* in Q 49:13 in a way that nothing, not even the most sacred things such as religion and homeland, has the right to deprive anyone of their dignity, let alone lead to racism, sectarianism, and differences in the level of civilization among nations.[95] Aṣ-Ṣadr regards the diversity as various windows to truth[96] or as fruits from different climates,[97] each with its own qualities, useful and irreplaceable in its place.

87 Ibid., 12: 134f.
88 For further reading on this topic, refer to: Rasoul Rasoulipour, "Imam Musa Sadr and Dialogical Religion," *Crosscultural Studies of Religion and Theology* 1 (2023): 2–29.
89 Aṣ-Ṣadr, "Falsafat al-Liqāʾ."
90 Aṣ-Ṣadr, *Masīra*, 10: 350.
91 Ibid., 2: 116.
92 Ibid., 10: 88.
93 Ibid., 5: 194.
94 Ibid., 12: 317.
95 Ibid., 1: 104.
96 Ibid., 2: 116.
97 Aṣ-Ṣadr, "Falsafat al-Liqāʾ."

Aṣ-Ṣadr believes that all human understandings are equal in one thing: their imperfection. This imperfection is the premise for openness and dialogue;[98] hence, for the advancement of society and service to humanity, dialogue must be an ongoing process. He has repeatedly stated that he established the Supreme Shia Council in Lebanon not for sectarianism, but rather to combat sectarianism and represent Shia interests for dialogue aimed at advancing Lebanese society.[99]

6.2 Social Cohesion and Community Building

Aṣ-Ṣadr, in line with his problem-solving approach and leadership position, interprets Qur'ānic injunctions not only as ethical and moral but also as social. When addressing Qur'ānic terminology and legal injunctions, he interprets them not in an essentialist and individualistic manner, but rather as guidelines aimed at the balanced growth of human societies. In fact, he draws inspiration from the Muslim Neoplatonic philosophers, advocating not only for virtuous individuals but for virtuous communities.[100] Aṣ-Ṣadr's social interpretations of legal injunctions diverge from the prevalent literalist and essentialist approaches in jurisprudence and theology. In this regard, he also resolves the conflict regarding legal injunctions between essentialism and teleologism (giving priority to sharia purposes [*maqāṣid ash-sharīʿa*]) by referring to the centrality of civil and social humanity. He does not overlook the methodology and established legal principles of jurisprudence, yet he does not confine himself to the achievements of the past and proposes innovative objectives in the philosophy of legal injunctions.

In this regard, aṣ-Ṣadr views the concept of spending in Q 2:195 as a means to revitalize the latent human talents among the underprivileged for the construction of a society, and considers neglecting it as leading to the collapse of society and the descent of all into the flames of discord and turmoil.[101] He regards jihad as a means to preserve national unity, dignity, and the welfare of the people.[102] He interprets fasting in Q 2:45 as a small exercise to surpass transient individual desires in favor of social interests.[103] He sees *al-qawl as-sadīd* (firm word) in Q 33:70 as a basis for creating possible connections and elevated social relationships.[104] He considers lying as a disruptor of the possibility of dialogue in society,[105] and in Q 49:12, he views backbiting as erasing the individual's social existence and making social participation impossible

98 Aṣ-Ṣadr, *Masīra*, 1: 291f.
99 Ibid., 2: 369.
100 Ibid., 11: 460–70.
101 Ibid., 11: 287.
102 Ibid., 8: 213.
103 Ibid., 10: 362.
104 Aṣ-Ṣadr, *Qulū*, 107.
105 Aṣ-Ṣadr, *Masīra*, 10: 137f.

for building a community.[106] The expression *naffāthāt fī l-ʿuqad* (blowing on knots) in Q 113:4 is seen as fanning the flames of rumors, discord, and mental and psychological entanglements to prevent constructive interaction and calculated movement of society.[107] Furthermore, as will be discussed in the following section, he considers polygyny in Q 4:3 as a conditional social necessity for the guardianship of war orphans, rather than an unconditional ruling.[108] Therefore, aṣ-Ṣadr considers the ultimate purpose of religion to be the improvement of the civil condition of humanity and consequently, human society.[109] He states that Islam is from human, for human, and towards human.[110]

6.3 Revisiting the Role and Status of Women

One of the key aspects of aṣ-Ṣadr's efforts to unravel the complexities of the diverse human society is the elimination of discrimination, including gender-based discrimination. This issue can be addressed through the establishment and empowerment of various institutions, some of which continue to exist today. The endeavor to eliminate gender-based discrimination is also a common theoretical concern among jurisprudence, interpretation, and society, which aṣ-Ṣadr has been involved in throughout his presence in Lebanon, engaging in discussions with Sunni scholars and even Christians. This issue encompasses matters such as civil marriage,[111] polygyny, temporary marriage, and paternal intervention in the marriage of daughters. Each of these questions is directly related to the interpretation of Qurʾānic verses, on the one hand, to the jurisprudential methodology, and on the other hand, to the culture and institutions of the pluralistic society under the leadership of aṣ-Ṣadr and the surrounding communities. In all these cases, aṣ-Ṣadr distances himself from the juridical consensus and traditional interpretative approaches, presenting fresh interpretations that clearly reflect his position as a social leader. Aṣ-Ṣadr consistently opposes civil marriage and, fundamentally, any removal of religious characteristics from personal sta-

106 Ibid., 11: 83.
107 Ibid., 10: 16.
108 Ibid., 3: 271–341.
109 Ibid., 2: 490.
110 Ibid., 11: 392f.
111 Although aṣ-Ṣadr always opposes the legalization of marriage between religions in Lebanon, he personally has read the marriage sermon between a Muslim woman and a Christian man in a few instances. Such marriages are considered impossible from the perspective of Islamic jurisprudence, including Shiite scholars, to this day. It seems that the jurisprudential basis of aṣ-Ṣadr's action can be traced back to his development of the concept of Islam in his mind. He considers the Qurʾānic term *muslim* to include followers of both Muḥammad and Jesus. These examples also demonstrate the distinction between individual agency and legislative authority in aṣ-Ṣadr's mind. e.g., aṣ-Ṣadr, *Masīra*, 12: 119.

tus laws.[112] Contrary to the consensus of all Islamic jurisprudential schools, he considers polygyny permissible only under exceptional circumstances, particularly when the relative abundance of orphans in society and justice towards female spouses are highlighted in Q 4:3.[113] He aligns with the prevailing Shiite jurisprudential trend by opposing the permanence of the prohibition of temporary marriage, which was permissible in early Islam as mentioned in Q 4:24,[114] and, based on Q 4:19, he views the role of a father in choosing a spouse for his daughter as advisory rather than obligatory.[115]

Drawing on the story of burying alive daughters in the pre-Islamic era and an allegorical interpretation of the phrase *qatl al-mawʾūda* (killing the little girl) in Q 81:8, he concludes that depriving daughters of leaving the house is a repetition of the experience of the pre-Islamic practice of killing daughters.[116] Furthermore, he contends that loving women, as attributed to Muḥammad in a hadith, signifies paying attention to women to empower them in society.[117] Aṣ-Ṣadr, in an interview, asserts that Islam has recognized women's rights, but Muslims have degraded women. When the interviewer expresses surprise at aṣ-Ṣadr's views and asks if he does not believe that the biggest obstacle to implementing these ideas are Muslims themselves, aṣ-Ṣadr responds, "I am aware of this issue, but God is greater."[118]

6.4 Identity

The belief in the diverse social identity of humans leads aṣ-Ṣadr to a rejection of the dissolution of diversity and the imposition of singular solutions, as well as a refusal to dissolve the individual within society. While denouncing sectarianism and sectarian political systems as absolute evils, aṣ-Ṣadr regards religious traditions and their diversity as sources of common good[119] and distinguishes between religious bias and religious faith, deeming the former as objectionable and the latter as necessary, both on an individual and societal level.[120] It is for this reason that aṣ-Ṣadr strongly opposes the imposition of secularism, understood as the eradication of diverse religious practi-

112 As-Sayyid Mūsā aṣ-Ṣadr, *Kalimatun huwa qāʾiluhā* (Beirut: Markaz al-Imām Mūsā ṣ-Ṣadr li-l-Abḥāth wa-d-Dirāsāt, 2019), 58.
113 Aṣ-Ṣadr, *Masīra*, 3: 271–341.
114 Ibid., 72.
115 Ibid., 1: 412.
116 Ibid., 11: 205.
117 Ibid., 5: 139.
118 Ibid., 2: 76.
119 Ibid., 3: 76.
120 Ibid., 12: 175.

ces, especially in the realm of civil law and personal status, believing it to be a form of bias.[121]

Aṣ-Ṣadr believes that governance should be heavenly rather than religious. By heavenly, he does not mean a reference to religious description. Rather, he means governance should be impartial, akin to the canopy of the sky and should be able to simultaneously support the majority and the minority.[122] The Arabic word *samā'* (heaven/sky) is derived from the root *S-M-W*, meaning to rise above, and aṣ-Ṣadr believes that we should overlook any characteristic other than humanity, and nothing, including religion, should be a source of discrimination or a barrier to human dialogue. For if even one person cannot share their unique divine attributes with society, it is the society that suffers the most damage.[123]

6.5 Defense and Struggle for Homeland

Aṣ-Ṣadr, in 1975/1395, while ascending to a transreligious and transnational social leadership, was confronted with the complex phenomenon of civil war. The multifaceted civil war in Lebanon had the potential to undermine all of aṣ-Ṣadr's theoretical and practical achievements. Following the historic defeat of the Arabs in the 1967/1387 war and the escalation of Israeli territorial aggressions into southern Lebanon at the beginning of the 1970s, aṣ-Ṣadr became aware of abstract concepts such as homeland and future. From the early 1970s, in his public speeches, he encouraged people to purchase and learn to use weapons.[124] His argument was that there was no comprehensive and capable government to protect southern Lebanon. This is despite the fact that in the first ten years of his presence in Lebanon, he had not even once talk about weapons. However, the concept of homeland drew him as a political and social leader towards new interpretations of the Qur'ān. For the first time in the early 1970s, he interpreted the word *zīna* (adornment) in Q 7:31 as a weapon, citing a tradition attributed to the Prophet, saying, "The weapon is the adornment of men." He repeated this interpretation in the following years and considered the purchase and learning to use weapons as equivalent to buying the Qur'ān and performing prayers.[125] Aṣ-Ṣadr, while encouraging armed resistance against occupying forces, vehemently opposed the onset of the civil war in 1975/1395. Just two months after the start of this war, he went on a hunger strike in a mosque in Beirut, declaring that he would not end his strike until the war was stopped.[126] Such a hunger strike is considered in the sharia as

121 Ibid., 2: 528–30.
122 Ibid., 2: 239f.
123 Ibid., 10: 6f.
124 Ibid., 2: 277.
125 Ibid., 5: 40.
126 Ibid., 6: 97–101.

a form of suicide, and is not permissible according to traditional jurists. However, aṣ-Ṣadr, as a mujtahid, considered it permissible for himself. Aṣ-Ṣadr states, "I want to prove with my hunger strike that there is a louder voice than the sound of bullets."[127] Ultimately, when aṣ-Ṣadr heard that some Shiites have surrounded several Christian villages, he broke his hunger strike and personally, along with several Christian officials, moved towards those villages, breaking the siege. In a television interview, he called the inhabitants of those Christian villages his own children.[128] Aṣ-Ṣadr's call to buy weapons for the defense of Lebanon's soil against external enemies and then his hunger strike to end the civil war demonstrate that the concept of homeland has gradually dominated all his hermeneutics, despite its diversities and divisions.

7 Conclusion

The Qur'ānic hermeneutics of a young aṣ-Ṣadr begins in Qom with the primary concern of deducing sharia. His hermeneutic develops gradually through dialogue with carriers of Qur'ānic approaches in the fields of culture, religion, economy, and politics, as a result of his two migrations to Iraq and Lebanon and his encounters with diverse communities. What supports aṣ-Ṣadr throughout his experience of assimilating pluralism and addressing his audience is the established mystical approach in the views and actions of his forefathers and some of his teachers. Each time, aṣ-Ṣadr, in response to his diverse audience, utilizes a pre-established interpretive approach to serve the purpose of alleviating human suffering. As a leader of an emerging diverse community, he endeavored to interpret the Qur'ān in a transreligious manner or in a way that he himself called heavenly, meaning anti-discriminatory.[129]

He believed that if such diversity did not exist in Lebanon, the Lebanese would have to create it to prove that coexistence between religions, and between East and West is possible.[130] At the peak of his successes in institution-building, dialogue, and the attainment of a new methodology in understanding religion, as a social leader, aṣ-Ṣadr suddenly found himself confronted with the immensely complex issue of civil war. In an effort to end the war, he resorted to strikes and sought to utilize all his accumulated knowledge and political acumen to resolve this problem that had gripped Lebanon. Finally, during a journey, one of his methods for increasing dialogue, promoting pluralism and solving homeland issues suddenly disappeared. To

127 Ibid., 6: 99.
128 Ibid., 6: 21.
129 As-Sayyid Mūsā aṣ-Ṣadr, *Qāba qawsayni aw adnā* (Beirut: Markaz al-Imām Mūsā ṣ-Ṣadr li-l-Abḥāth wa-d-Dirāsāt, 2023), 82.
130 Aṣ-Ṣadr, *Masīra*, 7: 131.

this day, there is no credible news about what happened to him after his enforced disappearance.

The significance of aṣ-Ṣadr's interpretation of Islam in the context of Lebanon and the increasingly multicultural world is hidden in a path that he paved for active, problem-solving, human-centered, and faith-based participation of Muslims. As an established leader, aṣ-Ṣadr utilized all his hermeneutical insights to open up a horizon beyond the boundaries between science and religion, reason and sharia, ethics and politics, East and West, Muslims and Christians, tradition and modernity, past and future, war and peace, love and struggle, resistance and dialogue, and innovation and authenticity. The horizon he sought to open was a human-centered horizon for the utilization of religious traditions and cultures towards alleviating human suffering.

8 Bibliography

8.1 Sources

Ganjī, ʿAlī. *Ayāme Taḥṣīl: Rawāyatī ʾaz Dwrāne Ṭalabegīe Sayyid Mūsā Ṣadr*. Tehran: Imam Moussa Sadr Cultural & Research Institute, 2023.

Al-Ḥillī, Al-Ḥasan b. Sulaymān. *Mukhtaṣar Baṣāʾir ad-darajāt*. Najaf: Manshūrāt al-Maṭbaʿa al-Ḥaydariyya fī an-Najaf, 1991.

Ibn ʿArabī, Muḥyī d-Dīn. *The Translator of Desires*, trans. Michael Sells. Princeton: Princeton University Press, 2021.

Kamāliyān, Moḥsen. *Mashāhir-i Khāndān-i Ṣadr*. Vol. 18. Qom: Rāze Neī, 2007.

Al-Khūʾī, as-Sayyid Abū l-Qāsim. *Al-Bayān fī tafsīr al-Qurʾān*. Dār ath-Thaqalayn, 2007.

Al-Kulaynī, Muḥammad b. Yaʿqūb. *Al-ʾÚsūl mīn al-Kāfī*. Ed. ʿAlī Akbar al-Ghaffārī, 8 vols. Qom: Dār al-Kutub al-Islāmiyya, 1986.

Rūmī, Jalāl ad-Dīn. *The Masnavi*, trans. Mojaddedi Jawid. Oxford: Oxford University Press, 2013.

Aṣ-Ṣadr, as-Sayyid Ṣadr ad-Dīn. *Al-Mahdī*. Beirut: Dār az-Zahrāʾ, 1978.

Aṣ-Ṣadr, as-Sayyid Mūsā. *Aḥādīth as-saḥar*. Beirut: Markaz al-Imām Mūsā ṣ-Ṣadr li-l-Abḥāth wa-d-Dirāsāt, 1999.

Aṣ-Ṣadr, as-Sayyid Mūsā. *Dirāsāt li-l-ḥayā*. Beirut: Markaz al-Imām Mūsā ṣ-Ṣadr li-l-Abḥāth wa-d-Dirāsāt, 1999.

Aṣ-Ṣadr, as-Sayyid Mūsā. *Nāy wa Ney: Justārhāyī darbāre-yi eslām, ensān, lubnān wa-muqāwamat*. Ed. ʿAlī Ḥujjatī Kermanī. Tehran: Imam Moussa Sadr Cultural & Research Institute, 2004.

Aṣ-Ṣadr, as-Sayyid Mūsā. "Az Pedaram Beguyam." *Ruznāme-ye Etelāʿāt* 24654 (2010): 6.

Aṣ-Ṣadr, as-Sayyid Mūsā. *Masīrat al-Imām as-Sayyid Mūsā ṣ-Ṣadr*. Ed. Yaʿqūb Ḥasan Ḍāhir, 12 vols. Beirut: Dār Bilāl, 2014.

Aṣ-Ṣadr, as-Sayyid Mūsā. *Qulū lā ilāha illā llāh tufliḥū*. Beirut: Markaz al-Imām Mūsā ṣ-Ṣadr li-l-Abḥāth wa-d-Dirāsāt, 2016.

Aṣ-Ṣadr, as-Sayyid Mūsā. *Kalimatun huwa qāʾiluhā*. Beirut: Markaz al-Imām Mūsā ṣ-Ṣadr li-l-Abḥāth wa-d-Dirāsāt, 2019.

Aṣ-Ṣadr, as-Sayyid Mūsā. *Qāba qawsayni aw adnā*. Beirut: Markaz al-Imām Mūsā ṣ-Ṣadr li-l-Abḥāth wa-d-Dirāsāt, 2023.

Aṣ-Ṣadr, as-Sayyid Mūsā. "*Falsafat al-Liqāʾ*." *imamsadr.net*. Accessed on February 20, 2024. https://imamsadr.net/Home/news-details.php/8167.

Aṣ-Ṣaduq, Muḥammad b. ʿAlī. *Kitāb al-Khiṣāl*. Qom: Jamāʿat al-Mudarrisīn fī l-Hawza al-ʿIlmiyya fī Qum al-Muqaddasa, 1983.
Aṭ-Ṭabāṭabāʾī, as-Sayyid Muḥammad Ḥusayn. *Al-Mīzān fī tafsīr al-Qurʾān*, 20 vols. Qom: Jamāʿat al-Mudarrisīn fī l-Ḥawza al-ʿIlmiyya fī Qum al-Muqaddasa, n.d.

8.2 Secondary Literature

Amirpur, Katajun. "The Changing Approach to the Text: Iranian Scholars and the Quran." *Middle Eastern Studies* 41, no. 3 (2005): 337–50.
Chittick, William C. *The Sufi Path of Knowledge: Ibn al-'Arabi's Metaphysics of Imagination*. State University of New York Press, 1989.
Chittick, William C. "Waḥdat al-Wujūd in India." *Ishraq: Islamic Philosophy Yearbook* 3 (2012): 29–40.
Culp, John. "Panentheism." In *The Stanford Encyclopedia of Philosophy* (Fall 2023 Edition), edited by Edward N. Zalta and Uri Nodelman. Accessed February 19, 2024. https://plato.stanford.edu/archives/fall2023/entries/panentheism/.
Hirtenstein, Stephen. "Human Dignity and Divine Chivalry: Rights, Respect, and Toleration According to Ibn 'Arabi." In *Secularization, Desecularization, and Toleration: Cross-Disciplinary Challenges to a Modern Myth*, edited by Vyacheslav Karpov, Manfred Svensson. Cham (Switzerland): Palgrave Macmillan, 2020: 59–79.
Kramer, Martin. "The Oracle of Hizbullah: Sayyid Muhammad Husayn Fadlallah." In *Spokesmen for the Despised: Fundamentalist Leaders in the Middle East*, edited by R. Scott Appleby. Chicago: University of Chicago Press, 1997: 83–181.
Mallat, Chibli. *The Renewal of Islamic Law: Muhammad Baqer as-Sadr, Najaf and the Shiʿi International*. Cambridge: Cambridge University Press, 1993.
March, Andrew F. "Taking People as they are: Islam as a "Realistic Utopia" in the Political Theory of Sayyid Qutb." *American Political Science Review* 104 (2010): 189–207.
Milani, Milad. *The Nature of Sufism: An Ontological Reading of the Mystical in Islam*. Routledge, 2022.
Pakatchi, Ahmad. "Āyāt al-aḥkām." In *Encyclopaedia Islamica*, edited by Farhad Daftary. Brill, 2015. Accessed on February 19, 2024. https://referenceworks.brillonline.com/entries/encyclopaedia-islamica/ayat-al-ahkam-COM_0319.
Rahnema, Ali. *Shiʿi Reformation in Iran: The Life and Theology of Shariʿat Sangelaji*. New York: Routledge, 2015.
Rasoulipour, Rasoul. "Imam Musa Sadr and Dialogical Religion." In *Crosscultural Studies of Religion and Theology* 1 (2023): 2–29.
Rustom, Mohammed. "Is Ibn al-ʿArabī's Ontology Pantheistic?" *Journal of Islamic Philosophy* 2 (2006): 53–67.
Saʿīdī, Farīdeh. "Ṣadr, Khāndān." In *Dāneshnāmeh-ye Jahān-e Eslām*, edited by Gholām-ʿAlī Haddād-ʿĀdel. Vol. 29. Tehran: Dāʾerat al Maʿāref-i Bozorg-i Eslāmī, 2021: 419–68.
Wielandt, Rotraud. "Main Trends of Islamic Theological Thought from the Late Nineteenth Century to Present Times." In *The Oxford Handbook of Islamic Theology*, edited by Sabine Schmidtke. Oxford: Oxford University Press, 2016: 707–64.

Taraneh R. Wilkinson

Süleyman Ateş

1 Introduction

Süleyman Ateş is a Turkish scholar of theology and public figure specializing in *tafsīr*, i.e. Qurʾānic exegesis. He is the author of a Turkish interpretation/translation of the Qurʾān, a major commentary on the Qurʾān, an encyclopedia of the Qurʾān, as well as various shorter exegetical works, articles, and other publications on topics related to the Qurʾān. He is widely known for his twelve-volume *tafsīr*, *Yüce Kur'an'ın Çağdaş Tefsiri* (A Contemporary Interpretation of the Sublime Qurʾān) and its subsequent abridgments. Trained initially according to madrasa-style Islamic education, he then pursued a career in the state-funded Turkish University İlahiyat faculty system and in the Turkish Diyanet, the government Directorate for Religious Affairs. As an academic and public figure, his method comprises both classical and modernist dimensions, boasting extensive and detailed engagement with a wide variety of classical *tafsīr* works, including Sufi *tafsīr* and modernist *tafsīr*, limited engagement with Orientalist commentary, and even comparisons with excerpts of Jewish and Christian scriptures. Despite his extensive classical training, he is generally considered to align with modernists in their approach to *tafsīr* and relies heavily on modernist sources. His hermeneutical approach stresses the centrality of monotheism (*tawḥīd*) and moral development in an attempt to reclaim the message of the Qurʾān through updated language that is both accessible to his contemporary Turkish readers and compatible with a modern scientific worldview. His influence in Turkish society saw a peak roughly parallel to the peak in the influence of Turkish İlahiyat faculties on Turkish Islam seen in the late 1990s. Turkish scholarly interest in his legacy, though often critical,[1] has seen some increase since the 2010s.

1 E.g., Ahmet Yazıcı, "Süleyman Ateş'in Görüşleri Bağlamında Kur'an Kıssaları," *İslâmî Araştırmalar* 22, no. 2 (2011): 106–17; Mustafa Hocaoğlu, "Süleyman Ateş'in Tefsirinde Bütünlük Sorunu: Ehli Kitab'ın Cennet'e Girmesi ve Tevrât'ın Tahrifi Meseleleri," *İslâmî Araştırmalar* 23, no. 1 (2012): 42–54; Gökhan Atmaca, "Süleyman Ateş'in Bakara Sûresi 62. Âyeti Tefsiri Üzerine Bir Değerlendirme," *Sakarya Üniversitesi İlahiyat Fakültesi Dergisi* 15, no. 28 (2013): 1–20; Mehmet Kaya, "'Kur'an Yolu' Tefsirinde Süleyman Ateş'e Yönelik Bazı Eleştirilere Dair Bir İnceleme," *Türk İslâm Medeniyeti Akademik Araştırmalar Dergisi* 13, no. 25 (2018): 123–48; Yunus Emre Gördük, "Elmalılı Hamdi Hz. Âdem'in Evrimle Yaratıldığını Mı Söylüyor?," *Süleyman Demirel Üniversitesi İlahiyat Fakültesi Dergisi* 42 (2019): 27–42.

https://doi.org/10.1515/9783111320069-014

2 Life, Works, Political and Intellectual Context

Born in Elazığ in 1933/1352[2] with the original surname Ataş, Süleyman Ateş, by his own account, memorized the Qur'ān in his home village of Tadım beginning at age five, becoming a *ḥāfiẓ* at about eight or nine years of age.[3] Ateş initially pursued madrasa-style education rather than attending a public school, studying with several members of the local ulema in Elazığ and Erzurum. In Elazığ, Hacı Muharrem Sırrı Hilmî Efendi taught him and served as his guide into the Naqshbandiyya Sufi order.[4] In Erzurum, he studied with Hacı Faruk Bey as well as the then mufti of Erzurum, Solakzade Sadık Efendi. After passing the state exam to obtain his elementary school diploma at the age of twenty, he then enrolled in the Elazığ İmam-Hatip high school in 1953/1372, graduating in 1960/1379. Although he was aware of the then newly opening Islamic Higher Institutions, which were more religiously conservative than the Ankara İlahiyat Faculty, he opted to pursue university studies. In order to go on to study at a state university he was required to pass the state high school exit exams and, in a few short months, he obtained his diploma from Elazığ Lisesi, earning a scholarship to study *tafsīr* at the Ankara University İlahiyat Faculty. During his undergraduate studies in Ankara, he also served as a state-sanctioned imam. His early education was marked by a tension and synthesis between tradition and modernizing state institutions that would continue to define his later work.

After finishing his initial studies at the Ankara University İlahiyat Faculty in 1964/1384, he returned in 1965/1385 to pursue doctoral studies. In 1968/1388 he completed his doctoral thesis on the Sufi *tafsīr* of Abū ʿAbd ar-Raḥmān as-Sulamī under the direction of M. Tayyip Okiç (1902–1977/1320–1397), an influential Bosnian scholar who migrated to Türkiye in 1945/1364 and founded the *tafsīr* and hadith departments at the Ankara University İlahiyat Faculty after being appointed a professorship there in 1950. Ateş also fostered rapport and worked with Hilmi Ziya Ülken (1901–1974/1319–1394) during his time as an ilahiyat student.[5] He then left the faculty to complete mandatory military service from 1969/1389 to 1971/1391. Upon his return to teaching duties

2 1938 is the date on his official birth registration. In a 1997 interview Ateş explains that his official birth date had been 1935 due to the delay in reporting and Ateş later changed the official date from 1935 to 1938 to be able to enroll in public school: Süleyman Ateş, "Prof. Dr. Süleyman Ateş ile Söyleşi," *Kur'ân Mesajı* 1, no. 1 (1997): 66.

3 Mustafa Öcal, *Diyanet İşleri Başkanları ve Hizmetleri* (Istanbul: Ensar, 2016), 359. Other accounts relate he had memorized the Qur'ān by the age of ten, e.g., Abdullah Takim, "The Turkish Qur'ānic Exegete Süleyman Ateş and His Contemporary Commentary of the Qur'ān," in *Unity and Diversity in Contemporary Muslim Thought*, ed. Abbas Poya and Farid Suleiman (Newcastle upon Tyne: Cambridge Scholars Publishing, 2017): 14.

4 Abdullah Takim, "The Turkish Qur'ānic Exegete Süleyman Ateş and His Contemporary Commentary of the Qur'ān," in *Unity and Diversity in Contemporary Muslim Thought*, ed. Abbas Poya and Farid Suleiman (Newcastle upon Tyne: Cambridge Scholars Publishing, 2017), 15.

5 Öcal, *Diyanet İşleri Başkanları ve Hizmetleri*, 377; Takim, "The Turkish Qur'ānic Exegete," 15.

at the faculty, he took up the study of English. Ateş did not stay in Ankara long before traveling to Iraq and Egypt in 1973/1393. Ateş went on to complete a second thesis on the Allegorical (*İşari*) School of *tafsīr*, i.e. Sufi *tafsīr* that looks for hidden inner meanings in the Qur'ān, before attaining to the equivalent of associate professor (*doçent*) in 1973/1393. From 1976/1396 to 1978/1398 Ateş served as Head of Religious Affairs (Diyanet İşler Başkanlığı). His relationship with the Diyanet has since had both positive and negative dimensions. Ateş praised the institution in an interview published in the journal *İslamiyat* in 2001/1421 and received a recognition from the Diyanet for his contributions to the *TDV İslam Ansiklopedisi* in 2014/1435.[6] However, the official Diyanet *Kur'an Yolu Türkçe Meâl ve Tefsir*, a Qur'ān translation and commentary published in 2016/1437, heavily criticized many of Ateş's positions in his major *tafsīr*.[7]

In 1979/1399 he became a full professor at the Ankara İlahiyat Faculty, after which he spent a period abroad, as a Guest Professor and German language student at Ruhr-University Bochum, and then taught in the Arab world, first at Imam Muhammad Ibn Saud Islamic University in Riyadh (1980–1982/1400–1402, 1984–1987/1404–1407) and subsequently at Emir Abdelkader University of Islamic Sciences in Algeria (1987–1988/1407–1408). During this time Ateş was working on his twelve-volume magnum opus *Yüce Kur'an'ın Çağdaş Tefsiri*, the first volumes of which appeared in 1988/1408. Upon his return to Türkiye, he did not resume a position at the Ankara University İlahiyat Faculty; instead, he taught at various ilahiyat faculties, first serving as the head of the Department of Fundamental Islamic Sciences at Samsun Ondokuz Mayıs University İlahiyat Faculty (1988–1995/1408–1415), then as faculty member at Istanbul University İlahiyat Faculty from 1995–1999/1415–1419, including some teaching at Marmara University İlahiyat Faculty, across the Bosphorus, in 1996/1416. His move to teach at other ilahiyat faculties in Türkiye occurred during a period when the Turkish government was opening new ilahiyat faculties, expanding the total from ten in 1982/1402 (most of which were newly converted into ilahiyat faculties as a result of the Constitution of 1982/1402) to twenty-four ilahiyat faculties by 1997/1417. From 1997/1417 to 2000/1420 he published a journal intended for a popular audience called *Kur'an Mesajı*, i.e. the Message of the Qur'ān.

Ateş officially retired in 1999/1419 but continued to remain active as a scholar and public figure, even traveling to the Netherlands (2001–2002/1422–1423) to teach at the Islamic University of Rotterdam and serve as a founding member at the now defunct Islamic University of Europe. As a public figure, Ateş wrote a daily column for the Turkish newspaper *Vatan Gazetesi*, from the paper's inception in 2002 to 2011/1422–1432. He has also created a personal website, www.suleymanates.com. In recent years, despite an overall decline in influence, his work has received increasing attention

6 TDV İslâm Ansiklopedisi, "SÜLEYMAN ATEŞ Müellifin özgeçmişi [Author's Resume]," last modified January 31, 2018, https://islamansiklopedisi.org.tr/muellif/suleyman-ates.
7 Kaya, "'Kur'an Yolu'."

from scholars and graduate students wishing to assess his contributions and legacy, from the 2000s onward.[8]

Süleyman Ateş has penned dozens of books and many more articles, both scholarly and popular. Several milestone publications dominate his academic trajectory and inform his more popular works. He published his doctoral thesis as *Sülemî ve Tasavvufî Tefsîri* (As-Sulamī and His Sufi Commentary on the Qur'ān) in 1969/1389. Five years later his second monograph *İşârî Tefsir Okulu,* (The Allegorical School of Qur'ānic Interpretation), came out in 1974/1394 and was later translated into Persian.[9] By 1975/1395 he had published an annotated Turkish translation of the Qur'ān, *Kur'ân-ı Kerîm ve Yüce Meâli,* republished in 1977/1397 with the Arabic text. In 1975/1395, the Turkish government accepted the translation for official use in İmam-Hatip high schools. Like his later *tafsīr,* this translation of the Qur'ān contains explanations and footnotes that attempt to integrate modern scientific knowledge about the world into the interpretive apparatus.[10]

His major *tafsīr* work *Yüce Kur'ân'ın Çağdaş Tefsîri* (1988–1992/1408–1412) established his authority as a Qur'ānic exegete, combining both classical sources and methods with some modern and scientific considerations. It has also received some criticism from religious authorities. Without contest this is his most important work, the culmination of twenty years of research and the first *tafsīr* to be completed by a Turkish academic in modern Türkiye.[11] Due to the success of his major *tafsīr,* it has since been abridged and reprinted in two major editions. The first abridgment was the six-volume *Kur'ân-ı Kerîm Tefsîri* in 1995/1415 printed by Milliyet. This six-volume abridgment was endorsed by a host of prominent ilahiyat professors of the late twentieth century, namely M. Sait Yazıcıoğlu, Mehmet S. Aydın, M. Hayri Kırbaşoğlu, including the then-current Head of Religious Affairs Mehmet Nuri Yılmaz, as well as two academics outside of ilahiyat faculties. These fifteen glowing reviews printed at the start of the work[12] epitomize the success of his *tafsīr* as it was received in the 1990s. A further abridgment of his work appeared in 2013 as a three-volume *Kur'ân-ı Kerîm Tefsîri*

8 For annotated bibliography and summary of graduate theses on Ateş up to 2014, see Ahmet Yazıcı, *Türkçe Tefsir Çalışmalarında Klasik-Modern Algısı: Ömer Nasuhi Bilmen-Süleyman Ateş Örneği* (Istanbul: Ravza Yayınları, 2014), 117f. There have been at least half a dozen graduate theses undertaken by ilahiyat students since 2014.

9 Takim, "The Turkish Qur'ānic Exegete," 16.

10 Cf. Mehmet Yolcu, Ayşe Tokay, and Ayşenur Bıçak, "Süleyman Ateş ve Abdullah Parlıyan'ın Meallerinin Karşılaştırılması," *e-Şarkiyat İlmi Araştırmalar Dergisi* 10, no. 4 (2018): 1328–53.

11 İsmail Cerrahoğlu and Talat Koçyiğit began jointly publishing a *tafsīr, Kur'an-ı Kerim Meal ve Tefsir,* in 1984, approximately four years before Süleyman Ateş began publishing his *tafsīr*; however, the former's publication was only completed by Koçyiğit in 2011 (Yazıcı, *Türkçe Tefsir,* fn. 108). A non-academic *tafsīr* was also published prior to Ateş's magnum opus which attempted to address in some measure relationship of the Qur'ān to contemporary scientific worldviews, Celal Yıldırım's (b. 1932/1350) *İlmin Işığında Asrın Kur'an Tefsiri* (1986).

12 Süleyman Ateş, *Kur'ân-ı Kerîm Tefsîri* (Istanbul: Milliyet, 1995), 1: 5–16.

through the publisher Yeni Ufuklar Neşriyat, a publishing house he has worked with for decades.

A final major work that Ateş completed in the wake of his *tafsīr* is his thirty-volume *Kur'an Ansiklopedisi* (1997–2003). This encyclopedia of the Qur'ān offers an alphabetically organized, term-by-term, topic-by-topic thematic guide to matters related to the Qur'ān and its exegesis. In this work, Ateş produces the relevant Qur'ānic verses according to their order of revelation under each topic heading, offers extant interpretations, and then concludes with his own views.

Some lesser works that contain Qur'ānic exegesis, summaries of his views as an exegete, or responses to criticisms of his major *tafsīr* include *İslâm'a İtirazlar, Kur'ân-ı Kerîm'den Cevaplar* (Objections to Islam, Answers from the Holy Qur'ān) (1966), *Kur'an Nizâmı* (The Order of the Qur'ān) (1979), *Kur'an-ı Kerim'in Evrensel Mesajına Çağrı* (A Call to the Universal Message of the Holy Qur'ān) (1990), *Gerçek Din Bu*, vols. 1–2 (This is True Religion) (1992), *Kur'an Prensipleri* (Principles of the Qur'ān) (1996), *Kur'an'da Nesh Meselesi* (Abrogation in the Qur'ān) (1996), *Yeni'den İslam'a* (Approaching Islam Anew), vols. 1–2 (1996/7), *99 Soruda Kur'ân-ı Kerîm* (The Holy Qur'ān in 99 Questions) (1997), *Namazda Okunan Sûrelerin Tefsiri* (A Commentary of Suras Read During Prayer) (1997), *Kur'ân'da Pergamberler Tarihi* (A History of Prophets in the Qur'ān) (2002).[13] Most of these monographs and booklets tend to address a broader, non-academic audience.

Ateş has undertaken translations from English and Arabic, including works by Toshihiko Izutsu (1914–1993/1332–1413), Sayyid Quṭb (1906–1966/1324–1386), and Abū 'Abd ar-Raḥmān as-Sulamī (937–1021/325–412). Additionally, Ateş is author of an autobiography *Bir Ömür Böyle Geçti* (A Life Went by like this) (2007). Some of his more influential articles will be discussed below.

3 Analysis of the Major Hermeneutical Works and Hermeneutical Method of Süleyman Ateş

With religious training in madrasa-style settings as well as state-funded institutions, as a *mufassir* Ateş makes use of both classical as well as more recent approaches to interpreting the Qur'ān. Despite his classical training and several stints spent in the Arab world, Turkish contemporaries tend to view his method as leaning more toward modernist than classical.[14] However, in an effort to distinguish his work from mere capitulation to Orientalist criticisms of Muslim exegetical practices, some of his more appreciative readers stress that his aim is not "to incorporate Islam into the modern era" but

13 For a more exhaustive list of Ateş's work and translations see Yazıcı, *Türkçe Tefsir*, 99–108.
14 Yazıcı, *Türkçe Tefsir*, 98.

"rather to islamicise modernity to a certain extent through the Qurʾān."[15] This modernist tendency in Ateş's *tafsīr* works is largely tied to the influence of his doctoral advisor M. Tayyip Okiç, who, himself a mixture of madrasa-style and European university pedagogy, exhibits tendencies popularized in the modernist[16] Islamic reform movements of the early twentieth century. Like his former mentor Okiç, Ateş stands under the influence of Muḥammad ʿAbduh (1849–1905/1266–1323) and M. Rashīd Riḍā (1865–1935/1282–1354), focusing strongly on the Qurʾān's ethical role and spiritual guidance to its reader.[17] Praised by many of his contemporaries for his erudition, he has nevertheless received criticism for some of his departures from classical positions.

His major *tafsīr*, *Yüce Kur'ân'ın Çağdaş Tefsîri,* is written in relatively accessible language, explicitly composed for the new generations of Turkish Muslims whose language was increasingly removed from the Ottoman of their forefathers.[18] It is thought that Ateş had his intended readers in mind when he called his *tafsīr* a so-called "contemporary" *tafsīr*, i.e. a *tafsīr* intelligible and accessible to Ateş's younger Turkish Muslim contemporaries.[19] His *tafsīr* continues to be hailed as one of the most significant efforts of its kind in Türkiye to apply modern scientific approaches to exegesis,[20] and in this spirit it "rejects superstition and advocates a Qurʾānic enlightenment."[21]

In terms of general layout and structure, his twelve-volume *tafsīr* opens with a fifty-five pages long introduction, explaining the author's methods and aims along with a general overview of Qurʾānic exegetical sciences. His commentary then follows the official ordering of suras. He first offers the Arabic text, then his Turkish translation, and proceeds to commentate. As he comments on the verse clusters he has just presented and translated, Ateş explains the context, including alternate sura names, relevant etymologies and the relation of individual verses to preceding and subsequent verses, any relevant vocabulary and etymology of key words, and then concludes each sura analysis with a summary of the general principles and lessons of the sura. After the final summary he also offers notes on the *qirāʾāt*, or alternate Qurʾānic readings, at the end of each sura. His commentary is occasionally interspersed with charts and science-related visuals. Most notably, his commentary on Q 75 features an anatomical visual of the various stages of development of the human zygote into a fetus.[22]

---

15 Takim, "The Turkish Qurʾānic Exegete," 23.

16 I use the term "modernist" with some reservation, since in the Turkish context scholars may use more specific terminology to designate the influence of modernist thinkers or trends, such as "the Manar school" (Manar okulu) or "the school of sociological *tafsīr*" (içtimai tefsir ekolü), e.g., Yazıcı, *Türkçe Tefsir*, 109, fn. 98.

17 Ibid., 109.

18 Süleyman Ateş, *Yüce Kur'ân'ın Çağdaş Tefsîri* (Istanbul: Yeni Ufuklar Neşriyat, 1988), 1: 59f.

19 Yazıcı, *Türkçe Tefsir*, 111.

20 Yolcu, Tokay, Bıçak, "Süleyman Ateş," 1335; Yazıcı, *Türkçe Tefsir*, 220.

21 Takim, "The Turkish Qurʾānic Exegete," 18.

22 Ateş, *Yüce Kur'ân'ın Çağdaş Tefsîri*, 10: 207–31.

His commentary prioritizes explaining Qur'ānic verses with other Qur'ānic verses, but also makes a point to bring up relevant *asbāb an-nuzūl* (occasions of revelation) drawing extensively on hadith. Ateş also lays out relevant disputed interpretations and debates among the ulema, and although he draws heavily on classical sources, he freely criticizes his sources and at times actively departs from them. He also compares what he deems relevant Qur'ānic verses to material in the Jewish Torah and Christian Gospels, including the spurious gospel of Barnabas.[23] Occasionally, he even uses poetry to elaborate complex principles, for example citing verses by Erzurumlu İbrahim Hakkı (1703–1780/1115–1194).[24] Footnotes are used to provide references to other *tafsīr* works, other Qur'ānic verses, and relevant non-*tafsīr* literature, including some contemporary sources. The twelve-volume *tafsīr* closes with a registry of Qur'ānic chapters and a sura-by-sura summary of thematic content. There is also a summary of suras and index at the end of each volume, for ease of navigation.

In terms of sources, he draws on the major hadith collections (Muḥammad b. Ismā'īl al-Bukhārī, Aḥmad b. Ḥanbal, Muslim b. al-Ḥajjāj, Abū Dāwūd as-Sijistānī, Abū 'Abd Allāh Muḥammad b. Mājah, Abū 'Īsā Muḥammad at-Tirmidhī, Abū Muḥammad ad-Dārimī, Yaḥyā b. Sharaf an-Nawawī, Abū Bakr al-Bayhaqī, Abū 'Abd ar-Raḥmān an-Nasā'ī, etc.), *sīra* and *ṭabaqāt* literature, and a variety of *tafsīr* sub-genres. In addition to classical *tafsīr* literature, Ateş heavily cites modernist *tafsīr* literature, such as Rashīd Riḍā's *Tafsīr al-manār* and late Ottoman Elmalılı Hamdi Yazır's (1878–1942/1295–1361) *Hak Dini Kur'ân Dili* (1935–1939/1355–1340). When it comes to *asbāb an-nuzūl*, Ateş lists more than one and does not hesitate to indicate preference, even rejecting some *asbāb*. His engagement with hadith and views on *asbāb an-nuzūl* are strongly influenced by Palestinian Qur'ānic exegete Muhammad İzzet Derveze's (1888–1984/1305–1404) *Tafsīr al-ḥadīth*.[25]

Apart from critically appraising hadith literature and commenting on classical discussions on occasions of revelation, he engages in legal (*fiqhī*) exegesis and what some call scientific (Turkish: *ilmi/bilimsel*) exegesis, in addition to Sufi exegesis, a topic he specialized in with his doctoral studies and in his early career. Ateş cites opinions of all four Sunni legal schools in contrast to his predecessor Elmalılı Hamdi Yazır, who was commissioned by the Diyanet to write his *tafsīr* in line with Ḥanafī *fiqh*.[26] Ateş has even occasionally cited Shiite sources in the context of legal exegesis. His own views are hard to place in any one legal school.[27]

23 Ibid., 9: 415.

24 Ibid., 1: 365.

25 Yazıcı, *Türkçe Tefsir*, 165.

26 Mustafa Bilgin, "Hak Dını Kur'an Dili," in *TDVIA*: https://islamansiklopedisi.org.tr/hak-dini-kuran-dili.

27 Yazıcı, *Türkçe Tefsir*, 206.

His approach has been characterized as a message of returning to the Qur'ān,[28] and in this sense he shares much in common with modernist contemporaries. Ateş has been most commonly criticized for being inconsistent in his views.[29] However, this charge of inconsistency is not directed at the sum of his fundamental views over time, which remain fairly consistent over time and across publications; rather, criticisms of Ateş are largely directed at the inconsistencies that arise as he applies his views to various Qur'ānic verses. Some of his most contested as well as influential exegetical moves involve a novel reading of abrogation, his specific application of historical criticism to Qur'ānic stories and hadith, his position on evolution, and his views on other religions and religious tolerance.

3.1 Abrogation

Scholars have criticized Ateş for denying *naskh*, or abrogation.[30] While it is true that Ateş's view largely departs from the classical conception of abrogation, he nonetheless retains the category of abrogation. Classical views of abrogation entail abrogation of legal rulings without the abrogation of the Qur'ānic proof text, abrogation of the Qur'ānic verse without the abrogation of any associated legal ruling, and abrogation of both verse and ruling. One verse of the Qur'ān may abrogate another verse, and legal schools and opinions differ on whether hadith may abrogate something in the Qur'ān. Ateş only acknowledges the abrogation of both Qur'ānic verse and ruling, meaning that for Ateş the Qur'ān does not contain abrogated verses. He further stresses the distinction between Qur'ānic revelation and hadith, placing the Qur'ān definitively above the sunna in terms of legal and theological authority. According to Ateş, any abrogated verse of the Qur'ān which God caused Muḥammad to forget never made it into the transmitted and recorded Qur'ān. In this view,[31] he follows ʿAbduh and Riḍā's reading of Q 8:6–7, where Muḥammad's forgetting is not as a result of any flaw but as a result of divine purpose.[32]

3.2 Historicity

As has been and continues to be the case with many exegetes influenced by modernist and reformist efforts to critically engage tradition in order to rediscover the essential

28 Takim, "The Turkish Qur'ānic Exegete," 22.
29 Kaya, "'Kur'an Yolu'," 127.
30 E.g., Sırrı Ataman, *Unutturulan Ayetler: Turan Dursun – Süleyman Ateş Tartışması* (Istanbul: Berfin Yayınları, 2008).
31 Ateş, *Yüce Kur'ân'ın Çağdaş Tefsîri*, 1: 217f., 5: 140.
32 Yazıcı, *Türkçe Tefsir*, 129f.

or authentic Qur'ānic message, Ateş sought to "to liberate [the Qur'ān] from the various interpretative additions and accretions of the exegetes and to present it authentically as it is."[33] This effort to rediscover the essential Qur'ānic message entails some measure of historical criticism of the Qur'ān's traditional reception, as well as a tendency to avoid exclusively engaging the Qur'ān on a literal level, preferring less literal interpretations in cases where a literal reading might contradict with historical or scientific fact.

While Ateş in no way denies the divine origin and perfection of the Qur'ān, he distinguishes between the Qur'ān and hadith literature, placing the Qur'ān firmly above the hadith. Distinct from the Qur'ān in Ateş's view, the hadith lies open to some measure of historical criticism. As far as Ateş is concerned, the statements of hadith were formally written and collected several generations after Muḥammad's lifetime and have likely been shaped by the cultures, concerns, and politics of subsequent generations.[34] Further, apart from when Muḥammad was communicating revelation to his people, Ateş considers Muḥammad's words the words of a normal person. Additionally, Ateş holds that Muḥammad never wrote his own words down so as to prevent confusion of his words with the word of God, and at most only a few companions recorded his sayings in his lifetime. With this reasoning in mind, Ateş regards it as scientifically unacceptable to consider all hadith nothing other than Muḥammad's words, and even in cases where Muḥammad's words have been faithfully recorded, it is still necessary to distinguish between Muḥammad's prophetic vs. human utterances.[35] Nevertheless, such concerns do not prevent Ateş from making use of hadith literature to support his Qur'ān commentary. This attempt to apply hadith to the Qur'ān in a manner that determines some hadith as unimportant or irrelevant and others as significant or relevant has been denounced as faulty and unduly innovative by some contemporaries.[36]

Moreover, Ateş does not read all Qur'ānic accounts as strictly historical accounts, and for this position he has also received criticism. For instance, Ateş has been faulted for casting doubt on the historical veracity of stories in the Qur'ān at the expense of giving an enlightened reading to his modern readers.[37] Like modernists such as Fazlur Rahman (1919–1988/1337–1408), who, in his time active in British, North American, and Pakistani contexts, has incidentally shaped Turkish discussions on approaches to

33 Takim, "The Turkish Qur'ānic Exegete," 19.

34 Ateş, *Yüce Kur'ân'ın Çağdaş Tefsîri*, 5: 111, 9: 104; cf. Yavuz Köktaş, "Yüce Kur'an'ın Çağdaş Tefsiri Adlı Eserde Hadis Olgusu ve Hadis Yorumları," *Dinbilimleri Akademik Araştırma Dergisi* 4, no. 1 (2004): 33.

35 Köktaş, "Yüce Kur'an'ın," 34f.

36 Ibid., 41.

37 Yazıcı, "Süleyman Ateş'in," 114.

the Qur'ān and hadith,[38] Ateş focuses on finding universal principles in the Qur'ān, at times actively reclaiming such principles from the Qur'ān's traditional reception history. This reclamation also entails reading the Qur'ān as revelation that attempts to relate to its audience and to its audience's worldview rather than as a strict log of facts and historical data.[39] Thus, Ateş tends to interpret tales of the prophets as stories (*öykü*), without assuming that all stories should be read literally or be primarily concerned with actual historical events. On this point, Ateş's approach has been likened to that of Muḥammad Aḥmad Khalaf Allah (1916–1997/1334–1417).[40] In espousing this view, Ateş has no intention to cast doubt on the divine provenance of the Qur'ān, he is simply more concerned with revelation's aim of encouraging actionable moral reflection and of cultivating a proper understanding of monotheism.[41]

For instance, Ateş suggested in a Vatan article from 2010/1431 that there might have been more than the one flood recounted in Genesis.[42] For comparison, in his twelve-volume *tafsīr*, Ateş stresses that the story of Noah and the Flood has been divinely tailored to address Muḥammad's Arab interlocutors so that they may draw parallels to their own time and grasp the application to their own situation.[43] When Ateş does bring up history and science in his exegesis, he is sure to affirm that the Qur'ān, properly understood, is compatible with scientific and historical fact. By contrast, he faults the Torah for historical inaccuracies in its recounting of creation and Flood narratives.[44]

3.3 Evolution

In a controversial article he published 1975/1395, entitled "Kuran-ı Kerim'e Göre Evrim Teorisi," i.e. "Evolutionary Theory According to the Qur'ān," Ateş addressed the question of evolutionary theory and its compatibility with the Qur'ān. He argued that evolution, if understood as gradual change according to natural laws made by God, was compatible with and supported by the Qur'ān. Ateş had prepared this article

---

38 For an appraisal and commendation of Fazlur Rahman's Qur'ānic approach see Mehmet S. Aydın's article "Fazlur Rahman ve İslâm Modernizmi," *İslamî Araştırmalar* 4, no. 4 (1990): 273–84, and for an analysis of Fazlur Rahman's model applied to a specific Turkish Qur'ānic exegete see Felix Körner's treatment of Mehmet Paçacı in *Revisionist Koran Hermeneutics in Contemporary Turkish University Theology: Rethinking Islam* (Würzburg: Ergon Verlag, 2005).

39 Yazıcı, *Türkçe Tefsir*, 184,188.

40 Ibid., 186.

41 Cf. Ateş, *Yüce Kur'ân'ın Çağdaş Tefsîri*, 5: 33.

42 Yazıcı, "Süleyman Ateş'in," 114.

43 Ateş, *Yüce Kur'ân'ın Çağdaş Tefsîri*, 10: 81f.

44 E.g., Süleyman Ateş, *Kur'an Nizâmı* (Istanbul: Yeni Ufuklar Neşriyat, [year not specified]. Original printing 1979), 23.

from a lecture he gave as part of the requirements to attain his associate professorship.[45] Its publication immediately sparked harsh criticisms.

Despite contemporary objections to his position on evolution, he nevertheless stands in the tradition already established by prior generations of Turkish religious scholars and holds the Qur'ān to be compatible with modern science. Like some members of the generation of state-supported religious scholars before him, Ateş, also relying heavily on Elmalılı Hamdi Yazır's *Hak Dini Kur'ân Dili* on this subject, did not see a contradiction between science and religion and even entertained the possibility of evolution as compatible with the Qur'ān. Another predecessor, Ahmet Hamdi Akseki (1887–1951/1304–1370), cited classical sources that cast doubt on the idea of Adam as the biological origin of humanity and did not think it necessary to believe that the world was created six thousand years ago, given that the Qur'ān does not specify the age of the earth.[46] Ateş takes a similar position and draws on Ahmet Hamdi Akseki's interpretation of Adam in Q 32:7, where the creation of human beings began from clay, but what happened between the beginning of humanity's creation and its fulfillment in Adam is left open-ended.[47] Ateş even specifies that to interpret the Qur'ānic reference to day in the context of creation as a normal human day would be incorrect – there is for him certainly a broader time scale at work.[48] In sum, from clay to Adam, the creation of an ensouled human being might involve intermediary steps over a long period of time. For Ateş, evolution, or *evrim*, as he referred to it in his 1975/1395 article, meant "gradual change and progress" and could be thought of in terms of three categories: inorganic evolution, organic evolution, and evolution of the soul.[49] There are definite parallels between his article and his *tafsīr*, although in the *tafsīr* his language is arguably more cautious regarding human evolution. Natural selection, for Ateş, is understood as "the ability of nature to select what is best for itself" and is nothing other than a "natural law made by God."[50] However, this does not mean that human beings came from apes; nevertheless, he does allow for the possibility that apes might have evolved from a not-yet-complete human prototype, aka a common ancestor.[51] Ateş even goes so far as to say that Muslims developed the theory of evolution before Darwin, citing Abū ʿUthmān al-Jāḥiẓ (d. 868/254), Abū ʿAlī Aḥmad b. Miskawayh (d. 1030/421) and even Erzurumlu İbrahim Hakkı (d. 1772/1186) as proffering salient antecedents to modern evolutionary theory. This position can also be

45 Takim, "The Turkish Qur'ānic Exegete," 16.

46 Veysel Kaya, "Can the Quran Support Darwin? An Evolutionist Approach by Two Turkish Scholars after the Foundation of the Turkish Republic," *MW* 102, no. 2 (2012): 362.

47 Ibid., 367.

48 Süleyman Ateş, "Kuran-ı Kerim'e Göre Evrim Teorisi," *Ankara Üniversitesi İlahiyat Fakültesi Dergisi* 20 (1975): 128.

49 Ibid., 127; Kaya, "Can the Quran Support Darwin?," 365.

50 Kaya, "Can the Quran Support Darwin?," 367.

51 Ateş, "Kuran-ı Kerim'e Göre Evrim Teorisi," 131.

found in his twelve-volume *tafsīr*, where he interprets Q 71 to speak of a kind of evolution (*tekâmül*) of life and of the human species, moving from plants, to animals, and finally humans; however, he stresses here that humans did not evolve from other animals and that only God knows the details of this evolution.[52] In addition to weathering criticisms for proposing a Qurʾānic view of evolution, Ateş has been accused of inconsistency in his interpretation of evolution in the Qurʾān. For instance, in the second volume of his *Yüce Kur'ân'ın Çağdaş Tefsîri* he reads Q 3:59 to mean Adam had no parents in much the same way as Jesus lacked a father, a position which seems to run against the idea of a gradual evolution of humankind.[53]

3.4 Other Religions and Paradise

By the publication of the first eight volumes of his *Yüce Kur'ân'ın Çağdaş Tefsîri*, Ateş had already received some criticism for his position on the People of the Book, particularly Jews and Christians, whose scriptures appear frequently in his exegesis. In 1989/1409, he published the article "Cennet Kimsenin Tekelinde Değildir" (Paradise is No One's Monopoly) as a defense of the position in his *tafsīr*, leading with the assertion that no religious group possessed a monopoly on heaven.[54] This article was challenged by hadith scholar Talât Koçyiğit, also a former student of M. Tayyip Okiç, in a response article entitled "Cennet Mü'minlerin Tekelindedir" (Paradise is the Monopoly of the Believers). Contrary to the stance of Ateş, Koçyiğit asserted that Muslims had a monopoly on heaven. Ateş followed up with a further rebuttal in 1990/1410, entitled "Cennet Tekelcisi mi? (Can We Monopolize The[sic] Paradise?). In this article Ateş argued that no one possesses a monopoly on heaven. For his inclusive position on the People of the Book, which he has maintained since that controversy, he has been likened to Wilfred Cantwell Smith (1916–2000/1334–1421), a Christian inclusivist and Orientalist.[55]

In his initial 1989/1409 article, Ateş interprets the Qurʾān as stipulating three conditions for entry into heaven: monotheism, unwavering belief in the afterlife, and good deeds, citing Q 2:62 and Q 5:69 as proof texts. He further draws on Q 2:111 and Q 4:123–24 to argue that Christians and Jews were mistaken when they thought they had a monopoly on heaven and that Muslims can fall into the same error. He stresses the

52 Ateş, *Yüce Kur'ân'ın Çağdaş Tefsîri*, 10: 80.
53 Gördük, "Elmalılı Hamdi Hz," 32.
54 Cafer S. Yaran, "Non-Exclusivist Attitudes Towards Other Religions in Recent Turkish Theology and Philosophy of Religion," in *Change and Essence: Dialectical Relations Between Change and Continuity in the Turkish Intellectual Tradition*, ed. Şinasi Gündüz and Cafer S. Yaran (Washington, DC: The Council for Research in Values and Philosophy, 2005), 8.
55 Muammer Erbaş, "Süleyman Ateş ve W. Cantwell Smith'e Göre Ehli Kitab (Hıristiyanlar)'ın ve Müslümanların Kurtuluşu," *Dokuz Eylül Üniversitesi İlahiyat Fakültesi Dergisi* 40 (2014): 33–68.

verses in the Qur'ān that depict the People of the Book as having at least some things correct and he extends this to assert that any religious community contains good and bad individuals. He goes so far as to claim that the term "Islam" has two meanings: that of the final and most complete religion and the name for all divine religions in human history up until Muḥammad, citing the Qur'ānic reference to Abraham in Q 22:78. Or in other words, Muḥammad did not invent a new religion but came to restore the purity of original monotheism, citing Q 46:9 and Q 6:90 as proof texts. In yet another controversial move, he reads Q 2:41 to suggest that the revelation of the People of the Book is not utterly abolished, arguing that the Qur'ān asks the People of the Book to purge their faith of beliefs that contradict monotheism, adding that such monotheists among the People of the Book may indeed go to heaven. Even though Ateş downplays the doctrine of *taḥrīf* or corruption of Jewish and Christian scriptures, he still holds that the Qur'ān requires some correction of Jewish and Christian beliefs. Koçyiğit in his turn criticizes Ateş for taking the verses he used to promote an inclusivist reading of the Qur'ān out of the initial context in which these verses were revealed, i.e. in the spirit of invitation to Islam not in the spirit of inclusion.[56] Moreover he berates Ateş for misreading Q 4:163 to say that Muḥammad and Noah received the same essential revelation when in fact it merely states that they received revelation in a similar manner.[57] Ateş's rebuttal in 1990/1410 ultimately stresses that God can let whomever He wishes into heaven, citing Q 17:54 to assert that God can allow polytheists into heaven if He so chooses. However, he does qualify his inclusivism to stipulate that the People of the Book who have heard of Muḥammad must then accept Muḥammad's prophethood. Finally, given the psychological improbability of all non-Muslim monotheists choosing to give up their faith and converting to Islam, despite being good and faithful human beings, he cites Q 38:28 to support the idea that God will likely have mercy on such non-Muslims who worship God and live righteously.

In 1999/1419 he reaffirmed this position in an English article entitled "The Attitude of the Koran Towards the Divine Religions," going so far as to claim that "the religions, which are revealed by the One God – Allah – are in their essence all equal. The Torah, the Gospel and the Koran are revelations which confirm themselves mutually."[58] He states: "The Koran recognizes all those, who believe in the oneness of God and who serve him alone, in (their) essence as Muslims – even though they are not called Muslims – and the Koran considers them as the people who will enter Paradise."[59] Again, similar to his previous work, he stresses that Islam is not only the name of a religion but also stands for the "monotheistic principle" common to all People of the Book.[60]

56 Yaran, "Non-Exclusivist Attitudes," 12.

57 Ibid., 13.

58 Süleyman Ateş, "The Attitude of the Koran towards the Divine Religions," *Kur'an Mesajı: İlmi Araştırmalar Dergisi* 2, no. 22, 23, 24 (1999): 51.

59 Ibid., 52.

60 Ibid., 58.

He claims that the essence of the scriptures of People of the Book and the Qur'ān is the same: devotion to the one God, belief in the afterlife, good deeds, and the pursuit of humanity.[61] Consistent with his previous Turkish-language articles on the subject, he affirms that the People of the Book may possibly enter Paradise provided that they recognize the Qur'ān as divine revelation and affirm Muḥammad as a prophet and that this can be done without conversion to Islam as long as such People of the Book correct their beliefs according to the injunctions of the Qur'ān.[62] How he imagines this corrective can be seen in his comparative analysis of Qur'ānic verses against excerpts from the Torah and New Testament followed by Sufi accounts of interactions with ascetics among the People of the Book in various passages of his major *tafsīr*.

While entertaining a more inclusive position than many of his critics, Ateş's inclusive reading of the Qur'ān's position on non-Muslim monotheists, as mentioned above, still carries Qur'ānic corrections and stipulations. Regardless, he continues to receive criticism for these views. For instance, the 2016/1437 official Diyanet Qur'ān commentary *Kur'an Yolu* criticizes Ateş for reading "*kitāb*" in Q 3:3 as the revelation given to Moses and Jesus,[63] a reading that reflects his views of Muslims and the People of the Book as sharing some common essential revelation.

This debate on the status of the People of the Book has shaped the positions of later generations of Turkish theologians. Those who have taken up Ateş's position tend to stress not only an inclusive reading of the Qur'ān but also warn against the correlation of exclusivism and violence.[64] At the same time, among the more recent theologians Ateş's interpretation has its share of critics.[65]

3.5 Tolerance

The question of tolerance, while related to Ateş's views on the People of the Book, extends also to intellectual criticisms of Islam. For Ateş, in Islam there is no intolerance, and this position came to the fore in Ateş's condemnation of the assassination of the atheist apostate, and former mufti of Sivas, Turan Dursun on September 4, 1990/Ṣafar 15, 1411 for the latter's critical views on Islam. Ateş addresses the issue of tolerance in the Qur'ān in his work *Gerçek Din Bu* (1992/1412), written in response to both criticisms leveled against Islam by Turan Dursun, who claimed that God was vengeful, and in response to Dursun's assassination. Ateş reads intolerance as incompatible

---

61 Ibid., 54f.
62 Ibid., 57.
63 Kaya, "'Kur'an Yolu'," 129.
64 Yaran, "Non-Exclusivist Attitudes," 17f. Those who have taken up Ateş's inclusive reading of the Qur'ān include Mehmet Okuyan, Mustafa Öztürk, and Hidayet Aydar, (p. 22), as well as Muammer Erbaş.
65 Hocaoğlu, "Süleyman Ateş'in."

with the essence of Islam and employs this principle in his reading of the Qur'ān to stress that Q 9:36, which contains an injunction to fight idolaters, does not in fact abrogate Qur'ānic verses like Q 88:21, Q 3:20, Q 16:82, or Q 42:48. Ateş reads these latter verses as divine counsel against the use of violence in seeking converts.[66] Thus while he may be blamed for inconsistent textual readings by his critics, his position on tolerance, for instance, aligns with his views on abrogation, i.e. since he does not recognize the Qur'ān to contain abrogated verses, a Qur'ānic verse that counsels fighting against non-Muslims does not abrogate verses which suggest more peaceful or tolerant approaches to seeking converts or interacting with non-Muslims.

4 Conclusion

While certainly erudite and informed by classical *tafsīr*, the exegetical approach Süleyman Ateş, in a more modernist vein, stresses the overarching principles of monotheism and moral development. These principles play out, for instance, in his views on tolerance, other religions, and his attitude towards scientific knowledge. One way of characterizing his legacy might be as the generational successor to Elmalılı Hamdi Yazır's *Hak Dini Kur'ân Dili*. While he has received much criticism for his views, as an intellectual product of the Ankara İlahiyat Faculty and important shaper of its legacy, Ateş stands in a wider tradition of Turkish Qur'ānic exegesis which seeks to combine traditional exegetical sciences with religious inclusivism and a moderate openness to contemporary scientific outlooks.

5 Bibliography

5.1 Sources

Ateş, Süleyman. *İslâm'a İtirazlar, Kur'ân-ı Kerîm'den Cevaplar*. Ankara: n.p., 1966.
Ateş, Süleyman. *Sülemî ve Tasavvufî Tefsîri*. Istanbul: Sönmez Neşriyat, 1969.
Ateş, Süleyman. *İşârî Tefsir Okulu*. Ankara: Ankara Üniversitesi İlahiyat Fakültesi Yayınları, 1974.
Ateş, Süleyman. *Kur'ân-ı Kerîm ve Yüce Meâli*. Ankara: Hayat Yayinlari, 2019. Originally published in 1974/5 and in 1977 with Arabic text.
Ateş, Süleyman. "Kuran-ı Kerim'e Göre Evrim Teorisi." *Ankara Üniversitesi İlahiyat Fakültesi Dergisi* 20 (1975): 127–46.
Ateş, Süleyman. *Kur'an Nizâmı*. Istanbul: Yeni Ufuklar Neşriyat, (year not specified). Original printing 1979.
Ateş, Süleyman. *Yüce Kur'ân'ın Çağdaş Tefsîri*, 12 vols. Istanbul: Yeni Ufuklar Neşriyat, 1988–1992.
Ateş, Süleyman. "Cennet Kimsenin Tekelinde Değildir." *İslâmî Araştırmalar* 3, no. 1 (1989): 7–24.
Ateş, Süleyman. "Cennet Tekelcisi mi?." *İslâmî Araştırmalar* 4, no. 1 (1990): 29–37.

66 Ataman, *Unutturulan Ayetler*, 26.

Ateş, Süleyman. *Kur'an-ı Kerim'in Evrensel Mesajına Çağrı*. Istanbul: Yeni Ufuklar Neşriyat, 1990.
Ateş, Süleyman. *Gerçek Din Bu*, 2 vols. Istanbul: Yeni Ufuklar Neşriyat, 1992.
Ateş, Süleyman. *Kur'ân-ı Kerîm Tefsîri*, 6 vols. Istanbul: Milliyet, 1995.
Ateş, Süleyman. *Kur'an'da Nesh Meselesi*. Istanbul: Yeni Ufuklar Neşriyat, 1996.
Ateş, Süleyman. *Kur'an Prensipleri*. Istanbul: Yeni Ufuklar Neşriyat, 1996.
Ateş, Süleyman. *Yeni'den İslam'a*, 2 vols. Istanbul: Yeni Ufuklar Yayınları, 1996/7.
Ateş, Süleyman. *Namazda Okunan Sûrelerin Tefsiri*. Istanbul: Yeni Ufuklar Neşriyat, 1997.
Ateş, Süleyman. "Prof. Dr. Süleyman Ateş ile Söyleşi." *Kur'ân Mesajı* 1, no. 1 (1997): 66–75.
Ateş, Süleyman. *99 Soruda Kur'ân-ı Kerîm*. Istanbul: Radikal Yayınları, 1997.
Ateş, Süleyman. *Kur'an Ansiklopedisi*, 30 vols. Istanbul: Kur'an Bilimleri Araştırma Vakfı, 1997–2003.
Ateş, Süleyman. *Die geistige Einheit der Offenbarungsreligionen*, trans. Abdullah Takim. Istanbul: Yeni Ufuklar, 1998.
Ateş, Süleyman. *Die Ziele des Korans*, trans. Abdullah Takım. Istanbul: Yeni Ufuklar, 1998.
Ateş, Süleyman. "The Attitude of the Koran towards the Divine Religions." *Kur'an Mesajı: İlmi Araştırmalar Dergisi* 2, no. 22, 23, 24 (1999): 51–72.
Ateş, Süleyman. "Soruşturma." *İslâmiyât* 4, no. 1 (2001): 113–15.
Ateş, Süleyman. *Kur'ân'da Pergamberler Tarihi*. Istanbul: Vatan Yayınları, 2002.
Ateş, Süleyman. *Maktab-i tafsīr-i Ishārī "İşârî Tefsir Okulu*, trans. Tawfīq H. Subḥānī. Tehran: Markaz-i Nashr-i Dānishgāhī, 2002.
Ateş, Süleyman. *Bir Ömür Böyle Geçti*, 2 vols. Istanbul: Yeni Ufuklar Neşriyat, 2007.
Ateş, Süleyman. "The Spiritual Unity of the Revealed Religions: İlâhi Dinlerin Ruhbirliği," trans. Abdullah Takım and Holger Brune. *Journal of Religious Culture* 117 (2008): 1–19.
Ateş, Süleyman. *Kur'ân-ı Kerîm Tefsîri*, 3 vols. Istanbul: Yeni Ufuklar Neşriyat, 2013.

5.2 Secondary Literature

Ataman, Sırrı. *Unutturulan Ayetler: Turan Dursun – Süleyman Ateş Tartışması*. Istanbul: Berfin Yayınları, 2008.
Atmaca, Gökhan. "Süleyman Ateş'in Bakara Sûresi 62. Âyeti Tefsiri Üzerine Bir Değerlendirme." *Sakarya Üniversitesi İlahiyat Fakültesi Dergisi* 15, no. 28 (2013): 1–20.
Aydın, Mehmet S. "Fazlur Rahman ve İslâm Modernizmi." *İslamî Araştırmalar* 4, no. 4 (1990): 273–84.
Bilgin, Mustafa. "Hak Dinı Kur'an Dili." In *TDVIA*, edited by TDV İslâm Araştırmaları Merkezi. Accessed March 11, 2024. https://islamansiklopedisi.org.tr/hak-dini-kuran-dili.
Erbaş, Muammer. "Süleyman Ateş ve W. Cantwell Smith'e Göre Ehli Kitab (Hıristiyanlar)'ın ve Müslümanların Kurtuluşu." *Dokuz Eylül Üniversitesi İlahiyat Fakültesi Dergisi* 40 (2014): 33–68.
Gördük, Yunus Emre. "Elmalılı Hamdi Hz. Âdem'in Evrimle Yaratıldığını Mı Söylüyor?" *Süleyman Demirel Üniversitesi İlahiyat Fakültesi Dergisi* 42 (2019): 27–42.
Hocaoğlu, Mustafa. "Süleyman Ateş'in Tefsirinde Bütünlük Sorunu: Ehli Kitab'ın Cennet'e Girmesi ve Tevrât'ın Tahrifi Meseleleri." *İslâmî Araştırmalar* 23, no. 1 (2012): 42–54.
Kaya, Mehmet. "'Kur'an Yolu' Tefsirinde Süleyman Ateş'e Yönelik Bazı Eleştirilere Dair Bir İnceleme." *Türk İslâm Medeniyeti Akademik Araştırmalar Dergisi* 13, no. 25 (2018): 123–48.
Kaya, Veysel. "Can the Quran Support Darwin? An Evolutionist Approach by Two Turkish Scholars after the Foundation of the Turkish Republic." *MW* 102, no. 2 (2012): 357–70.
Köktaş, Yavuz. "Yüce Kur'an'ın Çağdaş Tefsiri Adlı Eserde Hadis Olgusu ve Hadis Yorumları." *Dinbilimleri Akademik Araştırma Dergisi* 4, no. 1 (2004): 31–55.
Körner, Felix. *Revisionist Koran Hermeneutics in Contemporary Turkish University Theology: Rethinking Islam*. Würzburg: Ergon Verlag, 2005.
Öcal, Mustafa. *Diyanet İşleri Başkanları ve Hizmetleri*. Istanbul: Ensar, 2016.

Takim, Abdullah. *Koranexegese im 20. Jahrhundert: islamische Tradition und neue Ansätze in Süleyman Ateş's "Zeitgenössischem Korankommentar."* Istanbul: Yeni Ufuklar, 2007.

Takim, Abdullah. "The Turkish Qur'ānic Exegete Süleyman Ateş and His Contemporary Commentary of the Qur'ān." In *Unity and Diversity in Contemporary Muslim Thought*, edited by Abbas Poya and Farid Suleiman. Newcastle upon Tyne: Cambridge Scholars Publishing, 2017: 12–46.

TDV İslâm Ansiklopedisi. "SÜLEYMAN ATEŞ Müellifin özgeçmişi [Author's Resume]." Last modified January 31, 2018. https://islamansiklopedisi.org.tr/muellif/suleyman-ates.

Yaran, Cafer S. "Non-Exclusivist Attitudes Towards Other Religions in Recent Turkish Theology and Philosophy of Religion." In *Change and Essence: Dialectical Relations Between Change and Continuity in the Turkish Intellectual Tradition*, edited by Şinasi Gündüz and Cafer S. Yaran. Washington, DC: The Council for Research in Values and Philosophy, 2005: 7–23.

Yazıcı, Ahmet. "Süleyman Ateş'in Görüşleri Bağlamında Kur'an Kıssaları." *İslâmî Araştırmalar* 22, no. 2 (2011): 106–17.

Yazıcı, Ahmet. *Türkçe Tefsir Çalışmalarında Klasik-Modern Algısı: Ömer Nasuhi Bilmen-Süleyman Ateş Örneği.* Istanbul: Ravza Yayınları, 2014.

Yolcu, Mehmet, Ayşe Tokay, and Ayşenur Bıçak. "Süleyman Ateş ve Abdullah Parlıyan'ın Meallerinin Karşılaştırılması." *e-Şarkiyat İlmi Araştırmalar Dergisi* 10, no. 4 (2018): 1328–53.

Abbas Poya

Muḥammad Mujtahid Shabistarī

1 Introduction

In the last several decades, some noteworthy exegeses have been published in Iran. For instance, the well-known and widely circulated *al-Mīzān fī tafsīr al-Qur'ān* (The Balance in Interpretation of Qur'ān) by the theologian and philosopher Muḥammad Ḥusayn Ṭabāṭabā'ī (1892–1981/1310–1402)[1] enjoys great recognition among both traditional and progressive scholars. It is also worth mentioning here two other authors whose works, however, remain incomplete. These are Maḥmūd Ṭāliqānī (1914–1979/1332–1399),[2] a reform scholar and advocate of a contemporary understanding of Islam who wrote *Partuwī az Qur'ān* (A Ray of Light from the Qur'ān), and the religious intellectual and political activist Muḥammad Taqī Sharī'atī (1907–1987/1325–1407),[3] who compiled the *Tafsīr-i navīn* (Modern Exegesis). These two works are generally characterized as scientific, socio-critical, and reform-oriented. More traditional works, i.e., those which follow more conventional exegetical methodologies, are the *Tafsīr-i namūna* (Exemplary Exegesis) by Nāṣir Makārim Shīrāzī (b. 1924/1342)[4] and the *Mafāhīm al-Qur'ān* (Meanings of the Qur'ān) by Ja'far Subḥānī (b. 1930/1349).[5]

The traditional understanding of the Qur'ān, which also represents the dominant perception of the Qur'ān exegetes across all denominations and epochs, can be formulated as follows: God sent the text of the Qur'ān, both in word and content, in sections to Muḥammad over a period of almost twenty-three years through Archangel Gabriel. Muḥammad then verbally communicated to people what God – so to say – dictated to him. Some of his companions memorized the texts revealed on specific occasions and

1 On Ṭabāṭabā'ī see: Hamid Algar, "'Allāma Sayyid Muḥammad Ḥusayn Ṭabāṭabā'ī: Philosopher, Exegete, and Gnostic," *JIS* (2006): 1, accessed October 10, 2023, http://www.islamiclaws.com/marja/Allameh%20Tabatabai.pdf.

2 On Ṭāliqānī see: Hamid Dabashi, *Theology of Discontent: The Ideological Foundation of the Islamic Revolution in Iran* (New Jersey & London: Transaction Publishers, 2006), 216–72.

3 On Sharī'atī see: Ali Rahnema, "Ali Shariati: Teacher, Preacher, Rebel," in *Pioneers of Islamic Revival*, ed. Ali Rahnema (Kuala Lumpur: Strategic Information Research Development, 2008), 208–45 (208f.).

4 On Shīrāzī see: "Zindagīnāma," accessed October 10, 2023, https://makarem.ir/main.aspx?lid=0&typeinfo=22&catid=30399.

5 On Subḥānī see: "Zindagīnāma," accessed October 10, 2023, https://www.noorsoft.org/fa/Mostabser/View/63493.

Note: This chapter is a revised and updated version of the previously published article "How the Prophet Saw the World: On the Qur'ānic Exegesis of Moḥammad Mojtahed Shabistarī," in *Unity and Diversity in Contemporary Muslim Thought*, ed. Abbas Poya and Farid Suleiman (Newcastle upon Tyne: Cambridge Scholars Publishing, 2017), 208–27.

https://doi.org/10.1515/9783111320069-015

wrote them down in a coordinated plan after the death of Muḥammad. Thus, the Qur'ānic text emerges from an extraordinary, unearthly, and purely divine source. This assumption affirms the miraculous and sacral nature of the Qur'ān, thereby leaving it and its content beyond any critical reflection. Moreover, the traditional scholars assert that the Qur'ān is in heaven (*fī lawḥin maḥfūẓ*; Q 85:22). People simply must follow the Qur'ān; it is exempt from human intervention, changes in content, or language. This assumption is also supported by the exegetes mentioned above, who work with philosophical, scientific, and socio-critical approaches.

Nevertheless, some religious intellectuals and scholars are trying to challenge the traditional assumptions about the Qur'ān, understand it as a human phenomenon, and as a result desacralize it. Next to ʿAbdulkarīm Surūsh (b. 1945/1364) and Siddīqa Wasmaqī (b. 1961/1381), Muḥammad Mujtahid Shabistarī (b. 1936/1355) is the most powerful religious thinker who maintains this controversial position. Surūsh perceives Muḥammad as a narrator of prophetic dreams, hence the title of his series of articles, *Muḥammad rāwī-yi ru'yāhā-yi rasūlāna* (Muḥammad, the Narrator of Prophetic Dreams).[6] According to Surūsh, Muḥammad was a child of his time: He was shaped by his personal experiences and led by dreams which he communicated in the Qur'ān to other people. Wasmaqī understands revelation (*waḥy*) as the result of Muḥammad's own reflection. In this respect, *waḥy* is not something with which God inspires Muḥammad or communicates to him by means of an angel. Rather, God and the contents of revelation are what the human – in this case the prophet Muḥammad – discovers and articulates while searching for God. Thus, everything that Muḥammad announced and reported about God, His attributes and the creation was the result of his own perception, which was expressed in the Qur'ān.[7] Like Surūsh and Wasmaqī, Shabistarī considers the genesis of the Qur'ān to be closely related to both the personal character of Muḥammad and the historical and cultural contexts in which the text was produced. Shabistarī's aim is to remove the Qur'ān from heaven and introduce it as a worldly phenomenon.

Shabistarī sees the Qur'ān as a narrative of prophetic experiences that shaped Muḥammad's view of the world. In that respect, Muḥammad narrated his perception of the world in the Qur'ān. His experiences were like those of any other human being. The peculiarity, however, lies in his intense encounters with God. Therefore, when the Qur'ān talks about God, nature, and human being, it presents how Muḥammad saw them and not how they objectively were.[8]

---

6 Cf. ʿAbdulkarīm Surūsh, *Kalām-i Muḥammad, rū'yā-yi Muḥammad* (Online publication: Suqrāṭ, 2018), 85–237, accessed October 10, 2023, https://www.goodreads.com/book/show/53400859.

7 Cf. Siddīqa Wasmaqī, *Masīr-i payāmbarī*, (Teheran: s.n., 2021), accessed October 10, 2023, www.zeitoons.com/wp-content/uploads/2021/11/path-of-prophetood.pdf.

8 Cf. Shabistarī's series of articles "Qirā'at-i nabavī az jahān" (Prophetic Reading of the World), accessed October 10, 2023, http://mohammadmojtahedshabestari.com/مقالات-قرائت-نبوی-از-جهان-درآمدی-بر-پ/.

Although Shabistarī has not written a *tafsīr* work yet, he is concerned with promoting a new approach to studying the Qur'ān and challenges the conventional readings that take place in traditional educational institutions. Simultaneously, he deems the attempts made by Ṭāliqānī and Sharī'atī insufficient for an adequate understanding of the Qur'ān, and – by extension – of Islam. Shabistarī considers the "official reading" (*qirā'at-i rasmī*) of religion and the Qur'ān as too strongly oriented towards *fiqh* (jurisprudence). He argues that the current situation in the field is one of crisis and therefore pleads for an interpretive opening-up in understanding of religion and the Qur'ān. Shabistarī believes that religious texts can be read and understood in diverse ways, that all readings may be criticized, and that, through reflecting on these different readings, one may arrive at a suitable interpretation of the religion and its sacred texts.[9]

These assertions are part of a field that hardly any other 'religious intellectual' (*rūshanfikrān-i dīnī*) has approached. These so-called 'religious intellectuals'[10] are usually post-revolutionary Iranian scholars who argue against the official readings of Islam and instead advocate for formulating a religious understanding in harmony with human rights, democratic forms of government, and individual and social freedoms. In addition to Shabistarī, this group also includes the pre-revolutionary intellectual 'Alī Sharī'atī as well as both 'Abdulkarīm Surūsh and Muṣṭafā Malīkyān, who became better known after the founding of the Islamic Republic.[11] While Sharī'atī has written about the ideologization of Islam, Malīkyān has mostly examined the questions of ethics, Surūsh treated the epistemological themes, and Shabistarī has made the discussion surrounding new methodological approaches to interpreting the Qur'ān his primary field of inquiry. The point of departure for all these thinkers is the relation between religious prescriptions and the challenges of modernity.

2 Biography

Muḥammad Mujtahid Shabistarī was born in 1936/1354 in Shabistar in the Iranian province of East Azerbaijan.[12] In 1951/1370, he went to Qom to study classical Islamic disciplines. Over a period of eighteen years, he studied Arabic Language, Islamic Phi-

---

9 Moḥammad Mujtahid Shabistarī, *Naqdī bar qirā'at-i rasmī az dīn* (Tehran: Ṭarḥ-i Nū, 2000), 7.

10 For more on this phenomenon, see Abbas Poya, *Denken jenseits von Dichotomien. Iranisch-religiöse Diskurse im postkolonialen Kontext* (Bielefeld: transcript, 2014), 136–42.

11 On the three thinkers, see: Poya, *Denken jenseits von Dichotomien*, 95–198.

12 On Shabistarī's life, see the short autobiography on his website, "Zindagīnāma," accessed October 10, 2023, http://mohammadmojtahedshabestari.com/زندگی%e2%80%8cنامه/; cf. S. Mahmoud Sadri, "Sacral Defense of Secularism: The Political Theologies of Soroush, Shabestari, and Kadivar," *International Journal of Politics, Culture and Society* 15 (2001): esp. 260f.; Farzin Vahdat, "Post-revolutionary Islamic Modernity in Iran: The Intersubjective Hermeneutics of Mohamad Mojtahed Shabestari," in *Modern*

losophy, Theology, Jurisprudence, and Juridical Methodology. He moved to Hamburg in 1969/1389 to assume responsibility of the local Islamic center at the Imam Ali Mosque. During his nine-year stay in Germany, he grappled with different religious outlooks, especially those of Christianity, Judaism, and Buddhism. He participated in inter-faith dialogues, which distinguished him from other scholars of Islamic theology. He returned to Iran in 1978/1398 at the time of social and political turbulence caused by the founding of the Islamic Republic. He founded his own journal, *Andisha-yi islāmī* (*Islamic Thought*), which was discontinued after only a few issues due to financial constraints. Shabistarī appeared as a guest speaker and expert on numerous radio and television shows. Following the first elections after the Islamic Revolution, he became a member of the parliament. The increasingly rigorous official politics of the country, however, caused him to gradually distance himself from the political sphere. He later accepted an offer from the Ministry of Sciences (*Wizārat-i ʿUlūm*) to lecture at the University of Tehran in the fields of Philosophy and Islamic Theology. Later, he shifted to Religion and Mysticism. In 2006/1427, together with many other non-conformist lecturers and thinkers, Shabistarī was forced into early retirement. Nevertheless, he maintained contact with European academic institutions and was regularly invited as a visiting professor and guest speaker in Austria and Germany. Today, he is counted among the best-known intellectual critics of the official policies of his country and generally of Islamic rigorism.

Shabistarī was also intellectually active before the founding of the Islamic Republic, especially during his study in Qom. He was one of the few progressive scholars at the time and regularly contributed to the well-known and bountifully equipped journal *Maktab-i islām* (The School of Islam).[13] Along with many other academics, Shabistarī shared the dream of an ideal Islamic order. His ambitions at the time were geared toward new interpretations of traditional religious views and their reconciliation with the revolutionary ideas of the younger generation in order to guarantee freedom and justice.[14]

After breaking away from the official politics of his country, Shabistarī primarily concerned himself with the elaboration of new theories of Qurʾānic exegesis. His central focus has been to show that no Qurʾānic exegesis, no Islamic understanding, and, generally, no recognition or realization is formulated out of thin air. He instead suggests that any production of knowledge is always rooted in an individual's own personal and socio-political contexts and interests.[15]

Muslim Intellectuals and the Qurʾān, ed. S. Taji-Farouki (Oxford: Oxford University Press, 2004), 193–224, esp. 198–201.

13 Vahdat, "Post-Revolutionary," 199.

14 Muḥammad Manṣūr Hāshimī, *Dīn andishān-i mutajaddid: Rūshanfikrī-yi dīnī az Sharīʿatī tā Malīkyān* (Tehran: s.n., 2006), 235.

15 Muḥammad Mujtahid Shabistarī, *Hirminūtīk, kitāb va sunnat* (Tehran: Ṭarḥ-i Nū, 2000), 7–9.

This thesis applies equally to Shabistarī himself and his perception of the Qur'ān. Shabistarī's thoughts on Qur'ānic exegesis indeed stand in clear and direct relation to his views on political and ethical matters.[16] For this reason, the next section will briefly elaborate his position *vis-à-vis* democracy, freedom, and the plurality of religious interpretation, before moving on to the methodological aspects of his exegetical approach.

3 Democracy: The Best Possible Way of Governing

According to Shabistarī, the main concerns of religions – and therefore of Islam – are freedom and justice. Islam does not prescribe a specific pattern of government. Rather, it supports the ruling system that is most likely to guarantee freedom and justice. Shabistarī suggests that, in today's world, democracy is the only way of governing that guarantees freedom and justice, and as such should be supported.[17] However, Shabistarī is well-aware that democratic rules can be abused for non-democratic goals to maintain the *status quo* and avoid relinquishing power. For this reason, he declares that, on the one hand, believers can and should support democratic rule out of religious conviction. On the other hand, he emphasizes that presenting a majority does not grant believers the permission to insist on the maintenance of established sharia rules[18] and thereby on the implementation of discriminatory regulations that stand against human rights.[19] He goes a step further by stating that respecting the will of the majority is not in itself a guarantee of democratic rule. First and foremost, Shabistarī suggests, freedom and justice must be guaranteed to all members of society as basic human rights.[20] In order to avoid any possibility of the abuse of democracy, he questions concepts such as "religious democracy" or "democratic religious state."[21] It is with these kinds of expressions that the religious intellectual ʿAbdulkarīm Surūsh attempts to oppose the official political system of the *vilāyat-i faqīh* (rule of the juridi-

---

16 Sarkohi makes this connection in reverse, stating that Shabistarī's position on democracy and human rights is firmly rooted in his understanding of religion and its interpretation. Arash Sarkohi, *Der Demokratie- und Menschenrechtsdiskurs der religiösen Reformer in Iran und die Universalität der Menschenrechte* (Würzburg: Ergon, 2014), 72.

17 Muḥammad Mujtahid Shabistarī, *Ta'ammulātī dar qirā'at-i insānī az dīn* (Tehran: Ṭarḥ-i Nū, 2004), 147f.

18 Ibid., and Shabistarī, *Naqdī*, 116f.

19 Shabistarī, *Ta'ammulātī*, 150f., and 179.

20 Ibid., 179.

21 Ibid., 180.

cal scholar) in Iran. At the same time, Surūsh was trying to formulate an explicitly democratic framework for a religious society such as that of Iran.[22]

Shabistarī is not generally opposed to adding the word 'religious' (*dīnī*) if it indicates one's support of democracy based on religious convictions. However, one must uphold the essential preconditions of a democratic society: freedom, justice, and equality for all its citizens.[23] This is why Shabistarī is in favor of separating the political sphere from matters of religion. In an interview for the online journal *Qantara.de* in 2012/1433, he summarized his views regarding the separation of religion and state as follows:

> I am of the opinion that religious and political institutions are different institutions with different functions. That is the reason for their separation. But it does not follow that the religiosity of people cannot be the moral or ethical impulse for politics. This is entirely possible. Therefore, I do not say that politics and religion are separate from one another. Instead, I always say: Political and religious institutions should be separate. However, of course there can and should exist forms of cooperation between them.[24]

Shabistarī believes that there is no specifically Islamic pattern of government. In his view, it was not the task of the Prophet Muḥammad to introduce or implement a particular political system. Rather, he was a prophet, like all other messengers of God, whose goal was to save human beings from harm and teach them how to conduct their lives in a spiritual manner.[25] The political practice of Muḥammad and the first generation of Muslims was not determined by God, and, thus, was not holy. Instead, it was "human" (*insānī*), "secular" (*ʿurfī*), and historically/geographically limited – therefore, it is contestable and not applicable to all times and places.[26] Muḥammad operated in the contemporary social structures of his time and tried – within this framework – to actualize his moral and legal concepts. Thus, by way of example, Islam contributed to improving the situation of women at the time, but did not do away with the structures of patriarchy. For the same reason, in the political realm, Muḥammad did not develop his own mode of government, but rather contributed to improving the existing political relations of power. He was, primarily, a moral authority.[27]

---

22 On the concept of 'democratic religious state' by Surūsh, see: ʿAbdulkarīm Surūsh, "Ḥukūmat-i dimūkratīk-i dīnī," in Surūsh, *Farbihtar az id'ūlūzhī* (Teheran: Mu'assasa-yi Farhingī-yi Ṣirāṭ, 1994), 273–83. A German translation of the article is available: Poya, *Denken jenseits von Dichotomien*, 249–60.

23 Shabistarī, *Ta'ammulātī*, 145–52.

24 Qantara, "Warum Islam und Demokratie zusammen passen" (last modified March 25, 2016), accessed December 12, 2023, http://de.qantara.de/content/interview-mit-mohammad-mojtahed-shabestari-warum-islam-und-demokratie-zusammen-passen. [My translation]

25 Shabistarī, *Ta'ammulātī*, 155f.

26 Ibid., 156.

27 Ibid., 167f.; cf. Qantara, "Islam und Demokratie." For Shabistarī's views on democracy and how it relates to Islam in Iran, see Katajun Amirpur, "Islam und Demokratie – Die Geschichte einer Aneignung," *Blätter für deutsche und internationale Politik* 56, no. 11 (2011): 87–95. It cannot be overlooked that Shabis-

Shabistarī does not detect any opposition to democracy in Islam.[28] On the contrary, Islam is passionately in favor of democratic principles, since it sees freedom and justice as fundamental religious values that are best guaranteed in a democratic way of governing.[29] However, Shabistarī makes it clear that favoring democracy – or any other governmental system – is not a question of epistemology, but rather a concrete decision. It is a matter of positioning oneself, as a Muslim, in relation to supporting or opposing democracy:

> The question is not whether or not Islam and democracy are compatible. I have always said so. The correct formulation of the question is: Do Muslims want democracy or not? My answer is, if Muslims want democracy, then they will find an interpretation of Islam that is compatible with it. If they do not want democracy, then they will not find the said interpretation.[30]

4 Freedom and Plurality of Religious Readings

One of Shabistarī's primary concerns is the defense of civil liberties, human rights, and the freedom of belief.[31] From his point of view it is important that scholars in a community of believers should have no fear of crossing any 'red lines.' Rather, he says one must accord critics enough space to express their opinions freely. It is then up to the believer to work out how to suitably state and represent their convictions regarding the critique that has been advanced.[32]

In Shabistarī's thinking, religious freedom and, in general, the Universal Declaration of Human Rights, occupy an important place. He often emphasizes that only human rights can protect people from the poison of extremist, violent, and religiously motivated groups like IS (Islamic State). In Shabistarī's understanding, the traditional sharia lacks this ability. According to him, it can even create the following tendency:

---

tarī's view regarding Muhammad, his role as a prophet, and the question of whether he had a state concept is similar to the ideas of ʿAlī ʿAbd ar-Rāziq (1887–1966/1305–1386). The so-called 'spiritual father' of Islamic secularism advocated similar ideas in his book *al-Islām wa-uṣūl al-ḥukm: Baḥth fī l-khilāfa wa-l-ḥukūma fī l-islām*, first published in 1925. The book is available online under: https://mc.dlib.nyu.edu/files/books/nyu_aco000732/nyu_aco000732_hi.pdf (last accessed October 10, 2023). For an English translation of the book see: Ali ʿAbd ar-Raziq, *Islam and the Foundations of Political Power*, trans. Maryam Loutfi, ed. Abdou Filali-Ansary (Edinburgh, UK: Edinburgh University Press, 2012).

28 On Shabistarī's analysis of the relationship between religion/Islam and democracy, see his contribution "Demokratie und Religiösität," in *Unterwegs zu einem anderen Islam. Texte iranischer Denker*, ed. and trans. Katajun Amirpur (Freiburg im Breisgau: Herder, 2009), 25–36.

29 Shabistarī, *Ta'ammulātī*, 147.

30 Qantara, "Islam und Demokratie." [My translation]

31 On Shabistarī's conceptualization of the relationship between religion/Islam and human rights, see: "Die Menschenrechte und das Verständnis der Religion," in *Unterwegs zu einem anderen Islam: Texte iranischer Denker*, ed. and trans. Katajun Amirpur (Freiburg im Breisgau: Herder, 2009), 37–44.

32 Shabistarī, *Hirminūtīk, kitāb va sunnat*, 203.

> The 'correctness' of human rights in the contemporary world is based on their being the theoretical antidote. However, the theology and official legal doctrine of Islam are not only poor to counter those dangerous ideas [of Dāʿish], they are even the hotbed of the ISIS and the ISIS followers. Muslim theologians and jurists must acknowledge human rights and, through this historic step, make it possible for Islamic or Muslim culture to live and flourish in today's world.[33]

Therefore, Shabistarī attaches great importance to understanding religiosity as an open and dialogical process. His perception of religious devotion is one of ongoing discussion, a constant exchange between human beings and God. Consequently, religious scripture will always be ambivalent and should always be read, re-read, understood, and interpreted anew. If religious writings were definitive, according to Shabistarī, the conversation between God and human beings would have no meaning. He adds that some people, in refusing to tread the path of exegesis and modern understanding of religious writings, are keen to declare that the communication between God and human beings has already been terminated.[34]

In another one of his writings, Shabistarī arrives at the following conclusion: Since religious belief can only be developed in the state of freedom, external (political) freedom is a necessary precondition to internal (mental) one. Therefore, beliefs can develop disparately under different ways of governing, but their unfolding is generally not compatible with an understanding of religion that is exclusive and totalitarian. If such conditions prevail, belief will decay.[35]

Shabistarī basically sees thinking as a never-ending process: "Thought comes from thinking, and wherever thinking is restricted, this characteristic is lost. Thinking generally means that thoughts may never be restricted. Thinking can thus be equated with freedom."[36] Elsewhere, he writes:

> If we say that knowledge about God and the characteristics that we have ascribed to him represent the definitive definition of God and that nobody has the right to say anything different about God, then we are basically claiming that God is at our disposal. When we confine God, then God denies the human and takes away the freedom of human beings.[37]

According to Shabistarī, revelation is addressed to the free individual. The philosophy of revelation is that God acknowledges the human as a free being. For if God did not recognize human beings as free, He would not speak to them, but would, instead, coerce them.[38]

---

33 Muḥammad Mujtahid Shabistarī, "Mabnāhā-yi tiʾūrik-i Dāʿish" (Theoretic Principles of ISIS), accessed October 10, 2023, http://mohammadmojtahedshabestari.com/مبناهای-تئوریک-داعش/.

34 Shabistarī, *Hirminūtīk, kitāb va sunnat*, 302.

35 Muḥammad Mujtahid Shabistarī, *Imān va āzādī* (Tehran: Ṭarḥ-i Nū, 2000), 64–89.

36 Ibid., 23.

37 Ibid., 26.

38 Ibid., 28.

Shabistarī refuses to look at belief as something rigid or fixed. His argument is that because human beings live in multi-faceted social structures and institutions, their beliefs can also become institutionalized. However, with the institutionalization of religion there is the risk that human beings themselves be rejected by the institution. This happens when God, with whom human beings can freely communicate, is replaced by the institution of religion.[39]

Shabistarī also argues for a dynamic and comprehensive kind of *ijtihād* in order to do justice to the different forms of spirituality. He finds the conventional kind of *ijtihād* insufficient and ultimately doomed to fail. The decision to join a religion, according to Shabistarī, must be completely voluntary. To impose a religion via propaganda, or with force, is to betray that religion:

> If some people believe that belief is a commodity that can be forced onto people with propaganda, on television or through the radio, or that it is a law that can be imposed through the use of force, or that it is an ideology with which to infiltrate the young generation of a society, or that it is nothing more than a science about which books can be compiled, comparable to other disciplines, and necessitating the training of teachers for its transmission, then they are truly wrong. Rather, adopting a belief is the freest, most important and most dignified decision a human being makes. Every word and every deed that subverts this free decision and harms its dignity – even if in the name of religion – is a betrayal of belief itself.[40]

5 Hermeneutics of Scripture

Guided by his understanding of freedom, justice, and plurality of religious readings, Shabistarī endeavors to explain why a 'modern' hermeneutic approach to the Qur'ān is important in his book *Hirminūtīk, kitāb va sunnat* (Hermeneutics, Qur'ān, and Sunna).

The predominant mindset in traditional religious circles in Iran, Shabistarī finds, is that one can and should interpret the Qur'ān and sunna without preconception, interests, or expectations.[41] It is simply assumed that with a knowledge of the Arabic and methodology, one automatically reaches the right conclusions.[42] Thereby, however, the philosophical, theological, anthropological, sociological, or psychological backgrounds of exegetic or legal assumptions are disregarded.[43] Precisely to this point Shabistarī wishes to draw attention: The pre-conceptions, interests, and expectations

39 Ibid., 29.
40 Ibid, 42.
41 Shabistarī, *Hirminūtīk, kitāb va sunnat*, 7.
42 Ibid.
43 Ibid., 8.

of any individual exegete will affect their perspective, and consequently their interpretation of the Qur'ān.[44]

Accordingly, Shabistarī calls his concept of a modern Islamic theology *ilāhiyyāt-i tafahhumī* (hermeneutic theology). It assumes that religions and their prophets are concerned with understanding things and being understood by the human beings. This is an active exchange between the prophets and the people. The prophets understood the things in a certain way and communicated it to the people. The people then tried to understand what was communicated to them. This process of understanding is shaped by the particular circumstances in which people find themselves. "The understanding that comes about later is always shaped by previous understandings."[45]

According to Shabistarī, the starting point in analyzing this problem is establishing where the encounter of religious conviction and modernity occurs, as well as in modernity's conceptualizations of progress. Shabistarī confirms that there is a tense relationship between existing religious belief and modern life. He presents three established positions on the question of the relationship between modernity and the sharia.[46] One group contends, according to Shabistarī, that the sharia is complete and that it answers all possible questions. This group believes that the *ijtihād* of past times is entirely sufficient, the sharia is in no need of renewal, and the gate of *ijtihād* must remain firmly closed.[47] Consequently, possible reinterpretations of the sharia are viewed as dubious and destructive.[48] The second group – whose view Shabistarī describes as *ifrāṭī* (overdoing) – has nothing against Islamic belief, but fundamentally rejects the normative prescriptions of Islam and instead advocates the adoption of other ways of life.[49] Shabistarī is probably alluding to the left and Western-liberal positions within the internal Iranian discourse here, and he rejects it. The third group, which he seems to support, includes reformists and modernists. According to this group, the normative is an important part of Islam, but it must be subjected to contin-

44 Ibid., 9.

45 Muḥammad Mujtahid Shabistarī, *Naqd-i bunyāẓā-yi fiqh va kalām. Sukhanrānīhā, maqālāt, quftgūhā*, online publication, 2017, 342, accessed October 10, 2023, http://mohammadmojtahedshabestari.com/upload/book/Naqd%20Bonyan.pdf.

46 Shabistarī, *Hirminūtīk, kitāb va sunnat*, 92.

47 When speaking of "closing the door/gate of *ijtihād*" (*insidād bāb al-ijtihād*), the main concern is with the question of whether scholars – even after the establishment of the four Sunni legal schools – should continue to practice *ijtihād* as an independent methodology for legal/judicial inquiries. Scholars of Islamic studies and Muslim intellectuals differ on this matter, although recently the question is frequently answered affirmatively. On the issue, see: Abbas Poya, *Anerkennung des Iǧtihād – Legitimation der Toleranz. Möglichkeiten Innerer und äußerer Toleranz im Islam am Beispiel der Iǧtihād-Diskussion* (Berlin: Klaus Schwarz, 2003), 67–69.

48 Shabistarī, *Hirminūtīk, kitāb va sunnat*, 92.

49 Ibid.

uous *ijtihād*, which Shabistarī regards as examining one's theological and anthropological conceptualizations in light of continually evolving scientific knowledge.[50]

When it comes to questioning the relationship between religion and scientific or philosophical assumptions, Shabistarī finds that traditional scholars tend to have a distorted perception. For when there is talk of a contradiction between scientific and philosophical assumptions and religious convictions, many scholars perceive a fundamental contradiction between what a human being may rationally comprehend and that which God has decreed. Consequently, according to Shabistarī, these scholars claim that in the case of a contradiction between human assumption and divine utterance, one must subordinate oneself to God. Shabistarī finds both the question and the answer misleading. As he sees it, there is no contradiction between the human and God, but only between the assumptions of some human beings among themselves. For it is human beings who concern themselves with religious texts and announce their conclusions as so-called 'religious truths.' Ultimately, the contradiction lies between religious assumptions on the one hand, and scientific/philosophical assumptions on the other, in both cases being a matter of human endeavors.[51]

Shabistarī himself is concerned with finding a hermeneutic means of resolving this contradiction. He sees the proper function of a modern exegetical scholar in working out the connections between religious convictions, assumptions, and the interests and expectations of interpreters, for this is the meaning of hermeneutics. In Shabistarī's theory, the foundational assumption is that Muslims in today's world are inevitably influenced by transmitted knowledge, academic discoveries, and religious teachings through the recognition of God and his Prophet, as one structures one's own life according to Muḥammad's words. Consequently, it is not possible to look at the religious conviction or practice of Muslims independent of their embeddedness in the contemporary structures of knowledge.[52]

As a qualified legal scholar, Shabistarī attempts to show that, in order to apply *ijtihād* today, a scholar must have knowledge of a great deal more than the conventions of the traditional discipline of *uṣūl al-fiqh* (juridical methodology). The modern intellectual's knowledge base must surpass legal traditions and methodology; it must engage with hermeneutics and consider such a position in *ijtihād* in order to present a reading of Islam that is suited to our times. A scholar must be aware of the developments and changes in the world as well as of the historicity of religious concepts, i.e., it cannot be assumed that previous sharia rulings are eternally valid. He underpins his theories by basing them on the classical assumptions of the scholarly discipline of *uṣūl*, indicating that classical scholars were already engaged in discussions surrounding these assumptions and the importance of considering them. They had, for exam-

50 Ibid.
51 Ibid., 232f.
52 Ibid., 34–43.

ple, determined that a 'verbal sign' (*dalālat-i lafẓī*) is an 'indistinct sign' (*dalālat-i ẓannī*).[53] Shabistarī is clearly alluding to the fact that every understanding of the Qur'ān – even a legal interpretation of Qur'ānic statements – can be short-lived, and that no reading of textual sources can claim to be definitive.

Such a hermeneutical understanding also guides Shabistarī's approach to political questions. His theory is that the Qur'ān does not prescribe any clearly defined method for exercising authority, nor any particular way of governing. Instead, it simply advocates for just and fair governance. The role of religion is not to prescribe certain methods or patterns of government, but more importantly to determine the general values according to which people ought to be governed.[54] For this reason, the methods and patterns of governance may change. What must remain constant, however, is the maxim of ruling justly – to which all rulers must adhere.[55]

6 The Human Reading of Religion and the Qur'ān

Shabistarī attempts to integrate hermeneutical approaches to religious understanding or Qur'ān exegesis, while considering the role of the preconception, interests, and expectations of the subject. This invariably concords with his theory of a 'human reading of religion' (*qirā'at-i insānī az dīn*).

The exegetes of Qur'ān have long been concerned with finding the most reliable ways to comprehend God's intentions in scripture, be it through philological, historical, literary, or other means. These attempts are based on the assumption that God purposefully created an order of things that encompasses all aspects of life. Moreover, they assert that this intention was announced via Muḥammad in the Qur'ān and that, consequently, believers are obliged to understand and put into practice the meaning of these words as manifested in the Qur'ān.

Shabistarī inverts this thought-process and completely reformulates the question of exegesis. He differentiates between two exegetic perspectives, namely God's perspective of human beings and the human perspective of God. The former assumes that God wants human beings to be religious and that He has determined and specified what He expects of them. In contrast, the latter assumes that human beings voluntarily decide to pursue the path of faith, and that, as part of this process, they turn to God. Ultimately, they define the parameters and content of a religious life for themselves. Shabistarī himself adheres to the second perspective and defines it as the human way of reading religion.[56]

53 Ibid., 32.
54 Ibid., 60.
55 Ibid.
56 Shabistarī, *Ta'ammulātī*, 7.

He elaborates his theory by explaining that this human reading of religion does not reduce the individual to his religious dimension, but also takes into account all the other aspects of being, such as philosophical and legal facets. It is a spiritual positioning that a human being decides to adopt on their own accord, even if in monotheistic religions the exemplary guide is to be found in the lives of the prophets. In this approach, the human being is placed at the center of religious understanding; they are not defined as something abstract, but are instead seen as historical beings with all their experiences and impressions. In this reading, then, religion belongs to the human and not the other way around. In more traditional readings of religion, the individual is often reduced to a bundle of findings and regulations that are holy, transcendental, and far removed from human experience. Thereby, a clear separation between God and human beings is noted. Accordingly, the revelation (*waḥy*) is an extrasensory and sacred phenomenon that penetrates the human world, changing and overturning the individual and his universe. No realization is acknowledged unless it originates from a transcendental space.[57]

Overall, Shabistarī contrasts his "human reading of religion" (*qirā'at-i insānī az dīn*) with the "official reading of religion" (*qirā'at-i rasmī az dīn*), to which he dedicates an entire work.[58] By the "official reading of religion," he means the interpretation of religion and scripture that is heavily bound by sharia and fixated on the state (*fiqhī/ḥukūmatī*). It is a support system for official Iranian politics – which, in turn, supports the 'official reading' as well.[59] This reading uses the 'language of mandatory' (*zabān-i taklīf*) and is always concerned with prescribing what is permitted or prohibited. Such language is employed not only in the realm of rites, but also in connection with other themes and interpersonal political and social questions.[60] In contrast, Shabistarī expresses the need for a 'language of rights' (*zabān-i ḥaqq*). He believes that in today's world one should no longer speak of political, social, or economic questions in the 'language of mandatory,' as today one can choose alternatives. Here, it is not a matter of choice between the 'right' (*ḥaqq*) and the 'wrong' (*bāṭil*), but rather between various options – each of which has its own justifications.[61] In this day and age, there are different answers to questions in all realms of life: from how to enter into marriage or raise children, to ethical and political questions, such as the definition and practice of justice. Nobody can claim that one answer is definitively correct and that all the others are wrong. Thus, according to Shabistarī, a human reading of religion or scripture must consider this multiplicity of possibilities, even in exegetical questions.[62]

---

57 Ibid., 81–85.
58 Shabistarī, *Naqdī*.
59 Shabistarī, *Ta'ammulātī*, 60.
60 Ibid., 60f.
61 Ibid., 61–62.
62 Shabistarī, *Naqdī*, 7.

In addition, Shabistarī is convinced that every text can be read and interpreted in different ways. Since access to religion/Islam is exclusively via scripture, religion/Islam can also be read and interpreted differently. Historical experience supports this position. In every age – depending on preconception, interests, and the contemporary context – differing perceptions of religion and religious texts have been produced. Shabistarī, then, comes to the following conclusion: "The reading of religion is the reading of religious texts, and these can be read, understood, and interpreted in different ways. For this reason, there are multiple readings of religion. The history of religion is a clear testament to this fact."[63]

Consequently, Shabistarī finds it inadequate, given modern hermeneutic analysis, to declare any interpretation of a text as the only viable one. Islam, in all its possible formations, is an interpretation of the Qurʾān and the Sunna – and every understanding of the Qurʾān and Sunna, as long as it does not repudiate Muḥammad, is justifiable.[64]

Shabistarī, however, does not attempt to defend a kind of epistemological relativism. Without engaging in an epistemological discussion, he clarifies that correct interpretations are, indeed, possible and have specific characteristics: "The assumption that there exist different readings of religious texts does not mean that the readings and interpretations of religion may proceed under anarchic conditions, or that all readings are correct and that no discussion is necessary. [. . .] A reading is only acceptable if it is coherent and provable, but these can be numerous."[65]

7 The Qurʾān: How Muḥammad Saw the World

Shabistarī's central thesis is that although the Qurʾān has a divine origin, it concerns Muḥammad in both its wording and content. The claim of Muḥammad was that he was chosen as a prophet and spoke to people through an 'invisible support' (*imdād-i ghaybī*), which he described as revelation (*waḥy*). He did not claim that he received the words and content of the verses he expressed directly from God. Shabistarī suggests that if we read the Qurʾān as a historical text, we clearly see that it reflects a constant and intense dialogue between Muḥammad and his surroundings. In this dialogue, Muḥammad seeks to convince his tribe, the Christians, the Jews, and others of its message and belief in a single God. For this purpose, he uses all viable rhetorical skills, historical clues, and ethical reminders. However, a conversation can only take place if he himself is the creator of the spoken word. A word (*kalām*) that does not

63 Ibid.

64 Ibid., 114.

65 Shabistarī, *Naqdī*, 7 and 151.

come from a speaker is not a word at all and can hardly be understood by its interlocutors.[66]

At the same time, Shabistarī clarifies that, unlike many other scholars, he does not see Muḥammad as a 'mouthpiece' (*kānāl-i ṣawṭī*) who merely transmitted and disseminated words that were supplied to him without any active participation in the production of the text. On the contrary, Shabistarī emphasizes that one can only speak of 'human oration' (*kalām-i insānī*) if the words and content are in fact the product of a human being. The Qur'ān materialized over a period of more than twenty years, and until this day it maintains ongoing dialogue with humans and historical events.[67]

In reference to the German theologian and philosopher Albert Keller (1932–2010/1350–1431), Shabistarī defines language as a means of voicing forms of expression (*shaklā-yi iẓhārāt*) that consists of five elements: speakers, listeners/addressees, context, the community in which the language is spoken, and the designatum or the meaning that is being conveyed.[68] These five elements must all be present in order to discuss language. In the context of the Qur'ān, then, if we say that Muḥammad spoke and was taken seriously, then he must necessarily have been a participant in the coming into being of the spoken word.[69] If Muḥammad functioned merely as a 'mouthpiece' or 'reciter,' then one could not meaningfully comprehend the efforts of scholars of Qur'ānic exegesis to understand the text.[70] Regarding the meaning of revelation (*waḥy*), Shabistarī writes that it is the 'enabling of speaking' (*tawānāsāzi ba sukhan guftan*). To confirm this theory, he refers to the use of the expression in other contexts. In Q 16:68, the discussion is one of the revelations to the bees. Related to bees, *waḥy* means that God gave the bees the ability and competence to make abodes for themselves in mountains, trees, and trellises.[71] The Qur'ānic verses are the result of the *waḥy* but are not the *waḥy* itself. *Waḥy* can be understood as God's way of speaking to Muḥammad, but divine speaking is certainly not a human speaking, rather it is inconceivable. One can imagine this as God teaching Muḥammad.[72] Basically, Shabistarī understands the Qur'ān as a narrative of interpretation in which all natural events, history, and human destiny are understood as acts of God. Above all, this text wants to give the world a religious meaning.[73]

---

66 Cf. Shabistarī's series of articles, "Qirā'at-i nabavī az jahān," Part 1.
67 Ibid.
68 Cf. Albrecht Keller, *Sprachphilosophie* (Freiburg: Karl Alber, 1989), 45–48.
69 Cf. Shabistarī's series of articles "Qirā'at-i nabavī az jahān," Part 1.
70 Ibid.
71 Ibid.
72 Ibid.
73 Ibid.

In the second part of his article series on the *Prophetic Reading of the World*, Shabistarī stresses that he does not intend to explain the nature of revelation with his approach. Instead, he is primarily concerned with how the text of the Qurʾān can be understood. According to him, the Qur'ān is an Arabic text that anyone – Muslim or non-Muslim – can understand. It is human speech – speech of the human Muḥammad – and a monotheistic reading of the world based on the hermeneutic experience of the Prophet Muḥammad.[74] Shabistarī affirms that his approach is based on modern philosophy of language and linguistics and has hardly anything to do with the traditional theological discussions in Islam about the language, content, and origins of the Qurʾān.[75]

With reference to the modern philosophy of language, Shabistarī asserts that we can only understand a text expressed in the language of human beings when the speaker is a human being. The assumption that God sent down the Qurʾān to Muḥammad verbatim who then received it as a recording device and passed it on to the people makes it incomprehensible to both believers and unbelievers alike.[76] Again, the central point in Shabistarī's thesis is that everyone can understand the Qur'ān, a believer as well as an unbeliever. This is because, Shabistarī suggests, the Qur'ān is addressed to all people and communicates with them in a variety of contexts. Furthermore, all Muslim theologians, philosophers, and mystics are concerned with the metaphysical creation of the Qur'ān, no matter the school to which they are attributed. In this respect, it is not possible, given the differing approaches of all the text's interpreters, to arrive at a general understanding of the Qur'ān.[77]

Shabistarī's approach to understanding religion and religious texts is a clear plea for the renunciation of the conventional and dominant Qur'ān exegesis. At the same time, it is also a critique of some modern approaches, such as the so-called scientific exegesis (*tafsīr-i ʿilmī*, which is used by some modern scholars) to prove that Qur'ān antedates all possible scientific knowledge.[78]

Shabistarī's thesis that the Qur'ān is a historical text and the result of Muḥammad's experiences with his environment leads one to a thorough reconsideration of Islamic rules. Shabistarī therefore considers the religious norms formulated in the Qurʾān as spatio-temporal. The fact that a text is historical, according to Shabistarī, means that it originates under certain historical, social, economic, cultural, epistemo-

74 Cf. Shabistarī's series of articles "Qirāʾat-i nabavī az jahān," Part 2.

75 Ibid.

76 Ibid.

77 Ibid.

78 Generally, on 'modern' Qurʾānic exegesis, and specifically the 'scientific' Qurʾānic exegesis, see: Rotraud Wielandt, "Exegesis of the Qurʾān: Early Modern and Contemporary," in *EQ*, ed. Jane Dammen McAuliffe (Leiden/Boston: Brill, 2002), 2: 124–42, here 123ff.

logical, and linguistic conditions, i.e., it arises in a particular shape of possible human expressions. Thus, its meaning is limited to the conditions of that environment. If the environment or some of its conditions change, that meaning can no longer exist.[79] His perspective also calls into question the belief of certain scholars that Qur'ān resolves all possible normative problems and that the normative instructions contained in Qur'ān are valid for contemporary times. He argues against this, stating that the normative concerns addressed in Qur'ān have to do with the prophet's experiences and his specific historical and social contexts. Shabistarī believes that the normative determinants in the Qur'ān are the result of a contextually specific interpretation of contemporary social realities in the Hejaz, aimed at bringing those social realities into a harmonious equilibrium with the will of God. They should therefore not be understood as normative commands for all societies and all times.[80]

8 Conclusion

Shabistarī's thoughts regarding the Qur'ān and its interpretation ask for an understanding of the Qur'ān as the speech of a human being, namely Muḥammad. He additionally suggests that Qur'ān is a text in which Muḥammad imparts his experiences and perceptions of the world he inhabited. The central and immanent message of the Qur'ān, which simultaneously explains its divine origin, is the belief in a single God (*payām-i tavḥīdī*). Guided by this message, Muḥammad perceived the world, history, and humanity all in the ways explained in the Qur'ān. In this framework, and in the context of his immediate social surroundings, Muḥammad also developed normative provisions. Today's scholars can and must interpret the Qur'ān anew, in tune with contemporary living conditions of people and regarding the text's central message (belief in a single God). A particular way of governing is not to be inferred from either the Qur'ān or the actions of Muḥammad and his companions. The main concern of a religion, and thus of Islam, is the safeguarding of freedom, justice, and human rights. Since democratic rule is likely to protect these values, it can only be favored by Islam.

What is particularly striking in Shabistarī's work is the coherence in his thought. His work on freedom and politics is in alignment with his open Qur'ānic exegetical approach. His more recent view of the Qur'ān as a contextually bound narrative is the logical extension of his thesis that the Qur'ān is the speech of Muḥammad. This assumption, in turn, is directly related to his hermeneutic convictions that every text

79 Cf. Muḥammad Mujtahid Shabistarī, "Pishfahmhā-yi tafsīr-i āzād-i Qur'ān-i karīm" (Premises of the Open Understanding of the Honorable Qur'ān), accessed October 10, 2023, http://bamdadaiman.blogfa.com/post/74.
80 Ibid.

and oration is influenced by the preconception, interests, and expectations of the respective author or speaker.

Furthermore, Shabistarī's explanations of the interpretation of religion and religious texts should be understood as interreligious and non-denominational. Even if his starting point is Islam, Shabistarī's thoughts are applicable to the concept of religion in general, and can, therefore, be translate into one's own cultural and religious contexts.

9 Bibliography

9.1 Sources

ʿAbd ar-Rāziq, ʿAlī. *al-Islām wa-uṣūl al-ḥukm: Baḥth fī l-khilāfa wa-l-ḥukūma fī l-islām*. Accessed October 10, 2023. https://mc.dlib.nyu.edu/files/books/nyu_aco000732/nyu_aco000732_hi.pdf.

ʿAbd ar-Raziq, Ali. *Islam and the Foundations of Political Power*, trans. Maryam Loutfi. Ed Abdou Filali-Ansary. Edinburgh, UK: Edinburgh University Press, 2012.

Shabistarī, Muḥammad Mujtahid. "Demokratie und Religiösität." In *Unterwegs zu einem anderen Islam: Texte iranischer Denker*, trans. and edited by Katajun Amirpur. Freiburg im Breisgau: Herder, 2009: 25–36.

Shabistarī, Muḥammad Mujtahid. "Die Menschenrechte und das Verständnis der Religion." In *Unterwegs zu einem anderen Islam: Texte iranischer Denker*, trans. and edited by Katajun Amirpur. Freiburg im Breisgau: Herder, 2009: 37–44.

Shabistarī, Muḥammad Mujtahid. "Qirāʾat-i nabavī az jahān" (Prophetic Reading of the World). Accessed October 10, 2023. http://mohammadmojtahedshabestari.com/مقالات-قرائت-نبوی-از-جهان-در-آمدی-بر-پ/.

Shabistarī, Muḥammad Mujtahid. *Hirminūtīk, kitāb va sunnat*. Tehran: Ṭarḥ-i Nū, 2000.

Shabistarī, Muḥammad Mujtahid. *Imān va āzādī*. Tehran: Ṭarḥ-i Nū, 2000.

Shabistarī, Muḥammad Mujtahid. *Naqdī bar qirāʾat-i rasmī az dīn*. Tehran: Ṭarḥ-i Nū, 2000.

Shabistarī, Muḥammad Mujtahid. *Naqd-i bonyāẕā-ye fiqh va kalām. Sukhanrānīhā, maqālāt, quftgūhā*, online publication, 2017. Accessed October 10, 2023. http://mohammadmojtahedshabestari.com/upload/book/Naqd%20Bonyan.pdf.

Shabistarī, Muḥammad Mujtahid. *Taʾammolātī dar qirāʾat-i insānī az dīn*. Tehran: Ṭarḥ-i Nū, 2004.

Shabistarī, Muḥammad Mujtahid. "Mabnāhā-yi tiʾūrik-i Dāʿish" (Theoretic Principles of ISIS). Accessed October 10, 2023. http://mohammadmojtahedshabestari.com/مبناهایتئوریک-داعِش/.

Shabistarī, Muḥammad Mujtahid. "Pishfahmhā-yi tafsīr-i āzād-i Qurʾān-i karīm" (Premises of the Open understanding of the honorable Qurʾān). Accessed October 10, 2023. http://bamdadaiman.blogfa.com/post/74.

Surūsh, ʿAbdulkarīm. "Ḥukūmat-i dimūkratīk-i dīnī." In ʿAbdulkarīm Surūsh. *Farbihtar az idʾūlūzhī*. Teheran: Muʾassasa-yi Farhingī-yi ṣirāṭ, 1994: 273–83.

Surūsh, ʿAbdulkarīm. *Kalām-i Muḥammad, rūʾyā-yi Muḥammad* (Muhammad's Speech, Muhammad's Dreams). Online publication: Suqrāṭ, 2018. Accessed October 10, 2023. https://www.goodreads.com/book/show/53400859.

9.2 Secondary Literature

Algar, Hamid. "ʿAllāma Sayyid Muḥammad Ḥusayn Ṭabāṭabāʾī: Philosopher, Exegete, and Gnostic." Accessed October 10, 2023. http://www.islamiclaws.com/marja/Allameh%20Tabatabai.pdf.

Amirpur, Katajun, ed. and trans. *Unterwegs zu einem anderen Islam: Texte iranischer Denker*. Freiburg: Herder, 2009.

Amirpur, Katajun, "Islam und Demokratie – Die Geschichte einer Aneignung." *Blätter für deutsche und internationale Politik* 56, no. 11 (2011): 87–95.

Dabashi, Hamid. *Theology of Discontent: The Ideological Foundation of the Islamic Revolution in Iran*. New Jersey & London: Transaction Publishers, 2006.

Hāshimī, Muḥammad Manṣūr. *Dīnandishān-i mutajaddid: Rūshanfikrī-yi dīnī az Sharīʿatī tā Malīkyān*. Tehran: s.n., 2006.

Keller, Albrecht. *Sprachphilosophie*. Freiburg: Karl Alber, 1989.

Poya, Abbas. *Anerkennung des Iǧtihād – Legitimation der Toleranz. Möglichkeiten Innerer und äußerer Toleranz im Islam am Beispiel der Iǧtihād Diskussion*. Berlin: Klaus Schwarz, 2003.

Poya, Abbas. *Denken jenseits von Dichotomien. Iranisch-religiöse Diskurse im postkolonialen Kontext*. Bielefeld: transcript, 2014.

Qantara. "Warum Islam und Demokratie zusammen passen." Accessed October 10, 2023. http://de.qantara.de/content/interview-mit-mohammad-mojtahed-shabestari-warum-islam-und-demokratie-zusammen-passen?qt-nodes_popularity=2.

Rahnema, Ali. "Ali Shariati: Teacher, Preacher, Rebel." In *Pioneers of Islamic Revival*, edited by Ali Rahnema,. Kuala Lumpur: Strategic Information Research Development, 2008: 208–45.

Sadri, S. Mahmoud. "Sacral Defense of Secularism: The Political Theologies of Soroush, Shabestari, and Kadivar." *International Journal of Politics, Culture and Society* 15, no. 2 (2001): 257–70.

Sarkohi, Arash. *Der Demokratie- und Menschenrechtsdiskurs der religiösen Reformer in Iran und die Universalität der Menschenrechte*. Würzburg: Ergon, 2014.

Vahdat, Farzin. "Post-revolutionary Islamic Modernity in Iran: The Intersubjective Hermeneutics of Mohamad Mojtahed Shabestari." In *Modern Muslim Intellectuals and the Qurʾān*, edited by Suha Taji-Farouki. Oxford: Oxford University Press, 2004: 193–224 [Reprint 2010].

Wasmaqī, Siddīqa. *Masīr-i payāmbarī* (The Way of Prophecy). Teheran: s.n., 2021. Accessed October 10, 2023. www.zeitoons.com/wp-content/uploads/2021/11/path-of-prophetood.pdf.

Wielandt, Rotraud. "Exegesis of the Qurʾān: Early Modern and Contemporary." In *EQ*, edited by Jane Dammen McAuliffe. 6 vols. Leiden/Boston: Brill, vol. 2, 2002: 124–42.

"Zindagīnāma" (biography of Jaʿfar Subḥānī). Accessed October 10, 2023. https://www.noorsoft.org/fa/Mostabser/View/63493.

"Zindagīnāma" (biography of Muḥammad Mujtahid Shabistarī). Accessed October 10, 2023. http://mohammadmojtahedshabestari.com/زندگی%e2%80%8cنامه/.

"Zindagīnāma" (biography of Nāṣir Makārim Shīrāzī). Accessed October 10, 2023. https://makarem.ir/main.aspx?lid=0&typeinfo=22&catid=30399.

Ramy Abdin

Imran Nazar Hosein

1 Introduction

If Imran Hosein's hermeneutics were to be described in one sentence, it would the following: "He counts the stars and calls them all by name" (Psalm 147:4). Indeed, Hosein's Qur'ānic hermeneutics is unprecedented and creative in its approach. He regards the Qur'ān as the absolute truth and thus as the essence of Muslim epistemology. However, this very truth can only be obtained by believers who recite and reflect upon the Qur'ān, making them sensitive to divine guidance. Hosein provides not only a methodology for studying and understanding the Qur'ān but also a unique way of reciting the Qur'ān that is subservient to the divine system of time. In this chapter, I would like to examine his hermeneutics by discussing his methodology and mode of recitation, thereby presenting the ideas of an intriguing but also controversial Islamic scholar of the twenty-first century.

2 Biography

In 1942/1361, Imran Khalid Ahmad Nazar Hosein was born in the Caribbean nation of Trinidad and Tobago. As a descendent of Muslims from Hyderabad Deccan (India) who migrated to the "New World," he grew up in an educated family, whose heritage goes back to the historical region of Khorasan.[1] His father Ibrahim Nazar Hosein worked as a teacher in Trinidad, where he later became a school principal.[2] According to Imran Hosein, his father fell deeply in love with a Christian woman, though they never married due to religious differences over how to educate children. Although his father did not seem to be very fond of the traditional Islamic way of life, he held his religious identity and heritage in high regard, which is why he wanted his children to grow up as Muslims. His father therefore married a young Muslim widow from Bihar[3] who had lost her husband one year after marrying him. Imran Hosein de-

1 Khorasan is located in Central Asia, which includes the states of Iran, Afghanistan, Tajikistan, and Turkmenistan.

2 Because there is no extensive written biography of Imran Hosein, his biographical data comes from an interview in which he narrates his life and career in detail. The interview was published on April 29, 2021, on YouTube.

See Imran Hosein, "Flashback Zindagi of Sheikh Imran Hosein Complete Interview: 'The Story of My Life,'" YouTube video, posted by "Sheikh Imran Hosein," April 29, 2021, https://youtu.be/VLgvRVrQAkU?t=307.

3 Bihar is a state located in eastern India.

https://doi.org/10.1515/9783111320069-016

scribes their marriage as the greatest sacrifice his father made, as he was obviously heartbroken after learning that the Christian woman whom he had loved had died shortly after he married.[4]

When Imran Hosein reached high school age, his father sent him to Queen's Royal College; however, Hosein wanted a religious education and chose to attend a Roman Catholic college instead.[5] His first contact with Islamic religious education took place outside of formal education, namely through a Muslim youth organization, where he learned the basics of Islamic practice. Encouraged by his instructors at the organization, Imran Hosein discovered his talent for oration and lecturing, which may have paved the way for his scholarly activity as a lecturer.[6] After his father passed away, he had to start helping to provide for his family at age seventeen by working as a school teacher.[7]

When Imran Hosein turned eighteen, he met Indian Islamic scholar Muhammad Fazlur Rahman Ansari (d. 1974/1394), a shaykh from the Qādirī Sufi tradition and a former student of the renowned, celebrated Pakistani philosopher and poet Muhammad Iqbal (d. 1938/1357). At that time, Ansari was visiting Trinidad on a lecture tour. It was the first time that Hosein had met an Islamic scholar who was able to explain modern science through the lens of Islam and, in particular, through the Qur'ān.[8] The encounter with Ansari must have had a huge impact on Hosein's way of thinking, as he dedicated a whole book series to Ansari, who gave him the epistemological foundations for Qur'ānic understanding.[9] At the farewell ceremony for Ansari in Trinidad, the Indian scholar called on the audience to choose five young men to study the religious sciences of Islam. Inspired by Ansari, Hosein applied for a scholarship to study at the prestigious university of al-Azhar in Cairo, which he eventually obtained. He thus traveled to Egypt, where he began learning Arabic and the basic Islamic sciences.[10] However, Hosein was not content with his studies in Egypt, as he missed Ansari's contemporary, creative approach to religion.[11] Imran Hosein therefore interrupted his studies in Egypt and went to Pakistan to continue them when Ansari invited him to study at the Aleemiyah Institute of Islamic Studies in Karachi, which was, at that time, still in the process of being established. Hosein considers his decision to leave Cairo to study in Karachi as the wisest decision he ever made.[12]

---

4 Hosein, "The Story of My Life," https://youtu.be/VLgvRVrQAkU?t=911.

5 Ibid., https://youtu.be/VLgvRVrQAkU?t=1539.

6 Ibid., https://youtu.be/VLgvRVrQAkU?t=1705.

7 Ibid., https://youtu.be/VLgvRVrQAkU?t=1899.

8 Ibid., https://youtu.be/VLgvRVrQAkU?t=1915.

9 Imran Hosein, *An Introduction to Methodology for Study of the Qur'an* (San Fernando Trinidad and Tobago: Imran N. Hosein Publications, 2016), 104f.

10 Hosein, "The Story of My Life," https://youtu.be/VLgvRVrQAkU?t=2137.

11 Ibid., https://youtu.be/VLgvRVrQAkU?t=2563.

12 Ibid., https://youtu.be/VLgvRVrQAkU?t=2658.

Alongside their religious studies at the Aleemiyah Institute, Ansari directed his students to study at a public university as well. Consequently, Hosein participated in a study program at the University of Karachi, which included the subjects philosophy, psychology, and sociology.[13] However, Hosein had to interrupt his studies at the public university as he was suffering from health issues.[14] Hosein also had to pause his studies in Karachi when his mother fell ill in order to return to Trinidad, where he stayed for several months. He then embarked upon a journey to visit different cities around the world, such as London, Paris, Geneva, Istanbul, and Ankara, marking the beginning of his activity as a traveling scholar.[15] He eventually returned to Karachi in September 1968/Jumādā al-Ākhira 1388.[16] During this time, Hosein studied different subjects such as the philosophy and history of science under respected scholars and academics like Burhan Ahmed Farooqi (d. 1995/1415), Syed Sibte Nabi Naqvi,[17] and Yousaf Saleem Chishti (d. 1984/1404) as part of his religious studies at the Aleemyiah Institute. Imran Hosein finally graduated from the institute in 1971/1391. Shortly after graduating, he wrote his first book, entitled *Islam and Buddhism*, which was based on the answers to his final exam at Aleemiyah.[18]

At the age of twenty-nine, Imran Hosein returned to Trinidad to work at the Ministry of Foreign Affairs. Due to an apparent need for local diplomats, the ministry instructed Hosein to study international relations at the University of the West Indies. He then traveled to Switzerland to enroll in a postgraduate course in international economics at the University of Geneva. However, he never completed the final thesis for his economics degree. Hosein explained that the reason for discontinuing his studies was his lack of knowledge in the field of Islamic eschatology, which prevented him from finishing his thesis.[19]

Imran Hosein returned to Trinidad to pursue a career as a diplomat, even though Ansari expressed dissatisfaction with Hosein's chosen occupation. Hosein eventually gave up his work at the Ministry of Foreign Affairs in order to dedicate his life to teaching and lecturing on Islam. Hosein returned to Karachi to work at the Aleemiyah, where he took the position of the Principal of the institute when his former teacher Ansari died. However, Hosein did not agree with the institute's new affiliation with the Barelvi movement, which ultimately led to his departure.[20]

13 Ibid., https://youtu.be/VLgvRVrQAkU?t=3069.

14 Ibid., https://youtu.be/VLgvRVrQAkU?t=3243.

15 Ibid., https://youtu.be/VLgvRVrQAkU?t=3382.

16 Ibid., https://youtu.be/VLgvRVrQAkU?t=3809.

17 Although it was not possible to find any biographical data, he certainly lived in the twentieth century. His works include *Islam and Contemporary Science* and *The Evolution of Cosmology in the Hands of the Muslims Up to the Time of al-Biruni.*

18 "Hosein, The Story of My Life," https://youtu.be/VLgvRVrQAkU?t=3867.

19 Ibid., https://youtu.be/VLgvRVrQAkU?t=4476.

20 Ibid., https://youtu.be/VLgvRVrQAkU?t=5454.

Hosein subsequently moved to New York, where he began working in the Pakistani Islamic organization Tanzeem e-Islami, whose founder and leader at the time was Islamic scholar Israr Ahmad (d. 2010/1431). During this period, Imran Hosein devoted himself intensively to the subject of the end times, which became the focal point of his scholarly activity. He began writing his best-selling book *Jerusalem in the Quran*, which he published in 2001/1422. Due to religious differences with the Tanzeem e-Islami, and with Israr Ahmad in particular, he parted ways with the organization and left the United States after September 11, 2001. Since then, he has lectured extensively on the end times and his interpretations of major prophecies in countries such as Malaysia, Pakistan, Belgium, Serbia, Albania, and Great Britain.[21] He has authored more than thirty monographs, most of them on the end times, including *Surah Al-Kahf and The Modern Age*; *An Islamic View of Gog and Magog in the Modern World*; *Dajjal, the Qur'an, and Awwal Al-Zamaan: The Antichrist, the Holy Qur'an, and the Beginning of History*; and *Methodology for the Study of the Qur'an.*[22] Moreover, his online presence must be emphasized: besides his own website, where he sometimes publishes short essays, he consistently releases video lectures on the You Tube channel "Sheikh Imran Hosein." It can be argued that his online presence has contributed to his growing reputation. With over 400,000 followers, the amount of attention Hosein receives exceeds that of many Islamic scholars who are similarly active on YouTube.[23]

3 Imran Hosein's Hermeneutical Framework

The main focus of this section will be on Imran Hosein's method of interpreting the Qur'ān, which, due to his research interest, is necessarily related to his eschatological interpretations. However, it is crucial to understand some of his political and religious views, as they provide insights into his hermeneutical approach.

Hosein's primary political problem is the dominant political ideology of Western civilization, which is based on the idea of secularism, i.e., the separation of religion and politics. He argues that separating politics from the divine is a form of blasphemy against God, which inevitably leads to a godless way of living in human society based on sin and *shirk* (i.e., idolatry). Instead, human civilization should be built on the

21 Ibid., https://youtu.be/VLgvRVrQAkU?t=5867.

22 "Free E-Books," *imranhosein.org*, accessed February 29, 2024, http://imranhosein.org/n/e-books/.

23 Link to his YouTube channel, "Sheikh Imran Hosein," *YouTube,* accessed October 12, 2023, https://www.youtube.com/@SheikhImranHosein.

foundations of faith.[24] However, this foundation does not necessarily have to be Islamic,[25] as Hosein also supports Russia as an Orthodox Christian civilization and views it as a faith-based country.[26] Furthermore, in sura 19, *al-Kahf* (The Cave), Hosein establishes his view on the parable of Dhū l-Qarnayn, whom he calls the "Great Traveler." The Great Traveler demonstrates how a ruler is supposed to wield his (political) power, grounding his authority in faith and thus reigning with justice and empathy.[27]

Moreover, Imran Hosein describes the current global economic system as bogus and utterly *haram* (i.e., prohibited by Islamic law). He refuses to use paper money and condemns the banking system, which is based on usury (*ribā*), for being a sophisticated power network that exploits and enslaves, ultimately only enriching the ruling class. He therefore calls for the establishment of an alternative monetary system based on the use of gold and silver, a currency with intrinsic value that can be traced back to its material quality.[28]

Imran Hosein is a proponent of interreligious dialogue, in particular with followers of Orthodox Christianity, as he regards Orthodox Christians as believers and thus as allies of Islam in the battle against Western hegemony. He derives his view directly from the Qur'ān, arguing that God himself distinguished between true and false Christians:

> In all of the above the Qur'ān is severely critical of Rūm of the West, rather than Rūm of the East. As a consequence, it must be recognized that the Qur'ān does not treat all Christians (and Jews) the same way. Rather it recognizes that some Christians (and Jews) are indeed believers, while the rest are not.[29]

Thus, his regular travels to Orthodox Christian countries such as Serbia,[30] Armenia,[31] Bulgaria,[32] and Russia demonstrate his endeavors to build good relations between Islam and Orthodox Christianity. The pinnacle of his interreligious efforts so far was a

---

24 Imran Hosein, *Jerusalem in the Quran: An Islamic View of the Destiny of Jerusalem* (San Fernando: Masjid Jāmi'ah, 2011²), 179–96.

25 Imran Hosein, *Constantinople in the Quran* (San Fernando: Imran N. Hosein Publications, 2018), 79–84.

26 Imran Hosein, *The Quran the Great War and the West* (San Fernando: Imran N. Hosein Publications, 2018), 21–24.

27 For more details, see Ramy Abdin, "The Faces of Gog and Magog in Islam," in *Gog and Magog: Contributions Toward a World History of an Apocalyptic Motif*, ed. Georges Tamer, Andrew Mein, and Lutz Greisiger (Berlin/Boston: De Gruyter, 2023), 810–13.

28 Hosein, *Jerusalem in the Quran*, 210–33.

29 Hosein, *Constantinople in the Quran*, 80.

30 Imran Hosein, "Islamic Eschatology Faculty of Law, Belgrade Serbia By Sheikh Imran Hosein," YouTube video, posted by "Sheikh Imran Hosein," October 13, 2015, https://www.youtube.com/watch?v=sKzkgfpcUvo.

31 Imran Hosein, "Sheikh Imran Hosein Visits the Armenian Genocide Museum and Memorial in Yerevan, Armenia," YouTube video, posted by "Sheikh Imran Hosein," July 14, 2022, https://www.youtube.com/watch?v=x_7O7g42k9o.

32 Imran Hosein, "My Visit to Bulgaria – A Brief Report," YouTube video, posted by "Sheikh Imran Hosein," July 13, 2023, https://www.youtube.com/watch?v=R76HSr21cPI.

lecture he gave on Islamic eschatology at the Moscow State University during a conference organized by the Russian philosopher Alexander Dugin (b. 1962/1382).[33]

Without a doubt, Hosein's understanding functions within the framework of what he calls Islamic eschatology. For him, the alliance between Islam and Orthodox Christianity reveals the opposition between faith and the present-day, anti-Christian secular powers,[34] whose political and ideological impact will culminate in the appearance of the *Dajjāl*, the Great Deceiver (or Antichrist).[35] Muslims have a responsibility to acknowledge the reality that currently surrounds them by correctly understanding global developments, which only becomes possible if believers develop a spiritual sensibility for divine guidance. This guidance ought to be received through the Qur'ān itself, as Imran Hosein regards it as a manifestation of absolute truth. To recognize this truth, the believer must meet certain conditions, enabling them to penetrate the reality of the world with their enlightened spiritual gaze.[36]

4 Imran Hosein's Hermeneutics

4.1 Grasping the Stars: How to Study the Qur'ān

As previously mentioned, Imran Hosein regards the Qur'ān as absolute truth, not only in a theological sense but also as an epistemological basis for explaining actual world events: "The Qur'an has declared that its primary function is to explain all things: '. . . And We have sent down to thee (O Muhammad) the Book (i.e., the Qur'an) which explains all things . . .' (Qur'an, al-Nahl, 16:89)."[37] In his *Introduction to Methodology for Study of the Quran*, he states that the Qur'ān must be studied like the stars in the sky, as the locations of the stars point the traveler in the right direction:[38] "The Qur'ān, directs our attention to the stars above and informs us that they were placed in the sky as an adornment and in order to beautify the sky."[39] The Qur'ānic symbolism of the stars, which are depicted as lamps (*maṣābīḥ*) and as an adornment of the heavens (*zīnat al-kawākib*), is a simile that alludes to a way of studying and understanding the divine message:[40] "But it is part of the divine wisdom with which He [God] teaches

33 "Sheikh Imran Hosein with Prof. Alexander Dugin & Russian Intellectuals," YouTube video, posted by "Hamzah H," November 11, 2015, https://www.youtube.com/watch?v=lB8OUE_QUMs.

34 Hosein, *The Quran the Great War and the West*, 21–24.

35 Hosein, *Jerusalem in the Quran*, 93–106

36 Hosein, *An Introduction to Methodology for Study of the Qur'an*, 4–13.

37 Hosein, *Jerusalem in the Quran*, 3f. Please note that all translations of Qur'ānic verses as well as the parentheses in the verses have been taken from Imran Hosein's works.

38 Hosein, *An Introduction to Methodology for Study of the Qur'an*, 99–102.

39 Ibid., 99.

40 Ibid., 99, 105.

methodology for the study of the Qur'ān, that he has described those very stars in the sky above as lamps in his work."[41] Hosein continues: "The reason why He has described the stars as lamps is because they have an important additional function to perform, since Allah Most High has put them there so that they can be used to locate direction."[42] Moreover, Imran Hosein advocates a holistic reading and examination of the Qur'ān carried out in the same way that the astronomer studies the location of the stars in order to determine their constellations, which reveal one's own location as well as the right direction one must follow:

> In order for the stars to function as lamps with which we can locate direction for travel, or direction of the Qiblah (i.e., direction of prayer), it is necessary for us to first study how they are interconnected and how they combine to form a whole. We must carefully study the stars to grasp the "big picture," and only then can we read our present location and find the desired direction.[43]

The interconnectedness of the stars presented here must be understood as a reference to the hermeneutical approach of Imran Hosein's teacher, Fazlur Rahman Ansari. In his major work *The Qur'anic Foundations and Structure of Muslim Society*, Ansari points out that the Qur'ān has a semantic unity characterized by the interrelation and interconnection of every verse and chapter. This interrelation forms the harmonious totality of the divine message, which he calls the "system of meaning."[44] To be more specific, Qur'ānic passages and even whole chapters should not be read or studied in isolation. Rather, the believer must examine them holistically in order to grasp and locate an intelligible system, which all together is grounded in the logic of theoretical and religious consciousness.[45] Hosein links this system of meaning to the Qur'ānic symbolism of the stars in heaven. The perfect harmony depicted in the sky indicates the soundness of God's revelation, whose verses do not contain any inconsistencies or contradictions.[46] A student of the Qur'ān should therefore view the divine text as a unity in order to avoid the error of utilizing fragmentary exegesis, which leads to false conclusions, not only about the Qur'ān and its intended meanings but also about the given reality:

> Hence in studying the verses of the Qur'ān to discover the system of meaning of a subject, the student must always struggle to locate that explanation which harmoniously integrates as a perfect whole all the data in the Qur'ān on that subject. In the same way that no star is out of place in the sky above, so too no verse is out of place in the Qur'ān. No verse contradicts another verse, and no verse cancels another verse![47]

41 Ibid., 100.
42 Ibid., 100.
43 Ibid., 102.
44 Ibid., 104.
45 Ibid., 104.
46 Ibid., 105–7.
47 Ibid., 108.

In order to study the Qur'ān successfully, the student must live in accordance with Qur'ānic virtues: "No Muslim can possibly succeed in studying the Qur'ān while still borrowing or lending money on interest."[48] It can be noted that Hosein's hermeneutics begins with the (spiritual) state and religious practice of the reader, which together influence the implementation of hermeneutics. Moreover, proficiency in Arabic is necessary, otherwise it becomes impossible to understand the Qur'ān's deeper meaning; thus, any translation of the Qur'ān is insufficient to substitute the original.[49]

The Qur'ān as the source of absolute truth is supposed to dominate Muslim epistemology to such an extent that it even renders the hadith a subsidiary source.[50] For the collections of Prophetic tradition, in contrast to the Qur'ān, are to be regarded as a source of relative truth; in other words, the hadith is fallible, especially when it is used as an epistemological basis. That does not mean that Imran Hosein rejects the hadith as a religious source – he interprets several prophecies about the end times,[51] which can indeed be found in the collections of Prophetic tradition. However, the hadith must never contradict the Qur'ān. If there is a blatant contradiction between the Qur'ān and the hadith, the Qur'ān will always be given preference, regardless of the hadith's degree of authenticity.[52] Nevertheless, Hosein argues for a holistic reading of the sunna, similar to his Qur'ānic approach, as the Prophetic tradition has the capacity to elaborate and amplify God's revelation.[53] The student must use the Qur'ān as his or her primary source of interpretation. Hosein provides an example of the way that his methodology should be implemented by presenting an analysis of the release of Gog and Magog, peoples who will appear in the end of times to cause great corruption on earth. He states that in order to bring forth a proper study of Gog and Magog and their release, one must start by looking into the primary source – namely the Qur'ān – to detect any references to this issue.[54]

The Qur'ān contains exactly two references to Gog and Magog, namely in Q 18:94 and Q 21:95. As Hosein explains: "Our first effort must therefore be to discover the principle of unity that links these two [verses]."[55] The first verse depicts Gog and Magog as part of the story of Dhū l-Qarnayn, who is implored by the oppressed to protect them from the corruption of Gog and Magog. It reads:

> They said: O Dhul-Qarnain! Gog and Magog commit Fasad (corruption, awesome oppression and wickedness) in the Land. May we render you tribute (i.e., can we pay you something) in order

---

48 Ibid., 70.

49 Ibid., 63.

50 Ibid., 226–40.

51 For example, Hosein, *Jerusalem in the Quran.*

52 Hosein, *An Introduction to Methodology for Study of the Qur'an*, 226–40.

53 Hosein, *Jerusalem in the Quran*, 117–24.

54 Ibid., 117–19.

55 Ibid., 118.

> that you might erect a barrier between us and them (so that we can be protected from them)? (Qur'ān, al-Kahf, 18:94)[56]

Imran Hosein explains this verse by identifying Gog and Magog as descendants of Adam, and thus as responsible human beings who cannot be defeated in battle, as even an individual like Dhū l-Qarnayn, blessed by God with power and divine authority, would not seek to fight them.[57] Hosein aims to prove his thesis with the following hadith: "It will be under such conditions that Allah will reveal to Jesus ('alaihi as-Salām) these words: I have brought forth from among My servants such people against whom none will be able to fight [. . .]."[58] This is why Dhū l-Qarnayn was only able to imprison Gog and Magog by building a barrier to protect the other group of people.[59] However, the next verses proclaim that this barrier will come down when the warning (*wa'd*) of the Lord comes to pass.[60] Hosein states: "Which W'ad [sic] (warning) was he referring to? The answer is quite clearly revealed in the famous hadith in which the Prophet (sallalahu 'alaihi wa-sallam) narrated ten major Signs of the Last Day. Among the ten was the release of Gog and Magog."[61] Imran Hosein concludes that, if the barrier is destroyed, and Gog and Magog are consequently set free, it will be a major sign that the end times have begun, marking the commencement of the Last Age.[62]

The second reference to Gog and Magog in the Qur'ān gives a further indication of their emergence. It reads: "But there is a ban on a City which We have destroyed; that they (the residents) shall not return – until Gog and Magog are let through (their barrier) and they swiftly swarm down from every height (or spread in every direction). (Qur'ān, al-Anbiyā', 21:95–6)."[63] The verse links the release of Gog and Magog to the return of a people to a city. To know whether Gog and Magog were released or not, one must identify the town mentioned in the verse. Hosein points again to specific hadith traditions pertaining to Gog and Magog that help to geographically locate their activity.[64] In one tradition, Gog and Magog pass by Lake Tiberias, were they drink its water: "Allah will send Gog and Magog and they will swarm down from every slope. The first of them will pass the lake of Tiberias (i.e., the Sea of Galilee) and drink out of it. And when the last of them passes, he will say: There was once water there identified this town with Jerusalem."[65] Another geographical hint in the hadith

---

56 Ibid., 117, translated by Imran Hosein.
57 Ibid., 119f.
58 Ibid., 119, original emphasis.
59 Ibid., 120.
60 Ibid.
61 Ibid.
62 Ibid., 120.
63 Ibid., 123, translated by Imran Hosein.
64 Ibid., 123–25.
65 Ibid., 124, original emphasis.

literature is provided by their arrival at the mountain of aṭ-Ṭūr in the Holy Land.[66] A different tradition seems to provide an alternative name for the aforementioned mountain: the hadith states, “Gog and Magog would walk until they reach the mountain of al-Khamr, and it is a mountain of Bait al-Maqdis (i.e., Jerusalem).”[67] After collecting every piece of information about the geographic locations of Gog and Magog’s activity, Hosein deduces that the only town that comes into question is Jerusalem, since no other city is mentioned in the hadith literature.[68] Moreover, Jerusalem was actually destroyed twice in ancient history, which Hosein also uses as evidence to identify the town as Jerusalem, as the verse describes a town that God has previously destroyed. Hence, the prohibition against returning to this city can only pertain to the Jews, as they were the people who ruled over Jerusalem.[69] Looking at the given reality, Hosein explains,

> But the more important implication of the return of the Jews to the “city” (Jerusalem) and the establishment of the State of Israel is that Gog and Magog have now fulfilled that stage of their mission referred to in *Surah al-Anbiyah, 21:96,* i.e., they have descended from every height (or spread out in every direction) and have taken control of the world. The world order which brings the Jews back to the Holy Land is that of Gog and Magog![70]

Hosein views the founding of Israel as the realization of the Qur’ānic prophecy. Therefore, the Jews’ return to the Holy Land indicates the fulfillment of God’s promise, namely the release of Gog and Magog, who, Hosein says, stand for a corrupt world order based on blasphemy and exploitation, whose activity is located within Western civilization.[71]

In sum, Hosein’s method can be described as follows:

- Locating the principle of unity in the primary source: gathering every issue-specific piece of information from the Qur’ān by performing a holistic reading of its verses and chapters
- Locating the principle of unity in the secondary source: gathering every piece of issue-specific information from the hadith by performing a holistic reading of the particular traditions connected to the issue examined
- Verification: linking primary and secondary sources together in relation to the information gathered (i.e., examining analogies, correlations, and possible contradictions between the Qur’ān and the hadith)

---

66 Ibid.
67 Ibid, original emphasis.
68 Ibid., 124f.
69 Ibid., 123–25.
70 Ibid., 125, original emphasis.
71 Abdin, “The Faces of Gog and Magog in Islam,” 813f.

- Performing a *Zeitdiagnostik* (diagnosis of the times):[72] a multi-perspective observation of (global) changes throughout human history until the present age. This multi-perspectivity is based on approaches and insights gained not only from religious sources but also from a number of disciplines such as history, philosophy, the social and political sciences, economics, etc.
- Deduction: formalizing correlations between religious insights and time-diagnostic knowledge (i.e., insights into general changes in human civilizations as well as their ongoing development) to provide explanations of certain events and potentially identify the manifestation and fulfillment of religious prophesies

Hosein's methodology can be characterized as a multidisciplinary approach that is not restricted to religious sources; rather, due to its time-diagnostic aspect, it incorporates modern disciplines in order identify the Signs of the Last Age. However, Imran Hosein also discusses the application of what he regards as the wrong methodology, with which students will potentially draw false conclusions. In this respect, he reinforces the importance of the religious sources by emphasizing their correct hierarchy: "In order to determine the identity of the 'town' referred to above we examined all the material that exists in the Qur'an and Ahadith that pertain to Gog and Magog. We found only one town that is linked to Gog and Magog, and it is Jerusalem."[73] Hosein continues, "If we use the wrong methodology of commencing our study with the Hadīth, rather than the Qur'ān, we can be misled to misinterpret the Hadīth."[74]

Imran Hosein refers to the mistake of linking the release of Gog and Magog to the return of the Messiah, which apparently occurs in particular hadith traditions. One hadith states that Jesus will be sent by God to Damascus. The Messiah will then confront the Antichrist and kill him at the Gate of Lod in Jerusalem. After that, God will *send* Gog and Magog, who will swarm down from every height. This will occur after Jesus has completed his mission and taken his chosen followers to safety.[75]

If we start our analysis with the hadith while overlooking the references to Gog and Magog in the Qur'ān, we risk drawing false conclusions due to our fragmented observation of the subject at hand. Indeed, this tradition evokes the impression that Gog and Magog will be released after the appearance of the Messiah.[76] However, Hosein rejects this view, arguing that, in the hadith, the Prophet Muḥammad spoke

72 I discuss the term *islamische Zeitdiagnostik* in relation to an Islamic perspective to Modern technology in Ramy Abdin: "Die Auswüchse der Technik: Ein islamischer Weckruf," in *Normativität des Korans im Zeichen gesellschaftlichen Wandels: Theologische und religionspädagogische Perspektiven*, ed. Fahima Ulfat et al. (Baden-Baden: Nomos, 2022), 277–306.

73 Hosein, *Jerusalem in the Quran*, 30.

74 Hosein, *An Introduction to Methodology for Study of the Qur'an*, 76f.

75 Muslim b. al-Ḥajjāj, *Sahih Muslim*, 2937a (Book 54 [*Kitāb al-fitan wa-ashrāṭ as-sāʿah*], Chapter 20: ad-Dajjāl, Hadith 134), https://sunnah.com/muslim:2937a.

76 Hosein, *An Introduction to Methodology for Study of the Qur'an*, 81f.

about the sending of Gog and Magog (*yabʿath Allahu Yaʾğūğ wa-Maʾğūğ*) depicted in the Qurʾān, which is not the same as their release:

> Not only did [the] wrong methodology of ignoring the Qur'ān lead to an error concerning the time of the release of Gog and Magog into the world, and negligence in respect of locating the explanation in the Qur'ān for today's universal Fasād, but it also led to a misinterpretation of the Hadīth which spoke of Allah sending Gog and Magog rather than releasing them.[77]

Furthermore, Hosein points to the diminishing water level of Lake Tiberias, which confirms the hadith of Gog and Magog drinking and completely consuming its water. Hence, the drying up of Lake Tiberias indicates their release and activity. Hosein also presents a reference to the hadith of the Prophet Muḥammad having a vision of Gog and Magog.[78] In his vision, the Prophet witnessed them making a hole in the barrier built by Dhū l-Qarnayn, prompting him to warn the Arabs that a great evil was approaching: "This vision not only clearly indicated that the release of Gog and Magog commenced in the Prophet's life-time, but it also prophesied the eventual destruction of the Arabs by Gog and Magog at that time when "Khabath,"[79] or scum, prevails."[80]

4.2 Gazing to the Moon: How to Recite the Qurʾān

Imran Hosein regards the Qurʾān as a scripture which ought to be read constantly by the believer. The very meaning of the name Qurʾān (recitation) already indicates one of its most important purposes designated by divine wisdom, namely, to be a book of recitation. Moreover, one cannot study or understand the Qurʾān without continuously reciting it.[81] This being the case, there are certain conditions that must be met in order to recite the Qurʾān in the right manner. The first is to seek God's protection from Satan before reciting the Qurʾān (*al-istiʿadha*). Second, the believer must recite it in a melodious and pleasant way (*tartīl*). It is also obligatory to give one's undivided attention while listening to Qurʾānic recitation. Moreover, the reader should be firm in his or her belief that the Qurʾān is a book of *healing* (*shifāʾ*), as it was sent down to preserve the physical, spiritual, and moral health of believers.[82] Hosein adds that only those who have pure hearts can study and understand the Qurʾān. This purity is obtained by having faith in the one God as well as in the absolute validity of the

77 Ibid., 81.

78 Ibid., 82–85.

79 Derived from the Arabic term *khabath*, which means "impurity" or "filth." See Edward William Lane, *Madd al-qāmūs: An Arabic-English Lexicon* (Beirut: Libraire Du Liban, 1968), 2: 694, art. kh-b-th.

80 Hosein, *An Introduction to Methodology for Study of the Qur'an*, 84.

81 Imran Hosein, *The Qurʾān and the Moon: Divine Methodology for Monthly Recitation of the Qurʾān* (San Fernando: Imran N. Hosein Publications, 2020), 2f.

82 Ibid., 2–4.

Qur'ān.[83] In other words, only a believer has the foundation and therefore the capacity to understand the divine revelation. Aside from that, the believer must devote themselves to truth by standing up against oppression if necessary: "In order for someone to qualify to study the Qur'ān, his heart must not only have faith in the Qur'ān as the Word of the One God, but it must also be faithful to 'Truth'. This writer hastens to declare, as emphatically as he can, that Truth has zero tolerance for 'oppression.'"[84] For Hosein, the believer must be a representative of veracity and bravery who reinforces his or her integrity in the presence of corruption and evil. Therefore, his hermeneutics is grounded in spirituality (i.e., the spiritual state of the believer) as well as in moral conduct; the beginnings of Qur'ānic understanding lie within believers themselves.

The Qur'ān must be recited in the way that God himself recited it: "And when We have recited this Qur'ān, O Muhammad, you must follow that way of recitation! (Qur'ān, al-Qiyāmah, 75:18)"[85] Hosein understands this verse as a divine command to recite the Qur'ān in the right manner. According to the hadith, the Prophet Muḥammad reviewed the Qur'ān once a year in the month of Ramadan together with the Archangel Gabriel. Therefore, believers, too, must follow the practice of the Prophet, requiring them to likewise recite the Qur'ān at least once a year, at Ramadan.[86] However, Imran Hosein rejects arbitrary methods of recitation, as he believes that the order and partition of the suras was determined by God himself.[87] To identify the correct way to perform the monthly recitation, one must understand the very purpose of the suras and their arrangement.[88]

According to Edward William Lane, a sura is a "row of stones or bricks of a wall: any degree of a structure [. . .]. Hence its application in relation to the Ḳur-án [sic], [to signify a chapter thereof,] because each of what are thus called forms one degree, or step, distinct from another, or [leading] to another [. . .]."[89] For Hosein, the sura is a guiding line dividing the Qur'ān into thirty parts, meaning that one part can be recited on each day of the month. Nevertheless, he emphasizes that the correct partitioning of the Qur'ān must consider each sura a self-contained chapter, which should induce the reader to always finish reciting the chapter and thus to be faithful to the divine division and arrangement.[90] This is why Hosein does not accept the common partitioning of the Qur'ān, which divides each part of it (*juz'*) into twenty pages, regardless of whether the sura has ended or not. This common method of partitioning potentially

83 Ibid., 3.
84 Ibid.
85 Ibid., 5.
86 Ibid., 5–8.
87 Ibid., 23–26.
88 Ibid., 9f.
89 Lane, *Madd al-qāmūs: An Arabic-English Lexicon*, 4: 1465, art. s-w-r.
90 Hosein, *The Qur'ān and the* Moon, 9–21.

interrupts the Qur'ān in the middle of a sura, breaking it up into pieces,[91] which contravenes the divine arrangement: "It is now universally accepted that the first *Juz* of the *Qur'ān* ends at Verse 141 or 142 of *Sūrah al-Baqarah,* and that the second *Juz* ends at Verse 253 of *Sūrah al-Baqarah,* etc."[92] He adds, "Our readers would be horrified to realize that the *Qur'ān* was chopped 21 times in this arbitrary division of *Ajza.*"[93] Hosein goes even further by arguing for a prohibition against any subdivision of the suras. He aims to prove his opinion with the following verse: "Say to them O Muhammad: I am a warner who warns you, and I do so in a manner which is clear and without ambiguity – warning you of Divine anger which descends on the Muqtasimīn, who divide the Qur'ān arbitrarily into bits and pieces. (Qur'ān, al-Hijr, 15:89-93)."[94]

Hosein suggests arranging the Qur'ān in a different way, namely by dividing up its parts according to the length and arrangement of the suras, while avoiding any subdivision:[95] "Such recitation can be at *intervals* of daily recitation for the entire Qur'ān to be completed in one lunar month, and this is the way that Allah Most High recited the Qur'ān Himself (through the Angel Gabriel during the month of Ramadān)."[96] He argues that God placed the longer suras at the beginning of the Qur'ān in order to test believers:

> Allah Most Wise placed the longest *Sūrah* of all at the beginning of the *Qur'ān* for a specific reason and purpose, and our view is that He did so to *test us* and to force us to *think.*" [. . .] Allah Most High placed the longest *Sūrah* of the Qur'ān at the very beginning of the Qur'ān to test us to see whether we would respect the boundaries which He placed in the Qur'ān, and, as a consequence, would recognize the whole of *Sūrah al-Baqarah* as the first *Juz* to be recited [. . .].[97]

Hosein also proposes partitioning the Qur'ān into thirty parts for daily recitation, making it possible to complete the recitation within a lunar month while paying respect to the divine division.[98] He adheres to lunar time in order to connect the reciter with the system of time ordained by God. This connection is supposed to alter and correct temporal perception so that the human heart beats in harmony with the lunar cycle and thus with the divine system of time. In this regard, Imran Hosein emphasizes the hadith in which the Prophet Muḥammad prophesized that time will move faster in the end times. Hosein argues that most of humankind is already experiencing an acceleration of time, which is not only an indication of the fulfillment of this prophecy but also evidence of the beginning of the end times. Therefore, the adher-

---

91 Ibid., 23–27.
92 Ibid., 23, original emphasis.
93 Ibid., 26, original emphasis.
94 Ibid., 27, translated by Imran Hosein.
95 Ibid., 29f.
96 Ibid., 18, original emphasis.
97 Ibid., 18, original emphasis.
98 Ibid., 29.

ence to solar time distorts temporal perception, depriving the observer of any divine blessing in relation to temporal experience. Solar time is a temporal perception that was established by the activity of the Antichrist,[99] who Hosein believes is already at work influencing global developments in our given reality.[100]

So, if the Qur'ān is to be recited, the first chapter, sura 1, *al-Fātiḥa* (The Opening), which is the key to the Qur'ān, should be recited each time before the believer starts reading.[101] Hosein begins his division with sura 2, *al-Baqara* (The Cow); each of the first six parts (*ajzā'*) consists of one chapter:

- Day 1: sura 2, *al-Baqara*
- Day 2 sura 3, *Āl 'Imrān* (The House of 'Imrān)
- Day 3: sura 4, *an-Nisā'* (Women)
- Day 4: sura 5, *al-Mā'ida* (The Table Spread)
- Day 5: sura 6, *al-An'ām* (The Cattle)
- Day 6: sura 7, *al-A'rāf* (The Heights)[102]

The first six suras can be described as the section of long suras. On the seventh day, the reciter must combine the eighth and ninth suras – namely *al-Anfāl* and *at-Tawba* – and read them as one part, as the eighth is short compared to those that precede it, and, exceptionally, the ninth does not begin with the *basmala*,[103] which signifies a divine allusion to the fact that these two are supposed to be read together. On the seventh day, the middle section of Hosein's division begins. It is the section in which each part consists of two suras. The sixteenth chapter, *an-Naḥl* (The Bees), is the only exception in this section, as it is a long sura. More importantly, the following two suras, *al-Kahf* and *al-Isrā'*, should be recited together due to their divine interrelation:[104]

- Day 7: sura 8, *al-Anfāl* (The Spoils) – sura 9, *at-Tawba* (Repentance)
- Day 8: sura 10, *Yūnus* (Jonah) – sura 11, *Hūd*
- Day 9: sura 12, *Yūsuf* (Joseph) – sura 13, *ar-Ra'd* (The Thunder)
- Day 10: sura 14, *Ibrāhīm* (Abraham), – sura 15, *al-Ḥijr* (The Rocky Tract)
- Day 11: sura 16 an-Naḥl (The Bees)
- Day 12: sura 17, *al-Isrā'* (The Night Journey) – sura 18, *al-Kahf* (The Cave)
- Day 13: sura 19, *Maryam* (Mary) – sura 20, *Ṭā Hā*
- Day 14: sura 21, *al-Anbiyā'* (The Prophets) – sura 22, *al-Ḥajj* (The Pilgrimage)
- Day 15: sura 23, *al-Mu'minūn* (The Believers) – sura 24, *an-Nūr* (Light)[105]

99 Ibid., 45–70.
100 Hosein, *Jerusalem in the Quran*, 93–106.
101 Hosein, *The Qur'ān and the Moon*, 15–17.
102 See ibid., 30.
103 Ibid.
104 Ibid., 37.
105 Ibid., 37f.

On the fifteenth day, the reciter will be able to witness the full moon at night. Reciting sura an-Nūr (the Light) at full moon symbolizes the zenith of the lunar month, indicating the harmony between the recitation and the divine system of time. Moreover, the reciter is reminded of the commencement of the second half of the month, in which the moon begins to fade.[106]

The sixteenth day initiates the penultimate section, in which each part constitutes of two to five suras.

- Day 16: sura 25, *al-Furqān* (The Criterion) – sura 26, *ash-Shuʿarāʾ* (The Poets)
- Day 17: sura 27, *an-Naml* (The Ants) – sura 29, *al-ʿAnkabūt* (The Spider)
- Day 18: sura 30, *ar-Rūm* (The Byzantines) – sura 33, *al-Aḥzāb* (The Parties)
- Day 19: sura 34, *Sabaʾ* (Sheba) – sura 37, *aṣ-Ṣāffāt* (Those Ranged in Ranks)
- Day 20: sura 38 *Ṣād* – sura 40, *Ghāfir* (The Forgiver)
- Day 21: sura 41, *Fuṣṣilat* (Expounded) – sura 44, *ad-Dukhān* (Smoke)
- Day 22: sura 45 *al-Jāthiya* (Upon their Knees) – sura 49, *al-Ḥujurāt* (The Private Apartments)
- Day 23: sura 50, *Qāf* – sura 54, *al-Qamar* (The Moon)[107]

When the reciter reaches the twenty-first day, he or she enters the last third of the month. In the month of Ramaḍan, this is the stage when the Night of Decree (*laylat al-qadr*) occurs: "It would be on that blessed night that the angels and the *Rūh* (i.e., the *Rūh al-Quddus* or Holy Spirit, who is Gabriel) would descend from above in order to perform every errand assigned to them by Allah Most High."[108] Imran Hosein affirms this statement by referring to a verse from the sura *Fuṣṣilat*, which is actually recited on the twenty-first day: "But, behold, as for those who declare, 'Our Lord-God is Allah,' and then steadfastly pursue the right way – upon them do angels descend, (and this occurs on *Lailatul Qadr*) [saying:] 'Fear not and grieve not, but receive the glad tiding of that paradise which has been promised to you!' (Fussilāt, 41:30)"[109] In this regard, Hosein also points to a verse in sura 44, *ad-Dukhān*, which refers to the revelation of the Qurʾān in the Night of Decree: "Behold, from on high have We bestowed it on a blessed night: for, verily, We have always been warning [man]. On that [night] was made clear, in wisdom, the distinction between all things [good and evil] at a behest from Ourselves: for, verily, We have always been sending [Our messages of guidance]. (ad-Dukhān, 44:3-5)."[110]

The twenty-third day resembles the beginning of the last week of the lunar month, as well as the phase of the old moon. As the moon wanes, its light is succeeded by the glimmer of the stars that adorn the night sky during this period of time. Ac-

106 Ibid., 38.
107 Ibid., 38–41.
108 Ibid., 39, original emphasis.
109 Ibid.
110 Ibid., 40.

cordingly, the reciter reads the chapters "*an-Najm*" (the Stars) and "*al-Qamar*" (the Moon) on this day, symbolizing the moon being replaced by the stars.[111]

The last section consists of the short suras. By completing it, the reciter finishes his recitation of the Qur'ān:

- Day 24: sura 55, *ar-Raḥmān* (The Compassionate) – sura 58, *al-Mujādila* (She Who Disputes)
- Day 25: sura 59, *al-Ḥashr* (The Gathering) – sura 68, *al-Qalam* (The Pen)
- Day 26: sura 69, *al-Ḥāqqa* (The Undeniable Reality) – sura 74, *al-Muddaththir* (The Covered One)
- Day 27: sura 75, *al-Qiyāma* (The Resurrection) – sura 79, *an-Nāzi'āt* (The Wresters)
- Day 28: sura 80, *'Abasa* (He Frowned), – sura 88, *al-Ghāshiya* (The Overwhelming Event)
- Day 29: sura 89, *al-Fajr* (The Dawn) – sura 97, *al-Qadr* (The Decree)
- Day 30: sura 98, *al-Bayyina* (The Clear Proof) – sura 114, *an-Nās* (Mankind)[112]

Imran Hosein emphasizes the twenty-ninth day of Ramaḍan by identifying it as the Night of Decree. He argues that the twenty-ninth is the day in the holy month most neglected by Muslims, as their attention is generally drawn to the odd nights before. Thus, Hosein arranges the sura on the twenty-ninth day in such a way as to commemorate the greatest of all nights. The suras *al-Fajr* (the Dawn), *al-Layl* (the Night), and *al-Qadr* (The Decree) are therefore among the chapters that must be recited on this day. On the same day, the reciter must look out for the new moon in the night sky. If the moon becomes visible on the twenty-ninth day, the reciter ought to complete his recitation; otherwise, he delays his *khatm* (i.e., the completion of his recitation) until the thirtieth day.[113]

5 Conclusion

Hosein's hermeneutics is embedded within an eschatological framework. In this regard, his main emphasis lies on the Qur'ān, as it is a means of explaining everything, encompassing not only the field of interpretative exegesis but also the aspect of knowing and explaining (contemporary) world events. He argues for the application of holistic hermeneutics, as the Qur'ān has its own system of meaning that establishes an intelligible unity with interrelations between every single chapter and verse. In order to study and eventually understand the divine message, believers must be pious and virtuous in their conduct so that they become sensitive to divine guidance. Moreover,

111 Ibid., 41.
112 Ibid., 41–43.
113 Ibid., 42f.

the Qur'ān must be recited in a manner that is in harmony with the divinely ordained system of time. This system takes effect when believers recite the Qur'ān according to the correct division of the suras and, in this process, adhere to the Islamic calendar (i.e., lunar time). With this adjustment to the divine system of time, temporal experience changes and the exegete gains the epistemological ability to read God's signs, not only in the Qur'ān, but also in the given reality. Hosein's hermeneutics can be described as end times exegesis encompassed by a mystical concept of religious practice. The latter might be the result of the influence of his mentor Ansari, as his teacher was a shaykh of the Qādirī Sufi tradition. His interest in the end times was possibly evoked by Muhammad Iqbal, who wrote some poetic verses about Gog and Magog: "Set loose are all the hordes of Gog and Magog to the Muslim eye manifest is the meaning of the word yansilun (i.e., the two verses of the Qur'ān, al-Anbiyāh', 21:95–96, which end with the word yansiloon)."[114] Imran Hosein's Qur'ānic hermeneutics is definitely creative and unique, but also controversial in conventional Islamic scholarship, as he diverges from traditional Qur'ānic exegesis. He is a scholar who openly discusses his political views, which makes him a potential target of harsh criticism. However, it is precisely because of his political and economic interest that the element of diagnosing the times becomes a part of his hermeneutics, which makes his understanding of the Qur'ān and the hadith even more intriguing.

6 Bibliography

6.1 Sources

Hosein, Imran. *Jerusalem in the Quran: An Islamic View of the Destiny of Jerusalem*. San Fernando: Masjid Jāmi'ah, 2011^2.

Hosein, Imran, *An Islamic View of Gog and Magog in the Modern World*. San Fernando: Masjid Jāmi'ah, 2012^2.

Hosein, Imran. *An Introduction to Methodology for Study of the Qur'an*. San Fernando: Imran N. Hosein Publications, 2016.

Hosein, Imran. *Constantinople in the Quran*. San Fernando: Imran N. Hosein Publications, 2018.

Hosein, Imran. *The Qur'an, the Great War and the West*. San Fernando: Imran N. Hosein Publications, 2018.

Hosein, Imran. *The Qur'ān and the Moon: Divine Methodology for Monthly Recitation of the Qur'ān*. San Fernando: Imran N. Hosein Publications, 2020.

114 Imran Hosein, *An Islamic view of Gog and Magog in the Modern World* (San Fernando: Masjid Jāmi'ah, 2012^2), 186.

6.2 Online Sources

Hosein, Imran. "Islamic Eschatology Faculty of Law, Belgrade Serbia By Sheikh Imran Hosein." YouTube video. Posted by "Sheikh Imran Hosein." October 13, 2015. https://www.youtube.com/watch?v=sKzkgfpcUvo.

Hosein, Imran. "Flashback Zindagi of Sheikh Imran Hosein Complete Interview: "The Story of My Life." YouTube video. Posted by "Sheikh Imran Hosein." April 29, 2021. https://youtu.be/VLgvRVrQAkU?t=307.

Hosein, Imran. "Sheikh Imran Hosein visits the Armenian Genocide Museum and Memorial in Yerevan, Armenia." YouTube video. Posted by "Sheikh Imran Hosein." July 14, 2022. https://www.youtube.com/watch?v=x_7O7g42k9o.

Hosein, Imran. "MyVisit to Bulgaria – A Brief Report." YouTube video. Posted by "Sheikh Imran Hosein." July 13, 2023. https://www.youtube.com/watch?v=R76HSr21cPI.

"Sheikh Imran Hosein with Prof. Alexander Dugin & Russian Intellectuals." YouTube video. Posted by "Hamzah H." November 11, 2015. https://www.youtube.com/watch?v=lB8OUE_QUMs.

6.3 Secondary Literature

Abdin, Ramy. "Die Auswüchse der Technik: Ein islamischer Weckruf." In *Normativität des Korans im Zeichen gesellschaftlichen Wandels: Theologische und religionspädagogische Perspektiven*, edited by Fahima Ulfat et al. Baden-Baden: Nomos, 2022: 277–306.

Abdin, Ramy. "The Faces of Gog and Magog in Islam." In *Gog and Magog: Contributions Toward a World History of an Apocalyptic Motif*, edited by Georges Tamer, Andrew Mein, and Lutz Greisiger. Berlin/Boston: De Gruyter, 2023: 787–817.

Lane, Edward William. *Madd al-qāmūs: An Arabic-English Lexicon*. 8 Vols. Beirut: Libraire Du Liban, 1968.

Muslim b. al-Ḥajjāj. *Sahih Muslim*. Accessed November 22, 2023. https://sunnah.com/muslim:2937a.

Munirul Ikhwan

Muhammad Quraish Shihab

1 Introduction

No one in the history of modern Indonesia has been more prolific in Qur'ān-related works than Muhammad Quraish Shihab. He is one of a few Indonesian graduates of al-Azhar in Cairo who enjoy intellectual prestige and social fame at a national level. Holding a doctoral degree in Qur'ānic exegesis from al-Azhar, Shihab has built his intellectual project around Qur'ānic interpretation. He seems to have filled the gap that Hamka[1] left in the national scene when he moved to Jakarta in 1984/1404. Shihab has successfully maintained his networks with the capital's elites, scholars, and bureaucrats. The active role he plays in giving religious lectures and writing on Qur'ān-related topics, and his succession of appearances on some national television programs, have increased his public fame as an expert on the Qur'ān.

There have been some studies dealing with Shihab's exegetical works, ranging from examinations of the tradition, methods, and ideology upon which his interpretations are based,[2] to observations of his thematic (*mawḍūʿī*) method of interpretation[3] and discussions of his Qur'ānic interpretation of particular issues such as gender[4] and non-Muslims.[5] However, these studies have not comprehensively dealt with Shihab's philosophy of interpretation. This chapter examines Shihab's Qur'ānic hermeneutics, which has become the basis of his intellectual project to articulate Islam in a modern Muslim society. It contends that, although Shihab is attracted to the ideas of Muslim

1 An acronym for Haji Abdul Malik Karim Amrullah (1908–81/1326–1401), a prominent Islamic scholar, Qur'ānic exegete, politician, and writer.

2 Johanna Pink, *Sunnitischer Tafsīr in der modernen islamischen Welt* (Leiden: Brill, 2011); Johanna Pink, "Tradition and Ideology in Contemporary Sunnite Qur'ānic Exegesis: Qur'ānic Commentaries from the Arab World, Turkey and Indonesia and their Interpretation of Q 5:51," *WI* 50 (2010): 3–59.

3 Muhammadiyah Amin and Kusmana observe that Shihab bases his thematic interpretation on the linguistic and historical analysis of the Qur'ānic text, and that he reflects on contemporary Muslim realities through his understanding of the text. Muhammadiyah Amin and Kusmana, "Purposive Exegesis: A Study of Quraish Shihab's Thematic Interpretation of the Qur'an," in *Approaches to the Qur'an in Contemporary Indonesia*, ed. Abdullah Saeed (Oxford: Oxford University Press, 2005).

4 Anshori, "Penafsiran Ayat-Ayat Jender Dalam Tafsir al-Mishbah" (PhD dissertation, UIN Jakarta, 2006).

5 Eva F. Nisa, "Non-Muslims in the Qur'ān: A Critical Study on the Concept of Non-Muslims in Tafsir Al-Mishbāh of Muhammad Quraish-Shihab" (master's thesis, Leiden University, 2004).

Note: The basis of this chapter is my PhD thesis, "An Indonesian Initiative to Make the Quran Down-to-Earth: Muhammad Quraish Shihab and His School of Exegesis" (Freie Universität Berlin, 2015).

https://doi.org/10.1515/9783111320069-017

reformists, his Qur'ānic hermeneutics is still unable to free itself from Arabicate interpretive tradition.[6]

2 A Biographical Sketch of Muhammad Quraish Shihab

Muhammad Quraish Shihab's life and career have been described at length in various Indonesian academic and biographical works. Shihab is a leading Indonesian exegete. He brings together what most Indonesian Muslims consider to be the parameters of religious authority. Born on February 16, 1944/Ṣafar 22, 1363 in Rappang, South Sulawesi, into an educated family of the *'Alawī sāda,*[7] he was a passionate student of Islam from a very young age. After finishing his elementary schooling in 1956/1375, he was sent to Malang, East Java, to attend secondary school and to learn Islam at a *pesantren* (traditional religious school), Darul Hadits al-Faqihiyah, under the guidance of a captivating Tarim-born teacher, al-Ḥabīb ʿAbd al-Qadīr b. Aḥmad Bilfaqīh (1896–1962/1313–1381), whose charisma and spirituality impressed Shihab.

Shihab traveled to Cairo in 1958/1377 to continue his studies at al-Azhar. He started secondary school as al-Azhar was undergoing a reform project supervised by reformist Grand Imam Maḥmūd Shaltūt (1893–1963/1310–1382). Shihab writes that his greatest motivation for going to Cairo was to pursue studies in Qur'ānic exegesis. He was deeply impressed by the lectures his father gave on the Qur'ān during his childhood as well as by his father's open-mindedness.[8] His grandfather, Ali Shihab (1864–1915/1280–1333), was active in Jamʿiyyat al-Khayr, a reform-oriented Arab educational

6 By "Arabicate interpretive tradition," I mean a tradition of interpretation that claims that the truth of religious guidance is to be sought within the structures of the revealed text. In this regard, employing detailed grammatical and syntactic analysis becomes an important hermeneutical step toward understanding indications and instructions within the revealed text. Wael Hallaq employs the term "Arabicate hermeneutics" to designate this idea. Wael Hallaq, *Sharīʿa: Theory, Practice, Transformations* (Cambridge et al.: Cambridge University Press, 2009), 449, 502.

7 The *'Alawī sāda* (sing. *sayyid*) families claim to be the descendants of al-Imām Aḥmad b. ʿĪsā l-Muhājir (d. 956/345) who migrated from Basra with his family to Hadramawt in 952 AD. They claim their descent from the Prophet Muḥammad through his daughter Fāṭima (d. 632/11), who married ʿAlī b. Abī Ṭālib (d. 661/40), a cousin of the Prophet. In Hadramawt, the *'Alawī sāda* occupy the highest social stratum. In a diaspora community like Indonesia, they still enjoy this prestigious status and continue to practice endogamous marriage. Natalie Mobini-Kesheh, *The Hadrami Awakening* (Ithaca, NY: Southeast Asia Program Publication, 1999), 25; Frode Jacobsen, *Hadrami Arabs in Present-Day Indonesia: An Indonesia-Oriented Group with an Arab Signature* (Oxon/New York: Routledge, 2009), 23, 64.

8 His father, Abdurrahman Shihab, was a merchant, politician, preacher, and professor of Qur'ānic exegesis at the Muslim University of Indonesia and then at the State Islamic University (IAIN) of Makassar. His mother, Asma, was a sister of Sultan Rappang and of Bugis ethnicity.

institution and social organization based in Batavia (now Jakarta).[9] His father, Abdurrahman (1905–1986/1323–1406), was also educated at this institution and conveyed his religious knowledge to his children, which Quraish Shihab later realized was a reflection of the thought of Muslim reformist scholars like Muḥammad ʿAbduh (1849–1905/1265–1323), Muḥammad Iqbāl (1877–1938/1294–1357), and Abū l-Aʿlā Mawdūdī (1903–79/1321–99).[10] His family's reformist tendencies explain why Quraish Shihab was sent to Cairo for religious studies instead of Mecca or Hadramawt, which had become common destinations for the religious schooling of most children of the Hadrami diaspora.[11] It should also be noted that the majority of Indonesians saw Cairo in the twentieth century as a new learning destination offering a cosmopolitan milieu and modern experience.[12]

While studying at al-Azhar, Shihab stayed at Madīnat al-Buʿūth al-Islāmiyya, a dormitory for international students that was constructed between 1954/1373 and 1959/1378. The dormitory was part of President Gamal Abdel Nasser's strategy to include al-Azhar in his modernization and international agenda.[13] A small enrollment of international students allowed young Shihab to make personal contacts with some prominent Azharī scholars. He has stated that he had a very close relationship with Shaykh ʿAbd al-Ḥalīm Maḥmūd (1910–78/1328–98) – Dean of the Faculty of Theology at the time – with whom he engaged in rich discussions on various religious topics. Shihab was impressed by Maḥmūd's ability to reconcile Sufism and rationalism.[14] Moreover, the abundance of books in Cairo catered to Shihab's passion for reading. He was

---

9 Jamʿiyyat al-Khayr, which was founded in 1901 by an Arab community in Batavia (now Jakarta), but officially recognized by the Dutch colonial government in 1905/1323, is often called the first modern Islamic institution in the Dutch East Indies. Its founders were the nascent reformist Hadrami elite, primarily from the families of Shihāb and al-Mashhūr. The institution once invited teachers from around the Arab world to instruct its students. Aḥmad as-Sūrkatī (1872–1943/1289–1362), a reformist teacher from Sudan who was deeply influenced by the progressive thought of Muḥammad ʿAbduh and Muḥammad Rashīd Riḍā, was one of the invitees. He founded al-Irshād in 1915/1333 after a dispute with the Arab *sāda* of Jamʿiyyat al-Khayr regarding the position of *sāda* among non-*sāda* Arabs. See Mobini-Kesheh, *The Hadrami Awakening*, 36; Deliar Noer, *The Modernist Muslim Movement in Indonesia 1900–1942* (Singapore/New York: Oxford University Press, 1973), 58; Mona Abaza, *Indonesian Students in Cairo: Islamic Education, Perceptions and Exchanges* (Paris: Association Archipel, 1994), 41, 54f.; M.C. Ricklefs, *A History of Modern Indonesia since c. 1200* (Hampshire: Palgrave, 2001[3]), 215.

10 Mauluddin Anwar, Latief Siregar, and Mustofa Hadi, *Cahaya, Cinta dan Canda M. Quraish Shihab* (Ciputat: Lentera Hati, 2015), 12–16; Muhammad Quraish Shihab, *Membumikan Al-Quran: Fungsi dan Peran Wahyu dalam Kehidupan Masyarakat* (Bandung: Mizan, 2013), 20.

11 Abaza, *Indonesian Students in Cairo*, 40; Nico J.G. Kaptein, *Islam, Colonialism and the Modern Age in the Netherlands East Indies* (Leiden: Brill, 2014), 5.

12 Michael Laffan, *Islamic Nationhood and Colonial Indonesia* (London/New York: Roudledge Curzon, 2003), 127; Michael Laffan, "An Indonesian Community in Cairo: Continuity and Change in a Cosmopolitan Islamic Milieu," *Indonesia* 77 (2004): 1–26.

13 Abaza, *Indonesian Students in Cairo*, 115.

14 Anwar, Siregar, and Hadi, *Cahaya, Cinta Dan Canda M. Quraish Shihab*, 67.

interested in reading the works of ʿAbbās Maḥmūd al-ʿAqqād (1889–1964/1306–84), a prominent Egyptian journalist, poet, and literary critic, some of whose works deal with Islam and the Qurʾān.

In 1967/1387, Shihab completed his bachelor's degree in *tafsīr* (Qurʾānic exegesis), and two years later, he obtained his master's degree in the same field, with a thesis entitled *al-Iʿjāz at-tashrīʿī li-l-Qurʾān al-karīm* (The Legal Inimitability of the Qurʾān). After completing his master's degree, he returned to Makassar, where he was entrusted with a number of academic and public positions. In 1975/1395, he married Fatmawati Assegaf, who came from a family of Hadrami *sāda* living in Surakarta. In 1980/1400, he traveled to Cairo to pursue his doctorate at al-Azhar and eventually defended his thesis, *Naẓm ad-durar li-l-Biqāʿī: taḥqīq wa-dirāsa* (*Naẓm ad-durar* by al-Biqāʿī: Edition and Studies), which was awarded the grade *summa cum laude* (*mumtāz maʿa martabat ash-sharaf al-ūlā*).

Shihab launched his intellectual and public career when he moved to Jakarta in 1984/1404. Appointed a lecturer at the State Islamic University (IAIN) of Jakarta, he gradually became familiar to Muslim academics and state executives. He was co-chairperson of the Indonesian Council of Ulama (Majelis Ulama Indonesia, MUI). From 1992/1412 to 1998/1418, he was also Rector of IAIN Jakarta, before being appointed Minister of Religious Affairs by President Suharto in his last term in 1998/1418. Shihab's rising public career during the authoritarian regime of Suharto can be explained by the fact that his articulations of Islam never contradicted government policy, which sought to oppress any kind of "political Islam." Rather, he supported the idea of Islam as a foundation of personal piety and civic pluralism that could work within the government's developmentalist (*pembangunan*) project. Furthermore, we cannot neglect his position as a religious teacher to Suharto's Cendana family, which demonstrates the trust the family had in him when it came to religious matters. His appointment as a minister was short, as Suharto's government collapsed in May 1998/1418. He was then appointed Indonesian Ambassador to Egypt and Djibouti in Cairo from 1999/1419 to 2002/1422, a period in which he had many opportunities to consult more sources to complete his magnum opus, *Tafsir al-Misbah*.

After retiring from government positions, Shihab had more time to focus on cultural and intellectual projects. He founded the Center for Qurʾānic Studies (Pusat Studi al-Qur'an, PSQ) on September 18, 2004/Shaʿbān 4, 1425, which was designed as an incubator for the dissemination of his vision of "indigenizing Qurʾānic values within a pluralistic society."[15] By indigenizing, the institution aims to introduce programs that fit different segments of society from various cultural backgrounds. The center designs its curricula to produce new generations of (Indonesian) Qurʾānic exegetes, qualified religious teachers and preachers, and conveyers of *wasaṭiyyat al-Islām*, a perceived ideal and best practice of Islam. Shihab invites colleagues who

15 For further information about PSQ, see PSQ's profile on https://www.psq.or.id

share the same mission to develop strategies and to participate in the processes of teaching, training, and supervising the participants in its programs.

From 2010/1431 to 2017/1439 Shihab had been entrusted with leading the Indonesian Branch of the World Association for al-Azhar Graduates (WAAG). The foundation of the WAAG is al-Azhar's most recent strategic move to not only maintain its alumni networks but also to promote the orthodoxy of *wasaṭiyya,* the vision of an ideal Islam that rejects two opposing extremes: religious excessiveness and disobedience. Together with some international Muslim scholars and dignitaries, Shihab has been elected a member of the Muslim Council of Elders (Majlis al-Ḥukamāʾ), an international Islamic body founded on July 18, 2014/Ramaḍān 21, 1435. The Council is headed by the Grand Shaykh of al-Azhar, Aḥmad aṭ-Ṭayyib (b. 1946/1365), and aims to promote peace and oppose violence in the Islamic world and beyond.[16]

During the Islamic month of Ramadan, Shihab regularly appears on national television programs. He was a host of *Lentera Hati* on Metro TV, which featured Shihab's thematic interpretation of the Qurʾān (*tafsīr mawḍūʿī*). In 2004/1425, *Lentera Hati* was replaced by *Tafsir al-Misbah,* in which Shihab delivered a lecture based for the most part on his sequential verse-by-verse interpretation of the Qurʾān (*tafsīr taḥlīlī*). He has also appeared on SCTV and RCTI to deliver short religious sermons. The rise of digital media among Indonesian Muslims prompted Shihab and his daughter, Najwa Shihab, to launch and host the *Shihab & Shihab* talk show program, in which they discuss Islam and the everyday Muslim life. The program airs on Narasi TV, a digital television channel founded by Najwa in 2018/1439.

Shihab has authored more than fifty books (all in Indonesian). His works can be classified into five categories: The first deals with Qurʾānic exegesis and includes works such as *Wawasan al-Quran* (Qurʾānic Insights, 1996/1417) and *Tafsir al-Misbah* (2000/1421), which appeal to readers with religious education; and *Secercah Cahaya Ilahi* (A Glimpse of Divine Light, 2000/1421) and *Tafsir al-Lubab* (2012/1433), which address average readers. The second category comprises books on Qurʾānic hermeneutics, such as *Membumikan al-Quran* (Indigenizing the Qurʾān, 1992/1412) and *Kaidah Tafsir* (Principles of Interpretation, 2013/1434). The third includes works on theological topics, such as *Logika Agama* (The Logics of Religion, 2005/1425) and *Sunni-Syiah Bergandengan Tangan Mungkinkah?* (Sunni-Shiite Hand in Hand – Is It Possible? 2007/1427), which reflect Shihab's intention to reconcile religion and reason as well as Sunni and Shiite relations. The fourth category concerns Islamic legal issues and includes titles like *Jilbab: Pakaian Wanita Muslimah* (Headscarf: Muslim Female Dress, 2004/1424) and *Quraish Shihab Menjawab 1001 Soal Keislaman* (Quraish Shihab: Answering 1001 Questions on Islam, 2008/1428). Finally, the fifth consists of his books on

16 See M. Quraish Shihab, "Majelis al-Hukama'," accessed January 3, 2015, http://quraishshihab.com/akhlak/majlis-al-hukama/; Muslim Council of Elders "Who We Are," accessed December 13, 2019, https://www.muslim-elders.com/en/page/7/who-we-are.

broad Islamic topics, such as *Islam yang saya Anut* (Islam as I Adhere, 2018/1439), *Islam yang Saya Pahami* (Islam as I Understand, 2018/1439), and *Wasathiyyah* (The Middle Way, 2019/1440). This last category mainly reflects Shihab's response to current Islamic discourse.

3 Guidance-Oriented Interpretation

Shihab's vision of Qur'ānic interpretation generally reflects the spirit of reformist Muslim scholars, in particular Muḥammad ʿAbduh, who is often seen as the most prominent mastermind of Islamic modernism and, more importantly, as an Azharī scholar who laid down a theological basis that would later give younger Azharī scholars and graduates the courage to work on religious reform.[17] ʿAbduh used *tafsīr* as a basis for developing and disseminating his reformist thought.[18] Other Muslim scholars working toward Islamic reformism have also made the Qur'ān the basis for disseminating their reformist ideas to a wider public.[19] Walking in the footsteps of reformist scholars, Shihab has used Qur'ānic exegesis as an intellectual foundation for his refashioning of religious tradition in a modern, pluralistic Muslim society.

Following ʿAbduh, Shihab approaches the Qur'ān as a primary book of religious guidance – as it refers to itself as "guidance for all humans" (*hudan li-n-nās*).[20] Accordingly, Qur'ānic interpretation, he says, ought to address the actual problems that Muslim society encounters. But Shihab admits that Qur'ānic guidance cannot be easily grasped when its interpretation becomes an exclusively scholarly enterprise that extensively discusses Qur'ānic vocabulary, syntax, and principles of interpretation. Shihab's *Tafsir al-Qur'an al-Karim* (1997/1417), which consists of a collection of some twenty-four chapters arranged in chronological order, was intended as a more scholarly project, aiming to feature the "sequence of the divine guidance" given to the Prophet and his people.[21] However, he felt that this work did not fulfil the purpose of

17 ʿAbduh wrote a treatise on Islamic monotheism, *Risālat at-tawḥīd*, which he intended to serve as a theological basis for Islamic modernism. He injected Muʿtazilī rationalism into Ashʿarī theology, which was adhered to by al-Azhar and its scholars. See Nasr Hamid Abu Zayd, "The Dilemma of the Literary Approach to the Qur'an," *Alif: Journal of Comparative Poetics* 23 (2003): 8–47.

18 *GRIK*, 323.

19 See J.M.S. Baljon, *Modern Muslim Koran Interpretation* (Leiden: Brill, 1968).

20 Muhammad Quraish Shihab, *Tafsir Al-Misbah* (Ciputat: Lentera Hati, 2012), 1: ix; Shihab, *Membumikan Al-Quran*, 37. Muḥammad Rashīd Riḍā mentions that discussions of interpretive technicalities, Arabic grammar, and eloquence should not be the main work of an exegete. Qur'ānic interpretation, as Muḥammad ʿAbduh affirms, should be directed primarily toward revealing Qur'ānic guidance (*al-ihtidā' bi-l-Qur'ān*). See Muḥammad Rashīd Riḍā and Muḥammad ʿAbduh, *Tafsīr al-manār* (Cairo: Dār al-Manār, 1947^2), 1: 25.

21 I unfortunately was unable to consult this rare book.

effectively deploying Qur'ānic guidance to a wider audience due to its complexity.[22] This led him to write some less complicated exegetical works.

Shihab primarily discusses two major genres of Muslim interpretation in his *Membumikan al-Quran* (Making the Qur'ān Down-to-Earth) and *Kaidah Tafsir* (Principles of Interpretation): conventional, sequential, verse-by-verse interpretation (*taḥlīlī*, sometimes referred to as *tajzīʾī* or *musalsal*) and thematic interpretation (*mawḍūʿī*, or sometimes *tawḥīdī*). Shihab acknowledges that each has its own merits. *Taḥlīlī* interpretation provides detailed explanations of many things pertinent to a given verse: its context of revelation, its content, its relationship to other verses, and its literary quality. However, he believes that this kind of interpretation cannot provide direct guidance on specific issues as it has to discuss various aspects related to a particular verse.[23] In this regard, he regards *mawḍūʿī* interpretation as a more suitable method that can provide Muslims with a sense of "living" guidance. With this kind of interpretation, an exegete examines a real issue in Muslim society by collecting a number of verses that potentially deal with it in order to draw conclusions about what the Qur'ān may say about the issue in question. With *mawḍūʿī* interpretation, Shihab contends, Qur'ānic guidance becomes more visible.[24]

Shihab attributes credit for the *mawḍūʿī* method to his Egyptian Azharī colleague ʿAbd al-Ḥayy al-Farmāwī (b. 1942/1361), who presents the method in a rather concrete way. In Shihab's view, al-Farmāwī's method comprises several steps: determining the issue to be discussed, collecting verses related to it, arranging the verses in chronological order, examining the meaning and the correlation of those verses within the structure of each chapter (*sūra*), considering relevant prophetic traditions, and then investigating the verses comprehensively.[25] Shihab develops al-Farmāwī's method further with some additional emphases: First, the exegete must have a good understanding of contemporary problems that require theological guidance from the Qur'ān. Second, the development of Qur'ānic messages on particular issues must be taken seriously. Third, a semantic discussion of key Qur'ānic vocabulary is necessary in order to gain comprehensive understanding of certain words that may have several meanings. And fourth, the causes of revelation must be taken into account in order to discover the precise meaning of each verse in its historical context.[26] With such additional annotations, Shihab seems to suggest that those who want to interpret the

22 Shihab, *Tafsir Al-Misbah*, 1: xiii–xiv.

23 Shihab, *Membumikan Al-Quran*, 180–82.

24 Ibid., 180. The *mawḍūʿī* interpretation is often academically termed "purposive exegesis" as the exegete interprets a collection of verses to purposively address a particular contemporary issue. See Amin and Kusmana, "Purposive Exegesis: A Study of Quraish Shihab's Thematic Interpretation of the Qur'an"; Johanna Pink, "Striving for a New Exegesis of the Qur'ān," in *The Oxford Handbook of Islamic Theology*, ed. Sabine Schmidtke (Oxford: Oxford University Press, 2016), 785.

25 Shihab, *Membumikan Al-Quran*, 176.

26 Ibid., 177–79.

Qur'ān should not "push" its verses to address contemporary issues without reasonable precedent.

Shihab is widely known as the first scholar to introduce *mawḍūʿī* exegesis to Indonesia.[27] *Wawasan al-Quran* is an example of how Shihab approaches the Qur'ān thematically. He highlights particular contemporary issues such as the foundations of belief, human interactions, social and political questions, the sciences, and the arts by interpreting certain relevant verses. The book is a compilation of public sermons that he initially gave to governmental officials and middle-class Muslims at Istiqlal Mosque in Jakarta. Acknowledging that *Wawasan* is still too complicated for average readers, he published a more concise *tafsīr* work in 1999/1419, *Secercah Cahaya Ilahi*, which follows the principles of *mawḍūʿī* exegesis in many respects. Like *Wawasan*, *Secercah* is a collection of revised sermons that Shihab gave on various occasions as well as religious opinions that have been published in some print media.

Despite the advantages of *mawḍūʿī* exegesis in providing living guidance, Shihab does not abandon *taḥlīlī* interpretation, which is part of a time-honored Muslim exegetical tradition. Working with this model is a particular pride for a Muslim scholar. *Tafsir al-Misbah* demonstrates Shihab's approach to the Qur'ān as an integral text whose sections are uniquely interconnected in a holistic bond. *ʿIlm al-munāsabāt* (the science of interrelation) proves to be his core theory for addressing the seemingly unconnected parts of the Qur'ān and expounding upon its miraculous nature.[28] A commentary by Ibrāhīm b. ʿUmar al-Biqāʿī (d. 1480/885), a premodern Syrian scholar, titled *Naẓm ad-durar fī tanāsub al-āyāt wa-s-suwar*, which had been the subject of his doctoral thesis, is his primary source. Shihab is convinced that aspects of Qur'ānic eloquence (*balāgha*) that are missing in *mawḍūʿī* exegesis can be discussed in *taḥlīlī*.[29] With *taḥlīlī*, Shihab intends to achieve what cannot be achieved in *mawḍūʿī* interpretation. Nevertheless, he maintains what he calls a different style of *mawḍūʿī* exegesis, i.e., interpreting a Qur'ānic sura by shedding light on its principal purposes and exam-

---

27 Following Shihab's introduction of the method, thematic interpretation has become an important academic subject and exegetical method among students at Indonesian Islamic universities.

28 The idea of the interrelation of the Qur'ānic verses is not at all new. Abū Jaʿfar az-Zubayr (d. 1308/708), Badr ad-Dīn az-Zarkashī (d. 1392/794), and Jalāl ad-Dīn as-Suyūṭī (d. 1505/911) had discussed this topic in their works. Fakhr ad-Dīn ar-Rāzī (d. 1209/605) and Ibrāhīm b. ʿUmar al-Biqāʿī (d. 1480/885) attempted to adopt this concept in their commentaries. However, Mustansir Mir argues that *ʿilm al-munāsabāt* never acquired the status of mainstream exegetical thought in premodern times. It constitutes an important issue in modern Muslim exegesis as a response to Western criticism that the Qur'ān consists of disjointed verses collected in a bizarre way. Mustansir Mir, "The *Sūra* as a Unity: A Twentieth Century Development in Qur'ān Exegesis," in *Approaches to the Qur'an*, ed. G.R. Hawting and Abdul Kader A. Shareef (New York: Routledge, 1993), 211f., 218.

29 Shihab, *Tafsir Al-Misbah*, 1: xv; Ibrāhīm b. ʿUmar al-Biqāʿī, *Naẓm ad-durar fī tanāsub al-āyāt wa-s-suwar* (Cairo: Dār al-Kitāb al-Islāmī, 1992^2), 6.

ining its central topics.[30] Still, *Tafsir al-Misbah* is too difficult for average readers. He therefore went on to introduce a more concise work in this genre, *Tafsir al-Lubab.*

4 A Reformed Arabicate Hermeneutics

As a Muslim scholar, Shihab believes in the outstanding literary quality and superiority of the Qur'ān. He refers to the Qur'ān as *"bacaan sempurna"* (the perfect reading) as it stuns humans with its eloquent expression, its harmonious structure, its accurate choice of vocabulary, and its rich guidance.[31] For Muslims, the concept of *iʿjāz* (the miraculous nature of the Qur'ān) inspires both a sense of beauty and majesty. Like other Muslim theologians, Shihab contends that Qur'ānic *iʿjāz* becomes apparent in the congeniality of its vocabulary, the information it provides about the unseen (*ghayb*), and its coverage of some scientific findings. It not only confirms the truth about Muḥammad's mission but also sanctions the divine origin of the Qur'ān[32] by challenging its opponents to produce a text of its kind, or even just part of it.[33] The Muslim contention is that nobody has been able to meet this challenge.

Shihab's reformist approach is apparent in the way that he privileges the Qur'ānic text over other Islamic religious texts, including sunna (the prophetic tradition).[34] Indeed, Muslims consider the Qur'ān to be the most privileged text, but some often approach it through the "dictation" of the sunna, as the latter provides more concrete examples of Islamic practices. Shihab's argues that the Qur'ān is superior to the sunna on the basis of the Muslim idea of the Qur'ān as a divine, verbal revelation that enjoys the status of decisive authenticity (*qaṭʿī l-wurūd*). The Qur'ān is said to have been transmitted by a large number of Muslims over successive generations, and is implau-

30 Shihab observes that there are two styles of thematic interpretation. The first analyzes a collection of verses that may share a particular topic and elucidates the meaning of each in order to draw out the Qur'ānic insight into a particular issue as discussed above. Meanwhile, the second interprets a Qur'ānic chapter (*sūra*) by shedding light on its principal purposes and examining its central topics. The second model resembles *taḥlīlī* interpretation since it discusses fixed themes known from the internal structure of a chapter. See Shihab, *Membumikan Al-Quran*, 111–18.

31 Muhammad Quraish Shihab, *Mukjizat Al-Quran* (Bandung: Mizan, 2013), 115–58; Muhammad Quraish Shihab, *Wawasan Al-Quran: Tafsir Maudhu'i atas Pelbagai Persoalan Umat* (Bandung: Mizan, 2014), 3f.

32 Shihab, *Membumikan Al-Quran*, 37–44.

33 Shihab, *Mukjizat Al-Quran*, 46–48.

34 Muḥammad Rashīd Riḍā mentions that the Qur'ān is the origin (*aṣl*) and foundation (*asās*) of Islam. The Prophet is infallible in religious matters, while Muslims are entrusted to run their individual and collective worldly affairs as long as they do not violate the general principles of religion. Muḥammad Rashīd Riḍā, *Yusr al-Islām wa-uṣūl at-tashrīʿ al-ʿāmm* (Cairo: Maṭbaʿat al-Manār, 1928), 17f.

sible that they would have agreed to maintain the same lie (*tawātur*).[35] Meanwhile, the authenticity of the sunna is merely speculative (*ẓannī l-wurūd*). Most of its reports have not been transmitted in the *tawātur* way, and many do not even convey the original statements of the Prophet verbatim.[36]

There is a hermeneutical consequence of privileging the Qur'ānic text that impacts how other texts should be understood. The ultimate authority attached to the Qur'ān means that other texts, including the sunna, have to be understood within the framework created by the Qur'ān.[37] Shihab believes that the sunna has two roles in relation to the Qur'ān: confirmation (*bayān ta'kīd*) and explanation (*bayān tafsīr*). In response to whether the sunna can enact an independent ruling aside from the Qur'ān, Shihab follows in the footsteps of reformist scholars. According to him, if the sunna seems to contradict the Qur'ān, there is no other way but to defend the Qur'ān, whose authenticity is decisive, and to leave aside the sunna, whose authenticity is speculative.[38] As such, Shihab lays down a principle that turns the Qur'ān into a frame of reference within which the sunna and other Islamic texts must be dealt with.[39]

As far as the *i'jāz* is concerned, it is not surprising that Shihab emphasizes salient linguistic analysis when approaching the Qur'ān. Like most other Muslim exegetes, Shihab's approach to the Qur'ān is deeply rooted in what Wael Hallaq calls "Arabicate

35 In the Western scholarship of the Qur'ān, such notions of authenticity have been severely challenged, particularly by Western revisionist scholars who doubt that the origins of the Qur'ān can be traced back to the era of the Prophet Muḥammad in Arabia. Revisionist studies question the validity of the traditional Muslim accounts of the Qur'ān's origin and consider it a later phenomenon from outside Arabia. Such revisionist perspectives will never be accepted by Muslim theologians, for they have the potential to deconstruct the very Muslim notion of the Qur'ān's origin as God's verbal revelation to the Prophet Muḥammad. See Fred M. Donner, *Narratives of Islamic Origins: The Beginnings of Islamic Historical Writing* (Princeton: The Darwin Press, 1999²); Nicolai Sinai and Angelika Neuwirth, "Introduction," in *The Qur'ān in Context*, ed. Angelika Neuwirth, Nicolai Sinai, and Michael Marx (Leiden: Brill, 2010); John Wansbrough, *Qur'anic Studies: Sources and Methods of Scriptural Interpretation* (New York: Prometheus Books, 2004).

36 Shihab, *Membumikan Al-Quran*, 188.

37 See Peter Heath, "Creative Hermeneutics: A Comparative Analysis of Three Islamic Approaches," *AR* 36 (July 1989): 178.

38 Shihab, *Membumikan Al-Quran*, 191.

39 Shihab calls for a critical reading of the sunna by investigating its particular function and whether the Prophet said something in his position as the Messenger of God, a religious interpreter (*muftī*), a leader, a judge, or just as an ordinary man. He is convinced that all of the Prophet's interpretations are correct, but this does not necessarily imply that they are always relevant to all situations. The Prophet may interpret some Qur'ānic verses simply by reflecting on examples and practices unique to the context of his time. Shihab's position as such can be seen, for example, in his interpretation of the scolded (*al-maghḍūb 'alayhim*) and the misguided (*aḍ-ḍāllūn*) in Q 1:7. He rejects the interpretation that God's wrath is directed at only a particular group based on transmitted reports that explicitly mention that the scolded were Jews and the misguided were Christians. To him, God's wrath can befall anybody who violates God's teachings, including Muslims. Ibid., 192; Shihab, *Tafsir Al-Misbah*, 1: 86f.

hermeneutics."[40] Shihab's analysis mainly covers discussions of the basic meaning of key Qur'ānic terms, their semantic meanings within the structure of the Qur'ānic text, and some grammatical and rhetorical aspects of the Qur'ān. Shihab contends that the Arabic language has a rich vocabulary containing varied derivatives that can produce various meanings. Nevertheless, these meanings are still unified by a basic meaning. When included in a sentence, a single word may produce a new (conceptual) meaning that may shift from its basic meaning – but not in a way that is totally disconnected.[41]

For instance, Shihab holds that the word *kufr* essentially means "to cover" or "to hide"; it was originally used to designate a farmer who covers the seeds he plants with soil.[42] The word is used in the Qur'ān to convey various meanings, e.g., the denial of truth (Q 2:89) and God's existence (Q 3:10–12), and ingratitude (Q 14:7), depending on the context in the text.[43] Moreover, Shihab deals carefully with the Arabic preposition (*ḥarf al-jarr*) that follows a verb. According to him, *hadā* together with *ilā* merely means "to give information," while *hadā* without any preposition has a more powerful meaning, denoting not only "to show guidance" but also "to ensure that someone receives guidance." Shihab cites two Qur'ānic verses that speak about the Prophet in relation to the possibility of him giving guidance. Q 42:52, he says, demonstrates the Prophet's ability to convey divine guidance, while Q 28:56 discards the Prophet's ability to "grant" any guidance.[44] This implies that the task of the Prophet was only to convey divine guidance, whereas giving guidance remains the monopolized domain of God.

On his quest for meaning, Shihab adopts hermeneutical instruments originally rooted in Islamic legal theories (*uṣūl al-fiqh*), i.e., *qaṭ'ī d-dalāla* (definite evidence) and *ẓannī d-dalāla* (probable evidence). According to him, every text has two possible meanings. Certainly, for the author, it conveys only one "authentic meaning" (*dalāla ḥaqīqiyya*), which is his purpose as expressed in the text. For readers who come from other cultural and intellectual backgrounds, however, it may designate relative meaning (*dalāla nisbiyya*).[45] Here, Shihab is indebted to fourteenth-century Andalusian Mālikī legal scholar, Ibrāhīm b. Mūsā sh-Shāṭibī (d. 1388/790), who introduced what Wael Hallaq calls a "holistic theory" of interpretation. This theory suggests that a verse can-

---

40 Wael B. Hallaq, *Sharī'a: Theory, Practice, Transformations* (Cambridge et al.: Cambridge University Press, 2009), 449, 502.

41 Muhammad Quraish Shihab, *Kaidah Tafsir* (Ciputat: Lentera Hati, 2013), 37f.

42 Muhammad Quraish Shihab, *Lentera Hati* (Bandung: Mizan, 2007), 140; Shihab, *Tafsir Al-Misbah*, 6: 331.

43 Shihab summarizes that the various meanings of *kufr* lead to a negative religious mentality. *Kufr* is used in the Qur'ān to denote "a dirty soul" that prevents man from comprehending truth and divine guidance. Shihab, *Tafsir Al-Misbah*, 1: 117f.

44 Shihab, *Kaidah Tafsir*, 79f; Shihab, *Tafsir Al-Misbah*, 9: 622f. Q 42:52, "And indeed you [Muḥammad] guide to (*la-tahdī ilā*) a straight path"; Q 28:56, "Indeed you [Muḥammad] do not guide whom you like (*lā tahdī man aḥbabta*), but Allah guides whom He wills [. . .]."

45 Shihab, *Membumikan Al-Quran*, 213.

not be understood properly in isolation from other verses or the circumstances in which the Qur'ān was revealed.[46] This implies that most individual verses would belong to the category of *ẓannī*, or probable evidence, if they were to be read independently. The clarity of meaning may emerge when several pieces of *ẓannī* evidence, which appear to point in the same direction, are brought together, thereby producing a predominantly clear meaning. The Qur'ānic command regarding prayer, for instance, cannot inherently be understood as a religious obligation in the absence of corroborating attestations – such as those found in the sunna – which clearly mandate its observance.[47]

How can a fixed text that initially responded to religious and social questions in a particular historical context be regarded as the ultimate authority for addressing the modern realities of Muslim society? This question is becoming increasingly significant, especially in light of the way that the upheavals of modernity have deeply encroached upon the Muslim world and exposed contrasting values and ways of life. As a theologian, Shihab holds firmly to the notion of the universality of the Qur'ānic message across Muslim generations. This can be seen in the continuity in Muslim Qur'ānic interpretations over the course of history. However, Shihab denounces the notion of universality that suggests the necessity of returning everything back to the social setting in which the Qur'ān was initially revealed. Shihab believes that the Qur'ān must be interpreted in a way that advances its significance in new realities. The Qur'ān has to be constantly reread and interpreted in the new context according to the developments and changing conditions of Muslim society.[48]

In this regard, the role of the interpreter is pivotal in the construction of meaning. The text cannot be understood without the role of the interpreter, but the interpreter cannot produce meaning outside the text. The text must say something, and the interpreter – with his own intellectual and cultural capital – must construct meanings within the text. With regard to the unique interaction between the text and the interpreter, Shihab considers interpretation to be a human attempt to decode God's speech. With this definition he acknowledges that the act of interpretation is an intellectual enterprise that may generate disagreements among interpreters themselves. He argues that only God knows the exact meaning of the Qur'ān, simply because He is the Owner (*Pemilik*) of the text.[49] Interpretation therefore must be directed as much as possible toward expounding the purpose of the Owner of the text as indicated in the text itself.

Because the Qur'ān was not revealed in a social and historical vacuum, Shihab believes that knowledge of the Qur'ān's historical circumstances is imperative. Like

46 Wael B. Hallaq, "Law and the Qur'an," in *EQ*, ed. Jane Dammen McAuliffe (Leiden/Boston: Brill, 2003), 3: 164.

47 Shihab, *Membumikan Al-Quran*, 214–17.

48 Ibid., 132–41.

49 Ibid., 112f.

most other Muslim exegetes, his sources are the exegetical reports that narrate the revelation of a particular verse.[50] Those reports have been identified and compiled by some Muslim scholars into a separate genre, namely *asbāb an-nuzūl* (the causes of revelation), which has a particular literary character. A report is called *sabab* (pl. *asbāb*) if it talks about something or someone involved in some way in the life of the Prophet, connecting the situation with the revelation of a particular Qurʾānic text.[51] Shihab seems to take *asbāb an-nuzūl* for granted, without investigating the reliability of those reports any further, e.g., where they originated, how they were compiled, and how some of them appear to be in conflict with each other.

Muslim exegetes agree that *asbāb an-nuzūl* can be used as a tool for understanding the Qur'ān. But they differ on whether an interpreter must consider *al-ʿibra bi-ʿumūm al-lafẓ* (the generality of the wording) or *al-ʿibra bi-khuṣūṣ as-sabab* (the specificity of the cause) when interpreting the Qur'ān to address new realities. Shihab finds that most Muslim exegetes defend the supremacy of *ʿumūm al-lafẓ* on the ground that it corroborates the universal mission of Islam, drawing all Muslims to the message as dictated by the Qurʾānic wordings. The proponents of this approach argue that there is no fundamental difference between *al-ʿibra bi-ʿumūm al-lafẓ* and *al-ʿibra bi-khuṣūṣ as-sabab,* claiming that the difference only exists at the theoretical level because, in practice, both often produce similar interpretations.[52]

However, Shihab does not agree that both approaches always have the same implications, as the Qur'ān was not revealed *tabula rasa*. The Qurʾānic *sūra*s and verses

50 About the importance of the causes of revelation, ʿAlī b. Aḥmad al-Wāḥidī (d. 1075/468) said, "it is not possible to know the interpretation of a given verse without knowing its story and the causes of its revelation." Ibn Daqīq al-ʿĪd (d. 1302/702) also said, "exploring the cause of revelation is a firm way to understand the meanings of the Qur'ān." Ibn Taymiyya (d. 1328/728) has contended that "knowing the cause of revelation helps in understanding a given verse; therefore, knowing the cause leads to knowing the effect." Muslim scholars use *asbāb an-nuzūl* to identify the reason behind the divine legal ruling (*ḥukm*), to remove confusion regarding the meaning of the Qurʾānic text, to specify the general wording of Qurʾānic verses, or to generalize verses that initially addressed certain individuals. See Jalāl ad-Dīn as-Suyūṭī, *Lubāb an-nuqūl fī asbāb an-nuzūl* (Beirut: Mu'assasat al-Kutub ath-Thaqāfiyya, 2002), 7; Jalāl ad-Dīn as-Suyūṭī, *al-Itqān fī ʿulūm al-Qur'ān,* ed. Muḥammad Abū al-Faḍl Ibrāhīm (Cairo: Maktabat Dār at-Turāth, 1985), 1: 82f; Badr ad-Dīn az-Zarkashī, *Al-Burhān fī ʿulūm al-Qur'ān,* ed. Muḥammad Abū l-Faḍl Ibrāhīm (Cairo: Maktabat Dār at-Turāth, 2000), 1: 122–29..

51 Al-Wāḥidī is deemed to have been one of the earliest scholars to write a book of this genre. He collected as many reports as he could in his book *Asbāb an-nuzūl,* which is often seen as having firmly established this genre. Emerging at a time when the authority of tradition, rather than reason, was generally judged to be supreme within Islamic learning, al-Wāḥidī's work became a starting point for Muslim scholars in subsequent periods to assure that the causes of revelation, which have to be sought through transmitted reports, became an important key to Qurʾānic exegesis. Andrew Rippin, "Occasions of Revelation," in *EQ*, ed. Jane Dammen McAuliffe (Leiden/Boston: Brill, 2003), 3: 570.

52 Standing for the generality of the wording, Jalāl ad-Dīn as-Suyūṭī suggested that interpretations from two very different perspectives often produce similar results because there are always factors that may generalize a verse for all cases, or specify it for limited ones. Suyūṭī, *al-Itqān*, 196f.

were gradually revealed over a certain period of time, corresponding with the realities of seventh-century Arabia. Pointing out the dialectical relations between the text and its context, Shihab argues that the text-context relationship will not receive proper attention if approached through *al-ʿibra bi-ʿumūm al-lafẓ*.[53] Thus, he defends the minority opinion: the Qurʾān must be read by "considering the specificity of the cause," which can enrich and develop Qurʾānic exegesis when it comes to dealing with new issues.

Shihab contends that, when considering "the specificity of the cause," the Qurʾān must be interpreted through analogy (*qiyās*) in order to capture the essence of the Qurʾānic text as revealed within its initial cultural and social context, before articulating its significance in the new context.[54] However, Shihab affirms that the analogy in question is different from that frequently employed by traditional Muslim jurists, which is said to have been derived from formal Aristotelian logic.[55] He believes that this kind of analogy does not suffice to produce insightful interpretations as it merely attaches a new reality to an old one directly addressed by the Qurʾān due to the perceived similarity of their effective cause (*ʿilla*). In other words, the adoption of formal Aristotelian logic simply means relating new realities to conclusions that have already been drawn from the Qurʾānic text. In order to bridge the gap between the original context and new realities, Shihab proposes considering human interests (*al-maṣāliḥ al-mursala*) in the operation of analogy.[56] In doing so, Shihab seems to emphasize that Qurʾānic interpretation must be conducted by examining the Qurʾānic text and the collective social sensitivities of contemporary Muslim society.

Shihab mentions influential Sudanese intellectual and politician Ḥasan at-Turābī (d. 2016/1437) when he introduces *qiyās al-maṣāliḥ al-mursala*.[57] However, neither he

53 Shihab, *Membumikan Al-Quran*, 134.
54 Ibid., 135.
55 Wael B. Hallaq believes that Greek formal logic was accommodated by Sunni legal theories (*uṣūl al-fiqh*) some centuries after it was introduced to the intellectual landscape of Islam. Some elements of Greek logic can be traced in the writings of scholars like Abū Bakr al-Jaṣṣāṣ (d. 980/370), Imām al-Ḥaramayn al-Juwaynī (d. 1085/478), and Abū Ḥāmid al-Ghazālī (d. 1111/505) who, to a large extent, integrated Greek logic into Islamic legal theory. Wael B. Hallaq, "Logic, Formal Arguments and Formalization of Arguments in Sunnī Jurisprudence," *AR* 37, no. 3 (1990): 318; Wael B. Hallaq, *A History of Islamic Legal Theories* (Cambridge et al.: Cambridge University Press, 1997), 137.
56 Shihab, *Membumikan Al-Quran*, 135f.
57 Concerning the term, Shihab consults a secondary source, Yūsuf Kamāl's *al-ʿAṣriyyūn Muʿtazilat al-yaum* (The Modernists are the Muʿtazila of Today). The book was actually intended by its author as a response against twentieth-century modernist scholars, who, in his view, had gone too far in deconstructing the Islamic religious tradition with their stirring call for independent reasoning (*ijtihād*). Kamāl highlighted a characteristic shared by modernists who were content to argue using the Qurʾān, the sound *sunna*, and reason. Kamāl argued against the modernist opinion that considered the Prophetic tradition to be non-binding in human transactions; for, from the modernist point of view, they had to be examined from the perspective of general principles and the purposes of religion. Kamāl

nor at-Turābī elaborates on the term in any more detail.[58] Nevertheless, it is clear that Shihab is attempting to relate his interpretation to the path of Muslim utilitarianists, who have primarily developed their legal theory on the basis of *maṣlaḥa* (public interest).[59] Positing the Qur'ān as the cornerstone of religion, Shihab contends that the Qur'ān only provides detailed guidance on matters that are beyond human reason or matters that are not subject to development and change – such as the foundations of belief, ritual, and metaphysical doctrines. As for matters subject to development and change, the Qur'ān only provides general principles, which function as guidelines for the ability to adapt to those developments and changes.[60] He therefore differentiates between "purely religious matters" and "worldly social affairs." Demonstrating his reliance on ash-Shāṭibī, he contends that matters of worship in which reason has no place must be directly subject to the dictates of the revealed text, whereas human transactions (*muʿāmala*) must be derived from its essence and purpose.[61] Moreover, Shihab frequently cites Muḥammad aṭ-Ṭāhir b. ʿĀshūr's (d. 1973/1393) works, which offer a renewed theory of *maqāṣid ash-sharīʿa* (the purposes of Islamic jurisprudence) that closely corresponds to the contemporary challenges of Muslim society.[62]

Working on the basis of a principle of adaptability, Shihab proposes a harmonious and dynamic relation between the fixed Qur'ānic text and constantly changing circumstances. He believes that the text must be interpreted by looking carefully at its semantic meaning on the one hand and at the character, culture, and "positive" developments of society on the other.[63] In other words, the text must be understood within the framework of the contemporary context; at the same time, the changing reality must be steered according to general values derived from the text. Shihab argues that totally adopting the religious understandings of earlier Muslim generations is not an entirely suitable choice. Compelling a Muslim generation to thoroughly assume the religious understandings of a prior generation will only incur difficulties. Moreover, it contradicts the teachings of Islam, which, according to him, support every positive

called the modernists contemporary Muʿtazila due to their perceived sharing of "rationalistic heterodoxy." See Yūsuf Kamāl, *al-ʿAṣriyyūn Muʿtazilat al-yaum* (Mansoura: Dār al-Wafāʾ, 1986), 11, 14.

58 Hallaq, *A History of Islamic Legal Theories*, 228.

59 Modern Muslim utilitarianists relate to Muḥammad ʿAbduh's school of thought. They subscribe to a set of principles laid down by premodern jurists, particularly Najm ad-Dīn at-Ṭūfī (d. 1316/716) and Abū Isḥāq ash-Shāṭibī (d. 1388/790), but modify those principles according to the advantages of their era. Ibid., 214.

60 Shihab, *Wawasan Al-Quran*, 620.

61 Shihab, *Membumikan Al-Quran*, 120. In his *al-Muwāfaqāt*, ash-Shāṭibī says, "*Al-aṣl fī l-ʿibādāt bi-n-nisba ilā l-mukallaf at-taʿabbud, dūna al-iltifāt ilā l-maʿānī, wa-l-aṣl fī l-ʿādāt al-iltifāt ilā l-maʿānī.*" See Ibrāhīm b. Mūsā sh-Shāṭibī, *Al-Muwāfaqāt fī uṣūl ash-sharīʿa*, ed. Abdullāh Darrāz (Cairo: Al-Maktaba at-Tijāriyya al-Kubrā, 1975), 2: 300.

62 See also Pink, "Tradition and Ideology in Contemporary Sunnite Qur'ānic Exegesis," 8.

63 Shihab, *Membumikan Al-Quran*, 134.

development in society and the nature of social life, which is always subject to change.[64]

Shihab's interpretations of Q 24:31 and Q 33:59 concerning the etiquette of clothing for adult Muslim women is an example of how he takes different Muslim customs into account. Shihab differs from the majority of Muslim scholars, who opine that wearing the headscarf is obligatory for Muslim women by reflecting on Muslim women's clothing in the Prophet's time. He contends that the Qur'ān does not clearly determine the boundaries of "Islamic" female clothing. To him, the practices associated with female clothing mentioned in the sunna exclusively reflect the context of the Prophet's Arab time. He therefore believes that the notion of "respectable clothing" derived from the Qur'ān can be adjusted to reflect the customs of people whose women might not wear headscarves.[65]

However, Shihab's interpretation of the verses relating to the division of inheritance (*mīrāth*) does not reflect modern thinking about equal male-female sharing. Shihab maintains the Muslim majority opinion that the male heir should receive more shares than the female (double the amount) as stated explicitly in the wording of the Qur'ān. To him, this division reflects the principle of justice because the man bears more of the (financial) burden in his family.[66] Shihab's position can be explained by considering how he deals with what Muslim scholars have called *qaṭʿiyyāt* (texts perceived to bear one definite meaning) in Islam. The allocation of inheritance is mentioned explicitly in the Qur'ān for most heirs, and it seems to leave no other alternative interpretation. Moreover, this issue is closely related to the jurisprudence of marriage in Islam, which regulates rights and obligations between couples. Muslims may be willing to relinquish criminal law and many aspects of civil law to modern states, but not Islamic marital law.

5 Concluding Remarks

Our understanding of Muhammad Quraish Shihab's Qur'ānic hermeneutics must be placed within the context of the wave of Islamic reformism that attracted some Indonesian students to Cairo in the twentieth century. As far as Islamic reformism is concerned, the Qur'ān, which is understood as the basis of Muslim tradition, is playing a central role in a fresh, new articulation of Islam. As Islamic reformism cannot neglect

64 Ibid., 141.
65 Shihab, *Tafsir Al-Misbah*, 8: 528–34; Muhammad Quraish Shihab, *Jilbab Pakaian Wanita Muslimah* (Ciputat: Lentera Hati, 2012). For more on Shihab's interpretation of Muslim female clothing, see Munirul Ikhwan, "Interpreting the Qur'ān between Sharīʿa and Changing Custom: A Case of Women's Dress in Indonesia," in *New Trends in Qur'anic Studies: Text, Context, and Interpretation*, ed. Mun'im Sirry (Atlanta: Lockwood Press, 2019), 211–32.
66 Shihab, *Tafsir Al-Misbah*, 2: 445.

the "instructions" of the very basis of Muslim tradition, Muslim reformists are trying to transform the dysfunctionality of their tradition into a positive functionality in the face of a new, changing era. In this regard, tradition is seen as a potential asset in the formation of a unique modernity and an "indigenous" path toward social progress, instead of cultural setback.

In the hands of Shihab, the Qur'ān's function in a modern context cannot escape Arabicate hermeneutics. The centrality of Arabicate hermeneutics in the Muslim exegetical tradition is a consequence of the theological doctrine of *i'jāz*, which affirms the linguistic perfection and harmony of the Qur'ānic text. The space for introducing creative meanings can be found in what has been called *ẓanniyyāt* (probable evidences), since verses in this category are not seen to indicate any definite meanings. In Shihab's theory, the functionality of the Qur'ān is achieved by considering contemporary human interests and customs in the process of producing meaning.

6 Bibliography

6.1 Sources

Anwar, Mauluddin, Latief Siregar, and Mustofa Hadi. *Cahaya, Cinta Dan Canda M. Quraish Shihab*. Ciputat: Lentera Hati, 2015.

Muslim Council of Elders. "Who We Are." Accessed December 13, 2019. https://www.muslim-elders.com/en/page/7/who-we-are.

Shihab, Muhammad Quraish. *Lentera Hati*. Bandung: Mizan, 2007.

Shihab, Muhammad Quraish. *Jilbab Pakaian Wanita Muslimah*. Ciputat: Lentera Hati, 2012[6].

Shihab, Muhammad Quraish. *Tafsir Al-Misbah*. 15 vols. Ciputat: Lentera Hati, 2012[5].

Shihab, Muhammad Quraish. *Kaidah Tafsir*. Ciputat: Lentera Hati, 2013.

Shihab, Muhammad Quraish. *Mukjizat Al-Quran*. Bandung: Mizan, 2013.

Shihab, Muhammad Quraish. *Membumikan Al-Quran: Fungsi Dan Peran Wahyu Dalam Kehidupan Masyarakat*. Bandung: Mizan, 2013.

Shihab, Muhammad Quraish. *Wawasan Al-Quran: Tafsir Maudhu'i atas Pelbagai Persoalan Umat*. Bandung: Mizan, 2014.

Shihab, Muhammad Quraish. "Majelis al-Hukama'." Accessed January 3, 2015. http://quraishshihab.com/akhlak/majlis-al-hukama/.

6.2 Secondary Literature

Abaza, Mona. *Indonesian Students in Cairo: Islamic Education, Perceptions and Exchanges*: Paris: Association Archipel, 1994.

Abu Zayd, Nasr Hamid. "The Dilemma of the Literary Approach to the Qur'an." *Alif: Journal of Comparative Poetics* 23 (2003): 8–47.

Amin, Muhammadiyah, and Kang Kusmana. "Purposive Exegesis: A Study of Quraish Shihab's Thematic Interpretation of the Qur'an." In *Approaches to the Qur'an in Contemporary Indonesia*, edited by Abdullah Saeed. Oxford: Oxford University Press, 2005.

Anshori. “Penafsiran Ayat-Ayat Jender Dalam Tafsir al-Mishbah.” PhD diss., UIN Jakarta, 2006.
Baljon, J.M.S. *Modern Muslim Koran Interpretation*. Leiden: Brill, 1968.
Al-Biqāʿī, Ibrāhīm b. ʿUmar. *Naẓm ad-durar fī tanāsub al-āyāt wa-s-suwar*. Cairo: Dār al-Kitāb al-Islāmī, 1992².
Donner, Fred M. *Narratives of Islamic Origins: The Beginnings of Islamic Historical Writings*. Princeton: The Darwin Press, 1999².
Goldziher, Ignaz. *Die Richtungen der islamischen Koranauslegung*. Leiden: Brill, 1920.
Hallaq, Wael B. “Logic, Formal Arguments and Formalization of Arguments in Sunnī Jurisprudence.” *AR* 37, no. 3 (1990): 315–58.
Hallaq, Wael B. *A History of Islamic Legal Theories*. Cambridge: Cambridge University Press, 1997.
Hallaq, Wael B. “Law and the Qur’an.” In *EQ*, edited by Jane Dammen McAuliffe. 6 vols. Leiden/Boston: Brill, vol. 3, 2003: 149–72.
Hallaq, Wael B. Sharīʿa: *Theory, Practice, Transformations*. Cambridge: Cambridge University Press, 2009.
Heath, Peter. “Creative Hermeneutics: A Comparative Analysis of Three Islamic Approaches.” *AR* 36 (July 1989): 173–210.
Ikhwan, Munirul. “Interpreting the Qur’ān between Shariʿa and Changing Custom: A Case of Women’s Dress in Indonesia.” In *New Trends in Qur’anic Studies: Text, Context, and Interpretation*, edited by Mun’im Sirry. Atlanta: Lockwood Press, 2019: 211–32.
Jacobsen, Frode. *Hadrami Arabs in Present-Day Indonesia: An Indonesia-Oriented Group with an Arab Signature*. Oxon/New York: Routledge, 2009.
Kamāl, Yūsuf. *Al-ʿAṣriyyūn Muʿtazilat al-yaum*. Mansoura: Dār al-Wafāʾ, 1986.
Kaptein, Nico J.G. *Islam, Colonialism and the Modern Age in the Netherlands East Indies*. Leiden: Brill, 2014.
Laffan, Michael. *Islamic Nationhood and Colonial Indonesia*. London/New York: Routledge Curzon, 2003.
Laffan, Michael. “An Indonesian Community in Cairo: Continuity and Change in a Cosmopolitan Islamic Milieu.” *Indonesia* 77 (2004): 1–26.
Mir, Mustansir. “The Sūra as a Unity: A Twentieth Century Development in Qur’ān Exegesis.” In *Approaches to the Qur’an*, edited by G.R. Hawting and Abdul Kader A. Shareef. New York: Routledge, 1993: 211–24.
Mobini-Kesheh, Natalie. *The Hadrami Awakening*. Ithaca, NY: Southeast Asia Program Publication, 1999.
Nisa, Eva F. “Non-Muslims in the Qur’ān: A Critical Study on the Concept of Non-Muslims in Tafsir Al-Mishbāh of Muhammad Quraish-Shihab.” Master’s thesis, Leiden University, 2004.
Noer, Deliar. *The Modernist Muslim Movement in Indonesia 1900–1942*. Singapore/New York: Oxford University Press, 1973.
Pink, Johanna. *Sunnitischer Tafsīr in der modernen islamischen Welt*. Leiden: Brill, 2011.
Pink, Johanna. “Tradition and Ideology in Contemporary Sunnite Qurʾānic Exegesis: Qurʾānic Commentaries from the Arab World, Turkey and Indonesia and their Interpretation of Q 5:51.” *WI* 50 (2010): 3–59.
Pink, Johanna. “Striving for a New Exegesis of the Qurʾān.” In *The Oxford Handbook of Islamic Theology*, edited by Sabine Schmidtke. Oxford: Oxford University Press, 2016: 765–92.
Ricklefs, M.C. *A History of Modern Indonesia since c. 1200*. Hampshire: Palgrave, 2001³.
Riḍā, Muḥammad Rashīd. *Yusr al-Islām wa-uṣūl at-tashrīʿ al-ʿāmm*. Cairo: Maṭbaʿat al-Manār, 1928.
Riḍā, Muḥammad Rashīd, and Muḥammad ʿAbduh. *Tafsīr al-manār*. 12 vols. Cairo: Dār al-Manār, vol. 1, 1947².
Rippin, Andrew. “Occasions of Revelation.” In *EQ*, edited by Jane Dammen McAuliffe. 6 vols. Leiden/Boston: Brill, vol. 3, 2003: 569–73.
Ash-Shāṭibī, Ibrāhīm b. Mūsā. *Al-Muwāfaqāt fī uṣūl ash-sharīʿa*. Ed. Abdullāh Darrāz. 4 vols. Cairo: Al-Maktaba at-Tijāriyya al-Kubrā, vol. 2, 1975.
Sinai, Nicolai, and Angelika Neuwirth. “Introduction.” In *The Qurʾān in Context*, edited by Angelika Neuwirth, Nicolai Sinai, and Michael Marx. Leiden: Brill, 2010.
As-Suyūṭī, Jalāl ad-Dīn. *Lubāb an-nuqūl fī asbāb an-nuzūl*. Beirut: Muʾassasat al-Kutub ath-Thaqāfiyya, 2002.

As-Suyūṭī, Jalāl ad-Dīn. *Al-Itqān fī ʿulūm al-Qurʾān*. Ed. Muḥammad Abū l-Faḍl Ibrāhīm. 4 vols. Cairo: Maktabat Dār at-Turāth, 1985.

Wansbrough, John. *Qur'anic Studies: Sources and Methods of Scriptural Interpretation*. New York: Prometheus Books, 2004.

Az-Zarkashī, Badr ad-Dīn. *Al-Burhān fī ʿulūm al-Qurʾān*. Ed. Muḥammad Abū l-Faḍl Ibrāhīm. 4 vols. Cairo: Maktabat Dār at-Turāth, 2000.

Katajun Amirpur

ʿAbdulkarīm Surūsh

1 Introduction

ʿAbdulkarīm Surūsh (b. 1945/1364, commonly known in the West as Abdulkarim Soroush) is one of the world's leading Muslim intellectuals today. In 2005, *Time Magazine* included him in its list of the hundred most influential thinkers worldwide. His origin from a traditional religious background is evidenced by his given name Farajullāh Hāj Ḥusayn Dabbāgh. Born on Ashura, the anniversary of Imam Ḥusayn b. ʿAlī's death, he was named Ḥusayn. His chosen pseudonym is composed of the names he gave two of his sons. *Karīm* is traditionally one of the ninety-nine most beautiful names of God. The prefix *ʿabd* means a slave or servant, so *ʿAbdulkarīm* in combination translates as "servant of the bountiful." *Surūsh* is the angel of poetic inspiration.

Surūsh attended the *ʿAlaviyyi* school, an institution founded by religious *bazārīs* as an alternative to both secular public education and traditional religious colleges in the 1950s. It aimed to make its students proficient in both traditional and modern subjects and quickly turned into one of Iran's most influential educational institutions. Many intellectual revolutionary leaders graduated from it.

The foundations of Surūsh's later academic work were laid during his time in school. His Physics teacher drew his attention to a question then current with many prominent thinkers: leading Iranian intellectuals such as Mihdī Bāzargān (1907–1995/1325-1416), who would later become prime minister (January to November 1979/Ṣafar to Dhū l-Ḥijja 1399), and ʿAllāmeh Muḥammad Ḥusayn Ṭabāṭabāʾī (1903–1981/1321–1401), one of the clergy's recognised philosophers, tried to reconcile modern science with religion by showing that its findings were not in conflict with the Qurʾān and the tradition of the Prophet and the Imams.

The solution they offered did not satisfy Surūsh. Aged seventeen, he began studying Islamic philosophy and the classical disciplines of religious scholarship, *fiqh, uṣūl al-fiqh* and *tafsīr*, alongside his school education. Studying both the Qurʾānic interpretation of older scholars such as Fakhr ad-Dīn ar-Rāzī (1149–1209/544–606), Abū ʿAlī Faḍl b. Ḥusayn aṭ-Ṭabarsī (d. 1153/548), and Muḥsin Fayż Kāshānī (d. 1680/1091) and modern exegetes like Sayyid Quṭb, Maḥmūd Ṭāliqānī (1914–1979/1332–1399), and Ṭabāṭabāʾī, Surūsh was fascinated by the differences in their understanding of the Qurʾānic revelation and sought to understand the reason for this variety.

After graduation, Surūsh studied pharmacy and completed mandatory military service. He then moved to Bushehr to serve as director of the local laboratory for food and sanitary supplies. After fifteen months, he went on to briefly work in a Tehran laboratory for medical drug control. In 1973, he travelled to Britain to obtain a Master's degree in analytical chemistry from the University of London. Though enrolled in a scientific subject, he simultaneously studied philosophy and epistemology at Chel-

https://doi.org/10.1515/9783111320069-018

sea College's Faculty for History and Philosophy. Surūsh stayed in Britain for five and a half years and completed his studies by obtaining a doctorate in Chemistry.

At this point, tensions between the Shah's government and parts of the Iranian population were increasing. Iranian students in Europe and the United States, long a politically active and often pro-opposition constituency, were increasingly drawn into the conflict that embroiled their native country. Surūsh joined a Muslim community which he then used as his political platform. In the months leading up to the revolution, its premises became a focus of opposition activity and a meeting point for Iranian enemies of the Shah in Britain. Along with the then relatively unknown Surūsh, prominent intellectuals such as Ayatollah Murtażā Muṭahharī (1920–1979/1338–1399) and Ayatollah Muḥammad Ḥusayn Bihishtī (1925–1981/1343–1401), both close confidants of Ayatollah Rūḥullāh Khumaynī, lectured there.

2 From Court Ideologue to Dissident

After the revolution, Surūsh returned to Iran to teach at Tehran University's Faculty of Education. He was also frequently seen on television lecturing on the legal theory of the nascent Islamic Republic. In 1981/1401, he joined the faculty at the Society of Philosophy, an institute dedicated to the study of Western philosophy founded under the Shah, and the Cultural Revolution Headquarters, re-named the Supreme Cultural Revolution Council in 1984. This body was charged with establishing Islamic principles at universities – which had been closed for the duration – by developing new curricula and purging both faculty and student body of those who refused to conform to the new system.

After 1984/1404, Surūsh ended his political activism and dedicated himself entirely to his academic work. He taught at Tehran University as well as accepting visiting professorships at various non-government run universities, and even briefly lecturing at the theological college in Qom. Along with Islamic mysticism, his subjects included philosophy of religion, epistemology, comparative philosophy, and the philosophy of empirical sciences. From 1988 to 1994, he also preached in mosques and the sermons he delivered at the Imam Ṣādiq Mosque in Tehran became famous. His lectures on mysticism were even televised and served to spread his fame as an eloquent teacher and public speaker.

In his academic work, Surūsh engaged deeply with the scientific and intellectual development of the West or, as he himself put it, confronted the challenge of Western thought. He argued that scientific and philosophical traditions had to be studied and engaged with, a view that, he lamented, few traditional religious scholars shared. They would rather reflexively condemn all things Western as irreligious.

Surūsh applied Western methods in his research, studying the variety of Qurʾānic interpretations from an epistemological perspective. One explanation he adopted was drawn from the Western philosophy of science. A desire and open plea for a fruitful exchange with Western thought is central throughout Surūsh's work and fundamental to his attitude.

His stay in London also provided further insights: the West, Surūsh argued, had dedicated itself to happiness in this world since the Renaissance, while his own more mystically inclined culture taught a rejection of the world. Earthly happiness lost out to eternal salvation in Oriental thinking. Confronting the European attitude so antithetical to his own gave Surūsh a new impulse to grapple with the function of religion. In his research into the expectations the citizens of modernity had of their religion, it was the study of Christianity that confirmed his suspicions: the interpretation of religion was changeable over time. How else could any modern individual refer to themselves as an earnest Christian? This insight brought him to grips with the relationship between religion and the modern concept of reason, a field to which he dedicated years of study.

His criticism of the political system did not begin until the death of Khumaynī in 1989/1409, a fact that may be owed to Surūsh's great admiration for the Islamic Republic's founder. He greatly respected Khumaynī as a revolutionary leader who had succeeded in making religion – by nature silent – speak out and become a source of revolutionary activism. His critique of the ruling clergy and its claim to exclusivity in religious interpretation caused Surūsh's fall from grace in 1996. He was first attacked in the media, then targeted with death threats and physically assaulted in response to his article *Freedom and the Clergy* (*Ḥurrīyyat va rauḥānīyyat*). In it, Surūsh argued that the Shiite clergy had lost its freedom and morphed into a social stratum living at the expense of the faithful who were required to pay the *khums* ('fifth'). Their defining feature had become their source of funding, and they had thus automatically adopted all negative characteristics that exclusive social groups tended to develop:

> These advantages turn a group of individuals into a guild, a unitary body, and allow them to distinguish themselves from others. It is this pride of status that draws a distinction between one's own people and the rest, that establishes the Own and the Other, guards jealously the secrets of the trade, cultivates distinctive traditions and manners, internal gradations and ranks, speaks of a We, defends this We against outside threats, considers it sacred, important, the bearer of knowledge and the mysteries of the law, slowly finds its way into a ruling tradition and ideology, elects some great personalities as its sanctified leaders in matters practical and scholarly, follows set patterns of thought, holds them equal to truth and religion, and comes to see any violation of the tenets of this pride as an act of blasphemy.[1]

In short, Surūsh argues that public funding turns the clergy into a social interest group. This group claims a monopoly to defend its privilege, and this claim to exclusive insight leads to the ossification of religious thinking. He writes: "Any group that sees itself as the bearer of the sole valid interpretation of religion and seeks to base

1 ʿAbdulkarīm Surūsh, "Khadamāt va ḥasanāt-i dīn," *Kiyan* 5, no. 27 (1995): 3.

political power and material advantage on this claim must be rejected."[2] That was heady stuff in a theocratic state. Following numerous threats against his life, Surūsh first took to staying abroad sporadically, officially in his capacity as a visiting professor. For the past twenty years, he has lived in permanent exile. His academic work took him to such highly reputed institutions as the Wissenschaftskolleg in Berlin, the Leiden-based ISIM Institute (Institute for the Study of Islam in the Modern World), Harvard University, Princeton, Chicago, and Yale.

3 The Changeable Nature of Religious Knowledge

Surūsh's primary goal is to change the political and social conditions in Iran. That is why he studies religion. His interest is not individual teachings or definitions. Instead, he takes an epistemological approach. The epistemological theory of Karl Popper offers two elements that are central to his endeavour: It assumes both the possibility of an infinite growth in understanding and the fact that all understanding can merely approach but not fully encompass its object. Further, it views development as a process that can, under certain circumstances, be qualified as "wrong", but never as historically "worthless."

The term Surūsh uses to designate his primary epistemological theory is *kalām-i jadīd*. This is also the name of the book series in which his works are published by Sarat as well as of the lectures he gave at Qom and at various universities in Tehran. He himself renders it as *Modern Theology* in English, taking little issue with the fundamental incompatibility of the Islamic term *kalām* with the Christian concept of theology. In both cases, he argues, the issue is to speak of God.

The central tenet of Surūsh's position is that religious knowledge is changeable. As all human insight is changeable – depending on the time and the state of science – so is human knowledge of religion. Over time, new interpretations of faith emerge, appropriate to the conditions under which the interpreters themselves live. Surūsh easily dismisses the question whether religion can properly be subject to academic study the same way as any other aspect of reality is. His epistemological approach allows him to state that the traditional reading of Islam need not be devalued in the process. Neither is religion's claim to absolute truth contested. Centrally, he writes: "If there is a religious law of this absolute nature, then it lies solely with the lawgiver – with God."[3]

Surūsh's epistemological stance is thus that it is possible to infinitely increase understanding, but that such increase will only ever approach its object. Mere humans can never truly know what God expects of them. They can never understand what

2 ʿAbdulkarīm Surūsh, "Saqf-i maʿīshat bar sutūn-i sharīʿat," *Kiyan* 5, no. 26 (1995): 26.
3 ʿAbdulkarīm Surūsh, *Qabż va bast-i tiʾūrīk-i sharīʿat*. Teheran: Kalām-i jadīd: 1994[3], 305.

divine law really is or what it intends. God's intentions are unfathomable, and all assumed knowledge of them, Surūsh claims echoing Popper, is assumption. Epistemology teaches him that human understanding of the Qurʾān can never be taken as identical to the text itself and its true intent. Humans can only hope to recognise God's ultimate goal, and the goal of religion can never be in contradiction with human concepts.

The text of the Qurʾān, like all texts, is open and invites interpretation. Human understanding of the Qurʾān, like all human understanding, depends on the circumstances of its time and the state of science. Over time, Surūsh argues, new interpretations of faith arise that are adapted to the circumstances in which its interpreters live.

Surūsh does not try to pass this off as a novel insight. Quite the contrary, he regards the prevalent rigid interpretation as a phenomenon of modernity. Past scholars had always assumed that interpretation was changeable. In defence of this stance, Surūsh cites the concept of *ijtihād* and the multiplicity of historical interpretations of the Qurʾān. The fact that exegetes could arrive at so many different readings further proved, he claims, that there is a fundamental difference between the human understanding of religion and religion as such.

Thus, Surūsh differentiates between religion and religious understanding. He sees religious understanding as based on religion, but not identical with it. This opens space for reinterpretation. The differences between religion and religious understanding, he argues, clearly prove the need to make such a distinction. One example he cites is that religion, at least in the view of the religious, is free from contradiction while religious understanding, i.e. the interpretation of religious texts by scholars, frequently contains contradictions regardless of whether it is in the field of jurisprudence, theology, or exegesis.

A further argument Surūsh offers is that while religion itself is perfect, religious understanding clearly is not. By the first statement, he means that God has revealed all that is needed for right guidance in his Scripture. There is nothing lacking in religion. Religious understanding, on the other hand, is not perfect. If it were, the entire field of Qurʾānic exegesis would have become superfluous centuries ago.

Surūsh sees this differentiation between religion and religious understanding as the key formula to separate the changeable from the invariant, the holy and the mundane elements of religion. He also sees it as a point from which a revitalisation of religious thought can be undertaken. All other approaches of a modern interpretation are rejected. He is especially inimical to the so-called scientific interpretation *(tafsīr-i ʿilmī)* which, he claims, had only produced contradictions. Undertaking *tafsīr-i ʿilmī* had not solved the questions of the faithful, but led them to doubt the truth of their faith.

Exegetes embracing a scientific interpretation sought to prove that scientific insights such as the discovery of microbes or other modern ideas were not contrary to the Qurʾān, but could actually be derived from it. They argued that it was only now, with the knowledge of nineteenth and twentieth-century science spreading through the Muslim world, that the Qur'ān could be understood properly.

A leading proponent of the scientific interpretation in Iran is Ayatollah Maḥmūd Ṭāliqānī. His commentary *Partauī az Qur'ān* [*A Ray of Light from the Qur'ān*] proposes a modernist interpretation that was influential especially with Iranian leftist groups. Ṭāliqānī viewed the Qur'ān as a living document of great relevance for contemporary problems, and saw current issues such as social justice prefigured in it.

Surūsh lauds Ṭāliqānī for understanding that expectations of religion are shaped by the contemporary state of knowledge and current problems. Ṭāliqānī, too, had tried to reconcile religion and modernity. However, he criticises his conclusion and dismisses Ṭāliqānī's approach as misguided. Surūsh holds that there is no way to prove that modern scientific knowledge is prefigured in the Qur'ān. The existence of microbes cannot be logically deduced from its text. What the method did prove, he further argued, was that Ṭāliqānī, too, approached Qur'ānic exegesis with methods he had brought in from outside the sphere of religious scholarship. He had interpreted the Qur'ān through insights from political and social theory. The value of social justice was taught there, and Ṭāliqānī had used it as the basis for his exegesis. This fundamental characteristic of *tafsīr-i 'ilmī,* Surūsh argues, supports his thesis that there is a natural exchange of ideas among academic fields.

He supports this view by citing epistemology with its assumption that all human understanding is in perpetual dialogue. The views a theologian, an exegete, or a jurist hold of the Sunna and the Holy Book will of necessity change as their extra-religious understanding does. Since all understanding of religion is thus founded on knowledge outside the religious realm, it is necessarily in flow. This is also the argument Surūsh uses to explain the differences in the interpretations of prominent scholars that once started him on his own academic journey.

What makes his theory dangerous to Iran's religious establishment is that it calls into question a privileged body of knowledge. If humans can never fully fathom the will of God, different views of religion must be permitted and ideas that have not stood the test of time be rejected – regardless of who first formulated them. Surūsh undermines the claim to absolute obedience to the *valī-yi faqīh* (Guardian Jurist), Iran's supreme religious and political authority, when he argues that even his understanding of religion is human and thus fallible. Indeed, he openly accuses the current *valī-yi faqīh*, 'Alī Khāmini'ī (b. 1939/1358) of illegitimately claiming to be the sole religious authority.

According to Surūsh, there can be no definitive religious understanding, because the Qur'ān, like all texts, is open and invites interpretation. That is also why he argues that it can be interpreted in a way that does not contradict the tenets of democracy and human rights. This attitude was bound to cause him problems in Iran, where the position of Khumaynī still dominates public discourse. In the view of the Islamic Republic's founder, rights belonged solely to God. Humans had no rights as such, only duties towards God. It is possible for God or his deputy on earth, the *valī-yi faqīh*, to grant humans certain rights, but since these are not natural, they can be revoked as God or his representative may choose. According to Khumaynī, all humans must further be subject to the good of the public, the *umma*, and can therefore not claim any individual liberties before the government. This position is also held by the current Guardian Jurist Khāminiʾī. The Western claim to universality for human rights is countered with the argument that Muslims, owing to the historical and social developments specific to their culture, wished to honour the rights of God before those of man.

Surūsh has left the relativistic position of the leading clergy far behind. Neither is he interested in establishing that Islam is compatible with human rights, nor indeed that human rights stem from Islam, as some Muslim apologists claim. To him, human rights are simply the dictate of human reason. It is therefore impossible for them to contradict religion because God's will does not contradict reason. The fact that they originated in a context outside of religion does not stop Surūsh from considering their realisation in an Islamic system possible and indeed necessary. The rights of God would remain inviolate and human and divine values in full agreement. His position here is one rarely seen outside secularist circles; like them, he assumes that humans have rights that derive from outside the sphere of religion solely on the strength of being human. Unlike the secularists, though, he argues for this position from a religious stance.

If Surūsh's argument were to be followed to its logical conclusion, a number of laws in Islamic jurisprudence could no longer be applied. This especially concerns the class of *ḥadd* penalties, punishments for transgressions that include the amputation of hands or stoning to death. Surūsh argues that it is not necessary to follow Islamic law in detail. He distinguishes between values of a primary and secondary degree. Secondary values concern details of religious law such as dress codes and thus differed from religion to religion. All religions as well as human reason, though, are in full agreement on the significance of primary values such a justice and human dignity.

4 Whose Qurʾān?

In 2007/1428, Surūsh was accused of apostasy based on an interview he gave to a Dutch newspaper announcing the English translation of one of his books. The work had already been published in Iran in 1999/1420 titled *Qabż-i tajrubi-yi nabāvī* [*The Expansion of Prophetic Experience*]. Reactions to the announcement were harsh, probably motivated not least by the title the interview was published under: "The Word of Mohammad". In it, Surūsh had said:

> But the Prophet is also the creator of the revelation in another way. What he receives from God is the content of the revelation. This content, however, cannot be offered to the people as such,

> because it is beyond their understanding and even beyond words. It is formless and the activity of the person of the Prophet is to form the formless, so as to make it accessible. Like a poet again, the Prophet transmits the inspiration in the language he knows, the styles he masters and the images and knowledge he possesses.[4]

Above all, he clearly formulated the consequences that he saw emanating from the central role he believed the Prophet Muḥammad had played in the formation of the text:

> A human view of the Koran makes it possible to distinguish between the essential and the accidental aspects of religion. Some parts of religion are historically and culturally determined and no longer relevant today. That is the case, for instance, with the corporal punishments prescribed in the Koran. If the Prophet had lived in another cultural environment, those punishments would probably not have been part of his message.[5]

The publication of the book in Farsi in 1999/1420 had not led to a comparable outcry. The most likely explanation for this, along with the unfortunate title of the interview, is that the socio-political circumstances at the time had been far less restrictive than they were in 2007/08. Another reason may have been that the interview, published through the Iranian-registered website *aftab*-News,[6] brought Surūsh's view to the attention of a wider public. Ayatollah Jaʿfar Subḥānī of Qom learned of it through *aftab*-News on February 3, 2008/ Muḥarram 25, 1429 and was very incensed. In addition, Surūsh laid out his view much more pointedly in the offending interview than he had in his earlier essay, especially with regard to its concrete consequences. His earlier publication was phrased in far less direct and more cautious terms.

In *The Expansion of Prophetic Experience*, Surūsh criticises the traditional Islamic view that the Prophet Muḥammad was a mere messenger, receiving Revelation from God and passing it on with identical content and in identical words. Surūsh argues that this limits the role of the Prophet to that of a mindless instrument. He believes that Muḥammad received the Revelation, processed it, and taught it in his own words. He had been especially chosen and called by God for this task.

Surūsh writes that we owe some of the most interesting observations on the nature of Revelation and the prophetic experience to the fourteenth-century historian Walī d-Dīn ʿAbd ar-Raḥmān b. Muḥammad b. Khaldūn (1332–1406/732–808). We learn from his works that the Prophet was gradually able to bear Revelation better, which explains why the Meccan sura are short, the Medinan ones longer.

4 Michel Hoebink, "The Word of Mohammad. An Interview with Abdulkarim Soroush," in *The Expansion of Prophetic Experience. Essays on Historicity, Contingency and Plurality in Religion*, ed. Forough Jahanbakhsh (Leiden: Brill, 2009), 273.

5 Hoebink, "The Word of Mohammad," 275.

6 See https://aftabnews.ir.

In the sense of Ibn Khaldūn, Surūsh argues that the message of Revelation also changed according to its context. In Mecca, the Prophet wanted to rouse the people, to alarm them and draw their attention. To do so, he needed short, clear texts that addressed them emotionally and took unequivocal stances. In Medina, the task was to further the missionary effort and consolidate the teachings. This required legislation, detailed explanations, and a dialogue with the populace. As the Prophet was better able to bear Revelations and had gained in prophetic experience, the form and content of the message changed.

The longer an experience lasts, Surūsh believes, the more perfect it becomes. A poet will become a better poet the more poetry he writes, a speaker more eloquent the more he speaks. This is true for all experience even if the essence of what is experienced remain unchanged. The truth that is experienced remains invariant and its validity is not questioned. The innermost experience of the Prophet, though, was subject to such development and growth. With every passing day, he became more attuned to his position, his mission and purpose. He became more determined and clearer in carrying out his duties, better prepared to reach his goals, more successful in his work and more assured of its success.

Surūsh sees the Prophet's personality as his capital, both a recipient vessel and a generator, both subject and object of the religious experience of Revelation. As his personality developed, so did his experience – and vice versa, so that Revelation came to follow him rather than him following Revelation. In the language of a mystic: his proximity to God had brought the Prophet to a state where God was inherent in his sight.

For Surūsh, it follows that the responses of the Prophet would have become yet more fully complete had he lived longer. He believes that the Qurʾān could have been far larger and he attempts to illustrate this by posing a number of questions: if ʿĀʾisha bt. Abī Bakr had not been accused of adultery, would we have sura 24, *an-nūr* (The Light)? If there had been no war between the tribes, would sura 33, *al-aḥzāb* (The Parties) exist? If Abū Lahab had not existed or if his wife had not been hostile to the Prophet, would there be a sura 111, *al-masad* (The Palm Fiber). All of these were historical events of more or less significance whose traces are found in the Qurʾān.

Surūsh draws an analogy to the way that students influence their teacher, asking a question here, requesting explanations there. All of these questions and explanations are written into religion. That is what he means by saying religion is continuing and historical. He concludes that Islam was born in the context of this interaction and engagement. The Prophet did not bring a completed book to pass on to humanity. He did not say: see what you can do with this and act accordingly. The Qurʾān was revealed sequentially. Though the spirit of the revelation was always the same, the text was formed in response to events. The relationship between the Prophet and his people was dialogical. He spoke, and he listened, and then spoke of what he had heard. Surūsh agrees with Faḍl ar-Raḥmān (d. 1988/1408) here that everyday events played an important role in the creation of religion, and if something else had happened, Islam would have been created differently.

He further cites the fact that the Qur'ān only mentions events in Mecca and Arabia around the time of the Prophet as evidence that its environment shaped the text. There is no mention of the Iranian prophet Zoroaster simply because he was unknown there, but an ethnic group called the Magi which lived in Iran and was in contact with the Arabs is mentioned a few times. All the prophets mentioned in the Qur'ān were prophets of Israel of whom the Arabs had heard and whose names were more or less familiar. Surūsh explains that just as the Arabic language as it was spoken by and around the Prophet is reflected in the Qur'ān, so are the beliefs and views of his environment.

His view of the historical Muḥammad can be quite critical: though chosen as a Prophet, he was still human and subject to all the limitations of his time and place. Muḥammad's role as the messenger was to preach to humans a Revelation that was beyond the human capacity to grasp. He had to make them understand the incomprehensible, make them grasp the ineffable and infinite. This was a mission that of necessity could only be fulfilled through his mind and language, and these were historically located and burdened with the inadequacies of their time and place.

This fact, Surūsh writes, explains all contradictions and errors found in the Qur'ān. It reflects the Prophet's education, the extent of his knowledge and understanding. These are subject to error, but the Qur'ān also contains statements that concern fundamental aspects of religion and therefore must be inerrant and intended for eternity. Surūsh holds that these are all statements on the nature of God, on life after death, and the rules of prayer. These parts of the Qur'ān are identical with the true word of God. Errors and contradictions emerge in the statements, instructions and rules regarding historical events, other religions, and especially earthly life and human community. These passages reflect the state of cultural and mental development of the time and must be replaced where modern knowledge has superseded them.

While many see an epistemological turn as the opportunity to create an Islamic aggiornamento, Surūsh's critics claim he has abandoned his religion. This is the view of Ayatollah Jaʿfar Subḥānī who has engaged thoroughly with Surūsh's theses and written several rebuttals. The reformist cleric Muḥsin Kadīvar, (b. 1959/1378) unlike Subḥānī not likely to parrot the collective views of the Iranian clerical establishment, is also critical of Surūsh's approach. Though he agrees with Surūsh in many aspects of his political criticism, Kadīvar fears that Surūsh's approach to the Qur'ān would lead to the dissolution of new religious thought *(inhilāl-i nauandīshī-yi dīnī)* and that not much would be left of religion if it were carried to its logical conclusion. The moderate Grand Ayatollah Ḥusayn ʿAlī Muntaẓirī (1922–2009/1340–1430), too, questioned its validity.

Yet Surūsh's proposal on how to understand the nature of the word of God and its Revelation has led to this topic being seriously addressed in Iranian religious discourse again. Thanks to Surūsh, Iranian scholars are once more looking to answer this complex question. Ayatollah Subḥānī's response was serious and by no means polemical when he argued that Surūsh was caught up in his own contradictions: on the one hand he claims that Muḥammad was able to convey Revelation regarding life after death or the nature of God directly as he received it from God, yet on the other hand he declares that Muḥammad's historically conditioned knowledge made him prone to error with regard to earthly life. Why was he able to perfectly convey one thing, but not the other?

This is a counter-argument at least worth considering. In addition, Subḥānī points to numerous Qur'ānic passages that clearly state the Qur'ān is the word of God and Muḥammad conveys the Revelation. His argument is located with the text. It is true that the Qur'ān is called the word of God several times in the Qur'ān. The question is whether Surūsh ever cast doubt on that.

5 Conclusion

Surūsh exemplifies how life shapes one's theology and attitude towards religion and its role. He has come a long way, having started as the regime's ideologue. Today, he argues not only for the separation of 'church' and state but also for a complete rethinking of its ideological basis, the word of God.

6 Bibliography

6.1 Sources

Hoebink, Michel. "The Word of Mohammad. An Interview with Abdulkarim Soroush". In *The Expansion of Prophetic Experience. Essays on Historicity, Contingency and Plurality in Religion*, edited by Forough Jahanbakhsh. Leiden: Brill, 2009: 271–75.

Surūsh, ʿAbdulkarīm. "Saqf-i maʿīshat bar sutūn-i sharīʿat." *Kiyan* 5, no. 26 (1995): 25–31.

Surūsh, ʿAbdulkarīm. "Khadamāt va ḥasanāt-i dīn." *Kiyan* 5, no. 27 (1995): 2–17.

Surūsh, ʿAbdulkarīm, *Qabż va bast-i ti'ūrīk-i sharīʿat*. Teheran: Kalām-i jadīd: 1994[3].

Shehzad Saleem

Jāved Aḥmad Ghāmidī

1 Introduction

Jāved Aḥmad Ghāmidī was born in 1951/1370 in EB 15, a small village in Sāhīwāl, a district in the Punjab province of Pakistan. After graduating from a local school, he went to Lahore in 1967/1387. In 1972/1392, he graduated from the prestigious Government College, Lahore, with a degree in philosophy and English literature. During his studies, he continued his religious education in the traditional manner, with individual mentoring from various teachers and scholars. In 1973/1393, he came under the tutelage of Amīn Aḥsan Iṣlāḥī (d. 1997/1418), who was destined to have a significant impact on him. Ghāmidī was also associated with the famous scholar and revivalist Abū l-Aʿlā Mawdūdī (d. 1979/1399) for several years. In 2009/1430, Ghāmidī migrated to Malaysia due to security concerns and, in 2019/1440, settled in Dallas, Texas, in the United States.

Ghāmidī has written and lectured widely on the Qur'ān, Islamic law, and various other areas of Islam. He is the founder and President of the Al-Mawrid Institute of Islamic Sciences[1] as well as the Chief Editor of the Urdu monthly *Ishraq*[2] and the English monthly *Renaissance*.[3] He also founded the Musʿab School System.[4] He appears regularly on various TV channels to discuss Islam and contemporary issues as part of his campaign to educate people about Islam. His lectures and talks are available on his website.[5]

2 Works

Ghāmidī has drawn heavily on the Qur'ānic thought of his two illustrious predecessors, Ḥamīd ad-Dīn Farāhī (d. 1863/1279) and Amīn Aḥsan Iṣlāḥī, presenting many of their views in a more precise manner. Nevertheless, many of his contributions to Islamic thought are his own achievements. He is fluent in Arabic, Persian, Urdu, and English. However, he generally expresses his religious views in his mother tongue of Urdu.

---

1 Follow the link to the homepage of the institute: www.al-mawrid.org (accessed December 18, 2023).
2 See www.ghamidi.org (accessed December 18, 2023).
3 See www.monthly-renaissance.com (accessed December 18, 2023).
4 See www.musab.edu.pk (accessed December 18, 2023).
5 See www.javedahmadghamidi.com (accessed December 18, 2023).

https://doi.org/10.1515/9783111320069-019

His book *Mīzān* (English title: *Islam – A Comprehensive Introduction*) is based on his understanding of the contents of religion and the application of his hermeneutical principles, which will be discussed in the following paragraphs. He has also written an abridged edition of *Mīzān*, titled *Al-Islām* (English title: *Islam – A Concise Introduction*).

Furthermore, he has produced a five-volume, annotated translation of the Qur'ān, *Al-Bayān*. His translation closely reflects the brevity and terseness of the classical Arabic used in the Qur'ān, where ideas are conveyed using minimal words, and concepts that are understood are left unsaid. To achieve this brevity, the author employs various literary devices from classical Arabic that are not commonly found in other languages.

Burhān and *Maqāmāt* are two of Ghāmidī's other books on religious issues. The former is a treatise in which he critically analyzes contemporary religious thought, while the latter is a collection of his essays on contemporary views and issues that require *ijtihād*. *Khayāl-o Khāmah* is a collection of his poems. *Al-Bayan*, *Mīzān*, *Al-Islām* and most articles of *Burhān* and *Maqāmāt* have been translated into English. The articles of these two latter books have been compiled in an anthology titled *Selected Essays of Javed Aḥmad Ghāmidī*.

3 Background to Ghāmidī's Hermeneutical Approach

As previously mentioned, Ghāmidī draws heavily on the works and approaches of his two illustrious predecessors: Ḥamīd ad-Dīn Farāhī and Amīn Aḥsan Iṣlāḥī. In fact, it can be said that it is the former to whom the origins of Ghāmidī's distinct hermeneutical approach can be traced. Thus, it could rightly be deemed the Farāhīan hermeneutical approach. Farāhī was a cousin of the famous theologian and historian Shiblī Nuʿmānī (d. 1914/1332), from whom he learnt Arabic. He also studied Arabic literature under Fayḍ al-Ḥasan Sahāranpūrī (d. 1887/1304), a prodigious scholar of the language. After graduating from Allāhbād University, Farāhī taught at various institutions, including Aligarh and Dār al-ʿUlūm in Hyderabad, India.

For almost fifty years, Farāhī reflected on the Qur'ān, which remained his main interest and the focal point of all his writings. His greatest contribution was to redirect the attention of Muslim scholars to the Qur'ān as the basis and ultimate authority in all matters of religion. He emphasized that the Qur'ān should be considered the *mīzān* (the scale that weighs the truth) and the *furqān* (that which distinguishes between good and evil), as it itself claims. Thus, prophetic statements (*aḥādīth*) cannot change or modify the Qur'ān in any way. They should be interpreted in the light that emanates from this divine book and not vice versa. In light of the Qur'ān's status, he insisted on the univocity of the text and rejected the idea that variant readings should be considered part of the Qur'ān itself. It was his deep deliberations on the Qur'ān that led him to unfold its *naẓm* (coherence) in a unique way. By taking the three constituents of *naẓm* into consideration – order (*tartīb*), proportion (*tanāsub*), and unity (*waḥdāniyyah*) – he aimed to prove that a single interpretation of the Qur'ān was possible. This alone had the far-reaching consequence of the principle of Qur'ānic *naẓm*: serious differences in the interpretation of the Qur'ān that have given rise to the menace of religious sectarianism, he said, were actually the result of disregarding thematic and structural coherence in the arrangement of and relationships between various Qur'ānic verses and paragraphs. Each sect had adopted its own interpretation because isolating a verse from its context can produce multiple meanings. Farāhī argued that it was the coherence of the Qur'ān alone that, if considered, could produce a definite and integrated understanding of the Divine Message. Besides a host of other books authored mainly in Arabic, Farāhī wrote a commentary on fourteen suras of the Qur'ān titled *Tafsīr Niẓām al-Qur'ān* (Urdu translation: *Majmū'a tafāsīr-i Farāhī*).[6]

Amīn Aḥsan Iṣlāḥī[7] was born in 1904/1322 in a small village in Azamgarh (Uttar Pradesh), India. From 1925/1343 to 1930/1348, he stuck closely to Farāhī like a shadow. It was in this formative period that he developed a deep understanding of the Qur'ān and learned the principles of its interpretation from his mentor. In his *Tadabbur-i Qur'ān* (1958–80/1377–1400), he produced an impactful *tafsīr* spread out over nine volumes. It does not simply reflect the principles of his illustrious mentor, al-Farāhī, but also bears the stamp of originality. It is indeed a unique work that ushered in a new era in the field of scriptural interpretation. Iṣlāḥī used a Qur'ānic verse to validate that the Almighty divided the Qur'ān into seven chapters. Each of these chapters has a theme, and the suras are arranged in these chapters keeping this theme in view. Within each chapter, the suras themselves generally occur in pairs based on the subjects they discuss. Each sura also has a specific theme which is the most comprehensive statement of its contents.

4 Hermeneutical Principles

In his treatise, *Mīzān*, Ghāmidī mentions ten hermeneutical principles for understanding the Qur'ān:

1. Appreciation of Classical Arabic
2. Eloquence of language
3. Uniqueness of style
4. The Final authority

---

6 Ḥamīd ad-Dīn 'Abd al-Ḥamīd al-Farāhī, *Majmū'a tafāsīr-i Farāhī*, trans. Amīn Aḥsan Iṣlāḥī (Lahore: Faran Foundation, 1991).

7 See www.amin-ahsan-islahi.org (accessed January 5, 2024).

5. Parallel verses and constructions
6. The final book on religion
7. Subject matter of the Qur'ān
8. Coherence in the discourse
9. Arrangement of the qur'ān
10. Historical background

In order to provide a full picture of Ghāmidī's Qur'ānic hermeneutics, I will paraphrase Ghāmidī's position on each of these ten points in the following. The paraphrase will be based on my English translation of Ghāmidī's *Mīzān*.[8]

4.1 Appreciation of Classical Arabic

The Qur'ān was revealed in Classical Arabic, which was spoken by the tribe of Quraysh in Mecca (Q 19:97) during the age of *jāhiliyya*.[9] Although the Almighty has endowed it with inimitable eloquence and articulacy, its substance is no different from the language that was spoken by the Messenger of God and the people of Mecca at that time.

Consequently, correctly understanding this book depends on accurate knowledge and true appreciation of that language. It is essential that a person who wants to reflect on the Qur'ān and who attempts to interpret and explain is a very competent scholar of Classical Arabic. He or she should also be adept at appreciating its styles and linguistic features so that the language, at least, is not an impediment to him or her understanding the Qur'ān.

Another important fact about the Qur'ān, which every student of this divine book should be aware of, is that the Arabic it is written in it is not the Arabic in which authors like al-Ḥarīrī (d. 1122/516) and poets like al-Mutanabbī (d. 965/354) composed their masterpieces, nor is it the Arabic in which Abū l-Qāsim az-Zamakhsharī (d. 1144/538) and Fakhr ad-Dīn ar-Razī (d. 1209/606) wrote their Qur'ān commentaries. It is not the Arabic of the contemporary newspapers that are published in Arab countries, nor is it the Arabic of the prose or poetry written by today's Arab literati. There is no doubt that all of this is Arabic as well; however, it is very different from the Arabic of the Qur'ān, which can rightly be described as "Classical Arabic." Thus, the difference in the vocabulary, idiom, style, and construction of Classical Arabic and the Arabic spoken and written today is the same as the contrast, for example, between the Urdu and Persian of Mīrzā Ghālib (d. 1869/1286) and Mīr Taqī Mīr (d. 1810/1225), and Saʿdī Shīrāzī

8 See Javed Ahmad Ghamidi, *Mīzān* (*Islam: A Comprehensive Introduction*), trans. Shehzad Saleem (Al-Mawrid: Fine Printers, 2021^{5}).
9 Ibid., 19–25.

(d. 1292/691) and 'Umar Khayyām (d. 1131/525), and the Urdu and Persian of the newspapers and journals on the Indian subcontinent and in Iran. Similarly, this difference can be gauged by comparing the broad difference, for example, between the English of Shakespeare (d. 1616/1025) and Milton (d. 1674/1085) on the one hand and the language written and spoken today in Britain on the other. It is therefore essential to note that neither contemporary nor medieval Arabic plays a role in developing an appreciation of the language. In fact, becoming too involved in these forms of Arabic can be detrimental to one's understanding of the Qur'ān. Consequently, the very first thing that a reader must turn to in order to understand the language of the Qur'ān is the Qur'ān itself.

Moreover, it is an established belief that the astounding miracle of the language and literature of the Qur'ān has been transmitted to us unchanged. Thus, it is an acknowledged fact that the Qur'ān is not only the final and ultimate authority in all matters of religion but also represents the final criterion and standard for the language of its times.

After the Qur'ān, we find this language in the hadith of the Prophet and the traditions (*āthār*) of the companions. Of course, only a very small portion of them has been transmitted verbatim and is thus in a position to be presented as a criterion and as representative of Classical Arabic. Yet, whatever portion we have is a great treasure for students of this language. Its words, idioms, style, and constructions are the best examples of the language in which the Qur'ān was revealed. Since the original wording has been preserved in the supplications of the Prophet, in his conversations with his companions, and in the various parables that he used to explain some aspects of religion, the parallels of this language can be best observed in these three types of narratives. Thus, if students of the Qur'ān consult these sources, they can gather invaluable examples of Classical Arabic, which can help them to understand both difficult words as well as the background and occasions on which they are spoken in the Qur'ān.

Following these three sources, the greatest source of Classical Arabic is the classical literature of the Arabs. This literary corpus includes the works of celebrated poets like Imru' al-Qays (d. 544), Zuhayr (d. 609), 'Amr b. Kulthūm (d. 584), Labīd (d. 661/41), an-Nābigha (d. 604), Ṭarafa (d. 569), 'Antara (d. 601), al-A'shā (d. 629/7), and al-Ḥārith b. Ḥilliza (d. 580), as well as orators like Quss b. Sā'ida (d. 600). Scholars in this field know that a large part of this literature can be found in the anthologies of the poets and in *al-Aṣma'iyyāt, Mufaḍḍaliyyāt, al-Ḥamāsa, al-Mu'allaqāt as-sab'*, as well as in the works of writers like 'Amr b. Baḥr al-Jāḥiẓ (d. 869/255) and Abū l-'Abbās al-Mubarrad (d. 899/286). Many collections of poetic works from that era that had previously been inaccessible have now been published. Undoubtedly, much of the Arabic language has been transmitted to us through consensus and *tawātur*, and has been preserved in primary works like *at-Tahdhīb, al-Muḥkam, aṣ-Ṣiḥāḥ, al-Jamhara*, and *an-Nihāya*. However, it is also a fact that the greatest source of the language that has not been transmitted through *tawātur* is also the Classical Arabic literature of that age. Though some of it was composed later and attributed to that age, just as hadith scholars can distinguish between rightly and wrongly reported narratives, critics of the Arabic language can similarly distinguish between the original and the fabricated on the basis of objective standards of textual criticism.

4.2 Linguistic Eloquence

The Qur'ān was not just revealed in Arabic – it was revealed in eloquent Arabic.[10] The language is clear and cogent, and there is no vagueness in it. Every word is in the right place, and every style adopted is well known to its addressees (Q 26:193–95; Q 38:29).

This is an obvious reality of the Qur'ān. If this premise is accepted, then it must be conceded that none of the words used or styles adopted in the Qur'ān are rare or unknown (*shādh*). Its words and styles are well known to and conventionally understood by its addressees. No aspect of the language has any peculiarity or rarity to it. Consequently, when interpreting the Qur'ān, the conventionally understood and known meanings of the words should be taken into consideration. Aside from them, no interpretation is acceptable. Thus, in Q 55:6 "*wa-n-najmu wa-sh-shajaru yasjudān,*" the meaning of the word "*an-najm*" can only be "stars." In Q 22:52 "*wa-mā arsalnā min qablika min rasūlin wa-lā nabiyyin illā idhā tamannā alqā sh-shayṭānu fī umniyyatihi,*" the word "*tamannā*" can only mean "desire." In Q 88:17 "*a-fa-lā yanẓurūna illā l-ibili kayfa khuliqat,*" the word "*al-ibil*" has been used solely to denote "camel." The only meaning of the word "*bayḍ*" in Q 37:49 "*ka-annahunna bayḍun maknūn*" is "eggs." Finally, in Q 108:2, "*fa-ṣalli li-rabbika wa-nḥar,*" the word "*naḥr*" can only mean "sacrifice." They do not mean "plants," "recital," "clouds," "the hidden sheath of eggs," or "folding hands on the chest" respectively.

4.3 Uniqueness of Style

The Qur'ān has a unique style.[11] It has the simplicity and continuity of prose, yet it is not prose. It has the beat, rhythm, and poise of poetry, yet it is not poetry. It is not the kind of book that we are usually familiar with, where there are chapters and sections that deal with a specific topic or topics. The people of Arabia would sometimes call it poetry and sometimes likened it to the rhyming prose of the soothsayers, and it is that very uncertainty that shows that they were not satisfied with what they said about it. In reality, the Qur'ān's style makes it a unique book. It has the flow of tumultuous

10 Ibid., 25f.
11 Ibid., 26–28.

torrents and the vigor of pounding waves. Its sound reasoning has many variations that cannot be emulated; topics are connected to one another with subtle harmony. It cites stories and anecdotes; the discourse returns to its central theme every now and then. Verses that portray threats, intimidation, and punishment can be found in various styles, while other verses depict sorrow and longing. Emphatic expressions are another hallmark of its style. Similarly, we find verses that express intense emotions of disgust, indifference, and apathy. Instances that reflect warmth and affection are as refreshing as dew drops, and instances that convey wrath and rage are as fiery and resounding as thunder. The unique modes of address that it contains simply enchant the reader, putting him or her in a state of trance.

But what exactly is the genre of the Qur'ān? This is difficult to define precisely. The divine book bears some resemblance to oration, but cannot be strictly classified as such. As a student of the Qur'ān, it is important to keep the following points in mind: First, in order to understand the Qur'ān, its ambience should be studied; this means that the background, situation, and the requisites in which a sura was revealed must be determined. Nothing is required for this beyond deliberating on the Qur'ān itself, and the light of the Qur'ān suffices for this. When a person deliberates on the Qur'ān, concentrates on each and every one of its words, tries to understand the rhythm and beat of those words and the construction of the sentences, the occasions on which a discourse was uttered become fully clear. The extent of this clarity is such that they become evidence of themselves, and no external argument is required for any corroboration.

Second, the direction of the Qur'ān's address should be ascertained at every point. Its direction of address shifts a number of times at very short intervals and sometimes even within a single verse. In one instant, Muslims are the addressees and, in the next, the *mushrikūn*; similarly, the People of the Book are addressed in one verse, before the address suddenly shifts to Muslims. There are similar shifts between singular and plural addressees. Moreover, such changes occur in the speaker and the person being spoken to. The speaker might be God, before Gabriel then unexpectedly assumes the speaker's role. In another passage, the speaker is Gabriel, and then suddenly the discourse is emanating from the mouth of Muḥammad. In short, just as an orator shifts from one addressee to another by changing his or her tone, facial expressions, and the grandeur of the words used, in a similar manner, the address in the Qur'ān changes rapidly. Thus, it is essential that this aspect is given full consideration when interpreting and explaining the Qur'ān. It must be ascertained whether the *speaker* is, for example, God, Gabriel, the Prophet, or the people. Similarly, it should be determined whether the person being *spoken to* is God, the Prophet, or the people. It must also be ascertained whether the people are Muslims, Hypocrites, the People of the Book, or the Idolaters among the Ishmaelites, or whether two or three such groups or all of them are being addressed. Then, there may also be instances of ambiguity in the address. Sometimes, a verse seems to address the Prophet; however, in reality, the speech is directed at the Muslim *umma*. Similarly, what appears to be a verse addressing him is actually directed at the leadership of the Quraysh or at the People of the Book. Examples of such address abound in the Qur'ān. It is therefore essential that this distinction be made with full caution and that it be fully ascertained who the actual addressee is. Without taking this aspect into account, the true purport of the Qur'ān cannot be grasped.

Third, a differentiation must be made between verses that have a general meaning and those that have a specific meaning that is confined to a certain era. There are many places in the Qur'ān where the words are general; however, the context attests with full certainty that something specific is meant. The Qur'ān uses the word "*an-nās*" (people), but this does not refer to all the people of the world, and on many occasions, it does not even refer to all the people of Arabia. Rather, it refers to a group among them. The Qur'ān uses the expression "*'alā d-dīni kullihi*" (on all the religions), which does not refer to all religions of the world; similarly, it refers to "*al-mushrikūn*" (polytheists), but this does not mean all those who are guilty of polytheism. Similarly, the phrase "*wa-in min ahli l-kitāb*" (and from these People of the Book) does not refer to all the People of Book in the world. The Qur'ān also mentions "*al-insān*" (human beings), but this does not refer to mankind. This, then, is a common style in the Qur'ān, and if it is not taken into consideration when explaining and interpreting the divine book, the reader can end up misunderstanding the whole purport of the Qur'ān. Thus, it is of paramount importance that the interpretation of Qur'ānic words always remains subject to its context and usage.

4.4 The Final Authority

The Qur'ān is a *mīzān* (Q 42:17) (scale that tells good from evil) and a *furqān* (Q 25:1) (that which distinguishes between good and evil) on this Earth, and a *muhaymin* (Q 5:48) (guardian) over other divine scriptures.[12] The Qur'ān is *al-furqān* in the sense that it provides the final and absolute verdict in distinguishing truth from falsehood. This word also implies that this book is the standard by which everything must be judged and has the final say on matters relating to religion. Everyone must consult it alone to resolve differences of opinion. Nothing can judge it; it reigns supreme in the domain of religion and divine guidance, and no one may subordinate it to any other thing.

It is also *muhaymin* (Q 5:48) (dominant) over previous scripture, which means that the Qur'ān is the real, authentic, and trustworthy version of the Book of God. Thus, when the texts of other scriptures were lost to posterity and their translations were greatly tampered with, it was this Qur'ān that was given the authority to judge what was right and wrong in those scriptures. Whatever it judges to be right is right, and whatever it judges to be wrong is wrong and must necessarily be rejected.

12 Ibid., 28–47.

This is the status that the Qur'ān asserts for itself. Consequently, the following principles need to be adhered to on the basis of that status: First, no divine revelation extraneous to the Qur'ān – not even the Prophet to whom the Qur'ān was revealed – can specify a general directive of the Qur'ān or alter any of its directives. Everything must either be accepted as religion or rejected on the grounds that it does not exist on the basis of the Qur'ān. Everything accepted as religion shall be rigorously scrutinized in the light of this divine guidance. Thus, the basis of belief and faith must be directly derived from it.

Second, the text of the Qur'ān is univocal. The meaning conveyed by each of its words is definitive. Whatever it intends to say, it says with full certainty, and there is no ambiguity. In no issue is it unable to convey what it means to. The meanings of its words perfectly match those words, and the meanings do not in any way contradict what the words say. The only way to approach the Qur'ān in order to understand it is through its words. With full certainty, those words convey what they stand for, and there is no question of any doubt or ambiguity in this regard.

4.5 Parallel Verses

The Qur'ān presents its message in various ways and in a variety of styles.[13] As a result, it is unparalleled by any other works in explaining its own verses, which are inimitable and set in very concise diction. Thus, it introduces itself as "*kitāban mutashābihan*" (Q 39:23). It says, "This is a Book whose verses were first concise and then they were explained by Him Who is wise and all-knowing" (Q 11:1).

Thus, the style adopted was initially concise, brief, and succinct, and those succinct verses, which carried a world of meaning, were later explained. Hence, it must be accepted as a principle that the Qur'ān explains itself. This principle applies not only to the directives of the Qur'ān, the historical references it cites, and other allusions it makes but also to the invaluable treasure of the parallels of its own words and styles. Difficulties encountered in understanding them can be resolved through recourse to this treasure.

4.6 The Final Book on Religion

The Qur'ān is the final book of the religion it presents, not the first.[14] The history of this religion began when God created human beings on Earth, and the basic realities of religion were ingrained into their nature. These include, first, that they have a Cre-

13 Ibid., 47f.
14 Ibid., 48–52.

ator who created them; He alone is their Lord, and as a natural corollary to this, He alone should be worshipped by them. Second, human beings have been sent to this world to be tried and tested, and, for this reason, they have been given a clear awareness of good and evil; they have not only been given the freedom to exercise their will but have also been given sovereignty on this Earth. This trial of human beings will continue until their death. If they pass this trial, they will be granted access to the Kingdom of Heaven, where they will be free from the regrets of the past and fears of the future.

Third, the Almighty will keep sending His guidance at various times according to the needs of human beings. If they obey this guidance, they will not go astray, and if they evade it, they will be eternally doomed in the Hereafter. Consequently, the Almighty has fulfilled His promise and provided guidance to mankind by selecting people from among them and by delivering His guidance to mankind through them. This guidance contains both *al-ḥikma* (faith and morality) and *ash-sharīʿa* (law). The former obviously did not require any change, while the latter was revealed as per the needs of a people until the time of Abraham, when its directives crystallized in the form of a sunna for all mankind. In the time of Moses, when the formal state of the Israelites had been established, the Torah was revealed and the directives of the *sharīʿa* regarding the collectivity were also revealed. At that time, when certain aspects of *ḥikma* were not before the eyes of people, they were made evident to them through the Psalms and Gospels. When the original texts of those scriptures became extinct, the Almighty sent the last of His Messengers and gave him the Qurʾān (Q 5:48). This is the history of religion.

Thus, with this in mind, the following preceded the Qurʾān: 1) the Innate Guidance found in human nature, 2) the Tradition of the Religion of Abraham, and 3) the Scriptures of the Prophets. The first of these aspects relates to the basics of faith and morality. In the terminology of the Qurʾān, a major portion of this is called *maʿrūf* and *munkar*. The former refers to issues that are regarded as good by human nature, and the latter refers to things that are regarded as evil and therefore abhorred by it. The Qurʾān does not provide a comprehensive list of these aspects; however, it says that a person is innately aware of good and evil and is able to fully distinguish between the two on this basis. It thus demands that a person accept *maʿrūf* and shun *munkar* (Q 9:71).

The second of the above-mentioned aspects is called *millat-i ibrāhīm* in the Qurʾān. Prayer, fasting, pilgrimage, and *zakāh* are all directives of *millat-i ibrāhīm*. The addressees of the Qurʾān were fully aware of these directives and practiced them to a great extent. In the narrative that recounts Abū Dharr's acceptance of faith, he explicitly says that he used to pray diligently even before Muḥammad declared his prophethood.[15] It is established that Friday prayer was not unknown to the addressees of the Qur'ān.[16] They would also offer the funeral prayer[17] and would fast in the very manner we fast today.[18] *Zakāh*, too, was familiar to them as a specific proportion of their wealth, just as it is today (Q 70:24). Regarding the worship rituals of *ḥajj* and *'umra*, every knowledgeable person is aware that, though the Quraysh added some religious innovations, the rites of those worship rituals were virtually the same as they are today. In fact, it is evident from certain narratives that people were even aware of those innovations. Consequently, there is a narrative recorded by both al-Bukhārī and Muslim stating that the *ḥajj* offered by Muḥammad before his prophethood was conducted without the innovations of the Quraysh, in the very manner it had been offered since the time of Abraham.[19]

The same is true in the case of animal sacrifice, *i'tikāf* (the ceremonial stay at the mosque during Ramaḍān), and circumcision, as well as some other Islamic customs and aspects of etiquette. All these things were already known and specified, and the Arabs were aware of them as age-old traditions that had been transmitted from one generation to the next. Thus, there was no need for the Qur'ān to describe them in detail. The Arabs were fully aware of what the Arabic words that referred to them meant. If the Qur'ān asked the Arabs to pray, to fast, to offer *ḥajj*, or to pay *zakāh*, they knew the meaning of those terms. The Qur'ān was never the first directive on any of those matters. It only reformed and revived them, and explained some aspects – and that, too, to the extent that it was essential. This tradition of the religion of Abraham, which in religious parlance is called sunna, is regarded by the Qur'ān as the religion of God. And when the divine book (Q 16:123) asks the Prophet to follow the religion of Abraham, it indeed asks him to adhere to the sunna as well.

The third of these aspects is the divine scriptures which are present in the Bible in the form of the Torah, the Gospels, and the Psalms. Their recipients have lost parts of them to posterity and have also been guilty of making interpolations to them. However, a rich abundance of the sharia and *ḥikma* revealed by the Almighty is still present in them in its ancient, divine style. Students of the Qur'ān know that it has referred to them at various places, has made concise allusions to the prophetic tales mentioned in them, and has negated the interpolations of the Jews and the Christians and criticized the history presented in those scriptures. The Qur'ān has based its *itmām al-ḥujja* (communication of the truth to the extent that nobody has a pretext to

15 Abū l-Ḥusayn b. Ḥajjāj an-Nīsabūrī Muslim, *al-Jāmi' aṣ-ṣaḥīḥ* (Riyadh: Dār as-Salām, 2000^2), 1086–88, (no. 6359).

16 Muḥammad b. Mukarram b. Manẓūr, *Lisān al-'arab* (Beirut: Dār Ṣādir, n.d.), 8: 58.

17 Jawwād 'Alī, *al-Mufaṣṣal fī tārīkh al-'arab qabl al-islām* (Beirut: Dār al-'Ilm li-l-Malāyīn, 1986^2), 6: 338.

18 Abū 'Abd Allāh Muḥammad b. Ismā'īl al-Bukhārī, *al-Jāmi' aṣ-ṣaḥīḥ* (Riyadh: Dār as-Salām, 1999^2), 321 (no. 2002).

19 Ibid., 270 (no. 1664).

deny it) on those very scriptures, and it unequivocally declares that its fountainhead and origin is the same as that of those scriptures (Q 3:3–4; Q 4:163).

4.7 A Chronicle of Muḥammad's *Indhār*

With regard to its subject matter, the Qur'ān is a narrative of a messenger's *indhār* (warning).[20] The Qur'ān speaks of this reality. The reason for this is that the Qur'ān was revealed as an amalgamation of sharia and *ḥikma*, but was also revealed to become the real means of the Prophet's *indhār* to his people (Q 6:19).

It is well known that Muḥammad was not only a *nabī* (prophet) but also a *rasūl* (messenger). Prophets are personalities to whom the Almighty reveals divine guidance so that they can guide people. However, not every prophet is a messenger. Messengerhood is a status bestowed on only some prophets. According to some of the details provided by the Qur'ān, a *rasūl* decides the fate of his addressees if they intentionally deny the truth, passing the judgment of God on them in this very world. The Qur'ān informs us that this final phase in the preaching endeavor of a *rasūl* comes after it passes through the phases of *indhār* (warning), *indhār-i 'ām* (augmented warning), *itmām al-ḥujja* (conclusive communication of the truth), and *hijra wa-barā'a* (migration and acquittal). It is in this final phase that the divine court of justice is set up on Earth. Punishment is meted out to those who reject truth, while those who have accepted it are rewarded. In this way, a miniature Day of Judgement is witnessed on Earth.

The history of the preaching endeavors of the *rusūl* related in the Qur'ān shows that, at this stage, one of the following two situations generally arises: Either a *rasūl* has a few companions only and there is no place available for him to migrate to, or the *rasūl* has a substantial number of companions, and the Almighty also provides them with a place to which they can migrate, where they are bestowed with political authority. In both situations, the established practice of the Almighty manifests itself – the practice which the Qur'ān refers to in the following words: "For each community, there is a messenger. Then when their messenger comes, their fate is decided with justice, and they are not wronged" (Q 10:47).

When considering the subject matter of the Qur'ān, it is important for a student of the divine book to keep the following three aspects relating to its exegesis and interpretation in mind: First, after deliberating on the contents of a sura, the exact phase in which it was revealed should be determined. A reader is required to go so deep within this endeavor that he or she is able to very satisfactorily say that, for example, a sura has been revealed in the phase of *indhār*, in the phase of migration and

20 Ghāmidī, *Islam: A Comprehensive Introduction*, 52–54.

acquittal, or in the phase of reward and punishment. Moreover, each verse of a sura should be interpreted while keeping this distinction in mind.

Second, the addressees of each sura must be determined from among the people present at the time of the Qur'ān's revelation. They could be the idolaters, the People of the Book, the Hypocrites, the Prophet and his followers, or some specific group from among these religious denominations. It must also be determined whether parts of a sura address a secondary addressee besides the primary one. Consequently, the antecedent of every pronoun, the entity referred to in every defining article (*alif lām*), and the connotations of every term and expression should be determined in the light of the sura's addressees.

Third, in the case of directives that relate to *jihād*, the supremacy of the truth, and the political authority that derives from that supremacy, it must be specifically determined whether they are permanent directives of the sharia or if they specifically relate to the addressees of the prophetic times, as the directive cannot extend beyond those addressees.

4.8 Coherence in the Text

Each sura is a coherent collection of verses.[21] These verses are not disjointed or haphazardly placed in a sura. In fact, each sura has a theme, and all the verses are aptly placed with regard to this theme. When a sura is studied while keeping its theme in mind, and when its coherence becomes evident as a result of that study, it comes out as a well-knit unit. It is because of this coherence in the Qur'ān that, when it calls (Q 2:23; Q 11:13) upon its addressees to emulate it, it does not ask them to produce independent verses but to produce one or more suras like it.

The following points will help in understanding this coherence: First, each sura has a central theme around which its contents revolve, creating a unified whole. This theme serves as the most comprehensive statement of the sura's contents. The theme is to the sura what the soul is to the body. Second, the main text of the sura is accompanied by an introduction and a conclusion. In some cases, the content of the sura can be divided into sections and paragraphs, and in other cases into only paragraphs. Paragraphs represent small shifts in the subject, and sections represent greater shifts. The verses in the introduction and the conclusion may also be divided into paragraphs at times, according to the subject they discuss. Third, the paragraphs and sections relate to each other not through verse-to-verse linear connections but through various literary devices, such as parables, comparisons, or parallelisms, as well as through statements and passages that are conditional, parenthetical, inferential, modifying, or cyclic, or that signify corollaries, conclusions, questions, or answers. This, of

21 Ibid., 54–57.

course, is not an exhaustive list. Finally, the text of the sura progresses through these paragraphs and sections, gradually reaching its culmination. As a result, the sura assumes a distinct and unique form and shape, and becomes a complete and independent whole.

4.9 Arrangement of the Qur'ān

The suras of the Qur'ān are not haphazardly compiled as is generally thought.[22] They were arranged in a specific order by the Almighty, and, like the arrangement of the verses within a sura, the arrangement of the suras within the Qur'ān is very apt and meaningful regarding the topics they discuss. In short, according to this arrangement, the Qur'ān is divided into seven distinct chapters, and the suras within each chapter occur in pairs. The suras are paired on the basis of the topics discussed, and each member of a pair has a complementary relationship with the other. Some suras are an exception to this scheme, such as sura 1, *al-Fātiḥa* (The Opening), which functions as an introduction to the whole Qur'ān. Some other suras supplement or conclude a group. This scheme, with its seven sura chapters and its pairings, is described by the Qur'ān in the following words: "And We have bestowed upon you seven *mathānī*, which is this great Qur'ān" (Q 15:87). Each group in the Qur'ān begins with one or more Meccan suras and ends with one or more Medinan suras.

The following list is a brief summary of the seven Qur'ānic groups. It is primarily based on the interpretation of Ghāmidī's mentor, Amīn Aḥsan Iṣlāḥī.[23]

- Group I: Sura 1, *al-Fātiha* (The Opening) – Sura 5, al-*Mā'ida* (The Table)
  Meccan: 1
  Medinan: 2–5
- Group II: Sura 6, *al-An'ām* (The Cattle) – Sura 9, *at-Tawba* (Repentance)
  Meccan: 6, 7
  Medinan: 8, 9
- Group III: Sura 10, *Yūnus* (Jonah) – Sura 24, *an-Nūr* (Light)
  Meccan: 10–23
  Medinan: 24
- Group IV: Sura 25, *al-Furqān* (The Criterion) – Sura 33, *al-Aḥzāb* (The Parties)
  Meccan: 25–32
  Medinan: 33

22 Ibid., 57–60.

23 Amīn Aḥsan Iṣlāḥī. *Tadabbur-i Qur'ān*, 9 vols (Lahore: Faran Foundation, 1985). For more details on Iṣlāḥī's Qur'ānic hermeneutics, see Mustansir Mir's chapter on him in the fourth volume of this *Handbook*.

- Group V: Sura 34, *Saba'* (Sheba) – Sura 49, *al-Ḥujurāt* (The Private Apartments)
  Meccan: 34–46
  Medinan: 47–49
- Group VI: Sura 50, *Qāf* – Sura 66, *at-Taḥrīm* (Forbiddance)
  Meccan: 50–56
  Medinan: 57–66
- Group VII: Sura 67, *al-Mulk* (Sovereignty) – Sura 114, *an-Nās* (Mankind)
  Meccan: 67–112
  Medinan: 113–14
  Each group has a theme, and the suras within it are arranged according to that theme. The following are examples of the themes of the first two chapters:[24]
- Chapter I: (Sura 1, *al-Fātiḥa* – Sura 5, *al-Mā'ida*)
  The theme of the first chapter is communicating the truth to the Jews and Christians [of prophetic times] in a conclusive manner, instituting a new *umma* in their place, its *tazkiya* and *taṭhīr*,[25] and describing its final covenant with the Almighty.[26]
- Chapter II: (Sura 6, *al-An'ām* – Sura 9, *at-Tawba*)
  The theme of the second chapter is communicating the truth to the polytheists of Arabia in a conclusive manner, the *tazkiya* and *taṭhīr* of the believers, and describing the last miniature Day of Judgement witnessed in this world before the advent of the greater Day of Judgement.[27]

4.10 Historical Background

The Qur'ān must be understood in the context in which it was revealed.[28] According to conventional history, it was revealed to Muḥammad in Arabia in the seventh century. It is evident that Muḥammad explained the Qur'ān wherever and whenever need arose, as did the scholars among his companions, as well as the scholars and researchers who came after them.

This history of the Qur'ān is an established fact and demands the following from its students: First, they must be aware of the history of the period and place in which

---

24 The themes of each of the chapters have been translated from Ghāmidī's annotated translation of the Qur'ān: *Al-Bayān*. See Javed Aḥmad Ghāmidī, *Al-Bayān* (Lahore: Topical Printing Press, 2018).

25 Both *tazkiya* and *taṭhīr* are used by the author as synonyms to refer to the purification of believers' beliefs and practices; the word *taṭhīr* has the additional meaning of separating and organizing the believers from the hypocrites and the disbelievers. Whenever this additional meaning is implied, the author has specified it in the explanation of the sura.

26 Ghāmidī, *Al-Bayān*, 1: 15.

27 Ibid., 2: 15.

28 Gāmidī, *Islam: A Comprehensive Introduction*, 60f.

the Qur'ān was revealed. Second, full importance must be given to whatever is mentioned in the hadith and traditions (*āthār*) with reference to the Prophet and his companions. Traditions undoubtedly contain many spurious things, and because the original words have not always been preserved, and a narrator has relied on his own intellect to transmit those words, the original words have undergone many changes; however, this literature still contains much valuable information. Third, whatever scholars and exegetes have written must be given due consideration. It is only by benefiting from the works of previous scholars that new dimensions are added in such disciplines; such additions cannot be made by ignoring earlier scholarship. True knowledge does not come about through arrogance and haughtiness; it comes with humility and a sincere love for the truth. Thus, it is essential that students of the Qur'ān always study the primary exegetical works on the Qur'ān when they are forming an opinion or reflecting upon a verse and teaching it.

Before considering the works of leading scholars who have dealt with Qur'ānic exegesis in the Farāhī school, three primary works on the exegesis of the Qur'ān should be considered: *Jāmiʿ al-bayān* by aṭ-Ṭabarī, *at-Tafsīr al-kabīr* by ar-Rāzī, and *al-Kashshāf* by az-Zamakhsharī. The first of these is a compendium of the opinions of past authorities, the second mostly deals with theological issues, and the third with declensions and syntax. From these primary works, a student of the Qur'ān should only select that which is in harmony with the words of the Qur'ān, its coherence, and arrangement. However, it is important that these commentaries on the Qur'ān are not ignored.

5 Theological Influence

From 1973/1393 to 1994/1415, Ghāmidī remained a student of Islāḥī as he studied the Qur'ān and hadith under him. This included regular, periodic visits to Islāḥī's residence in a village near Sheikhupura, a city almost thirty miles from Lahore. Islāḥī later moved to Lahore, meaning that student-teacher meetings became more frequent.

Ghāmidī has been regularly teaching various student cohorts the Qur'ān and his approach since 1970/1390, both at a personal level and under the auspices of al-Mawrid, a Foundation for Islamic Research and Education, which he established in Lahore in 1983/1403. Moreover, he has also held weekly Qur'ān sessions for the masses, where he has also embarked upon teaching his own *tafsīr*. His work on the Qur'ān has been instrumental in revisiting and persuasively revising many previous interpretations of various Qur'ānic verses. While his two predecessors, Farāhī and Iṣlāḥī, confined themselves to exegetical explanations, Ghāmidī's work has not only added new dimensions to the commentaries of those two scholars but has also produced a comprehensive exposition of religious matters under the title *Mīzān* (referred to earlier). He has shown that the Qur'ān itself divides religious matters into two categories: *al-ḥikma* and *ash-sharī'a*. Whereas the former refers to topics in the philosophy of religion, the latter includes those that pertain to divine law. Ghāmidī further classifies these two categories into subcategories. The former comprises two subcategories, faith and ethics, and the latter ten subcategories: the sharia of worship rituals, social sharia, political sharia, economic sharia, the sharia of preaching, the sharia of *Jihād*, penal sharia, dietary sharia, Islamic customs and etiquette, and oaths and their atonement.

In each of these categories, Ghāmidī has made unique contributions to the interpretation of the directives of the Qur'ān. Examples include his views on variant readings of the Qur'ān, the subject matter of the Qur'ān, the specific nature of the preaching mission of Abraham's progeny, the punishment of apostasy, the testimony and *diyat* (penalty for murder) of women, the etiquette of gender interactions, slavery in Islam, the system of government envisaged by Islam, the requisites of citizenship, inheritance laws, Islamic punishments and the general and specific directives of *jihad*, and the second advent of Jesus.

Another of Ghāmidī's original contributions is his concept and definition of the word sunna. He categorizes it as something distinct from hadith, laying down certain principles to precisely determine its corpus. By applying these principles, he has actually come up with a list of the sunna's contents.

Ghāmidī also presents an integrated framework for the concepts and terms of Islam in his essay *Dīn-i Ḥaqq* (The True Religion). This framework in itself is representative of a complete interpretation of Islam, contrasting with the two other prevailing interpretations of Islam in the Muslim *umma*: *taṣawwuf*-based interpretation and *jihād*-based interpretation.

One noteworthy view held by Ghāmidī is that there is no religious basis for *ijmā'* (consensus of Muslim scholars on an interpretation).[29] He has shown that the Qur'ānic verse and the hadith on which it purportedly stands have been grossly misinterpreted.

6 Reception

Ghāmidī has been appearing regularly on all major television channels in his home country since 2005/1426. His expertise is highly valued by television anchors, social and political leaders, intellectuals, and the educated masses. Amid the general religious culture of intellectual stagnation, Ghāmidī is a voice of reason who stands for continuous deliberation on the texts of the scriptures.

29 For details, see Shehzad Saleem, *Selected Essays of Javed Ahmad Ghamidi* (Lahore: BPH Printers, 2019[2]), 49–54.

In the last four decades, some scholars in his home country have criticized his views. Most of these views relate to his interpretations of the Qur'ān and the hadith. In fact, his critics have circulated a list of the views they find objectionable, including:

1. Jesus is dead.
2. Al-Mahdī is not going to arrive near the Day of Judgement.
3. No one except God has the right to declare anyone a disbeliever (*kāfir*).
4. Having a beard is not a religious directive.
5. Hadith cannot add any new beliefs or deeds to the religious corpus.
6. The sunnas of the Prophet Muḥammad are twenty-seven in number.
7. The concept of consensus is a religious innovation.
8. Apostasy was only punishable by death in Prophetic times.
9. The punishment of death by stoning and the prohibition against drinking liquor are not divinely ordained.
10. There is only one legitimate reading of the Qur'ān.
11. There are no restrictions on *ijtihād* (exercising independent judgement in matters not covered by the Qur'ān or the sunna). Even an ordinary person can engage in it.
12. Waging jihad to achieve the supremacy of Islam has no basis.
13. Sufism (*taṣawwuf*) is a parallel religion to Islam.
14. There is no difference between the testimony of man and woman, Muslim and non-Muslim.
15. Music is not forbidden in Islam.
16. All portraits except those that were or can be worshipped are permissible.
17. Women can lead men in prayer.

In March 2020/1441, Ghāmidī began recording his detailed responses to these objections on a weekly basis. It is a work in progress and will take a few years to complete.

7 Bibliography

7.1 Sources

Al-Farāhī, Ḥamīd ad-Dīn ʿAbd al-Ḥamīd. *Majmūʿa tafāsīr-i Farāhī*. Translated by Amīn Aḥsan Iṣlāḥī. Lahore: Faran Foundation, 1991.

Ghāmidī, Javed Aḥmad. *Al-Bayān*. 5 vols. Lahore: Topical Printing Press, 2018.

Ghāmidī, Javed Aḥmad. *Mīzān*. Lahore: Topical Printing Press, 2018.

Ghāmidī, Javed Aḥmad. *Burhān*. Lahore: Topical Printing Press, 2018[10].

Ghāmidī, Javed Aḥmad. *Maqāmāt*. Lahore: Topical Printing Press, 2017[4].

Ghāmidī, Javed Aḥmad. *Khayāl u Khāmah*. Lahore: Topical Printing Press, 2018[4].

Ghāmidī, Javed Aḥmad. *Al-Islām*. Lahore: Fine Printers, 2019[7].

Ghāmidī, Javed Ahmad. *Mīzān (Islam: A Comprehensive Introduction)*, trans. Shehzad Saleem. Lahore: Fine Printers, 2021[5].

Ibn Manẓūr, Muḥammad b. Mukarram. *Lisān al-ʿarab*. 15 vols. Beirut: Dār Ṣādir, n.d.
Iṣlāḥī, Amīn Aḥsan. *Tadabbur-i Qurʾān*. 9 vols. Lahore: Faran Foundation, 1985.
Al-Farāhī, Ḥamīd ad-Dīn ʿAbd al-Ḥamīd. *Majmūʿa tafāsīr-i Farāhī*, trans. Amīn Aḥsan Iṣlāḥī. Lahore: Faran Foundation, 1991.
Saleem, Shehzad, *Selected Essays of Javed Ahmad Ghamidi*. Lahore: BPH Printers, 2019².

7.2 Secondary Sources

ʿAlī, Jawwād. *Al-Mufaṣṣal fī tārīkh al-ʿarab qabl al-islām*. 10 vols. Beirut: Dār al-ʿIlm li-l-Malāyīn, 1986².
Al-Aṣmaʿī, Abū Saʿīd ʿAbd al-Mālik b. Qurayb. *Al-Aṣmaʿīyāt ikhtiyār al-Aṣmaʿī*. Ed. Aḥmad Muḥammad Shākir and ʿAbd as-Salām Muḥammad Hārūn. Cairo: Dār al-Maʿārif, 1993.
Al-Azdī, Abū Bakr Muḥammad b. Durayd. *Al-Jamhara fī l-lugha*. Ed. Ramzī Munīr Baʿlabakkī, 3 vols. Beirut: Dār al-ʿIlm li-l-Malāyīn, 1987.
Al-Azharī, Abū Manṣūr Muḥammad b. Aḥmad. *At-Tahdhīb fī l-lugha*. Ed. Muḥammad ʿAwaḍ, 15 vols. Beirut: Dār Iḥyāʾ at-Turāth al-ʿArabī, 2001.
Al-Bukhārī, Abū ʿAbd Allāh Muḥammad b. Ismāʿīl. *Al-Jāmiʿ aṣ-ṣaḥīḥ*. Riyadh: Dār as-Salām, 1999².
Aḍ-Ḍabbī, al-Mufaḍḍal b. Muḥammad b. Yaʿlā b. ʿĀmir b. Sālim. *Mufaḍḍaliyāt*. Ed. Aḥmad Muḥammad Shākir and ʿAbd as-Salām Muḥammad Harūn. Beirut: n.d.
Al-Jawharī, Abū Naṣr Ismāʿīl. *Tāj al-lugha wa-ṣiḥaḥ al-ʿarabiyya*. 6 vols. Beirut: Dār al-ʿIlm li-l-Malāyīn, 1987.
Al-Jazarī, Abū s-Saʿdāt al-Mubārak b. Muḥammad b. al-Athīr. *An-Nihāya fī gharīb al-ḥadīth wa-l-āthār*. Ed. Ṭāhir Aḥmad az-Zāwī and Maḥmūd Muḥammad aṭ-Ṭannāḥī, 5 vols. Beirut: al-Maktaba al-ʿIlmiyya, 1979.
Muslim, Abū l-Ḥusayn b. Ḥajjāj an-Nīsabūrī. *Al-Jāmiʿ aṣ-ṣaḥīḥ*. Riyadh: Dār as-Salām, 2000².
Al-Qurashī, Abū Zayd Muḥammad b. Abī l-Khaṭṭāb. *Jamharat ashʿār al-ʿarab*. Beirut: Dār al-Qalam, n.d.
Ar-Rāzī, Fakhr ad-Dīn. *At-Tafsīr al-kabīr*. 32 vols. Beirut: Dār al-Kutub al-ʿIlmiyya, 2000/1421.
As-Suhaylī, Abū l-Qāsim ʿAbd ar-Raḥmān b. ʿAbd Allāh. *Ar-Rawḍ al-unuf fī tafsīr as-sīra*. 7 vols. Beirut: Dār Iḥyāʾ at-Turāth al-ʿArabī, 2000.
Aṭ-Ṭāʾī, Abū Tammām Ḥabīb b. Aws. *Dīwān al-ḥamāsa*. 2 vols. Beirut: Dār al-Qalam, n.d.
Aṭ-Ṭabarī, Abū Jaʿfar Muḥammad b. Jarīr. *Jāmiʿ al-bayān ʿan tāʾwīl āy al-Qurʾān*. 30 vols. Beirut: Dār Iḥyāʾ at-Turāth al-ʿArabī, 2001.
Az-Zamakhsharī, Abū l-Qāsim Muḥammad b. ʿUmar. *Al-Kashshāf ʿan ḥaqāʾiq at-tanzīl wa-ʿuyūn al-aqāwīl fī wujūh at-taʾwīl*. Ed. ʿAbd al-Razzāq al-Mahdī, 4 vols. Beirut: Dār al-Kitāb al-ʿArabī, n.d.
Az-Zubaydī, Muḥammad Murtaḍā. *Tāj al-ʿarūs min jawāhir al-qāmūs*. 40 vols. N.p.: Dār al-Hidāya, n.d.

Margherita Picchi

amina wadud

1 Introduction

amina wadud (born 1952/1371) is a Muslim African-American theologian, Professor Emerita of Islamic Studies, and activist. Awarded the Danish Democracy Prize in 2007/1428, wadud[1] is the first Islamic scholar to have systematically employed the category of gender as an analytical tool for Qurʾānic hermeneutics. On this basis, she is recognized as a pioneer of what came to be known as Islamic feminism, which she defines as a knowledge-building project that "aligns with other feminisms in maintaining the radical idea that women are human beings,"[2] and bases its call for gender equality and reciprocity on a rigorous interpretation of Islamic primary sources, especially the Qurʾān.[3]

She became internationally known as 'The Lady Imam' after leading a gender-mixed congregation in Friday prayer in New York on March 18, 2005/Ṣafar 8, 1426, which drew both acclaim and criticism from Muslims worldwide and has been more widely discussed in mainstream media than her scholarly work. Nevertheless, wadud's intellectual work is extensive, including two monographs, the seminal book *Qur'an and Woman* (1992/1412) and the semi-autobiographical *Inside the Gender Jihad* (2006/1427), several journal articles and chapters in edited books, public lectures, sermons, blog posts, and numerous interviews.

Her work can best be summarized as a "radical theological engagement similar to liberation theology."[4] Although it began in the context of gender, 'wadudian hermeneutics'[5] is more generally concerned with promoting social justice and contrasting oppression in all forms, including racism, homotransphobia, ableism, religious exclusivism, environmental injustice, and economic exploitation. In emphasizing the importance of reclaiming female agency in interpreting Islamic primary sources, wadud calls for a "democratization of authority where the experience of believers comes into

1 The name is written in lowercase following wadud's own preference.
2 amina wadud, "*Musawah:* Gender Equity through Qur'anic discourse," in *Communities of the Qur'an: Dialogue, debate and diversity in the twenty-first century*, eds. Emran El-Badawi and Paula Sanders (London: Oneworld, 2019), 81–97, 88.
3 The term 'Islamic feminism' is highly controversial and still has no clear definition: the one used here is amina wadud's. The heated debate over the definition of 'Islamic feminism' is perhaps best captured in the well-known discussion between Margot Badran and Asma Barlas, published in Anytta Kynsilehto, ed., *Islamic Feminism: Current Perspectives,* Tampere Peace Research Institute Occasional Paper No. 96 (Tampere: Tampere Peace Research Institute, 2008).
4 Kecia Ali and amina wadud, "The Making of the 'Lady Imam': An interview with amina wadud," *Journal of Feminist Studies in Religion* 35, no. 1 (2019): 67–79, 72.
5 The expression is borrowed from Hibba Abudigeri, "The Renewed Woman of American Islam: Shifting Lenses Toward 'gender *jihad*'?," *MW* 91 (2001): 1–18.

https://doi.org/10.1515/9783111320069-020

dialogue with policies and practices,"[6] and a redefinition of religious leadership that challenges the monopoly of traditional scholars and institutions in interpreting religious sources.

Based on wadud's acknowledgment that the reader's life experience inevitably affects their interpretation of a text, this chapter will retrace specific milestones in her biography. Subsequently, it will examine her major hermeneutical works, analyzing the features of her exegetical method and several of the hermeneutical keys she relies on, with a particular focus on what she calls 'the tawhidic paradigm.' Finally, as it is impossible to give here a full picture of the various subjects that wadud explores in her works, this article will focus on two fundamental issues that are particularly representative of her theological views that have proven influential for later scholars of Islamic feminism: the exegesis of the Qur'ānic account of the creation of humankind, and the interpretation of the Qur'ānic verse 4:34.

2 amina wadud's Life and Work

amina wadud was born Mary Teasley in a poor, Christian African-American household in Silver Spring (Maryland) on September 25, 1952/Muḥarram 6, 1372.

In her "personal transition, most often called conversion,"[7] wadud became Buddhist before embracing Islam. From that transitional experience, she retained the practice of meditation as well as the understanding that "the bearded white man in the clouds on the throne was one conception of God but not necessarily the sum total."[8]

Her transition to Islam occurred in the broader context of the 'Black Power' political and cultural movement that emerged in the years following the death of Malcolm X. This movement "was shaped by and connected to Islam on a deep cultural level."[9] Although she was "very conscious" of its struggle,[10] wadud was not directly involved in the Black Muslim Movement, and she describes her formal entry into Islam as "partially an accident."[11] On Thanksgiving Day, 1972/1392, she entered a mosque in Washington DC to inquire about Islam and ended up pronouncing the *shahāda:* "Because I

6 wadud, "*Musawah*," 95.

7 amina wadud, *Inside the Gender Jihad* (Oxford: Oneworld, 2006), 2.

8 Ali and wadud, "The Making of the 'Lady Imam'," 76.

9 Patrick Bowen, *A History of Conversion to Islam in the United States*, vol. 2, *The African-American Islamic Renaissance, 1920– 1975* (Leiden/Boston: Brill, 2017), 553. For a focus on African American women converts to Islam, see Carolyn Rouse, *Engaged Surrender: African American Women and Islam* (Berkeley: University of California Press, 2004.

10 *The noble struggle of amina wadud*, directed by Elli Safari (New York: Women Make Movies, 2007).

11 amina wadud, "Conversion Story #643 (not a real number)," *Religion Dispatches* May 5, 2011, https://religiondispatches.org/conversion-story-643-not-a-real-number/.

had my hair covered and was wearing a long skirt, I think the brothers assumed I was already there."[12]

The turning point in her transition occurred four months later when she was given a copy of the Qur'ān, and "fell in love." She began studying Arabic to understand the sacred text in its primary language. Later, she embarked on a journey that included moving to Libya (1975–1977/1395–1397) and then to Egypt (1981–1982/1401–1402), where she began studying *tafsīr* with a professor at Cairo University. However, she was unsatisfied with her teacher's interpretative methodology, or better by his lack of transparency about the method he used to arrive at a particular meaning: "And because he did not give any indications how he had gotten there, I was left to take it for granted that he 'knew'."[13]

By the time she returned to her home university, she had attained enough fluency in Arabic and confidence in Islamic studies to engage in *tafsīr* by herself. She combined what she had learned in Islamic institutions with Western methodologies of textual analysis: "linguistics introduced me to literature on hermeneutics and postmodernist analysis, offering a framework to begin considering interpretation as an intentional process of textual evaluation toward meaning."[14] She chose to focus her PhD dissertation on analyzing the Qur'ānic view on women and gender relations. This work was later published as her first book *Qur'an and Woman*, although the extent of its importance was not evident at the time. Despite earning her PhD at the University of Michigan (1988/1408), wadud was unable to find a job in US Academia. In 1989/1409 she joined the International Islamic University in Malaysia, where she stayed until 1992/1412, as the first PhD female Assistant Professor in the Department of Revealed Knowledge.

The years spent in Malaysia had a tremendous influence on her work: in her words, "I was given the opportunity to link my theories about Islamic theology with active social involvement *in the name of Islam*."[15] Soon after her arrival, she was invited to join a group of women who had recently decided to found a collective dedicated to the reading of gender in the Qur'ān, but who lacked the expertise to engage in *tafsīr* directly.[16] wadud became the eighth founding member of the group, which eventually took the name Sisters in Islam (SIS). She helped establish the methodological foundations for an egalitarian reading of the Qur'ān through weekly study sessions, and directly contributed to the writing of the group's first two pamphlets: "Are

12 amina wadud, "Al-Qur'an and Me," *Religion Dispatches*, September 8, 2010, https://religiondispatches.org/al-quran-and-me/.

13 Ibid.

14 amina wadud, "Qur'ān, Gender and Interpretive Possibilities," *Hawwa* 2, no. 3 (2004): 316–36, 328.

15 wadud, *Inside the Gender Jihad*, 258.

16 The encounter between wadud and SIS is narrated by wadud herself in *Inside the gender Jihad*, 113–17; see also Zainah Anwar and Rose Ismail, "Amina Wadud and Sisters in Islam: a Journey toward Empowerment," in *A Jihad for Justice: Honoring the work and life of Amina Wadud*, eds. Kecia Ali, Julianne Hammer and Laury Silvers (USA: 48HrsBooks, 2012), 63–72.

Women and Men Equal before Allah?" and "Are Muslim Men Allowed to Beat Their Wives?".[17]

Together with other members of SIS, wadud participated in the World Conference on Women held in Beijing in 1995/1416; there, "two polarized perspectives emerged that had dominated academic research, political action and international interventions on Muslim women for decades,"[18] namely secular feminism and Islamism. These apparently opposite approaches were paradoxically similar insomuch as they argued that the values of religion and gender equality are incompatible and that one must be chosen against the other. Unsatisfied with the either/or decision imposed upon them, "a third group on the fringe would grow to become more coherent in the following decade. This third group was the origin of Islamic feminism."[19]

While wadud's "tendency toward theory was put to the test of actual practices" through her involvement with SIS,[20] the first edition of *Qur'an and Woman* was published in Kuala Lumpur (1992/1412). The book was soon acknowledged as a groundbreaking work that brought a fresh new gender perspective in the discipline of *tafsīr*. In the following years, it was translated into several languages and re-published with a new preface and added glossary by Oxford University Press in 1999/1419. It was then widely used as a textbook for courses related to gender and Islam, particularly in Western universities.

In the meantime, wadud had returned to the US to take the position of professor of Islamic Studies at Virginia Commonwealth University. She later made her first headlines by delivering a famous and still controversial sermon (*khuṭba*)[21] before a gender-mixed audience at Claremont Main Road Mosque (Cape Town, South Africa) on August 12, 1994/Ṣafar 4, 1415.[22] The event was welcomed by progressive Muslims, but also provoked

17 Sisters in Islam, *Are Women and Men Equal before Allah?* (Petaling Jaya: Sisters in Islam 1991); Sisters in Islam, *Are Muslim Men Allowed to Beat their Wives?* (Petaling Jaya: Sisters in Islam 1991).

18 amina wadud, "The Ethics of *Tawhid* over the Ethics of *Qiwamah*," in *Men in Charge? Rethinking Authority in Muslim Legal Tradition*, eds. Ziba Mir-Hosseini, Mulki Al-Sharmani, and Jana Rumminger (Oxford: Oneworld, 2015), 256–74, 262.

19 wadud, "*Musawah*," 87.

20 wadud, *Inside the Gender Jihad*, 117.

21 It should be noted that wadud's speech was formally defined as a 'pre-*khuṭba* lecture,' not an actual sermon, in order to "allay traditional sensitivities while attempting to challenge other aspects of the status quo" (wadud, *Inside the Gender Jihad*, 178). The official label notwithstanding, this article follows wadud's own definition of the lecture as a *khuṭba*, supported by South African liberation theologian Farid Esack, who attended the event and later commented that what she delivered "was, for all intent and purpose, a sermon" (Farid Esack, *Qur'ān, Liberation and Pluralism* [Oxford: Oneworld 1997], 246). See also Abdulkader Tayob, "The pre-khutba, khutba and Islamic change," *Al-Qalam*, August 15, 1994.

22 The event is narrated by wadud herself in *Inside the Gender Jihad*, 158–86; see also Esack, *Qur'ān, Liberation and Pluralism*, 246–48; the articles by Sa'diyya Shaikh, "Engaging surrender: the intimacy and power of the gender jihad," in Ali, Hammer and Silvers, *A Jihad for Justice*, 213–16; Fatima Seedat, "An intellectual mentor," in Ali, Hammers and Silvers, *A Jihad for Justice*, 191–95.

what the mosque's imam Rashied Omar described as "a massive overreaction on the part of a community not used to having their horizons changed."[23] In commenting on the uproar following the sermon, Omar used the term 'gender jihad,' which wadud has since adopted to identify her "struggle to establish gender justice in Muslim thought and praxis."[24]

In the long term, the event made history. However, the immediate aftermath "was a painful and traumatic time"[25] for the organizers as well as for wadud, who chose to refuse further requests for delivering sermons or leading the ritual prayer in public occasions.

She returned to the spotlight more than ten years later. On March 18, 2005/Ṣafar 8, 1426, she stood in a hall of the Cathedral of St. John the Divine in New York City in front of a gender-mixed congregation, fulfilling the role of imam in leading the Friday prayer. The event was broadly publicized and received significant media coverage, with hundreds of articles published on the prayer and the 'quiet heretic' who led it.[26] wadud gained international fame as 'the Lady Imam' (a label she only recently assumed for herself), for the prayer soon proved to be a historical event that "forced and inspired a fruitful discussion on the nature of woman's place in the mosque that continues to this day."[27]

In 2006/1427, wadud published her second book, *Inside the Gender Jihad – Women's Reform in Islam*. It is a semi-autobiographical work that builds upon the theological reflection in *Qur'an and Woman*, contextualizing it in the author's personal experience as an academic, a twice-divorced mother of five who raised her children mostly as a single parent, and a transnational activist engaged in affirming gender equality in different Muslim contexts.

wadud retired from her position at Virginia Commonwealth University in 2008/1429 shortly after being promoted to Full Professor, "because it honestly did not resolve my deep spiritual and political yearnings, and I did not see it moving any closer as the time went, so I tried something else."[28]

Her retirement from academia did not signify a retreat into the private sphere; on the contrary, to this day wadud remains earnestly involved both in activism and

23 Quoted in Fahmi Gamieldien, *The History of the Claremont Main Road Mosque, its people and their contribution to Islam in South Africa* (Claremont: Claremont Main Road Mosque, 2004), 89.

24 wadud, *Inside the Gender Jihad*, 10.

25 Jaamia Galant, "Engaged Surrender: Reflecting on 25 years of the Gender Jihad" (pre-khuṭba lecture, Claremont Main Road Mosque, Friday August 30, 2019).

26 For a thoughtful analysis on the mediatic impact of the prayer, see Juliane Hammer, "Performing Gender Justice: the 2005 woman-led prayer in New York," *Contemporary Islam* 4 (2010): 91–116.

27 Ahmed Elewa and Laury Silvers, "'I *am* one of the people:' a survey and legal analysis on woman-led prayer in Islam," *Journal of Law and Religion* 26, no. 1 (2010–11): 141–71.

28 amina wadud, "To blog or not to blog? Is that even a question . . .?," *Feminism and Religion*, July, 2013, https://feminismandreligion.com/2013/07/18/to-blog-or-not-to-blog-is-that-even-a-question-by-amina-wadud/.

research. Among her various activities, it is worth mentioning her contribution as a resource person to the transnational network Musawah ("equality"), launched by Sisters in Islam in 2009/1430. wadud is part of the coordinating team of Musawah's course *Islam & Gender Equality and Justice* (I-nGEJ),[29] and authored a chapter in the widely praised anthology *Men in Charge? Rethinking Authority in Muslim Legal Tradition.*[30] In 2020 she moved from the United States to Indonesia, where she is currently visiting Professor for the Indonesian Consortium for Religious Studies.[31]

With regard to research, wadud has recently concluded a three-year project on sexual diversity and human dignity in Islamic primary sources, which reflects her recent engagement in support for LGBTQI+ Muslim rights.[32] Although the findings of this research are not yet published, wadud has shared internationally some of the findings in a number of closed workshops, the recordings of which are available on her Patreon channel.[33]

3 Wadudian Hermeneutics

3.1 The Qurʾān and its Exegesis

The Qurʾān lies at the center of wadud's theology: although she acknowledges the sunna as the second primary source of Islam, she "place[s] greater significance on the Qur'an"[34] and only seldom takes the hadith literature into analysis.

It should also be noted that wadudian hermeneutics does not approach the Qurʾān as the only manifestation of God's will. Quoting from Q 31:27, wadud reminds that the Qurʾān "did not and, as it says directly, could not exhaust the knowledge of Allah through language alone"[35] and criticizes any equation between God and the Qurʾān. According to wadud, the Qurʾān is the most exquisite example of God's guidance to human-

29 "Islam & Gender Equality and Justice (I-nGEJ)," Musawah, accessed January 23, 2024], https://www.musawah.org/capacity-building/ingej/.

30 Ziba Mir-Hosseini, Mulki Al-Sharmani, Jana Rumminger, eds., *Men in Charge? Rethinking Authority in Muslim Legal Tradition* (Oxford: Oneworld, 2015).

31 https://www.icrs.or.id/visiting-professor, accessed on January 23, 2024).

32 amina wadud, "LGBTQI Muslims and International Movements for Empowerment," *Feminism and Religion*, September 20, 2014, https://feminismandreligion.com/2014/09/20/lgbtqi-muslims-and-international-movements-for-empowerment-by-amina-wadud/.

33 "The Lady Imam," Patreon, accessed November 5 2020 https://www.patreon.com/TheLadyImam/posts.

34 amina wadud, *Qur'an and Woman: Rereading the Sacred Text from a Woman's Perspective* (New York: Oxford University Press, 1999), xvii.

35 wadud, *Inside the Gender Jihad*, 212.

ity, but all signs and symbols (*ayāt*) "are metaphors pointing to that ultimate reality that transcends all symbols and signs, be they in language of revelation or in the creation or nature."[36]

wadud's interest in interpretative methodologies has grown throughout her career; rather than adopting a rigid methodology with a single, conclusive definition, wadudian hermeneutics is better defined as an ongoing process involving discussion and dialogue with activists and fellow scholars, reflective thinking, and continual revision of its tenets.

This continuous process of revision and self-examination is perhaps best exemplified by wadud's dynamic positionality concerning feminism and the label 'feminist' that has been applied to her work since the first edition of *Qur'an and Woman.* In order to respond to this label, wadud chose to use the preface of the 1999/1420 edition of the same book to clearly state that hers was not a feminist perspective and defining her as such without her consent made her feel "as though I do not count as a human being."[37] Seven years later, in her second book *Inside the Gender Jihad,* wadud positioned herself as 'pro-faith' – 'pro-feminist'. She explained that "despite how others may categorize me, my work is certainly feminist, but I still refuse to self-designate as feminist, even with 'Muslim' put in front of it, because my emphasis on faith and the sacred prioritize my motivations in feminist methodologies;" also, "as an African-American, the original feminist paradigms were not intended to include me, as all the works on Womanism have soundly elucidated."[38] Finally, in her blog article *The F* word: feminism in Islam,* wadud consciously made an epistemological shift, reclaiming the label 'Islamic feminist' that was initially imposed on her by others. In the end, "there really never was a conflict in the first place; just a set of circumstances that caused the confusion that one had to choose between one and the other, instead of living both."[39]

This tireless process of revision notwithstanding, wadudian hermeneutics has always remained consistent in its adamant refutation of any association between God and *ẓulm* (oppression, injustice), since "God does not oppress,"[40] and in its affirmation of a radical theological engagement for the sake of justice and reciprocity among humans. wadud's stance is that oppression exists because of what humans do, and the whole point of being a Muslim is to be consciously engaged in dismantling it.

36 wadud, *Qur'ān, Gender and Interpretive Possibilities,* 319.
37 wadud, *Qur'an and Woman,* xviii.
38 wadud, *Inside the Gender Jihad,* 79–80.
39 amina wadud, "The F* word: feminism in Islam," *Religion Dispatches,* May 5, 2011, https://religiondispatches.org/the-f-word-feminism-in-islam/.
40 amina wadud, "Towards a Qur'anic Hermeneutics of Social Justice: Race, Class and Gender," *Journal of Law and Religion* 12, no. 1 (1995–96): 37–50, 46.

3.2 A Hermeneutical Model of Interpretation

In giving the Qur'ānic text absolute precedence over any other sources, wadudian hermeneutics is rooted in the classical exegetical principle that the best interpreter of the Qur'ān is the Qur'ān itself (*tafsīr al-Qur'ān bi-l-Qur'ān*). Nevertheless, she reproaches traditional interpreters for having reduced this tool "to using a single term or verse to interpret another term or verse," so that "intra-Qur'anic interpretation is atomistic and seriously limited."[41]

In wadud's view, traditional works of *tafsīr* are rendered more problematic as they have all been written exclusively by men. Female voices have been marginalized, invalidating women's experiences as insignificant to the understanding of the divine message. Moreover, classical Islamic scholars molded their conception of gender relations on the notion of just rulership, not of equality or reciprocity. This all-male traditional exegesis has been nearly sacralized by later scholars, to the extent that the distinction between the Qur'ān and its exegesis has become blurred and rendered almost indistinguishable to many modern Muslims. As a consequence, traditional patriarchal interpretations that highlight the superiority of men over women are considered to epitomize the true spirit of Islam.

wadud argues that correcting this unbalance must be a moral imperative; it is fundamental to reclaim the idea that "women's experiences matter,"[42] and to "read for gender in the Qur'an," which means "to read before the text, behind the text as well as **through** the text to determine rhetorical devices for hegemony and their power assertions along gender lines."[43]

In order to overcome the limits of the traditional approach to Qur'ānic exegesis, wadud proposes a hermeneutical model of interpretation "concerned with three aspects of the text: 1. The context in which the text was written (in the case of the Qur'an, in which it was revealed); 2. The grammatical composition of the text (how it says what it says); 3. The whole text, its *weltanschauung* or worldview."[44]

The Qur'ān was revealed in a specific historical context, and it is necessary to acknowledge the contextual nature of its contents in order to search for their profound and universal meaning. In quoting the words of the theologian Fazlur Rahman (1919–1988/1337–1408), who profoundly influenced her hermeneutics, wadud argues that the Qur'ān is "God's response through Muhammad's mind [. . .] to a historical situation"[45] and although the moral guide it offers is universal, the particulars of its solutions are historically situated in the context of the 7th-century Arabian Peninsula. It would be a

41 wadud, "Towards a Qur'anic Hermeneutics," 43–44.
42 wadud, "*Musawah*," 95.
43 Ibid., 89. Emphasis in original
44 wadud, *Qur'an and Woman*, 3.
45 Fazlur Rahman, *Islam and Modernity: Transformation of an Intellectual Tradition* (London: University of Chicago Press, 1982), 8.

deplorable anachronism to consider them irrevocable in our days, or worse, it would imply reducing God to the patriarchal context in which the Qur'ān was revealed. In wadud's view, the construction of priorities based on universals rather than particulars is "the most under-considered aspect of holistic intra-Qur'ānic interpretation."[46] It represents a necessary tool for finding the most appropriate application of a Qur'ānic principle to the social context of today.

wadud is particularly concerned with the linguistic constraints in which the Qur'ānic message is articulated. In order for the Qur'ān to be understood by the people who lived in the specific historical context of its revelation, the divine message had to express itself in their human language, namely Arabic, adapting its truth to its specific epistemology and particular linguistic and grammatical rules. For example, Arabic is a grammatically gendered language, with each verb, noun, and adjective assigned either a male or female case. The masculine gender is considered the unmarked form, meaning that masculine words or pronouns are used for gender-indefinite or inclusive reference. God in the Qur'ān is referred to with the grammatically male pronoun *huwa*. This is due to the limits of the Arabic language and the patriarchal context of its revelation, but it would be a grave mistake to equate God with the imperfect signifiers for God, as the Qur'ān is also clear that God transcends gender: "He is not male. She is not female."[47]

"The Qur'an is constrained"[48] by this and all other limits of the Arabic language, and in order to get at its spirit "[e]ach term should therefore be examined on the basis of its language act, syntactical structures, and textual contexts in order to more fully determine the parameters of meaning."[49]

In summary, wadud maintains that "the Qur'an is universal in intent, but not in its contextual epistemology;"[50] it is "a window to look *through*, a doorway with a threshold one must pass *over* toward the infinite possibilities that point humanity toward a continuum of spiritual and social development."[51] In order to derive general ethical principles from contextual assertions found in the Qur'ān, one should read the text holistically, as a discursive unity, taking into consideration both the historical context and the linguistic construction of its teaching. This theological work should not be detached from reality, for "theory alone is insufficient to bring an end to patriarchy and gender asymmetry. There is a crucial interplay between belief in certain ideas and the practical implementation of gender justice in the context of present global circumstances."[52]

---

46 wadud, "Towards a Qur'anic Hermeneutics," 44.
47 wadud, "The Ethics of *Tawhid*," 426.
48 wadud, *Inside the Gender Jihad*, 212.
49 wadud, *Qur'an and Woman*, xiii.
50 wadud, *Inside the Gender Jihad*, 153.
51 Ibid., 197.
52 Ibid., 42.

3.3 Hermeneutical Keys

3.3.1 *Tawḥīd* and the Tawhidic Paradigm

Wadudian hermeneutics is based on an egalitarian idealized cosmology grounded on the principle of *tawḥīd*. Theologically, *tawḥīd* refers to God's unity, the fundamental principle of Islam that any Muslim acknowledges through the *shahāda*, the profession of faith. "There is no Islam without *tawḥīd*,"[53] and its opposite, *shirk*, is defined in the Qur'ān as the only unforgivable sin which inevitably dooms those who commit it to Hell. At a cosmological level, "*tawḥīd* is an umbrella term that encompasses the world,"[54] for God is not only One but *unites* all things. Finally, "as an ethical term, *tawḥīd* relates to relationships and developments within the social and political realm, emphasizing the unity of all human creatures beneath one Creator."[55]

Building upon the concept of *tawḥīd*, throughout her career wadud progressively elaborated what she defines 'the tawhidic paradigm,' her key interpretative method, and the cornerstone of her entire theology.

In her first book *Qur'an and Woman*, wadud declares her goal to be the building of a "hermeneutics of *tawḥīd*" and defines the latter as "the unicity of Allah, harmony in and unity of all creation under a single Creator."[56] In discussing the story of human creation in the Qur'ān, wadud argues that the unity of God, as radically opposed to the 'paired' nature of all created things, "establishes a special link between the Creator, Allah, and the created, humankind,"[57] but she does not further develop this concept.

The earlier elaboration of the tawhidic paradigm can be found in the opening chapter of *Inside the Gender Jihad*, where wadud defines it as "unequivocal and fundamental to my work on gender."[58] Indeed, as *tawḥīd* operates as a principle of equilibrium and harmony, at a social level it must translate into a principle of equality and reciprocity among humans, who are all equal under the Creator. Relying on the I-Thou-God model as elaborated by the Jewish philosopher and theologian Martin Buber (1878–1956),[59] wadud argues that although dualism appears to be the fundamental feature of this world, "there is no basis for schism in the duality of 'self' and 'other', since ontologically they are co-dependent according to the statement '*min kulli shay' khalaqnaa zawjayn*' (and from all things we have created a pair (51:49)."[60] While the traditional model of

53 wadud, "The Ethics of *Tawhid*," 414.

54 wadud, "*Musawah*," 91.

55 wadud, *Inside the Gender Jihad*, 28.

56 wadud, *Qur'an and Woman*, xii and xxvi.

57 Ibid., 26.

58 wadud, *Inside the Gender Jihad*, 24.

59 Martin Buber, *I and Thou*, trans R.G. Smith (London: Bloomsbury, 2013) [or. ed. *Ich und Du* 1923].

60 wadud, *Inside the Gender Jihad*, 29.

gender relations based on complementarity "has an unequal power dimension"[61] and produces a model of human relations organized with a hierarchical axis with God at the top, men as mediators in the middle, and women at the bottom, the tawhidic paradigm creates a new horizontal axis of reciprocity between humans. Thus, it

> becomes the inspiration for removing gender stratification from all levels of social interaction: public and private, ritual and political. Not only does it mean the I and Thou are equal, but also it means that I and Thou *are one within the oneness of Allah.* Social, liturgical, and political functions become determined by the capacity of both women and men in a larger realm of education, dedication, and contribution with no arbitrary exclusion of women from performing any of these functions.[62]

In her early publications on the topic, wadud graphically represented the tawhidic paradigm as an equilateral triangle, with Allah at the vertex and "man" and "woman" at the vertices of the base.

This graphic was intended to visually demonstrate how the tawhidic paradigm "maintains the metaphysics of Allah's supremacy" but "removes the female's dependency on the male and returns her on a direct line with the divine." As a result, "the human-to-human relationship can only be constructed horizontally, as one of mutual support and reciprocity."[63]

In later articles and lectures, however, wadud has adjusted this graphic representation, reacting to what she perceived as problematic responses by some readers who misinterpret the triangle as literal.[64] She clarified that the nature of *tawḥīd* is not a triad, but rather a circle, "the center of which is everywhere, the circumference of which is nowhere," quoting St. Augustine's description of God.[65] According to wadud, this description of God is best visually rendered with the Mandelbrot set, the video projection of a mathematical fractal.

Zooming into the Mandelbrot set effectively renders how God is present "in the most mimetic things, and the largest" and was useful for her to have "a deeper understanding of how the binaries work and at the same time how the binaries have limited our capacity to embrace the dignity and full humanity of people who don't fit into the binary."[66] This applies not only to gender, as wadud's interest in the concept of 'pairs' (*zawj*), in which "all things were created" (Q 51:49), has become increasingly intersectional over time. She now considers gender to be an interdependent category,

61 Ibid., 28.
62 Ibid., 33.
63 wadud, "The Ethics of *Tawhid*," 271.
64 amina wadud, "Can One Critique Cancel All Previous Efforts?," *Journal of Feminist Studies in Religion* 32, no. 2 (2016): 130–34.
65 Augustine of Hippo, *The City of God*, trans. Henry Bettenson (London: Penguin, 2003) [or.ed. *De civitate dei*, 413–426 CE].
66 amina wadud, "The tawhidic paradigm", available on Youtube, https://www.youtube.com/watch?v=P7FNrsXWrv4, accessed February 13, 2024.

deeply intertwined with race and class. In wadud's view, all these binaries should be understood as archetypical, not literal, for "there is always the manifestation of the other within each."[67]

3.3.2 Islam as *Engaged Surrender*

The Arabic word "Islam" is most commonly translated in English as "submission [to God's will]," and a "Muslim" is "one who submits." wadud prefers to translate Islam as *engaged surrender,* explaining that while "submission is enforcement situated completely outside of the one who submits," "surrender is mutually accepting human volition while recognizing Allah's ultimate authority."[68] Engaged surrender requires human agency and "involves an active consciousness in participation in our social lives, community lives, economic lives, and political lives, by the heart which is always open to the will of Allah, and which always gives precedent to Allah's will."[69] wadud reclaims this definition as most significant to her identity and work as a Muslim woman, Islamic scholar, and gender activist; and it is worth noting that "Islam as engaged surrender" was the title of the *khuṭba* she delivered on August 12, 1994/Rabīʿ al-Awwal 5, 1415, at the Claremont Main Road Mosque in Cape Town. In her sermon, which she still considers one of the best lectures of her career,[70] wadud explored the long-debated topic of what 'Islam' means by adopting a distinct woman perspective, discussing specifically female experiences – pregnancy and childbirth – as an important site of immanent relationship with the Divine, and a universal center for reflecting on God's nature as *ar-Raḥmān, ar-Raḥīm;* names that come from the same root word as *raḥm,* the womb.[71]

3.3.3 Khilāfa

Embracing Islam as engaged surrender with one's full moral agency is the purpose of human creation. In the Qurʾānic story of the origins of humankind, God announces to the angels that "Indeed I will create on earth a *khalīfa*" (Q 2:30), a term wadud translates with *trustee* or *moral agent.* According to her interpretation, this verse explains that humans are *meant* to be on earth (there is no Fall in the Qurʾān) and that they are meant to be there as God's trustees: something that implies moral responsibility

67 Ibid.
68 wadud, *Inside the Gender Jihad,* 23–24.
69 Ibid., 158.
70 Personal interview with amina wadud, March 12, 2020.
71 For the sermon's full text, see wadud, *Inside the Gender Jihad,* 158–63.

"of each human being to establish social justice, as a representative of the divine will or cosmic harmony."[72]

The Qur'ān does not pose any restrictions on the capacity of *khilāfa* on the basis of race, class, or gender; however, "the patriarchal paradigm of public agency corrupts moral excellence"[73] and illegitimately excludes women from the primordial contract between humans and their Creator.

In wadud's view, reclaiming the full capacity of women as *khalīfa* is an essential instrument for gender activists, and it offers a faith-based legitimation for reforming all laws that restrict women's full citizenship in modern Muslim-majority nation-states.[74]

3.3.4 Taqwā

Although all human beings are equal in relation to their One Creator, and they all have the capacity for *khilāfa*, the Qur'ān does make distinctions among people.

> The value of the distinctions between humankind on earth can be clearly summed up by the Qur'anic statement in sura *Al-Ḥujarāt* (49:13): 'We created you male and female and have made you nations and tribes that you may know one another. [. . .] Indeed the most noble of you from Allah's perspective is whoever (he or she) has the most *taqwa.*'[75]

Often translated as piety, wadud instead defines *taqwā* as "consciousness of Allah"[76] or "moral consciousness in the trustee of Allah,"[77] which is a concept that is closely connected to *khilāfa.* She considers Q 49:13 to be self-explanatory: God does not distinguish people based on gender, class, or race, as only *taqwā* matters, and all actions matter in determining one's *taqwā.*

wadud emphasizes the importance of integrating the domestic parameters of *taqwā* as a basis for public policy. Historically, ethical and moral paradigms have focused on men's actions (and power) in the public space, disconnecting *taqwā* from intimate family relationships. As a consequence, the value of caretaking responsibilities in determining one's value has been minimized, and "men and sons are encouraged to develop ideas of manhood as unlimited empowerment."[78] By quoting feminist

72 wadud, *Inside the Gender Jihad,* 35.
73 wadud, *Qur'ān, Gender and Interpretive Possibilities,* 325.
74 amina wadud, "Citizenship and Faith," in *Women and Citizenship,* ed. Marylin Friedman (New York: Oxford University Press, 2005), 170–87.
75 wadud, *Qur'an and Woman,* 36.
76 Ibid., 37.
77 wadud, *Inside the Gender Jihad,* 40.
78 Ibid., 41.

thinkers such as Joan Tronto and Sharon Welch, wadud suggests that politicizing the ethics of care based on *tawḥīd* is "essential to comprehensive social justice."[79]

3.3.5 ʿAdl

wadud sees *ʿadl*, or justice, as a concept with a deep cosmological meaning that transcends jurisprudence. *ʿAdl* is the balance of forces in God's creation: day and night, heaven and earth, permanence and change, male and female, and so on. *ʿAdl* is also the purpose of sharia, something upon which all Islamic scholars have unanimously agreed. However, contemporary conceptions of justice differ significantly from those of classical Islamic jurisprudence (*fiqh*), which still influence the dominant interpretation of sharia in Muslim contexts.[80]

In defining justice, traditional exegetes encoded "fundamentally different notions of what it means to be a human being, notions in which some human beings are not fully equal, as in the case of slavery and its practice by Muslims, or in which women are considered to be under the dominion of men."[81] In other words, they have not ignored the Qurʾānic appeal to apply justice but instead interpreted it based on their own idea of justice, which is elaborated in the context of a strongly hierarchical and patriarchal society.

On the contrary, a concept of justice built upon the tawhidic paradigm is based on the principle of equality and reciprocity among all human beings, and it becomes a necessary tool for the creation of a political system that allows every individual to fulfill their *khilāfa*.

3.4 Qurʾān and Gender Justice: Major Themes

3.4.1 The Creation of Humankind

In her book *Qur'an and Woman*, wadud examines the Qurʾānic account of the origin of humankind through a distinct linguistic approach. She focuses primarily on Q 4:1, which she translates as follows: "and *min* His *ayāt* (is this:) that He created You (humankind) *min a* single *nafs*, and created *min* (that *nafs*) its *zawj*, and from these two He spread (through the earth) countless men and women (4:1)."[82]

---

79 Ibid., 45.

80 For a short overview of the distinction between sharia and *fiqh*, see amina wadud, "Shari'ah is not the Law," *Religion Dispatches*, May 5, 2011, https://religiondispatches.org/shariah-is-not-the-law/.

81 wadud, "*Musawah*," 94.

82 wadud, *Qur'an and Woman*, 17.

She draws four key terms from this verse (*min, āyāt, nafs, zawj*) to explain how "the Qur'an does not make a distinction between men and women in this creation."[83] *Min* is a particle that can be used for the English preposition *from* but can also imply that two things are of the same nature; while the word *āyāt* refers to God's 'signs', which can be implicit (empirical signs that are found in nature and humans can perceive) or explicit (verbal symbols or words that can only be known by revelation). As for *nafs*, the Arabic word for the 'soul' is grammatically feminine but conceptually neutral: the verse does not state that the creation of humankind began with a man, Adam, but instead points at the united origin of all humans from a 'single soul.' The word *zawj* confirms this reading, for it is grammatically masculine but conceptually neutral: the verse indicates that like all created things, humans are made in 'pairs', but share a single origin in God so that no hierarchy can exist between the two components. The Qur'ānic story of the creation of humankind is meant to explain that "just as we have one point of origin, so do we also have one destination: from one to the many and back to one again. What remains to be seen then is the Qur'anic treatment of the dynamics involved in the interactions between the many."[84]

3.4.2 Q 4:34

Perhaps one of the most successful and influential arguments made by wadud in her books pertains to her interpretation of the Qur'ānic verse 4:34. This verse – one of the most heavily debated in modern times – appears to establish a husband's dominance over his wife:

> Husbands are *qawwāmūna 'alā* (responsible for) their wives, with [the bounties] God has given to some more than others and with what they spend out of their money. Righteous wives are devout and guard what God would have them guard in their husbands' absence. If you fear *nushūz* (high-handiness) from your wives, remind them [of the teaching of God], then ignore them in bed, then *iḍribūhunna* (hit them). If they obey you, you have no right to act against them: God is most high and great.

This verse is the basis on which traditional Islamic jurists have built what Islamic feminism scholars defined as 'the *qiwāma* postulate,' or the 'maintenance-obedience model,' an asymmetrical conception of marital relations where a wife's obedience is provided to the husband in exchange for maintenance.[85]

amina wadud was one of the early scholars to deconstruct the traditional exegesis of Q 4:34 and to suggest alternative readings that could enhance gender equality. In

83 Ibid., 15.

84 Ibid., 26–27.

85 Mulki Al-Sharmani, "Marriage in the Islamic Interpretive Tradition: Revisiting the Legal and the Ethical," *Journal of Islamic Ethics* 2 (2018): 76–96.

this sense, her interpretation has shrewdly evolved during her career. In her book *Qur'an and Woman,* wadud examined this verse through a linguistic approach, focusing mainly on the words *qawwāmūna, nushūz* and *iḍribūhunna.* According to her analysis, *qawwāmūna* should be not be understood as a privilege but rather a responsibility, for this verse "establishes an ideal obligation for men [who don't bear children] with regard to women to create a balanced and shared society."[86] *Nushūz* cannot mean wifely disobedience to the husband, as traditional scholars have unanimously concluded, "since the Qur'an uses *nushūz* for both male and the female."[87] The word should be intended as referring to abusive domestic behavior, regardless of the gender of the person committing it. Finally, the ascending scale of disciplinary measures that culminates with *iḍribūhunna* (hit them) should be read not as permission but rather "as a severe restriction of existing practices,"[88] considering the Qur'ān was revealed in a context where abusing one's wife was considered socially and culturally acceptable. Moreover, the verb *ḍaraba* does not refer to severe beating but should be intended as a purely symbolic act to be used in extreme circumstances only.

This argument was later criticized as apologetic and unrealistic: "if someone reads or hears *iḍribūhunna* in a village in Egypt, will they think 'tap lightly with a feather as a symbolic act of displeasure?' No, they won't. Case closed."[89]

In response to this criticism and as a consequence of her direct involvement in activism, with her second book wadud decided to "accept the critiques of moving beyond my apologetics"[90] and substantially changed her approach. She shifted from a qualified 'no' that rationalizes men's violence to an unequivocal 'no' that condemns any kind of disciplinary role of the husband over his wife:

> There is no getting around this one, even though I have tried through different methods for two decades. I simply do not and cannot condone permission for a man to 'scourge' or apply *any kind* of strike to a woman. [. . .] This leads me to clarify how I have finally come to say 'no' outright to the literal implementation of this passage.[91]

Saying 'no' to what has become historically unacceptable does not mean to reject the Qur'ān or to deny its divine nature, but rather to acknowledge that God "cannot be contained or constrained by text- let alone by the search for meanings of that text by any human being, with all his or her flaws and inadequacies."[92]

86 wadud, *Qur'an and Woman*, 73.
87 Ibid., 75.
88 Ibid., 76.
89 Sarah Eltantawi, "Finding 'Yes' by Saying 'No': A Tribute to Amina Wadud," in Ali, Hammer and Silvers, *A Jihad for Justice*, 155–58.
90 wadud, *Inside the Gender Jihad*, 188.
91 Ibid., 200.
92 Ibid., 197.

Moreover, wadud argues that appointing men as superiors of women is to confer upon men a role that belongs to God exclusively, and thus constitutes an act of *shirk*. When a person places him or herself above another, he or she is falling into *istikbār* (arrogance), the same misplaced sense of value that pushed Satan to disobey God and to refuse to bow to Adam on the claim that "I am better than Him," thus causing his fall.[93]

Because of the dominance of the *qiwāma* paradigm, "Muslim women are **not** experiencing justice and well-being in their lives, despite adhering to what has heretofore been dubbed 'divine'."[94] The 'way forward' from the Satanic logic that supports the *qiwāma* ideal is to embrace the tawhidic paradigm of reciprocity and equality, and to reclaim the full capacity of women as moral agents (*khalīfa*) before God.

4 Conclusion

amina wadud is a polarizing figure. Due to her radical theological engagement, her supporters consider her a liberation hero and one of the most important theologians of our time. Margot Badran, for instance, argues that "her breakthrough Islamic thinking is at the beginning of the 21st century what [Muhammad] Abduh's work was at the beginning of the 20th."[95] On the other side of the spectrum, traditional Islamic scholars denounce her as a heretic. On several occasion, the mere announcement of her presence as a lecturer at certain events has led to controversy. In this regard, it should be noted that most of her critics – and even some of her supporters – only seldomly take her academic work into consideration: it is the symbolic power of her action and her gender that sets her at the center of the debate.

Nonetheless, her theological work is broad. Throughout her career, she has developed a hermeneutical method that builds upon Fazlur Rahman's work – his 'double movement' theory in particular – but grounds the theory in the lived reality of women and of all other categories of people who experience oppression in the name of Islam.

The pillar of wadudian hermeneutics is the tawhidic paradigm, a model for human relations that affirms the unconditional and unmediated relationship between God and every human being. The purpose of human creation is to be engaged in surrender to *tawḥīd* and to become God's moral trustee (*khalīfa*) on earth, with the aim of maintaining harmony and justice (*ʿadl*) between the Self and the Other. Capacity for *khilāfa* cannot be restricted on the basis of race, class or gender: only moral con-

93 amina wadud, "I am better than him," *Religion Dispatches*, August 30, 2010, https://religiondispatches.org/i-am-better-than-him/.

94 wadud, "*Musawah*," 96. Emphasis in original.

95 Margot Badran, "Seismic Shift from Patriarchy to Equality: Amina Wadud on Reading the Qur'an and Revolution," in Ali, Hammer and Silvers, *A Jihad for Justice*, 149–51, 151.

sciousness (*taqwā*) matters in the determination of one's value in the eyes of God. According to wadud, "God does not oppress," as *ẓulm* exists because of human action, and the whole point of being a Muslim is to be consciously engaged in dismantling it.

It would be premature to evaluate wadud's impact on the discipline of *tafsīr*. However, it is undeniable that her work has been a precious contribution to the development of the interpretive possibilities that contrast unjust socio-cultural norms and legal provisions that impede gender equality. These norms are based on traditional authoritarian readings of the Islamic primary sources that remain nearly sacralized by contemporary Muslims to this day.

5 Bibliography

5.1 Selected Works by amina wadud

wadud, amina. "On belonging as a Muslim Woman." In *My Soul is a Witness: African-American Women's spirituality*, edited by Gloria Wade-Gayles. Boston: Beacon Press, 1995.

wadud, amina. "Towards a Qur'anic Hermeneutics of Social Justice: Race, Class and Gender." *Journal of Law and Religion* 12, no. 1 (1995–96): 37–50.

wadud, amina. "Alternative Qur'anic interpretation and the status of Muslim women." In *Windows of Faith: Muslim Women Scholar-Activists in North America*, edited by Gisela Webb. New York: Syracuse University Press, 2000: 3–21.

wadud, amina. *Qur'an and Woman: Rereading the Sacred Text from a Woman's Perspective*. New York: Oxford University Press, 1999.

wadud, amina. "American Muslim Identity: Race and Ethnicity in Progressive Islam." In *Progressive Muslims: On Justice, Gender and Pluralism*, edited by Omid Safi. Oxford: Oneworld, 2003: 270–85.

wadud, amina. "Qur'ān, Gender and Interpretive Possibilities." *Hawwa* 2, no. 3 (2004): 316–36.

wadud, amina. "Citizenship and faith." In *Women and Citizenship*, edited by Marylin Friedman. New York: Oxford University Press, 2005: 170–87.

wadud, amina. *Inside the Gender Jihad*. Oxford: Oneworld, 2006.

wadud, amina. "The Ethics of *Tawhid* over the Ethics of *Qiwamah*." In *Men in Charge? Rethinking Authority in Muslim Legal Tradition*, edited by Ziba Mir-Hosseini, Mulki Al-Sharmani, Jana Rumminger. Oxford: Oneworld, 2015: 256–74.

wadud, amina. "Can One Critique Cancel All Previous Efforts?." *Journal of Feminist Studies in Religion* 32, no. 2 (2016): 130–34.

wadud, amina. "*Musawah*: Gender Equity through Qur'anic discourse." In *Communities of the Qur'an: Dialogue, debate and diversity in the twenty-first century*, edited by Emran El-Badawi and Paula Sanders. London: Oneworld, 2019: 81–97.

5.2 Online Sources

Patreon. "The Lady Imam." September 20, 2014. https://www.patreon.com/TheLadyImam/posts.

wadud, amina. "I am better than him." *Religion Dispatches*. August 30, 2010. https://religiondispatches.org/i-am-better-than-him/.

wadud, amina. "Al-Qur'an and Me." *Religion Dispatches*, September 8, 2010. https://religiondispatches.org/al-quran-and-me/.
wadud, amina. "Conversion Story #643 (not a real number)." *Religion Dispatches*, May 5, 2011. https://religiondispatches.org/conversion-story-643-not-a-real-number/.
wadud, amina. "Shari'ah is not the Law." *Religion Dispatches*, May 5, 2011. https://religiondispatches.org/shariah-is-not-the-law/.
wadud, amina. "The F* word: feminism in Islam." *Religion Dispatches*, May 5, 2011. https://religiondispatches.org/the-f-word-feminism-in-islam/.
wadud, amina. "To blog or not to blog? Is that even a question . . .?." *Feminism and Religion*, July, 2013. https://feminismandreligion.com/2013/07/18/to-blog-or-not-to-blog-is-that-even-a-question-by-amina-wadud/.
wadud, amina. "LGBTQI Muslims and International Movements for Empowerment." *Feminism and Religion*, September 20, 2014.
wadud, amina. "The tawhidic paradigm 1." Video recording of zoom lecture on wadud's Patreon channel. April 23, 2020.
wadud, amina. "The tawhidic paradigm 2.1." Video recording of zoom lecture, posted on wadud's Patreon channel. April 28, 2020.

5.3 Secondary Literature and Other Works

Abudigeri, Hibba. "The Renewed Woman of American Islam: Shifting Lenses Toward 'gender *jihad*'?." *MW* 91 (2001): 1–18.
Ali Kecia, Hammer Julianne and Silvers Laury. *A Jihad for Justice: Honoring the Work and Life of Amina Wadud*. USA: 48HrsBooks, 2012.
Ali, Kecia, and wadud amina. "The Making of the 'Lady Imam': An interview with amina wadud." *Journal of Feminist Studies in Religion* 35, no. 1 (2019): 67–79.
Al-Sharmani, Mulki. "Marriage in the Islamic Interpretive Tradition: Revisiting the Legal and the Ethical." *Journal of Islamic Ethics* 2 (2018): 76–96.
Anwar, Zainah, and Rose Ismail. "Amina Wadud and Sisters in Islam: a Journey toward Empowerment." In *A Jihad for Justice: Honoring the Work and Life of Amina Wadud*, edited by Kecia Ali, Julianne Hammer and Laury Silvers. USA: 48HrsBooks 2012: 63–72.
Augustine of Hippo. *The City of God*, trans. Henry Bettenson. London: Penguin, 2003.
Badran, Margot. "Seismic Shift from Patriarchy to Equality: Amina Wadud on Reading the Qur'an and Revolution." In *A Jihad for Justice: Honoring the Work and Life of Amina Wadud*, edited by Kecia Ali, Julianne Hammer and Laury Silvers. USA: 48HrsBooks 2012: 149–51.
Barlas, Asma. "Amina Wadud's hermeneutics of the Qur'an: women rereading sacred texts." In *Modern Muslim Intellectuals and the Qur'an*, edited by Suha Taji-Faruki. New York: Oxford University Press, 2006: 97–124.
Bowen, Patrick. *The African-American Islamic Renaissance, 1920–1975*. Vol. 2, *A History of Conversion to Islam in the United States*, 2 vols. Leiden/Boston: Brill, 2017.
Buber, Martin. *I and Thou*, trans. R.G. Smith. London: Bloomsbury, 2013.
Elewa, Ahmed and Silvers Laury. "'I *am* one of the people:' a survey and legal analysis on woman-led prayer in Islam." *Journal of Law and Religion* 26, no. 1 (2010–11): 141–71.
Eltantawi, Sarah. "Finding 'Yes' by Saying 'No': A Tribute to Amina Wadud." In *A Jihad for Justice: Honoring the Work and Life of Amina Wadud*, edited by Kecia Ali, Julianne Hammer and Laury Silvers. USA: 48HrsBooks 2012: 155–58.
Esack, Farid. *Qur'ān, Liberation and Pluralism*. Oxford: Oneworld 1997.

Galant, Jaamia. “Engaged Surrender: Reflecting on 25 years of the Gender Jihad.” Pre-khuṭba lecture delivered at Claremont Main Road Mosque, Friday August 30, 2019.
Gamieldien, Fahmi. *The History of the Claremont Main Road Mosque, its people and their contribution to Islam in South Africa*. Claremont: Claremont Main Road Mosque, 2004.
Hammer, Juliane. “Performing Gender Justice: the 2005 woman-led prayer in New York.” *Contemporary Islam* 4 (2010): 91–116.
Hidayatullah, Aysha A. *Feminist Edges of the Qur'an*. New York: Oxford University Press, 2014.
Kynsilehto, Anytta, ed. *Islamic Feminism: Current Perspectives*. Tampere Peace Research Institute Occasional Paper No. 96 (Tampere: Tampere Peace Research Institute, 2008).
Mir-Hosseini, Ziba, Mulki Al-Sharmani and Jana Rumminger, eds. *Men in Charge? Rethinking Authority in Muslim Legal Tradition*. Oxford: Oneworld, 2015.
Musawah. “Islam & Gender Equality and Justice (I-nGEJ).” Accessed January 23 2024. https://www.musawah.org/capacity-building/ingej/.
Rahemtullah, Shadaab. *Qur'an of the Oppressed: Liberation Theology and Gender Justice in Islam*. New York: Oxford University Press, 2017.
Rahman, Fazlur. *Islam and Modernity: Transformation of an Intellectual Tradition*. London: University of Chicago Press, 1982.
Rouse, Carolyn. *Engaged Surrender: African American Women and Islam*. Berkeley: University of California Press, 2004.
Safari, Elli, dir. *The noble struggle of amina wadud*. New York: Women Make Movies, 2007.
Seedat, Fatima. “An intellectual mentor.” In *A Jihad for Justice: Honoring the Work and Life of Amina Wadud*, edited by Kecia Ali, Julianne Hammer and Laury Silvers. USA: 48HrsBooks 2012: 191–95.
Shaikh, Sa'diyya. “Engaging surrender: the intimacy and power of the gender jihad.” In *A Jihad for Justice: Honoring the Work and Life of Amina Wadud*, edited by Kecia Ali, Julianne Hammer and Laury Silvers. USA: 48HrsBooks 2012: 213–16.
Sisters In Islam. *Are Muslim Men Allowed to Beat their Wives?* Petaling Jaya: Sisters in Islam 1991.
Sisters In Islam. *Are Women and Men Equal before Allah?* Petaling Jaya: Sisters in Islam 1991.
Tayob, Abdulkader. “The pre-khutba, khutba and Islamic change.” *Al-Qalam*, August 15, 1994.

Nimet Şeker

Qur'ānic Hermeneutics by and for Women

1 The Rise of Female Exegesis in the Twentieth Century

According to a tradition in Abū Jaʿfar Muḥammad b. Jarīr aṭ-Ṭabarī's encyclopedic commentary on the Qur'ān *Jāmiʿ al-bayān*, the wife and companion of the Prophet ʿĀ'isha bt. Abī Bakr once explained the following verses 8–11 in sura 53, *an-Najm* (The Star): "And then [he] approached – coming down until he was two bow-length away or even closer – and revealed to God's servant what He revealed."[1] This ambiguous passage raises the – theologically highly relevant question – who or what the prophet met. The verse *fa-awḥā ilā ʿabdihi mā awḥā* can at least be understood as a statement about the event of revelation, according to which God delivered his servant, the prophet, a verse from the Qur'ān. But according to this understanding, it is God again, who "then approached [Muḥammad] – coming down until he was two bow-length away or even closer" (*thumma danā fa-tadallā fa-kāna qāba qawsayni aw adnā*). ʿĀ'isha interpreted this verse as the description of an encounter between Gabriel and the prophet, at which Muḥammad saw the angel at the horizon. In her exegesis she explained that the angel used to show himself to the prophet in human shape. This time though, he came in his true form, completely covering the horizon.[2] She thus denied that the prophet had a vision of the divine (*ru'yat Allāh*). ʿĀ'isha denies the vision of God even more clearly in her exegesis with reference to the verse Q 6:103 ("No vision can take Him," *lā tudrikuhu al-abṣār*). Later scholars such as Fakhr ad-Dīn ar-Rāzī (d. 1210/606) made use of this interpretation. In ar-Rāzī's theological commentary on the Qur'ān, his interpretation is not only based on ʿĀ'isha's exegesis, he also cites her explanation as an authoritative support for his own interpretation.[3] For him, ʿĀ'isha certainly possessed authority and reputation in the field of exegesis.

Nevertheless, in Sunni *tafsīr* literature, with a few exceptions such as ʿĀ'isha, female exegetes were not perceived as part of the "transhistorical exegetical communities."[4] Only in the twentieth century did ʿĀ'isha's tradition of female exegesis come

1 *Thumma danā fa-tadallā fa-kāna qāba qawsayn*[i] *aw adnā fa-awḥā ilā ʿabdihi mā awḥā*, M.A.S. Abdel Haleem, *The Qur'ān* (Oxford: Oxford University Press, 2008, reprinted with corrections 2010). Citations refer to the reprint edition.

2 Muḥammad b. Jarīr aṭ-Ṭabarī, *Jāmiʿ al-bayān ʿan ta'wīl āy al-Qur'ān. Tafsīr aṭ-Ṭabarī*, ed. ʿAbd Allāh b. ʿAbd al-Muḥsin at-Turkī (Cairo: Dār Hijr, 2001), 22: 18f.

3 Fakhr ad-Dīn ar-Rāzī, *Tafsīr al-Fakhr ar-Rāzī al-mushtahir bi-t-tafsīr al-kabīr wa-mafātīḥ al-ghayb* (Beirut: Dār al-Fikr, 1981), 13: 133.

4 Aisha Geissinger, *Gender and Muslim Constructions of Exegetical Authority. A Rereading of the Classical Genre of Qur'ān Commentary* (Boston/Leiden: Brill, 2015), 280.

https://doi.org/10.1515/9783111320069-021

back to life, now in the form of written commentary. These texts can be divided into three categories: commentaries that follow the traditional system and explain the entire Qur'ān according to the order of the canonical text; commentaries that attempt the application of new methods and focus on the interpretation of particular verses and suras; and commentary works which, out of a certain thematic interest, are devoted to the question of what the Qur'ān says about the position of women. For various reasons, the last category in particular has received a lot of attention and growth in publications in recent decades.

Nonetheless, works such as *Qur'ān and Woman* by Amina Wadud or *Believing Women in Islam* by Asma Barlas are based on a hermeneutics of the Qur'ān that can be described as Qur'ān-focused, which is specifically built for this type of commentary. Joining a general trend in modern Islamic thought, according to which the entire Islamic religious practice should be reformed, they turn their attention to the Qur'ān as the only normative text.[5] In gender-egalitarian readings of the Qur'ān, additional reasons for adopting such a *solo corano* approach are relevant. There is a generally accepted notion that the exegetical scholarly tradition, as well as the text corpus of hadiths, have shaped and constructed a hierarchical world in which the male is the norm and the female is a deficient derivative of the norm. In contrast, it is believed that the Qur'ānic text, as the unadulterated word of God, is exempt from any misogyny and gender hierarchy. The Qur'ān guarantees women a catalog of rights, puts them ontologically on the same level as men and credits them with free will and freedom of choice. With the Qur'ān at the center of a theology of women's liberation, Qur'ānic hermeneutics becomes a means for establishing this liberation. For this reason, gender-egalitarian hermeneutics has produced theologizing forms of exegesis[6] that not only aim at explaining the Qur'ānic text, but also at discussing the implications of exegesis for further gender-related issues of theological anthropology, the conception of God, or religiously constructed power relations in Muslim communities. This thematic reading not only has little in common with the classical *tafsīr*, it also demar-

5 The spread of the Qur'ān-exclusive and thus hadith-skeptic epistemology in modern Islamic reformism begins with the writings of Sayyid Aḥmad Khān (d. 1898/1315) in India as well as Muḥammad ʿAbduh (d. 1905/1323), Muḥammad Rashīd Riḍā (d. 1935/1354) and Sayyid Quṭb (d. 1966/1386) in Egypt and had an impact on many reform thinkers in the twentieth century. Amina Wadud also receives the works of Sayyid Quṭb and Abū l-Aʿlā Mawdūdī (d. 1979/1399) in *Qur'ān and Woman*. Daniel Brown, *Rethinking Tradition in Modern Islamic Thought* (Cambridge: Cambridge University Press, 1999); Yudian Wahyudi, "The Slogan 'Back to the Qur'ān and the Sunna': A Comparative Study of the Responses of Ḥasan Ḥanafī, Muḥammad ʿĀbid al-Jābirī and Nurcholish Madjid" (PhD diss., McGill University Montreal, 2002); Aisha Y. Musa, "The Qur'ānists," *Religion Compass* 4, no. 1 (2010): 12–21; Johanna Pink, "Striving for a New Exegesis of the Qur'ān," *The Oxford Handbook of Islamic Theology*, ed. Sabine Schmidtke, (Oxford: Oxford University Press, 2016): 787f.

6 Walid A. Saleh, "Contemporary *Tafsīr*: The Rise of Scriptural Theology," in *The Oxford Handbook of Qur'ānic Studies*, ed. Mustafa Shah and Muhammad Abdel Haleem (Oxford: Oxford University Press, 2020): 693–703.

cates itself from precisely this tradition for the reasons mentioned above. Apart from the work of ʿĀʾisha ʿAbd ar-Raḥmān, it is influenced by the teachings of the Pakistani reformist Fazlur Rahman (d. 1988/1408) and tries to implement his *double-movement* method, by taking ethical premises such as gender equity and gender equality as normative, divinely ordained principles in the Qurʾān.

Although there is a large number of Qurʾān commentaries by female scholars in various languages such as Arabic, Persian and Turkish, the best-known and most widely received works in the field of Qurʾānic exegesis by and for women are written in English. This chapter is therefore intended to provide an overview of important contributions to exegesis by women in different languages, and then present the hermeneutical foundations of gender-egalitarian exegesis from the pens of its best-known representatives.

2 Women's Exegesis of the Qurʾān

That the Islamic tradition did not produce Qurʾān commentaries by female scholars until the twentieth century is a widespread view that is also regularly expressed in academic circles. Yet until today, there are no historical studies of the actual contribution of women to the field exegesis (*tafsīr*)[7] and hermeneutics of the Qurʾān. While it is correct that the Egyptian scholar ʿĀʾisha ʿAbd ar-Raḥmān, known by her pen name Bint ash-Shāṭiʾ (d. 1998/1418) is one of the most important and well-known representatives of modern Qurʾānic exegesis, contrary to what is often assumed, she is not the first female author of a written commentary on the Qurʾān. In bibliographical and bio-bibliographical sources, for example, a number of works by female scholars are named without it being known whether a manuscript or a printed edition thereof exists. This category includes, for example, a commentary on the Qurʾān[8] attributed to Jānā Bīgūm (d. 1660/1070),[9] and a book entitled *Zīb at-tafsīr fī tafāsīr al-Qurʾān*[10] attributed to the Mughal princess Zīb an-Nisāʾ al-Hindiyya (d. 1702/1113).[11]

7 In this chapter only written comments on the Qurʾān are considered. Exegetical traditions from early Islamic commentators such as ʿĀʾisha bt. Abī Bakr, Ḥafsa bt. ʿUmar, and Umm Salama are not counted as part of the written tradition of *tafsīr*.

8 ʿAbd al-Ḥayy b. Fakhr ad-Dīn al-Ḥusnī, *al-Iʿlām bi-man fī tārīkh al-hind min al-aʿlām* (*Nuzhat al-khawāṭir wa-bahjat al-masāmiʿ wa-n-nawāẓir*) (Beirut: Dār Ibn Ḥazm, 1999), 1: 517f.

9 Jānā Bīgūm (d. 1660/1070) was the daughter of the commander-in-chief of the Mughal army, ʿAbd ar-Raḥīm b. Bayram Khān (d. 1627/1036).

10 Ismāʿīl Bāshā al-Baghdādī (Bağdatlı İsmail Paşa), *Hadiyyat al-ʿārifīn asmāʾ al-muʾallafīn wa-āthār al-muṣannafīn* (Istanbul: Millî Eğitim Bakanlığı, 1951), 1: 375.

11 Zīb an-Nisāʾ was the eldest daughter of the Mughal emperor Aurangzeb (d. 1707/1119) and is best known for her divan poetry.

The first printed *tafsīr* work by a female author, which is also accessible today, seems to come from the Shiite theologian Nuṣrat Bīgum Amīn (d. 1983/1403).[12] The commentary *Makhzan al-ʿirfān dar tafsīr-i Qurʾān* in Persian explains the entire Qurʾān from a multi-method perspective and comprises fifteen volumes.[13] Nuṣrat Amīn was born in Isfahan in 1886/1303 and studied philosophy in addition to her traditional Islamic education. She made numerous contributions in Arabic and Persian to the fields of hadith, *fiqh* (Islamic jurisprudence), *kalam* (scholastic theology), *akhlāq* (morality), mysticism and exegesis. She received *ijāzas* (authorization) in *ijtihād* (legal reasoning) and *riwāya* (transmission) from a number of Ayatollahs, including Grand Ayatollah ʿAbd al-Karīm Ḥaʾirī Yazdī (d. 1937/1356). She also gave *ijāzas* in *ijtihād* and *riwāya* to her students. Nuṣrat Amīn is still considered a *mujtahida* (authoritative interpreter of Islamic Law) and enjoys a high reputation as a theological authority in Shiite circles.[14] Her voluminous Qurʾān commentary *Makhzan al-ʿirfān* is listed as an integral part of the pre-revolutionary curricula in Islamic theology at Tehran University.[15] Contrary to her expansive work and her deep interest in theological topics, with the exception of a practical handbook on femininity and motherhood, which is addressed to a general rather than scholarly audience,[16] she did not write on women's issues, nor did she develop a Qurʾān exegesis from a female perspective. Nonetheless, it is astonishing that her work, despite her high reputation as a theologian, is little known today.

The first Muslim woman to develop a reputation for herself as an exegete with an independent methodology that is characteristic of her known is the Egyptian theologian and Arabist ʿĀʾisha ʿAbd ar-Raḥmān (d. 1998/1418), known under her pen name Bint ash-Shāṭiʾ. Born in 1931/1349 into a religious family of scholars in the Egyptian provincial town of Dumyāṭ, her education, contrary to her father's wishes, was supported by her mother and grandfather, a theologian at al-Azhar University.[17] At that time, formal religious education for girls and women was not yet possible, nor was al-Azhar University accessible to women. She then trained as a teacher, and gained admission to Cairo University through part-time distance learning courses. Under the

---

12 Mirjam Künkler and Roja Fazaeli, "The Life of Two Mujtahidahs. Female Religious Authority in Twentieth-Century Iran," in *Women, Leadership and Mosques. Changes in Contemporary Islamic Authority*, ed. Hilary E. Kalmbach and Masooda Bano (Leiden: Brill, 2012), 127–60; Maryam Rutner, "Religious Authority, Gendered Recognition, and Instrumentalization of Nusrat Amin in Life and after Death," *Journal of Middle East Women's Studies* 11, no. 1 (2015): 24–41.

13 Nuṣrat Bīgum Amīn, *Makhzan al-ʿirfān dar tafsīr-i Qurʾān* (Isfahan: Kitābfurūsh-i Saqāfī, 1956).

14 For detailed information on her biography and writings see Künkler and Fazeli, "The Life of Two Mujtahidahs," 132–41.

15 Künkler and Fazeli, "The Life of Two Mujtahidahs," 140.

16 Nuṣrat Bīgum Amīn, *Kitāb ravish-i khūshbakhtī wa-tavṣiyyah bih khāharān-i imānī* (The Ways to Happiness and Advice for Sisters in Faith) (Isfahan: Saghāfī, 1968–69); Ladan Rahbari, "Women's Ijtihad and Lady Amin's Islamic Ethics on Womanhood and Motherhood," *Religions* 11, no. 2 (2020): 88.

17 Shuruq Naguib, "Bint al-Shāṭiʾs Approach to *tafsīr*: An Egyptian Exegete's Journey from Hermeneutics to Humanity," *JQS* 17, no. 1 (2015): 49.

supervision of Ṭāhā Ḥusayn, she received her doctorate in 1952/1371 with an outstanding thesis on Abū l-ʿAlā l-Maʿarrī. She continued her academic career, which she had already started as an MA student in the Arabic department of the Faculty of Philology, as professor of Arabic language and literature at ʿAyn Shams University from 1962/1381, and was finally professor of Qur'ānic exegesis at Qarawiyyīn University until the 1990s. She was invited to teaching in Umm Durmān and Kharṭūm, Algiers, Beirut, in the Arab Emirates and in Riyadh.

ʿĀ'isha ʿAbd ar-Raḥmān's academic writings can be grouped into the following three subjects: textual studies of the Qur'ān and hadith; Arabic language and literature; and biographical works on the female members of the prophetic family. In the field of exegesis her literary and linguistic commentary on the early Meccan suras *at-Tafsīr al-bayānī li-l-Qur'ān al-karīm* (The Literary Exegesis of the Qur'ān), in which she practically applies the methodology of Amīn al-Khūlī,[18] and *al-Iʿjāz al-bayānī li-l-Qur'ān al-karīm wa-masā'il Nafiʿ b. al-Azraq* (The Qur'ān's Rhetorical Inimitability and the Questions of Nafīʿ b. al-Azraq) are significant contributions.

But even in her exegetical works, we do not find any interpretation of the Qur'ān from a specifically female perspective. On the other hand, in her shorter treatises such as *Maqāl fī l-insān* (A Treatise on the Human Being),[19] *al-Mar'a al-muslima ams wa-l-yawm* (The Muslim Women: Yesterday and Today)[20] and *al-Mafhūm al-islāmī li-taḥrīr al-mar'a* (The Islamic Concept of Women's Liberation),[21] she expresses her concern for Muslim women in modern times. Interestingly, even in these shorter contributions, which are not titled as *tafsīr*, she undertakes interpretations in terms of women's emancipation and gender equality.[22] Today she is not considered a feminist,[23] but in her work she clearly shows a strong awareness of issues such as gender equality and the needs of modern Arab-Islamic women. Thus, in the field of religious studies, she can undoubtedly be seen as a pioneer of women's rights from an Islamic perspective. Many

18 ʿĀ'isha ʿAbd ar-Raḥmān, *at-Tafsīr al-bayānī li-l-Qur'ān al-karīm* (Cairo: Dār al-Maʿārif, 1990[7]), 1: 10–14. Naguib notes that a comprehensive study of the implementation of al-Khūlī's method in her work is pending and that this aspect should not obscure the peculiar nature of her work, Naguib, "Bint al-Shāṭi's Approach to *tafsīr*," 52.

19 ʿĀ'isha ʿAbd ar-Raḥmān, *Maqāl fī l-insān. Dirāsa qur'āniyya* (Cairo: Dār al-Maʿārif, 1969[2]).

20 ʿĀ'isha ʿAbd ar-Raḥmān, "al-Mar'a al-muslima ams wa-l-yawm," in *al-Islām, al-yawm wa-ghadan*, ed. Ibrāhīm Salāma et al. (Cairo: Dār Iḥyā' al-Kutub al-ʿArabiyya, 1960), 67–94.

21 ʿĀ'isha ʿAbd ar-Raḥmān, *al-Mafhūm al-islāmī li-taḥrīr al-mar'a* (Umm Durmān: Jāmiʿat Umm Durmān, 1966–67); Aisha Abdul Rahman, "The Islamic Conception of Women's Liberation," trans. Nazih Khater, *Al-Raida* 125 (Spring 2009): 37–43.

22 Naguib, "Bint al-Shāṭi's Approach to *tafsīr*," 51.

23 Ruth Roded, "Bint Al-Shati's *Wives of the Prophet*. Feminist or Feminine?," *British Journal of Middle Eastern Studies* 33, no. 1 (2006): 51–66; Roxanne D. Marcotte, "Bint al-Shāṭi' on Women's Emancipation," in *Coming to Terms with the Qur'ān. A Volume in Honour of Professor Issa Boullata*, ed. Khaleel Mohammed and Andrew Rippin (North Haledon: Islamic Publications International, 2007), 200–2; Naguib, "Bint al-Shāṭi's Approach to *tafsīr*," 70f.

of the ideas and interpretations of later gender-egalitarian exegesis actually go back to ideas that ʿAbd ar-Raḥmān presents in a concise form in her shorter treatises, especially in *al-Mafhūm al-islāmī li-taḥrīr al-marʾa* (1967). It is noteworthy that, despite the striking similarity to later feminist-exegetical positions, her work is not cited in later writings.[24]

Other contributions to Qurʾānic exegesis by female scholars are *Naẓarāt fī kitāb Allāh* (Reflections on the Book of God)[25] by the Egyptian activist Zaynab al-Ghazālī (d. 2005/1426), the sixteen-volume work *al-Mubaṣṣir li-nūr al-Qurʾān* (The Beholder of the Light of the Qurʾān)[26] by Nāʾila Hāshim Ṣabrī, the four-volume work *Taysīr at-tafsīr* (The Accessible Exegesis)[27] by Fawqiyya ash-Shirbīnī, and *Tafsīr gharīb al-Qurʾān* (Exegesis of Unusual Terms in the Qurʾān)[28] by Kāmila bt. Muḥammad al-Kawwārī. In addition, the thirteen-volume work *Okuyucu Tefsiri* (Reader's Exegesis)[29] by Semra Kürün Çekmegil and the twelve-volume work *Kur'an Tahlili* (Analysis of the Qurʾān)[30] by Necla Yasdıman are available in Turkish. The work *Yüce Kur'an-ı Kerim ve Meali* (The Glorious Qurʾān and its Commentary)[31] by Ayşa Zeynep Abdullah was also recently published in Turkish.

In any case, there are numerous internationally known exegetical works in English from the twentieth and twenty-first centuries that are aimed at establishing a gender-egalitarian exegesis. This form of thematic exegesis will be discussed below.

3 Women's Re-readings of the Qurʾān

3.1 An Egalitarian Ontology of Gender

Almost all contributions to the gender-egalitarian interpretation of the Qurʾān are based on a moral foundation that is derived from constitutive statements on the nature and purpose of man in the Qurʾān. For this purpose, the creation narrative is read as constituting an ontological equality of man and woman. This idea is then connected to the human trusteeship of God on earth (*khilāfa*), and the central belief in the unity of

---

24 Apart from Nimat Hafez Barazangi's critical preoccupation with ʿĀʾisha ʿAbd ar-Raḥmān's work, only one reference could be found in a contribution by Azizah al-Hibri: Nimat Hafez Barazangi, *Woman's Identity and the Qurʾān. A New Reading* (Gainesville: University Press of Florida, 2004), 42, 44, 70–79, 130; Azizah al-Hibri, "Islam, Law and Custom: Redefining Muslim Women's Rights," *American University International Law Review* 12, no. 1 (1997): 20.

25 Zaynab al-Ghazālī, *Naẓarāt fī kitāb Allāh*, vol. 1 (Cairo: Dār ash-Shurūq, 1994).

26 Nāʾila Hāshim Ṣabrī, *al-Mubaṣṣir li-nūr al-Qurʾān* (Jerusalem: Maṭbaʿat ar-Risāla, 1998–2003).

27 Fawqiyya ash-Shirbīnī, *Taysīr at-tafsīr* (Cairo: Maktabat al-Īmān, 2008).

28 Kāmila al-Kawwārī, *Tafsīr gharīb al-Qurʾān* (Beirut: Dār Ibn Ḥazm, 2008).

29 Semra Kürün Çekmegil, *Okuyucu Tefsiri* (Malatya: Nida Dergisi Yayınları, 2006).

30 Necla Yasdıman, *Kur'an Tahlili* (Izmir: Cemre, 2006).

31 Ayşa Zeynep Abdullah, *İndirilme Sırasına Göre Yüce Kur'an-ı Kerim ve Meali* (Istanbul: Hermes, 2019).

God (*tawḥīd*). Fundamental to this is a deconstruction of the traditional reading of verse 1 of sura 4, *an-Nisā'* (Women), according to which Eve was created from Adam's rib and woman had a subordinate position to man in the order of creation.[32] Contrary to the traditional reading based on hadiths, "a borrowed foreign concept,"[33] egalitarian interpretations consider only the wording of the verse: by re-reading expressions such as *min nafsin wāḥidatin* and *zawjahā* in Q 4:1, all of humanity, man and woman, is regarded as coming from a single, original soul. Thus, *nafs* is understood as an ungendered,[34] pre-created soul, which stands for the egalitarian origin of sexually and culturally differentiated individuals. It is further stated that there is no evidence in the wording of the verse that men are superior to women in creation. The terms *nafs* and *zawj* do not denote Adam and Eve.[35] The emphasis here is on the common origin of man and woman from *nafs wāḥida*. This reading is further connected to the central dogma of divine unity (*tawḥīd*), stating that it establishes equality of all creatures before God.[36] Likewise, the creation of all things as equal pairs (*azwāj*)[37] is understood as their otherness to the absolute unity of God (Q 42:11).[38] This means that there is no gender hierarchy in the origin of mankind. Woman does not act as a descendant of man, nor as his opposite or binary pole.[39] Rather, man and woman are created by their common origin to coexist in love and mercy (*mawadda wa-raḥma*) for one another (Q 30:20).[40] The destiny of mankind, and thus of woman and man, is the fulfilment of the trustee-

32 For a summary of traditional and modern readings of the creation story and the role of Eve in it, see Yvonne Y. Haddad, "Eve: Islamic Image of Woman," *Women's Studies International Forum*, 5, no. 2 (1982): 135–44; Hatice Arpaguş, "The Position of Woman in the Creation. A Qur'ānic Perspective," in *Muslima Theology. The Voices of Muslim Women Theologians*, ed. Marcia Hermansen, Ednan Aslan, and Elif Medeni (Frankfurt a.M.: Peter Lang, 2013), 113–32.

33 Abdul Rahman, "The Islamic Conception of Women's Liberation," 38.

34 Jerusha Tanner Lamptey, *Divine Words, Female Voices. Muslima Explorations in Comparative Feminist Theology* (New York: Oxford University Press, 2018), 49, 159, 162.

35 Riffat Hassan, "Woman and Man's 'Fall'. A Qur'ānic Theological Perspective," in *Muslima Theology. The Voices of Muslim Women Theologians*, ed. Marcia Hermansen, Ednan Aslan and Elif Medeni (Frankfurt a.M.: Peter Lang, 2013), 101–13.

36 Azizah al-Hibri, "An Introduction to Muslim Women's Rights," in *Windows of Faith: Muslim Woman Scholar-Activists in North America*, ed. Gisela Webb (Syracuse: Syracuse University Press, 2000), 50f.; Asma Barlas: *"Believing Women" in Islam: Unreading Patriarchal Interpretations of the Qur'ān* (Austin: University of Texas Press, 2002), 13, 102f., 106–8; Amina Wadud, *Inside the Gender Jihad. Women's Reform in Islam* (London: Oneworld, 2006), 24–32.

37 *Hūd*/Q 11:40, *Fāṭir* (The Originator)/Q 35:11, *Yāsīn*/Q 36:36, *az-Zukhruf* (Gold Ornaments)/Q 43:12, *adh-Dhāriyāt* (The Scatterers)/Q 51:49, *an-Najm* (The Star)/Q 53:45, *ar-Raḥmān* (The Compassionate)/Q 55:52; Wadud, *Qur'ān and Woman*, 20–23.

38 Wadud, *Qur'ān and Woman*, 26.

39 Lamptey, *Divine Words, Female Voices*, 162.

40 Riffat Hassan, "An Islamic Perspective," in *Women, Religion, and Sexuality. Studies on the Impact of Religious Teachings on Women*, ed. Jeanne Becher (Geneva: WCC Publications, 1990), 101; Amina Wadud, *Qur'ān and Woman. Rereading the Sacred Text from a Woman's Perspective* (New York, Oxford: Oneworld, 1999[2]), 19–22; al-Hibri, "Islam, Law, and Custom," 26f.; Asma Barlas, *Believing Women*, 132–35.

ship of God on earth (*khilāfa*, Q 2:30 and Q 33:72), as Amina Wadud states:[41] "Being *khalifah* is equivalent to fulfilling one's human destiny as a moral agent, whose responsibility is to participate in upholding the harmony of the universe. With respect to society, harmony means working for justice."[42] The meaning of this quote becomes clearer when the word "human" is replaced with "woman." With this, Wadud states that women have been given free will to follow their moral agency on earth.

It is ʿAbd ar-Raḥmān who is the first exegete to present this reading of the human condition (*insāniyya*)[43] of women, thus preparing a discussion that generations of female scholars would take up after her. Her reading of the term "human" (*al-insān*) in the Qurʾān leads to the idea that the human being – and thus women as ontologically equal beings – is a creature who is distinguished by God from all other creatures through a specific mandate and obligation. This includes the interpretation of normative scriptures (*bayān*[44] or *ijtihād*[45]), the fulfilment of religious precepts and of the *khilāfa*, the doing of good, and the prevention of bad.[46] All of this falls under the moral obligation of *khilāfa*, which God has imposed on man and woman alike: "The freedom of the Muslim woman finds its source in this determined origin. In the Islamic view, this freedom is the perfection of the woman's humanity with all the rights and consequences related to this humanity."[47] From this theological foundation, women's rights as "human rights"[48] are derived as "God-given rights"[49]: these include the right to education,[50] the right to financial independence[51] and the right to divorce.[52] The interpretation of Q 4:1 implying ontological equality leads to the demand that women have equal access to the exegesis of normative texts. At the same time, this demand is implemented by the female exegetes themselves through a critique of the androcentric tradition and the reformulation of exegesis on gender relations: "If we accept verse 4:1 [. . .] as *relevant* to understanding the Islamic stance on women, *a Muslim women who identifies with the Qurʾān becomes an essential partner in the interpretation process of the Qurʾān*, just as her access and partic-

41 Al-Hibri, "An Introduction to Muslim Women's Rights," 53; Wadud, *Inside the Gender Jihad*, 32–37.
42 Wadud, *Inside the Gender Jihad*, 34.
43 ʿAbd ar-Raḥmān, *at-Tafsīr al-bayānī*, 1: 177.
44 Ibid.; Abdul Rahman, "The Islamic Conception of Women's Liberation," 39.
45 Al-Hibri, "Redefining Muslim Women's Rights," 40, 43; al-Hibri, "An Introduction to Muslim Women's Rights," 57.
46 ʿAbd ar-Raḥmān, *at-Tafsīr al-bayānī*, 1: 177; ʿAbd ar-Raḥmān, *Maqāl fī l-insān*, 45–48; Naguib, "Bint al-Shāṭi's Approach to *tafsīr*," 61f., 69f.
47 Abdul Rahman, "The Islamic Conception of Women's Liberation," 38.
48 Ibid, 39.
49 Al-Hibri, "An Introduction to Muslim Women's Rights," 58.
50 Abdul Rahman, "The Islamic Conception of Women's Liberation," 39.
51 Al-Hibri, "An Introduction to Muslim Women's Rights," 57.
52 Ibid., 58.

ipation in the interpretation process is a must for her self-identity," says Nimat Barazangi.[53]

3.2 The Premise of Gender Equality

Notwithstanding the widely shared view that the Qur'ān regards women and men as essentially equal beings, the discussion about how to define gender equity and gender equality remains vague. Rather, it seems that two separate theoretical concepts are always considered together, presupposing them as normative, ahistorical and universalistic core messages of the Qur'ān.[54] Despite the absence of a theological reflection of justice and equality, some aspects of this idea can be inferred from the works.

At first, equality seems to be thought of on a triadic model. On the one hand, there is the direct, immediate relationship of the individual to God and the responsibility of the individual before God, and on the other hand, the hierarchy-free relationship between man and woman in an equal partnership.[55] In this tripartite model, the status of the individual is not determined by gender, but rather through individual *taqwā* (God consciousness).[56]

The Qur'ānic statements on otherworldly reward and punishment are also interpreted in terms of equality. Thus, the verses Q 3:195[57] and Q 33:35[58] are understood as promising a just reward – also between man and woman – in the hereafter. Amina Wadud shows how verses emphasize the principle of equity in questions of death, resurrection, last judgment, reward and punishment after death (Q 3:185–86). In doing so, she interprets the recurring terms *nafs* (soul, being) and *zawj*, plural *azwāj* (part-

53 Barazangi, *Woman's Identity and the Qur'ān*, 43, italics in original. "Bint al-Shāti' is not simply asserting women's equal humanity, nor their right to education, but their original right to understand, interpret, and explain divine discourse, and to contest the meanings established by male interpreters," Naguib, "Bint al-Shāṭi's Approach to *tafsīr*," 72.

54 Raja Rhouni, *Secular and Islamic Feminist Critiques in the Work of Fatima Mernissi* (Leiden/Boston: Brill, 2010), 252; Kecia Ali, *Sexual Ethics and Islam. Feminist Reflections on Qur'an, Hadith, and Jurisprudence* (London: Oneworld, 2008), 154f.

55 Wadud, *Inside the Gender Jihad*, 28–31; Amina Wadud, "The Ethics of *Tawhid* and the Ethics of *Qiwamah*," in *Men in Charge? Rethinking Authority in Muslim Legal Tradition*, ed. Ziba Mir-Hosseini, Mulki Al-Sharmani, and Jana Rumminger (Oxford: Oneworld, 2015), 271f.; Aysha A. Hidayatullah, *Feminist Edges of the Qur'ān* (Oxford: Oxford University Press, 2014), 168f.; Lamptey, *Divine Words, Female Voices*, 158f.

56 Abdul Rahman, "The Islamic Conception of Women's Liberation," 40.

57 Wadud, *Qur'ān and Woman*, 51; al-Hibri, "An Introduction to Muslim Women's Rights," 52; Lamptey, *Divine Words, Female Voices*, 46.

58 Wadud, *Qur'ān and Woman*, 99; Barlas, *Believing Women in Islam*, 143; Ali, *Sexual Ethics*, 114f., 123f.

ner) in Q 81:7 and Q 21:47[59] as gender-neutral expressions, concluding that their use does not denote gender differentiation. According to Wadud, other verses such as Q 2:284 and Q 40:40[60] point to the just, therefore equal, treatment of men and women in questions of the hereafter.[61] In respect to paradise and hell, too, with the exception of the virgins of paradise (*ḥūr ʿīn*), there is no distinction between the punishment and the reward of men and women: "Recompense is *aquired* not through gender, but through actions performed by the individual before death [. . .] (45:1–2)."[62]

With regard to verses that contradict the principle of gender equality against the background of contemporary values, many exegetes apply a hermeneutic of historical contextualization.[63] According to Wadud, the description of the large-eyed paradise maidens in the Meccan suras addresses a specific male audience in the hierarchical society of the Qurʾān's context of revelation, especially certain Meccan patriarchs, and their innermost desires and needs. In the Medinan suras, however, the Qurʾān only uses the gender-neutral expression *zawj, azwāj*.[64] Wadud also interprets the expression *azwāj muṭahhara* (pure spouses) in the verses Q 2:25, Q 3:15 and Q 4:57 to mean "female partner" or "wife from this world." The verses emphasize the "partnership, friendship and harmony" in paradise.[65] This explanation runs contrary to the traditional reading, according to which *azwāj muṭahhara* means pure women of paradise.[66] But there is no discussion in Wadud's exegesis of a possible reading of *azwāj muṭahhara* as addressing male desires in the Medinan context, as is clear from the counting in Q 3:14 of *an-Nisāʾ* among the coveted things (*ash-shahawāt*), and from the reference to heavenly *azwāj muṭahhara* in the following verse Q 3:15.[67] Indeed, the

59 *Wa-idhā n-nufūsu zuwwijat*, "when souls are sorted into classes"; *wa-naḍaʿu l-mawāzīna l-qisṭa li-yawmi l-qiyāmati fa-lā tuẓlamu nafsun shayʾan*, "We will set up scales of justice for the Day of Resurrection so that no one can be wronged in the least," Abdel Haleem, *The Qurʾān*, 411, 205.

60 *Man ʿamila sayyiʾatan fa-lā yujzā illā mithlahā wa-man ʿamila ṣāliḥan min dhakarin aw unthā wa-huwa muʾminun fa-ūlāʾika yadkhulūna l-jannata yurzaqūna fīhā bi-ghayri ḥisābin*, "Whoever does evil will be repaid with its like; whoever does good and believes, be it a man or a woman, will enter Paradise and be provided for without measure," Abdel Haleem, *The Qurʾān*, 304.

61 Wadud, *Qurʾān and Woman*, 46–49, italics in original.

62 Ibid., 50.

63 Hidayatullah, *Feminist Edges of the Qurʾān*, 65–86; Nimet Seker, *Koran und Gender. Exegetische und hermeneutische Studien zum Geschlechterverhältnis im Koran* (Hamburg: Editio Gryphus, 2020), 117–45.

64 Wadud, *Qurʾān and Woman*, 54–56.

65 Ibid., 56f.

66 These women of paradise are described by the exegetes as being free of any worldly physical imperfection of women such as menstruation, and birth is also counted among them, aṭ-Ṭabarī, *Jāmiʿ al-bayān*, 1: 419–22; Abū Manṣūr Muḥammad al-Māturīdī, *Taʾwīlāt ahl as-sunna. Tafsīr al-Māturīdī*, ed. Majdī Bāsallūm (Beirut: Dār al-Kutub al-ʿIlmiyya, 2005), 1: 405; Abū Iṣḥāq Aḥmad ath-Thaʿlabī an-Nīsābūrī, *al-Kashf wa-l-bayān. Tafsīr ath-Thaʿlabī*, ed. ʿAlī b. ʿĀshūr and Naẓīr as-Sāʿidī (Beirut: Dār Iḥyāʾ at-Turāth al-ʿArabī, 2001), 1: 181.

67 *Zuyyina li-n-nāsi ḥubbu sh-shahawāti min an-nisāʾi wa-l-banīna wa-l-qanāṭīri l-muqanṭarati min adh-dhahabi wa-l-fiḍḍati wa-l-khayli l-musawwamati wa-l-anʿāmi wa-l-ḥarthi dhālika matāʿu l-ḥayāti d-

"selectiveness" of Wadud's reading is described as a conscious departure from the androcentric aspects of the otherworldly reward.[68]

A second aspect of the principles of equality and justice is seen in verses like Q 30:21 and Q 9:71, which speak of a reciprocal relationship between man and woman, indicating mutual responsibility, respect and love in male-female relationships. According to this notion, the idea of existential complementarity[69] of men and women is not just based on a relationship. It is rather derived from the same essence of both genders, then transferred to the relationship between all people. Men and women "need each other to realize their full existence. Husband and wife are founding partners of one social cell, and companions in the journey of life"[70] writes ʿAbd ar-Raḥmān, addressing the men:

> So walk with her on this path, for she is your daughter, your sister, and your partner for life. She is the mother of your children and maker of your future, so remember God's *aya* on her behalf: "And among His signs is this, that He created for you mates among yourselves, that ye may dwell in tranquility with them, and He has put love and mercy between your (hearts): verily in that are signs for those who reflect" (30:21).[71]

Nonetheless, this does not mean that men and women are interchangeable in their social functions.[72] Even if it sounds paradoxical, the recognition of equality does not cancel out the difference between men and women; rather, equality applies in relation to moral and spiritual issues, whereas difference applies in social functions.[73] Asma Barlas says, "difference does not always imply inequality."[74] It is only through

---

dunyā wa-llāhu indahū ḥusnu l-maʾābi. Qul a-unabbiʾukum bi-khayrin min dhālikum li-lladhīna ttaqaw ʿinda rabbihim jannātun tajrī min taḥtihā l-anhāru khālidīna fīhā wa-azwājun muṭahharatun wa-riḍwānun min Allāhi wa-llāhu baṣīrun bi-l-ʿibādi, "The love of desirable things is made alluring for men – women, children, gold and silver treasures piled up high, horses with fine markings, livestock, and farmland – these may be the joys of this life, but God has the best place to return to. [Prophet], say, 'Would you like me to tell you of things that are better than all of these? Their Lord will give those who are mindful of God Gardens graced with flowing streams, where they will stay with pure spouses and God's good leasure – God is fully aware of His servants," Abdel Haleem, *The Qur'ān*, 35.

68 Karen Bauer, "'The Male Is Not Like the Female' (Q 3:36): The Question of Gender Egalitarianism in the Qur'ān," *Religion Compass* 3, no. 4 (2009), 650. In *Inside the Gender Jihad*, Wadud recognizes the androcentric character of the paradise virgins in the Qur'ān, but does not deviate from her reading of *zawj, azwāj*, Wadud, *Inside the Gender Jihad*, 193.

69 Hassan, "An Islamic Perspective," 109; Wadud, *Qur'ān and Woman*, 21; al-Hibri, "Islam, Law, and Custom," 32, 38; ʿAisha ʿAbdul Rahman, "Islam and the New Woman," trans. Anthony Calderbank, *Alif: Journal of Comparative Poetics* 19 (1999), 202; al-Hibri, "An Introduction to Muslim Women's Rights," 53; Azizah al-Hibri, "Qur'ānic Foundations of Rights of Muslim Women in the 21st Century," in *Women in Indonesian Society: Access, Empowerment and Opportunity*, ed. M. Atho Mudzhar et al. (Yogyakarta: Sunan Kalijaga Press 2002), 5, 20.

70 Abdul Rahman, "The Islamic Conception of Women's Liberation," 40.

71 Ibid., 42, italics in original.

72 Wadud, *Inside the Gender Jihad*, 155.

73 Wadud, *Qur'ān and Woman*, 102.

74 Barlas, *Believing Women in Islam*, 199, 145.

the difference between men and women and the associated social gender roles that women's rights emerge.[75] Aysha Hidayatullah formulates a fundamental criticism on this point, according to which the term "gender" is not defined in gender-egalitarian hermeneutics and is used in an undifferentiated manner to denote "women." She notes the lack of a discussion of academic theories about essentialist gender understandings, gender binaries and cultural gender constructions.[76]

3.3 The Discussion about Gender Roles

With regard to Qur'ānic ordinances or ideals about gender roles, the exegetes make contradicting statements. On the one hand, there is the statement that the Qur'ān does not prescribe any gender roles or division of labor between men and women.[77] The only role ascribed to women can be found in pregnancy and childbirth, and this is based on female biology. But beyond that, the (social) role of motherhood is not derived from the Qur'ān as an obligation.[78] Still, gender-egalitarian readings positively reinterpret the traditional hierarchical exegesis of *qiwāma* (*ar-rijālu qawwāmūna ʿalā n-nisāʾi*, Q 4:34) as male guardianship or duty of care and thus as a social responsibility. ʿĀʾisha ʿAbd ar-Raḥmān even understands *qiwāma* ("guardianship") as a male right. Yet, it is neither an absolute nor an unconditional right:

> It is about time for men to understand that their legitimate right of guardianship over us is neither absolute nor is it for all men in general over all women. It is a conditional right "because Allah has given the one more (strength) that the other, and because they support them from their means" (4:34), and if a man could not meet this condition, he loses his right to guardianship.[79]

Unlike the North American exegetes, ʿAbd ar-Raḥmān reads the first part of the verse Q 4:34 – like many classical scholars[80] – in relation to the term *daraja* (rank) from Q 2:228 (*wa-lahunna mithlu llādhī ʿalayhinna bi-l-maʿrūfi wa-li-r-rijāli ʿalayhinna daraja-*

75 Ibid., 199.

76 Hidayatullah, *Feminist Edges of the Qur'ān*, 129. She also contradicts the argument that a relationship between man and woman that is characterized by "reciprocity" or "reciprocity" annuls the principle of hierarchy. According to her understanding, androcentric verses do not exclude a partnership based on love and reciprocity: "In the premodern context of the Qur'ān's revelation, equality may not have been understood as a premise of loving, caring relationships; relationships of male-female dominance may not have necessarily offended ideals of mercy and tranquility," Hidayatullah, *Feminist Edges of the Qur'ān*, 165.

77 Wadud, *Qur'ān and Woman*, 8, 20, 104.

78 Ibid., 20, 22; Barlas, *Believing Women in Islam*, 179.

79 Abdul Rahman, "The Islamic Conception of Women's Liberation," 41.

80 Ayesha S. Chaudhry, "Wife-Beating in Pre-Modern Islamic Tradition. An Inter-Disciplinary Study of Hadith, Qur'ānic Exegesis and Islamic Jurisprudence" (PhD diss., New York University, 2009), 318–20; Karen Bauer, "Room for Interpretation. Qur'ānic Exegesis and Gender" (PhD diss., Princeton University, 2008), 61f., 84, 96, 98, 102.

tun). Be that as it may, she derives the principle of equality from Q 2:228 and interprets it as a bonus of social manhood (*rujūla*) and not as a feature of biological masculinity (*dhukūra*). Although she does not explain what she means by the two terms,[81] her differentiation suggests that she does not regard *daraja* and *qiwāma* as natural-essential, but rather conditional features. This interpretation of *qiwāma* as a responsibility and not as a male power privilege can also be found in later exegesis. Azizah al-Hibri sees *qiwāma* as an obligation (*taklīf*) and not as a male privilege (*tashrīf*).[82] For Barlas though, *qiwāma* is to be brought into harmony with the principle of equality that is expressed in Q 9:71 (*wa-l-muʾminūna wa-l-muʾminātu baʿḍuhum awliyāʾu baʿḍin*). The man's obligation is therefore seen in the financial and material provision of the woman, derived from the addition *wa-bi-mā anfaqū min amwālihim* (Q 4:34).[83] The guardianship, which is positively applied in this way, is further expanded by Wadud and related to the biological function of women to bear children: "Ideally, *everything* she needs to fulfil her primary responsibility [i.e. childbearing] comfortably should be supplied in society, by the male: this means physical protection as well as material sustenance."[84] In this way, the premise that the Qur'ān does not assign roles and functions to men and women is further modified. The interpretive endeavors of the female exegetes clearly amount to interpreting wording and verses in terms of spiritual or ethical-legal hierarchy, according to the principle of gender equality.

4 Interpretation of Verses that Grant Privileges to Men

4.1 *Daraja* (Q 2:228)

Revisionist and novel interpretations have been put forward with regard to a number of other verses that Muslims associate with the ontological and legal superiority of men over women.

It is argued that verse Q 2:228,[85] which speaks of a higher "rank" (*daraja*) of the man over the woman (*wa-li-r-rijāli ʿalayhinna darajatun*), first establishes equality of

81 Hosn Abboud, "The Arab Women Discourses on Feminism and Islam: Fear of an Oxymoron," *Al-Raida* 148–49 (2015/2016), 36.

82 Al-Hibri, "Qur'ānic Foundations," 19.

83 Barlas, *Believing Women in Islam*, 186f.

84 Wadud, *Qur'ān and Woman*, 73, italics in original.

85 "*Wa-l-muṭallaqātu yatarabbaṣna bi-anfusihinna thalāthata qurūʾin wa-lā yaḥillu lahunna an yaktumna mā khalaqa llāhu fī arḥāmihinna in kunna yuʾminna bi-llāhi wa-l-yawmi l-ākhiri wa-buʿūlatuhunna aḥaqqu bi-raddihinna fī dhālika in arādū iṣlāḥan wa-lahunna mithlu lladī ʿalayhinna bi-l-maʿrūfi wa-li-r-rijāli ʿalayhinna darajatun wa-llāhu ʿazīzun ḥakīmun*," "Divorced women must wait for three monthly periods before remarrying, and, if they really believe in God and the Last Day, it is not lawful for them to conceal what

men and women in legal issues of divorce (*wa-lahunna mithlu llādhī ʿalayhinna bi-l-maʿrūfi*).[86] It is argued that the sentence *wa-li-r-rijāli ʿalayhinna darajatun* is limited only to the statement made in the verse about divorce and the waiting period (*ʿidda*) after it.[87] The higher "rank" or "advantage" of the husband over his wife is seen in his right to reverse the divorce during the waiting period; this right is not granted to wives.[88] Another reading of *daraja* relates to the male prerogative to pronounce divorce without consultation and without a waiting period, while the woman's right to divorce is subject to restrictions such as the three-month waiting period.[89] In addition, *daraja* is interpreted as a male privilege with regard to the possibility of remarriage without a waiting period.[90] Through interpretations like these, the common association of *qiwāma* with *daraja* is declared to be invalid.

4.2 Polygyny (Q 4:3)

When interpreting verse Q 4:3 on the question of whether polygyny is allowed for men, an intra-textual reading is carried out. From a joint consideration with the preceding verse Q 4:2, the permission for the marriage of up to four female orphans is interpreted as a restriction of an existing practice. It is assumed that the verse intends to restore justice in dealing with orphans and their property managed by their male guardians. Furthermore, Wadud reflects on the verse with regard to a fair financial and material provision for the orphans and tries to create a historical context by introducing the idea of marriage as a hierarchical relationship ("marriage of subjugation") in the context of revelation. She states that not only women, but especially wealthy orphans have found themselves in a relationship of dependency on a male guardian or husband and were exposed to the improper marriage with or through the guardian. The verse exhorts to treat orphans fairly as potential spouses. With respect to Q 4:129, justice in marriage and the fair treatment of women in a polygynous marriage are also reflected upon.[91] When it comes to the question of whether the

---

God has created in their wombs: their husbands would do better to take them back during this period, provided they wish to put things right. [Divorced] women have [rights] similar to their obligations, according to what is fair, and [ex-]husbands have a degree [of right] over them: [both should remember that] God is almighty and wise," Abdel Haleem, *The Qurʾān*, 25f.

86 Abdul Rahman, "The Islamic Conception of Women's Liberation," 41; Wadud, *Qurʾān and Woman*, 69; Wadud, *Inside the Gender Jihad*, 25.

87 Barlas, *Believing Woman in Islam*, 196.

88 Hassan, "An Islamic Perspective," 113f.

89 Wadud, *Qurʾān and Woman*, 68; Wadud, *Inside the Gender Jihad*, 25.

90 Hassan, "An Islamic Perspective," 114.

91 Wadud, *Qurʾān and Woman*, 83f.; al-Hibri, "An Introduction to Muslim Women's Rights," 68; Barlas, *Believing Women in Islam*, 190f.

Qur'ān permits marriage with up to four women, there is a strong tendency to negate this.[92] Interestingly enough, the interpretation of verse Q 4:3 contains key ideas that have already been expressed in a well-known tradition by ʿĀ'isha bt. Abī Bakr.[93] The lack of reference to this hadith is probably due to the overall rejection of the hadith material in many works on gender-egalitarian exegesis.[94]

4.3 *Wa-ḍribūhunna* (Q 4:34)

The second part of verse Q 4:34 continues to be the subject of controversy. In addition to the question of what is meant by the female *nushūz*, the controversy also concerns the recommended measures, formulated as imperatives, that should follow a female *nushūz*[95] (*fa-ʿiẓūhunna wa-hjurūhunna fī-l-maḍājiʿi wa-ḍribūhunna*). The implications of *ḍaraba* (to beat) in Q 4:34 were discussed extensively in premodern exegesis.[96] Through a semantic analysis of the word in the Qur'ān, Amina Wadud tries to find a new meaning for *ḍaraba*. According to her, the verb *ḍaraba* can also mean "setting an example," as in *ḍaraba llāhu mathalan*[97] ("Allah gives or sets as an example [. . .]."[98]). She states that it is also used in the sense of "turning away" and "setting off on the path" (*ḍaraba fī l-arḍ*[99]). Wadud does not formulate it explicitly at this point, but suggests that *wa-ḍribūhunna* does not mean "hit them," but "turn away from them" or "set an example."[100] She then emphasizes that even if *ḍaraba* means "to beat," in view of the Qur'ānic condemnation of female infanticide and the violence used by companions of the prophet against women, it should be read not as a permission, but rather as a

92 Al-Hibri, "A Study of Islamic Herstory: Or How Did We Ever Get Into This Mess?," *Women's Studies International Forum* 5, no. 2 (1982): 216; Barlas, *Believing Women in Islam*, 191.

93 Aṭ-Ṭabarī, *Jāmiʿ al-bayān*, 6: 358–60.

94 Hassan, "An Islamic Perspective," 94f.; Barlas, *Believing Women in Islam*, 42–50.

95 For the debate about the term *nushūz* see Vardit Rispler-Chaim, "Nušūz between Medieval and Contemporary Islamic Law: The Human Rights Aspect," *AR* 39, no. 3 (1992), 315–27; Wadud, *Qur'ān and Woman*, 74f.; Ali, *Sexual Ethics*, 120f.; Ayesha S. Chaudhry, "Marital Discord in Qur'ānic Exegesis: A Lexical Analysis of Husbandly and Wifely *Nushūz* in Q. 4:34 and Q. 4:128," in *The Meaning of the Word. Lexicology and Qur'ānic Exegesis*, ed. Stephen R. Burge (Oxford: Oxford University Press, 2015), 325–49; Seker, *Koran und Gender*, 78f.

96 Bauer, *Room for Interpretation*, 158–61; Chaudhry, *Wife-Beating in Pre-Modern Islamic Tradition*, 262–306.

97 This usage can be found in the verses Q 2:26, 18:32, 18:45, 30:28, 39:29, 43:57.

98 Wadud, *Qur'ān and Woman*, 76.

99 This usage can be found in the verses Q 2:273, 3:156, 4:101, 5:106, 73:20.

100 Wadud, *Qur'ān and Woman*, 76. A similar interpretation can be found in Mehmet Okuyan, "Kadına Yönelik Şiddete Kur'an'ın Bakışı," (The Qur'anic View on Violence Against Women) *Ondokuz Mayıs Üniversitesi İlahiyat Fakültesi Dergisi* (*Journal of the Faculty of Divinity of Ondokuz Mayıs University*) 23 (2007): 122. Okuyan finds numerous references to the variety of meanings of the verb *ḍaraba* in the Qur'ān.

prohibition and restriction of violence against women.[101] This reading requires further discussion, among other things because the interpretation "turn away from them/set an example" is already implied in the second imperative *wa-hjurūhunna fī l-maḍājiʿ* ("stay away from them in bed"). This raises the question of why the statement from the second imperative should be repeated in the third imperative. In the academic context, the discussion about the meaning of *wa-ḍribūhunna* is an ongoing endeavor and attempts are constantly being made to find new ways of interpreting it. The number of new publications on the topic is proportionally high.[102]

5 Responses and Criticism

Contemporary efforts to read the Qurʾān on a gender-egalitarian basis have become a constantly evolving endeavor, with this discussion being carried forward by a younger generation of feminist academics. In the academic world, feminist and gender-egalitarian Qurʾānic hermeneutics have received great recognition and reception, but also critique. On certain points, gender-egalitarian exegesis has attested methodical and hermeneutical inconsistency and selectiveness, and has been criticized for producing anachronistic and apologetic interpretations.

The obvious paradox of reading the Qurʾān as a timeless text with a universal message and at the same time applying a historicizing approach to selected hierarchical verses[103] can be rated as a narrowed application of the *double movement*-method of Fazlur Rahman. Behind this method, however, lies an overarching hermeneutical problem that is not reflected theoretically at any point: the historicizing of verses that seem incompatible with the principle of justice, while at the same time pursuing a universalis-

101 Wadud, *Qurʾān and Woman*, 76; Barlas, *Believing Women in Islam*, 188.

102 Maysam J. al-Faruqi, "Women's Self-Identity in the Qurʾān and Islamic Law," in *Windows of Faith: Muslim Women Scholar-Activists in North America*, ed. Gisela Webb (Syracuse: Syracuse University Press, 2000), 72–101; Hadia Mubarak, "Breaking the Interpretive Monopoly. A Re-Examination of Verse 4:34," *Hawwa* 2, no. 3 (2004): 261–89; Laleh Bakhtiar, "*The Sublime Quran*: The Misinterpretation of Chapter 4 Verse 34," *European Journal of Women's Studies* 18, no. 4 (2011): 431–39; Shahbaz Ahmad Cheema, "The Concept of Qawama: A Study of Interpretive Tensions," *Hawwa* 11, no. 2–3 (2014): 235–51; Nimet Seker, "Ermahnt sie, meidet sie im Bett und schlagt sie! Zur Frage der Geschlechtergewalt in Q 4:34," in *Gewalt in den Heiligen Schriften von Islam und Christentum*, ed. Hamideh Mohagheghi, Klaus von Stosch (Paderborn: Schönigh, 2014), 135–43; Nafiseh Ghafournia, "Towards a New Interpretation of Quran 4:34," *Hawwa* 15, no. 3 (2017): 279–92; Hadia Mubarak, "Change Through Continuity: A Case Study of Q. 4:34 in Ibn ʿĀshūr's *al-Taḥrīr wa'l-tanwīr*," *JQS* 20, no. 1 (2018): 1–27; Nevin Reda, "The Qur'an and Domestic Violence: An Islamic Feminist, Spiritually Integrative Reading of Verse 4:34," *International Journal of Practical Theology* 23, no. 2 (2019): 257–73.

103 Ali, *Sexual Ethics and Islam*, 153; Rhouni, *Secular and Islamic Feminist Critiques*, 14; Karen Bauer, "In Defense of Historical-Critical Analysis of the Qurʾān," *Journal of Feminist Studies in Religion* 32, no. 2 (2016): 127f.

tic approach. In addition, the lack of a theoretical discussion on equality and justice as categories of analysis is criticized.[104] In this way, contemporary ideas of equality and justice are believed to be congruent with the Qur'ānic construction of justice.[105]

With regard to methodology – contrary to the claims to contextualize the text with the help of historical sources – no form of source criticism and text criticism is being consistently applied. Rather, the employed method of historical contextualization can be summed up as making generalized assumptions about the historical context of Qur'ānic revelation.[106]

Another point of criticism is the truth claim of egalitarian readings, while they at the same time state that the exegetical tradition has so far misinterpreted the Qur'ān. Feminist academics have questioned this claim and appealed to the authors to view their interpretations as equally subjective constructions and individual exegetical choices.[107]

6 Conclusion

With this refined level of discussion, there are many indications that Qur'ānic hermeneutics from a female perspective still offers numerous opportunities for further development.

Approaches to the interpretation of the Qur'ān by and for women have significantly influenced academic and non-academic engagements with the Qur'ān since the mid-twentieth century. These approaches not only address questions specific to women but also contribute to broader discussions on exegetical and hermeneutical matters concerning the Qur'ān. Today, a diverse array of research topics and methodological approaches characterizes this field.

For a brief period, academic discourse on gender issues in the Qur'ān and gender-egalitarian exegesis was thought to have reached its conclusion. It appeared that all pertinent points had been addressed, leading to repetitive discussions. However, this perception has been challenged by recent research contributions from the 2020s and ongoing scholarly efforts in the field.[108] Additionally, the methodological advancements in addressing these issues indicate a dynamic evolution of thought. Importantly, numerous female exegetes from Arabic- and Turkish-speaking regions, hitherto unknown, await discovery by researchers.

---

104 Hidayatullah, *Feminist Edges of the Qur'ān*, 130f.

105 Ali, *Sexual Ethics and Islam*, 154f.; Rhouni, *Secular and Islamic Feminist Critiques*, 266.

106 Seker, *Koran und Gender*, 117–45.

107 Ali, *Sexual Ethics and Islam*, 153.

108 E.g. Celene Ibrahim, *Women and Gender in the Qur'an* (Oxford: Oxford University Press, 2020); Hadia Mubarak, *Rebellious Wives, Neglectful Husbands. Controversies in Modern Qur'anic Commentaries* (Oxford: Oxford University Press, 2022).

7 Bibliography

7.1 Sources

Abdel Haleem, Muhammed A.S. *The Qur'ān*. Oxford: Oxford University Press, 2008. Reprinted with corrections 2010.

ʿAbd ar-Raḥmān, ʿĀʾisha. "Al-Marʾa al-muslima ams wa-l-yawm." In *al-Islām, al-yawm wa-ghadan*, edited by Ibrāhīm Salāma et al. Cairo: Dār Iḥyāʾ al-Kutub al-ʿArabiyya, 1960: 67–94.

ʿAbd ar-Raḥmān, ʿĀʾisha. *Al-Mafhūm al-islāmī li-taḥrīr al-marʾa*. Umm Durmān: Jāmiʿat Umm Durmān, 1966–67.

ʿAbd ar-Raḥmān, ʿĀʾisha. *Maqāl fī l-insān. Dirāsa qurʾāniyya*. Cairo: Dār al-Maʿārif, 1969².

ʿAbd ar-Raḥmān, ʿĀʾisha. *At-Tafsīr al-bayānī li-l-Qurʾān al-karīm*. 2 vols. Cairo: Dār al-Maʿārif, 1990⁷.

ʿAbdul Rahman, ʿAisha. "Islam and the New Woman," trans. Anthony Calderbank. *Alif: Journal of Comparative Poetics* 19 (1999): 194–202.

Abdul Rahman, Aisha. "The Islamic Conception of Women's Liberation," trans. Nazih Khater. *Al-Raida* 125 (Spring 2009): 37–43.

Abdullah, Ayşa Zeynep. *İndirilme Sırasına Göre Yüce Kur'an-ı Kerim ve Meali*. Istanbul: Hermes, 2019.

Ali, Kecia. *Sexual Ethics and Islam. Feminist Reflections on Qur'an, Hadith, and Jurisprudence*. London: Oneworld, 2008.

Amīn, Nuṣrat Bīgum. *Makhzan al-ʿirfān dar tafsīr-i Qur'ān*. Isfahan: Kitābfurūsh-i Saqāfī, 1956.

Amīn, Nuṣrat Bīgum. *Kitāb ravish-i khūshbakhtī wa-tavṣiyyah bih khāharān-i imānī*. Isfahan: Saghāfī, 1968–69.

Arpaguş, Hatice. "The Position of Woman in the Creation. A Qur'ānic Perspective." In *Muslima Theology. The Voices of Muslim Women Theologians*, edited by Marcia Hermansen, Ednan Aslan, and Elif Medeni. Frankfurt a.M.: Peter Lang, 2013: 113–32.

Al-Baghdādī, Ismāʿīl Bāshā (Bağdatlı İsmail Paşa). *Hadiyyat al-ʿārifīn asmāʾ al-muʾallafīn wa-āthār al-muṣannafīn*. 2 vols. Istanbul: Millî Eğitim Bakanlığı, 1951.

Bakhtiar, Laleh. "*The Sublime Quran*: The Misinterpretation of Chapter 4 Verse 34." *European Journal of Women's Studies* 18, no. 4 (2011): 431–39.

Barazangi, Nimat Hafez. *Woman's Identity and the Qur'ān. A New Reading*. Gainesville: University Press of Florida, 2004.

Barlas, Asma. *"Believing Women" in Islam: Unreading Patriarchal Interpretations of the Qur'ān*. Austin: University of Texas Press, 2002.

Çekmegil, Semra Kürün. *Okuyucu Tefsiri*. 13 vols. Malatya: Nida Dergisi Yayınları, 2006.

Cheema, Shahbaz Ahmad. "The Concept of Qawama: A Study of Interpretive Tensions." *Hawwa* 11, no. 2–3 (2014): 235–51.

Al-Faruqi, Maysam J. "Women's Self-Identity in the Qur'ān and Islamic Law." In *Windows of Faith: Muslim Women Scholar-Activists in North America*, edited by Gisela Webb. Syracuse: Syracuse University Press, 2000: 72–101.

Ghafournia, Nafiseh: "Towards a New Interpretation of Quran 4:34." *Hawwa* 15, no. 3 (2017): 279–92.

Al-Ghazālī, Zaynab. *Naẓarāt fī kitāb Allāh*. Cairo: Dār ash-Shurūq, 1994.

Hassan, Riffat. "An Islamic Perspective." In *Women, Religion, and Sexuality. Studies on the Impact of Religious Teachings on Women*, edited by Jeanne Becher. Geneva: WCC Publications, 1990: 93–128.

Hassan, Riffat. "Woman and Man's 'Fall'. A Qur'ānic Theological Perspective." In *Muslima Theology. The Voices of Muslim Women Theologians*, edited by Marcia Hermansen, Ednan Aslan, and Elif Medeni. Frankfurt a.M.: Peter Lang, 2013: 101–13.

Al-Hibri, Azizah. "A Study of Islamic Herstory: Or How Did We Ever Get Into This Mess?," *Women's Studies International Forum* 5, no. 2 (1982): 207–19.

Al-Hibri, Azizah. "Islam, Law and Custom: Redefining Muslim Women's Rights." *American University International Law Review* 12, no. 1 (1997): 1–44.

Al-Hibri, Azizah. "An Introduction to Muslim Women's Rights." In *Windows of Faith: Muslim Woman Scholar-Activists in North America*, edited by Gisela Webb. Syracuse: Syracuse University Press, 2000: 50–71.

Al-Hibri, Azizah: "Qur'ānic Foundations of Rights of Muslim Women in the 21st Century." In *Women in Indonesian Society: Access, Empowerment and Opportunity*, edited by M. Atho Mudzhar et al. Yogyakarta: Sunan Kalijaga Press, 2002: 3–26.

Al-Ḥusnī, ʿAbd al-Ḥayy b. Fakhr ad-Dīn. *Al-Iʿlām bi-man fī tārīkh al-hind min al-aʿlām (Nuzhat al-khawāṭir wa-bahjat al-masāmiʿ wa-n-nawāẓir)*. 8 vols. Beirut: Dār Ibn Ḥazm, 1999.

Ibrahim, Celene. *Women and Gender in the Qur'an*. Oxford: Oxford University Press, 2020.

Al-Kawwārī, Kāmila. *Tafsīr gharīb al-Qur'ān*. Beirut: Dār Ibn Ḥazm, 2008.

Lamptey, Jerusha Tanner. *Divine Words, Female Voices. Muslima Explorations in Comparative Feminist Theology*. New York: Oxford University Press, 2018.

Al-Māturīdī, Abū Manṣūr Muḥammad. *Ta'wīlāt ahl as-sunna. Tafsīr al-Māturīdī*. Ed. Majdī Bāsallūm, 10 vols. Beirut: Dār al-Kutub al-ʿIlmiyya, 2005.

Mubarak, Hadia. "Breaking the Interpretive Monopoly. A Re-Examination of Verse 4:34." *Hawwa* 2, no. 3 (2004): 261–89.

Mubarak, Hadia. *Rebellious Wives, Neglectful Husbands. Controversies in Modern Qur'anic Commentaries*. Oxford: Oxford University Press, 2022.

Okuyan, Mehmet: "Kadına Yönelik Şiddete Kur'an'ın Bakışı" (The Qur'ānic View on Violence Against Women). In *Ondokuz Mayıs Üniversitesi İlahiyat Fakültesi Dergisi* (*Journal of the Faculty of Divinity of Ondokuz Mayıs University*) 23 (2007): 94–133.

Ar-Rāzī, Fakhr ad-Dīn. *Tafsīr al-Fakhr ar-Rāzī al-mushtahir bi-t-tafsīr al-kabīr wa-mafātīḥ al-ghayb*. 32 vols. Beirut: Dār al-Fikr, 1981.

Reda, Nevin. "The Qur'an and Domestic Violence: An Islamic Feminist, Spiritually Integrative Reading of Verse 4:34." *International Journal of Practical Theology* 23, no. 2 (2019): 257–73.

Ṣabrī Hāshim, Nā'ila. *Al-Mubaṣṣir li-nūr al-Qur'ān*. 16 vols. Jerusalem: Maṭbaʿat ar-Risāla, 1998–2003.

Seker, Nimet. "Ermahnt sie, meidet sie im Bett und schlagt sie! Zur Frage der Geschlechtergewalt in Q 4:34." In *Gewalt in den Heiligen Schriften von Islam und Christentum*, edited by Hamideh Mohagheghi and Klaus von Stosch. Beiträge zur Komparativen Theologie 10. Paderborn: Schönigh, 2014: 135–43.

Seker, Nimet. *Koran und Gender. Exegetische und hermeneutische Studien zum Geschlechterverhältnis im Koran*. Hamburg: Editio Gryphus, 2020.

Ash-Shirbīnī, Fawqiyya. *Taysīr at-tafsīr*. 4 vols. Cairo: Maktabat al-Īmān, 2008.

Aṭ-Ṭabarī, Abū Jaʿfar Muḥammad b. Jarīr. *Jāmiʿ al-bayān ʿan ta'wīl āy al-Qur'ān. Tafsīr aṭ-Ṭabarī*. Ed. ʿAbd Allāh b. ʿAbd al-Muḥsin at-Turkī, 26 vols. Cairo: Dār Hijr, 2001.

Ath-Thaʿlabī an-Nīsābūrī, Abū Iṣḥāq Aḥmad. *Al-Kashf wa-l-bayān. Tafsīr ath-Thaʿlabī*. Ed. ʿAlī b. ʿĀshūr and Naẓīr as-Sāʿidī. Beirut: Dār Iḥyā' at-Turāth al-ʿArabī, 2001.

Wadud, Amina. *Qur'ān and Woman. Rereading the Sacred Text from a Woman's Perspective*. New York, Oxford: Oneworld, 1999[2].

Wadud, Amina. *Inside the Gender Jihad. Women's Reform in Islam*. London: Oneworld, 2006.

Wadud, Amina. "The Ethics of *Tawhid* and the Ethics of *Qiwamah*." In *Men in Charge? Rethinking Authority in Muslim Legal Tradition*, edited by Ziba Mir-Hosseini, Mulki Al-Sharmani, and Jana Rumminger. Oxford: Oneworld, 2015: 256–74.

Yasdıman, Necla. *Kur'an Tahlili*. 7 vols. Izmir: Cemre, 2006.

7.2 Secondary Literature

Abboud, Hosn. "The Arab Women Discourses on Feminism and Islam: Fear of an Oxymoron." *Al Raida* 148–49 (2015/2016): 34–52.

Bauer, Karen. "Room for Interpretation. Qur'ānic Exegesis and Gender." Phd diss., Princeton University, 2008.

Bauer, Karen. "'The Male Is Not Like the Female' (Q 3:36): The Question of Gender Egalitarianism in the Qur'ān." *Religion Compass* 3, no. 4 (2009): 637–54.

Bauer, Karen. "In Defense of Historical-Critical Analysis of the Qur'ān." *Journal of Feminist Studies in Religion* 32, no. 2 (2016): 127–28.

Brown, Daniel. *Rethinking Tradition in Modern Islamic Thought.* Cambridge: Cambridge University Press, 1999.

Chaudhry, Ayesha S. "Wife-Beating in Pre-Modern Islamic Tradition. An Inter-Disciplinary Study of Hadith, Qur'ānic Exegesis and Islamic Jurisprudence." PhD diss., New York University, 2009.

Chaudhry, Ayesha S. "Marital Discord in Qur'ānic Exegesis: A Lexical Analysis of Husbandly and Wifely *Nushūz* in Q. 4:34 and Q. 4:128." In *The Meaning of the Word. Lexicology and Qur'ānic Exegesis*, edited by Stephen R. Burge. Oxford: Oxford University Press, 2015: 325–49.

Geissinger, Aisha. *Gender and Muslim Constructions of Exegetical Authority. A Rereading of the Classical Genre of Qur'ān Commentary.* Boston/Leiden: Brill, 2015.

Haddad, Yvonne Y. "Eve: Islamic Image of Woman." *Women's Studies International Forum*, 5, no. 2 (1982): 135–44.

Hidayatullah, Aysha A. *Feminist Edges of the Qur'ān.* Oxford: Oxford University Press, 2014.

Künkler, Mirjam and Roja Fazaeli. "The Life of Two Mujtahidahs. Female Religious Authority in Twentieth-Century Iran." In *Women, Leadership and Mosques. Changes in Contemporary Islamic Authority*, edited by Hilary E. Kalmbach and Masooda Bano. Leiden: Brill 2012: 127–60.

Marcotte, Roxanne D. "Bint al-Shāṭi' on Women's Emancipation." In *Coming to Terms with the Qur'ān. A Volume in Honour of Professor Issa Boullata*, edited by Khaleel Mohammed and Andrew Rippin. North Haledon: Islamic Publications International, 2007: 179–208.

Mubarak, Hadia. "Change Through Continuity: A Case Study of Q. 4:34 in Ibn ʿĀshūr's *al-Taḥrīr wa'l-tanwīr.*" *JQS* 20, no. 1 (2018): 1–27.

Musa, Aisha Y. "The Qur'ānists." *Religion Compass* 4, no. 1 (2010): 12–21.

Naguib, Shuruq. "Bint al-Shāṭi's Approach to *tafsīr*: An Egyptian Exegete's Journey from Hermeneutics to Humanity." *JQS* 17, no. 1 (2015): 45–84.

Pink, Johanna. "Striving for a New Exegesis of the Qur'ān." In *The Oxford Handbook of Islamic Theology*, edited by Sabine Schmidtke. Oxford: Oxford University Press, 2016: 787–88.

Rahbari, Ladan. "Women's Ijtihad and Lady Amin's Islamic Ethics on Womanhood and Motherhood." *Religions* 11, no. 2 (2020): 88.

Rhouni, Raja. *Secular and Islamic Feminist Critiques in the Work of Fatima Mernissi.* Leiden/Boston: Brill, 2010.

Rispler-Chaim, Vardit. "Nušūz between Medieval and Contemporary Islamic Law: The Human Rights Aspect." *AR* 39, no. 3 (1992): 315–27.

Roded, Ruth. "Bint Al-Shati's *Wives of the Prophet.* Feminist or Feminine?." *British Journal of Middle Eastern Studies* 33, no. 1 (2006): 51–66.

Rutner, Maryam. "Religious Authority, Gendered Recognition, and Instrumentalization of Nusrat Amin in Life and after Death." *Journal of Middle East Women's Studies* 11, no. 1 (2015): 24–41.

Saleh, Walid A. "Contemporary *Tafsīr*: The Rise of Scriptural Theology." *The Oxford Handbook of Qur'ānic Studies*, edited by Mustafa Shah and Muhammad Abdel Haleem. Oxford: Oxford University Press, 2020: 693–703.

Wahyudi, Yudian. "The Slogan 'Back to the Qur'ān and the Sunna': A Comparative Study of the Responses of Ḥasan Ḥanafī, Muḥammad ʿĀbid al-Jābirī and Nurcholish Madjid." PhD diss., McGill University Montreal, 2002.

Joas Wagemakers

Fundamentalist Qur'ānic Hermeneutics

1 Introduction

If one googles the term "fundamentalism," originally coined to label certain American Christians in 1920/1338,[1] one definition one will find is "a form of a religion, especially Islam or Protestant Christianity, that upholds belief in the strict, literal interpretation of scripture."[2] This definition suggests a uniform approach to – in the case of Islam – the Qur'ān that is belied by the much more diverse reality of Islamic fundamentalism, as this chapter will show. A broader definition of the phenomenon of fundamentalism can be found in the volumes edited by Marty and Appleby as part of their "Fundamentalism Project": "[a] strategy, or set of strategies, by which beleaguered believers attempt to preserve their distinctive identity as a people or group. Feeling this identity to be at risk, fundamentalists fortify it by a selective retrieval of doctrines, beliefs, and practices from a sacred past."[3]

This definition stresses the idea that fundamentalists consider themselves to be under some sort of threat ("beleaguered believers," "at risk") against which they need to guard themselves ("preserve their distinctive identity," "fortify"). With regard to Islamic fundamentalists, it is important to realize that there is no agreement on what kind of threat believers are under (or what should be done in defense). Shepard, whose points of reference include famous fundamentalist scholars like the Pakistani Abū l-Aʿlā Mawdūdī (1903–1979/1321–1399), the Egyptian Sayyid Quṭb (1906–1966/1324–1386) and the Iranian Ruhollah Khomeini (1902–1989/1320–1409), describes fundamentalism – or "radical Islamism," as he labels it – as a trend that embraces progress through Islam. This is done by viewing Islam as applicable to "all aspects of social as well as personal life"[4] and, in response to both Westernization and the lack of Islam in Muslim societies, urging the application of the sharia (Islamic law) in public life.[5] This description of fundamentalism is thus based on the perception of what one might call a *political* threat. Indeed, Choueiri defines fundamentalism as "a certain intellectual stance that claims to derive political principles from a timeless, divine

1 Michael Cook, *Ancient Religions, Modern Politics: The Islamic Case in Comparative Perspective* (Princeton, NJ: Princeton University Press, 2014), 371–72.

2 See Lexico, "fundamentalism", accessed on January 13, 2020, www.lexico.com/en/definition/fundamentalism.

3 Martin E. Marty and R. Scott Appleby, "Introduction," in *Accounting for Fundamentalisms: The Dynamic Character of Movements*, ed. Martin E. Marty and R. Scott Appleby (Chicago/London: The University of Chicago Press, 1994), 1.

4 William E. Shepard, "Islam and Ideology: Towards a Typology," *IJMES* 19, no. 3 (1987): 314.

5 Ibid., 315. See also Andrew Rippin, *Muslims – Their Religious Beliefs and Practices*, vol. 2, *The Contemporary Period* (London/New York, 1993), 39–40, who draws heavily on Shepard.

https://doi.org/10.1515/9783111320069-022

text."[6] As such, this way of approaching fundamentalism is in line with other terms used to describe this phenomenon, such as "political Islam" and "Islamism".

In contrast to the political threat discerned by Shepard, Cook, in his treatment of Islamic fundamentalism, describes how some Muslims – like the original American Christian fundamentalists – perceive what may be labelled as a *doctrinal* threat from outsiders to the "purity" of their beliefs. He sees two ways of dealing with this: flexibly and inflexibly. While the former typifies modernists, the latter can be applied by holding on to a religious tradition as it was handed down throughout the ages, which Cook refers to as "conservatism," or by focusing on the source of one's religion instead, which he labels "fundamentalism." Such fundamentalists "identify one component of their religious tradition as its foundation while the rest is superstructure; [. . .] locate authority in the foundation rather than the superstructure; and [. . .] take the authority of the foundation seriously in a substantial way."[7]

Both approaches to fundamentalism, whether focused on a political or a doctrinal threat, are relevant for this chapter. Although fundamentalism is a broad phenomenon, I limit myself to two types of twentieth- and twenty-first-century Islamic fundamentalists, namely the Muslim Brotherhood and (Jihādī-)Salafīs. As we will see below, the Qur'ānic hermeneutics of the former can be seen as showing the belief in a political threat; the way the latter explain the Qur'ān exhibits the perception of a doctrinal threat. I will first deal with the Muslim Brotherhood and (Jihādī-)Salafism and some of the major hermeneutical works they rely on. I then analyze the aims, methods and results of their engagement with the Qur'ānic text, before ending with a conclusion. This way, it becomes clear that Islamic fundamentalists do not uniformly adhere to a single "strict, literalist interpretation of scripture" and that, in fact, even "literalist interpretation[s]" are not always alike.

2 Major Muslim Brotherhood and (Jihādī-)Salafī Works

Before mentioning some of the major works used by Muslim Brothers and (Jihādī-)Salafīs, we need to focus briefly on what both groups of fundamentalists believe in and stand for. As we will see, their historical background and development helps explain what type of threat they discern (political or doctrinal) and how this is reflected in their method of interpreting the Qur'ānic text.

6 Youssef M. Choueiri, *Islamic Fundamentalism: The Story of Islamist Movements* (London/New York: Continuum, 2010[3]), 7.
7 Cook, *Ancient*, 373.

2.1 Threats as Perceived by the Muslim Brotherhood and (Jihādī-) Salafīs

The Society of the Muslim Brotherhood (*Jamā'at al-Ikhwān al-Muslimīn*) was founded in Egypt by the school teacher Ḥasan al-Bannā in 1928/1346. Through his outreach activities and his growing number of followers, this lay organization spread throughout Egypt in the 1930s and 1940s and, in later years, also to other countries in the Middle East and beyond. Al-Bannā's message was a simple and populist one, claiming that "Islam is the solution" (*al-Islām huwa al-ḥall*).[8] This vague slogan betrayed a lack of doctrinal specificity, which was underlined by al-Bannā's own words that the Muslim Brotherhood was "a Salafī call (*da'wa Salafiyya*) [. . .], a Sunni way [. . .], a Sufi truth [. . .], a political organization [. . .], a sports group [. . .], a league of cultural knowledge [. . .], an economic company [. . .] and a social idea."[9]

The early Muslim Brotherhood thus espoused a broad ideology that could accommodate different Islamic points of view. As such, the threat the organization saw was not a doctrinal one. Instead, its members staunchly opposed British colonialism and even contributed to Arab efforts to "liberate" Palestine in the 1940s.[10] They also believed the problems in their own society could be solved by applying the sharia at the state level, a goal they have long tried to work towards through participation in Egypt's political system, as well as those of other Muslim countries. The early Brotherhood's perceived threat is thus a political one formed by both outside forces as well as those in the country itself.[11]

Salafīs, by contrast, perceive the danger threatening them quite differently. The word *salaf* ("predecessors," often used to refer to the first three generations of Muslims) as an Islamic religious term seems to have been used in the phrase *madhhab as-salaf* (the doctrine of the predecessors) from the tenth century onward. This phrase

8 For more on the early Muslim Brotherhood in Egypt, see Brynjar Lia, *The Society of the Muslim Brothers in Egypt: The Rise of an Islamic Mass Movement, 1928– 1942* (Reading, UK: Ithaca Press, 1998); Richard P. Mitchell, *The Society of the Muslim Brothers* (Oxford, etc.: Oxford University Press, 1969. Reprinted 1993).

9 Ḥasan al-Bannā, *Majmū'at rasā'il al-imām ash-shahīd Ḥasan al-Bannā* (N.p.: Dār at-Tawzī' wa-n-Nashr al-Islāmiyya, 1992), 122–23.

10 Abd al-Fattah Muhammad El-Awaisi, *The Muslim Brothers and the Palestine Question: 1928–1947* (New York: I.B. Tauris, 1998), 135–99.

11 A more recent account of the Brotherhood's relationship with the West is Martyn Frampton, *The Muslim Brotherhood and the West: A History of Enmity and Engagement* (Cambridge, MA/London: Belknap/Harvard University Press, 2018). The organization's approach towards their own political regimes has greatly evolved, but is broadly aimed at trying to establish a civil Islamic state. See, for example, Chris Harnisch and Quinn Mecham, "Democratic Ideology in Islamist Opposition? The Muslim Brotherhood's 'Civil State'," *MES* 45, no. 2 (2009): 189–205; Annette Ranko, *The Muslim Brotherhood and Its Quest for Hegemony in Egypt: State Discourse and Islamist Counter-Discourse* (Wiesbaden: Springer, 2015); Bruce K. Rutherford, *Egypt After Mubarak: Liberalism, Islam, and Democracy in the Arab World* (Princeton, NJ: Princeton University Press, 2008).

referred to a fideist reading of Scripture that subjugated reason (*ʿaql*) to reliance on the transmitted text of the Qurʾān (*naql*), particularly with regard to the important theological issue of God's names and attributes (*asmāʾ Allāh wa-ṣifātuhu*). In mediaeval times, a "*salafī*"[12] was thus someone who espoused these specific theological views.[13] This approach was taken in opposition to rationalism, speculative theology (*kalām*) and the negation (*taʿṭīl*), distortion (*taḥrīf*), anthropomorphic explanation (*tashbīh*) or metaphorical interpretation (*taʾwīl*) of the texts advocated by other groups and trends.[14]

This fideist theological approach still forms what may be called the core of modern-day Salafism, which took shape in the twentieth century, but several other elements have often accompanied this way of reading scripture among both medieval *salafīs* as well as adherents to today's Salafī trend. Among them is the rejection of the following of legal precedents (*taqlīd*) of the schools of Islamic law (*madhāhib*, sing. *madhhab*) in favor of independent interpretation of the Qurʾān (*ijtihād*), which scholars revered by today's Salafīs, such as Taqī d-Dīn Aḥmad b. Taymiyya (1263–1328/661–728) and especially Ibn Qayyim al-Jawziyya (1292–1350/691–751), advocated.[15] They also include a skepticism of (certain forms of) Sufism,[16] which can also be discerned – though perhaps in a more nuanced way – in the writings of the same medieval *salafī* scholars.[17] Other elements in today's Salafism include an emphasis on in-group communal solidarity (and a rejection of outsiders and their beliefs), particularly through the concept of *al-walāʾ wa-l-barāʾ* (loyalty and disavowal),[18] which – in a rudimentary form – was also advocated by Ibn Taymiyya[19]

---

12 I use the italicized, uncapitalized "*salafī*" to refer to the term's medieval meaning and the Roman, capitalized "Salafī" to denote the modern-day adherents to Salafism.

13 Henri Lauzière, *The Making of Salafism: Islamic Reform in the Twentieth Century* (New York: Columbia University Press, 2016), 28–29.

14 Nader El-Bizri, "God: Essence and Attributes," in *The Cambridge Companion to Classical Islamic Theology*, ed. Tim Winter (Cambridge, etc.: Cambridge University Press, 2008), 124–27.

15 Bernard Haykel, "On the Nature of Salafi Thought and Action," in *Global Salafism: Islam's New Religious Movement*, ed. Roel Meijer (London: Hurst & Co., 2009), 43–44.

16 Joas Wagemakers, "Why Salafis Have Anti-Sufi Attitutes," *Oasis Newsletter* 6 (2017): https://www.oasiscenter.eu/en/why-salafis-have-anti-sufi-attitudes (accessed 6 February 2024).

17 Ovamir Anjum, "Sufism without Mysticism? Ibn Qayyim al-Ǧawziyyah's Objectives in *Madāriǧ al-Sālikīn*," *OM* 90, no. 1 (2010): 173–75; Arjan Post, "A Glimpse of Sufism from the Circle of Ibn Taymiyya," *Journal of Sufi Studies* 5 (2016): 159–63.

18 Sabine Damir-Geilsdorf, Mira Menzfeld and Yasmina Hedider, "Interpretations of al-walaʾ wa-l-baraʾ in Everyday Lives of Salafis in Germany," *Religions* 10, no. 2 (2019): 1–18; Uriya Shavit, "Can Muslims Befriend Non-Muslims? Debating *al-Walaʾ wa-l-Baraʾ* (Loyalty and Disavowal) in Theory and Practice," *Islam and Christian-Muslim Relations* 25, no. 1 (2014): 67–88; Joas Wagemakers, "Framing the 'Threat to Islam': *Al-Walaʾ wa-l-Baraʾ* in Salafi Discourse," *ASQ* 30, no. 4 (2008): 1–22.

19 Joas Wagemakers, "The Transformation of a Radical Concept: *Al-Walaʾ wa-l-Baraʾ* in the Ideology of Abu Muhammad al-Maqdisi," in *Global Salafism: Islam's New Religious Movement*, ed. Roel Meijer (London: Hurst & Co., 2009), 85–87.

and Ibn Qayyim al-Jawziyya,[20] but has been developed further by Wahhabi[21] scholars since the eighteenth century.[22] A final element in today's Salafism that also has deep roots in various pious branches of Islam is the focus on ritual purity, which includes strictly following the Prophet Muḥammad and the *salaf* in their manner of prayer, dress, speech and behavior.[23] From these different elements of Salafism, we cannot just conclude that Salafis claim to emulate the *salaf* as strictly and in as many spheres of life as possible, but also that the threat they guard themselves against is clearly a doctrinal one.

Despite Salafis' perception of a doctrinal threat, they are not all alike in this respect. Apart from politically quietist Salafis, who do perceive the threat they need to counter as only doctrinal in nature, there are also political Salafis and, more importantly for this chapter, Jihādī-Salafis. Both have directly or indirectly been influenced by more politicized groups, such as the Muslim Brotherhood, and believe that – apart from a doctrinal threat – there is also a political threat in the form of rulers who are unwilling to apply the sharia. Political Salafis therefore employ a clearly political discourse and engage in various types of activism, while Jihādī-Salafis advocate the overthrow of regimes in the Muslim world through jihad.[24] As such, Jihādī-Salafis, organizationally represented by groups like al-Qāʿida and the Islamic State (IS), combine the two types of threat discerned in this chapter by perceiving a doctrinal threat to the purity of their beliefs, but also a political one to fully Islamic rule, represented by the supposedly apostate leaders of Muslim countries and Western powers involved in the Muslim world.

2.2 Major Muslim Brotherhood and (Jihādī-)Salafī Works of Qur'ānic Hermeneutics

The Muslim Brotherhood, precisely because it is a layperson's organization that does not deal much with doctrinal issues, is less focused on Qur'ānic hermeneutics than other groups and movements and has produced relatively little in the way of works systematically explaining scripture. It has, however, been influenced by major works of Qur'ānic hermeneutics by several scholars who are somehow related to the Muslim

20 Joas Wagemakers, "Revisiting Wiktorowicz: Categorising and Defining the Branches of Salafism," in *Salafism After the Arab Awakening: Contending with People's Power*, eds. Francesco Cavatorta and Fabio Merone (London: Hurst & Co., 2016), 20.

21 I define Wahhabism as the Najdī (i.e., Central Arabian) version of Salafism. See Joas Wagemakers, *Salafism in Jordan: Political Islam in a Quietist Community* (Cambridge, etc.: Cambridge University Press, 2016), 33–34.

22 Joas Wagemakers, "The Enduring Legacy of the Second Saudi State: Quietist and Radical Wahhabi Contestation of *al-Walāʾ wa-l-Barāʾ*," *IJMES* 44, no. 1 (2012): 93–110.

23 Samir Amghar, *Le salafisme d'aujourd'hui: Mouvements sectaires en Occident* (Paris: Michalon Éditions, 2011), 139–49; Wagemakers, *Salafism*, 170–74.

24 For more on the exact differences between the different branches of Salafism, see Wagemakers, "Revisiting."

Brotherhood. The first of these is the reformist scholar Muḥammad Rashīd Riḍā (1865–1935/1282–1354), whose work of Qur'ānic exegesis (*tafsīr*) was originally published in his famous journal *Al-Manār* (The Lighthouse), which strongly influenced Ḥasan al-Bannā and of which the latter became the editor after Riḍā's death,[25] and was later published as a twelve-volume exegetical work.[26]

A second work that influenced the Muslim Brotherhood is *Tafhīm al-Qur'ān* (Instructing the Qur'ān),[27] by the aforementioned Pakistani scholar Mawdūdī. His thinking – particularly his politicized interpretations of terms such as *jāhiliyya* (the pre-Islamic age of ignorance) and *ḥākimiyya* (sovereignty) – had an important impact on some Brothers imprisoned in Egypt, especially Sayyid Quṭb.[28] The latter himself, perhaps the leading thinker within Muslim Brotherhood ranks in the 1960s, also wrote an important work of Qur'ānic hermeneutics entitled *Fī ẓilāl al-Qur'ān* (In the Shade of the Qur'ān),[29] which remains one of the few significant works of exegesis produced by a member of the Muslim Brotherhood.

For today's Salafīs, major classical books of *tafsīr* include aṭ-Ṭabarī's (839–923/224–311) work on the Qur'ān,[30] which is said to have been the one recommended most by Ibn Taymiyya.[31] The work that modern-day Salafīs value above all others, however, is most probably the exegesis by Ibn Kathīr (1301–1373/701–775), who – like Ibn Qayyim al-Jawziyya – was another one of Ibn Taymiyya's students who followed in his footsteps and authored *Tafsīr al-Qur'ān al-ʿaẓīm* (Exegesis of the Great Qur'ān).[32] Other, later and less prominent examples of *salafī* exegesis include the one by Jamāl ad-Dīn al-Qāsimī (1866–1914/1283–1332).[33] Modern-day Salafī works of *tafsīr* include those by ʿAbd ar-Raḥmān b. Nāṣir as-Saʿdī (1889–1956/1307–1376),[34] Muḥammad al-Amīn ash-Shinqīṭī

25 Mehdi Sajid, "A Reappraisal of the Role of Muḥibb al-Dīn al-Khaṭīb and the YMMA in the Rise of the Muslim Brotherhood," *Islam and Christian-Muslim Relations* 29, no. 2 (2018): 193–95.

26 Muḥammad Rashīd Riḍā, *Tafsīr al-Qur'ān al-ḥakīm* (Cairo: Dār al-Manār, 1934–35).

27 Abū l-Aʿlā Mawdūdī, *Towards Understanding the Qur'ān*, ed. and transl. Zafar Ishaq Ansari (Leicester, UK: Islamic Foundation, 1988).

28 Adnan A. Musallam, *From Secularism to Jihad: Sayyid Qutb and the Foundations of Radical Islamism* (Westport, CT/London: Praeger, 2005), 151f.

29 Sayyid Quṭb, *Fī ẓilāl al-Qur'ān* (Cairo: Dār ash-Shurūq, 1980/81).

30 Ibn Jarīr aṭ-Ṭabarī, *Jāmiʿ al-bayān fī ta'wīl al-Qur'ān* (Beirut: Dār al-Kutub al-ʿIlmiyya, 2005).

31 Izza Rohman, "Salafi Tafsirs: Textualist and Authoritarian?," *Journal of Qur'ān and Ḥadīth Studies* 1, no. 2 (2012): 200.

32 ʿImād ad-Dīn Abū l-Fidā' Ismāʿīl b. Kathīr al-Qurashī d-Dimashqī, *Tafsīr al-Qur'ān al-ʿaẓīm*, ed. Aḥmad ʿAbd as-Salām az-Zuʿbī (Beirut: Sharikat Dār al-Arqam b. Abī l-Arqam, n.d.).

33 Muḥammad Jamāl ad-Dīn al-Qāsimī, *Tafsīr al-Qāsimī l-musammā Maḥāsīn at-ta'wīl*, ed. Muḥammad Fu'ād ʿAbd al-Bāqī (Cairo: n.p., n.d.).

34 ʿAbd ar-Raḥmān b. Nāṣir as-Saʿdī, *Taysīr al-karīm ar-raḥmān fī tafsīr al-kalām al-mannān* (Cairo: Dār al-Ḥadīth, 2005).

(1907–1973/1325–1393),[35] Muḥammad b. Ṣāliḥ al-ʿUthaymīn (1925–2001/1344–1422)[36] and Abū Bakr Jābir al-Jazāʾirī (1921–2018/1340–1440).[37]

3 Aims, Methods and Results of Fundamentalist Qur'ānic Hermeneutics

Muslim Brothers and (Jihādī-)Salafīs, as different types of fundamentalists, have translated the threats they perceive – either political or doctrinal (or both) – into specific aims and methods of reading and explaining the Qur'ān. In the case of the Muslim Brotherhood, this has resulted in a political reading of Scripture, while Salafīs have geared their *tafsīr* towards maintaining the doctrinal "purity" of their beliefs, and Jihādī-Salafīs have combined the two.

3.1 The Muslim Brotherhood's Political Threat in Qur'ānic Hermeneutics

As al-Bannā once stated, the Muslim Brotherhood believes that "the rulings and teachings of Islam are all-encompassing (*shāmila*), organizing people's affairs in the world and in the hereafter." This includes various spheres of life, among them "religion and state (*dīn wa-dawla*)."[38] Such an approach to Islam, which explicitly encompasses politics and strives to translate religion into a political platform, has remained part of the Brotherhood's approach ever since, including in Quṭb's *tafsīr*, which, as Wild writes, "fights two major enemies: the powerful but spiritually bankrupt anti-Islamic West and the apostate Muslim societies and individuals, who are no better than pagan societies."[39]

This aim of using the Qur'ān to make the case that Islam is confronted with a political threat – be it from outside Western forces or from those inside the Muslim world – has its methodological roots in modern readings of Muslim Scripture. While pre-modern *tafsīr* seldom strayed outside the realm of religion as such, this changed with the advent of nineteenth–century Western colonialism, when Muslims were con-

35 Muḥammad al-Amīn ash-Shinqīṭī, *Aḍwāʾ al-bayān fī īḍāḥ al-Qurʾān bi-l-Qurʾān* (Beirut: Dār Iḥyāʾ at-Turāth al-ʿArabī, n.d.).
36 Muḥammad b. Ṣāliḥ al-ʿUthaymīn, *Tafsīr al-Qurʾān al-karīm*, ed. Fahd b. Nāṣir as-Sulaymān (Riyadh: Dār ath-Thurayyā, 2002).
37 Abū Bakr Jābir al-Jazāʾirī, *Aysar at-tafāsīr li-kalām al-ʿalī l-kabīr* (Medina: Maktabat al-ʿUlūm wa-l-Ḥikam, 2003).
38 Al-Bannā, *Majmūʿat*, 119.
39 Stefan Wild, "Political Interpretation of the Qur'ān," in *The Cambridge Companion to the Qur'ān*, ed. Jane Dammen McAuliffe (Cambridge, etc.: Cambridge University Press, 2006), 282.

fronted with large-scale non-Muslim rule of their lands for the first time. This led to new methods of explaining the Qur'ān, including treating *tafsīr* as related to contemporary issues in society. One of the scholars who was in the forefront of applying this method was Rashīd Riḍā.[40] As mentioned above, al-Bannā had been strongly influenced by Riḍā's journal *Al-Manār*, but the Brotherhood as a whole had also been affected by him, especially his Qur'ānic readings on Islamic statehood and how to apply these in modern times.[41]

This new method resulted in a highly politicized reading of Qur'ānic texts by Brotherhood leaders. With regard to the Western threat, for example, the founder of the group's Syrian branch, Muṣṭafā s-Sibāʿī (1915–1964/1333–1384), focused on Western treatment of women, which he clearly presents as having been one of inequality, submission and dependence throughout history.[42] As-Sibāʿī portrays Islam, however, as a religion that provides women with not just more rights than they had in pre-Islamic times, but also with equality in humanity, social affairs and legal rights[43] – although he does acknowledge differences between men and women[44] – with copious references to the Qur'ān.[45]

Al-Bannā sought to counter the political threat he believed Muslims were confronted with by following Riḍā and presenting a framework for an Islamic state on the basis of the Qur'ān, prefacing his views on "Government in Islam" by quoting Q 5:49: "And judge between them (*uḥkum baynahum*) according to what God has sent down, and do not follow their caprices (*ahwā'ahum*), and beware of them lest they tempt thee away from any of what God has sent down to thee [. . .]."[46] He also incorporates the will of the people into this system by pointing to the concept of consultation (*shūrā*) with reference to Q 3:159 ("take counsel with them (*wa-shāwirhum*) in the affair") and Q 42:38 ("their affair being counsel between them (*shūrā baynahum*)").[47]

Quṭb, finally, perpetuated this method of relating the Qur'ān to contemporary issues, relying on classical scholars to only a very limited extent.[48] This appears to re-

40 Ibid., 276–80.

41 Marion Boulby, *The Muslim Brotherhood and the Kings of Jordan, 1945–1993* (Atlanta, GA: Scholars Press, 1999), 126–27.

42 Muṣṭafā s-Sibāʿī, *Al-Mar'a bayna al-fiqh wa-l-qānūn* (Beirut: Dār al-Warrāq, 1999), 13–20.

43 Ibid., 26–27.

44 Ibid., 27–35.

45 Ibid., 23–26.

46 Al-Bannā', *Majmūʿat*, 317. These and subsequent translations from the Qur'ān are taken from A.J. Arberry, ed. and trans., *The Koran Interpreted* (New York: Touchstone, 1955. Reprint 1996]), although I do not use his verse-numbering and instead rely on the more common verse-numbering found in other Qur'āns.

47 Al-Bannā, *Majmūʿat*, 319.

48 Neguin Yavari, "*Tafsīr* and the Mythology of Islamic Fundamentalism," in *Tafsīr and Islamic Intellectual History: Exploring the Boundaries of a Genre*, eds. Andreas Görke and Johanna Pink (Oxford, etc.: Oxford University Press, 2014), 293.

flect not only his and others' new methodology of Qur'ānic hermeneutics, but also that neither Quṭb nor Brotherhood leaders like al-Bannā were religious scholars. To Quṭb, Muslims lived in a situation of *jāhiliyya*, which had resulted from Western influence on a cultural, political and economic level. To counter this, Quṭb believed an Islamic society should be brought about by a vanguard (*ʿuṣba* or *ṭalīʿa*) of true believers to lead Muslims back to the right path, which should be followed by setting up an Islamic state.[49] The latter centered on his interpretation of Q 5:44–47, which states that "[. . .] whoso judges not (*wa-man lam yaḥkum*) according to what God has sent down – they are the unbelievers (*al-kāfirūn*) [. . .] the evildoers (*aẓ-ẓālimūn*) [. . . and] the ungodly (*al-fāsiqūn*)." Quṭb explicitly ties these verses to (political) authority (*sulṭān*) in his *tafsīr*.[50] He thus seems to interpret the word "judges" as "rules" or "governs", giving it a more political flavor, thereby providing the basis for the excommunication (*takfīr*) of and jihad against supposedly apostate Muslim rulers who fail to apply Islamic law, as well as for his idea of a sharia-based Islamic state.[51]

The idea of seeing Western influence and "apostate" Muslim rule as a political threat that should be countered by a politicized reading of the Qur'ān, particularly in Quṭb's work, has had a major influence on other fundamentalist scholars, ranging from Shiite revolutionaries in Iran and the Lebanese Hizbullah to the Palestinian Hamas in the West Bank and the Gaza Strip.[52] More specifically, such views have had an important impact on the Syrian Muslim Brother Saʿīd Ḥawwā (1935–1989/1354–1410), whose own *tafsīr*[53] was strongly influenced by Quṭb's hermeneutical work, although it has been less popular than his.[54] The same applies to Muḥammad Abū Fāris (1938–2015/1357–1436), a prolific author and member of the Jordanian Muslim Brotherhood. The writings of al-Bannā and Quṭb had a major impact on him,[55] including with regard to the political reading of the Qur'ān by the latter, who he sometimes refers to simply as "the author of (*ṣāḥib*) *aẓ-Ẓilāl*", in reference to Quṭb's *tafsīr*.[56] Finally, although less clear than his impact on other Muslim Brothers, Quṭb's radical reading of the Qur'ān has also influenced Jihādī-Salafīs, to whom we must now turn.

---

49 Ibid., 296–300.

50 Quṭb, *Ẓilāl*, 2: 889.

51 Yvonne Yazbeck Haddad, "The Qur'anic Justification for an Islamic Revolution: The View of Sayyid Quṭb," *MEJ* 37, no. 1 (1983): 27f.; Wild, "Political," 283.

52 Wild, "Political," 282f.

53 Saʿīd Ḥawwā, *Al-Asās fī t-tafsīr* (Cairo: Dār as-Salām li-ṭ-Ṭabāʿa wa-n-Nashr wa-t-Tawzīʿ wa-t-Tarjama, 1985).

54 Johanna Pink, *Sunnitischer Tasfīr in der modernen islamischen Welt: Akademische Traditionen, Popularisierung und nationalstaatliche Interessen* (Leiden: Brill, 2011), 105–7.

55 Muḥammad ʿAbd al-Qādir Abū Fāris, *Manhaj at-taghyīr ʿinda ash-shahīdayn Ḥasan al-Bannā wa-Sayyid Quṭb* (Amman: Dār ʿAmmār, 1997).

56 See, for example, Muḥammad ʿAbd al-Qādir Abū Fāris, *an-Niẓām as-siyāsī fī l-Islām* (Amman: Dār al-Furqān li-n-Nashr wa-t-Tawzīʿ, 1980), 34, 90.

3.2 (Jihādī-)Salafīs' Doctrinal Threat in Qur'ānic Hermeneutics

Unlike the Muslim Brotherhood, Salafīs mostly seem to aim their *tafsīr* at countering doctrinal threats. The groundwork for this was laid by Ibn Taymiyya, perhaps the most important *salafī* scholar to today's Salafīs, although he never authored a full *tafsīr* himself.[57] He discerns what he considers faulty readings of the Qur'ān and concentrates on those who superimpose their own meanings on the text, irrespective of what the words actually say, by either depriving it of its true meanings or by adding meanings it does not have.[58] Ibn Taymiyya attributes such readings to "groups among the people of religious innovations (*ahl al-bida'*) who believe in a doctrine (*madhhaban*) that differs from the truth that is adhered to by [the followers of] the middle [position], who do not agree on an error (*ḍalāla*), like the predecessors of the community (*salaf al-umma*) and its imams."[59] He names the groups that espouse such allegedly deviant Qur'ānic readings as "the Khawārij, the Rawāfiḍ, the Jahmiyya, the Mu'tazila, the Qadariyya, the Murji'a and others" and criticizes the Mu'tazila in particular for deriving faulty doctrines from such readings.[60]

Ibn Taymiyya's method of countering such doctrinal threats is rooted in his belief of where things went wrong. He claims that the Prophet Muḥammad himself explained and interpreted the Qur'ān to his followers and that serious differences in Qur'ānic readings did not occur until after the first three generations of Muslims. Because of the influence of reason, rather than the transmission of tradition, the faulty readings mentioned above crept into *tafsīr*.[61] Ibn Taymiyya therefore argues that reading the Qur'ān should be done through the prism of those Muslim generations who did not engage in mistaken readings themselves: the *salaf*. By excluding other sources of exegesis, like reason and considered opinion (*ra'y*), and explaining Qur'ānic texts solely through other Qur'ānic texts and through the interpretations of the Prophet Muḥammad and the *salaf*, the true meaning of Scripture could be reached.[62] This method of *tafsīr* was partly adopted by the aforementioned Ibn Kathīr[63] and also seems to have largely been embraced by twentieth-century Salafī exegetes.[64]

---

57 Younus Y. Mirza, "Ibn Taymiyya as Exegete: Moses' Father-in-Law and the Messengers in *Sūrat Yā Sīn*," *JQS* 19, no. 1 (2017): 40–41; Walid A. Saleh, "Ibn Taymiyya and the Rise of Radical Hermeneutics: An Analysis of *An Introduction to the Foundations of Qur'anic Exegesis*," in *Ibn Taymiyya and His Times*, eds. Yossef Rapoport and Shahab Ahmed (Oxford, etc.: Oxford University Press, 2010), 148–54.

58 Taqī d-Dīn Aḥmad b. 'Abd al-Ḥalīm b. Taymiyya, *Muqaddima fī uṣūl at-tafsīr*, ed. 'Adnān Zarzūr (Kuwait: Dār al-Qur'ān al-Karīm, 1971. Reprint 1972), 81–82. For an extensive analysis of this document, see Saleh, "Ibn Taymiyya," 128–48. The passage referred to here is dealt with on page 140.

59 Ibn Taymiyya, *Muqaddima*, 82.

60 Ibid., 82–85.

61 Saleh, "Ibn Taymiyya," 126–39.

62 Ibid., 148–50.

63 Ibid., 152–54.

64 Rohman, "Salafi," 203.

Ibn Taymiyya's legacy in this respect has resulted in a modern-day Salafī hermeneutical method that is characterized by several aspects. Firstly, and as mentioned before, the influence of reason is excluded or strictly limited and subordinated to following revelation;[65] secondly, the Qur'ānic text is seen as fixed in meaning, uninfluenced by the role of the interpreter, clear in and of itself and, as such, uniform in its true interpretation;[66] thirdly, the text of the Qur'ān is read in a decontextualized way, ignoring the historical, cultural and economic background of Scripture,[67] although there is some room for taking the reasons of revelation (*asbāb an-nuzūl*) into account;[68] fourthly, Salafīs have a tendency to explain the Qur'ān in a segmented way, rather than focusing on the text's internal coherence, while they nevertheless generalize and universalize the hermeneutical conclusions based on this approach;[69] and, finally, they downplay or de-emphasize the underlying purposes (*maqāṣid*) or general interest (*maṣlaḥa*) found in Qur'ānic discourse.[70]

The result of this hermeneutical method is a literal, textualist and doctrinally "pure" *tafsīr* that avoids the supposed mistakes of other exegetes. This is expressed, for example, in the way the names and attributes of God are viewed. As mentioned above, this issue was basically what decided whether or not one was a *salafī* in medieval times and has been adopted by modern-day Salafīs. Thus, while other scholars view references to God's attributes metaphorically or anthropomorphically, Salafīs ranging from the Saudi pro-regime Ibn Ṣāliḥ al-ʿUthaymīn[71] to the Jordanian Jihādī-Salafī Abū Muḥammad al-Maqdisī[72] read them literally, but – with reference to Q 42:11 ("like Him there is naught") – without ascribing any definitive form to them through extra-textual means such as speculation.[73]

Ironically, the decontextualized and generalized nature of Salafī Qur'ānic hermeneutics results in readings that are perhaps particularly susceptible to interpretations that ascribe meanings to words that they may not have had in the original context, which is precisely what Ibn Taymiyya was so strongly against. Ibn Kathīr's reading of Q 60:1–9, for example, decontextualizes these verses by essentially ignoring the Muslims' seventh-century conflict with polytheistic Meccans that plays in the background

65 Adis Duderija, "Neo-Traditional Salafi Qur'an-Sunnah Hermeneutic and the Construction of a Normative *Muslimah* Image," *Hawwa* 5, nos. 2–3 (2007): 313–15; Duderija, "Neo-Traditionalist Salafi Qur'an-Sunna Hermeneutics and Its Interpretational Implications," *Religion Compass* 5, no. 7 (2011): 322.

66 Duderija, "Neo-Traditional," 300–4; Duderija, "Neo-Traditionalist," 319–21; Rohman, "Salafi," 202–5.

67 Duderija, "Neo-Traditional," 305f.

68 Ibid., 306; Duderija, "Neo-Traditionalist," 321; Rohman, "Salafi," 206, 207.

69 Duderija, "Neo-Traditional," 307f.; Duderija, "Neo-Traditionalist," 322f.

70 Ibid., "Neo-Traditional," 309f.; Duderija, "Neo-Traditionalist," 323.

71 Mohammad Gharaibeh, "Zur Glaubenslehre des Salafismus," in *Salafismus: Auf der Suche nach dem wahren Islam*, eds. Behnam T. Said and Hazim Fouad (Freiburg, etc.: Herder, 2014), 113–23.

72 Abū Muḥammad al-Maqdisī, "Hādhihi ʿaqīdatunā," accessed January 24, 2020, www.ilmway.com/site/maqdis/MS_35456.html. In the older Word-version (no longer available), this topic was discussed on page 5.

73 Rohman, "Salafi," 203.

here and generalizes the enemies mentioned in this sura in such a way that they include a much broader group of unbelievers.[74] This, in turn, allowed later Salafī scholars to apply such verses much more broadly and, in the case of al-Maqdisī, even facilitates his application of this sura to political rulers of the Muslim world.[75]

The latter example shows that Jihādī-Salafīs use their *tafsīr* to counter a doctrinal threat *and* a political threat, thereby combining the two forms of fundamentalism described earlier. Al-Maqdisī's writings also show that his Qur'ānic hermeneutics, though generally decontextualized, sometimes do make use of *asbāb an-nuzūl*. This is particularly evident in his reading of a passage that was also extremely important to Quṭb: Q 5:44. The Syrian a-political Salafī scholar Muḥammad Nāṣir ad-Dīn al-Albānī (1914–1999/1332–1420) points out that, in contrast to Quṭb's reading, the "unbelievers" in this verse who do not judge "according to what God has sent down" should not be equated with modern-day Muslim rulers. The reason for this is, he writes, that this verse was originally revealed with regard to Jews, who presumably rejected Islamic laws in their hearts because of their different religion, which cannot be said of today's leaders of the Muslim world. Al-Maqdisī agrees about the *asbāb an-nuzūl* of this verse, but states that the text does not mention what Jews believed in their hearts. If that means that this is no precondition for *takfīr*, then why not also apply it to Muslim rulers? If, on the other hand, the Jews' unbelief could be deduced from their actions, should the same not apply to leaders of Muslim countries, too?[76]

Decontextualization therefore does not mean completely ignoring context altogether, nor does reading texts literally, which Jihādī-Salafīs are often said to do,[77] mean that they agree in their Qur'ānic hermeneutics. This is the case with regard to the question of to what extent people can be held responsible for their sins if no messenger has been sent to warn them (based on verses such as Q 17:15, 20:134 and 28:47), which ultimately leads to diverging views among Jihādī-Salafīs on the fate of people who vote in democratic elections.[78] It becomes even clearer with regard to Q 47:4 ("When you meet the unbelievers, smite their necks (*fa-ḍarba r-riqāb*)"). Ḥusayn b. Maḥmūd, a writer affiliated with IS, states that this verse commands Muslims to decapitate their unbelieving

74 Joas Wagemakers, "Defining the Enemy: Abū Muḥammad al-Maqdisī's Radical Reading of Sūrat al-Mumtaḥana," *WI* 48, nos. 3–4 (2008): 361–63.

75 Ibid., 364–69.

76 Wagemakers, "Transformation," 100f.

77 Rosalind W. Gwynne, "Usama bin Ladin, the Qur'an and Jihad," *Religion* 36 (2006): 61–90; Donald Holbrook, "Using the Qur'an to Justify Terrorist Violence: Analysing Selective Application of the Qur'an in English-Language Militant Islamist Discourse," *Perspectives on Terrorism* 4, no. 3 (2010): 15–28; Amritha Venkatraman, "Religious Basis for Islamic Terrorism: The Quran and Its Interpretations," *Studies in Conflict & Terrorism* 30, no. 3 (2007): 229–48.

78 Joas Wagemakers, "An Inquiry into Ignorance: A Jihādī-Salafī Debate on *Jahl* as an Obstacle to *Takfīr*," in *The Transmission and Dynamics of the Textual Sources of Islam: Essays in Honour of Harald Motzki*, eds. Nicolet Boekhoff-van der Voort, Kees Versteegh and Joas Wagemakers (Leiden: Brill, 2011), 305–24.

enemies, including the American journalist James Foley, who was beheaded by IS in 2014. Yet fellow Jihādī-Salafī Abū Maḥmūd al-Filasṭīnī, who sympathizes with al-Qāʿida, disagrees. Using the Qur'ān itself to contextualize this verse, he points to Q 21:107 ("We have not sent thee, save as a mercy (*raḥmatan*) unto all beings") as well as to the *basmala* ("In the Name of God, the Merciful, the Compassionate"), which precedes almost every sura, to stress that mercy is central to the Qur'ān. As such, mercy should also be applied in killing someone. While beheading someone with a sharp sword is the least painful (and hence most merciful) way of doing so, al-Filasṭīnī claims that IS does precisely the opposite: it does not "smite" people's necks, but uses "slaughter" (*dhabḥ*) to kill them in a painful and merciless way, thereby deviating from the very verse it claims legitimizes this practice.[79]

Thus, while it is true that (Jihādī-)Salafīs generally employ a strongly fideist, textualist, decontextualized, segmented (but generalized) and literalist reading of the Qur'ān, this still leaves some room for differences and certainly does not always yield the same results. This is particularly the case with Jihādī-Salafīs, who – unlike a-political Salafīs – employ a fundamentalist reading of the Qur'ān that not only seeks to defend them from a doctrinal threat, but also from a political one. This leads them to stretch their generalization of segmentally read passages from Scripture even further, causing greater divergence, despite their literalist approach.

4 Conclusion

Fundamentalism may be seen as the strategy through which believers seek to counter a perceived threat to their identity by selectively retaining or emphasizing aspects of a past they see as sacred. In Islam, we generally find two different perceptions of this threat: a political and a doctrinal one. While the ideologically broad and vague early Muslim Brotherhood clearly discerned a political threat coming from the West and their own regimes, Salafīs were (and are) far more interested in theological matters and therefore see their own beliefs as challenged by a doctrinal threat. Jihādī-Salafīs, a radical, revolutionary and often violent sub-set of Salafism, see the threat as both political and doctrinal.

Fundamentalists' perception of this threat is translated into how they read the Qur'ān: Muslim Brothers explain Scripture through the prism of (or at least taking into account) contemporary societal and political issues, resulting in a reading that is critical towards Western influence and "un-Islamic" rule in Muslim countries; Salafīs, by contrast, interpret the Qur'ān in a fideist, textualist, decontextualized, segmented (yet nevertheless generalized) and literalist way so as to block out all supposed doctrinal deviations, thereby trying to retain the "purity" of their beliefs. Jihādī-Salafīs

79 Joas Wagemakers, "Salafi Source Readings Between Al-Qaeda and IS," *Oasis* 12, no. 23 (2016): 59–62.

combine both methods of interpreting the Qur'ān. Unlike what might have been expected, fundamentalists in Islam thus do not have a single, "literal" way of reading, explaining and applying Scripture, but actually do so in divergent ways. In fact, even those (Jihādī-)Salafīs for whom literalism is, indeed, important sometimes differ greatly in their *tafsīr*. This shows that even among a group that is often treated as homogeneous, diverse Qur'ānic hermeneutics clearly exists.

5 Bibliography

5.1 Sources

Abū Fāris, Muḥammad ʿAbd al-Qādir. *An-Niẓām as-siyāsī fī l-islām*. Amman: Dār al-Furqān li-n-Nashr wa-t-Tawzīʿ, 1980.

Abū Fāris, Muḥammad ʿAbd al-Qādir. *Manhaj at-taghyīr ʿinda ash-shahīdayn Ḥasan al-Bannā wa-Sayyid Quṭb*. Amman: Dār ʿAmmār, 1997.

Al-Bannā, Ḥasan. *Majmūʿat rasāʾil al-imām ash-shahīd Ḥasan al-Bannā*. N.p.: Dār at-Tawzīʿ wa-n-Nashr al-Islāmiyya, 1992.

Ḥawwā, Saʿīd. *Al-Asās fī t-tafsīr*. Cairo: Dār as-Salām li-ṭ-Ṭabāʿa wa-n-Nashr wa-t-Tawzīʿ wa-t-Tarjama, 1985.

Ibn Kathīr al-Qurashī ad-Dimashqī, ʿImād ad-Dīn Abū l-Fidāʾ Ismāʿīl. *Tafsīr al-Qurʾān al-ʿaẓīm*. Ed. Aḥmad ʿAbd as-Salām az-Zuʿbī. 4 volumes. Beirut: Sharikat Dār al-Arqam b. abī l-Arqam, n.d.

Ibn Ṣāliḥ al-ʿUthaymīn, Muḥammad. *Tafsīr al-Qurʾān al-karīm*. Ed. Fahd b. Nāṣir as-Sulaymān. Riyadh: Dār ath-Thurayyā, 2002.

Ibn Taymiyya, Taqī d-Dīn Aḥmad b. ʿAbd al-Ḥalīm. *Muqaddima fī uṣūl at-tafsīr*. Ed. ʿAdnān Zarzūr. Kuwait: Dār al-Qurʾān al-Karīm, 1971. Reprinted 1972.

Al-Jazāʾirī, Abū Bakr Jābir. *Aysar at-tafāsīr li-kalām al-ʿalī l-kabīr*. Medina: Maktabat al-ʿUlūm wa-l-Ḥikam, 2003.

Al-Maqdisī, Abū Muḥammad. "Hādhihi ʿaqīdatunā." (1997) Accessed January 24, 2020. www.ilmway.com/site/maqdis/MS_35456.html.

Mawdūdī, Abū l-Aʿlā. *Towards Understanding the Qurʾān*, ed. and trans. Zafar Ishaq Ansari. Leicester, UK: Islamic Foundation, 1988–.

Al-Qāsimī, Muḥammad Jamāl ad-Dīn. *Tafsīr al-Qāsimī l-musammā Maḥāsīn at-taʾwīl*. Ed. Muḥammad Fuʾād ʿAbd al-Bāqī. 9 volumes. Cairo: No publisher, n.d.

Quṭb, Sayyid. *Fī ẓilāl al-Qurʾān*. Cairo: Dār ash-Shurūq, 1980/1981.

Riḍā, Muḥammad Rashīd. *Tafsīr al-Qurʾān al-ḥakīm*. 12 vols. Cairo: Dār al-Manār, 1934–1935.

As-Saʿdī, ʿAbd ar-Raḥmān b. Nāṣir. *Taysīr al-karīm ar-raḥmān fī tafsīr al-kalām al-mannān*. Cairo: Dār al-Ḥadīth, 2005.

Ash-Shinqīṭī, Muḥammad al-Amīn. *Adwāʾ al-bayān fī īḍāḥ al-Qurʾān bi-l-Qurʾān*. Beirut: Dār Iḥyāʾ at-Turāth al-ʿArabī, n.d.

As-Sibāʿī, Muṣṭafā. *Al-Marʾa bayna al-fiqh wa-l-qānūn*. Beirut: Dār al-Warrāq, 1999.

Aṭ-Ṭabarī, Ibn Jarīr. *Jāmiʿ al-bayān fī taʾwīl al-Qurʾān*. Beirut: Dār al-Kutub al-ʿIlmiyya, 2005.

The Koran Interpreted, ed. and trans. A.J. Arberry. New York: Touchstone, 1955. Reprint 1996.

5.2 Secondary Literature

Amghar, Samir. *Le salafisme d'aujourd'hui: Mouvements sectaires en Occident*. Paris: Michalon Éditions, 2011.

Anjum, Ovamir. "Sufism without Mysticism? Ibn Qayyim al-Ǧawziyyah's Objectives in *Madāriǧ al-Sālikīn*." *OM* 90, no. 1 (2010): 153–80.

Boulby, Marion. *The Muslim Brotherhood and the Kings of Jordan, 1945–1993*. Atlanta, GA: Scholars Press, 1999.

Choueiri, Youssef M. *Islamic Fundamentalism: The Story of Islamist Movements*. London/New York: Continuum, 2010^3.

Cook, Michael. *Ancient Religions, Modern Politics: The Islamic Case in Comparative Perspective*. Princeton, NJ: Princeton University Press, 2014.

Damir-Geilsdorf, Sabine, Mira Menzfeld and Yasmina Hedider. "Interpretations of al-wala' wa-l-bara' in Everyday Lives of Salafis in Germany." *Religions* 10, no. 2 (2019): 1–18.

Duderija, Adis. "Neo-Traditional Salafi Qur'an-Sunnah Hermeneutic and the Construction of a Normative *Muslimah* Image." *Hawwa* 5, nos. 2–3 (2007): 289–323.

Duderija, Adis. "Neo-Traditionalist Salafi Qur'an-Sunna Hermeneutics and Its Interpretational Implications." *Religion Compass* 5, no. 7 (2011): 313–25.

El-Awaisi, Abd al-Fattah Muhammad. *The Muslim Brothers and the Palestine Question: 1928–1947*. New York: I.B. Tauris, 1998.

El-Bizri, Nader. "God: Essence and Attributes." In *The Cambridge Companion to Classical Islamic Theology*, edited by Tim Winter. Cambridge, etc.: Cambridge University Press, 2008: 121–40.

Frampton, Martyn. *The Muslim Brotherhood and the West: A History of Enmity and Engagement*. Cambridge, MA/London: Belknap/Harvard University Press, 2018.

Gharaibeh, Mohammad. "Zur Glaubenslehre des Salafismus." In *Salafismus: Auf der Suche nach dem wahren Islam*, edited by Behnam T. Said and Hazim Fouad. Freiburg, etc.: Herder, 2014: 106–31.

Gwynne, Rosalind W. "Usama bin Ladin, the Qur'an and Jihad." *Religion* 36 (2006): 61–90.

Harnisch, Chris and Quinn Mecham. "Democratic Ideology in Islamist Opposition? The Muslim Brotherhood's 'Civil State'." *MES* 45, no. 2 (2009): 189–205.

Haykel, Bernard. "On the Nature of Salafi Thought and Action." In *Global Salafism: Islam's New Religious Movement*, edited by Roel Meijer. London: Hurst & Co., 2009: 33–57.

Holbrook, Donald. "Using the Qur'an to Justify Terrorist Violence: Analysing Selective Application of the Qur'an in English-Language Militant Islamist Discourse." *Perspectives on Terrorism* 4, no. 3 (2010): 15–28.

Lauzière, Henri. *The Making of Salafism: Islamic Reform in the Twentieth Century*. New York: Columbia University Press, 2016.

Lexico. "fundamentalism." Accessed on January 13, 2020. www.lexico.com/en/definition/fundamentalism.

Lia, Brynjar. *The Society of the Muslim Brothers in Egypt: The Rise of an Islamic Mass Movement, 1928–1942*. Reading, UK: Ithaca Press, 1998.

Marty, Martin E. and R. Scott Appleby. "Introduction." In *Accounting for Fundamentalisms: The Dynamic Character of Movements*, edited by Martin E. Marty and R. Scott Appleby. Chicago/London: The University of Chicago Press, 1994: 1–9.

Mirza, Younus Y. "Ibn Taymiyya as Exegete: Moses' Father-in-Law and the Messengers in *Sūrat Yā Sīn*." *JQS* 19, no. 1 (2017): 39–71.

Mitchell, Richard P. *The Society of the Muslim Brothers*. Oxford, etc.: Oxford University Press, 1969. Reprinted 1993.

Musallam, Adnan A. *From Secularism to Jihad: Sayyid Qutb and the Foundations of Radical Islamism*. Westport, CT/London: Praeger, 2005.

Pink, Johanna. *Sunnitischer Tasfīr in der modernen islamischen Welt: Akademische Traditionen, Popularisierung und nationalstaatliche Interessen*. Leiden: Brill, 2011.

Post, Arjan. "A Glimpse of Sufism from the Circle of Ibn Taymiyya." *Journal of Sufi Studies* 5 (2016): 156–87.
Ranko, Annette. *The Muslim Brotherhood and Its Quest for Hegemony in Egypt: State Discourse and Islamist Counter-Discourse*. Wiesbaden: Springer, 2015.
Rippin, Andrew. *The Contemporary Period*. Vol. 2, *Muslims – Their Religious Beliefs and Practices*, 2 vols. London/New York, 1993.
Rohman, Izza. "Salafi Tafsirs: Textualist and Authoritarian?," *Journal of Qur'ān and Ḥadīth Studies* 1, no. 2 (2012): 197–213.
Rutherford, Bruce K. *Egypt After Mubarak: Liberalism, Islam, and Democracy in the Arab World*. Princeton, NJ: Princeton University Press, 2008.
Sajid, Mehdi. "A Reappraisal of the Role of Muḥibb al-Dīn al-Khaṭīb and the YMMA in the Rise of the Muslim Brotherhood." *Islam and Christian-Muslim Relations* 29, no. 2 (2018): 193–213.
Saleh, Walid A. "Ibn Taymiyya and the Rise of Radical Hermeneutics: An Analysis of *An Introduction to the Foundations of Qur'anic Exegesis*." In *Ibn Taymiyya and His Times*, edited by Yossef Rapoport and Shahab Ahmed. Oxford, etc.: Oxford University Press, 2010: 123–62.
Shavit, Uriya. "Can Muslims Befriend Non-Muslims? Debating *al-Wala' wa-l-Bara'* (Loyalty and Disavowal) in Theory and Practice." *Islam and Christian-Muslim Relations* 25, no. 1 (2014): 67–88.
Shepard, William E. "Islam and Ideology: Towards a Typology." *IJMES* 19, no. 3 (1987): 307–36.
Venkatraman, Amritha. "Religious Basis for Islamic Terrorism: The Quran and Its Interpretations." *Studies in Conflict & Terrorism* 30, no. 3 (2007): 229–48.
Wagemakers, Joas. "Defining the Enemy: Abū Muḥammad al-Maqdisī's Radical Reading of Sūrat al-Mumtaḥana." *WI* 48, nos. 3–4 (2008): 348–71.
Wagemakers, Joas. "Framing the 'Threat to Islam': *Al-Wala' wa-l-Bara'* in Salafi Discourse." *ASQ* 30, no. 4 (2008): 1–22.
Wagemakers, Joas. "The Transformation of a Radical Concept: *Al-Wala' wa-l-Bara'* in the Ideology of Abu Muhammad al-Maqdisi." In *Global Salafism: Islam's New Religious Movement*, edited by Roel Meijer. London: Hurst & Co., 2009: 81–106.
Wagemakers, Joas. "An Inquiry into Ignorance: A Jihādī-Salafī Debate on *Jahl* as an Obstacle to *Takfīr*." In *The Transmission and Dynamics of the Textual Sources of Islam: Essays in Honour of Harald Motzki*, edited by Nicolet Boekhoff-van der Voort, Kees Versteegh and Joas Wagemakers. Leiden: Brill, 2011: 301–27.
Wagemakers, Joas. "The Enduring Legacy of the Second Saudi State: Quietist and Radical Wahhabi Contestation of *al-Walā' wa-l-Barā'*." *IJMES* 44, no. 1 (2012): 93–110.
Wagemakers, Joas. "Revisiting Wiktorowicz: Categorising and Defining the Branches of Salafism." In *Salafism After the Arab Awakening: Contending with People's Power*, edited by Francesco Cavatorta and Fabio Merone. London: Hurst & Co., 2016: 7–24, 241–48.
Wagemakers, Joas. *Salafism in Jordan: Political Islam in a Quietist Community*. Cambridge, etc.: Cambridge University Press, 2016.
Wagemakers, Joas. "Salafi Source Readings Between Al-Qaeda and IS." *Oasis* 12, no. 23 (2016): 55–63.
Wagemakers, Joas. "Why Salafis Have Anti-Sufi Attitutes." *Oasis Newsletter* 6 (2017) https://www.oasiscenter.eu/en/why-salafis-have-anti-sufi-attitudes (accessed 6 February 2024).
Wild, Stefan. "Political Interpretation of the Qur'ān." In *The Cambridge Companion to the Qur'ān*, edited by Jane Dammen McAuliffe. Cambridge, etc.: Cambridge University Press, 2006: 273–89.
Yavari, Neguin. "*Tafsīr* and the Mythology of Islamic Fundamentalism." In *Tafsīr and Islamic Intellectual History: Exploring the Boundaries of a Genre*, edited by Andreas Görke and Johanna Pink. Oxford, etc.: Oxford University Press, 2014: 289–319.
Yazbeck Haddad, Yvonne. "The Qur'anic Justification for an Islamic Revolution: The View of Sayyid Quṭb." *MEJ* 37, no. 1 (1983): 14–29.

Rahel Fischbach

History and Transcendence in Recent Muslim Qur'ānic Hermeneutics

1 Introduction

The relation between the transcendent and contingent in the Qur'ān occupied Muslim dialectic theologians (*mutakallimūn*) early on. While their metaphysical arguments about the Qur'ān may seem dated or obscure today, related discussions have resumed in the face of post-Enlightenment epistemologies that bracket or dispense with the divine. Far from theoretical, such debates carry socio-political weight. Proponents of historical method often view it as crucial for opening the Qur'ān to new readings and adapting Islam to modernity – a conviction that has entered public and political discourse.[1] In contrast, Muslims have long contested historical approaches that upset the belief in the Qur'ān as the direct word of God.[2]

Critics mainly situated in post-colonial studies fear that historicizing non-Western religions implicates epistemic violence, i.e., the process by which Western forms of knowing preclude or annul local forms of knowledge.[3] Saba Mahmood even contests viewing "liberal moderate Muslims" who approach the Qur'ān historically as an "indigenous social formation." She argues that in aligning Islam with the Western ideal type of privatized religion, they are complicit in displacing those forms of being religious that conflict with the hegemonic secular worldview. Mahmood's argument rests on the insight that secularism is not just the separation of church and state. Rather, as an organizing principle of society and an ideology secularism aims "to shape religion

1 Government and non-government entities alike have promoted this idea. See Rahel Fischbach, "Beyond Normative Secularity: Politics of Interfaith Dialogue in Lebanon," in *The Interfaith Movement: Mobilising Religious Diversity in the 21st Century*, eds. John Fahy and Jan-Jonathan Bock (New York/London: Routledge 2019): 158–62; Saba Mahmood, "Secularism, Hermeneutics, and Empire: The Politics of Islamic Reformation," *Public Culture* 18.2 (2006): 323–25; Katharina Völker, *Quran and Reform: Rahman, Arkoun, Abu Zayd* (Frankfurt a.M.: Peter Lang 2017), 23–27; Rotraud Wielandt, *Offenbarung und Geschichte im Denken moderner Muslime* (Wiesbaden: F. Steiner, 1971), 2f.; Jürgen Habermas, "Die Dialektik der Säkularisierung," *Blätter für deutsche und internationale Politik* 4 (2008): 33–46.

2 Rotraud Wielandt, "Wurzeln der Schwierigkeit innerislamischen Gesprächs über neue hermeneutische Zugänge zum Koran," in *The Qur'an as Text*, ed. Stefan Wild (Leiden: Brill, 1996), 6.

3 See Gayatri Spivak, "Can the Subaltern Speak?," in *Marxism and the Interpretation of Culture*, eds. Cary Nelson and Lawrence Grossberg (Urbana, IL: University of Illinois Press, 1998), 280–83. Michel Foucault defines an "*episteme* as the 'apparatus' which makes possible the separation not of the true from the false, but of what may not be characterized as scientific." See Michel Foucault, *Power/Knowledge: Selected Interviews and Other Writings 1972–77*, trans. Colin Gordon et al. (New York: Pantheon, 1980), 82.

https://doi.org/10.1515/9783111320069-023

and religious subjectivities according to ideals and suppositions" that predicate a particular relation to history.[4]

Examining the ways by which recent Muslim thinkers mediate history and transcendence in their Qur'ān hermeneutics qualifies the claims of both those who view secular history as the only means to open the Qur'ān to polyvalence and rethink its revelation and those who view it as alien to Islam. On the one hand, Mahmood's caveat against epistemic violence corrects naïve convictions about cultural transfer, which is always embedded in social and political practices within a wider field of knowledge and power. In short, Western forms of knowing gained widespread legitimacy, because they were produced by groups in positions of power.[5] On the other hand, overdrawing the Western-indigenous dichotomy brushes over the diversity within Islam and the West, and leaves little room "for space of intervention emerging in the cultural interstices that introduces creative invention into existence."[6]

The main part of this article explores the theories of five contemporary thinkers about the revelation, ontology, and interpretation of the Qur'ān. They all consider the relation between history and the Qur'ān, aim to open it to new meanings and engage critically with the Islamic exegetical tradition. Yet, their trajectories for reaching these aims differ. Three of them, subsequently referred to as modernists,[7] advocate history-based hermeneutics as the *sine qua non* for overcoming the alleged stasis of modern Islamic thought, namely, Naṣr Ḥāmid Abū Zayd (d. 2010/1431), Abdelkarim Soroush (b. 1945/1365), and Wajīh Qānṣū. The other two, Shafīq Jarādī and Nayla Ṭabbāra, consciously seek to entail the divine in their concepts of history.

To appreciate the problem of linking historical hermeneutics to Muslim "progress," the following section clarifies the correlation between secularism, history, and imperialism. Section three scrutinizes the core features of the approaches mentioned above in relation to those held by pre-colonial scholars and by thinkers often labeled

4 Mahmood, "Secularity," 330–35, 343. Quote from p. 335. Recent scholarship demonstrates that all modern states, European ones included, enact some form of politics toward religion, producing various "secularisms." See James Beckford, *Social Theory and Religion* (Cambridge/New York: Cambridge University Press, 2003), 44.

5 James Blaut, *Colonizer's Model of the World: Geographical Diffusionism and Eurocentric History* (New York: Guilford Press, 1993).

6 Homi K. Bhabha, *The Location of Culture* (New York: Routledge, 1994), 12.

7 Leirvik Oddbjørn calls these thinkers *reformist Muslims*. See Leirvik Oddbjørn, "*Waḥy* and *tanzīl*: Modern Islamic approaches to divine inspiration, progressive revelation, and human text," *Studia Theologica* 69, no.2 (2015): 103, 110. For him, the term "refers to thinkers with a declared commitment to certain 'progressive' values such as rationalism, anti-authoritarianism, democracy, gender equality, and pluralism (103)." They are also "interested in finding ways to escape the revealed text's imprisonment in ahistorical categories (110)." However, to me, the term modernist implies certain epistemological presuppositions linked to secular historical concepts of time and space more clearly than the term reformist. There exist Muslim reform thinkers who would accept most of Leirvik's reformist commitments but reject other features of a modernist agenda, such as empiricism, materialist rationalism, an uncompromising commitment to natural (i.e., Newtonian) law, individualism, etc.

traditional or Islamist. I focus on three assumptions about Muslim scholars who pursue historical Qur'ān hermeneutics, namely: They 1) consider the Qur'ān a merely human product, 2) evacuate the transcendent from the human sphere, and 3) dismiss tradition.[8] Finally, I situate their hermeneutical convictions within their socio-political context to critique labeling Muslim scholars as Western, indigenous, or orthodox. Once we grant that history is also part of Muslim public and intellectual discourse yet remain open to concepts of history that function not primarily without God(s), the preconceived notions of modern/traditional, Islamic/Western, etc. become blurry. As we will see, the ways by which Muslim authors mediate the transcendent in their Qur'ān hermeneutics vary, as they grapple with different theological and political concerns.

2 History, Secularism, and Imperialism

Historical consciousness is regularly claimed as a hallmark of modernity – a measure of and vehicle for progress. Per this rationale, the formation of modern individuals presupposes historicizing their faith, past, and identity.[9] Neutralizing "dangerous passages" in scripture through historical critique forms part of secularism's origin myth, i.e., the narrative of how Europe escaped endless "religious wars" when the secular state stepped in to relegate religion to its "proper sphere" of faith and conscience. Though historically incorrect, this myth entered Western political philosophy as an important argument in favor of secularism. Viewed as self-critical, unbiased, rational, and modern, secularism came henceforth to be construed as religion's non-dogmatic opposite.[10] However, the mode of thinking the past in terms of secular history has a hard time tolerating other forms of temporality. Based on an idea of time as empty, homogeneous, and linear, historical discourse "speaks of things temporal and terrestrial in a human, fallible voice, while staking its claim to authority on rigorous critical practice."[11]

8 A fourth point can only be addressed in passing, namely, that they owe their "allegiance to the sovereign rule of the state rather than structures of traditional authority." See Charles Hirschkind, "Heresy or Hermeneutics: The Case of Nasr Hamid Abu Zayd," *Stanford Humanities Review* 5, no. 1 (1996): 473. Quote from Mahmood, "Secularism," 340.

9 See, e.g., Ninian Smart, "Editorial Introduction," in *Nineteenth Century Religious Thought in the West*, ed. Smart et al. (Cambridge: Cambridge University Press, 1985), 1: 11; Paul Ricoeur, *Memory, History, Forgetting* (London/Chicago: University of Chicago Press, 2006), 63, 165.

10 See William T. Cavanaugh, *The Myth of Religious Violence: Secular Ideologies and the Roots of Modern Conflict* (Oxford: Oxford University Press, 2009). See also Wendy Brown, "Introduction," in *Is Critique Secular?*, ed. Talal Asad et al. (New York: Fordham University Press, 2013), vii–viii.

11 Bruce Lincoln, "Theses on Method," *Method and Theory in the Study of Religion* 8 (1996): 225. For the idea of history as linear, homogeneous, empty time, see Walter Benjamin, "Theses on the Philosophy of History," in Benjamin, *Illuminations*, trans. Harry Zohn (New York: Fontana, 1982), 263–66.

In suspending the foundations of practices and texts in which humans are not the only intentional agents, these precepts collide with the idea of the Qur'ān as God's word. They can also subvert native epistemologies and non-secular ways of being in the world. As Ashis Nandy argues, history cannot "be dealt with from outside of history," since it aligns with a secular definition of reality that excludes other experiences of the "real."[12] As a "model of a higher, overarching language" history renders other forms of relating to the past (e.g., ritual) inconceivable, translating them into its secular vocabulary.[13] Modernity in this case denotes not just a certain period, in which all people share equally, but conveys a spatial and temporary hierarchy that posits Europe as the desired endpoint of humanity.

Linking modernity to historical thinking while claiming the attribute modern for Europe aided in marking "them" from "us" and lent imperial endeavors intellectual support; for example, by discrediting other religions based on their "faulty scriptures" or "naïve" beliefs. Identifying a-historical forms of thinking as irrational or childlike, liberal and Marxist thinkers alike argued that non-European people needed guidance from outside, i.e., colonialism.[14] In tandem with philology, history also aided colonial administrators in arrogating the socially embedded authority of clerics to legal codes and the state. Colonial philologists located the essence of religion in its written sources. Only historical-philological method, they insisted, could grant direct access to the authentic religious truths formerly obscured by clerics and popular culture. In claiming to understand a religion better than its local practitioners, they helped establish and maintain colonial control over native traditions and populations.[15]

Therefore, at first sight, Mahmood's argument that Muslim thinkers who pursue historical Qur'ān hermeneutics have succumbed to the hegemonic (Western) secular paradigm may look convincing.[16] European theories indeed impacted Muslim attempts to historicize the Qur'ān. However, viewing their theories as breaches of tradition incommensurable with an allegedly less Western orthodoxy erects a false dichotomy. We may first note that Mahmood's argument seems to imply the "non-European" as *prima facie* counterhegemonic. Yet, contests over philological power are not new. The texts we approach to grasp a tradition are always already products of such contests. Forms of knowing that elude textuality, exercised by those without philological power, remain

12 Ashis Nandy, *The Romance of the State and the Fate of Dissent in the Tropics* (New Delhi: Oxford University Press, 2003), 84–90.

13 Dipesh Chakrabarty, *Provincializing Europe: Postcolonial Thought and Historical Difference* (Princeton/London: Princeton University Press, 2000), 72–75.

14 See Rasiah S. Sugirtharajah, *Exploring Postcolonial Biblical Criticism: History, Method, Practice* (Chichester: Wiley-Blackwell, 2012), 39–44; Spivak, "Subaltern," 280–83; Chakrabarty, *Provincializing*, 5–8, 40; Uday Singh Mehta, *Liberalism and Empire: A Study in Nineteenth Century British Liberal Thought* (London/Chicago: University of Chicago Press, 1999), 81.

15 Siraj Ahmed, *Archeology of Babel: The Colonial Foundation of the Humanities* (Stanford: Stanford University Press, 2018), 72–95.

16 Mahmood, "Secularism," 343.

invisible. Although crucial in forming Muslim selves, they were never fixed in writing. Ironically, despite its promise to emancipate humanity from clerical authority, historical-philological method extended its reign by investing religious authority in standardized texts, which impacted religious traditions worldwide.[17]

Second, measuring a tradition's "indigenousness" by its distance to secular-historical practice repeats the orientalist logic, on which the modern-tradition binary rests. On the one hand, this view feeds into older convictions about cultures as self-contained enclosures, portraying pre-colonial, non-Western traditions as static and homogeneous. On the other, it reinforces the idea of the "European miracle." Accordingly, modern Western thought formed and was globally diffused with no impact from non-Europe in return.[18] This narrative ignores that modern European thought developed to a large degree in the colonial encounter and downplays pre-modern cross-cultural interplay.[19]

Third, pitting historical Qur'ān hermeneutics against "orthodox" Islam is problematic, if orthodox is meant to signal closer proximity to "right" or "traditional" Islam. As Talal Asad notes, rather than "a mere body of opinions," orthodoxy indicates a "distinctive relationship [. . .] of power."[20] As such, the term does not necessarily imply cohesion with pre-modern Islamic discourses. Using the categories orthodox and secular-historical as objective, analytical opposites risks humoring certain socio-political structures of dominance and the polemics that sustain them, projecting a false coherence of current lived Islam while ignoring the diversity of pre-modern Islamic knowledge production.

The intent of what follows is not to argue that modernists are either more or less "Islamic" than their Muslim critics based on how closely each of them follows a textual blueprint or style of reasoning. Rather, my point is that under new metaphysical and epistemological conditions Muslim modernists breach at the same time as they

---

17 Ahmed, *Archeology*, 17–20, 39f.

18 Proponents of this view often credit Europe's historical place to "some special quality of race, culture, environment, mind or spirit, which gives this human community a permanent superiority over all other communities, at all times." As James Blaut argues, they ignore the impact colonialism had on Europe's economy, which "led to the selective rise of Europe, the modernization or development of Europe (and Europeanized areas [. . .]), and the underdevelopment of Asia, Africa, and Latin America." See Blaut, *Model*, 1–32. See also Chakrabarty, *Provincializing*, 7–9.

19 As Sebastian Conrad states, Enlightenment thought was "a product of, and response to, global conjectures; [. . .] the work of many authors in different parts of the world. [. . .] The non-European world was always present in eighteenth- century intellectual discussions." See Sebastian Conrad, "Enlightenment in Global History: A Historiographical Critique," *American Historical Review* 117, no. 4 (2012): 999–1027, esp. 1009. Ironically, critical Bible studies perhaps emerged in contact with Islamic exegesis. See Daniel Frank, *Search Scripture Well: Karaite Exegetes and the Origins of the Jewish Bible Commentary in the Islamic East* (Leiden: Brill, 2004), 251–57. See also Ahmed, *Archeology*, 18, 73.

20 As he explains: "Wherever Muslims have the power to regulate, uphold, require or adjust correct practices, and to condemn, exclude, undermine, or replace incorrect ones, there is the domain of orthodoxy." See Talal Asad, "The Idea of an Anthropology of Islam," *Qui Parle* 17, no. 2 (2009): 15.

bear continuities with strands of Islamic reasoning to different degrees and for various reasons. Especially troubling to their critics is the claim that the Qur'ān is a "human text," its form and content shaped, if not determined, by socio-cultural context rather than being trans-historical, eternal, and divine. To avoid misconceptions about the ostensive novelty of their ideas, the following section examines some pre-modern theories about Qur'ānic nature and revelation.

3 Divine Discourse and Human Comprehension

Many academics and modern self-identified orthodox Muslims imagine a nearly uniform "traditional" Islamic view of revelation that resembles the Christian idea of verbal inspiration.[21] According to this allegedly axiomatic view, the Prophet passively received a dictated message from God via Gabriel that he transmitted *verbatim* to the people. Abdullah Saeed chides this idea for portraying Muḥammad as a mere "bystander"[22] without any role in the process – comparable to "an audio-recording instrument such as a tape recorder."[23] Rotraud Wielandt thinks that ignoring the human aspect of the Qur'ān and its language led to exegetical monovalence. Other scholars similarly think that the belief in the Qur'ān as God's direct word results in literalism and aversion to new interpretations.[24] It needs to be noted upfront that there never existed one unified epistemic framework for apprehending the multivalent dimensions of the Qur'ān or its revelation.[25]

---

21 Anzaruddin Ahmad, "Applying Hermeneutics to the Qur'an: A Recritique," in *Critical Perspectives on Literature and Culture in the New World Order*, ed. Noritah Omar et al. (Newcastle: Cambridge Scholars Publishing, 2010), 94.

22 Abdullah Saeed, *Interpreting the Qur'ān: Towards a Contemporary Approach* (New York: Routledge, 2006), 37.

23 Abdullah Saeed, "Fazlur Rahman: a Framework for Interpreting the Ethico-legal Content of the Qur'an," in *Modern Muslim Intellectuals and the Qur'an*, ed. Suha Taji-Farouki (Oxford: Oxford University Press, 2006), 45.

24 Wielandt, *Offenbarung*, 1. See also, e.g., Ali Akbar, *Contemporary Perspectives on Revelation and Qur'ānic Hermeneutics: An Analysis of Four Discourses* (Edinburgh: Edinburgh University Press, 2020), 2, 19f.

25 Muḥammad ʿĀbid al-Jābirī (d. 2010/1431) identified three epistemic paradigms: 1) *bayān* is the foundation for ethical-legal and linguistic interpretation. Emphasizing the Qur'ān's linguistic constitution presupposes that it was sent as a clear, intelligible divine message; 2) *ʿirfān* ("gnosis") mainly refers to mystical approaches, based on experience and the imaginative faculty; 3) *burhān* (apodictic proof) is predicated on Aristotelian philosophy, reason, and demonstrative proof. The different knowledge regimes – again internally highly diverse – gave way to alternate concepts of human comprehension and interpretation. See Muḥammad ʿĀbid al-Jābirī, *Bunyat al-ʿaql al-ʿarabī: Dirāsa taḥlīliyya naqdiyya li-naẓm al-maʿrifa fī th-thaqāfa al-ʿarabiyya* (Beirut: Markaz Dirāsāt al-Waḥda al-ʿArabiyya, 2013), 13–15, 251–55, 383f.; Nimet Şeker, *Der Koran als Rede und Text: Hermeneutik sunnitischer Koranexegeten zwischen Textkohärenz und Offenbarungskontext* (Berlin: EB-Verlag, 2019), 126–32.

The Qurʾān is God's word. In it, God speaks. However, Muslim scholars differed over what exactly that meant, and in which form Muḥammad acquired *qurʾān* (reciting).[26] Many *mutakallimūn* affirmed that the Qurʾānic language was human and finite. They accepted that God spoke in the Qurʾān but denied that He had a voice. That the very words and letters were uncreated and eternally determined, as the Ḥanbalites argued, was a contested minority opinion.[27] For example, Abū Muḥammad b. Kullāb (d. 854/280), perhaps the first to explicitly posit the Qurʾān as eternal (*qadīm*), located its timeless meaning in God's essence. When descending upon prophets it would take on various forms in different texts (e.g., Gospel, Thora, Qurʾān) that were God's word(s) metaphorically.[28] Similarly, Abū Manṣūr al-Māturīdī (d. 944/333) argued that God's Speech as one of His attributes differed from its contingent manifestations: Qurʾānic content was eternal, but the physical means of transmitting it were not.[29]

Many pre-modern Muslim theories of revelation can be subsumed under the "Principle of Accommodation." Accordingly, God adapts His speech to the people for whom it is intended. The Qurʾān uses "elements of familiar and natural order to transcend them," which is "itself a property of [human] language" that "operates through analogies and metaphors."[30] A central verse in this respect states that God speaks to prophets through *waḥy*, from behind a veil, or through a messenger. In the Qurʾān, *waḥy* connotes a mysterious kind of communication – an inspiration, signaling, disclosing, whispering from afar, or transmitting a secret message. *Waḥy* can be verbal, non-verbal, visual, directed at humans or animals, and even (though rarely) refer to non-divine communication.[31] Another mode of revelation is *tanzīl* (sending down), which the Qurʾān uses solely for a divine address to humans or other acts issuing from God.

In Stefan Wild's view, the spatial dimension of *tanzīl* (a downward movement) and the mediating figure of Gabriel are meant to highlight the distance between God

---

26 For example, they debated whether he received God's word *verbatim* (*al-lafẓ*), together with its meaning (*al-maʿnā*) or only its meaning (*al-muḥtawā*), which he perhaps himself put into Arabic. The "*Kayfiyyāt al-inzāl wa-maʿnāhu*" (sciences of the kinds of descending and its meaning) debated the various ways by which *qurʾān* reached Gabriel and then Muḥammad. See Naṣr Ḥāmid Abū Zayd, *Mafhūm an-naṣṣ: Dirāsa fī ʿulūm al-Qurʾān* (Cairo: al-Hayʾa al-Miṣriyya al-ʿĀmma li-l-Kitāb, 1990), 47.

27 See Josef van Ess, "Verbal Inspiration?," in *The Qur'an as Text*, ed. Stefan Wild (Leiden/New York/Köln: Brill, 1996), 177f., 181–84, 190–92; Josef van Ess, *TG* (Berlin/New York: De Gruyter, 1991), 4: 615–17.

28 Wajīh Qānṣū, *an-Naṣṣ ad-dīnī fī l-Islām min at-tafsīr ilā t-talaqqī* (Beirut: Dār al-Fārābī, 2011), 74 f.

29 Philip Dorroll, "The Doctrine of the Nature of the Qurʾān in the Māturīdī Tradition," in *Mâtürîdî Düşünce ve Mâtürîdîlik Literatürü*, eds. M. Raşit Akpınar, Recep Tuzcu, and Aslı Menekşe (Istanbul: Selçuk Üniversitesi İmâm Mâtürîdî Uygulama ve Araştırma Merkezi, 2018): 129f.

30 See Amos Funkenstein, *Theology and the Scientific Imagination from the Middle Ages to the Seventeenth Century* (Princeton: Princeton University Press, 1986), 213–21.

31 See Q 42:51 for the three modes of revelation. Oddbjørn, "*Waḥy*," 104. Islamic tradition mostly identifies the angel Gabriel as God's agent who imparted *qurʾān* to Muḥammad. Yet, the Qurʾān mentions him only once in this capacity. In other verses, it calls this intermediary figure *rūḥ al-qudus*, i.e., the Holy Spirit (Q 16:102), and *rūḥ al-amīn*, i.e., the Trustworthy Spirit (Q 26:192–95).

and humans. Some scholars therefore think that revelation is a misleading translation in the Islamic context since the term entails that God unveils Himself by becoming visible, sensible, or audible to humans. In Wild's words, this "concept [is] foreign to the Qur'ān."[32] However, Wild's notion of revelation originates from within modernity itself and from its metaphysical framework rather than from scholastic Christian theology. Thomas Aquinas, for example, perceived *revelatio* not as the unveiling of God but of the human gaze, when it was divinely enabled to perceive "divine truths surpassing reason."[33] In a similar vein, for Muḥyī d-Dīn b. ʿArabī (d. 1240/638), revelation and interpretation meant actualizing the meaning of symbols, by which "the external form [is united] to the innermost ground (or 'heart') of the saint."[34] Projecting the modern apologetic view of revelation onto the entire Islamic tradition contorts the nuances within the diverse Islamic imaginaries regarding human-divine interaction.

A more fluid idea of revelation, on the other hand, softens the trope that, traditionally, Muḥammad was seen as a mere channel for a divine, monodirectional thou-shalt message.[35] Rejecting the historian's notion of the Qur'ān as a human product need not imply denying the Prophet any role in the process.[36] The mode of Qur'ānic revelation cannot be separated from its functions, such as *hudā* (guidance), *amr Allāh* (God's command), or *dhikr* (reminder). Such relational terms entail some form of participation and transformation on the side of the addressee.[37] Many Islamic scholars assumed a participatory role of language itself. For them, apprehension of divine truths, including the Prophet's, implied intimacy and union with God in the very perception of that truth, which they sought to reenact in their dealings with the Qur'ān. Rather than passively receiving an "object," they saw revelation as an event, in which humans are grasped by a higher reality that they also come to embody.[38]

---

32 Stefan Wild, "We have sent down to thee the book with the truth: Spatial and temporal implications of the Qur'anic concepts of *nuzūl, tanzīl,* and *inzāl,*" in *The Qur'an as Text*, ed. Stefan Wild (Leiden/New York/Köln: Brill, 1996), 137–42. Note that according to Josef van Ess, a few scholars among the Ḥanbalites did assume that Muḥammad had met God face to face. See van Ess, "Verbal Inspiration?," 186–88. For the mainstream view, see Toshihiko Izutsu, *God and Man in the Qur'an: Semantics of the Qur'anic Weltanschauung* (Tokyo: Keyo University, 2008), 174–78.

33 See John Montag, "Revelation: The False Legacy of Suarez," in *Radical Orthodoxy: A New Theology*, ed. John Milbank et al. (London/New York: Routledge, 1999), 40–44.

34 See Robert J. Dobie, *Logos and Revelation: Ibn 'Arabi, Meister Eckhart, and Mystical Hermeneutics* (Washington DC: Catholic University of America Press: 2010), 25–56. Ibn ʿArabī's full name was Muḥyī d-Dīn abū ʿAbd Allāh Muḥammad b. ʿAlī b. Muḥammad b. al-ʿArabī l-Ḥātimī ṭ-Ṭāʾī l-Andalusī l-Mursī d-Dimashqī.

35 For this view, see, e.g., Akbar, *Perspectives*, 18f.

36 The Qur'ān itself denies that a human produced any part of it. See, e.g., Q 10:38, 11:13–14, 17:88, 52:29–30.

37 See, e.g., Q 2:2, 4:105, 4:113, 7:2, 7:52, 14:1, 16:64, 16:89, 18:1, 57:25, 43:3–4.

38 Dobie, *Logos*, 47. Thus, ʿĀʾisha's *ḥadīth* equating the Qur'ān with Muḥammad's character does not mean that he produced the Qur'ān, but the opposite. See p. 55.

The Qur'ān's context-related descent reinforces its imminent, dialogical character. While two verses suggest that it was sent down all at once (*jumlatan wāḥidatan*),[39] many more imply that the Prophet received *qur'ān* piecemeal (*tadrījan*), for twenty-three years.[40] Its character as both oral speech and a written, linear text took shape in two alternate arrangements of verses and suras: one according to the order of their descent (*tartīb an-nuzūl*), the other according to their recitation (*tartīb at-tilāwa*) in the *muṣḥaf* (canonical codex).[41] The Qur'ān is transcribed speech; it does not record the non-linguistic clues of the speech act itself. In other words, the Qur'ān entails what is said, but not in which context, for which reason, to whom, or in which tone. Hence, various Islamic sciences attach great hermeneutical value to situating verses and suras spatially and temporally, using extra-textual references, such as *asbāb an-nuzūl* (occasions of revelation), *sīra* (biography of the Prophet), *ḥadith* (reports about the Prophet's words and deeds), and pre-Islamic poetry as well as context-related methods like *naskh* (abrogation). These exegetical devices presuppose an idea of the Qur'ān as an event in time addressing situations in the life of the early Muslim community.

One can argue that especially scholars like Taqī d-Dīn Aḥmad b. Taymiyya (d. 1328/729) or Abū ʿAbdallāh Muḥammad b. Idrīs ash-Shāfiʿī (d. 820/205), often called "literalists," pursued context-oriented exegesis. As Nimet Şeker explains, in viewing the first listeners of *qur'ān*, i.e., the Prophet and his companions (*ṣaḥāba*), as having had a direct grasp of its meaning that later generations lacked, they recognized the hermeneutical distance between God's word and its understanding.[42]

Yet, context-oriented exegesis did not mean for Muslim scholars that accidental contingencies (read: history) brought forth the Qur'ān. They generally did not see a conflict between gradual Qur'ānic descent and the *muṣḥaf* as the divinely intended textual form, even if they differed over how and by whom the Qur'ān received its final order.[43] While some aimed at limiting the range of its meanings, many *mufassirūn* strove to discover the Qur'ān's "richness and transforming possibilities." For them, exegesis was an "unending process."[44] Aware that understanding is an intricate affair, especially when

39 See Q 2:185, 97:1. A common explanation is that the entire Qur'ān existed, even pre-historically, on the Well-Preserved Tablet (*al-lawḥ al-maḥfūẓ*, see Q 85:21–22), before it descended in Ramadan in the Night of Destiny (*laylat al-qadr*, see Q 97:1). The Qur'ān does not specify where it was sent. Some scholars assume that it descended to the lowest heaven (*bayt al-ʿizza*), from where Gabriel delivered it piecemeal to Muḥammad. Others argue that it was sent all at once to his heart directly. See Şeker, *Koran*, 32f.

40 See, e.g., Q 17:106, 25:32–33, 69:38–43, etc. Muslim scholars differ as to whether his mission lasted 20, 23, or 25 years. See Şeker, *Koran*, 29f.

41 Muslim scholars offered creative explanations for its gradual revelation. Some thought its purpose was pedagogical, i.e., to train the Prophet to receive and memorize longer passages or to gradually familiarize the first Muslims with new laws, faith, and ethics, while answering their immediate questions.

42 See Şeker, *Koran*, 19–23, 133–92.

43 Şeker, *Koran*, 15–18, 25–52, 70f., 204–15.

44 Kristin Zahra Sands, *Ṣūfī Commentaries on the Qur'ān in Classical Islam* (London/New York: Routledge, 2006), 3–7, 32, 41. Walid Saleh contends that *tafsīr* was anthological, and "a polyvalent Qur'ān

the speaker is God and the addressee human, Muslim thinkers developed elaborate linguistic theories that left room for ambiguity and human error.[45]

It is in any case difficult to establish a correlation between literalism and the belief in the Qur'ān as God's word.[46] While Muslim scholars aimed to avoid conflating God – infinite, transcendent, unfathomable – and the finite, human, knowable, they balanced their views between *tanzīh* (God's total otherness) and *tashbīh* (His likeness) differently. Yet, they generally conceived of God's activity as inscribed into the fabric of the world, regardless of theological quarrels over the details of this relationship. It is precisely this balance where post-Cartesian, post-Kantian epistemic frameworks diverge. In Asad's words, "Secularism and its attendant historical worldview [. . .] removed and transferred God from the material world to a transcendental realm."[47]

In addition, a philosophical realism of linguistic representation de-ontologized language and posited "the experiment [. . .] opposite language as that which guaranteed and verified the latter."[48] In placing God outside of nature and rendering language historical, i.e., provisional and "rigidly restricted to representation," the revelation process became less intelligible and its linguistic message unreliable in conveying truth.[49] As we will see, however, most of the following authors adopt models of revelation that can be read as modified versions of accommodation theory. Yet, they entail the transcendent differently in their deliberations.

3.1 Naṣr Ḥāmid Abū Zayd: Can God Speak a Word to Man?

Egyptian literature professor Naṣr Ḥāmid Abū Zayd caused some furor with his claim that the Qur'ān was a "human text (*naṣṣ insānī*)" and a "cultural product."[50] Capitalizing on the idea of *waḥy* as a primarily immaterial communication between God and humans, he emphasized the human-divine dialectic of revelation. In his view, the Qur'ān in its material form was a linguistically encoded translation of divine *waḥy*

was seen as the norm. [. . .] A verse could have conflicting interpretations, each of which could be adduced as part of the meaning of the word of God without disrupting the notion of [its] clarity." See Walid Saleh, *The Formation of the Classical Tafsīr Tradition: The Qur'ān Commentary of al-Tha'labī (d. 427/1035)* (Leiden/Boston: Brill, 2004), 18, 20.

45 See, e.g., Şeker, *Koran*, 132–91; Thomas Bauer, *Die Kultur der Ambiguität: Eine andere Geschichte des Islam* (Berlin: Verlag der Weltreligionen, 2011), 143–91.

46 Bauer, *Kultur*, 116–19.

47 Talal Asad, *Formations of the Secular: Christianity, Islam, Modernity* (Stanford, CA: Stanford University Press, 2003), 23–27, 37–43, 211–15, 235f.

48 Michel de Certeau SJ, *The Mystic Fable*, vol. 1, *The Sixteenth and Seventeenth Centuries*, trans. Michael B. Smith (Chicago: University of Chicago Press, 1992), 123.

49 See Montag, "Revelation," 51–56.

50 Naṣr Ḥāmid Abū Zayd, *Naqd al-khiṭāb ad-dīnī* (Cairo: Sīnā li-n-Nashr, 1992), 93; Abū Zayd, *Mafhūm*, 9–12, 18.

into a human text, which entailed interpretation insofar as language is context-dependent and embodies its culture *structurally*.[51] Against the objection that he had banned God from the Qur'ān,[52] however, Abū Zayd's model of dispatcher (God) and receiver (Muḥammad) retains the idea of the Qur'ān as God's Word. In describing the Qur'ān as a message "from heaven to earth" that "God sent down upon the Prophet," Abū Zayd also maintained the spatial model of *tanzīl*. In other words, he did not reduce *waḥy* to an inner, mental state or experience.

Instead, he offered a kind of accommodation theory in that God "chose" to adapt His speech to human capacity by using the "cultural and linguistic signification system (*niẓām dalālī*) [. . .] of the first receiver."[53] Hence, for Abū Zayd, the initiative for the Qur'ān's shape and content derived from God. His view of history largely functions without God, as Charles Hirschkind rightly notes. However, Hirschkind's conclusion that Abū Zayd's epistemology denies metaphysical statements, as "the divine never enters human experience at all [but] remains outside of knowledge," needs qualification.[54] Rejecting metaphysical speculation about the Qur'ān before its descent can also be read as adhering to the theological premise of *tanzīh* and God's oneness (*tawḥīd*). Since God's speech is part of His essence, it cannot "descend" or be "inlibrated" in the Qur'ān. Otherwise, so Abū Zayd, Muslims would do to the Qur'ān what Christians did to Jesus, that is, deifying something other than God (*shirk*).[55]

---

51 See Abū Zayd, *Mafhūm*, 21, 27, 40–42. His account of different Islamic scholarly views regarding *waḥy* partly serves him to prove the sheer variety of opinions among Muslim scholars on the matter, and hence their relativity. He states further: "The text was transmitted from the first moment on – i.e. when the Prophet recited it when it was revealed to him – from its being as a divine text (*naṣṣ ilāhī*) to something cognizable, a human text (*naṣṣ insānī*) because it mutated from descent to interpretation. The comprehension of the texts by the Prophet constituted the first stage of the movement that springs from the link between text and human reason." See Abū Zayd, *Naqd*, 93.

52 "Munāẓarat ad-duktūr Muḥammad ʿAmāra, al-mufakkir al-islāmī l-kabīr, maʿa Naṣr Abū Zayd, ash-shuyūʿī, ʿan riddatihi" (Debate between Dr. Muḥammad ʿAmāra, the great Islamic thinker and Naṣr Abū Zayd, the Marxist, on his apostasy), Youtube Video, Qatar Television December 31, 1996, posted by "video max," November 9, 2013, accessed June 24, 2016, https://www.youtube.com/watch?v=KWSIQGLMOcc. See also Hirschkind, "Heresy," 38.

53 The full quotation reads: The Qur'ān "is a message from heaven to earth, but it is not a message that is independent of the rules of reality with all its structures which are embedded in this reality; and the most important of which is the cultural texture. The Absolute sheds light on itself (*yakshifu ʿan nafsihi*) through speech to humans, it 'condescends' (*yatanazzalu*) to them by using their cultural and linguistic frame of reference (*niẓām dalālī*)." See Abū Zayd, *Mafhūm*, 28, 47f., 64f. Quote from p. 64. See also Navid Kermani, *Offenbarung als Kommunikation: Das Konzept waḥy in Naṣr Ḥāmid Abū Zayds Mafhūm al-Naṣṣ* (Frankfurt a.M.: Lang, 1996), 40–44.

54 Hirschkind, "Heresy," 43, 45–47.

55 Abū Zayd, *Naqd*, 99. According to Christian dogma, the second person of the Trinity, the *logos*, became flesh. Jesus was hence fully divine and fully human, which the Qur'ān and Muslims reject. In Abū Zayd's view, declaring the Qur'ān literally the divine Word of God repeats this Christian mistake, see p. 196. Harry A. Wolfson first introduced the term inlibration ("enbookment") as an equivalent to the incarnation of the *logos* in Christianity. Following this analogy, while in Christianity the word be-

As we have seen, many *mutakallimūn* similarly argued that God's essence, including His speech, was inaccessible to humans.[56] Suspicion of metaphysical claims about God due to human limitation is also common in modern Muslim discourse. For example, Sayyid Quṭb (d. 1966/1386) – generally weary of metaphysical speculation – dedicated part of his commentary to the constraints of human thinking in matters divine. As he asserted, God's Word was "divine knowledge that knows no bounds, [humans] can neither grasp nor comprehend, much less emulate it."[57] Grand Ayatollah Muḥammad Ḥusayn Faḍlallāh (d. 2010/1431) similarly highlighted human cogitative limitations regarding God, whose essence transcended human frameworks.[58] Abū Zayd's epistemological premises doubtlessly differ from these thinkers. Yet, they all share reservations regarding strong metaphysical assertions. Meanwhile, Abū Zayd also affirmed that the Qur'ān conveyed God's character, will, and intention, albeit in human language, filtered through human interpretation.

Rather than settling Qur'ānic ontology, Abū Zayd sought new ways to interpret the Qur'ān. His strong insistence on its human side owed much to his conviction that belief in the Qur'ān as literally divine closed it to new readings. He had in mind those approaches that reduced its meaning to the opinions of the early Muslims. In his view, this position amounted to *shirk* because it conflated the human and divine by ignoring the historical contingency of human grasp and language. On the contrary, Abū Zayd denied the *tafsīr* of the early Muslims or Muḥammad privileged exegetical status.[59] Instead, he championed a literary-hermeneutical approach. Accordingly, meaning was not something out there, waiting to be discovered but the result of interpretation, in which the horizons of text and reader fuse to create ever new meanings.[60]

In prioritizing a view of the Qur'ān as a linear text, Abū Zayd downplayed the context-oriented character of *waḥy* as a communication process, for whose interpretation

---

comes flesh, in Islam, the word becomes text (or book). See Harry A. Wolfson, *The Philosophy of the Kalam* (Cambridge/Massachusetts: Harvard University Press, 1993), 50.

56 For al-Māturīdī, the letters and sounds of the Qur'ān are "indicators" (*dalālāt*) of the real speech of God, i.e., the physical means by which we recite and deduce the content from the Qur'ān. See Dorroll, "Doctrine," 130–36.

57 See Hannelies Koloska, *Offenbarung, Ästhetik und Koranexegese: Zwei Studien zu Sure 18 (al-Kahf)* (Wiesbaden: Harrassowitz, 2015), 200–2. Quṭb views *kalām*'s metaphysical arguments about the Qur'ān as superfluous. In his work, God seems as removed from the world as in other post-Cartesian works. While he preserves the transcendent otherness of God and the vocabulary of *tanzīl*, he eclipses its pre-modern rich multiplicity of meaning, seemingly reducing it to its deictic function. See Sayyid Quṭb, *Milestones: Ma'alim fi'l-tareeq*, trans. A.B. al-Mehri (Birmingham: Maktabah Booksellers and Publishers, 2006), 49–57. His theological convictions still need further study.

58 Muḥammad Ḥusayn Faḍlallāh, *'An sanawāt wa-mawāqif wa-shakhṣiyyāt: Hākadhā taḥaddatha – hākadhā qāla. Ḥāwarathu Munā Sukariyya* (Beirut: Dār an-Nahār, 2007), 56.

59 Abū Zayd, *Naqd*, 60f., 93f. In his view, Qur'ānic studies should follow an ascending dialectic (*dīyāliktīk ṣā'id*), which, in contrast to a descending dialectic (*dīyāliktīk hābiṭ*), starts from the particular and tangible, i.e., language and culture, to arrive at "the hidden." See Abū Zayd, *Mafhūm*, 3, 27–29.

60 Abū Zayd, *Mafhūm*, 27–30, 251–53; Abū Zayd *Naqd*, 93.

the opinions of its first addressees could, in fact, help. Perhaps his emphasis on the Qur'ān as a speech act primarily served to establish its historicity. Yet, the tensions inherent in viewing the Qur'ān as both a coherent literary text and speech act(s) also mark the Islamic exegetical tradition. His insistence on hermeneutic subjectivity has parallels in pre-modern learned discourse as well. Islamic scholars from various disciplines warned against determining God's intention with certainty. In *fiqh* (jurisprudence), doubt came to be one of the guiding legal principles (*qawāʿid fiqhiyya*).[61] Jurists also developed sophisticated theories about deriving probable precepts from Qur'ān and Sunna, factoring in dissent, textual ambiguity, and human error. In many ways, their linguistic and epistemological theories prefigured modern reader-response theory.[62]

Moreover, several scholars affirmed that the *ṣaḥāba* at times disagreed and even erred.[63] Abū Zayd deviated from their position in that he distrusted the sources and viewed the Prophet as fallible, too.[64] The idea that classical *tafsīr* was shaped by socio-political circumstances and cannot satisfy the Qur'ān's claim to transcend time and space is not unusual today.[65] Thus, Faḍlallāh argued that Qur'ānic readings changed because of power struggles and shifting social milieus.[66] Muḥammad Bāqir aṣ-Ṣadr empha-

---

61 See Intisar Rabb, *Doubt in Islamic Law: A History of Legal Maxims, Interpretation, and Islamic Criminal Law* (New York: Cambridge University Press, 2015). *Fiqh* means discernment, ability to judge, or comprehension of the *sharīʿa*, often referred to as Islamic jurisprudence or Islamic law.

62 Gregor Schwarb, "Capturing the Meaning of His Speech: The Relevance of *uṣūl al-fiqh* to an understanding of *uṣūl al-tafsīr* in Jewish and Muslim *kalām*," in *A Word Fitly Spoken: Studies in Mediaeval Exegesis of the Hebrew Bible and the Qur'ān*, eds. Meir Bar-Asher et al. (Jerusalem: The Ben-Zwi Institute, 2007), 145f. Schwarb notes that their "deliberations comprised a set of fundamental questions about the ways in which meanings are produced as well as about the conditions to be fulfilled for their (proper) understanding. In Islam of the classical period, such hermeneutic reflections came to be operative in such strictly religious disciplines as *uṣūl al-dīn*, *uṣūl al-fiqh*, *uṣūl al-ḥadīth*, or (*uṣūl*) *tafsīr al-Qur'ān*, represented in each case by a rich variety of doctrinal affiliations (within *kalām*, *fiqh*, etc.). [There were also] miscellaneous types of hermeneutic reflections in para- and extra-religious fields, such as fundamental linguistics (*uṣūl al-naḥw*), literary theory (*ʿilm al-balāgha*), literary criticism (*naqd al-shiʿr*), logic (*ʿilm al-manṭiq*), philosophy and medicine (*ṭibb*), natural sciences (*ʿulūm ṭabīʿiyya*) and so forth," p. 112. See also Bauer, *Kultur*, 143–91.

63 Abū Jaʿfar aṭ-Ṭabarī rejected their opinions if they contradicted the Qur'ān. He also reported divergences in their opinion. Agreeing with him, Najm ad-Dīn aṭ-Ṭufī (d. 716/1316) included a list of reasons why the *tafsīr* of the *ṣahābā* was *ijtihād*, open to disagreement and error, and not complete. See Şeker, *Koran*, 141f., 157–59.

64 Abū Zayd, *Naqd*, 87.

65 See, e.g., Faḍlallāh, *ʿAn sanawāt*, 57; Muḥammad Ḥusayn Faḍlallāh, *Dunyā l-marʾa* (Beirut: Dār al-Malāk li-ṭ-Ṭibāʿa wa-n-Nashr wa-t-Tawzīʿ, 1997), 18–33; Shafīq Jarādī, "Naqd an-naṣṣ ad-dīnī," in *Ḥiwār al-ḥaqīqa min ajli al-ḥayāt maʿan*, vol. 2, *Aʿmāl muntadā l-fikr al-masīḥī l-islāmī fī kulliyyat al-lāhūt li-sh-sharq al-adnā*, ed. George Sabra (Beirut: Manshūrāt Kulliyyat al-Lāhūt li-sh-Sharq al-Adnā, 2008), 97f.; Muḥammad aṭ-Ṭāhir b. ʿĀshūr, *aṭ-Ṭaḥrīr wa-t-tanwīr* vol. 1 (Tunis: ad-Dār at-Tūnisiyya li-n-Nashr, 1984), 29.

66 Muḥammad Ḥusayn Faḍlallāh, "al-ʿAlāqa al-islāmiyya al-masīḥiyya: Qirāʾāt islāmiyya fī r-rāhin wa-l-mustaqbal," in *al-ʿAlāqāt al-islāmiyya al-masīḥiyya: Qirāʾāt marjaʿiyya fī t-tārīkh wa-l-ḥāḍir wa-l-mustaqbal* (Beirut: Markaz ad-Dirāsāt al-Istrātījiyya wa-l-Buḥūth wa-t-Tawthīq, 1994), 18–33.

sized that Qurʾān exegesis, dialectic in nature, had to start from "reality" and "human experience." The Qurʾān's "ability to offer something new" depended on human engagement with it. While the Qurʾān was "inexhaustible, [in that] God's words are infinite, [. . .] linguistic *tafsīr*" could not exhaust its meaning, because language was finite and had "only limited capacities."[67]

3.2 Abdelkarim Soroush

While Abū Zayd retained the idea of the Qurʾān as God's Word – divine in origin, expressed in human language – Iranian intellectual Abdelkarim Soroush seems to suspend this idea altogether. His argument rests on thoroughly historicized grounds, despite its mystical phrasing. For Soroush, *waḥy* belongs to the phenomenological category of "religious experience," i.e., the sensing of phenomena that occur in this world but remain overwhelming and unnamable, i.e., transcendent. He uses the term transcendence in an already secularized way since such an experience can but need not be theocentric. *Waḥy* is a wholly human, hence, subjective, "frail," and "blighted" experience that humans can convey in only limited form, determined by language and culture. Over the reified outcome, i.e., the Qurʾān, God seems to have no sway.[68]

In the theological debate over the tension between the transcendent and immanent, Soroush sides with Immanuel Kant's notion of the *Ding-an-sich* (thing in itself), which is a *noumenal* unknown, not with Hegel's phenomenological Spirit in history that can somewhat be discerned.[69] Historically, Muslim thinkers held that God is both phenomenal and noumenal, balancing their views of God between *tashbīh* and *tanzīh*. God can be known phenomenally through the Qurʾān. Yet, Muslims also strove to assert the ontological gap between humans and God, Who is noumenally outside the grasp of human knowledge. Soroush fully accepts the noumenal-phenomenal distinction. Since humans are always situated in a specific context and understand the world through a particular lens, they have no context-independent avenue by which to perceive or articulate things.

This ontological situation makes epistemological claims about the *Ding-an-sich* impossible without using concepts and language acquired in the phenomenal realm. While assumptions about the noumenal may be offered, these claims to noumenal re-

67 As-Sayyid Muḥammad Bāqir aṣ-Ṣadr, *al-Madrasa al-Qurʾāniyya* (Beirut: Dār at-Taʿāruf, 1981), 22f.

68 See Abdolkarim Soroush, *The Expansion of Prophetic Experience: Essays on Historicity, Contingency, and Plurality in Religion*, ed. Forough Jahanbakhsh, trans. Nilou Mobasser (Leiden/Boston: Brill, 2009), 4, 9, 17, 74–76; Soroush, "Islamic Revival and Reform: Theological Approaches," in *Reason, Freedom, and Democracy: Essential Writings of Abdolkarim Soroush*, ed. Mahmoud and Ahmad Sadri (Oxford/New York: Oxford University Press, 2000), 34–37.

69 See Immanuel Kant, *Religion Within the Bounds of Bare Reason*, trans. Werner S. Pluhar (Indianapolis: Cambridge University Press, 2009).

ality cannot be independently assessed. In other words, the noumenal is truly transcendent – entirely inaccessible to human perception and knowledge. In the end, it is unclear whether God speaks at all in Soroush's idea of *waḥy*. While he affirms that Muḥammad's "mystical experience" of transcendence resulted from a mediated engagement with the divine,[70] religious experience is not confined to the Prophet. Other people can have it, too, like the Buddha, who was yet unaware that he experienced a revelation from God.[71]

On the one hand, acknowledging the potential truth in diverse experiences of transcendence has an equalizing effect. On the other, this inclusivist reading of *waḥy* leaves little room for particularity. In Soroush's view, individuals can participate in the "religious experience" that is *waḥy*. However, to achieve this participatory possibility, he first generalizes it to such an extent that its relevance as divine revelation risks losing its meaning. His theory of the essentials and accidentals of religion illustrates this tendency. Essentials are the universals of a religion. Accidentals could have been different. For example, the Qur'ān could have been longer or revealed in another language, "if the Prophet had lived longer."[72] In positing an unchanging Islamic core clothed in contingent "garments" (e.g., language, values, etc.), shaped by time, geography, and culture, Soroush achieves a "minimalistic form" of Islam that is limited to its spiritual aspects.[73] Therefore, passages and injunctions identified as accidentals lose their non-current binding force.[74]

3.3 Wajīh Qānṣū: One Structure of Generality – Many Meanings

Remarkably, in contrast to Abū Zayd, Soroush, and much of Islamic tradition, Lebanese philosophy professor Wahjīh Qānṣū argues that God revealed *Himself* to Muḥam-

70 For him, "Religious experience" equals that of "transcendence," occurring "as a dream, as the sensing of a scent or a bellow, the seeing of a figure or a color, a feeling of connectedness to a boundless vastness, a sense of contraction and gloom, [. . .] a love for an unseen beloved, the sensing of a spiritual presence, a oneness with a person or an object, a detachment from the self and suspension in nothingness, the comprehension of a mystery or the discovery of a secret, a tiring of earthly attachments and flight toward the eternal, being drawn to some gravitational force, a thirst, a vacuum, a flash, a sense of awe, [. . .] ecstasy." See Soroush, *Expansion*, 123–26.

71 Soroush, *Expansion*, 4–12, 26f., 45. As he states, "Viewing divine discourse as *nothing but* the Prophetic discourse is the best way of resolving the theological problem of how God speaks. [. . .] In this experience, it seems to the prophet *as if* someone comes to him and proclaims messages." P. 12. Emphasis mine.

72 Soroush, *Expansion*, 63.

73 Soroush, *Expansion*, 144f. He argues that "religion becomes human and historical when it enters history, [. . .] subjected to people's theoretical and practical intrusions; [. . .] dulled and dimmed by the passage of time and the veils that impede understanding; [. . .] much is gradually taken away from it and added on to it."

74 Soroush, *Expansion*, 63–68.

mad. While the initiative comes from God, *waḥy* occurred at times as a direct dialogue between them. *Waḥy* is the conjunction (*ittiṣāl*) of transcendence and contingence, "entailing both *descended speech* and the experience of prophethood."[75] Qānṣū emphasizes the life-changing aspect of this event that transformed the Prophet into a "person, in whose core the experience of *waḥy* intensified and poured forth in his words and deeds." Since Muḥammad was the "speaking Qur'ān [. . .] that walked among us,"[76] a clear demarcation between Qur'ān and Prophet is non-sensical, introduced only after Qur'ān and sunna were fixed in writing.[77]

By keeping historical categories open with a view to the event of *waḥy* Qānṣū admits "cracks in the structure of the homogeneity of historical time."[78] In other words, *waḥy* constitutes a divine breach in the system of representation of secular time. However, his theory of interpretation and the Qur'ān *as text* forecloses analogous disruptions of historical regularity. The uniqueness of *waḥy* also renders it incommensurable. Qānṣū affirms that divine Speech – adapted to human capacity – descended on the Prophet; but this *qur'ān* is not easily accessible anymore. It went through several stages influenced by language, politics, and culture, before it materialized in its canonized form.[79] Taking its dialogical style at face value, Qānṣū regards the material Qur'ān as the historical document about the "event of the Qur'ān" that describes the already "linguistically edited message" of the Prophet just as it documents human speech, angel's speech, and even the devil's. All these participate in the Qur'ānic discourse, and God is only one of them.[80]

Like Abū Zayd and Soroush, Qānṣū asserts that Qur'ānic meaning results from the fusion of horizons of text and reader. Yet, he deliberates in greater detail what the Qur'ān's materiality meant for its first recipients. Unlike them, we read the Qur'ān as a written text, fully redacted and nicely arranged. We know how the story went and what its outcome was – a situation that differs fundamentally from that of its first recipients. The effect the Qur'ān had on Muḥammad and the first Muslims can hence not be re-enacted. The Qur'ān was directed first at the Prophet and second at the people around him, whose questions, concerns, and needs it reflects; it does not address the questions of today's reader directly, who is always only a second-order recipient. The Qur'ān can take one to that experience but is not *waḥy* or God's word

75 Qānṣū, *Naṣṣ*, 53f. Emphasis mine. Based on reports detailing the strong effects of *waḥy* on the Prophet as "*qawl thaqīl*" (heavy speech), Qānṣū concludes that the words and revelatory (transforming and existential) process of *waḥy* cannot be seen as distinct.

76 The Qur'ān was "of blood and flesh and walked among the people and called to God." See Wahjīh Qānṣū, "an-Naṣṣ ad-dīnī wa-rihānāt al-ma'nā s-sīyāsī," *Qaḍāya Islāmiyya Mu'āṣira* 19, no. 61–62 (2015): 92.

77 In Qānṣū's view, nothing indicates that the Qur'ān in its exact wording, shape, or form existed eternally. In the view of Muḥammad Ḥusayn aṭ-Ṭabaṭabā'ī and Ibn 'Arabī, *waḥy* need not be verbal but can occur in other ways. See Qānṣū, *Naṣṣ*, 68–75.

78 For this phrase see Benjamin, "Theses," 265.

79 Qānṣū, *Naṣṣ*, 8, 51f., 80f., 218. The phases that led to the final redaction of the historical "religious text" were *waḥy*, *tablīgh* (promulgating *waḥy*), and documenting these processes in writing (*tadwīn*).

80 Qānṣū, *Naṣṣ*, 11–20; Qānṣū, "Rihānāt," 95–100.

in the literal sense.[81] In the end, history creates a corridor between the Qurʾān and modern readers in Qānṣū's account by detaching them from Muḥammad's revelatory experience.

Qānṣū wants to liberate reading from tradition and the restraints that produce "generic readers," trained to feel and think within certain confines.[82] In common with Abū Zayd and Soroush, he aims to open the Qurʾān to polyvalence envisioning an interpretative process that yields endless new meaningful readings. This interpretive multiplicity, however, comes at the cost of other history-independent avenues to the Qurʾān.[83] To put this differently, in aligning their hermeneutics to a secular definition of the real these thinkers risk excluding other experiences of the real and diminishing epistemological diversity. For Abū Zayd, Soroush, and Qānṣū, meaning inheres in the Qurʾān's linguistically structured text, not in experiencing it through non-rational means within a framework that embeds one's understanding in the practices of a specific community.[84] Reliving the Qurʾān's origin moment or partaking in its mystery through spiritual practices become improbable.

3.4 Thinking History with God I: Shafīq Jarādī

The reductive element of critical hermeneutics has been recognized by Shiite scholar Shafīq Jarādī, head of the *Maʿhad al-Maʿārif al-Ḥikmiyya* (The Sapiental Knowledge Institute), a philosophical teaching and research institute in Beirut. His deliberations help scrutinize the heuristic value of the modern-tradition binary, in our case: the alleged dichotomy between the "traditional" idea of Qurʾānic revelation as verbal inspiration said to disregard the human element in the revelatory process, on the one hand, and history believed to be capable of salvaging the human role and interpretive

81 Qānṣū, *Naṣṣ*, 11–21.

82 Qānṣū, *Naṣṣ*, 525–27; Wajīh Qānṣū, "at-Taṣāluḥ al-mujtamaʿī bayna ad-dīn wa-ʿalmaniyyat ad-dawla," *Al-Maḥajja* 11 (2010): 61f.

83 Qānṣū, "Rihānāt," 95–102; Qānṣū, *Naṣṣ*, 11–20.

84 Qānṣū expresses this perspective perhaps most clearly. For him, the problem with traditional approaches to the Qurʾān is that they place us "before a religious text that does not come to us freely, and we are not free to receive it however we want either. The text is not free to pour forth its meanings over us but comes to us from within a space of consciousness (*waʿy*), existential worldviews, and a network of scholarly a priori truths that condition our thinking about the text and [its] meaning. [It comes to us with] an etiquette and rules of conduct that condition its deliberation, a specific view of the foundational history of Islam (specifically the times of the prophet, and the "rightly-guided caliphs"), and a system (*niẓām*) of a specific relation between the text and the network of the interpreters who received it and embarked upon its understanding." See Qānṣū, *Naṣṣ*, 8f. Saba Mahmood argues that their dismissal of ritual and Islamic reading practices echoes a post-Kantian view of "primitive religion," indicated by ritual and "outward religiosity" as opposed to sincere, authentic spirituality that Kant saw as indicative of "human progress." See Mahmood, "Secularism," 342; Kant, *Religion*, xxvi, 135–37.

multiplicity, on the other. Jarādī wants to think a plurality of temporalities that remain contemporaneous "with us" by rehabilitating those pasts and presents that the scientific-historical paradigm marginalized.[85] His critique of historical hermeneutics is not a romantic longing for the past or a defense of the Qur'ān's divinity in the sense of saving the magic of some out-of-space occurrence. In fact, he criticizes the *mutakallimūn* and Islamic philosophers for neglecting the human role in revelation.[86]

However, he does not let secular history define that role either. Aiming to account for history in the Qur'ān without reducing it to that history, his theory of revelation presents a version of accommodation theory based on Mullā Ṣadrā's (d. 1640/1050) ontological epistemology.[87] Jarādī affirms that the language of the Qur'ān is human and reflects a certain time and culture. Language makes the Qur'ān intelligible, but it does not make it human. Jarādī's idea of *waḥy* and his hermeneutics rest on the premise that creation and revelation are not discrete phenomena but analogical and complementary in nature, because all existence emanates from God and remains rooted in Him. Language, humans, and nature are innately capable of transcending themselves.

Rather than "supernatural" information infused into a passive vessel, *waḥy* describes a relational process between the human and divine – a descending movement that answers the ascending movement of a longing self. By reminding a person of her embeddedness in larger orders and moving her ontologically closer to God, *waḥy* invites the human to read her self, whose natural disposition strives toward divination.[88] Humans are thus ontologically and epistemologically not entirely detached from God or confined to their historical contingency. Jarādī's thought reveals that historicizing the Qur'ān goes to the heart of human nature itself. He faults secular hermeneutics for their reductive, negative vision of humans, who are left without a way to transcend themselves, at least not ontologically.[89]

---

85 Shafīq Jarādī, "ʿArsh ar-rūḥ wa-insān ad-dahr," *Al-Maḥajja* 26 (2013): 5.

86 Shafīq Jarādī, "al-Waḥy min manẓūr ilāhiyyāt al-maʿrifa," *Al-Maḥajja* 25 (2012): 25.

87 Mullā Ṣadrā, i.e., Muḥammad b. Ibrāhīm b. Yaḥyā l-Qawāmī sh-Shirāzī, was one of the most prominent post-Avicennan figures in Islamic philosophy in the Safavid period, striving to achieve a synthesis of philosophical and Sufi thought. His school of thought is called transcendental philosophy or "al-Ḥikma al-Mutaʿāliya."

88 See Jarādī, "Waḥy," 16f., 28, 30. He explains: The "human carries in herself a true divinizing aspect while the knowledge that comes through revelation also resembles the knowledge of God (28) [. . .] *Waḥy* is not the human and is also not his opposite. [*Waḥy*] is ultimate perfection, not an essence separate from the human but the luminous emanation (*fayḍ*) that evokes in him hidden treasures, illuminating before him the courses of question, contemplation, and answer (18)." *Nafs* can refer to both soul and self. For a more detailed discussion of human nature, see Jarādī, "ʿArsh," 11–26; Shafīq Jarādī, "al-Imān ṣabghat Allāh: Wa-man aḥsan min Allāh ṣabghat," *Al-Maḥajja* 24 (2012): 13–30. He explains further that the first revelation, God's command "be" (*kun*), ushers in the world. The second revelation occurs through the imperative "recite" (*iqrāʾ*), issuing in the Qur'ān. See Jarādī, "Waḥy," 18f.

89 Shafīq Jarādī, "Death as Awakening" (unpublished article), 7–9.

As we have seen, Qānṣū grants the Prophet such ontological openness. But, as in the case of Abū Zayd and Soroush, history interrupts any further immediacy with the Qur'ān as revelation. Their hermeneutics presuppose a metaphysical separation of the world into things in themselves and their meaning, where the individual bestows meaning, structure, and significance upon things. Historical hermeneutics hence entail not just a restructuring of time but also of space that comes to be seen as something apart from the subject who interprets and makes sense of it.[90] Jarādī's hermeneutics defy such a binary conception of reality. His concepts of language and human nature collide with a Cartesian view of the cosmos that predicates the separation of the human from the divine, just as he rejects the Spinozian identification of nature with God.[91]

Instead, Jarādī conceptualizes knowing in line with Ibn ʿArabī and Mullā Ṣadrā. Their idea of *ījād* (God's bestowal of existence upon contingent beings) posits that all things have an inherent meaning because God created them for a specific purpose to begin with.[92] They can be known (or become apparent to the mind) through presence, witnessing (*shuhūd*), and unveiling (*kashf*), i.e., the "disclosure of [the meaning of] existence and intelligibility" through "removing the barriers of material embodiment and ignorance."[93] Yet, more explicitly than Ṣadrā, Jarādī considers the impact of time, place, and culture on the Qur'ān.

Like the modernists, he also asserts that we need new methods to approach it since classical *tafsīr* – human contingent reflection – can only yield so many meanings.[94] Jarādī views humans as multi-dimensional, which has two implications: First, since the Qur'ān addresses different human dimensions, we need multiple techniques and practices to approach it. Second, understanding the Qur'ān in turn also requires the multidimensionality of humans and relies on different and opposing opinions, as others may grasp Qur'ānic meanings to which we remain blind. Jarādī thus embraces a form of *tafsīr* that is open to religious and epistemological pluralism.[95] Other contemporary Mus-

---

90 As Timothy Mitchell argues, historical hermeneutics presuppose a kind of metaphysics, in which "the world is experienced in terms of an ontological distinction between physical reality and its representation – in language, culture, or other forms of meaning. Reality is material, inert, and without inherent meaning, and representation is the non-material, non-physical dimension of intelligibility." See Timothy Mitchell, *Colonising Egypt* (New York: Cambridge University Press, 1988), xiii.

91 Jarādī, "Waḥy," 39.

92 See Ibrahim Kalin, *Knowledge in Later Islamic Philosophy: Mulla Sadra on Existence, Intellect, and Intuition* (New York/Oxford: Oxford University Press, 2010), 223–29. As Kalin explains, "Ṣadrā unites the argument of 'what-ness' (*mā-huwa*) with the argument of 'why-ness' (*limā-huwa*) because to be able to say properly what a thing is, is to be able to say why it is and acknowledge its source. The unassailable relationship between the existence of something is established by the intrinsically intelligible reality of existence." See p. 229.

93 Ibid., 222. See also William Chittick, *The Heart of Islamic Philosophy: The Quest for Self-Knowledge in Afḍal al-Dīn Kashānī* (New York: Oxford University Press, 2001), 174.

94 Jarādī, "Naqd an-naṣṣ," 97f.

95 See Fischbach, "Normative Secularity," 169f.

lim thinkers, like Faḍlallāh, also ascribe to a multilayered concept of truth, albeit from within a different epistemological framework. Their thought demonstrates that the partiality of human knowledge and the humanness of Qur'ānic language, need not entail relativism or the privatization of truth claims, as Mahmood and Hirschkind imply.[96]

3.5 Thinking History with God II: Nayla Ṭabbāra

The work of Lebanese Qur'ān scholar and interfaith activist Nayla Ṭabbāra elucidates further that history can be thought of with God, and that pluralist or polyvalent readings do not require the Qur'ān's disenchantment. Ṭabbāra rejects the idea of an infallible *tafsīr* but affirms that there is a sacred text.[97] To revise the Islamic view of non-Muslims, she reads the Qur'ān diachronically, situating its verses according to their chronological order, and analyzes its stance on Christians in view of dogma, ethics, and politics. Drawing on the Islamic-hermeneutic principle of *naskh*, Ṭabbāra concludes that the Qur'ānic dogmatic standpoint remained consistently positive toward Christians, while its stance on political configurations changed, ushering in what she calls "unity in diversity." This qualified religious pluralism is rooted in the idea of a common origin and return to God and asserts both Islam's truth and religious differences as divinely willed.[98]

Her work portends a certain irrelevance of discussions over the metaphysical status of the Qur'ān for hermeneutics. She affirms a form of progressive revelation oriented toward the socio-political circumstances of its addressees. Yet, her approach does not deny the idea of the Qur'ān as God's word. As she states, "The Qur'ānic verses depict the changeable (*mutaḥawwil*). They treat, oppose, and interact with the changing circumstances [. . .] over a period of 23 years."[99] By insisting that verses depict, i.e., reference instead of reflect, a specific reality, she avoids reducing them to their historical context, leaving room for divine providence. In other words, she does not conceptualize history as secular, linear, homogenous time.

Notably, her point of departure is not the historical, circumstantial but the "absolute unchangeable." She thus follows what Abū Zayd criticized as a descending dialectic that starts from God, proceeds to the Prophet, and finally arrives at the contingent, historical.[100] Her reasoning is firmly Qur'ān-centered with some explanations taken

---

96 Mahmood, "Secularism;" Hirschkind, "Heresy," 467. For Faḍlallāh's deliberations regarding this topic, see Faḍlallāh, *'An sanawāt*, 50–63; Faḍlallāh, *al-'Alāqa*, 29, 167, 191f.

97 Rahel Fischbach, Interview with Nayla Ṭabbāra, December 17, 2012.

98 Nayla Ṭabbāra and Fādī Ḍaw, *ar-Raḥāba al-ilāhiyya: Lāhūt al-ākhir fi-l-Masīḥiyya wa-l-Islām* (Jounieh: al-Maktaba al-Būlusiyya, 2011), 157, 166f. Their book has since then been translated into French, English, and German.

99 Ṭabbāra, *Raḥāba*, 44.

100 Compare Abū Zayd, *Mafhūm*, 29.

from hadith, historical-critical scholarship, *tafsīr*, linguistics, semantics, and rhetorical analysis. She rejects prevalent readings but does so from within the rationalities provided by the pre-modern Islamic sciences. Her analysis thus remains faithful to the classical view of the Qur'ān as both a speech act and a linear, coherent text.

4 Modernity and Polyvalence

Without being exhaustive, this section contextualizes the Qur'ān hermeneutics of the previous thinkers to assess the epistemic rupture modernist approaches are said to imply in comparison to others seen as more congruous with "tradition." In general, reducing the Qur'ān and other Islamic "texts" to their linguistic structure detaches them from the discursive practices that previously produced their multidimensional meaning. Claiming direct access to these texts by means of linguistic method undercuts clerical authority and obscures the ontologies of texts that were typical of manuscript cultures. Yet, as Siraj Ahmed notes, such restructuring of religious authority and knowledge is not confined to historical hermeneutics but characterizes post-colonial religious developments generally. The now common view of Islam as text-based, standardized, and philologically defined is itself an epistemic rupture, as it "impos[es] analytic and bureaucratic order on multilingual terrains."[101]

Abū Zayd's criticism of *tafsīr* and contemporary Egyptian Islamic discourse did not entail a general rejection of tradition. Notably, his main target was those forms of modern Islam that focus on its origins to construe the normatively Islamic. He faulted them precisely not with conservatism but with simplifying, manipulating, and bypassing tradition. While advocating historical consciousness as the basis for reading the Qur'ān, he was more concerned with retrieving the diversity and the "lost dimensions" within the Islamic tradition.[102] Later in his life, he even revoked an exclusively linguistic-philological approach as being too narrow.[103] Qānṣū's vision of "re-structuring the epistemological and religious memory of our present consciousness and thoughts to renew the current interpretative space"[104] similarly addresses the lack of contemporary hermeneutic flexibility. After working at the institution of Faḍlallāh, he turned his focus toward the philosophy of religion, embracing Mohammed Arkoun's call for rethinking the areas of Islamic

101 Ahmed, *Archeology*, 39.

102 Abū Zayd, *Mafhūm*, 12f., 22. Quote from p. 12. He emphasized repeatedly that he merely strove to retrieve material that already existed in the Islamic tradition. See also Kermani, *Offenbarung*, 114.

103 Naṣr Ḥāmid Abū Zayd, "Den Koran neu denken: Für eine humanistische Hermeneutik," in *Gottes Menschenwort: Für ein humanistisches Verständnis des Koran*, ed. and trans. Thomas Hildebrandt (Freiburg: Herder, 2008): 159–228.

104 Qānṣū, *Naṣṣ*, 9.

thought that had been silenced by "external pressure and censorship."[105] The modernists' emphasis on the *Sitz im Leben* of text and reader can be understood as counteracting those who suppressed other opinions by claiming to speak for God and conflating the Qur'ān with their own interpretation.[106]

Today, Ibn Taymiyya's *ad fontes* approach has become popular. But as Walid Saleh contends, lifting the exegetical opinions of the Prophet, *ṣaḥāba*, and the generation following them (*tābiʿūn*) to the level of "prophetic knowledge" broke with much of *tafsīr* convention and only gained widespread acceptance in the latter half of the twentieth century. Yet, despite obvious differences, Ibn Taymiyya shared the modernists' aspiration to sever divine intention from human subjective opinion and considered linguistics an objective science, reliable in deriving truth, too.[107]

Conspicuously, rather than "traditionalism," Abū Zayd identified modernity's drive to clarity and univocity as a main factor for the loss of Qur'ānic meaning and a narrowing conception of Islam. He blamed the state for this situation[108] – an assessment that various studies confirm.

For example, Gregory Starrett demonstrated that in its state-building process, Egypt "secured for Islam an integral part in the governance of the nation," with which it followed the example of its former British colonizers.[109] Serving the needs of the modern nation-state, increasing government control of Islamic institutions and teaching had a lasting impact on Egyptian religious subjectivities. One result was a general trend to functionalize Islam, i.e., to reorient the metaphysical aims of Islamic practice toward socially useful goals. Another was Islam's objectification, i.e., the mode of representing it as a unified, univocal whole.[110] In other words, the "dominant religious discourse" has not been left epistemically unruptured, but was itself affected by new philology, colonialism, and secularism in the form of state administration of religious affairs. Modern sovereign states generally aim to shape religions within their borders and are suspicious of life forms perceived as challenging state legitimacy or national culture.[111]

In Iran, Ruholla Khomeini originally envisioned an Islamic Republic that would entail Islam's pluralism, so that "*fiqh* and interpretive ideas on different topics, *no*

105 For this demand in detail see Mohammed Arkoun, *Rethinking Islam Today* (Washington DC: Center for Contemporary and Arab Studies, Georgetown University, 1987), 19f. Qānṣū takes Arkoun's demand seriously, referencing him frequently.
106 E.g., Abū Zayd, *Naqd*, 60f., 93f.; Abū Zayd, *Mafhūm*, 120–22.
107 Walid A. Saleh, "The Rise of Radical Hermeneutics," in *Ibn Taymiyya and His Times*, eds. Shahab Ahmed and Yossef Rapoport (Karachi/New York: Oxford University Press, 2010): 123–62.
108 Abū Zayd, *Mafhūm*, 13, 25.
109 Gregory Starrett, *Putting Islam to Work: Education, Politics, and Religious Transformation in Egypt* (Berkley: University of California Press, 1998), 6.
110 Starrett, *Putting Islam to Work*, 9, 53, 129, 142, 229.
111 Brown, "Introduction," x.

matter how diverse, [could] engage with one another freely."[112] If this statement contrasts with the censorship prevailing after 1979, the reason was not that Khomeini suddenly denied Islam's polyvalence. Yet, modern nation-states, secular ones included, only allow for so much legal and epistemic flexibility, and declaring Qur'ānic meaning contingent on time privatizes its normative range. Many religious authorities in Iran feared losing political authority, i.e., the prerogative of making binding decisions based on Qur'ān and sunna to determine people's public and private lives.[113] The Revolution placed Islam at the center of Iranian politics and identity construction, rendering approaches that questioned the state version of Islam political as well. The regime feared especially those theories that undermined the truth claims that lent it legitimacy.

In the late 1980s, when Soroush became an influential Shiite thinker for those aiming to found institutions of civil society in Iran, he still deemed the Qur'ān flawless and eternal, believed in the possibility of spiritual infallibility, and shared Khomeini's love of Mullā Ṣadrā. As we have seen in Jarādī's approach, Ṣadrā's optimistic view of humans ascribed to them an innate potential for perfection, capable of gaining religious knowledge that was not "always time-bound."[114] After the Revolution, Soroush's view of humans darkened. Based on his experience of authoritarian Iranian politics, he perhaps feared that the idea of human perfectibility lent itself to totalizing political ideologies.[115] With his perspective on politics, his epistemology changed, too.

There is a general tendency among self-identified secularists, Muslim practitioners, and scholars alike to view Islamic legal discourse as the normatively Islamic. All thinkers presented in this article question the priority status of *fiqh*, albeit to different degrees. For example, Ṭabbāra claims that the privileged status of law in modern times silenced the Qur'ān. Narrowing it to one of its many facets reduced lived Islam by marginalizing alternative, normative Islamic discourses. Before calling their critique anti-traditional, one should note that recent studies have challenged the often-assumed privileged status of *fiqh* in pre-colonial times. Importantly, the rise of the modern nation-state, whose members are significantly more defined by law and its mechanisms than pre-modern people, augmented the authority claim of *fiqh*.[116]

---

112 Quoted in Behrooz Ghamari-Tabrizi, *Islam and Dissent in Post-revolutionary Iran: Abdolkarim Soroush, Religious Politics and Democratic Reform* (London/New York: Palgrave Macmillan, 2008), 145. Emphasis mine.

113 Ghamari-Tabrizi, *Islam*, 188, 241. As Ghamari-Tabrizi affirms, for "the new generation of intellectuals," hermeneutical Islam was a "deeply political issue [. . .] rather than [. . .] an epistemological or cognitive matter." Ibid., 271.

114 Soroush, "Islamic Revival," 31–33. For his earlier thought, see Ghamari-Tabrizi, *Islam*, 93–94, 188, 241.

115 Mahmoud Sadri and Ahmad Sadri, "Introduction," in *Reason, Freedom and Democracy: Essential Writings of Abdolkarim Soroush*, ed. and trans. Mahmoud Sadri and Ahmad Sadri (Oxford: Oxford University Press, 2000), xiii–xix.

116 Shahab Ahmed, *What is Islam? The Importance of Being Islamic* (Princeton/Oxford: Princeton University Press, 2016), 117–25.

The call for a move away from *fiqh* needs therefore not be seen as aiming to fit Islam into a secular framework but as an attempt to correct developments produced by (secular) political modernity. At the same time, Islamic legal practice also took Qurʾānic ambiguity and human shortcomings into account. Classically trained jurists today still follow the praxis of distinguishing between God's eternal intention or *lex aeterna* (*sharīʿa*) and human efforts to discern it.[117] These nuances are often lost in contemporary popular discourse and polemics.

Ṭabbāra and Qānṣū's works need to be read against the chaotic political upheavals in Lebanon over the past fifty years. At the latest since the Civil War (1975–90), the actions of religious militias led to the fragmentation of the country into separate enclaves, organized along confessional identity. In part, their work aims to reclaim religion from those groups that co-opted it for political, sectarian, and military purposes. In Ṭabbāra's view, rather than just "threatening stability and peace" they also transformed religion into an ideological identity giver and undermined Islam's spiritual side.[118] Yet, she does not reduce religion to the internal, private sphere. On the contrary, asserting an important place for religion in politics and the public sphere, she thinks religion will lead people to embrace diversity and impact society positively.[119]

5 Conclusion

The work of Abū Zayd, Soroush, and Qānṣū illustrates the epistemic and political implications of adopting secular history as a unitizing paradigm for reading the Qurʾān. Their hermeneutics entail a transformation of Muslim religious subjectivities by privileging history over other forms of memory construction, and by envisioning a privatized form of Islam that fits a political system "based on religious pluralism and respect for various faiths and convictions."[120] They also champion a model of governance, in which a secular state mediates between its citizens. Viewing scholars who advocate historical Qurʾān hermeneutics as social actors intent on changing the political configurations in which they operate helps understand the socio-religious views

117 See, e.g., Muḥammad Ḥusayn Faḍlallāh, *Fiqh ash-sharīʿa* (Beirut: Dār al-Malāk li-ṭ-Ṭibāʿa wa-n-Nashr wa-t-Tawzīʿ, 2001), 7; Muḥammad Ḥusayn Faḍlallāh, *Fī afāq al-ḥiwār al-islāmī l-masīḥī* (Beirut: Dār al-Malāk li-ṭ-Ṭibāʿa wa-n-Nashr wa-t-Tawzīʿ, 1994), 103.

118 Ṭabbāra, *ar-Raḥāba*, 8f. She explains that the conflation of religion and politics in Lebanon led even non-religious people to identify strongly with their confession and hampered the ability to perceive "others" from a perspective of faith instead of politics.

119 Rahel Fischbach, Interview with Nayla Ṭabbāra, Skype conversation, August 23, 2013.

120 Wahjīh Qānṣū, "at-Taṣāluḥ al-mujtamaʿī bayna ad-dīn wa-ʿalmaniyyat ad-dawla," *Al-Maḥajja* 11 (2010): 57. See also Abdolkarim Soroush, "Intellectual Autobiography: An Interview," in *Reason, Freedom and Democracy: Essential Writings of Abdolkarim Soroush*, eds. Mahmoud Sadri and Ahmad Sadri (Oxford: Oxford University Press, 2000), 20.

expressed in their works. At the same time, it would be simplistic to regard their work as mere reactions to political exigencies. Rather, they strive to rethink the Qur'ān within conceptual frameworks that fit their epistemological and metaphysical convictions – the world they live in.

The binary categories of secular vs. religious, indigenous vs. western, and modern vs. traditional are of only limited use to characterize their work, as their Qur'ān hermeneutics represent negotiations about the place of religion in a post-colonial world, in which the nation-state with its secular practices is already in place. But these categories are also problematic regarding their metaphysical and methodological convictions. The modern-traditional binary implies a certain coherence of the "pre-modern" that sets it clearly apart from the "modern." Most scholars acknowledge epistemic and hermeneutical ruptures in the pre-modern past. Yet, they often assume that those ruptures were not as radical or drastic as the modern one with its perception of the world as historical, natural, and materialistic. However, we should entertain the idea that the modern rupture is only one among others and that it is more porous than the tradition-modernity binary suggests.

The genealogy of regarding methodological innovation as the main indicator of "ruptures" is itself deeply rooted in Enlightenment assumptions about tradition and modernity.[121] The insistence on the radical novelty of the secular-historical worldview identified as modern, and its declared incompatibility with all pre-modern ones, owes much of its cogency to modernist ideology itself.

We need to recognize that these terms are value-laden, often used polemically and with normative intent. As Manfred Sing reminds us, authors are situated in complex and multifold discourses (nationalism, anti-state activism, interfaith dialogue, academia, et al.) that produce tensions and contradictions. Often, authors hold positions that can only be identified in relation to other authors who oppose them. In this respect, it is questionable to categorize authors by reducing them to certain characteristics. These ascriptions are already part of the struggles between different positions about legitimacy and Islamicity. The attributes of traditional-modern perhaps make sense in reference to other dimensions (political stance, pluralism, acceptance of other ways of being in the world, gender, etc.). Yet, the positions of similarly labeled persons on these different themes can vary.[122] Taking "historical consciousness" as the benchmark for modernity loses its heuristic value as soon as one scrutinizes an author's metaphysical, hermeneutical, and epistemological convictions in detail.

The late Protestant anti-critique contended that even the biblical Masoretic text was divinely inspired. The attempt to absolutize the authority of holy scripture was a

121 See Hadia Mubarak, "Change through Continuity: A Case Study of Q 4:34 in Ibn ʿĀshūr," *JQS* 20, no. 1 (2018): 3.

122 Manfred Sing, *Progressiver Islam in Theorie und Praxis: Die interne Kritik am hegemonialen Diskurs durch den „roten Scheich" ʿAbdallāh al-ʿAlāyilī* (Würzburg: Ergon, 2007).

reaction to the new historical criticism rather than a clinging to tradition.[123] One may cautiously claim that a similar development has also taken place in Islam, in which the Qur'ān has increasingly come to be seen as eternally fixed in form and interpretation. A comprehensive examination of the metaphysical convictions, operative in modern self-acclaimed traditional works on the Qur'ān could shed more light on this assumption.

6 Bibliography

6.1 Sources

Abū Zayd, Naṣr Ḥāmid. *Mafhūm an-naṣṣ: Dirāsa fī ʿulūm al-Qur'ān*. Cairo: al-Hay'a al-Miṣriyya al-ʿĀmma li-l-Kitāb, 1990.

Abū Zayd, Naṣr Ḥāmid. *Naqd al-khiṭāb ad-dīnī*. Cairo: Sīnā li-n-Nashr, 1992.

Abū Zayd, Naṣr Ḥāmid. "Den Koran neu denken: Für eine humanistische Hermeneutik." In *Gottes Menschenwort: Für ein humanistisches Verständnis des Koran*. Ed. and trans. Thomas Hildebrandt. Freiburg: Herder, 2008: 159–228.

Faḍlallāh, Muḥammad Ḥusayn. "Al-ʿAlāqa al-islāmiyya al-masīḥiyya: Qirā'āt islāmiyya fī r-rāhin wa-l-mustaqbal." In *al-ʿAlāqāt al-islāmiyya al-masīḥiyya: Qirā'āt marjaʿiyya fī t-tārīkh wa-l-ḥāḍir wa-l-mustaqbal*. Beirut: Markaz ad-Dirāsāt al-Istrātījiyya wa-l-Buḥūth wa-t-Tawthīq, 1994: 18–33.

Faḍlallāh, Muḥammad Ḥusayn. *Fī āfāq al-ḥiwār al-islāmī l-masīḥī*. Beirut: Dār al-Malāk li-ṭ-Ṭibāʿa wa-n-Nashr wa-t-Tawzīʿ, 1994.

Faḍlallāh, Muḥammad Ḥusayn. *Dunyā al-mar'a*. Beirut: Dār al-Malāk li-ṭ-Ṭibāʿa wa-n-Nashr wa-t-Tawzīʿ, 1997.

Faḍlallāh, Muḥammad Ḥusayn. *Fiqh ash-sharīʿa*. Beirut: Dār al-Malāk li-ṭ-Ṭibāʿa wa-n-Nashr wa-t-Tawzīʿ, 2001.

Faḍlallāh, Muḥammad Ḥusayn. *ʿAn sanawāt wa-mawāqif wa-shakhṣiyyāt: Hākadhā taḥaddatha – hākadhā qāla. Ḥāwarathu Munā Sukariyya*. Beirut: Dār an-Nahār, 2007.

Ibn ʿĀshūr, Muḥammad aṭ-Ṭāhir. *Aṭ-Taḥrīr wa-t-tanwīr*. 30 vols. Tunis: ad-Dār at-Tūnisiyya li-n-Nashr, 1984.

Jarādī, Shafīq. "Naqd an-naṣṣ ad-dīnī." In *Aʿmāl muntadā l-fikr al-masīḥī l-islāmī fī kulliyyat al-lāhūt li-sh-sharq al-adnā*. Vol. 2, *Ḥiwār al-ḥaqīqa min ajli al-ḥayāt maʿan*, edited by George Sabra. 4 vols. Beirut: Manshūrat Kulliyyat al-Lāhūt li-sh-Sharq al-Adnā, 2008: 95–103.

Jarādī, Shafīq. "Al-Imān ṣabghat Allāh: Wa-man aḥsan min Allāh ṣabghatan." *Al-Maḥajja* 24 (2012): 13–30.

Jarādī, Shafīq. "Al-Waḥy min manẓūr ilāhiyyāt al-maʿrifa." *Al-Maḥajja* 25 (2012): 15–43.

Jarādī, Shafīq. "ʿArsh ar-rūḥ wa-insān ad-dahr." *Al-Maḥajja* 26 (2013): 11–26.

Jarādī, Shafīq. "Death as Awakening." (unpublished article).

Qānṣū, Wajīh. "At-Taṣāluḥ al-mujtamaʿī bayna ad-dīn wa-ʿalmaniyyat ad-dawla." *Al-Maḥajja* 11 (2010): 41–66.

Qānṣū, Wajīh. *An-Naṣṣ ad-dīnī fī-l-Islām min at-tafsīr ilā t-talaqqī*. Beirut: Dār al-Fārābī, 2011.

Qānṣū, Wajīh. "An-Naṣṣ ad-dīnī wa-rihānāt al-maʿnā s-sīyāsī." *Qaḍāya Islāmiyya Muʿāṣira* 19, no 61–62 (2015): 87–127.

Quṭb, Sayyid. *Milestones: Ma'alim fi'l-tareeq*, trans. A.B. al-Mehri. Birmingham: Maktabah Booksellers and Publishers, 2006.

Aṣ-Ṣadr, as-Sayyid Muḥammad Bāqir. *Al-Madrasa al-Qur'āniyya*. Beirut: Dār at-Taʿāruf, 1981.

123 Marius Reiser, *Bibelkritik und Auslegung der Heiligen Schrift* (Tübingen: Mohr Siebeck, 2007), 259.

Soroush, Abdolkarim. "Intellectual Autobiography: An Interview." In *Reason, Freedom and Democracy: Essential Writings of Abdolkarim Soroush*. Eds. Mahmoud Sadri and Ahmad Sadri. Oxford/New York: Oxford University Press, 2000: 3–25.
Soroush, Abdolkarim. "Islamic Revival and Reform: Theological Approaches." In *Reason, Freedom, and Democracy: Essential Writings of Abdolkarim Soroush*. Ed. Mahmoud Sadri and Ahmad Sadri. Oxford/New York: Oxford University Press, 2000: 26–38.
Soroush, Abdolkarim. *The Expansion of Prophetic Experience: Essays on Historicity, Contingency, and Plurality in Religion*. Ed. Forough Jahanbakhsh, trans. Nilou Mobasser. Leiden/Boston: Brill, 2009.
Ṭabbāra, Naylā, and Fādī Ḍaw. *Ar-Raḥāba al-ilāhiyya: Lāhūt al-ākhar fi l-Masīḥiyya wa-l-Islām*. Jounieh: al-Maktaba al-Būlusiyya, 2011.

6.2 Secondary Literature

Ahmad, Anzaruddin. "Applying Hermeneutics to the Qur'an: A Recritique." In *Critical Perspectives on Literature and Culture in the New World Order*, edited by Noritah Omar et al. Newcastle: Cambridge Scholars Publishing, 2010: 81–100.
Ahmed, Shahab. *What is Islam? The Importance of Being Islamic*. Princeton/Oxford: Princeton University Press, 2016.
Ahmed, Siraj. *Archeology of Babel: The Colonial Foundation of the Humanities*. Stanford: Stanford University Press, 2018.
Akbar, Ali. *Contemporary Perspectives on Revelation and Qur'anic Hermeneutics: An Analysis of Four Discourses*. Edinburgh: Edinburgh University Press, 2020.
Arkoun, Mohammed. *Rethinking Islam Today*. Washington DC: Center for Contemporary and Arab Studies, Georgetown University, 1987.
Asad, Talal. *Formations of the Secular: Christianity, Islam, Modernity*. Stanford, CA: Stanford University Press, 2003.
Asad, Talal. "The Idea of an Anthropology of Islam," *Qui Parle* 17, no. 2 (2009): 1–30.
Bauer, Thomas. *Die Kultur der Ambiguität: Eine andere Geschichte des Islam*. Berlin: Verlag der Weltreligionen, 2011.
Beckford, James. *Social Theory and Religion*. Cambridge/New York: Cambridge University Press, 2003.
Benjamin, Walter. *Illuminations*, trans. Harry Zohn. New York: Fontana, 1982.
Bhabha, Homi K. *The Location of Culture*. New York: Routledge, 1994.
Blaut, James. *Colonizer's Model of the World: Geographical Diffusionism and Eurocentric History*. New York: Guilford Press, 1993.
Brown, Wendy. "Introduction." In *Is Critique Secular?*, edited by Talal Asad et al. New York: Fordham University Press, 2013: 7–20.
Cavanaugh, William T. *The Myth of Religious Violence: Secular Ideologies and the Roots of Modern Conflict*. Oxford: Oxford University Press, 2009.
Chakrabarty, Dipesh. *Provincializing Europe: Postcolonial Thought and Historical Difference*. Princeton/London: Princeton University Press, 2000.
Chittick, William. *The Heart of Islamic Philosophy: The Quest for Self-Knowledge in Afḍal al-Dīn Kashānī*. New York: Oxford University Press, 2001.
Conrad, Sebastian. "Enlightenment in Global History: A Historiographical Critique." *American Historical Review* 117, no. 4 (2012): 999–1027.
De Certeau SJ, Michel. *The Sixteenth and Seventeenth Centuries*, trans. Michael B. Smith. Vol. 1, *The Mystic Fable*, 2 vols. Chicago: University of Chicago Press, 1992.

Dobie, Robert J. *Logos and Revelation: Ibn 'Arabi, Meister Eckhart, and Mystical Hermeneutics*. Washington DC: Catholic University of America Press: 2010.
Dorroll, Philip. "The Doctrine of the Nature of the Qur'ān in the Māturīdī Tradition." In *Mâtürîdî Düşünce ve Mâtürîdîlik Literatürü*, edited by M. Raşit Akpınar, Recep Tuzcu, and Aslı Menekşe. Istanbul: Selçuk Üniversitesi İmâm Mâtürîdî Uygulama ve Araştırma Merkezi, 2018: 123–40.
Fischbach, Rahel. Interview with Naylā Ṭabbāra. December 17, 2012.
Fischbach, Rahel. Interview with Naylā Ṭabbāra. Skype conversation. August 23, 2013.
Fischbach, Rahel. "Beyond Normative Secularity: Politics of Interfaith Dialogue in Lebanon." In *The Interfaith Movement: Mobilising Religious Diversity in the 21st Century*, edited by John Fahy and Jan-Jonathan Bock. New York/London: Routledge 2019: 158–78.
Foucault, Michel. *Power/Knowledge: Selected Interviews and Other Writings 1972–77*, trans. Colin Gordon. New York: Pantheon, 1980.
Frank, Daniel. *Search Scripture Well: Karaite Exegetes and the Origins of the Jewish Bible Commentary in the Islamic East*. Leiden: Brill, 2004.
Funkenstein, Amos. *Theology and the Scientific Imagination from the Middle Ages to the Seventeenth Century*. Princeton: Princeton University Press, 1986.
Ghamari-Tabrizi, Behrooz. *Islam and Dissent in Post-revolutionary Iran: Abdolkarim Soroush, Religious Politics and Democratic Reform*. London/New York: Palgrave Macmillan, 2008.
Habermas, Jürgen. "Die Dialektik der Säkularisierung." *Blätter für deutsche und internationale Politik* 4 (2008): 33–46.
Hirschkind, Charles. "Heresy or Hermeneutics: The Case of Nasr Hamid Abu Zayd." *Stanford Humanities Review* 5, no. 1. (1996): 35–50.
Izutsu, Toshihiko. *God and Man in the Qur'an: Semantics of the Qur'anic Weltanschauung*. Tokyo: Keyo University, 2008.
Al-Jābirī, Muḥammad ʿĀbid. *Bunyat al-ʿaql al-ʿarabī: Dirāsa taḥlīliyya naqdiyya li-naẓm al-maʿrifa fī th-thaqāfa al-ʿarabiyya*. Beirut: Markaz Dirāsāt al-Waḥda al-ʿArabiyya, 2013.
Kalin, Ibrahim. *Knowledge in Later Islamic Philosophy: Mulla Sadra on Existence, Intellect, and Intuition*. New York/Oxford: Oxford University Press, 2010.
Kant, Immanuel. *Religion Within the Bounds of Bare Reason*, trans. Werner S. Pluhar. Indianapolis: Cambridge University Press, 2009.
Kermani, Navid. *Offenbarung als Kommunikation: Das Konzept waḥy in Naṣr Ḥāmid Abū Zayds Mafhūm al-Naṣṣ*. Frankfurt a.M.: Lang, 1996.
Koloska, Hannelies. *Offenbarung, Ästhetik und Koranexegese: Zwei Studien zu Sure 18 (al-Kahf)*. Wiesbaden: Harrassowitz, 2015.
Lincoln, Bruce. "Theses on Method." *Method and Theory in the Study of Religion* 8 (1996): 225–27.
Mahmood, Saba. "Secularism, Hermeneutics, and Empire: The Politics of Islamic Reformation." *Public Culture* 18, no. 2 (2006): 323–47.
Mehta, Uday Singh. *Liberalism and Empire: A Study in Nineteenth-Century British Liberal Thought*. London/Chicago: The University of Chicago Press, 1999.
Mitchell, Timothy. *Colonising Egypt*. New York: Cambridge University Press, 1988.
Montag, John. "Revelation: The False Legacy of Suarez." In *Radical Orthodoxy: A New Theology*, edited by John Milbank, Catherine Pickstock, and Graham Ward. London/New York: Routledge, 1999: 38–63.
Mubarak, Hadia. "Change through Continuity: A Case Study of Q 4:34 in Ibn ʿĀshūr." *JQS* 20, no. 1 (2018): 1–27.
"Munāẓarat ad-duktūr Muḥammad ʿAmāra, al-mufakkir al-islāmī l-kabīr, maʿa Naṣr Abū Zayd, ash-shuyūʿī, ʿan riddatihi" (Debate between Dr. Muḥammad ʿAmāra, the great Islamic thinker and Naṣr Abū Zayd, the Marxist, on his apostasy). Youtube Video. Qatar Television December 31, 1996. Posted by "video max." November 9, 2013. Accessed June 24, 2016. https://www.youtube.com/watch?v=KWSIQGLMOcc.

Nandy, Ashis. *The Romance of the State and the Fate of Dissent in the Tropics*. New Delhi: Oxford University Press, 2003.
Oddbjørn, Leirvik. "*Waḥy* and *tanzīl*: Modern Islamic approaches to divine inspiration, progressive revelation, and human text." *Studia Theologica* 69, no. 2 (2015): 101–21.
Rabb, Intisar. *Doubt in Islamic Law: A History of Legal Maxims, Interpretation, and Islamic Criminal Law*. New York: Cambridge University Press, 2015.
Reiser, Marius. *Bibelkritik und Auslegung der Heiligen Schrift*. Tübingen: Mohr Siebeck, 2007.
Ricoeur, Paul. *Memory, History, Forgetting*. London/Chicago: University of Chicago Press, 2006.
Sadri, Mahmoud, and Ahmad Sadri. "Introduction." In *Reason, Freedom, and Democracy: Essential Writings of Abdolkarim Soroush*. Ed. Mahmoud Sadri and Ahmad Sadri. Oxford/New York: Oxford University Press, 2000: ix–xix.
Saeed, Abdullah. "Fazlur Rahman: a Framework for Interpreting the Ethico-legal Content of the Qur'an." In *Modern Muslim Intellectuals and the Qur'an*, edited by Suha Taji-Farouki. Oxford: Oxford University Press, 2006: 37–66.
Saeed, Abdullah. *Interpreting the Qurʾān: Towards a Contemporary Approach*. New York: Routledge, 2006.
Saleh, Walid A. *The Formation of the Classical Tafsīr Tradition: The Qurʾān Commentary of al- Thaʿlabī (d. 427/1035)*. Leiden/Boston: Brill, 2004.
Saleh, Walid A. "The Rise of Radical Hermeneutics." In *Ibn Taymiyya and His Times*, edited by Shahab Ahmed and Yossef Rapoport. Karachi/New York: Oxford University Press, 2010: 123–62.
Sands, Kristin Zahra. *Ṣūfī Commentaries on the Qurʾān in Classical Islam*. London/New York: Routledge, 2006.
Schwarb, Gregor. "Capturing the Meaning of His Speech: The Relevance of *uṣūl al-fiqh* to an understanding of *uṣūl al-tafsīr* in Jewish and Muslim *kalām*." In *A Word Fitly Spoken: Studies in Mediaeval Exegesis of the Hebrew Bible and the Qur'ān*, edited by Meir Bar-Asher et al. Jerusalem: The Ben-Zwi Institute, 2007: 111–56.
Şeker, Nimet. *Der Koran als Rede und Text: Hermeneutik sunnitischer Koranexegeten zwischen Textkohärenz und Offenbarungskontext*. Berlin: EB-Verlag, 2019.
Sing, Manfred. *Progressiver Islam in Theorie und Praxis: Die interne Kritik am hegemonialen Diskurs durch den „roten Scheich" ʿAbdallāh al-ʿAlāyilī*. Würzburg: Ergon, 2007.
Smart, Ninian. "Editorial Introduction." In *Nineteenth Century Religious Thought in the West*, edited by Smart et al. 3 vols. Cambridge: Cambridge University Press, vol. 1, 1985.
Spivak, Gayatri. "Can the Subaltern Speak?." In *Marxism and the Interpretation of Culture*, edited by Cary Nelson and Lawrence Grossberg. Urbana, IL: University of Illinois Press, 1998: 280–83.
Starrett, Gregory. *Putting Islam to Work: Education, Politics, and Religious Transformation in Egypt*. Berkley: University of California Press, 1998.
Sugirtharajah, Rasiah S. *Exploring Postcolonial Biblical Criticism: History, Method, Practice*. Chichester: Wiley-Blackwell, 2012.
Van Ess, Josef. *TG*. 6 vols. Berlin/New York: De Gruyter, 1991.
Van Ess, Josef. "Verbal Inspiration?" In *The Qur'an as Text*, edited by Stefan Wild. Leiden/New York/Köln: Brill, 1996: 177–95.
Völker, Katharina. *Quran and Reform: Rahman, Arkoun, Abu Zayd*. Frankfurt a.M.: Peter Lang, 2017.
Wielandt, Rotraud. *Offenbarung und Geschichte im Denken moderner Muslime*. Wiesbaden: F. Steiner, 1971.
Wielandt, Rotraud. "Wurzeln der Schwierigkeit innerislamischen Gesprächs über neue hermeneutische Zugänge zum Koran." In *The Qur'an as Text*, edited by Stefan Wild. Leiden: Brill, 1996: 257–82.
Wild, Stefan. "We have sent down to thee the book with the truth. Spatial and temporal implications of the Qur'anic concepts of *nuzūl*, *tanzīl* and *inzāl*." In *The Qur'an as Text*, edited by Stefan Wild. Leiden/New York/Köln: Brill, 1996: 137–41.
Wolfson, Harry A. *The Philosophy of the Kalam*. Cambridge/Massachusets: Harvard University Press, 1993.

Felix Körner S.J.

Modern Qur'ānic Hermeneutics in Turkey

1 Terminology and Delimitation

Before we really start, let us clarify three conceptual issues raised by the chapter's headline. "Modern" means here that late twentieth-century and early twenty-first-century reflections will be presented.[1] The term "Qur'ānic hermeneutics" works on several levels. Three distinctions may be helpful. First, between 'explicit' and 'implicit' hermeneutics. What the interpreters *say* they do when interpreting is 'explicit' hermeneutics. It may be quite different from what they actually *do*: their implicit hermeneutics. A second distinction is that between the Qur'ān as 'object' and as 'subject.' The 'object' branch of Qur'ānic hermeneutics reviews what *interpreters* of the Qur'ān do and claim they do. The 'subject' branch of Qur'ānic hermeneutics studies how the *Qur'ān* interprets itself. Again, the 'implicit/explicit' distinction is helpful.[2] A third dis-

1 Cf., on this and the following themes: Akif Koç and İsmail Albayrak, eds., *Tefsire Akademik* Yaklaşımlar (Ankara: Otto, 2013); and by the present author: Felix Körner, *Revisionist Koran Hermeneutics in Contemporary Turkish University Theology. Rethinking Islam* (Würzburg: Ergon, 2005) (used here throughout the article); Körner, "Turkish Theology Meets European Philosophy. Emilio Betti, Hans-Georg Gadamer and Paul Ricœur in Muslim Thinking," *Revista Portuguesa de Filosofia* 62 (2006): 805–9; Körner, "When Islam Receives Criticism. Historical Koran Exegesis in Today's Turkey," *Rosenzweig Jahrbuch* 2 (2007): 153–69; Körner, "Bibelhermeneutik. Islamwissenschaftlich," "Geschichte und Geschichtlichkeit. Islamwissenschaftlich," and "Historizität. Islamwissenschaftlich," in *Lexikon der Bibelhermeneutik. Begriffe – Methoden – Theorien – Konzepte*, ed. Oda Wischmeyer (Berlin: de Gruyter, 2013), 209–10, 266, 306–7. Further literature in the following footnotes.

2 Since this is less widely seen, we need to explicate it briefly. Studying the Qur'ān's own ('subject') *implicit* hermeneutics is to uncover the Qur'ān's *de facto* intertextual strategies. Two examples may be adduced. First, how does the text later *explain* an unclear expression used previously? A marker at the beginning of such explicatory passages can be *yas'alūnaka 'an*, "they ask you about . . ." or the didactic question *mā adrāka mā*, "how can you know what . . . is?" Second, how does the text later *reinterpret* motifs which were used differently before (*interpretative Fortschreibung*, Nicolai Sinai). For example, how is the understanding of the figure of Ibrāhīm updated in light of the hijra? On this, cf., Angelika Neuwirth, *Koranforschung – eine politische Philologie? Bibel, Koran und Islamentstehung im Spiegel spätantiker Textpolitik und moderner Philologie* (Berlin: de Gruyter, 2014), and already her dictum that the exegesis of the Qur'ān begins while the genesis of the Qur'ān is still on the way: Neuwirth, "Koran," in *Grundriß der arabischen Philologie*, vol. 2, *Literaturwissenschaft*, ed. Hartmut Gätje (Wiesbaden: Ludwig Reichert, 1987), 96–135, 121; further: Nicolai Sinai, *Fortschreibung und Auslegung. Studien zur frühen Koraninterpretation* (Wiesbaden: Harrassowitz, 2009), 59. The Qur'ān's own ('subject') *explicit* hermeneutics is what the Qur'ānic text itself says about its usage. We might call that the metatextual rules: clarification is up to God (Q 75:19); and interpretation is not up to human opinion, Q 3:7:

Note: The present chapter was written during my fellowship at Wissenschaftskolleg zu Berlin. I am profoundly grateful for the help provided by the institution and its staff.

https://doi.org/10.1515/9783111320069-024

tinction needs to be applied when different 'object' hermeneutics are being studied: namely, the difference between 'descriptive' and 'normative' hermeneutics needs to be seen. Descriptive Qur'ānic hermeneutics *portrays* endeavors of other interpreters, e.g., exegetes who are considered classical. Normative Qur'ānic hermeneutics *proposes* models of understanding the Qur'ān today. The restriction to Turkey means that we will present authors who live and work in the country thus called. Sometimes, however, the linguistically 'Turkic' world, or the Ottoman empire, or emigrant authors with a primarily Turkish academic career, will also be taken into consideration.

With these clarification in mind, we can now draw this chapter's delimitation: it focuses on "explicit," "object," "normative" Qur'ānic hermeneutics in Turkey. For the other areas brief indications may suffice:

"Descriptive" Qur'ānic hermeneutics is quite common, also among Turkish authors.[3]

The "subject" branch of Qur'ānic hermeneutics is, in Turkey, a new area of research.[4]

"Implicit" Turkish Qur'ān hermeneutics can be observed, e.g., in the various Tefsir/ Meâl works or in popularizing modernists.[5]

2 Historical, Socio-political, Intellectual, and Academic Contexts

In Turkey, most producers of explicit Qur'ānic hermeneutics are professional academics; they may be working at centers for Qur'ānic studies – e.g., Kur'an Araştırmaları Merkezi (KURAMER), İslam Araştırmaları Merkezi (İSAM) – or, typically, at the universities' divinity schools. The paradoxical history of the establishment of these schools deserves attention. They are simultaneously traditional – and they are a novelty; They continue Ottoman lines, and take up Western structures; their foundation puts into action a specifically Kemalist agenda – and may be relevant for global Islam.

"Some of its verses are definite in meaning (*muḥkam*), these are the mother of the Scripture – and others are ambiguous (*mutashābih*). The perverse at heart eagerly pursue the ambiguities in their attempt to make trouble and to pin down a meaning (*ta'wīl*); but only God knows the true meaning."

3 İsmail Cerrahoğlu, *Tefsir Tarihi* (Ankara: Fecr Yayınevi, 2009); Cerrahoğlu, *Tefsir Usûlü* (Ankara: Türkiye Diyanet Vakfı, 2016); Muhsin Demirci, *Tefsir Tarihi* (Istanbul: MÜİFAV, 2014); Demirci, *Tefsir Usulü* (Istanbul: MÜİFAV, 2016).

4 Esra Gözeler, "Surelerin Mekkīliği ve Medenīliği. 22/el-Ḥacc Suresi Örneği," *Ankara Üniversitesi İlahiyat Fakültesi Dergisi* 58 (2017): 69–108.

5 Cf., on Salih Akdemir: Felix Körner, ed. and trans., *Glaube in Gesellschaft. Ankaraner Stimmen zur Politischen Theologie. Ausgewählte Texte übersetzt und kommentiert von Felix Körner SJ* (Freiburg: Herder, 2019): 229–63; on Yaşar Nuri Öztürk: Felix Körner, ed. and trans., *Alter Text – Neuer Kontext. Koranhermeneutik in der Türkei heute. Ausgewählte Texte, übersetzt und kommentiert von Felix Körner SJ* (Freiburg: Herder, 2006, 2015²), 209–43, and Kathrin Eith, *Koranexegese bei Yaşar Nuri Öztürk. Ein traditionskritischer Entwurf des Islams in der Türkei* (Würzburg: Ergon, 2013).

2.1 From Istanbul to Ankara

The Ottoman Ministry of Education, established in 1847/1263, inaugurated a European-style educational system designed to replace the traditional Islamic school (medrese).[6] In 1900/1318, a Department of Theology (*'ulūm-i 'āliye-i dīnīye shu'besi*) was created at the University (*Dār ül-fünun*, Istanbul). In 1910/1328, a twelve-year pre-university course of theology was established. It included only traditional subjects and textbooks, with one exception: in the seventh grade, "modern physics" (*ḥikmet-i cedīde*) was to be studied. In 1914/1332, the university department was replaced with a specialists' school at Istanbul's Süleymānīye Mosque. The school consisted of the three traditional departments: Tefsīr and Ḥadīth; Fiqh; Kelām and Ḥikmet (the latter covered the classically philosophical disciplines from cosmology to ethics). A law from 1917/1335 added a fourth department, namely, Literature.

In 1923/1341, the Republic of Turkey was founded. Mustafa Kemal (Atatürk) was the key figure in the process. He epitomized his project programmatically with the declaration, "We will turn ourselves into the most modern nation." As a consequence, all medreses were brought under the authority of the Ministry of Education, which ultimately led to their closure.

In 1924/1342, the Great National Assembly (in Ankara) made three decisions relevant to our topic. First, the caliphate was abolished. Second, the foundation of *İmām Ḫatīp Mektepleri* was decreed: schools for prayer leaders and Friday preachers. Soon, Turkey saw twenty-nine such schools. The Istanbul school's director had an "enthusiastic belief in the possibilities of reforming Islam to become a vital and inspiring force in the life of new Turkey."[7] But with a continuous decline in attendance, by 1932/1351 all *İmām Ḫatīp* (from 1928/1346 on, spelled "İmam Hatip") schools were closed. The only surviving, indeed thriving, type of junior theological education was the *dār ül-qurrā'* (school for Qur'ān recitation).

Third, the establishment of a Faculty of Theology at the University (of Istanbul) was decided upon. It was to be structured according to "modern scientific principles." But its students' numbers were decreasing rapidly, together with those of the İmam Hatip schools. In 1929/1347, a decree by the Ministry of Education abolished Arabic and Persian classes from the high school curriculum. Subsequently, knowledge of these languages was no longer a prerequisite for admission to the Faculty of Theology. In 1933/1352 and 1934/1353, laws regulated a reform of the University. The Faculty of Theology was turned into the "Institute of Islam Researches" (Islâm Tedkikleri Enstitüsü) and was now considered a department of the Literature Faculty's Oriental Institute.

6 Cf., Gotthard Jäschke, "Der Islam in der neuen Türkei. Eine rechtsgeschichtliche Untersuchung," *WI* 1 (1951): 1–174; corrections and additions: ibid. 2 (1953): 278–87; Howard A. Reed, "The Faculty of Divinity at Ankara. Part I," *MW* 46 (1956): 295–312; Reed, "The Faculty of Divinity at Ankara. Part II," *MW* 47 (1957): 22–35; Körner, *Revisionist Koran Hermeneutics*, 48–62.

7 Jäschke, "Der Islam in der neuen Türkei," 122.

In 1938/1357, Atatürk died. In 1947/1366, Turkey became a multi-party democracy. The demand for better training of Islamic religious staff was voiced. Private seminaries were opened, but students had to graduate from a secular school before being admitted to them. Less than a year later, however, parliamentarians demanded a re-opening of state-run preachers' schools. The atmosphere was heated. Two contemporary voices[8] illustrate the argument and the tensions:

> We have to save from extinction, or rather, to train anew, a generation of religious staff, both knowledgeable (bilgili) and patriotic (vatansever). (M. Tuncer)

> Not in the least shall we verge from the principle of laicism. Religion however is not only the bond between the individual and God. It is also a social fact, a matter of the community (cemaat işi). In the countryside, there are often no more religious representatives (hocalar) to bury the dead. If we do not educate religious staff, superstition will prevail. Already, we hear about heavenly letters (semavî mektublar) passing from hand to hand. Religious fanaticism is bad. But fanatic irreligion does not match with our tolerant society, either. It was not in order to ban Islam from Turkey that we accepted the principle of laicism. Against the Soviet advance, we have to utilize the force of the Islamic Belt, stretching from Java to the Atlantic Ocean. (Cihat Baban)

In 1948/1367, the reinstitution of İmam Hatip schools was decreed. Unlile before, they would no longer be under the jurisdiction of the Presidency of Religious Affairs (Diyanet), but under the Ministry of Education. This decision aimed to prevent a resurgence of the dualism of religious and secular education, of medreseli/mektebli mentality "still existent in oriental countries," as noted by Minister of Education Hasan Tahsin Banguoğlu (1904–1989/1322–1409).

Also in 1948/1367, the People's Party decided to establish a School of Islamic Theology in Ankara; since Turkish university structures follow German models, it was not named "School" but "Faculty (fakülte)". In 1949/1368, the Great National Assembly issued the relevant law. It was Banguoğlu again who spoke programmatically at the occasion. He demanded that the Faculty should counter reactionary movements: "It is a torch which the fabulists flee like bats." He suggested that on top of the classical canon, the Faculty should offer courses in psychology, sociology, general and comparative religious studies. In addition to Arabic and Persian, European languages should be taught; and alongside teachers from the former Istanbul Faculty, some lecturers from other Islamic countries should be employed, e.g., from Pakistan. Perhaps, Banguoğlu said, it was due to the revolutionary atmosphere that the old Faculty had been unable to find a sufficient number of students; with more than one hundred young men following the İmam Hatip courses in 1949/1368, things seemed to have changed. When Ankara University's president Hikmet Birand (d. 1972/1392) welcomed the new Faculty to the 1949/50/1368/69 winter term, he said: "The re-

8 Unless otherwise indicated, the following quotes are from Jäschke, "Der Islam in der neuen Türkei", 125 and 124f., my translation from Jäschke's German. He marks out the mention of the Atlantic Ocean with: "[!]".

ligious spirit can be a great power for Enlightenment (aydınlatıcı). [. . .] Religious reforms accomplished in past centuries by individuals are today being brought about by life itself, by society. To observe and record these events and, if necessary, take the leading role is the task of an alert institute of religion."

The Ankara Faculty saw immediate resistance in the country, both from the traditionalist and the secularist/revolutionary wings of the (ruling) People's Party. The traditionalists lamented subordination to the government. The secularists wanted to see religion confined to the private realm. We can identify those tensions also in light of the Ankara Faculty's name: İlahiyat Fakültesi. The designation *ilāhiyyāt* appears in many classical works of theology. But Arabic does not use the word for the whole discipline of 'theology'; it can, however, be used to mean parts of it: 'metaphysics,' 'doctrine of God,' or 'speculative theology.' In Turkish, by contrast, the name itself was used for the whole of theology, and as a marker of the new agenda – already for the short-lived 1924/1342 Istanbul institution. The 1949/1368 project explicitly wanted to avoid extremes. Neither did the founders have in mind another *kulliyyat ash-sharī'a* in a closed-in *madrasa* atmosphere, nor a mere philosophical, let alone sociological reflection on religion. Their intention was, rather, explicitly "Garptaki örneklerine benzer bir ilâhiyat fakültesinin kurulmasını – to establish a Divinity School similar to models in the West";[9] that is, a Faculty of Theology, just like a "Katolik Fakültesi, Protestan Fakültesi."[10] That meant that it needed to be "ilmî": scholarly, scientific. What is "scientific" theology? The pedagogue and writer İsmail Hakkı Baltacıoğlu (d. 1978/1368) explicated this in parliament: "Faculties, because they are houses of science, strive for comparison, empirical knowledge, and eventually, if possible, for explanation. So, as much as the [medrese] is subjective, the [faculty] is objective. This is the government's notable will: to create a Faculty of scientific dignity, of scientific character."[11]

9 İsmail (İsmayıl) Hakkı Baltacıoğlu published several titles in the field of religion, too; most notably, an article "Towards Religion," in which he argues that religion is not only the source of social institutions but also lives within these institutions, and life is therefore as impossible without religion as it is without blood ("Dine Doğru," *Ankara Üniversitesi İlahiyat Fakültesi Dergisi* 6 (1957): 44–59. The above quote is from what he said in Turkish Parliament on June 4, 1949: *T.B.M.M. Tutanak Dergisi*, Dönem: VIII, Cilt: 20, Toplantı: 3, 101. Birleşim (4. VI. 1949): 279, accessed online on April 24, 2020, https://www.tbmm.gov.tr/tutanaklar/TUTANAK/TBMM/d08/c020/tbmm08020101.pdf. Baltacıoğlu was in this very sentence quoting the justification (gerekçe) of the bill that proposed the foundation of an "İslâm İlahiyatı Fakültesi," as he called it (278).

10 Ibid., 278.

11 Ibid., 278. To give a live impression of the era's thinking in sharp contrast, a fuller quote is given here: The intention (Maksat) to establish a Divinity Faculty "is not to resurrect the medrese. For the difference between Faculty and medrese is quite fundamental. It is this: medrese studies are text-positivist and a priori. Faculties, because they are houses of science, strive for comparison, empirical knowledge and eventually, if possible, for explanation. So, as much as the first is subjective, the second is objective. This is the government's notable will: to create a faculty of scientific dignity, of scientific character. – Medreseyi diriltmek değildir. Çünkü fakülte ile medreseyi ayıran çok temelli bir karakter vardır. O da şudur: Medresenin çalışması nassi, apriyori, kablettecrübidir. Fakülteler, ilim evleri olduğundan, bunlar

2.2 From Ankara to Turkey and to Europe

Ankara Üniversitesi İlahiyat Fakültesi (AÜİF, Ankara Divinity School) became the model for many analogous foundations in Turkey. Between 1959/1378 and 1982/1402, eight Higher Islam Institutes were established to train religious education teachers and employees of the Presidency of Religious Affairs. In 1982/1402, seven of the eight institutes were converted into İlahiyat Faculties. Beginning in 1987/1407, many more Faculties of Theology were founded in Turkey. By 2005/1426, there were twenty-five such schools, all at public universities. Under the rule of the AK Party, in 2008/1429 private universities were also allowed to establish İlahiyat Faculties. By 2020/1441, more than 110 such schools had been founded in Turkey.

Kemal Atatürk's project was nothing less than a cultural revolution. Present-day Turkey seems to be undergoing a second such revolution. Especially after the 2016/1437 coup d'état attempt, Turkish state action and social atmosphere interfere with academic freedom. Theologians, like other academics, perceive the transformation of Turkey as having led them into "a society of fear." Not only for that reason, several leading figures in Qur'ānic hermeneutics have moved to European universities. This helps spread the previously rather local impact of Turkish theological approaches internationally. Ideas on Qur'ānic hermeneutics developed by Turkish authors are, however, also disseminated through other channels:

a. participation in international conferences;
b. creation of specialized theological degree programs for students of Turkish descent who have grown up in countries such as Belgium, Germany, the Netherlands, or the U.S.;
c. invitation of foreign interlocutors to the Turkish Faculties;
d. publication in other languages (especially English and Arabic);
e. involvement in multinational research projects (e.g., those initiated by Eugen-Biser-Stiftung);
f. academic visits to other places of learning (e.g., the AÜİF – Pontifical Gregorian University scholarly exchange program, which goes back to 1986/1406).

3 Intellectual influences

The various approaches and authors draw from different sources, but ten factors can be seen as fairly common.

---

mukayeseye, müşahadeye ve en sonunda da, mümkün olursa, izaha çalışmaktadırlar. Yani birinci sübjektif olduğu kadar, ikincisi objektiftir. O halde Hükümetin şayanı dikkat olan isteği şudur; bütün mânasiyle ilim haysiyeti ve ilim karakteri taşıyan bir fakülte meydana getirmektir."

a. Inner-Turkish exchange, debate, mutual challenging and cross-fertilization are, of course, a key factor.
b. The Ankara Faculty was from its beginning a place where internationally renowned scholars from different backgrounds taught together, thus creating an atmosphere of free research and exchange.[12]
c. Many of today's leading authors in Turkish Qur'ān hermeneutics mention as their own masters two Ankara theologians, namely Hüseyin Atay (1930–2023/1349–1445)[13] and Mehmed Said Hatiboğlu (born 1930/1349).[14]
d. The self-understanding of Turkish religious thought was already classically more open towards adaptation and philosophical reflection, thus contrasting itself against the Ḥanbalī legal school and Ashʿarī *kalām*, both of which were considered more typically "Arab." The Ottoman empire's majority legal school was, conversely, Ḥanafī, and its theology was inspired by the elegant syntheses of Abū Manṣūr al-Māturīdī (893–941/280–329). Being an Uzbek, he is happily accepted as a "Turk."[15]
e. Today's hermeneutically interested theologians can draw from an autochthonous Turkish tradition of Islamic thought responsive to modernity. Notably the Ottoman Empire's last decades had seen several reform thinkers;[16] above all, Filibeli Ahmed Hilmi (1865–1914/1282–1332 from Plovdiv/Philippopolis, now Bulgaria), İzmirli İsmail Hakkı (1869–1946/1286–1365 from İzmir/Smyrna, Anatolia), and Musa Carullah Bigiyef (1875–1949/1292–1368, a Kazan Tatar).[17]
f. Authors writing in Arabic are systematically read and received by today's Turkish theologians: both the classical exegetes and twentieth century figures (notably, Amīn al-Khūlī, ʿĀ'isha ʿAbd ar-Raḥmān/Bint ash-Shāṭi', Ḥasan Ḥanafī, Naṣr Ḥāmid Abū Zayd).

---

12 Among the first professors were Muhammed Hamidullah (d. 2002/1423) from India, M. Tayyib Okiç (d. 1977/1397) from what is today Bosnia-Herzegovina, Annemarie Schimmel (d. 2003/1423) from Germany, and Muhammed Tanci (aṭ-Ṭanjī, d. 1974/1394) from Morocco.

13 Şaban Ali Düzgün, "Hüseyin Atay," in*Çağdaş İslâm Düşüncürleri*, ed. Cağfer Karadaş (Istanbul: Ensar, 2009), 265–300; German translation and commentary in: Körner, *Glaube in Gesellschaft*, 23–80; Hüseyin Atay, *Kur'ana Göre Araştırmalar*, 7 vols. (Ankara: Atay ve Atay, 2013).

14 İhsan Işık, "Mehmed Said Hatiboğlu Hoca," in *Türkiye Edebiyatçılar ve Kültür Adamları Ansiklopedisi*, ed. İhsan Işık (Istanbul: Elvan Yayınları, 2007²); Mehmed Said Hatiboğlu, *Müslüman Kültürü Üzerine* (Ankara: Kitâbiyât, 2004); Hatiboğlu, *Hz. Peygamber ve Kur'an Dışı Vahiy* (Istanbul: Otto, 2009).

15 Many books celebrate him as "the great Turkish thinker," e.g. Ahmet Vehbi Ecer, *Büyük Türk Alimi Mâtürîdi* (Istanbul: Yesevi Yayıncılık, 2006), or the six-hundred-page congress volume, İsmail Kurt and Seyit Ali Tüz, eds., *Büyük Türk Bilgini İmâm Mâtürîdî ve Mâtürîdîlik* (Istanbul: Marmara Üniversitesi İlâhiyat Fakültesi Vakfı Yayınları, 2012). A selection of Māturīdī's work has been translated into Turkish by the AÜİF theologian Sönmez Kutlu, *İmam Mâturîdî ve Maturidilik* (Ankara: Kitabiyat, 2011; Otto, 2018⁸).

16 Seminal is Rotraud Wielandt, "Main Trends of Islamic Theological Thought from the Late Nineteenth Century to Present Times," in *The Oxford Handbook of Islamic Theology*, ed. Sabine Schmidtke (Oxford: University Press, 2016): 707–64, especially: 727–32.

17 The *Ankara Okulu Yayınları* (cf., next section) provide 10 volumes of his work in modern Turkish.

g. A marking influence was left among Turkish theologians by Fazlur Rahman, who visited Turkey for the first time in 1977/1397.
h. Research periods abroad are also regularly mentioned as important inspirations by the authors, especially in Austria, France, Germany, the U.K. and the U.S., but also in Arab countries, Indonesia, or Malaysia.
i. Western European scholars of Islam can have effects on Turkish theologians simply because their books are read in Turkey (in the original languages or in translation). So, a scholar of the Qur'ān such as Rudi Paret is sometimes mentioned as an inspiring challenge.
j. Twentieth/twenty-first-century Western philosophers are being used by Turkish specialists in Qur'ānic hermeneutics. Names frequently mentioned in this context are Hans-Georg Gadamer,[18] Paul Ricœur,[19] and Jürgen Habermas.[20]

4 The School of Ankara?

In Western perception, several of the authors to be discussed here are members of the "Ankara School" – if such a school can be said to exist. The name is now widely used, but it has also led to misunderstandings and rejections. An Ankara Ekolü did actually exist; however, that was a sociological movement of the early 1940s. The hermeneutical approach of the 1990s was never sufficiently homogenous to understand itself as one school; but the exegetes to be presented here were founding members of the books series named Ankara Okulu Yayınları: "Publications of the Ankara School." The series has flourished since 2002/1423. It now numbers well over four-hundred volumes. So, is there a theological Ankara school after all? It was the publisher who chose the name of the series; and he was not thinking of a "school" in the sense of a tradition of thought (Turkish, *ekol*). He rather used the word *okul*, which (just like "school" in English) can also mean "faculty" in the sense of "authorized university department." Furthermore, the members of Ankara University's İlahiyat Fakültesi (i.e., Divinity School) never comprised "the" Ankara school. Rather, like any lively academic institution, its members have different, even opposing presuppositions, agendas, approaches, and intuitions.

18 Burhanettin Tatar, "Kur'an'ı Yorumlama Sorunu," in *Kur'an ve Dil. Dilbilim ve Hermenötik Sempozyumu, 17–18 Mayıs 2001*, ed. Bekir Tileklioğlu (Van: Yüzüncü Yıl Üniversitesi, 2001), 493–508, German translation and commentary in: Körner, *Alter Text – Neuer Kontext*, 103–28.

19 Ömer Özsoy, "Kur'an hitabının tarihselliği ve tarihsel hitabın özgün anlamı sorunum," in Ömer Özsoy, *Kur'an ve tarihsellik yazıları* (Ankara: Kitâbiyât, 2004), 77–101.

20 Tahsin Görgün, *Anlam ve Yorum. Dini Metinlerin Anlaşılması ve Yorumlanması* (Istanbul: Gelenek Yayıncılık, 2003; Külliyat Yayınları, 2019²); Görgün, *İlahi Sözün Gücü. Varlık ve Bilgi Kaynağı Olarak Kur'an* (Istanbul: Gelenek Yayıncılık, 2004; Külliyat Yayınları, 2013²).

Beyond the question of its name, however, there is something like a hermeneutical Ankara School or, more precisely, there used to be. The authors designated themselves the "İslâmiyât Circle," after the theological journal they founded in 1998/1419: *İslâmiyât.* In 1999/1420, some of the circle's protagonists were still referring to themselves as a tradition of modernism.[21] Soon, they would no longer use that self-designation; and ten years later, in July 2009/1430, they closed their platform, the *İslâmiyât* journal.[22] Some authors have now parted ways in terms of habitation, profession, and opinion.

There are many presently active Turkish authors worth discussing, for example a faithful transmitter of Fazlur Rahman's methodology: Adil Çiftçi (born 1963/1383), sociologist of religion by training;[23] or the profound and original exegete of the Qur'ān Halis Albayrak (born 1955/1374).[24] His Qur'ānic hermeneutics situates the process of understanding within a dynamic triangle: the Qur'ān, its addressee, and. All three are interwoven, each making its presence felt within the other – creating a dynamic of "witnessing" that allows the present-day reader to engage in "perceiving" the Qur'ān.

In what follows, however, we will study only the hermeneutical theories of four authors whose approaches are, each in his own way, marked by a reception of the Islamic exegetical tradition and of twentieth-century philosophy.

4.1 Mehmet Paçacı (born 1959/1378): Back to History

Mehmet Paçacı's agenda might be condensed in the slogan "back to history."[25] It should be taken, in Paçacı's case, as a triple imperative:

1 Hark back to classical Islamic receptions of the Qur'ān: there, the recipients' own historical situation played a highly relevant – and reflected – role.

---

21 Mehmet Paçacı and Yasin Aktay, "75 Years of Higher Religious Education in Modern Turkey," *MW* 89 (1999): 398–413, 412.

22 A nine volume (7412 pp.) reprint makes the journal's lively history accessible again: *İslâmiyât Dergisi. Koleksiyon* (Ankara: Otto, 2013).

23 Körner, *Revisionist Koran Hermeneutics*, 109–34.

24 Halis Albayrak, *Tarihin İçinden Kur'an'ı Algılamak* (Istanbul: Şule, 2011).

25 The present author has provided translations and evaluations of five articles by Paçacı in German:

a. *The Qur'ān and I: how historical are we?:* "Kur'an ve Ben Ne Kadar Tarihseliz?," originally in: *İslâmî Araştırmalar* 9 (1996): 119–34, edited version in the book of the same title: Mehmet Paçacı, *Kur'an ve Ben Ne Kadar Tarihseliz?* (Ankara: Ankara Okulu Yayınları, 2000), 53–79; German translation and commentary in: Körner, *Alter Text – Neuer Kontext*, 31–76.

b. *What happened to the Qur'ān and its exegesis in modernity?:* "Çağdaş Dönemde Kur'an'a ve Tefsire Ne Oldu?," *İslâmiyât* 6, no. 4 (2003): 85–104; German translation and commentary in: Körner, *Alter Text – Neuer Kontext*, 129–63.

c. *An exegetical essay on Sura 110 from the point of view of the Semitic religious tradition:* "De Ki Allah 'Bir'dir': *ahad/œhâd.* Sami Dini Geleneği Perspektifinden İhlas Sûresi'nin Bir Tefsiri Denemesi," *İslâmiyat* 1 (1996): 49–71, edited version in the book, *Kur'an ve Ben Ne Kadar Tarihseliz?*, 155–84; German translation and commentary in: Körner, *Alter Text – Neuer Kontext*, 165–207.

2 Similarly today, Islam should discern the present historical situation and its role in it, in order to become the historic agent it used to be.
3 For that, use historical methodology in reading the Qur'ān, so that you can distinguish what is contingent (contextually conditioned) in it from its absolute (ever-valid) message.

Mehmet Paçacı's approach is influenced by Hans-Georg Gadamer's program of "rehabilitating tradition," and his analysis of the "historically effected consciousness" (taken by Paçacı as interest in the text's effects and efficacy throughout history). Also influential for Paçacı was Fazlur Rahman's description of the hermeneutical process as transposing a concrete Qur'ānic regulation – abstracted as a general rule – into a new situation. Like Fazlur Rahman, he wants to see that method not as something specifically modern but as something which classical Islamic authors have done, too. Paçacı, however, develops Fazlur Rahman's relatively conventional hermeneutics in three noteworthy directions. More than Fazlur Rahman,
1 he wants to see the contingent part of each Qur'ānic regulation as still meaningful: it is the Qur'ān's way of providing, alongside each revealed norm, a rule of how to apply it;
2 he is interested in the history of the Qur'ān's *de facto* reception, therefore studying especially its role in the decisions of judges (*quḍāt*): of all people, an exegete of the Qur'ān claims that rather than canonical Arabic *tafsīr*, the place that shows us how to bring back the Qur'ān to world relevance, is the Ottoman legal court;
3 he treasures Islam's classical eras (especially the Ottoman Empire) as models for today: of how Islam can prove to be the dynamic alternative to Western modernity.

Paçacı deploys his approach against three adversaries: European Christianity (copying Islam's historical consciousness, though claiming to be its creator), Western Orientalism (depriving Islam of its power to shape history), and the Qur'ānic revisionisms of Muḥammad ʿAbduh and his disciples (modernizing Islam by jettisoning tradition).

Problematic in Paçacı's approach may be
1 his tendency to contrast two blocks – Islam and the West – in a "clash of civilizations" as two clear-cut identities, two given entities in rivalry, rather than seeing their interaction as an entangled history of collaboration, interpenetration, and development;

---

d. *What was classical Qur'ān exegesis?*: "Klasik Tefsir Neydi?," in *Klasiği yeniden düşünmek Sempozyumu*, ed. Halit Özkan (İstanbul: Klasik, 2006); German translation and commentary in: *Münchener Theologische Zeitschrift* 58 (2007): 127–45.

e. *Critique of the modernist interpretation of "the woman in the Qur'ān:* "Çağdaşçı 'Kur'an'da Kadın' Yorumunun Eleştirisi," in *Tarihten Günümüze Kur'an'a Yaklaşımlar*, eds. Bilal Gökkır et al. (Istanbul: İlim Yayma Vakfı, 2010), 559–85; German translation and commentary in: Körner, *Glaube in Gesellschaft*, 178–227.

2 his inclination to idealize life in past empires more than to ponder the problems of pre-modern social realities;
3 his undertone of providing arguments for an interventionist policy to recreate an Ottoman identity, rather than describing (and perhaps relying on) processes of historical growth.

The strengths of Mehmet Paçacı's work in Qur'ānic hermeneutics are
1 his interdisciplinary approach of broad technical, linguistic, philosophical, and historical scope;
2 his capacity to incorporate – felicitously deviating from ideologies of contrast – methods and insights from various eras and backgrounds;
3 his innovative look at the hermeneutical problem, re-contextualizing it in Muslims' present-day responsibility to shape the world.

4.2 Ömer Özsoy (born 1963/1383): Qur'ān as Interaction

Ömer Özsoy's approach marks a new step in historicizing the Qur'ān.[26] His core insight is that the Qur'ān cannot be read as a book. A book should, to some extent, be
1 complete in itself,
2 arranged according to a meaningful structure,
3 comprehensible in itself.

By contrast, Özsoy says, the Qur'ān can be understood only by a listener who is aware of witnessing a communication between God and the first Muslims, the verses being only one half of that communication. The other half is what the first Muslims – Muḥammad and his adherents – said (asked, criticized, etc.) and did (rightly or wrongly). In short, the Qur'ān is an event of interaction. Özsoy's term for that is "discourse" (hitap, söylem). Since the Qur'ān is an interactive discourse between God and human beings,

26 Since 2004, many of Özsoy's publications are in German. The present author has provided translations and evaluations of two earlier articles by Özsoy in German:
a. *Present-day Muslims' problems of renewal, and the Qur'ān:* "Çağdaş Müslümanların Yenilenme Sorunları ve Kur'an," in Ömer Özsoy, *Kur'an ve tarihsellik yazıları* (Ankara: Kitâbiyât, 2004), 87–95, originally published as "Kur'an ve Müslümanların Yenilenme Sorunu," in *Değişim Sürecinde İslam Sempozyumu. 20–22 Nisan 1996, Ankara* (Ankara: Turkey Diyanet Vakfı, 1997), 7–14; German translation and commentary in: Körner, *Alter Text – Neuer Kontext*, 15–30.
b. *The historicity of the Qur'ānic discourse, and the problem of a historical discourse's original meaning*: "Kur'an hitabının tarihselliği ve tarihsel hitabın özgün anlamı sorunu," in Ömer Özsoy, *Kur'an ve tarihsellik yazıları* (Ankara: Kitâbiyât, 2004), 53–66, which is a heavily edited version of "Kur'an hitabının tarihselliği ve tarihsel hitabın nesnel anlamı sorunu," *İslâmî Araştırmalar* 9 (1996): 135–43; German translation and commentary in: Körner, *Alter Text – Neuer Kontext*, 77–101.

1 each verse needs to be heard together with its original historical situation;
2 the message for today is not in one verse or all verses but is the result of the Qur'ān's reaction to the first Muslims' behavior: what is criticized and what is approved of; and therefore,
3 the first Muslims' behavior that went uncriticized by the Qur'ān is also a message for today: a model to be followed;
4 a reading of the Qur'ān today, thus understood, can become a new interaction with God and a believer, who draws new conclusions for new situations.

Özsoy's approach is indebted to Paul Ricœur's conceptual proposal (le discours), to orality/literacy research, and to the scholarship of Mehmed Said Hatiboğlu. Three problems in his approach might be noted:
1 it is a methodological proposal that, so far, has not produced many concrete exegetical results;
2 it is highly speculative when it claims that what is not mentioned in the Qur'ān but can be reconstructed as practiced by Muḥammad and/or his followers can be taken to be divinely approved behavior;
3 while the Qur'ānic text is an objective given,[27] we have hardly any access to the first Muslims' actions beyond the Islamic traditions. They are historically dubitable. Critical research provides a fairly reliable account of what happened, but at the same time reminds us of its uncertainty.

These reservations do not thwart Özsoy's project; they rather shed a different light on it. In fact, his approach is interactive also in the sense that it is an invitation to scholars from several disciplines – history, philology, *tafsīr*, and hadith studies – to collaborate in reconstructing the second half of the Qur'ānic event: the history of the first community.

Inspired by Fazlur Rahman, Özsoy and Paçacı – each in his own way – look for the Qur'ān's meaning in its history. The following author holds a contrary position.

4.3 Burhanettin Tatar (born 1965/1385): God's Guidance on the Way

The Qur'ān's promise is to offer God's guidance (*hudā*), and that is what Muslims expect from the Qur'ān. So how to get it? By interpreting it; but according to Tatar,[28] interpre-

27 Özsoy is well aware that such a statement is an exaggeration. There is much debate on the heterogeneity of the textual traditions and on the genesis of the *muṣḥaf*. Ömer Özsoy has in fact written his own book on this problematic "An introductory essay to the history of the making of Qur'ān's text": *Kur'an'ın Metinleşme Tarihi. Bir Giriş Denemesi* (Ankara: İlâhiyât, 2002).
28 The present author has provided a translation and commentary of an article by Tatar on "The problem of interpreting the Qur'ān": "Kur'an'ı Yorumlama Sorunu," in *Kur'an ve Dil: Dilbilim ve Her-*

tation of the Qur'ān cannot presuppose that each particular Qur'ānic word has a particular meaning. Nor does its meaning subsist in relation to other Qur'ānic words; nor in its relation to a historical event. With this, Tatar rejects a Fazlur Rahman-like hermeneutics that can be sensed even behind Paçacı's and Özsoy's approaches. What does the Qur'ān mean then? "Where" is its meaning, if neither in itself, nor in its historical context, nor in its classical receptions? Tatar observes what Muslims actually do. He finds that the Qur'ān is their field of consensus, their space of communication. For the Qur'ān is not an objective text, dualistically in front of its interpreters. Rather, the Qur'ān is a "symbolic text." It serves as a mirror: in it, Muslims see what is actually in themselves. Meaning is, according to Tatar, in the interpreters' understanding; consequently, interpretation is not the discovery of meaning but its creation. As the word *hudā* implies, what the Qur'ān provides is guidance; and guidance happens dynamically, on the way. We need to transform our relation with the Qur'ān from hierarchy to partnership, from ideology to interaction, from violence to dialogue.

Tatar's hermeneutical proposal is a radical application of a metaphysics of relation (rather than of substance) to the interpretative process. The objection that one might raise is whether his relational hermeneutic is not relativistic. If Tatar's view is correct, then it would be impossible to misunderstand the Qur'ān. Might Muslims not have good reasons for expecting from the Qur'ān more than a mirror; to find in it an alterity; to discover in the Qur'ān meaning the readers are precisely not producing by themselves?

Tatar is, however to be credited for his originality and openness; and what is manifest in his work is his intimate knowledge of Gadamer. It should also be noticed that more than a decade after his proposal, Tatar found a late ally from the other end of the world. Shahab Ahmed's definition of what it means to be a Muslim, or indeed to be Islamic, strikes the same chord when he famously says, Islamic is what is meaningful through hermeneutical engagement with Revelation to Muḥammad as Pre-Text, Text, and Con-Text.[29]

4.4 Mustafa Öztürk (born 1965/1385): The Three Qur'ānic Universals

Mustafa Öztürk[30] sees two camps in Qur'ānic hermeneutics, "universalism" and "historicism." Historicists allow the Qur'ān to refer only to that particular time and cul-

menötik Sempozyumu, 17–18 Mayıs 2001 (Van: Yüzüncü Yıl Üniversitesi, 2001), 493–508; German translation and commentary in: Körner, *Alter Text – Neuer Kontext*, 103–28.

29 Shahab Ahmed, *What Is Islam? The Importance of Being Islamic* (Princeton: Princeton University Press, 2016), 364.

30 The present author has provided a translation and commentary of an article by Tatar on "The Qur'ān's character as historical discourse": Felix Körner, "Türkisch-islamische Theologie im Aufbruch. Mustafa Öztürk, 'Der Koran als geschichtliche Rede'," *Lebendiges Zeugnis* 63, no. 2 (2008): 84–108;

ture whose expressive means it uses. Universalists, by contrast, claim that the Qur'ān addresses all people, civilizations, and conditions. Rather than opting for one or the other camp, Öztürk offers a new road. The Qur'ān is for seventh-century Arabs; but it is relevant today, too. How? Rather than accepting a mere either/or, he says that the Qur'ān's material needs to be sifted. Much of it is "historical" (tarihsel, bound to the past, obsolete); but some of it is "superhistorical" (tarih üstü): its proclamation of the oneness of God, the Day of Judgement, and the ethics of virtue. These three are the Qur'ān's universal (evrensel) messages. They coincide with what other prophets and sages proclaimed in world history; which proves that the three messages are universal.

Öztürk thus uses "universality" in a double sense: what is *valid* for all cultures, and what is *present* in all cultures. His position gives immediate rise to an objection. If the only Qur'ānic message valid today can be found elsewhere as well, what do we need the Qur'ān for? That objection, however, springs to mind only if one studies the Qur'ān in a non-Muslim atmosphere. Öztürk, by contrast, has only Muslim readers in mind. He does not want to convince non-Muslims to read the Qur'ān. He rather writes for Muslims, whose fundamental religious reference point is the Qur'ān anyway. He simply wants to liberate them from the influence of those who preach seventh-century rules as if they had to be followed today.

5 Evaluation and Perspective

The structural, cultural, and personal constellations in Turkey have proven to be fertile ground for innovative, sometimes impressive approaches to the challenges of Qur'ānic hermeneutics. In conclusion, three questions should be discussed, namely, whether modern Qur'ānic hermeneutics in Turkey is historical-critical, whether it is influential beyond Turkey, and what future developments of the theories studied above would be desirable.

First, then, are endeavors in Qur'ānic hermeneutics in Turkey historical-critical? The question is often posed as if historical criticism were the one and only key for Islam's "opening up" towards modernity. Much depends on what is meant by 'historical-critical.' If only that interpretation is considered historical-critical which presupposes that God does not act in history, hardly any of the hermeneutics studied here can be seen as historical-critical. If, on the other hand, consideration of the historical context of a text suffices for an approach to be called historical-critical, all Islamic Qur'ān exegesis is such.

---

in Turkey, that publication caused much debate; Öztürk's seminal article is also discussed in Felix Körner, "Modernistische Koranexegese in der Türkei. Eine Diskussion mit Mustafa Öztürk," in *Im Dienst der Versöhnung. Für einen authentischen Dialog zwischen Christen und Muslimen*, ed. Peter Hünseler (Regensburg: Friedrich Pustet, 2008), 14–22.

Serious historical-criticism is today also critical of its own past excesses.[31] A list of nine criteria of historical criticism would be this:[32] an interpretation of a text is historical-critical in so far as the interpreter is aware that

(1) the text refers to a particular historical context
(2) the text is limited by it;
(3) there is a distance between text and interpreter;
(4) the interpreter is also conditioned by a particular consciousness, and
(5) the text had an influence on that consciousness;
(6) text and interpreter principally refer to the same reality;
(7) the distance between text and interpreter can be productive;
(8) historical research can to some degree establish the credibility or spuriousness of the text's claims;
(9) the text needs to be interpreted within the whole of history and interpretations of history.

If evaluated according to such a catalogue, Turkish exegetes go quite far in what may be called historical criticism.

The second question can be answered with one sentence: are Turkish approaches to Qur'ānic hermeneutics influential beyond Turkey? Yes, in the German-speaking world, several of the aforementioned authors are well-known, while in other parts of the world, reception depends often on the reputation that Turkish (religious) culture has.

Our final question is the invitation to present desirable future developments of Turkish hermeneutical theories. Since Qur'ānic hermeneutics is no longer only an intra-Islamic theme, but a point of exchange and debate also in non-Muslim contexts, themes that seemed self-evident before need new attention and new arguments.

1 Is the Qur'ān God's revelation? Before, Muslim exegetes either simply presupposed this, or they thought they could prove it in view of its formal characteristics (its "miraculous character, perfection"). Where such arguments are rejected as arbitrary and thus inconlcusive, Qur'ānic hermeneutics needs to offer today new, convincing justifications why this of all texts should be accepted as God's revelation to humanity. After all, the Qur'ān sometimes transmits ethically problematic particulars and some-

31 Samuel Vollenweider, "Die historisch-kritische Methode – Erfolgsmodell mit Schattenseiten. Überlegungen im Anschluss an Gerhard Ebeling," *Zeitschrift für Theologie und Kirche* 114 (2017): 243–59, and Knut Backhaus, "Aufgegeben? Historische Kritik als Kapitulation und Kapital von Theologie," *Zeitschrift für Theologie und Kirche* 114 (2017): 260–88.

32 Cf., by the present author: "Historisch-kritische Koranexegese? Hermeneutische Neuansätze in der Türkei," in *Krieg der Zeichen? Zur Interaktion von Religion, Politik und Kultur*, eds. Görge K Hasselhoff and Michael Meyer-Blanck (Würzburg: Ergon, 2006), 57–74, 59, and "Kritik der historischen Kritik. Eine neue Debatte in der islamischen Theologie," in *Der Islam in Europa. Zwischen Weltpolitik und Alltag*, eds. Urs Altermatt, Mariano Delgado, and Guido Vergauwen (Stuttgart: Kohlhammer, 2006), 127–42.

times conveys convictions which can also be otherwise known (the "universal contents").

2 Another point that has so far been taken for granted and is now seen as worth a dispute is the claim that the Qur'ān's core message refers to what people should do and which moral attitude they should take. A tendency – which can also be identified in liberal Christian theology[33] – becomes visible: an "ethical reductionism."[34] By contrast, it would certainly be worth refocusing the Qur'ān's elucidation of the meaning of life, its originality in presenting and representing God, its debates with Judaism and Christianity, and its poetic power.

3 Finally, the claim that religion loses its identity if it does not take its foundational text "literally" should be taken seriously. One can argue from the Qur'ān itself for the possibility and necessity of interpretation: according to the Qur'ān, instead of simply using force to bring about the state of affairs God wants, God is communicating with human beings linguistically and is thus indeed calling humanity to contemplate and reflect (Q 45:13).

6 Bibliography

6.1 Sources

Albayrak, Halis. *Tarihin İçinden Kur'an'ı Algılamak*. Istanbul: Şule, 2011.

Atay, Hüseyin. *Kur'ana Göre Araştırmalar*, 7 vols. Ankara: Atay ve Atay, 2013.

Bayındır, Abdülaziz. *Kur'an Işığında Doğru Bildiğimiz Yanlışlar*. Istanbul: Süleymaniye Vakfı Yayınları, 2017.

Cerrahoğlu, İsmail. *Tefsir Usûlü*. Ankara: Türkiye Diyanet Vakfı, 2016.

Demirci, Muhsin. *Tefsir Usulü*. Istanbul: MÜİFAV, 2016.

Görgün, Tahsin. *İlahi Sözün Gücü. Varlık ve Bilgi Kaynağı Olarak Kur'an*. Istanbul: Gelenek Yayıncılık, 2004; Külliyat Yayınları, 2013.

Görgün, Tahsin. *Anlam ve Yorum. Dini Metinlerin Anlaşılması ve Yorumlanması*. Istanbul: Gelenek Yayıncılık, 2003; Külliyat Yayınları, 2019.

Gözeler, Esra. "Surelerin Mekkīliği ve Medenīliği. 22/el-Ḥacc Suresi Örneği." *Ankara Üniversitesi İlahiyat Fakültesi Dergisi* 58 (2017): 69–108.

Hatiboğlu, Mehmed Said. *Hz. Peygamber ve Kur'an Dışı Vahiy*. Istanbul: Otto, 2009.

İslâmiyât Dergisi. Koleksiyon. Ankara: Otto, 2013.

İslamoğlu, Mustafa. *Kur'an'ı Anlama Yöntemi: Tefsir Usulü*. Istanbul: Düşün Yayıncılık, 2015.

Özsoy, Ömer. "Kur'an hitabının tarihselliği ve tarihsel hitabın nesnel anlamı sorunu." *İslâmî Araştırmalar* 9 (1996): 77–101.

Özsoy, Ömer. "Kur'an ve Müslümanların Yenilenme Sorunu." In *Değişim Sürecinde İslam Sempozyumu. 20–22 Nisan 1996, Ankara*, Ankara: TDV, 1997: 7–14.

Özsoy, Ömer. *Kur'an'ın Metinleşme Tarihi. Bir Giriş Denemesi*. Ankara: İlâhiyât, 2002.

33 Cf., Adolf von Harnack, *Das Wesen des Christentums* (Leipzig: J.C. Hinrichs, 1900).

34 Körner, *Revisionists Koran Hermeneutics*, 84.

Özsoy, Ömer. "Çağdaş Müslümanların Yenilenme Sorunları ve Kur'an." In Ömer Özsoy, *Kur'an ve tarihsellik yazıları*. Ankara: Kitâbiyât, 2004a: 87–95.
Özsoy, Ömer. "Kur'an hitabının tarihselliği ve tarihsel hitabın özgün anlamı sorunum." In Ömer Özsoy, *Kur'an ve tarihsellik yazıları*. Ankara: Kitâbiyât, 2004b: 77–101.
Paçacı, Mehmet. "De Ki Allah 'Bir'dir': *ahad/æhâd*. Sami Dini Geleneği Perspektifinden İhlas Sûresi'nin Bir Tefsiri Denemesi." *İslâmiyat* 1 (1996a): 49–71.
Paçacı, Mehmet. "Kur'an ve Ben Ne Kadar Tarihseliz?" *İslâmî Araştırmalar* 9 (1996b): 119–34.
Paçacı, Mehmet. *Kur'an ve Ben Ne Kadar Tarihseliz?* Ankara: Ankara Okulu Yayınları, 2000.
Paçacı, Mehmet. "Çağdaş Dönemde Kur'an'a ve Tefsire Ne Oldu?" *İslâmiyât* 6, no. 4 (2003): 85–104.
Paçacı, Mehmet. "Klasik Tefsir Neydi?" In *Klasiği yeniden düşünmek Sempozyumu*, edited by Halit Özkan. İstanbul: Klasik, 2006.
Paçacı, Mehmet. "Klassische Koranexegese – was war das?" *Münchener Theologische Zeitschrift* 58, trans. Felix Körner. İstanbul: Klasik, 2007: 127–45
Paçacı, Mehmet. "Çağdaşçı 'Kur'an'da Kadın' Yorumunun Eleştirisi." In *Tarihten Günümüze Kur'an'a Yaklaşımlar*, edited by Bilal Gökkır et al. Istanbul: İlim Yayma Vakfı, 2010: 559–85.
Tatar, Burhanettin. "Kur'an'ı Yorumlama Sorunu." In *Kur'an ve Dil. Dilbilim ve Hermenötik Sempozyumu, 17–18 Mayıs 2001*, edited by Bekir Tileklioğlu. Van: Yüzüncü Yıl Üniversitesi, 2001: 493–508.

6.2 Secondary Literature

Ahmed, Shahab. *What Is Islam? The Importance of Being Islamic*. Princeton: Princeton University Press, 2016.
Bacik, Gokhan. "Hermeneutics in Contemporary Turkey: An Analysis of Turkish Historicists." *Religions* 12, no. 11, 1027 (2021). https://doi.org/10.3390/rel12111027.
Backhaus, Knut. "Aufgegeben? Historische Kritik als Kapitulation und Kapital von Theologie." *Zeitschrift für Theologie und Kirche* 114 (2017): 260–88.
Baltacıoğlu, İsmail Hakkı. "Dine Doğru." *Ankara Üniversitesi İlahiyat Fakültesi Dergisi* 6 (1957): 44–59.
Cerrahoğlu, İsmail. *Tefsir Tarihi*. Ankara: Fecr Yayınevi, 2009.
Demirci, Muhsin. *Tefsir Tarihi*. Istanbul: MÜİFAV, 2014.
Dorroll, Philip. *Islamic Theology in the Turkish Republic*. Edinburgh: University Press, 2023.
Düzgün, Şaban Ali. "Hüseyin Atay." In *Çağdaş İslâm Düşünürleri*, edited by Cağfer Karadaş. Istanbul: Ensar, 2009: 265–300.
Ecer, Ahmet Vehbi. *Büyük Türk Alimi Mâtürîdi*. Istanbul: Yesevi Yayıncılık, 2006.
Eith, Kathrin. *Koranexegese bei Yaşar Nuri Öztürk. Ein traditionskritischer Entwurf des Islams in der Türkei*. Würzburg: Ergon, 2013.
Harnack, Adolf von. *Das Wesen des Christentums*. Leipzig: J.C. Hinrichs, 1900.
Hatiboğlu, Mehmed Said. *Müslüman Kültürü Üzerine*. Ankara: Kitâbiyât, 2004.
Işık, İhsan. "Mehmed Said Hatiboğlu Hoca." In *Türkiye Edebiyatçılar ve Kültür Adamları Ansiklopedisi*, edited by İhsan Işık. Istanbul: Elvan Yayınları, 2007.
Jäschke, Gotthard. "Der Islam in der neuen Türkei. Eine rechtsgeschichtliche Untersuchung." *WI* 1 (1951): 1–174; 2 (1953): 278–87; 6 (1961): 185–202; 8 (1963): 252–59; 9 (1964): 260–64; 10 (1965): 181–92; 12 (1969): 125–6; 13 (1971): 145–61; 18 (1977): 1–18; 20 (1980): 200–3.
Koç, Akif, and İsmail Albayrak, eds. *Tefsire Akademik Yaklaşımlar*. Ankara: Otto, 2013.
Körner, Felix. *Revisionist Koran Hermeneutics in Contemporary Turkish University Theology. Rethinking Islam*. Würzburg: Ergon, 2005.
Körner, Felix. trans. and ed., *Alter Text – Neuer Kontext. Koranhermeneutik in der Türkei heute. Ausgewählte Texte, übersetzt und kommentiert von Felix Körner SJ*. Freiburg: Herder, 2006a, 2015.

Körner, Felix. "Historisch-kritische Koranexegese? Hermeneutische Neuansätze in der Türkei." In *Krieg der Zeichen? Zur Interaktion von Religion, Politik und Kultur*, edited by Görge K. Hasselhoff and Michael Meyer-Blanck. Würzburg: Ergon, 2006b: 57–74.

Körner, Felix. "Kritik der historischen Kritik. Eine neue Debatte in der islamischen Theologie." In *Der Islam in Europa. Zwischen Weltpolitik und Alltag*, edited by Urs Altermatt, Mariano Delgado, and Guido Vergauwen. Stuttgart: Kohlhammer, 2006c: 127–42.

Körner, Felix. "Turkish Theology Meets European Philosophy. Emilio Betti, Hans-Georg Gadamer and Paul Ricœur in Muslim Thinking." *Revista Portuguesa de Filosofia* 62 (2006d): 805–9.

Körner, Felix. "When Islam Receives Criticism. Historical Koran Exegesis in Today's Turkey." *Rosenzweig Jahrbuch* 2 (2007): 153–69.

Körner, Felix. "Modernistische Koranexegese in der Türkei. Eine Diskussion mit Mustafa Öztürk." In *Im Dienst der Versöhnung. Für einen authentischen Dialog zwischen Christen und Muslime*, edited by Peter Hünseler. Regensburg: Friedrich Pustet, 2008a: 14–22.

Körner, Felix. "Türkisch-islamische Theologie im Aufbruch. Mustafa Öztürk, 'Der Koran als geschichtliche Rede'." *Lebendiges Zeugnis* 63, no. 2 (2008b): 84–108.

Körner, Felix. "Bibelhermeneutik. Islamwissenschaftlich." In *Lexikon der Bibelhermeneutik. Begriffe – Methoden – Theorien – Konzepte*, edited by Oda Wischmeyer. Berlin: de Gruyter, 2013a: 209–10.

Körner, Felix. "Geschichte und Geschichtlichkeit. Islamwissenschaftlich." In *Lexikon der Bibelhermeneutik. Begriffe – Methoden – Theorien – Konzepte*, edited by Oda Wischmeyer. Berlin: de Gruyter, 2013b: 266.

Körner, Felix. "Historizität. Islamwissenschaftlich." In *Lexikon der Bibelhermeneutik. Begriffe – Methoden – Theorien – Konzepte*, edited by Oda Wischmeyer. Berlin: de Gruyter, 2013c: 306–7.

Körner, Felix. trans. and eds., *Glaube in Gesellschaft. Ankaraner Stimmen zur Politischen Theologie. Ausgewählte Texte übersetzt und kommentiert von Felix Körner SJ*. Freiburg: Herder, 2019.

Kurt, İsmail, and Seyit Ali Tüz, eds. *Büyük Türk Bilgini İmâm Mâtürîdî ve Mâtürîdîlik*. Istanbul: Marmara Üniversitesi İlâhiyat Fakültesi Vakfı Yayınları, 2012.

Kutlu, Sönmez. *İmam Mâturîdî ve Maturidilik*. Ankara: Kitabiyat, 2011; Otto, 2018.

Neuwirth, Angelika. "Koran." In *Grundriß der arabischen Philologie*, vol. 2, *Literaturwissenschaft*, edited by Hartmut Gätje. Wiesbaden: Ludwig Reichert, 1987: 96–135.

Neuwirth, Angelika. *Koranforschung – eine politische Philologie? Bibel, Koran und Islamentstehung im Spiegel spätantiker Textpolitik und moderner Philologie*. Berlin: de Gruyter, 2014.

Paçacı, Mehmet and Yasin Aktay. "75 Years of Higher Religious Education in Modern Turkey." *MW* 89 (1999): 398–413.

Reed, Howard A. "The Faculty of Divinity at Ankara. Part I." *MW* 46 (1956): 295–312.

Reed, Howard A. "The Faculty of Divinity at Ankara. Part II." *MW* 47 (1957): 22–35.

Serinsu, Ahmet Nedim. *Tefsir Tarihi Atlası ve Uygulama Haritaları*. Ankara: Grafiker Yayınları, 2019.

Sinai, Nicolai. *Fortschreibung und Auslegung. Studien zur frühen Koraninterpretation*. Wiesbaden: Harrassowitz, 2009.

T.B.M.M. Tutanak Dergisi, Dönem: VIII, Cilt: 20, Toplantı: 3, 101. Birleşim (4. VI. 1949): 279, Accessed April 24, 2020. https://www.tbmm.gov.tr/tutanaklar/TUTANAK/TBMM/d08/c020/tbmm08020101.pdf.

Vollenweider, Samuel. "Die historisch-kritische Methode – Erfolgsmodell mit Schattenseiten. Überlegungen im Anschluss an Gerhard Ebeling." *Zeitschrift für Theologie und Kirche* 114 (2017): 243–59.

Wielandt, Rotraud. "Exegesis of the Qur'ān: Early Modern and Contemporary." In *Encyclopaedia of the Qur'ān*, vol. 2. Brill, edited by Jane D. McAuliffe. vol. 2. Brill, Leiden, 2002: 124–42.

Wielandt, Rotraud. "Main Trends of Islamic Theological Thought from the Late Nineteenth Century to Present Times." In *The Oxford Handbook of Islamic Theology*, edited by Sabine Schmidtke. Oxford: University Press, 2016: 707–64.

Wilkinson, Taraneh. *Dialectical encounters: Contemporary Turkish Muslim Thought in Dialogue*. Edinburgh: Edinburgh University Press, 2020.

Derya Buğur

Qur'ānic Hermeneutics in Turkish Translations of the Qur'ān

1 Introduction

As early as during the time of the ancient Uyghurs,[1] Turks began translating a large number of sacred texts from their respective religions into Turkish, especially those from the Uyghur eras of Buddhism and Manichaeism.[2] From the tenth century and the adoption of Islam through to the classical period of the Ottoman Empire, a desire arose for Turkish translations to convey the meaning and message of the "divine words" to Turkish-speaking people for better comprehension.[3] According to Abdülkadir Inan, the Qur'ān was first translated into Persian and was only published as a state work after it had been verified by fatwas (*fatwā, fatāwā*; legal opinions) on its permissibility.[4] Within academic discourse, we encounter scholarly viewpoints concerning the first Turkish translation of the Qur'ān, which was produced between the tenth and eleventh centuries.[5] It is believed that this translation was made as an interlinear translation (literal translation) of the Qur'ān by translating Abū Jaʿfar aṭ-Ṭabarī's commentary (*Tafsīr aṭ-Ṭabarī*) into Persian and abbreviating it.[6] The oldest known copy of this translation of the Qur'ān originated in his *tafsīr*, no. 73. It dates back to 1333/734 and is on display in the Museum of Turkish and Islamic Art (*Türk ve İslâm Eserleri Müzesi*) in Istanbul.[7]

---

1 Largest group of Turkic, Muslim people living in China. Ablet Kamalov, "Uyghur Historiography," in *Oxford Research Encyclopedia of Asian History*, accessed October 26, 2023, https://oxfordre.com/asianhistory/view/10.1093/acrefore/9780190277727.001.0001/acrefore-9780190277727-e-637.

2 Suat Ünlü, "Kur'an-ı Kerimin Türkçeye Çevrilmesi ve ilk Türkçe Kur'an Tercümeleri," *Dini Araştırmalar Dergisi* 9 (2007): 15.

3 Mustafa Öztürk, "II. Meşrutiyet'ten Günümüze Kur'an Mealleri," *Türkiye Araştırmaları Literatür Dergisi* 18 (2011): 174.

4 Abdülkadir Inan, "Eski Kuran Tercümelerinin Dili Meselesi," *Türk Dili: Aylık Fikir ve Edebiyat Dergisi* 1, no. 7 (1952): 160.

5 Gülden Sağol, *An Interlinear Translation of the Qur'ān into Khwarazm Turkish: Introduction, Text Glossary and Facsimile; Introduction and Text. 1* (Cambridge: Harvard University, Department of Near Eastern Languages and Civilizations, 1993), 23; Abdülkadir Inan, *Kur'ân-ı Kerîm'in Türkçe tercemeleri üzerinde bir inceleme* (Ankara: Diyanet İşleri Başkanlığı, 1961), 8.

6 Esra Karabacak, "Tarihi Seyrinden Yola Çıkarak Kur'an Tercümeleri Üzerine Bir Çalışma: Prof. Dr. Nihad M. Çetin'e Armağan," *Ankara Edebiyat Fakültesi Basımevi* (1999): 184.

7 Abdülkadir Erdoğan, "Kur'an Tercemelerinin Dil Bakımından Değerleri," *Vakıflar Dergisi* 1 (1938): 47; Zeki Velidi Togan, "Londra ve Tahran'daki İslamî Yazmalardan Bazılarına Dâir," *İslam Tetkikleri Enstitüsü Dergisi* 3 (1959–1960): 135.

https://doi.org/10.1515/9783111320069-025

During the Ottoman era, several commentaries were translated from Persian to Turkish, which encouraged interest in Turkish translations of the Qur'ān. Around the Tanzimat period (1839–76/1255–93), translations of certain suras and their interpretations started emerging in Turkish, which elevated the subject of a Turkish Qur'ān translation to special significance during the second Ottoman constitutional period (1908/1326) through to the founding of the Turkish Republic (1923/1342). According to Mustafa Öztürk, the Young Turks (*Jön Türkler*)[8] fanned the flames of Turkish national pride, which inevitably gave Turkish translations political and ideological connotations.[9] In addition to the first Turkish translations of the Qur'ān and the commentaries that were written over time, more translations were produced after the establishment of the Turkish Republic by various translators, such as Mehmet Akif Ersoy (d. 1938/1357), Muhammed Hamdi Yazır (also called Elmalı Hamdi) (d. 1942/1361), and Izmirli Ismail Hakkı (d. 1946/1365). Moreover, the works of Abū l-Aʿlā Mawdūdī (d. 1979/1399) and Sayyid Quṭb (d. 1966/1386) were translated into Turkish. In particular, the Turkish translation of the English Qur'ān translation "The Message of The Qur'ān" by Muhammad Asad (d. 1992/1413) drew considerable interest.[10] Although there have been longstanding debates about the translation of the Qur'ān, numerous translations have emerged over time, each differing from the other in terms of translation techniques. Above all since the 2000s, readers' expectations of translations have increased, which has played a central role in the emerging diversity and growing number of translations. According to estimates by Hasan Elik and Muhammed Coşkun, this number has now grown to around three hundred works,[11] which reveal the differences and variety in the methodologies used. Moreover, Turkish translations of the Qur'ān in particular stand out as they testify to the changes that have occurred in language, politics, society, and culture.

2 Major Works

Over the last ten centuries, since the arrival of the Turks in Anatolia, numerous Qur'ān translations and commentaries have been written. However, their creation and publication cannot be understood independently of political and cultural developments. When the Turks officially accepted Islam as their main religion in the tenth

8 Political opposition against the sultanate that advocated the establishment of a constitutional government. E.J. Zürcher, "Yeñi ʿOthmānlīlar," in *EI*², ed. P. Bearman et al., accessed January 9, 2024, http://dx.doi.org/10.1163/1573-3912_islam_SIM_8013.

9 Öztürk, "II. Meşrutiyet'ten Günümüze Kur'an Mealleri," 149f.

10 Hidayet Aydar, "Yaygınlık Bakımından Meallerimiz ve Mealler Üzerinde İstatistiksel Bir Değerlendirme," *FSM İlmi Araştırmalar İnsan ve Toplum Bilimleri Dergisi* 5 (2015): 119–20.

11 Hasan Elik, Muhammed Coşkun, "İndirildiği Dönemin Işığında Kur'an Tefsiri (Tevhit Mesajı) M.Ü. İlahiyat Fakültesi Vakfı Yayınları (2013): 15–19.

century, it was essential to translate the Qur'ān into Turkish in order to understand its meaning, principles, and rules in an appropriate manner and to teach the people.[12] Some believe that the chapters and verses of the Qur'ān were at least orally translated into Turkish in the tenth century, potentially even earlier, before the entire Qur'ān was translated into Qarākhānid Turkish in writing. Additionally, translations of individual Qur'ānic verses can be found in *Kutadgu Bilig*, a doctrinal poem written in 1069/461 by Yūsuf Khāṣṣ Ḥājib.[13] According to Zeki Velidi Togan, the first written translation of the Qur'ān emerged during the time of the Sāmānids (819–999/203–389), a Muslim dynasty that was originally Persian. During the reign of Manṣūr b. Nūḥ I (961–76/350–65), a committee of scholars translated the Qur'ān and the original Arabic commentary "*Tafsīr aṭ-Ṭabarī*" into Persian. The commentary was abridged during the translation process, resulting in an adequate interlinear Persian translation based on the Qur'ānic text.[14]

According to Togan, a Turkish translation would be written alongside the Persian version, as both translations would show similarities based upon analysis of the text, and the translation committee would also include Turkish scholars. He bases his finding on a comparative textual analysis of earlier Persian and Turkish copies and points out similarities in the translations.[15] Moreover, he argues that the Sāmānids strategically used the Turkish translation of the Qur'ān as a means of propaganda to spread Islam among the Turks. This was because the Turks constituted the majority of the empire's military power, meaning the military command was also dependent on Turkish officers. Aside from disseminating Islam by translating the Qur'ān, the Sāmānid rulers also aimed to gain the officers' trust and to strengthen the military's loyalty to the Sāmānid dynasty. A further explanation for the simultaneous translations is that most members of the translation committee came from Turkish areas, meaning that the production of a Turkish version alongside the Persian translation was a likely consequence.[16] Muhammet Abay explains in his work that both Fuad Köprülü (d.1966/1386) and Wilhelm Barthold (d.1930/1349) support this theory, using linguistic studies to argue that the Turkish Qur'ān was written between the tenth and eleventh centuries.[17]

In opposition to this, Inan argues that the first Turkish translations of the Qur'ān only emerged after Sāmānid rule and were written by the Muslim Qarākhānids in the eleventh century after they captured Bukhara (a city in what is now Uzbekistan) in

12 Ünlü, "Kur'an-ı Kerimin Türkçeye Çevrilmesi," 15.

13 Eleazar Birnbaum, "On Some Turkish Interlinear Translations of the Koran," *JT* 14 (1990): 113–14.

14 Gülden Sağol, *An Interlinear Translation of the Qur'ān*, 23.

15 Zeki Velidi Togan, "The Earliest Translation of the Qur'ān into Turkish," *Islâm Tetkikleri Enstitüsü Dergisi* 4, no. 1–2 (1964): 3–17.

16 Togan, "Londra ve Tahran'daki İslamî Yazmalardan," 135; Togan, *The Earliest Translation*, 3.

17 Muhammet Abay, "Türkçedeki Kur'an Meâllerinin Tarihi ve Kronolojik Bibliyografyası," *Türkiye Araştırmaları Literatür Dergisi* 10, no. 19–20 (2012): 235.

the tenth century. Thus, with the onset of the cultural flourishing of Turkic intellectuals, the first translation would have evolved through literary language.[18] Eleazar Birnbaum notes that this first "interlinear translation" (a literal translation with the Persian wording written beneath the Arabic text), along with the summary of Ṭabarī's commentary, served as a basis for later translations of the Qur'ān.[19] Further, Hidayet Aydar claims that the oldest Turkish translation of the Qur'ān most likely originated in such a translation.[20] In the early fifteenth century, a new tradition emerged, in which Turkish commentaries on the entire Qur'ān were composed on the basis of translations of earlier Arabic commentaries. In addition to the first Qur'ān commentary, the most significant commentaries in the Turkish language include *Tercüme-i Tefsir-i Tibyân* translated by Ayıntabi Mehmed Efendi, *Mevâkib Tefsiri* by Ismail Ferruh Efendi, and *Tafsilu'l-Beyan* translated by Süleyman Tevfik el-Hüseyni.[21] For the most part, the increase in Turkish translations of the entire Qur'ān started in the Second Constitutional Period (*II Meşrutiyet dönemi*) of the Ottoman Empire and with the establishment of the Turkish Republic. The Second Constitutional Period was an epoch in which, parallel to the growing ideology of nationalism and Turkishness, the idea of nationalizing religion took hold, which consequently raised demands for a Turkish identity of Islam.[22]

The rise of the Young Turk movement led to the Islamic call to prayer as well as Islamic prayer and worship being conducted in Turkish, which was widely criticized. Moreover, religious services were also practiced in Turkish in the early years of the Turkish Republic.[23] There was a frequent emphasis on the need to present the Qur'ān in clear language, free of superstition. During this time, the government of the Turkish Republic aimed to eliminate the ulema (*ʿulamāʾ*; religious scholars) acting as intermediaries between the Qur'ān and the people. Their reasoning was that it would prevent the dissemination of possible misinterpretations in public discourse, guided by the ulema's own individual agendas. Removing them would allow the Qur'ān to speak directly to the people.[24] Demands for Turkish nationalism became even more evident in the voices of various officials growing louder as they demanded a "Turkish Qur'ān" for Turks and also rejected the need to learn Arabic to understand the Islamic faith. These demands subsequently expanded to include the translation of the hadith and the *Khuṭba* (Friday Sermon) into any language to increase accessibility.[25] By the

18 Inan, *Kur'ân-ı Kerîm'in Türkçe*, 8, 13, 20.

19 Abay, "Türkçedeki Kur'an," 233–34; Birnbaum, *On Some Turkish Interlinear Translations*, 113.

20 Hidayet Aydar, "Kur'an Çalışmaları," *Istanbul Üniversitesi İlahiyat Fakültesi Dergisi, Darulfunun İlahiyat* 1 (1999): 166.

21 O. Karacan and B. Yaşar, "II. Meşrutiyet'ten Cumhuriyet'e Kur'an Tercümelerine Osmanlı Aydınlarının Yaklaşımı," *Birey ve Toplum Sosyal Bilimler Dergisi* 9 (2019): 96f.

22 Ibid., 93–116.

23 Öztürk, "II. Meşrutiyet'ten Günümüze Kur'an Mealleri," 174.

24 Abay, "Türkçedeki Kur'an," 246.

25 Karacan and Yaşar, "II. Meşrutiyet'ten Cumhuriyet'e Kur'an," 96–100.

time the Turkish Republic was founded in 1923/1341, several Turkish Qur'ān translations had been written in Ottoman letters. Moreover, Qur'ān schools were eventually abolished and the teaching of Arabic in schools discontinued, leading to a decrease in the number of people who could read the Qur'ān in Arabic and leaving room for forged translations. To avoid any possible misinterpretations in the translation of the Qur'ān, the Presidency of Religious Affairs in Türkiye (T.C. Diyanet İşleri Başkanlığı)[26] declared on February 21, 1925/Rajab 27, 1341 that it would not be possible to translate the Qur'ān into any language without commentary.

Thus, Elmalı Muhammed Hamdi Yazır (d. 1942/1361) was commissioned by the state authority to develop a Turkish Qur'ān translation with an additional commentary, the first volume of which was published in 1935/1355 under the title *Hak Dini Kur'ân Dili.* The translation was completed in a total of nine volumes by 1939/1358 and printed in several editions in 1960/1379, 1971/1391, and 1982/1402.[27] Since this translation was completed, it has become the most widely published and used Turkish translation.[28] Although Elmalı's translation is generally perceived as the first Turkish translation, the actual first Qur'ān translation, *Tanrı Buyruğu* by Ömer Rıza Doğrul, was written even earlier, in 1934/1353. This translation was exclusively carried out by private individuals and was therefore free from government grants, arrangements, or influences. Since this translation was published privately, it did not enjoy the same popularity as the one by Elmalı.[29]

During the 1930s and 1940s, the state was undergoing an increasing process of Turkification, which also took place on a religious level, leading to numerous controversies and debates. Switching the call to prayer from Arabic to Turkish was one of the republic's most controversial decisions, with the solely Turkish religious practice causing an uproar. In addition, debates arose about the most accurate, idiomatic, and meaningful way to translate the Qur'ān into Turkish. Elmalı in particular did not support this development, rejecting the term "Turkish Qur'ān" (*Türkçe Kur'an*) and arguing that such a translation could only be seen as subordinate to the original Arabic text. Due to the political and social unrest sparked by the process of nationalization, religious publishing was interrupted for the most part until 1949.

When the opposition Democratic Party came to power in 1950/1370, the previous governing party, the Cumhuriyet Halk Partisi (Republican People's Party), founded by Atatürk, was sidelined. The new government's erosion of secularism led to the reinstatement of the original Arabic call to prayer. At the same time, the population was

26 The religious authority established by Mustafa Kemal Atatürk after the foundation of the Turkish Republic.

27 Ishak Özgel, "Elmalılı Muhammed Hamdi Yazırın Tefsiri ve Hakkında Yapılmış Çalışmalar Bibliyografyası," *Türkiye Araştırmaları Literatür Dergisi* 19 (2012): 335–38.

28 Ibid., 326.

29 Ali Akpinar, "Tanrı Buyruğu," in *TDVIA*, accessed October 1, 2023, https://islamansiklopedisi.org.tr/tanri-buyrugu.

increasingly turning toward Islam. In addition to numerous religious publications, Turkish translations of the Qur'ān began being produced once more. The translations by Hasan Basri Çantay (d. 1964/1384) and Süleyman Ateş (b. 1933/1352) were among the most popular. In addition, the *tafsīr* by Sayyid Qutub, considered the "spiritual father"[30] of later Islamist movements, was also translated into Turkish.[31] The outbreak of the Islamic Revolution in Iran in 1979/1399, together with the religious policy of the new ruling party of Türkiye, resulted in the Turkish population's increasing religiosity, especially after the military coup of September 12, 1980/Dhū l-Qaʿda 2, 1400. A rapid increase in the number of Qur'ān translations being produced can be observed in connection with the ideology emerging from the Islamic movement of the 1970s. During this period, translations were produced by Ali Bulaç (b. 1951/1371), whose general sympathies can be traced to the important Turkish-Islamist Milli Görüş[32] (National Outlook) movement.[33] This movement generally had a strong influence in Türkiye until the end of the 1990s.[34] It should be noted that Necmettin Erbakan (d. 2011/1432), the founder of the movement, is regarded as the mentor of Türkiye's current head of state, Recep Tayyip Erdoğan. Since he himself was an integral member of this movement, his initial political activities took place in the context of Erbakan's parties (Fazilet Parti, Refah Parti).[35] In addition, in the 1980s, Celal Yıldırım (d. 2019/1441) translated the commentaries of Ibn Kathīr and Mawdūdī into Turkish,[36] whose works

---

30 Bassam Tibi, *Fundamentalismus im Islam: Eine Gefahr für den Weltfrieden?* (Darmstadt: Wissenschaftliche Buchgesellschaft, 2002), XXI.

31 Ibid., 2–3, 81; Abay, "Türkçedeki Kur'an," 251–57.

32 Milli Görüş, founded by Necmettin Erbakan. It simultaneously mobilized at the political level, first as the Milli Nizam (National Order) Party (1970–71) and then in 1972 as the Milli Selamet (National Salvation) Party (1972–80). Both parties were banned due to their religious fundamentalist attitudes. In the following years, Erbakan's political activities continued to intensify with the establishment of the Refah Parti (Welfare Party) in 1983, and he had managed to become Prime Minister by the time of the 1997 coup. Cf. Ahmet T. Kuru, "Changing Perspectives on Islamism and Secularism in Turkey: The Gülen Movement and the AK Party," in *Contributions of the Gülen Movement: Muslim World in Transition*, ed. Louis Cantori, Marcia Hermansen, David Capes, et al. (London: Leeds Metropolitan University Press, 2007), 143f.

33 Islami Analiz, "Bulaç: Bugüne kadar oyumu Milli Görüş'e verdim," October 22, 2015, accessed November 12, 2023, https://islamianaliz.com/haber/7450752/bulac-bugune-kadar-oyumu-milli-goruse-verdim-sandiga-gidersem-oyum-yine-saadet-partisine; T24, "Ali Bulaç: Sandığa gidersem oyum Saadet Partisi'ne . . ., " October 22, 2015, accessed November 12, 2023, https://t24.com.tr/haber/ali-bulac-sandiga-gidersem-oyum-saadet-partisine,313864.

34 Ahmet T. Kuru, "Assertive Secularism and the Multiculturalist Challenge (1989–2008)," in *Secularism and State Policies toward Religion: The United States, France, and Turkey* (Cambridge: Cambridge University Press, 2009), 178.

35 Shadi Hamid, "From Erbakan to Erdoğan: The Evolution of Turkish Foreign Policy," *Insight Turkey* 6, no. 1 (2004): 113–18.

36 Abay, "Türkçedeki Kur'an," 257.

are favored by fundamentalist movements.[37] According to the latest data from 2015/1436, Celal Yıldırım's Qur'ān translation is another of Türkiye's most popular works, which also include the Qur'ān translation by Suat Yıldırım (b. 1943/1362),[38] reflecting a sympathy for the Gülen movement.[39]

The analysis of a survey from 2015/1436 conducted by the University of Istanbul shows which Qur'ān translations students prefer.[40] These include the translations by Mustafa Öztürk (b. 1965/1385), Suat Yıldırım, Elmalı Muhammed Hamdi Yazır, Muhammad Asad, and Hasan Tahsin Feyizli. A similar survey was published that same year, this time examining the Qur'ān translations that are most widely read by the family members and neighbors of students all over Türkiye.[41] The survey interviewed a total of 2,580 families. The results showed that Elmalı Muhammed Hamdi Yazır's translation was being used in 51 percent (1,311 families) of households, while other translations, such as Suat Yıldırım's, were read in 17 percent (426 families) and Ali Bulaç's in 10 percent (266 families) of households. The second most favored translation (at 30 percent, equating to 781 families), was published by the Presidency of Religious Affairs in Türkiye (T.C. Diyanet İşleri Başkanlığı). The translation by the Religious Affairs Foundation (Türkiye Diyanet Vakfı) was preferred by 779 families (30 percent) and therefore took third place in the survey. Among the families interviewed, 17 percent chose Suat Yıldırım's translation, which, on the other hand, came second among students. Translators critical of the government like Ihsan Eliaçık (b. 1961/1381),[42] whose translation of the Qur'ān was banned by order of a Turkish court,[43] and Kemalist translator Yaşar Nuri Öztürk (d. 2016/1438)[44] did not make it onto the students' list and only scored two to three percent points in the family evaluations.[45]

---

37 Birgitt Krawietz, "Ibn Taymiyya, Vater des islamischen Fundamentalismus?" in *Salafismus in Deutschland: Ursprünge und Gefahren einer islamisch-fundamentalistischen Bewegung*, ed. Thorsten Gerald Schneiders (Bielefeld: Transcript, 2014), 75.

38 Aydar, "Yaygınlık Bakımından Meallerimiz," 123.

39 Salih Yucel, "Fethullah Gülen: Spiritual Leader in a Global Islamic Context," *Journal of Religion and Society* 12 (2010): 2; Mücahit Bilici, "The Fethullah Gülen Movement and Its Politics of Representation in Turkey," *The Muslim World* 96 (2006): 16.

40 Aydar, "Yaygınlık Bakımından Meallerimiz," 109–39.

41 Ibid.

42 Ceyda Nurtsch, "Interview with Turkish Theologian Ihsan Eliacik: The Koran and Social Justice," *Qantara*, January 15, 2014, accessed November 12, 2023, https://qantara.de/en/article/interview-turkish-theologian-ihsan-eliacik-koran-and-social-justice.

43 Karar, "Diyanet istedi mahkeme meal yasakladı," Febuary 21, 2023, accessed November 12, 2023, https://www.karar.com/guncel-haberler/diyanet-istedi-mahkeme-meal-yasakladi-1729935.

44 Memurlar, "Yaşar Nuri'den şaşırtan Erdoğan çıkışı," accessed November 12, 2023, https://www.memurlar.net/haber/551307/yasar-nuri-den-sasirtan-erdogan-cikisi.html.

45 Aydar, "Yaygınlık Bakımından Meallerimiz," 132.

3 Method

The word *meâl* (Arab. *maʾāl*) is derived from an Arabic root (*a-w-l/ā-la*). It has various meanings such as "to return," "to lead to," and "to arrive,"[46] and is defined as "meaning" and "concept" in Turkish dictionaries of the Qurʾān.[47] However, it is not mentioned in Arabic dictionaries in connection with the translation of the Qurʾān. In Turkish, the word *meâl* explains what a word or phrase signifies. Thus, according to Zülfikar Durmuş, it does not have the same meaning as the original Arabic word, but offers a meaningful and explanatory interpretation of the revelation.[48] Ihsan Eliaçık, on the other hand, criticizes the incorrect use of the term *meâl* in Turkish.[49] Most Qurʾān translations in Türkiye are titled *Kur'an-ı Kerim ve Yüce Meali*, which means "The Qurʾān and its Glorious Interpretation." In this context, there is the potential for the misunderstanding that a translation represents or reproduces the actual Qurʾānic text, which is not anywhere near possible without a commentary due to linguistic barriers and the ambiguity of the text.

This misunderstanding has to do with the fact that, for years, translators were in the habit of translating the original text literally, speckled with their own interpretations, which only appeared only in the footnotes. According to Eliaçık, because every *meâl* already means "the interpreted," it is necessary to embed commentary in the translation by considering the historical context, the explanation of the terms, and the cause of the revelation.[50] Without these aspects and without explicitly stating a specific purpose, it would amount to nothing but the manipulation of the reader. It is therefore debatable how far it can be considered correct to use the word *meâl* in respect of Qurʾān translations. While translation implies an objective approach, the word *meâl* refers to adding interpretations of a term or text. To be more precise, it refers to an interpretive form of translation.[51] Today's Turkish Qurʾān translations usually begin with an introduction and an index. The Qurʾānic text is presented either in the form of *muṣḥaf meâl, nuzul meâl*, or thematic *meâl*, which also differ in their translation methods. A distinction is made between "interlinear"/"literal" (*kelime/satır arası; harfî tercüme*) and "interpretive" (*tafsīrī*) translation. This also includes the Eastern Turkish translations, which were based on the principle of the earlier Persian

46 Hans Wehr, *Arabisches Wörterbuch für Schriftsprache der Gegenwart: Arabisch-Deutsch* (Wiesbaden: Harrassowitz 1985), 54.

47 Muhammed Şerif Kahraman, „Kur'ân'ın harfî ve tefsirî tercümesi" Siirt Üniversitesi Ilahiyat Fakültesi Dergisi 3:2 (2016): 152.

48 Zülfikar Durmuş, *Kur'an'ın Türkçe Tercümeleri* (Istanbul: Çıra Akademi, 2020), 42–52.

49 R. Ihsan Eliaçık, *Yaşayan Kur'an: Nüzul sırasına göre Türkçe meal – tefsir* (Istanbul: Inşa Yayınları, 2007), 11.

50 Ibid., 11–12.

51 Ibid., 11–13.

translations of the Qur'ān.[52] The first Turkish translations of the Qur'ān applied the aforementioned technique of interlinear translation,[53] according to which the Turkish words are written in a smaller script below the Arabic text. The fact that Arabic syntax differs somewhat from Turkish syntax prevents a proper understanding of the meaning in such translations.[54] Some scholars such as Hamdi Aktaş,[55] Halil Uysal,[56] and Hace Ahmet Didin[57] have limited themselves to a literal translation so as not to interfere with the divine wording, while others have attempted to convey the meaning of the Qur'ān through explanations and footnotes in the interpretive translation. This is possibly one of the reasons for the widespread use of interpretive translations, which allow translators to excessively interfere with the original text.[58]

3.1 Sequence of the Qur'ān

3.1.1 Muṣḥaf

Most translations (*meâl*) of the Qur'ān in Türkiye today follow the *muṣḥaf* order, starting with Sura 1, *al-Fātiḥa* (The Opening), and ending with Sura 114, *an-Nās* (Mankind).[59] The written version of the Qur'ān is written in Ottoman calligraphy. The art of calligraphy used to write out the Qur'ān in Turkish has given rise to the expression "The Qur'ān was revealed in Mecca, read in Cairo, and written down in Istanbul."[60] Aside from the artistic design of the layout, in the early years of the Turkish Republic, there were translations without the *muṣḥaf*, in which only the translation of the Qur'ān was reproduced without the Arabic original. However, these translations had the disadvantage of not being comparable with the original text due to the absence of the Arabic text. One reason why the translation would not contain the original was to allow readers to access the Qur'ān anytime and anywhere without the need for ritual washing. Some works were first published without the *muṣḥaf* and later re-published with the text inserted. An example of this type of translation is Süleyman Ateş's (b. 1933/1352)

52 Birnbaum, *On Some Turkish Interlinear Translations*, 113; Janos Eckmann, *Eastern Turkic Translations of the Koran* (Budapest: Akademiai Kiado, 1971), 151.
53 Öztürk, "II: Meşrutiyet'ten Günümüze Kur'an Mealleri," 185.
54 Abay, "Türkçedeki Kur'an," 299f.
55 "Kur'an-ı Kerim Meal ve Kelime Meali"; cf. Hikmet Koçyiğit, "Günümüz Türkçe Kur'an Meâlleri Üzerine Bir Tasnif Denemesi," *Istanbul Üniversitesi İlahiyat Fakültesi Dergisi* 37 (2017): 99.
56 "Kur'ân-ı Kerim Türk.e Anlamı ve Kelime Meali," cf. ibid., 99.
57 "Kur'an-ı Kerim Alt Yazılı Renkli Kelime Meali, İrab-i Kaide ve Belagat Uygulamalı," cf. ibid., 99.
58 Ibid., 97–100.
59 Ibid., 85.
60 Muhammet Abay, "Kur'an ve Tefsire Dair Başvuru Kaynakları," *Türkiye Araştırmaları Literatür Dergisi* 10, no. 19–20 (2012): 402.

1975/1395 translation, which was initially published without the Arabic text and later republished with it included.[61]

3.1.2 Asbāb an-Nuzūl

The category *Asbāb an-nuzūl* (occasions of revelation/Turkish: *nuzūl sırası*) describes the order of revealed verses that were sent down in sequence. Additionally, it depicts the historical occasions that describe the reasons for the revelation. Turkish Qur'ān translations written in order of the *nuzūl* have been published since 1994/1415 and are used to better understand the Qur'ān based on its preserved original order. These intensive efforts to comprehend the core of the Qur'ān, especially at present, have led to the approach of writing the texts and commentaries in order of the *asbāb* (occasions). Some authors believe that commentaries based on the *asbāb an-nuzūl* are important for achieving an authentic and holistic understanding of the Qur'ān, which fosters efforts to build Muslim identity. After the year 2000/1421, there was an explosion in the number of Qur'ān versions being published with the chapters arranged according to the occasions of revelation. They have therefore been met with great interest. Explanations of the occasions on which the Qur'ānic verses were revealed as well as glossaries have been added to some of these translations; others have been enriched with commentaries/explanations.[62]

3.1.3 Thematic Sequence

Thematic translations contain an index that categorizes the individual revelations of the Qur'ān by topic. These translations can thus be considered better-structured versions of Qur'ān commentaries. Such an index or directory allows the reader to engage with the translation in a more systematic manner. However, it is essential to note that such a subject index is not the same thing as a word index. A word index only provides the meanings of individual words and explains them. A subject index helps the reader to find verses on certain subjects in a summarized form. Nevertheless, there is also a subjective component to thematic organization. According to some scholars, translations categorized by subject cannot be considered complete translations as they may not include all verses of the Qur'ān.[63]

61 Koçyiğit, "Günümüz Türkçe Kur'an," 112.
62 Ibid., 88
63 Ibid., 87.

3.2 Translation Techniques

3.2.1 Interlinear Translation (*Kelime / Satır Arası / Harfî Tercüme*)

Literally translating the Qur'ān means conveying it from one language into another while preserving its verses and order (i.e., the original Arabic syntax). This approach aims to retain the original meaning of all words by substituting the equivalent word in the target language or using terms from other languages that have the same meaning.[64] Such translations must fully correspond to the original and reflect its meaning, method, style, and rhetoric.[65] As the meaning of words can change over time, it is particularly difficult to decipher their original meanings and translate them into the target language with appropriate words. It is therefore more a matter of interpretation based on the historical context than an issue of literal translation. This also applies to the Qur'ān, where literal translation is not possible because each word in the source language is replaced by the corresponding word in the target language. Furthermore, the harmony of the verses and their sequence takes precedence over the coherence of the meaning.[66] However, out of concern that human interpretations could potentially be added to the divine wording, most Turkish translations have been produced literally.[67] But when the idiomatic phrases of the Qur'ān are not assigned equivalent counterparts, and their original stories are not reflected in the translation, their meaning fades and deforms into simple speech.[68] According to İsmail Cerrahoğlu, there is consensus (*ijmā'*) among Islamic scholars that a literal translation of the Qur'ān cannot be achieved.[69] For literal translation, the original and the translation must be identical. The translation has to accurately reproduce the signification of the original. "This is out of the question for the Qur'ān, which is a revelation from God."[70]

3.2.2 Interpretive Translation (*Tafsīrī*)

Interpretive translation aims to explain the meaning of a phrase in another language without following the original wording and order of the verse being translated, as in

64 Muhsin Demirci, *Kur'an Tarihi* (Istanbul: M. Ü. İlahiyat Fakültesi Vakfı Yayınları, 2014), 190.

65 Faruk Gürbüz, *Tercüme Problemleri ve Mealler* (Istanbul: Insan Yayınları, 2004), 77.

66 Demirci, *Kur'an Tarihi*, 190.

67 Mustafa Öztürk, *Cumhuriyet Türkiyesi'nde Meal ve Tefsirin Serencamı* (Ankara: Ankara Okulu Yayınları, 2012), 72f.

68 Gürbüz, *Tercüme Problemleri ve Mealler*, 54–56.

69 Original citation: "Kur'ânın harfi tercemesinin yap ılamıyacağına islam alimleri icma' ile ittifak halindedirler. Çünkü harfi terceme bir ibdalden ibarettir. Asl ile bedel birbirinin ayn ı olmalıdır. Bedel, aslın kapsadığı şeyleri aynen içine almalıdır. Böyle bir şey vahy-i Ilahi olan Kur'ân için bahis konusu olamaz." İsmail Cerrahoğlu, *Tefsir Usûlü* (Ankara: Türkiye Vakfı Yayınları, 1995), 213.

70 Ibid.

literal translation. The main concern is to capture the intended meaning of the revelation and to translate it into the target language through the appropriate words. Therefore, *tafsīri* translation does not claim to be a literal rendering of the source text.[71]

4 Chosen Exegetes

For this chapter, I have selected sura 4, *an-Nisāʾ* (The Women), verse 34, in order to compare the translations by İhsan Eliaçık and the Diyanet (T.C. Diyanet İşleri Başkanlığı). The reason for my selection is the publication ban on Eliaçık's Qurʾān translation *Yaşayan Kuran Meâl Tesfir'i* (*Living Qurʾān*), which was imposed in February 2023/ Rajab 1444 as the result of a court case initiated by the Diyanet. According to the Diyanet, Eliaçık's translation contained "offensive elements." The court agreed, deciding to confiscate Eliaçık's work and prohibit its publication.[72] The ban has since been lifted, with a ruling that the Presidency of Religious Affairs does not have the authority to order the confiscation of books.[73]

I would like to emphasize the different methodologies and analyses in the two works in order to illustrate how the same text can be the subject of completely different translations and interpretations depending on the translator.

4.1 Diyanet İşleri Başkanlığı (The Presidency of Religious Affairs in Türkiye)

The year after the Turkish Republic was proclaimed in 1923/1341, the Presidency of Religious Affairs in Türkiye was established to regulate the religious affairs of society, starting with the conveyance of Islam in Turkish. Since then, it has been considered the highest religious authority, and it answers to the Turkish Prime Ministry. Further, it was intended to disrupt dependence on scholars to prevent religious abuse. Thus, the Diyanet was tasked with implementing all religious measures in Turkish, which

71 Naif Yaşar, *Kur'ân Tercüme Teknikleri* (N.p.: LAP LAMBERT Academic Publishing, 2017), 124f.

72 Duvar English, "Turkish Court Lifts Ban on Theologian's Quran Translation," *Duvar English*, March 14, 2023, accessed November 11, 2023, https://www.duvarenglish.com/turkish-court-lifts-ban-on-theologian-ihsan-eliaciks-quran-translation-news-62016; Karar, "Diyanet istedi mahkeme meal yasakladı," accessed November 11, 2023, https://www.karar.com/guncel-haberler/diyanet-istedi-mahkeme-meal-yasakladi-1729935; Cumhuriyet, "İhsan Eliaçık'ın kitabına 'İslam dini açısından sakıncalı' gerekçesiyle toplatma kararı," accessed November 11, 2023. https://www.cumhuriyet.com.tr/turkiye/ihsan-eliacikin-kitabina-islam-dini-acisindan-sakincali-gerekcesiyle-toplatma-karari-2054245.

73 Gazete Memur, "İhsan Eliaçık'ın Kuran meali hakkında toplatma kararı kaldırıldı," accessed November 11, 2023, https://gazetememur.com/gundem/ihsan-eliacikin-kuran-meali-hakkinda-toplatma-karari-kaldirildi,uI-oE4qs4km6Zau_2pHh-Q.

included translating the call to prayer from Arabic to Turkish (in 1932/1351) and publishing a Turkish translation of the Qur'ān (1938/1357).[74] The translation of the Qur'ān published by the Presidency of Religious Affairs is one of the most important institutionalized translations in Türkiye. At the same time, the authority answers to the president and therefore to the government of the day, meaning that it is always subject to the political agenda of the respective regime. For this reason, the Diyanet has been criticized in particular for both reflecting and being affected by official state ideology.[75]

Following the 2017/1438 referendum, the presidential system was introduced to Türkiye in 2018/1439, with Recep Tayyip Erdoğan becoming Türkiye's head of state. The resulting abolition of parliamentary checks and balances gave him sole power over the executive branch. He is also entitled to control the judiciary by individually appointing the most important members of the judicial apparatus.[76] Erdoğan's Justice and Development Party[77] rejects its categorization as a Muslim democratic party and instead exhibits the structure aimed at the targeted re-Islamization of the Turkish Republic. It utilizes certain means, such as the mass media, the film industry, and state religious scholars,[78] to not only glorify the Ottoman Empire and the Turkish nation but also to call for a return to Islamic traditions, thereby boosting Turkish nationalism.[79] The Diyanet currently answers to Recep Tayyip Erdoğan, meaning that he also directs the Diyanet's current ideology.[80]

4.2 Ihsan Eliaçık

Eliaçık's work was written in the context of theological studies and social movements, which makes him one of the most well-known political and theological representatives of the current era. He is considered contemporary Türkiye's main civil theologian, and his numerous works deal with social justice in Islam. His Qur'ān commentary, *Living Qur'ān* (2007/1428), was, as mentioned above, temporarily subject to a

74 Şayan Ulusan, "Atatürk Dönemi Din Uygulamaları (1923–1938) (Yorumlar, Eleştiriler, Değerlendirm," *Cappadocia Journal of History and Social Sciences* 11 (2018): 306–17.

75 Koçyiğit, "Günümüz Türkçe Kur'an," 93f.

76 Ismail Küpeli, "Drohende Alleinherrschaft in der Türkei: Per Verfassungsreferendum soll die Bevölkerung die Demokratie abschaffen," *RLS-Standpunkte* 5 (2017): 3.

77 Adalet ve Kalkınma Partisi (AKP), under the leadership of Recep Tayyip Erdoğan.

78 Soyalp Tamçelik, "Türkiye'de Popüler Kültürün Muhafazakârlaşması ve Eğlence Kültürünün Siyasallaşması," *JT* 11, no. 2 (2016): 1147.

79 Fatma Murat, "Türk Dizilerinin Tarih Kurgusu: Muhteşem Yüzyıl, Payitaht Abdülhamid, Kuruluş Osman," *Dergi Park Akademik* 8 (2022): 130–50; Umut Uzer, "Conservative Narrative: Contemporary Neo-Ottomanist Approaches in Turkish Politics," *Middle East Critique* 29 (2020): 275–90.

80 Sözcü, "Cumhurbaşkanı Erdoğan'dan Ali Erbaş mesajı," accessed November 11, 2023, https://www.sozcu.com.tr/cumhurbaskani-erdogandan-ayasofya-aciklamasi-wp5964783.

publication ban after a complaint was made by the Office of Religious Affairs.[81] In his view, the democratic and moral values of Islam have faded due to religious scholars joining autocratic governments. In this context, he criticizes above all the sultanate's rule as blasphemous, as it required unconditional obedience to the ruler and elevated the political leader to the rank of God.[82] At the same time, he opposes the rigid secular approach, which considers any religious expression in politics to be illegitimate and relegates religion to the private sphere.[83]

Eliaçık emphasizes that there is no place for coercion in Islam and criticizes Kemalist-liberal advocates for previous governments' banning of headscarves in government agencies and universities as well as religious-conservative advocates for oppressing people who are less religious than them.[84] Eliaçık's criticism of current policy is mainly directed against political Islam in Türkiye. He often criticizes Erdoğan for politicizing Islam,[85] thus rejecting the state's promoting or influencing of a particular religion. He criticizes the Diyanet for shaping the religious sphere of Türkiye on behalf of the government by underpinning state action with religion. Also, the Diyanet, he says, does not protest against the state's intervention into religious issues, as it represents the ideology of the government. Eliaçık condemns all clerics who serve the government's objectives as well as the chairman of the Diyanet, Ali Erbaş. Moreover, he endorses the complete abolition of the institution of the Diyanet in order to liberate religion from state control so that Islam can be implemented independently in the sphere of civil society.[86] Through his work and actions, Ihsan Eliaçık is actively involved in educating Turkish civil society about Islamic traditions. He has also supported local anti-capitalist movements, in particular the Anti-Capitalist Muslim Movement, which was actively involved in the 2013/1434 Gezi protests (*Gezi Parkı Protestoları*).[87] The movement was also part of the preparatory team for the Democratic Islamic Conferences initiated by the Kurdish movement. The frequency of Eliaçık's appearances on the national stage increased significantly during the Gezi protests in the summer of 2013/1434, where he opposed the

81 Cf. 4. "Chosen Exegetes."

82 R. Ihsan Eliaçık, *Demokratik özgürlükçü İslam* (Istanbul: Tekin Yayınevi, 2014), 15.

83 Ibid., 245.

84 Ibid., 203–9.

85 Susanne Güsten, "Türkei Islamisten gegen Erdogan," *Tagesspiegel*, April 13, 2018, accessed November 11, 2023, https://www.tagesspiegel.de/politik/islamisten-gegen-erdogan-5284864.html.

86 Ekin Karaca, "İhsan Eliaçık: Diyanet'in görevi devletin icraatlarına dinî kılıf bulmak," *Bianet*, accessed November 11, 2023, https://bianet.org/haber/ihsan-eliacik-diyanet-in-gorevi-devletin-icraatlarina-dini-kilif-bulmak-137669.

87 The protest of May 2013 was directed against the urban development plan for Istanbul's Taksim Gezi Park. Arising from outrage over the violent eviction of a sit-in in the park, the protests turned against the leading government of Recep Tayyip Erdoğan. Cf. Burcu Karakas, "Ten Years After Turkey's Gezi Protests Some Say Vote a Chance for Justice," *Reuters*, May 26, 2023, accessed November 11, 2023, https://www.reuters.com/world/middle-east/ten-years-after-turkeys-gezi-protests-some-say-vote-chance-justice-2023-05-26/.

populist discourse of the then-Prime Minister Erdoğan, who portrayed the protesters as secular elites trying to dominate the religious majority.[88] Eliaçık's sermon in Gezi Park was an important example of civil religion emphasizing solidarity with people of different identities and lifestyles and sharply criticizing the government for its pursuit of economic profit.[89]

5 Analysis of Q 4:34

One of the central disputes in gender relations in Islam is the issue of physical violence against and male superiority over women. This issue has not only been discussed at the international level by various scholars and ideologues, such as Asma Barlas, Khaled M. Abou El Fadl, and Saʿid Hawwa, but is also a controversial issue at a national level in Türkiye. Scholars like Mustafa Öztürk, Sonia Cihangir, Yaşar Nuri Öztürk (d. 2016/1437), and Yunus Vehbi Yavuz share the opinion that Islam rejects violence against women.[90] AKP Member of Parliament Mehmet Demir stated in a Facebook post on International Women's Day that, according to the Prophet Muḥammad, women can be beaten softly[91] and intimidated. This statement caused turmoil within Turkish society. Afterward he took the statement down from the Diyanet's website.[92] In 2020/1441, the website also advised women who had been affected by violence in marriage "to not react, be patient, walk away," which likewise led to considerable criticism.[93] Still, domestic violence in Islamic society is related to the understanding of Q 4:34. This verse is generally used as a justification for husbands to exercise violence against their wives.[94] In the following, I will examine the interpretation of Q 4:34 by discussing the understandings of two different exegetes.

88 Ateş Altınordu, "Occupy Gezi, Beyond the Religious-Secular Cleavage", *The Immanent Frame*, accessed November 11, 2023, https://tif.ssrc.org/2013/06/10/occupy-gezi-beyond-the-religious-secular-cleavage/.

89 Habertürk, "Gezi Parkı eylemlerinde son durum," accessed November 12, 2023, https://www.haberturk.com/gundem/haber/850301-gezi-parki-eylemlerinde-son-durum.

90 Yunus Vehbi Yavuz, *Kur'an'da Kadın Hak ve Özgürlüğü* (Istanbul: Bayrak Yayınları, 1999), 16; Mustafa Öztürk, "Kur'an yorumu ve şiddet meselesi," *Bilimname* 32 (2016): 12f.; Sonia Cihangir, *Kadına Dair: Hurafeler ve Gerçekler* (Istanbul: Ozan Yayincılık, 2016), 57–68; Yaşar Nuri Öztürk, *Der verfälschte Islam* (Düsseldorf: Grupello Verlag, 2007), 104–8.

91 Dünya, "Çaresiz kalındığında hafifçe dövün, Diyanet`ten açıklama," accessed November 12, 2023, https://www.dunya.com/gundem/quotcaresiz-kalindiginda-hafifce-dovunquote-diyanetten-haberi-203770.

92 Ibid.

93 Sözcü, "Şiddet gören kadına Diyanet'ten tavsiye," accessed November 12, 2023, https://www.sozcu.com.tr/2020/gundem/siddet-goren-kadina-diyanetten-tavsiye-tepki-verme-sabret-uzaklas-5644351/.

94 Sami Aldeeb, *Schlagt Die Frauen: Interpretation des Koranverses 92/4:34 durch die Jahrhunderte (Kurzversion Ohne Exegesen in Arabisch)* (Charleston: Amazon, 2022), 5f.

5.1 Diyanet İşleri Başkanlığı (Presidency of Religious Affairs in Türkiye)

5.1.1 Translation into Turkish and English

> Allah'ın, (iki cinse) birbirinden farklı özellik ve lütuflar bahşetmesi ve mallarından harcama yapmaları sebebiyle erkekler kadınların yöneticisi ve koruyucusudurlar. Sâliha kadınlar Allah'a itaatkârdır; Allah'ın korumasına uygun olarak, kimsenin görmediği durumlarda da kendilerini korurlar. (Evlilik hukukuna) baş kaldırmasından endişe ettiğiniz kadınlara öğüt verin, onları yataklarda yalnız bırakın ve onları dövün. Eğer size itaat ederlerse artık onların aleyhine başka bir yol aramayın; çünkü Allah yücedir, büyüktür.[95]

> God has endowed (both genders) with different qualities and favors, and by spending their wealth, men are the leaders and protectors of women. Righteous women are obedient to God and guarding themselves according to what God has guarded, even in situations when no one sees them. Advise those women whom you fear may rebel (against the law of marriage), leave them alone in their beds and beat them. And if they obey you, then seek no other way against them, because God is the Exalted, the Great.[96]

5.1.2 Method

The Diyanet's translation of the Qur'ān falls into the category of "institutional translation," which consists of one or more people coming together to prepare and produce a translation. The translators are not named, so it is not clear who or how many translators were involved in translating the Qur'ān.[97] This Qur'ān translation follows the original arrangement of the *muṣḥaf*, meaning from Q 1 to Q 114. Moreover, the *muṣḥaf* has been added to the Turkish translation of Q 4:34, so the original Arabic text precedes the Turkish translation and is followed by a comprehensive interpretation (*tafsīri*). Furthermore, the scholars draw on the traditions of the Prophet (*ḥadīth*) to explain their translation and commentary. The *tafsīri* method of translation has been used to translate the sura. To clarify its meaning, short explanations have been given in brackets at some points in the text.[98] This example of annotated translation can be seen at the beginning of the examined verse – "(iki cinse)" ("[both genders]") – and near the end of the sura – "(*Evlilik hukukuna*)" ("[against the law of marriage]"). The words in brackets provide additional information that is intended to make the sen-

95 Diyanet İşleri Başkanlığı Kur'an-I Kerim, "Nisâ Suresi – 34. 'Meal (Kur'an Yolu),'" accessed December 19, 2023, https://kuran.diyanet.gov.tr/tefsir/Nisâ-suresi/527/34-ayet-tefsiri.
96 From the original Turkish translation by the Presidency of Religious Affairs in Türkiye; my English translation.
97 Koçyiğit, "Günümüz Türkçe Kur'an," 93f.
98 Ibid., 84–99.

tence easier to understand.[99] The comparative (*karşılaştırmalı*) method of translation has been chosen here by reproducing some of the hadiths that explain the meaning of the verses with their sources to provide additional explanations of the given verse.[100]

5.1.3 Analysis and Interpretation

As already mentioned, an explanatory bracket – "(both genders)" – is used at the beginning of the verse to clarify that God's speech is about both genders. The Diyanet then translates the following sentence to provide that men are "protectors and leaders" of women because they spend their wealth on them: "God has endowed [both genders] with different qualities and favors, and by spending their wealth, men are the rulers and protectors of women." The key word in the original Arabic verse is *qawwām* (Turk.: *kavvâm*), the singular of *qawwāmūn*,[101] and has been translated by the Diyanet as "leader" and "protector." To justify their translation, they cite Islamic jurists (*Müctehidler*)[102] and Qur'ān commentators (*Müfessirler*)[103] who have used not only the wording of verses and hadiths but also the customs and traditions of a society to determine this meaning.[104] They have defined the term *kavvâm*as as "the one who stands above something, the one who dominates, the one who is heavily involved with him/her."[105] These scholars have also attributed additional meanings to the term, including "head, leadership, education, protection, defense, construction, acquisition and production."[106] The Diyanet suggests that more men than women possess these qualities. Additionally, the commentators have stated that *aile reisliği* ("head of the family") is now the predominant meaning of *kavvâm*. Regarding this section, it should be noted that the religious authority does not provide any verifiable sources for its claims. It is unclear which historical statements by the cited jurists and Qur'ān commentators they are relying on. Furthermore, they do not provide any grammatical explanation of the root, stem, or morphology of the term *kavvâm*. Further, they fail to specify which dictionary they are referring to in the "dictionary of definitions" they cite.[107]

---

99 5.1.1. Translation: Diyanet İşleri Başkanlığı, Nisâ Suresi – 34.

100 Koçyiğit, "Günümüz Türkçe Kur'an," 104.

101 Diyanet, "Ayet," first sentence.

102 Yunus H. Apaydın, "Müctehid," in *TDVIA*, accessed December 25, 2023, https://islamansiklopedisi.org.tr/muctehid.

103 Abdulhamit Birişik, "Müfessir," in *TDVIA*, accessed December 25, 2023, https://islamansiklopedisi.org.tr/mufessir.

104 Diyanet, "Nisâ Suresi – 34, 'Tefsir (Kur'an Yolu),'" accessed December 19, 2023, https://kuran.diyanet.gov.tr/tefsir/Nisâ-suresi/527/34-ayet-tefsiri.

105 My translation of the Turkish original. Diyanet, "Nisâ Suresi – 34" Tefsir (Kur'an Yolu) 2. paragraph: "bir şeyin üzerinde duran, hâkim olan, özen gösteren, onunla yakından ilgilenen."

106 My translation, ibid., "reislik, yöneticilik, eğitim, koruma, savunma, ıslah, kazanma, üretme."

107 Ibid.

In its interpretation, the Diyanet explains why it chose the terms "leader" and "protector" by making reference to two distinct aspects, arguing that not only husbands but all men should generally be considered as such. This is because "a) God has given some people superior abilities over others, and men are more suitable to protect and lead than women. b) Man has the responsibility to financially provide for his family."[108] Certain "Qur'ān commentators" regard the first motive (a) as an inherent aspect of human nature. In terms of protection (*koruma*) and leadership (*yönetme*), men are considered to possess more favorable attributes than women. Men, they say, typically prioritize reason and logic, whereas women tend to be more emotional. "In general, reason and logic remain in the foreground for men, while emotions predominate the actions of women."[109] First, it is unclear which "Qur'ān commentators" the Diyanet is referring to when it claims that men, unlike women, are naturally endowed with reason and logic. This statement is problematic as it implies the existence of gender-specific differences in mental endowment, which indicates favoritism by God: women cannot compete with men on any level due to their natural deficiencies; they are considered second-class humans, as they are not only physically but also mentally inferior to men, which renders women dependent on their spouses. The Diyanet claims that men's superiority is signified by their role as the protectors and leaders of women. However, it is important to note that the Diyanet has not standardized its interpretation of gender roles. For instance, its view about man's designation as a "protector" relies solely on the male anatomy, which by nature is physically stronger than the female anatomy. However, the text does not evaluate the position of the "leader."

By making reference to the wording of the Qur'ān – "and by spending their wealth, men are the rulers and protectors of women"[110] – the religious authority justifies the role of the man as woman's "leader," without providing any interpretation that explains this statement of the Qur'ān in more detail. It merely refers to men's financial responsibility for their families to legitimize their leadership role, although it also mentions that Islamic law places sole responsibility for the family's livelihood and all other financial obligations on men.[111] Accordingly, the core of the divine commandment is providing for the family, for which the man is responsible, and not the man's position of power over woman as the "leader," legitimized by his spending on her. It is therefore not clear from the Diyanet's statement how the idea of leading women is derived from man's obligation as the breadwinner of the family.

---

108 Diyanet, "Nisâ Suresi – 34," Tefsir (Kur'an Yolu) 1. Paragraph: a) "Allah insanların bir kısmına diğerlerinden üstün kabiliyetler vermiştir, bu cümleden olarak koruma ve yönetme bakımından erkekler, kadınlardan daha uygun özelliklerle donatılmışlardır. b) Erkekler aile geçimini ve diğer malî yükümlülükleri üstlenmişlerdir."

109 Ibid., b) "genel olarak erkeklerde akıl ve mantık ön plandadır, kadınlarda ise duygu öne çıkar." My translation.

110 Cf. 5.1.1.

111 Diyanet, "Nisâ Suresi – 34, 'Tefsir (Kur'an Yolu),'" 1. Paragraph.

The verse continues with the translation "Righteous women are obedient to God."[112] Again, it should be noted that there is neither a literal translation of the Arabic word *sâliha* into Turkish nor an examination of the grammatical composition of the term. In the Diyanet interpretation, the commentators refer to the rule that a wife must be obedient in the context of marriage, which therefore classifies her as *sâliha* or *nâshiza*. The term *sâliha* is described by the Diyanet as a woman who properly fulfills her duties and responsibilities as a wife both in the presence and absence of her husband and family. She always acts according to the rules set by God, in which case she is accepted by society for maintaining a family. At this point, there is neither any explanation about the exact rules that are explicitly "set" by God, nor are there are any further details about "the society" that, according to the Diyanet, has accepted these rules by mutual agreement. It is unclear whether these obligations are legal, cultural, or traditional, as societies change, meaning such rules cannot be assumed to apply to every society. The women who are described as *nâshiza* are the opposite of *sâliha*; *nâshiza* thus signifies a woman who rebels against Islamic family law. In this case, God gives the man (*aile reisi*) the responsibility to restore order and rebuke the "rebellious" wife. As the verse above shows, the woman's obedience to God is hardly mentioned; instead, her obedience to her husband and family is presented in terms of her duties.[113]

Amina Wadud argues that the obedience (*itaatkâr*) mentioned in the verse pertains to the obedience of all believers to God, rather than the wife's obedience to her husband.[114] Riffat Hassan sees another term for obedience in the word *ṣāliḥāt* (righteous), which is related to the term *ṣālāḥiyya*, which can be translated as "ability" or "possibility." The term is associated with an ability attributed only to women. This could refer to pregnancy[115] as well as to the phrase from the verse "they protect themselves even in situations that no one sees."[116] At this point, a more detailed explanation from the Diyanet on the wording "situations that no one sees" seems necessary, as the commentary is unclear about which "situation" or "circumstance" is signified. God's call to woman to protect herself even in situations that the public is not aware of could also refer to pregnancy. In that case, the wife would have to take responsibility for protecting herself and her child as long as her pregnancy is not noticed by society. At this point, obedience is directed toward God, who has entrusted her with the responsibility of bearing a child. The beginning of the verse leads to the possible interpretation that women should not take on the role of being providers in addition to the task of preg-

112 Diyanet, "Nisâ Suresi – 34, 'Meal (Kur'an Yolu),'" 34: "Sâliha kadınlar Allah'a itaatkârdır."
113 Diyanet, "Nisâ Suresi – 34, 'Tefsir (Kur'an Yolu),'" 3. Paragraph.
114 Asma Barlas, *Women's Readings of the Qur'ān: The Cambridge Companion to the Qur'ān* (Cambridge: Cambridge University Press, 2007), 263.
115 Yasmin Masidi, *Are Muslim Men Allowed to Beat their Wives?* (Malaysia: Sisters in Islam, 2009), 12.
116 Diyanet, "Nisâ Suresi – 34, 'Meal (Kur'an Yolu),'" 34: "[. . .] kimsenin görmediği durumlarda da kendilerini korurlar."

nancy. One potential meaning is therefore that each gender should fulfill their purpose assigned by God without being in a superior situation. However, the Diyanet is more concerned with bringing obedience and disobedience down to the human level, using the marriage relationship as an example. At the beginning of the passage on the rebelliousness of women – "Advise those women whom you fear may rebel (against the law of marriage)" – a bracketed phrase, "marriage law" (Evlilik hukuku), is added, clarifying the context.[117] The key term here is the Arabic word *nushūz*, which has been translated into Turkish as "to rebel" (Turk.: *baş kaldırmak*). The translation is not accompanied by any the historical context or clarification of Qur'ānic language. This would have allowed for a better understanding of the context of the actions and could have also clarified the meaning of rebellion and its context. Instead, the measures to be taken to deal with such a situation are presented by making a judgment about the woman's "unruliness." Before the final judgement, she should first be "advised" (*öğüt vermek*) in order to end the conflict. At this point, it is not about a clarifying conversation with the woman, but about the man instructing the woman. This means that the genders are no longer at eye level; rather, the male is placed above the female. Women should be left alone in their beds and beaten ("onları yataklarda yalnız bırakın ve onları dövün"). For all three measures, the imperative form has been used in the Turkish translation, based on the original. No explanatory bracket has been used for this sentence. This gives the impression that, in situations of marital conflict, only the husband is authorized to mediate and may resort to physical violence to resolve the problem. With this last translation "beat her" (*onları dövün*), the Diyanet has opted for a negative connotation of the Arabic word *ḍaraba* and this time uses the historical context to explain it.[118]

They rely on several prophetic traditions (hadiths) that explain the conditions, the manner, as well as the extent of beating, though the Prophet opposed violence in marriage. Based on the Diyanet's interpretation, it is possible to discern changing attitude to violence, from rejection to justification. The absolute prohibition on beating mentioned at the beginning takes a more relaxed form toward the end, so that it can be justified under certain conditions.[119] This development is illustrated by starting with a tradition in which the Prophet explicitly forbids the beating of the spouse:

> In these hadiths, which are more authentic and well-founded compared to the traditions to the contrary, the Prophet forbids the beating of wives, as he refers to those who beat their wives as "unwholesome." He also draws attention to the fact that this behavior is incompatible with sharing the same house and bed and is inhumane and immoral. (Bukhari, "Nikâh," 93)[120]

117 Diyanet, "Nisâ Suresi – 34, 'Meal (Kur'an Yolu),'" 34.

118 Ibid.

119 Diyanet, "Nisâ Suresi – 34, 'Tefsir (Kur'an Yolu),'" fourth to eighth paragraphs.

120 My translation of the original citation, Diyanet, "Nisâ Suresi – 34, 'Tefsir (Kur'an Yolu),'" paragraph 4.

This is followed by a consideration of legal opinions and statements from various scholars, ranging from the prohibition on beating to the prohibition on causing any visible signs of injury by beating. While some scholars explain the conditions and the extent of the violence and point out the parts of the body that should not be hit,[121] other scholars reject hitting and see *ḍaraba* more as a metaphorical allusion. To counter this argument, the Diyanet cites a tradition that refutes it, describing the act of beating as *ibâha* (not obligatory or forbidden) in the Qur'ān. The hadiths that prohibit violence against women have only proven to be *kerâhet* (frowned upon and unattractive).[122] Through historical contextualization, the commentators explain that women can only be beaten in societies where beating does not violate social and legal norms, and is not considered shameful and abnormal. It serves as a last resort to make the woman aware of the man's anger; hence, violence is only permitted in such situations. They also refer to a tradition from the second caliph, Omar, who complained about the dominance of women who no longer obeyed men due to a change in society. According to the Diyanet, this verse is an essential measure that must be applied when a woman violates family law. In applying these measures, place, time, and society have been considered, including the physical punishment of the wife by the husband.[123]

Again, the Diyanet argues that this punishment is condemned by the Prophet; however his condemnation does not invalidate the verse, as the hadith only provides additional information on the issue. At this point, it should be noted that the Diyanet is convinced that the Qur'ān legitimizes physical action against women under certain conditions, not least due to the indication that the sunna is not given any decision-making authority here. Based on the interpretation of Q 4:34 provided by the Diyanet, it becomes apparent that about half of the entire commentary deals with the topic of beating. In addition, as previously mentioned, there are only references (hadiths) to beating. A historical contextualization of this topic is provided, but it does not contain any references either. It can therefore be assumed that this verse is of particular importance to the Diyanet. In contrast to the other verses, the Diyanet intended to give a precise discussion of the issue of physical action against women. This approach is particularly evident in the selection of single hadiths, which the commentators discuss in greater detail. They also clarify the difference between what is forbidden and what is frowned upon by referring to the Prophet's negative position on the issue of violence toward women. Here it is explained that actions condemned by the Prophet do not necessarily always illustrate a sin. They are neither halal nor haram, so it is up to the individual to decide whether to commit or refrain from the condemned act. The Diya-

121 Diyanet, "Nisâ Suresi – 34, 'Tefsir (Kur'an Yolu),'" 5. Paragraph.

122 Ibid., sixth paragraph.

123 Ibid., seventh paragraph.

net therefore believes that the Prophet's sayings (sunna) on this point cannot override the validity of this verse.[124]

5.2 Ihsan Eliaçık

5.2.1 Translation into Turkish and English

> Erkekler, kadınlar üzerine titrer; onları koruyup kollarlar. Bu, Allah'ın onlardan kimine kiminden fazla vermesi ve erkeklerin mallarını harcamalarından dolayı böyledir. İyi, güzel ve doğru olan kadınlar, Allah'ın korunmasını buyurduğu mahremiyeti koruyan ve O'na saygıda kusur etmeyen kadınlardır. Şiddetli geçimsizlik yaşadığınız eşlerinizle önce oturup konuşun, olmazsa yataklarında yalnız bırakın, yine olmazsa bir müddet ayrılın. Barışıp anlaşırsa, hâlâ işi yokuşa sürüp bahaneler aramayın. Yücelik ve büyüklük Allah'a mahsustur.[125]

> Men care for women and protect and guard them. This is because God has given more to some of them than others and because men spend their wealth. The good, beautiful, and righteous women are those who protect the intimacy that God has commanded to be protected and do not fail to respect Him. When you have harsh disagreements with your wives, first sit down and talk to them, if that fails, leave them alone in their beds, and if that fails, separate from them for a while. If they reconcile and compromise, do not still drag it out and look for excuses. Glory and greatness belong to God.[126]

5.2.2 Method

Ihsan Eliaçık is the single author of this *meâl*. He has followed the method of *asbāb an-nuzūl* (Turk.: *nuzul sırası*), meaning that the arrangement of the verses is based on the order of their revelation. Therefore, the arrangement begins with sura 96, *al-ʿAlaq (al-alak)* (The Blood Clot), and ends with sura 110, *an-Naṣr (al-nasr)* (Help).[127] He chooses to translate without reproducing the original Arabic text (*metinsiz meâllller*), meaning that only the Turkish *meâl* is displayed. In particular, Eliaçık takes the approach of annotated translation (*tafsīri meâl*)[128] in which he provides his translation of Arabic terms and their explanation in the footnotes. To do so, he examines the morphological structure of a word, looking first at the root and the corresponding verbal noun (*maṣdar*). Then, he presents a few different terms from the same root as well as their definitions, showing various possible translations of the term in question. Thus, he first demonstrates the ambiguity of the word and its potential forms to better ex-

124 Diyanet, "Nisâ Suresi – 34, 'Tefsir (Kur'an Yolu),'" eighth paragraph.
125 Eliaçık, *Yaşayan Kur'an*, 890–94.
126 My translation based on Eliaçık's Turkish translation of Q 4:34.
127 Eliaçık, *Yaşayan Kur'an*, 3–5.
128 Cf. above 3.1.1 Interpretive translation (*tafsīri*).

plain its meaning. This is followed by a historical contextualization of the use of the word, its different forms, and its meanings in each period. Some of this is confirmed by references to the hadith. He also offers the reader synonyms for better understanding. This approach becomes evident in the first phrase of Q 4:34. The explanations and translations of the terms for each Qur'ānic verse, which extend over several pages, illustrate his consistent aim of providing a meaningful and comprehensible translation of the revelation.

5.2.3 Interpretation

Eliaçık derives the term "*GAVVAM*"[129] from the root word "*G-V-M*" and translates its verbal noun as "stand up" ("ayağa kalkmak"). The translation of the first sentence of the verse thus expresses that men assume both a caring and a protective function for women. It is initially unclear which transcription he used, as he neither adheres to the usual transcription standards (e.g., Ijmes, DMG, Isnad) nor gives any information about the guidelines for his transcription. The explanation of the word stem mentions that a dictionary was used, but does not explicitly state which one. After translating the verbal noun, he explains that the word is used as an exaggeration in verse, which therefore means "the one who stands up often" ("*çokça ayağa kalkan*"). He interprets this literal translation as "who is very caring/interested" ("*çok ilgili olan*"). The exaggerated meaning explains the extraordinary care men take for their wives. The "one who stands up often" could therefore describe a man who is always interested in his wife's concerns and takes care of her affairs. The subsequent sentence accordingly clarifies the roles of men and women without emphasizing any superior position.[130] The word "*FADL*"[131] is explained below· (Arab.: *faḍl*); one of its many meanings can be translated as "surplus"/"superiority" and therefore justifies the assumption of male superiority in this verse:[132] "God has given more to some of them than others."[133]

However, according to Eliaçık's explanation, the verse instead describes the different characteristics of women and men, who, like all people who are born as members of the same species, share fundamental similarities. However, they differ in their individual characteristics, such as skin color, size, gender, etc. For example, some people have darker skin, while others are taller. The latter does not indicate superiority; it is merely one of the differences among human beings. This logic also applies to the two genders. For example, the man has a "surplus" over the woman, since he carries the "seed" responsible for continuing the human species, which cannot be produced

129 According to the translator's spelling of the word.
130 Eliaçık, *Yaşayan Kur'an*, 890f. (footnote 32: "*GAVVAM*").
131 According to the translator's spelling of the word.
132 Eliaçık, *Yaşayan Kur'an*, 891 (footnote 33: "*FADL*").
133 Eliaçık, *Yaşayan Kur'an*, 890, "Allah'ın onlardan kimine kiminden fazla vermesi."

in the female body. The same logic applies to the woman, who has "more" than the man with her womb. Ultimately, however, the "surplus" qualities of the two genders rely on each other to benefit. This is an allusion to the creation of a living being, which is only possible when two genders work together. Therefore, the focus here is on the heterogeneity of both genders rather than male superiority.[134] Building on this explanation, Eliaçık turns to the term "*NUŞUZ*"[135] (Arab.: *nushūz*), which he translates from Arabic as "to rise," "to lift," "to rebel," and so on. The Turkish meaning would be "marital unrest"[136] ("When you have harsh disagreements with your wives . . ."), to which the Qurʾān draws particular attention.[137] At this point, the Qurʾān focuses on the serious conflicts in marriage caused by the woman. The Qurʾān uses patriarchal language corresponding to society's ideas and proposes different solutions. In general, the verse Q 4:34 would concern fundamental problems in marriage rather than a woman's rebellion against her husband, as such problems affect not only specific societies, but the whole of humanity. Above all, by recommending certain steps in the Qur'an for solving marital problems that did not involve violence, this verse was aimed to civilise the Arab society of that time (time of the revelation of the verse), which was prone to violence. Eliaçık emphasizes the "universal and moral messages" conveyed to the general public by the Qurʾān, whose verses should first be perceived on a human level rather than a purely Muslim one.[138]

Concerning the recommendation mentioned in the case of marital conflict, Eliaçık translates his explanation in the following way: "[F]irst sit down and talk to them, if that fails, leave them alone in their beds, and if that fails, separate from them for a while."[139] To explain this approach, he refers to a hadith in which many women complained to the Prophet about the physical violence committed against them by their husbands. According to the hadith, men who committed such acts were despised and not portrayed as "blessed" (*hayırlı*) men. It would therefore be extremely illogical if, despite the many complaints, a revelation continued to call for the beating of women. Furthermore, it would have been obvious that this verse was seen as a measure to civilize Arab society at a time when, Eliaçık says, women were exclusively subject to childbearing and physical violence. In the absence of a reference to the hadith quoted above, the accuracy of this statement cannot be verified. Nor is his subsequent assertion about Arab society at the time supported by any source.[140] The term in question

134 Eliaçık, *Yaşayan Kur'an*, 891–92 (footnote 33: "*FADL*").
135 According to the translator's spelling of the word.
136 Eliaçık, *Yaşayan Kur'an*, 892 (footnote 35: "*NUŞUZ*").
137 Ibid., 892, "Şiddetli geçimsizlik yaşadığınız eşlerinizle . . ."
138 Ibid., 892–95 (footnote 36: "*DARB*").
139 Ibid., 892f. "önce oturup konuşun, olmazsa yataklarında yalnız bırakın, yine olmazsa bir müddet ayrılın."
140 Ibid., 893.

is "*DARB*"[141] (*ḍaraba*), which has many different meanings and also functions as a universal term with different grammatical forms and uses. Eliaçık illustrates this diversity of terms with examples like "speculator" (*müdârib*), "pitching a tent" (*darabe heyme*), and "turmoil," "crisis" (*idtırâb*). *DARB* is derived from the root word "*D-R-B*," the verbal noun of which is translated as "to beat," "to beat up," "to make," "to leave," etc." There are also other meanings such as "to go away," "to leave a place for a while," "to open," "to separate." From this, Eliacik infers that the verse in question advocates for the temporary separation of the spouses, deciding against the use of "to beat" and in favor of "to separate" in his translation of *darb*. If the previous measures of communicating ("first sit down and talk to them") and avoiding the marital bed ("leave them alone in their beds") do not solve the marital conflict, the last option, temporary separation ("separate yourself from them for a while"), should be used.[142]

6 Conclusion

This chapter has given an overview of the development of Turkish translations of the Qur'ān, which changed over time due to political, social, and cultural conditions. From the very beginning of the adoption of Islam by the Turkish people, different methods of interpreting the Qur'ān were developed. From the initial intention of the first translations to properly understand and follow the faith, the goal of translation changed due to political interests. With the collapse of the Ottoman Empire and the establishment of the Republic of Türkiye, two opposing forms of government emerged, influencing the development and continued evolution of contemporary Turkish translations. Since the establishment of the Republic, Türkiye has been governed by different party-political ideologies, and this influence can be seen in Turkish translations of the Qur'ān over time. Depending on the ideology of the political leadership at a given time, Islam has moved to the forefront or the background, and translations have transitioned from a Kemalist to an Islamist style. This has resulted in many different translations of the Qur'ān, including translations by sympathizers of the political rulers. It is striking that the Qur'ān translations by Kemalist theologians such as Yaşar Nuri Öztürk have received little attention in Türkiye, while translations by sympathizers of Islamist movements like the Gülen movement enjoy a much better reputation.[143] Since the hermeneutical interpretation of texts results in various statements and contexts of meaning, which can be justified primarily by the *ijtihād*[144] (individual interpretation), there is the possibility for populist movements as well as fundamentalist currents to justify their

141 According to the translator's spelling of the word.

142 Eliaçık, *Yaşayan Kur'an*, 892f (footnote 36: "*DARB*").

143 Cf. 2. Major Works.

144 Lutz Berger, *Islamische Theologie* (Vienna: Facultas.wuv, 2010), 116.

interpretations by referring to this method, since a person's subjectivity can never be completely excluded. This is currently demonstrated in the Turkish translations of Qur'ānic verse Q 4:34 prepared by the Diyanet and Islamic theologian Ihsan Eliaçık. Although both deal with the same text, there are clear differences in both the interpretations and the translations. The analysis of the two interpretations shows that, on the one hand, the "superiority" of man is used to legitimize a relationship of domination and power over woman. On the other hand, his "superiority" is used to merely define a protective and caring function toward the woman. Ultimately, this shows that interpretation depends on the particular exegete and his or her intentions, as the same text can have very different meanings. In this context, it is also important to consider whether the exegetes are emphasizing the ambiguity of certain terms and deploying them in the sense of a meaningful explanation, as in the case of the terms *qānitāt* or *ḍaraba,* which do not necessarily indicate the unequal appreciation of genders.

The principle of male superiority and female subordination to men was established on the basis of prevailing readings of the Qur'ān. Barlas has explained this using the example of the term *qawwamūn.*[145] According to some authors, men are superior to women due to their supposed physical and intellectual advantages. This statement, he says, is not found anywhere in the verse. Furthermore, it is clear from the verse that some men are superior to women only in certain areas and that there is no general statement about the natural superiority of men.[146] According to Barlas, religious scholars (*ʿulamā'*) have succeeded in weakening the impulse for gender equality in various areas of tradition. Over time, the "egalitarianism associated with the Qur'ān" has lost its "subversive connotation."[147] The growing social conservatism and political quietism of Muslims are also reflected in the general exegesis of the Qur'ān, which has become increasingly misogynistic over time. Given that a divine text was created under specific social, political, geographical, and cultural conditions, it is impossible to understand the text without knowing its historical context.[148]

7 Bibliography

7.1 Sources

Diyanet Isleri Baskanligi. Kur'an-i Kerim. "Meal (Kur'an Yolu) Nisâ Suresi – 34." Accessed November 25, 2023. https://kuran.diyanet.gov.tr/tefsir/Nisâ-suresi/527/34-ayet-tefsiri.
Eliaçık, R. Ihsan. *Yaşayan Kur'an: Nüzul sırasına göre Türkçe meal – tefsir*. Istanbul: Inşa Yayınları, 2007.
Eliaçık, R. Ihsan. *Demokratik özgürlükçü İslam*. Istanbul: Tekin Yayınevi, 2014.

145 Barlas, *Women's Reading of the Qur'ān*, 262–63
146 Ibid., 262f.
147 Ibid., 256–57.
148 Ibid., 256–69.

Hasan Elik, Muhammed Coşkun. *İndirildiği Dönemin Işığında Kur'an Tefsiri (Tevhit Mesajı)*. Istanbul: M.Ü. İlahiyat Fakültesi Vakfı Yayınları, 2013.
Inan, Abdülkadir. "Eski Kuran Tercümelerinin Dili Meselesi." *Türk Dili: Aylık Fikir ve Edebiyat Dergisi* 1, no. 7 (1952): 19–22.
Wehr, Hans. *Arabisches Wörterbuch für Schriftsprache der Gegenwart: Arabisch-Deutsch*. Wiesbaden: Harrassowitz 1985.

7.2 Secondary Literature

Abay, Muhammet. "Kur'an ve Tefsire Dair Başvuru Kaynakları." *Türkiye Araştırmaları Literatür Dergisi* 10, no. 19–20 (2012): 389–435.
Abay, Muhammet. "Türkçedeki Kur'an Meâllerinin Tarihi ve Kronolojik Bibliyografyası." *Türkiye Araştırmaları Literatür Dergisi* 10, no. 19–20 (2012): 231–303.
Akpınar, Ali. "Tanrı Buyruğu." In *TDVIA*, edited by TDV İslâm Araştırmaları Merkezi. Accessed October 1, 2023. https://islamansiklopedisi.org.tr/tanri-buyrugu.
Aldeeb, Sami. *Schlagt die Frauen: Interpretation des Koranverses 92/4:34 durch die Jahrhunderte (Kurzversion Ohne Exegesen in Arabisch)*. Charleston: Amazon, 2022.
Apaydın, H. Yunus. "Müctehid." In *TDVIA*, edited by TDV İslâm Araştırmaları Merkezi. Accessed December 25, 2023. https://islamansiklopedisi.org.tr/muctehid.
Aydar, Hidayet. "Kur'an Çalışmaları." *Istanbul Üniversitesi İlahiyat Fakültesi Dergisi, Darulfunun İlahiyat* 1 (1999): 159–235.
Aydar, Hidayet. "Yaygınlık Bakımından Meallerimiz ve Mealler Üzerinde İstatistiksel Bir Değerlendirme." *FSM İlmi Araştırmalar İnsan ve Toplum Bilimleri Dergisi* 5 (2015): 109–47.
Barlas, Asma. *Women's Readings of the Qur'ān: The Cambridge Companion to the Qur'ān*. Cambridge: Cambridge University Press, 2007.
Berger, Lutz. *Islamische Theologie*. Vienna: Facultas, 2010.
Bilici, Mücahit. "The Fethullah Gülen Movement and Its Politics of Representation in Turkey." *The Muslim World* 96 (2006): 1–20.
Birişik, Abdulhamit. "Müfessir." In *TDVIA*, edited by TDV İslâm Araştırmaları Merkezi. Accessed December 25, 2023. https://islamansiklopedisi.org.tr/mufessir.
Birnbaum, Eleazar. "On Some Turkish Interlinear Translations of the Koran." *JT* 14 (1990): 113–38.
Cerrahoğlu, İsmail. *Tefsir Usûlü*. Ankara: Türkiye Vakfı Yayınları, 1995.
Cihangir, Sonia. *Kadına Dair: Hurafeler ve Gerçekler*. Istanbul: Ozan Yayincılık, 2016.
Demirci, Muhsin. *Kur'an Tarihi*. Istanbul: M. Ü. İlahiyat Fakültesi Vakfı Yayınları, 2014.
Durmuş, Zülfikar. *Kur'an'ın Türkçe Tercümeleri*. Istanbul: Çıra Akademi, 2020.
Eckmann, Janos. *Eastern Turkic Translations of the Koran*. Budapest: Akademiai Kiado, 1971.
Erdoğan, Abdülkadir. "Kur'an Tercemelerinin Dil Bakımından Değerleri." *Vakıflar Dergisi* 1 (1938): 47–51.
Fatma, Murat. "Türk Dizilerinin Tarih Kurgusu: Muhteşem Yüzyıl, Payitaht Abdülhamid, Kuruluş Osman." *Dergi Park Akademik* 8 (2022): 130–50.
Gürbüz, Faruk. *Tercüme Problemleri ve Mealler*. Istanbul: Insan Yayınları, 2004.
Hamid, Shadi. "From Erbakan to Erdoğan: The Evolution of Turkish Foreign Policy." *Insight Turkey* 6, no. 1 (2004): 113–18.
Inan, Abdülkadir, *Kur'ân-ı Kerîm'in Türkçe tercemeleri üzerinde bir inceleme*. Ankara: Diyanet İşleri Başkanlığı, 1961.
Kahraman, Muhammed Şerif "Kur'ân'ın harfî ve tefsirî tercümesi" *Siirt Üniversitesi Ilahiyat Fakültesi Dergisi* 3:2 (2016): 145–175.

Karabacak, Esra. "Tarihi Seyrinden Yola Çıkarak Kur'an Tercümeleri Üzerine Bir Çalışma: Prof. Dr. Nihad M. Çetin'e Armağan." *Ankara Edebiyat Fakültesi Basımevi* (1999): 183–89.
Karacan, O., and B. Yaşar. "II: Meşrutiyet'ten Cumhuriyet'e Kur'an Tercümelerine Osmanlı Aydınlarının Yaklaşımı." *Birey ve Toplum Sosyal Bilimler Dergisi* 9 (2019): 91–122.
Koçyiğit, Hikmet. "Günümüz Türkçe Kur'an Meâlleri Üzerine Bir Tasnif Denemesi." *Istanbul Üniversitesi İlahiyat Fakültesi Dergisi* 37 (2017): 79–121.
Krawietz, Birgitt. "Ibn Taymiyya, Vater des islamischen Fundamentalismus?" In *Salafismus in Deutschland: Ursprünge und Gefahren einer islamisch-fundamentalistischen Bewegung*, edited by Thorsten Gerald Schneiders. Bielefeld: Transcript, 2014: 67–88.
Kuru, Ahmet T. "Changing Perspectives on Islamism and Secularism in Turkey: The Gülen Movement and the AK Party." In *Muslim World in Transition: Contributions of the Gulen Movement*, edited by Louis Cantori, Marcia Hermansen, David Capes, et al. London: Leeds Metropolitan University Press, 2007.
Kuru, Ahmet T. "Assertive Secularism and the Multiculturalist Challenge (1989–2008)." *Secularism and State Policies toward Religion: The United States, France, and Turkey*. Cambridge University Press, 2009: 161–201.
Küpeli, Ismail. "Drohende Alleinherrschaft in der Türkei: Per Verfassungsreferendum soll die Bevölkerung die Demokratie abschaffen." *RLS-Standpunkte* 5 (2017):1–4.
Özgel, Ishak. "Elmalılı Muhammed Hamdi Yazırın Tefsiri ve Hakkında Yapılmış Çalışmalar Bibliyografyası." *Türkiye Araştırmaları Literatür Dergisi* 19 (2012): 327–62.
Öztürk, Mustafa. *Cumhuriyet Türkiyesi'nde Meal ve Tefsirin Serencamı*. Ankara: Ankara Okulu Yayınları, 2012.
Öztürk, Yaşar Nuri. *Der verfälschte Islam*. Düsseldorf: Grupello Verlag, 2007.
Öztürk, Mustafa. "II. Meşrutiyet'ten Günümüze Kur'an Mealleri." In *Türkiye Araştırmaları Literatür Dergisi* 18 (2011): 140–91.
Öztürk, Mustafa. "Kur'an yorumu ve şiddet meselesi." *Bilimname* 32 (2016): 7–40.
Masidi, Yasmin. *Are Muslim Men Allowed to Beat their Wives?* Malaysia: Sisters in Islam, 2009.
Rippin, Andrew. "Occasions of Revelation." In *EQ*, edited by Jane Dammen McAuliffe. Brill, 2001–2006. Article published online 2016. Accessed January 17, 2024. http://dx.doi.org/10.1163/1875-3922_q3_EQSIM_00305.
Sağol, Gülden. *An Interlinear Translation of the Qur'ān into Khwarazm Turkish: Introduction, Text Glossary and Facsimile. Introduction and Text; 1*. Cambridge, MA: Harvard University, Department of Near Eastern Languages and Civilizations, 1993.
Tamçelik, Soyalp. "Türkiye'de Popüler Kültürün Muhafazakârlaşması ve Eğlence Kültürünün Siyasallaşması." *JT* 11, no 2 (2016): 1139–52.
Tibi, Bassam. *Fundamentalismus im Islam: Eine Gefahr für den Weltfrieden?* Darmstadt: Wissenschaftliche Buchgesellschaft, 2002.
Togan, Zeki Velidi. "Londra ve Tahran'daki İslamî Yazmalardan Bazılarına Dâir." *İslam Tetkikleri Enstitüsü Dergisi* 3 (1959–1960): 133–60.
Togan, Zeki Velidi. "The Earliest Translation of the Qur'ān into Turkish." *Islâm Tetkikleri Enstitüsü Dergisi* 4, no. 1–2 (1964): 1–19.
Ulusan, Şayan. "Atatürk Dönemi Din Uygulamaları (1923–1938) (Yorumlar, Eleştiriler, Değerlendirm." *Cappadocia Journal of History and Social Sciences* 11 (2018): 306–20.
Uzer, Umut. "Conservative Narrative: Contemporary Neo-Ottomanist Approaches in Turkish Politics." *Middle East Critique* 29 (2020): 275–90.
Ünlü, Suat. "Kur'an-ı Kerimin Türkçeye Çevrilmesi ve ilk Türkçe Kur'an Tercümeleri." *Dini Araştırmalar Dergisi* 9 (2007): 9–57.
Yaşar, Naif. *Kur'ân Tercüme Teknikleri*. N.p.: LAP LAMBERT Academic Publishing, 2017.
Yavuz, Yunus Vehbi. *Kur'an'da Kadın Hak ve Özgürlüğü*. Istanbul: Bayrak Yayınları, 1999.

Yucel, Salih. “Fethullah Gülen: Spiritual Leader in a Global Islamic Context.” *Journal of Religion and Society* 12 (2010): 1–19.
Zürcher, E.J. “Yeñi ʿOthmānlïlar.” In *EI*², edited by P. Bearman et al. Brill, 1954–2005. Article published online 2012. Accessed January 9, 2024. http://dx.doi.org/10.1163/1573-3912_islam_SIM_8013.

7.3 Online Sources

Altınordu, Ateş. “Occupy Gezi, Beyond the Religious-Secular Cleavage.” *The Immanent Frame*. Accessed November 11, 2023. https://tif.ssrc.org/2013/06/10/occupy-gezi-beyond-the-religious-secular-cleavage/.
Cumhuriyet. “İhsan Eliaçık'ın kitabına 'İslam dini açısından sakıncalı' gerekçesiyle toplatma kararı.” Accessed November 11, 2023. https://www.cumhuriyet.com.tr/turkiye/ihsan-eliacikin-kitabina-islam-dini-acisindan-sakincali-gerekcesiyle-toplatma-karari-2054245.
Duvar English. “Turkish Court Lifts Ban on Theologian's Quran Translation.” *Duvar English*. March 14, 2023. Accessed November 11, 2023. https://www.duvarenglish.com/turkish-court-lifts-ban-on-theologian-ihsan-eliaciks-quran-translation-news-62016.
Dünya, “Çaresiz kalındığında hafifçe dövün, Diyanet'ten açıklama.” Accessed November 12, 2023. https://www.dunya.com/gundem/quotcaresiz-kalindiginda-hafifce-dovunquote-diyanetten-haberi-203770.
Gazete Memur. “İhsan Eliaçık'ın Kuran meali hakkında toplatma kararı kaldırıldı.” Accessed November 11, 2023. https://gazetememur.com/gundem/ihsan-eliacikin-kuran-meali-hakkinda-toplatma-karari-kaldirildi,uI-oE4qs4km6Zau_2pHh-Q.
Güsten, Susanne. “Türkei Islamisten gegen Erdogan.” *Tagesspiegel*. April 13, 2018. Accessed November 11, 2023. https://www.tagesspiegel.de/politik/islamisten-gegen-erdogan-5284864.html.
Islami Analiz. “Bulaç: Bugüne kadar oyumu Milli Görüş'e veridi.” October 22, 2015. Accessed November 12, 2023. https://islamianaliz.com/haber/7450752/bulac-bugune-kadar-oyumu-milli-goruse-verdim-sandiga-gidersem-oyum-yine-saadet-partisine.
Kamalov, Ablet. “Uyghur Historiography.” In *Oxford Research Encyclopedia of Asian History*. October 29, 2021. Accessed October 26, 2023. https://oxfordre.com/asianhistory/view/10.1093/acrefore/9780190277727.001.0001/acrefore-9780190277727-e-637.
Karaca, Ekin. “İhsan Eliaçık: Diyanet'in görevi devletin icraatlarına dinî kılıf bulmak.” *Bianet*. Accessed November 11, 2023. http://bianet.org/bianet/ifade-ozgurlugu/137669-ihsan-eliacik-diyanetin-gorevi-devletinicraatlarina-dini-kilif-bulmak.
Karakas, Burcu. “Ten Years after Turkey's Gezi Protests.” *Reuters*. Accessed November 12, 2023. https://www.reuters.com/world/middle-east/ten-years-after-turkeys-gezi-protests-some-say-vote-chance-justice-2023-05-26/.
Karar. “Diyanet istedi mahkeme meal yasakladı.” February 21, 2023. Accessed November 12, 2023. https://www.karar.com/guncel-haberler/diyanet-istedi-mahkeme-meal-yasakladi-1729935.
Memurlar. “Yaşar Nuri'den şaşırtan Erdoğan çıkışı.” Accessed November 12, 2023. https://www.memurlar.net/haber/551307/yasar-nuri-den-sasirtan-erdogan-cikisi.html.
Nurtsch, Ceyda. “Interview with Turkish Theologian Ihsan Eliacik: The Koran and Social Justice.” *Qantara*. January 15, 2014. Accessed November 12, 2023. https://qantara.de/en/article/interview-turkish-theologian-ihsan-eliacik-koran-and-social-justice.
Sözcü. “Cumhurbaşkanı Erdoğan'dan Ali Erbaş mesajı.” Accessed November 11, 2023. https://www.sozcu.com.tr/2020/gundem/cumhurbaskani-erdogandan-ayasofya-aciklamasi-5964783/.
Sözcü. “Şiddet gören kadına Diyanet'ten tavsiye.” Accessed November 11, 2023. https://www.sozcu.com.tr/2020/gundem/siddet-goren-kadina-diyanetten-tavsiye-tepki-verme-sabret-uzaklas-5644351/.
T24. “Ali Bulaç: Sandığa gidersem oyum Saadet Partisi'ne . . .”. October 22, 2015. Accessed November 12, 2023. https://t24.com.tr/haber/ali-bulac-sandiga-gidersem-oyum-saadet-partisine,313864.

Ervan Nurtawab and Peter G Riddell

Modern Qur'ānic Hermeneutics in Indonesia and Malaysia

1 Introduction

Although located at a great distance from the traditional Islamic heartlands of the Middle East, the Southeast Asian countries of Indonesia and Malaysia represent dynamic sites for Qur'ānic hermeneutical activity. Commentary writing on the Qur'ān in both countries increased exponentially during the twentieth century, and the exegetical output represents a blend of elements from other parts of the Muslim world and creative hermeneutical reflection by Southeast Asian exegetes. In the field of Qur'ānic hermeneutics, as in all fields of Islamic learning, it is no longer valid to talk of center and periphery of the Muslim world. Rather, the different centers of Islam nourish each other in a repeating cycle of cross-fertilization, with each benefiting from the creative genius of Islamic scholarly communities elsewhere.

2 Major Works in the Field that Engage with the Qur'ān

2.1 Indonesia

Modern *tafsīrs* are very much connected to the emergence of the reformist-modernist movement of the nineteenth and early twentieth century, greatly influenced by the Egyptian scholars Muḥammad ʿAbduh (d.1905/1323) and Rashīd Riḍā (d. 1935/1354). This movement encouraged Muslims to engage directly with the Qur'ān and the prophetic traditions and to modernize elements of Islamic education. Anthony H. Johns notes that the early twentieth century Indonesian *tafsīrs* increasingly removed both *isrāʾīlīyat* elements and discussion of the variant reading systems of the Qur'ān, following the increased use of the 1924/1342 Qur'ānic Egyptian standard edition that follows the reading system transmitted by Ḥafṣ b. Sulaymān (d.796/180).[1] However, some Indonesian modernist commentaries are identified as using some *isrāʾīlīyat* narrative

1 Anthony H. Johns, "Quranic Exegesis in the Malay world: in Search of a Profile," in *Approaches to the History of the Interpretation of the Qur'ān*, ed. Andrew Rippin (New York: Oxford University Press, 1988), 274–86.

https://doi.org/10.1515/9783111320069-026

elements to explain the verses of the Qur'ān.[2] In addition, the use of Biblical materials as a trend in global modern Qur'ānic exegetical activities[3] is also found in several Indonesian modern commentaries. In this regard, one obvious example is found in Zainuddin Hamidy (d.1957/1376) and Fachruddin HS's *Tafsir Quran*, and albeit in a small number, the use of Biblical materials for Qur'ānic interpretation can still be found in A. Hassan's (d.1958/1377) *Tafsir al-Furqan* (The Interpretation of The Criterion). Trends in modern *tafsīr* writing do not automatically replace the pre-modern exegetical activities that were so heavily influenced by *Tafsīr al-Jalālayn*. This work was translated and commented upon in local languages and continues to be used in both present traditional and modern educational milieus.[4]

Johanna Pink notes that while prioritizing the use of reason rather than referring to the existing authoritative commentaries, modernist readings of the Qur'ān greatly consider the historical context of the period of the Qur'ānic revelation in seventh-century Arabia, and then look for its significance and ultimate goals *(maqāṣid al-Qur'ān)*.[5] In the Indonesian context, early twentieth century modernism required ready access to a comprehensive understanding of the Qur'ān among urban communities. Mahmud Yunus (d.1982/1402) responded to this situation by composing a work that was both translation and interpretation of the whole Qur'ān in a brief but clear presentation.[6] In the final section of his work, he included a summary guide for readers to understand the content of the whole Qur'ān. The second half of the twentieth century witnessed an increase in the numbers of highly-educated members of the Indonesian public. Commonly based in urban areas, they sought to deepen their understanding of Islamic issues, yet they lacked knowledge of Arabic. For them, Islamic

---

2 Some examples are the commentaries of the Qur'ānic verses on the stories of Dhū l-Qarnayn and Gog and Magog in *Sūrat al-Kahf* found in Mahmud Yunus' *Tafsir Quran Karim*. See Ervan Nurtawab, "*Tafsīr al-Jalālayn* at the Crossroads: Interpreting the Qur'ān in Modern Indonesia," *Australian Journal of Islamic Studies* 6, no. 4 (2021): 4–24. More general on the story of Gog and Magog in Malay-Indonesian commentaries, see Majid Daneshgar, "Gog and Magog in Malay-Indonesian Islamic Exegetical Works," in *Gog and Magog: Contributions toward a World History of an Apocalyptic Motif*, ed. Georges Tamer, Andrew Mein and Lutz Greisiger (Berlin: De Gruyter, 2023): 597–615.

3 On the trend of quoting Biblical materials in the global modern Qur'ānic commentaries, see Samuel J. Ross, *Qur'ān Commentary and the Biblical Turn: A History of Muslim Exegetical Engagement with the Biblical Text* (Berlin: De Gruyter, 2023).

4 See Martin van Bruinessen, "Kitab kuning: Books in Arabic Script used in the Pesantren Milieu: Comments on a New Collection in the KITLV Library," *BKI* 146 (1990): 253–54; Peter G. Riddell, *Islam and the Malay-Indonesian World: Transmission and Responses* (Singapore: Horizon Books, 2001), 147; Nashruddin Baidan, *Perkembangan Tafsir Al-Qur'an di Indonesia* (Yogyakarta: Tiga Serangkai, 2003), 96–98; and Ervan Nurtawab, "Jalalayn Pedagogical Practice: Styles of Qur'an and Tafsir Learning in contemporary Indonesia" (PhD diss., Monash University, 2018).

5 Johanna Pink, *Muslim Qur'ānic Interpretation Today: Media, Genealogies and Interpretive Communities* (Sheffield: Equinox, 2019), 125.

6 Based on his statement found on the front cover of his 1993 printed edition. See Mahmud Yunus, *Tafsir Quran Karim Bahasa Indonesia* (Jakarta: Hidakarya Agung, 1993).

thought as articulated by Western scholars was subject to further investigation because such perspectives were based on Westernized-scientific points of view, not the correct *'aqīda* (Islamic creed). The composition of the *Tafsier Al-Qurânul Madjied 'An-Nur'* (The Interpretation of the Noble Qur'ān 'The Light') of Muhammad Hasbi Ash-Shiddieqy (d.1975/1395), for example, reflected this concern.

Late twentieth-century and early twenty-first-century Indonesia witnessed an increasing number of what Howard Federspiel calls 'enriched commentaries.'[7] Pioneered by the launch of three full commentaries, i.e. *Tafsier Al-Qurânul Madjied 'An-Nur'* by Ash-Shiddieqy, *Tafsir al-Azhar* by Buya Hamka (d.1981/1401), and a ten-volume commentary by a team from the Indonesian Ministry of Religious Affairs, the development of specifically Indonesian modern hermeneutics, generally in a thematically-presented work, can be seen in attempts to address the following three main issues: the Qur'ān and science, feminist or gender studies, and relations between Islam and nationhood or citizenship. This period also witnessed the activities of publishing Indonesian translations of a pre-modern Qur'ānic commentary, that is Ibn Kathīr's commentary, to segment the targeted market among the reformist Muslim groups.[8] More recently, Indonesian Muslims also witnessed increasing exegetical activity through websites and social media.

Trends in Qur'ānic scientific interpretation *(tafsīr 'ilmī)* can be seen from the translation of Maurice Bucaille's *La Bible, le Coran et la Science* by M. Rasjidi (d. 2001/1422), a renowned Indonesian intellectual graduate from the Sorbonne. The book first appeared in 1978/1398 and, as Johns notes, gained a fair degree of popularity.[9] From the 1980s to the 1990s, two Indonesian scholars, Nazwar Syamsu and Ahmad Baiquni (d.1998/1419), gained popularity through their published works on the Qur'ān and science. From 29 August – 1 September 1994/22 to 25 Rabīʿ al-awwal 1415 the fourth international conference on the miracles of the Qur'ān and the prophetic tradition on modern science and technology was held in Bandung. Supported by *Rābiṭat al-'Ālam al-Islāmī* (The Muslim World League) and some Indonesian Islamic organizations including IAIN (now UIN) Syarif Hidayatullah Jakarta, the conference was officially opened by the Indonesian Vice President, Try Sutrisno. Muslim intellectuals from fifteen countries, including Maurice Bucaille, contributed papers. The famous Indonesian Islamic publisher, Gema Insani Press, then published the conference proceeding in two volumes in 1995/1416.

Despite a decreasing trend during the 2000s, the second decade of the twenty-first-century in Indonesia witnessed significant attempts at drawing upon scientific perspec-

7 Howard M. Federspiel, *Popular Indonesian Literature of the Qur'an* (Ithaca, NY: Cornell Modern Indonesian Project, 1994), 72.

8 Johanna Pink, "Eight Shades of Ibn Kathīr: The Afterlives of a Premodern Qur'ānic Commentary in Contemporary Indonesian Translations," in *Malay-Indonesian Islamic Studies: A Festschrift in Honor of Peter G. Riddell*, ed. Majid Daneshgar and Ervan Nurtawab (Leiden: Brill, 2023), 109–33.

9 Johns, "Quranic Exegesis," 284.

tives in Qur'ānic exegetical activity. The government efforts in upgrading the organizational status of some state institutes of Islamic studies (*Institut Agama Islam Negeri*, or IAIN) into state Islamic universities (*Universitas Islam Negeri*, or UIN) since the early 2000s have attracted more Muslim scholars' attention to the integration of religious studies and sciences in the Islamic university curricula. The Indonesian government, through the Central Bureau of Qur'ān Affairs (*Lajnah Pentashihan Mushaf Alquran*, or LPMQ) of the Ministry of Religious Affairs in collaboration with the Indonesian Institute of Sciences (*Lembaga Ilmu Pengetahuan Indonesia*, or LIPI), also worked on the projects of writing Qur'ānic scientific commentaries where the results can be accessible online on their official website. Aside from that, there is the individual effort by Agus Purwanto (b. 1964/1384), a Muhammadiyah organization member who was granted a doctoral degree in physics from Hiroshima University in 2002/1423, to approach the Qur'ān and the prophetic traditions through his expertise in science. He published several books on related subjects, especially those with Mizan publisher in 2012–15. He also initiated the formation of Islamic high schools that he calls 'trensains', abbreviated from 'pesantren sains (the Islamic boarding school of sciences)'.

With regard to gender studies-based commentary writing, a trend has developed since the 1990s, especially through the formation of NGOs specializing in women issues. Two figures who deserve mention here are Kyai Husein Muhammad (b. 1953/1372)[10] and his disciple, Faqihuddin Abdul Kodir (b. 1971/1391). Abdul Kodir developed a method that he calls *qirā'a mubādala* (gender relation-based reading). His suggestion to consider gender relations in looking at religious texts has attracted a wide audience among the scholarly faculty of Islamic higher educational institutions (IAIN/UIN) and Muslim groups, especially those affiliated with the Nahdlatul Ulama organization. In addition, some scholars have addressed gender studies in doctoral research, such as Nasaruddin Umar (b. 1959/1379) in 1998/1419 at IAIN (now, UIN) Syarif Hidayatullah. The development of this genre was also endorsed by projects translating works of prominent feminist scholars, like Amina Wadud's *Rereading the Sacred Text from a Woman's Perspective*[11] and Asma Barlas' *Believing Women in Islam: Unreading Patriarchal Interpretations of the Qur'ān*.[12]

Regarding the discourse on Islam and nationhood, scholars deserving mention are Nurcholish Madjid (d. 2005/1426) and M. Quraish Shihab (b. 1944/1363). The following analysis shows how Nurcholish Madjid's thoughts on neo-modernism and pluralism stimulated debates among Muslims and were in line with the New Order govern-

10 For further on this figure, see Yusuf Rahman, "Feminist Kyai, KH. Husein Muhammad: The Feminist Interpretation on Gendered Verses and the Qur'an-based Activism," *Al-Jami'ah: Journal of Islamic Studies* 55, no. 2 (2017): 293–326.

11 Amina Wadud, *Rereading the Sacred Text from a Woman's Perspective* (New York: Oxford University Press, 1999).

12 Asma Barlas, *Believing Women in Islam: Unreading Patriarchal Interpretations of the Qur'ān* (Texas: The University of Texas Press, 2004).

ment's attempt to promote religious values that supported government developmental policies and ethnic differences.

Meanwhile, M. Quraish Shihab[13] should be given a privileged place in Indonesian modern Qur'ānic exegesis. Granted a doctoral degree in Qur'ānic exegesis from al-Azhar University in 1982/1402, Quraish Shihab soon became an iconic figure in this discipline. Since the 1980s, his works have attracted a wide readership among Indonesian Muslims, especially those who live in urban areas. His Qur'ānic understanding generally aims to motivate Muslims and, more recently, to support moderation. While he regularly produced works on thematic interpretation and the Qur'ānic sciences *('ilm at-tafsīr)*, he then decided to prepare a commentary on the whole Qur'ān, carrying the title *Tafsir al-Mishbah: Pesan, Kesan dan Keserasian Al-Qur'an* (The Interpretation of the Light: Message, Impression and Harmony of the Qur'ān). Its first volume appeared in 2000/1421, and the entire work was published in 2003/1424.

The existence of state Islamic universities has also paved the way for the reception of Western tradition in Qur'ānic studies and various approaches to the Qur'ān proposed by modern Muslim thinkers. The reception of these two points has significantly boosted the development of Qur'ānic exegetical discourses in the higher educational environment.[14] More currently, some Indonesian scholars with a lecturer position in Islamic theology faculties have been intensively promoting a variety of modern Qur'ānic interpretive methods. Sahiron Syamsuddin (b.1968/1388), for example, develops an interpretive method that he calls *ma'nā-cum-maghzā* (meaning with significance) where a reader of the Qur'ān tries to understand the historical context of the related texts as understood by the first Muslim generation, and then to establish its significance for the contemporary contexts. Many young scholars followed his lectures through which they have been trained to use this method to look at different aspects in the Qur'ān. Meanwhile, a number of Islamic scholars attempt to conceptualize an approach called *tafsīr maqāṣidī* (purposive reading) that emphasizes the ultimate goals in understanding the verses of the Qur'ān. Aksin Wijaya and Shofiyullah

13 See the contribution on Shihab in this volume of the *Handbook*

14 The Qur'ānic exegetical activities that limitedly circulate among the academics remind us of Pink's proposal for the typology of contemporary commentaries where she classifies the scholars' commentaries as the most conventional form. See Johanna Pink, "Tradition, Authority and Innovation in Contemporary *Sunnī Tafsīr*: Towards a Typology of Qur'ānic Commentaries from the Arab World, Indonesia and Turkey," *Journal of Qur'anic Studies* 12 (2010): 60–62. For the controversy about the acceptance of Western Qur'ānic discourse in the Indonesian Islamic universities and review of related literature of the Qur'ān, see Izza Rohman, "The Pursuit of New Interpretive Approaches to the Qur'ān in Contemporary Indonesia," in *The Qur'ān in the Malay-Indonesian World: Context and Interpretation*, ed. Majid Daneshgar, Peter G. Riddell and Andrew Rippin (London: Routledge, 2016): 83–136.

Muzammil, for example, offer a detailed concept of this interpretive method.[15] More practically, Abdul Mustaqim utilizes this approach to formulate the concept of Islamic moderation currently promoted by the Indonesian government.[16]

Some of the aforementioned figures also use YouTube and social media to gain a wider audience.[17] Quraish Shihab's lectures, for example, are generally uploaded to YouTube. Aside from that, Fadhli Lukman has identified three figures who actively use Facebook in disseminating their thoughts on the Qur'ān; i.e. Salman Harun, Buya Gusrizal Gazahar, and Irena Handono.[18] It is also worth mentioning Nadirsyah Hosen (b. 1973/1393) who regularly writes short articles on Qur'ānic commentary on his Facebook and Twitter account. Many of his online writings are then compiled and published with the title *Tafsir Al-Qur'an di Medsos: Mengkaji Makna dan Rahasia Ayat Suci pada Era Media Sosial* (Qur'ānic Interpretation on Social Media: Understanding the Meaning and Secret of the Holy Texts in the Social Media Era).[19]

2.2 Malaysia

While there is only limited evidence of Qur'ānic exegetical activity in the area of present-day Malaysia prior to 1900/1318, there was a significant surge in activity during the twentieth century. This is in large part due to the influence of Islamic reformist thinking brought back from the Middle East by Malays who had gone to study in Mecca and Cairo. At least fifty commentaries on the Qur'ān, in part or in entirety, were produced by Malaysian scholars during the twentieth century. The pace has picked up even further during the early twenty-first century due to the facilities provided by the digital age.

Commentary writing in Malaysia since the turn of the twentieth century has been carried out by wide-ranging scholars, both prominent and not so prominent. Famous

---

15 Aksin Wijaya and Shofiyullah Muzammil, "*Maqāṣidī Tafsīr*: Uncovering and Presenting *Maqāṣid Ilāhī-Qur'ānī* into Contemporary Context," *Al-Jami'ah: Journal of Islamic Studies* 59, no. 2 (2021): 449–78.

16 Abdul Mustaqim, "Argumentasi Keniscayaan Tafsir Maqashidi sebagai Basis Moderasi Islam." Pidato Pengukuhan Guru Besar dalam Bidang Ulumul Qur'ān disampaikan di Hadapan Rapat Senat Terbuka Universitas Islam Negeri Sunan Kalijaga Yogyakarta, 16 Desember 2019).

17 How the visual media and internet significantly changed authorship, audience, topics and styles in Qur'ānic commentaries, see Johanna Pink, *Muslim Qur'ānic Interpretation Today*, 81–103. More detailed in the case of the Indonesian YouTube channels, see U. Fikriyati and A. Fawaid, "Pop-Tafsir on Indonesian YouTube Channel: Emergence, Discourses and Contestations," *Proceedings of the 19th-Annual International Conference on Islamic Studies, AICIS 2019, 1–4 October 2019*, Jakarta-Indonesia.

18 Fadhli Lukman, "Digital Hermeneutics and A New Face of the Qur'an Commentary: The Qur'an in Indonesian's Facebook," *Al-Jami'ah: Journal of Islamic Studies* 56, no. 1 (2018): 95–120.

19 Nadirsyah Hosen, *Tafsir Al-Qur'an di Medsos: Mengkaji Makna dan Rahasia Ayat Suci pada Era Media Sosial* (Yogyakarta: Bunyan, 2017).

figures who produced commentaries included Muhammad Sa'id b. Umar (c.1854–1932/1270–1351), Syed Syeikh al-Hadi (1867–1934/1284–1353), Syeikh Uthman Jalaluddin al-Kelantani (1880–1952/1297–1371), Syeikh Muhamad Idris al-Marbawi (1893–1989/1311–1410), Nik Abdul Aziz Nik Mat (1931–2015/1350–1436), and Abdul Hadi Awang (b.1947/1366). Also noteworthy is the direct involvement of the Malaysian Government in supporting and publishing some Qur'ānic commentaries.

Some exegetical works produced during the twentieth and early twenty-first centuries sought to comment on a wide selection of Qur'ān suras. Examples include the famous *Tafsīr Nūr al-Iḥsān* (The Interpretation of the Light of Benevolence) by Muhammad Sa'id b. Umar, which served as a bridge between classical Malay exegesis and modern works. Also included in the category of general commentaries were *Syarah tafsir al-Jalalan Arab, Melayu yang dinama anwar al-huda wa amtar an-nada* (An Explanation of the *Tafsīr al-Jalālayn* in Arabic and Malay, entitled *anwar al-huda wa amtar an-nada*) by Syeikh Uthman Jalaluddin al-Kelantani; *Tafsir al-Qur'an al-Hakim bahasa Melayu* by Mustafa Abdul Rahman; *Terjemahan Tafsir Abdullah Yusuf Ali*, a translation into Malay with supplementary commentary of Abdullah Yusuf Ali's Qur'ān rendering, by Muhammad Uthman Elmuhammady; *Tafsir Al-Quran*, a multi-volume work by Nik Abdul Aziz; and *Tafsir Al-Wasit* (The Interpretation of the Intermediary), based on radio lectures by the prominent Syrian scholar Wahbah az-Zuhaili and published in 2019 by the Islamic Development Department of Malaysia (JAKIM), a department of the Malaysian Government.

By contrast, some exegetical works focused on a particular sura or suras, such as *Tafsir Juz' 'Amma* and *Tafsir Surat al-Fatihah* by Syed Sheikh al-Hadi; *Tafsir Surah Yasin* and *Tafsir Qur'an Marbawi Juzuk Alif Lam Mim* by Syeikh Muhamad Idris al-Marbawi; *Intisari Tafsir Juz' 'Amma* (A Summary Interpretation of *juz' 'amma*) by Syeikh Abu Bakar Ashaari; *Tafsir al-Qur'an al-Karim: Juzu 'Amma* by Muhammad Fauzi bin Haji Awang; and *Tafsir Ramuan Rapi Dari Arti Surah al-Kahfi* (A Concise Interpretation of the Meaning of *Sura al-Kahf*) by Muhammad Noor Ibrahim.

While a majority of the commentaries produced represented original exegesis in Malay, albeit drawing significantly on Arabic originals, some commentaries were translations into Malay of Arabic commentaries, such as a rendering into Malay of *Tafsīr al-Khāzin* by Tok Kenali (which has not survived), and the translation into Malay of Sayyid Quṭb's famous commentary *Fī ẓilāl al-Qur'ān* by Dato' Yusoff Zaky b. Yacob.

Some of the commentaries produced by Malaysian scholars represented dedicated publications, while others were compilations by students of notes on sermons and lectures, such as *Khulasah al-Quran*, representing notes on the sermons of Maulana Abdullah Awang Nuh (1905–47/1323–1366) by his student Wan Nawang Ismail.

Some scholars who produced exegetical works attained Malaysia-wide reputations and in some cases became known beyond Malaysia, such as Muhammad Sa'id b. Umar. Others were primarily influential in the State where they were based, such as Mustafa Abdul Rahman (in Perak) and Syeikh Abu Bakar Ashaari (in Perlis).

3 Analysis of the Aims, Methods and Results of the Treatment of the Qur'ān

3.1 Indonesia

Tafsir Qurän Karim by Mahmud Yunus is a major work coming from the first half of the twentieth century. Yunus divided his work into three sections, by which he intended to present a succinct commentary that was free from *isrāʾīliyyāt* elements which are considered irrational by the educated people of the modern time.[20] The first two sections were set out in two columns, with the Arabic text of the Qur'ān on the right and its translation on the left. The third section contains the explanation that serves as his commentary, placed underneath both the Qur'ān and translation and separated with a horizontal line. In an attempt to provide his readers with rapid access to an understanding of the entire Qur'ān, Yunus summarized the content of each chapter at the end of his work, including aspects of religion, scientific development and social life.

The second half of the twentieth century in Indonesia witnessed the composition of more detailed commentaries. One example is *Tafsir An-Nuur* by Ash-Shiddieqy. His intention in writing this commentary was to provide access to exegesis for Indonesian Muslims who were not proficient in Arabic, some of whom were not able to recite any part of the Qur'ān. Because of this, he also included the Arabic Qur'ān in Latin script.

The next development in Qur'ānic exegesis in late twentieth century Indonesia onward drew on more specific approaches. The first relates to commentators' attempts to respond to modern scientific knowledge. There are different orientations in Indonesian Qur'ānic scientific interpretations. One prioritizes the development of Islamic science while another has followed modern sciences for the purpose of understanding the meaning of the Qur'ān and deepening religious piety. An example of the former is Agus Purwanto's user-friendly books that have targeted various levels of Indonesian readers. In his *Ayat-ayat Semesta: Sisi-sisi al-Quran yang Terlupakan* (The Verses on the Universe: Elements in the Qur'ānic Verses that Have Been Forgotten), the first section of the book functions as a manual to consider the 800 indexed Qur'ānic verses on the related subject. Purwanto states that science – among other human-made things – truly represents the worldview of the creator, and that up to the present time there is only one dominant type of science, that is a Westernized science which supports materialism and negates the existence of God.[21] He therefore believes that science should base itself on the principles of Oneness and Qur'ānic revelation. He nevertheless admits that no civilization nowadays can develop a new science

20 Yunus, *Tafsir*, vi.

21 Agus Purwanto, *Ayat-ayat Semesta: Sisi-sisi al-Quran yang Terlupakan* (Bandung: Mizan, 2015), 186.

from zero. Purwanto places stress on certain Qur'ānic statements that encourage Muslims to undertake more scientific empirical investigations and at the same time to believe in God's existence.[22]

An example of the second approach is seen in the Indonesian government's project-based Qur'ānic commentaries in the early decades of the twenty-first century published by a team under the Central Bureau of the Qur'ān Affairs of The Ministry of Religious Affairs, in collaboration with the Indonesian Institute of Sciences, the *Lembaga Penerbangan dan Antariksa Nasional* and Bosscha Observatorium of Bandung Technological Institute. Thematically presented, these works explore various subjects in science, for example food and drink, time, microorganism, ocean and so on. For the purpose of analysis, we choose one commentary that discusses the subject of time. In the Minister of Religious Affairs' Preface, he points to the function of this work as a means of exploring the universe via proven scientific developments for the sake of strengthening the belief in Oneness. The content is very much dedicated to the elaboration of key vocabularies, showing that the Qur'ān utilizes seventeen words to indicate the existence of time for different purposes, especially for worshiping God in Islam. Many other pages are allotted to exploring the history of how human civilizations have developed the concept of time, the geological periodization on the creation of the planet, and calendar systems that are then applied in the modern period.

As for feminist studies-based commentaries, Umar's and Abdul Kodir's works are representative samples. In his *Argumen kesetaaan gender perspektif Al-Qur'ān* (Argument for the Gender Equality Based on the Qur'ānic Perspective), Umar encourages contextual understanding of the Qur'ānic verses dealing with gender relationships, arguing that those verses always have particular social contexts. Taking into account Western theories in gender relations and biological anatomy, Umar's focus is on the elaboration of some key vocabularies used in the Qur'ān that indicate sex and gender distinctions. He then reconsiders creation of humans and connects the subject to principles of gender equality and the Arabic grammatical elements that have in many ways legitimized male domination. Another commentary on gender-based relations is the *Qirā'ah Mubādalah* of Abdul Kodir. At the outset, Abdul Kodir tried to show the ways in which the Qur'ān and the prophetic traditions have fully recognized the equal positions between man and woman. They strengthen each other and neither has hegemony over the other.[23]

Nurcholish Madjid and Quraish Shihab are representative of those scholars who seek to connect the understanding of the Qur'ān to Indonesian public life. Madjid generally devoted himself to the study of the established Islamic doctrines and comparative religions, while promoting neo-modernism. One example of this is in the way he

---

22 See also Agus Purwanto *Nalar Ayat-ayat Semesta: Menjadikan Al-Quran Sebagai Basis Konstruksi Ilmu Pengetahuan* (Bandung: Mizan, 2015).

23 Faqihuddin Abdul Kodir, *Qirā'ah Mubādalah: Tafsir Progresif Untuk Keadilan Gender Dalam Islam* (Yogyakarta: IRCiSoD, 2019), 50.

conceptualizes intention as a basic commitment of any human action, in order to redefine the concept of *iḥsān* as optimalization of a Muslim work ethic. Madjid notes that as a Muslim-majority country, Indonesia is now suffering from a poor work ethic compared to people in countries where Confucianism predominates. He states that the ethical system of Confucius contributed to the development of those countries, and that Islam needs to be understood in a way that can create the highest standards of achievement among its followers.[24]

Two well-known thematic commentaries by Quraish Shihab to consider here are *Membumikan Al-Quran: Fungsi dan Peran Wahyu Dalam Kehidupan Masyarakat* (Make the Qurʾān Down-to-Earth: The Function and Role of Revelation in the Life of Society) first published in 1992/1413 and *Wawasan Al-Quran: Tafsir Maudhu'i atas Pelbagai Persoalan Umat* (Insights from the Qurʾān: A Thematic Interpretation of the Various Problems of *Umma*) first published in 1996/1417. As in Madjid's works, these two works by Shihab were originally articles presented for a limited circle. Categorized as among the scholars who continue the mainstream *tafsīr* tradition,[25] Shihab is known for his wide attention to the elaboration of Arabic key vocabularies and their harmony with the whole Qurʾānic text as a gateway to understanding the meaning of the Qurʾānic verses. He then connects these elaborations to many aspects of Muslim public life. His masterpiece, the fifteen-volume commentary of the whole Qurʾān entitled *Tafsir al-Mishbah* confirms his great interest in the elaboration of Qurʾānic vocabularies and the harmony of its verses and chapters.

Islamic moderation has become a very popular term in the current Indonesian Islamic context and is used for the government's diplomacy to counter the issues of radicalism and intolerance. Corresponding to the understanding of the official concept of moderation, some Indonesian scholars look for a legitimate basis for this concept in interpreting the Qurʾān. Mustaqim's work on purposive interpretation mainly responds to this issue. He states that being moderate in interpretation means that this interpretive method is accommodative of both textualism and contextualism.[26] Originally developed to construct principles in Islamic law, purposive interpretation can be used to understand many aspects in the Qurʾān. Here, the act of interpreting the Qurʾān needs to extend to the ways in which the text can provide contextual understanding and its significance in relation to the aspects of human life. In the understanding of such stories in the Qurʾān as the Prophet Nūḥ in Q 11:40 and Q 23:27, the purposive interpretation of both verses can capture an ecological aspect that everyone is responsible for protecting the ecosystem. In the verse mentioning spouse relationships (i.e. Q 2:187), this approach can extend to the understanding of equality be-

---

24 Nurcholish Madjid, *Islam, Doktrin dan Peradaban: Sebuah Telaah Kritis tentang Masalah Keimanan, Kemanusiaan dan Kemodernan* (Jakarta: Paramadina, 2000), 409f.

25 Pink, *Muslim Qurʾānic Interpretation Today*, 40–47.

26 Mustaqim, "Argumentasi," 6.

tween husband and wife, and that they have a duty of accepting their own spouse's shortcomings.[27]

Online writings and videos are nowadays increasingly in demand as a means of disseminating religious thinking. With its capability to reach audiences more effectively, the use of social media, however, can adversely affect the society due to the widespread use of hate speech and disinformation. In this connection, Hosen actively produced online writings on the Qur'ān as a quick response to Indonesian socio-political problems. An example of this is in the way Qur'ānic verses had been commodified during the Indonesian regional elections; namely, the widespread interpretation of Q 4:139 to attach the label of hypocrites to the supporters of Basuki Tjahaya Purnama or *Ahok*, a minority Chinese Christian who stood as the incumbent in the 2017/1438 Jakarta governor election. His response was that this verse cannot be used for such political purposes.[28] More scholars are now more aware of the importance of the internet in disseminating their commentaries by creating their own official websites such as that of Quraish Shihab, https://quraishshihab.com/. Current trends in the study of the Qur'ān and how it is understood in the Indonesian public sphere might provide more complex phenomena than we have highlighted above. More work in this field can definitely open interesting new windows into this area of research.

3.2 Malaysia

The prolific activity in Qur'ānic exegesis by Malay scholars during the twentieth century discussed above is notable for a variety of key features. The first that deserves consideration relates to how the exegetes in question drew on commentary writing from other regions, especially the Arab world.

The very first commentary in Malay upon the whole Qur'ān by ʿAbd ar-Raʾūf of Singkel (d. c. 1693/1104) had served as a vehicle for the introduction to the region of three Arabic language commentaries in particular: the *Tafsīr al-Jalālayn*, *Anwār at-tanzīl wa-asrār at-ta'wīl* (The Lights of Revelation and the Secrets of Interpretation) by ʿAbd Allāh b. ʿUmar al-Bayḍāwī and *Lubāb at-ta'wīl fī maʿānī t-tanzīl* (The Core of Interpretation in the Meanings of Revelation) by ʿAlī b. Muḥammad al-Khāzin al-Baghdādī.[29] The evident circulation of these works during the formative centuries of Islam in the Southeast Asian region meant that increasingly Malay students of Islam were being exposed to a variety of hermeneutical methods. *Tafsīr al-Jalālayn* provides a very effective and popular introduction to the science of exegesis, training students at an introductory level in word-by-word and phrase-by-phrase exegesis, and drawing

27 Mustaqim, "Argumentasi," 12–14.
28 Hosen, *Tafsir*, 113–18.
29 Peter G. Riddell, "The Sources of `Abd al-Ra'ūf's Tarjumān al-Mustafīd," *JMBRAS* 57, no. 2 (1984): 113–18.

upon hadith references, grammatical details, the *qirā'āt*, some philosophical perspectives as well as narrative elements. There are few better works to provide such a broad perspective on exegetical method, which explains why the *Tafsīr al-Jalālayn* continues to be widely used in Malaysia and surrounding regions today. The other two Arabic sources mentioned above, by al-Bayḍāwī and al-Khāzin, also spoke through the pages of ʿAbd ar-Raʾūf's *Tarjumān al-mustafīd* (The Interpreter of the Beneficial), educating Malay students of *tafsīr* by exposing them to further details of grammar, philosophical perspectives and, in the case of al-Khāzin, rich narrative detail.

These three Arabic language *tafāsīr* also played crucial roles in the shaping of later exegetical works produced by Malay scholars during the twentieth century. Haji Mohd Yusof Ahmad (1868–1933/1285–1352), better known as Tok Kenali, returned to Malaya after studying in the Arab world and transmitted his learning through the system of Islamic schools (*pondoks*) that proliferated in Malaya. He could be seen as an early step in the wave of twentieth century *tafsīr* in Malaysia. He encouraged the study of *tafsīr*, taught the *Tafsīr al-Jalālayn* to his students, and set out to translate al-Khāzin's commentary into Malay, a task that unfortunately did not reach fruition. Another unrealized goal on his part was to render *Tafsīr Ibn Kathīr* into Malay, which also points to the nascent influence of that work's *tafsīr bi-l-ma'thūr* approach in the region.

However, those who followed Tok Kenali in pursuing exegetical goals were not just interested in rendering Arabic works into Malay. Many sought to produce their own contribution to exegetical reflection. *Tafsīr Nūr al-Iḥsān* by Muhammad Sa'id b. Umar stayed close to the classical approach of ʿAbd ar-Raʾūf, choosing the *Tafsīr al-Jalālayn* as its foundation.[30] Nevertheless, this work carries distinctly Sufi resonances in parts, pointing both to his Naqshbandī affiliation and to his determination to speak into his Malay context.

For firm evidence of reformist influences, we turn to the exegetical work of Syed Syeikh al-Hadi. He sat at the feet of Shaykh Muḥammad ʿAbduh at al-Azhar and drew on his lessons in *tafsīr* in composing *Tafsir Juz' 'Amma* (Pulau Pinang 1345) and *Tafsir Surat al-Fatihah* (Pulau Pinang 1928/1346). His works were not cases of pure transmission but rather adaptation as well, shaping the commentary to appeal to a Malay readership. He used his *tafsīr* platform to target what he saw as imperfections in the practice of Islam by some Malays, such as the use of Qur'ān verses as magic to ward off the effects of evil spirits.[31] Apart from his dedicated works of *tafsīr*, Syed Syeikh al-Hadi

30 Mohd Sholeh Sheh Yusuff and Mohd Nizam Sahad, "Bacaan Intertekstual Terhadap Tafsīr Nur al-Ihsan: Satu Kajian Menurut Kaedah Parallel," *Labuan e-Journal of Muamalat and Society* 7 (2013): 48.

31 M. Abdullah, S. Arifin, and K. Ahmad, "The Influence of Egyptian Reformists and its Impact on the Development of the Literature of Quranic Exegesis Manuscripts in the Malay Archipelago," *Arts and Social Sciences Journal* (2012): 3.

worked with others to transmit the new reformist (*iṣlāḥ*) teachings through newspapers and magazines, such as *Tunas Melayu* (1913/1331) and *al-Ikhwan* (1926/1344).

As the twentieth century proceeded, commentary writing in Malaysia drew on an ever wider base of sources than the traditional triumvirate of Jalālayn-Bayḍāwī-Khāzin. In his *Tafsir al-Qur'an al-Hakim Bahasa Melayu* (1949/1368) which commented on the first twenty-seven *ajzā'* of the Qur'ān, Mustafa Abdul Rahman creatively engaged with the Qur'ān and its meanings. He laid the foundations for his work from *Tafsīr al-manār* and *Tafsīr al-Marāghī*, drawing on his own reformist educational formation at the *Maahad Il Ihya Assyarif* Islamic school in Perak.

In 1959/1379 the government school system in the newly independent Federation of Malaya incorporated *tafsīr* within its Islamic Studies curriculum, providing a further stimulus to the study of Qur'ānic exegesis.[32] During the 1950s, five new commentaries had appeared. A further eight appeared in the 1960s, with exponential growth in a context of Islamic resurgence that was fueled by both Malaysian government policy and international political and societal developments.

In his *Tafsir Surah Yasin* (1966/1386), Syeikh Muhamad Idris al-Marbawi gives space to the philosophical perspectives of Fakhr ad-Dīn ar-Rāzī among other commentators.[33] Like Syed Syeikh al-Hadi, he did not depend on hadith in his exegesis as many classical exegetes had done, preferring a Qur'ān-on-Qur'ān approach, reflecting the emphasis in the Meccan context where he was born and grew up.

The increasing trend in the latter part of the twentieth century for Malaysian students to study in the Middle East led to greater Arabic language proficiency and a greater facility with original Arabic commentaries among the Malaysian Muslim scholarly community.

With the surge in exegetical output in the twentieth century, the styles of exegetical works moved beyond the traditional *at-tafsīr al-ijmālī* (general exegesis) approach that had been the hallmark of the early Malay commentaries. The influence of *Tafsīr al-Jalālayn*, itself a work of *at-tafsīr al-ijmālī*, was profound in the early Malay commentaries, both from the seventeenth century (*Tarjumān al-mustafīd*) and the early twentieth century, such as *Syarah tafsir al-Jalalain Arab, Melayu yang dinama anwar al-huda wa amtar an-nada* by Syeikh Uthman Jalaluddin al-Kelantani (1880–1952/1297–1371).

However, as the twentieth century proceeded, exegetes increasingly opted for more detailed styles: *at-tafsīr at-taḥlīlī* (comprehensive exegesis) and *at-tafsīr al-mawḍū'ī* (thematic/topical exegesis).[34] Mustafa Abdul Rahman's *Tafsir al-Qur'an al-*

32 Haziyah Hussin et al. "The Trend of Malay Quranic Commentary Writing in Malaysia in the 20th Century," *Journal of Applied Sciences Research* 8 (2012): 4346.

33 Mustaffa Abdullah, *Khazanah Tafsir di Nusantara* (Kuala Lumpur: University of Malaya Press, 2011), 54.

34 Fahd b. 'Abd ar-Raḥmān b. Sulaymān ar-Rūmī, *Buḥūth fī uṣūl at-tafsīr wa-manāhijihi* [Investigation into the Foundations and Methods of Quranic Exegesis] (Riyadh: Maktabat at-Tawba, 1996).

Hakim Bahasa Melayu was an early example of *at-tafsīr at-taḥlīlī*, moving beyond general exegesis to focus on a detailed exposé in pursuing his reformist exegetical goals.

Changing styles were facilitated by changing modes of transmission. The electronic and digital revolution has had no less impact on the field of *tafsīr al-Qur'ān* than on other fields. Any study of modern exegesis in Malaysia must of necessity move beyond the printed page to consider also broadcast sermons, Youtube films and various other means of exegetical exposition.

Two prominent Malaysian Islamic scholars whose works have benefited from the electronic and digital revolution are the late Dato' Bentara Setia Haji Nik Abdul Aziz bin Nik Mat (1931–2015/1350–1436) and Dato' Seri Tuan Guru Haji Abdul Hadi Awang (b. 1947/1366), both associated with the Islamic Party of Malaysia. Nik Abdul Aziz delivered Friday sermons at his mosque in Kelantan over many years, with these sermons recorded on audio cassette. At the time of writing, these exegetical sermons are being compiled into a multi-volume printed edition of *Tafsir Al-Quran*. Similarly, Abdul Hadi Awang's sermons over many years are being published in individual volumes by sura by Harakah Publisher.[35]

The exegetical output by Nik Abdul Aziz and Abdul Hadi Awang illustrates the growth of *at-tafsīr al-mawḍūʿī* (thematic/topical exegesis), placing a focus on a particular theme or sura. Several studies of Malaysian *tafsīr* output in the twenty-first century have revealed the progressive dominance of comprehensive and thematic exegesis. Of sixty-six works of *tafsīr* produced in Malaysia between 2001–2015, 36% were works of *at-tafsīr at-taḥlīlī* while 27% represented *at-tafsīr al-mawḍūʿī*.[36] The Malaysian Government has itself contributed to this trend by preparing ten volumes representing the style of *at-tafsīr al-mawḍūʿī*.[37]

Nevertheless, attention to direct translation of Arabic *tafsīr* works has not diminished in quantity or quality. Indeed, between 2001–2015, some 36% of new works of *tafsīr* produced in Malaysia were translations into Malay of commentaries from other languages, principally Arabic. These include the 2007/1428 translation into Malay of the *Tafsīr al-munīr* by Mustaffa Abdullah, Zakaria Omar, Abd Karim Ali et al.[38]

The increased attention given to the field of *tafsīr al-Qur'ān* by both the Malaysian Government and the broader scholarly community has led to the growth of centers of learning where *tafsīr* studies flourish. A key example is the Academy of Islamic Studies of the University of Malaya, a well-resourced facility that produces many gradu-

35 Haji Abdul Hadi Awang, *Tafsir Surah Al Baqarah Ayat 83–160* (Kuala Lumpur: Penerbitan Harakah, 2015).

36 Sahlawati Abu Baka, Haziyah Hussin, and Wan Nasyrudin Wan Abdulla, "Analisis Perkembangan Penulisan Tafsir Maudu'i di Malaysia," *Al-Irsyad: Journal of Islamic and Contemporary Issues* 4 (2019): 147–49.

37 Mohammad Zaini Yahaya et al., *Tafsir Maudhu'iy Ibadah* (Kuala Lumpur: Dewan Bahasa dan Pustaka, 2014).

38 Abu Baka et al., "Analisis."

ates of the Islamic sciences, including those who write important dissertations on *tafsīr al-Qur'ān*. Output of such study centers also includes academic journals, such as *Al-Bayan: Journal of Qur'an and Hadith Studies* produced by the Academy of Islamic Studies and published by Brill Publishers.

Yet the heart of modern *tafsīr* activity in Malaysia for the masses rests in the mosques and pondoks, where Qur'ānic exegesis forms the backbone of regular worship and community gathering. Many mosque-based scholars generate a significant following through online sermons in which they bring their interpretation of particular themes to the Muslim faithful in Malaysia.

4 Conclusion

The discussion in this article has highlighted the healthy state of Qur'ānic exegesis in modern Indonesia and Malaysia. *Tafsīr* has developed in multiple directions in these two Southeast Asian nations during the twentieth and twenty-first centuries. First the sheer volume of works of Qur'ānis exegesis has exploded, from modest beginnings in earlier centuries to an ever-expanding dialogue between scholars in interpreting the Qur'ān. Second, the styles of exegesis have increased in complexity, with traditional chain-commentary styles continuing but being supplemented by thematic approaches which engage not only with matters of doctrinal detail but also with dynamic developments in the modern world. Furthermore, Qur'ānic exegesis has moved from being the preserve of the educated elite to being accessible to the masses, through modern digital communications and social media.

While Indonesia and Malaysia draw on the hermeneutical riches of other parts of the Muslim world, they also make a significant contribution to Islamic hermeneutics within their region and beyond. Two-way students and scholarly exchanges between Southeast Asia and other Muslim countries and regions ensure that the engagement with and interpretation of the Qur'ān in Indonesia and Malaysia will continue to flourish throughout the twenty-first century.

5 Bibliography

5.1 Sources

Abdul Aziz bin Nik Mat, Haji Nik. *Tafsir al-Quran surah-surah pilihan: An-Naba', At-Takwir, At-Thaariq, al-Ghasiyyah, As-Syams, Al-Lail*. Sg. Buloh, Selangor: Kemilau Publika, 2012.

Abdul Aziz bin Nik Mat, Haji Nik. *Tafsir Surah Yunus*. Kota Bharu, Kelantan: Maahad Ad-Dakwah Wal-Imamah, 1998.

Abdul Hadi Awang, Haji. *Tafsir Surah Al Baqarah Ayat 83–160*. Kuala Lumpur: Penerbitan Harakah, 2015.

Abdul Hadi Awang, Haji. *Tafsir Surah Yaasin: Menyelami Lubuk Hati Al-Quran*. Kuala Lumpur: Penerbitan Muttaqin, 1999.
Abdul Kodir, Faqihuddin. *Qirā'ah Mubādalah: Tafsir Progresif Untuk Keadilan Gender Dalam Islam*. Yogyakarta: IRCiSoD, 2019.
Abdul Rahman Mahmud, Mustafa. *Tafsir al-Qur'an al-Hakim bahasa Melayu*. Pulau Pinang: Persama Press, 1949.
Al-Hadi, Syed Sheikh. *Tafsir al-Fatihah*. Pulau Pinang: Jelutung Press, 1928.
Al-Kelantani, Uthman Jalaluddin. *Syarah tafsir al-Jalalain Arab, Melayu yang dinama anwar al-huda wa amtar an-nada*. Pulau Pinang: Persama Press, 1934.
Al-Qari bin Haji Salleh, Abdullah. *Tafsir Surah al-Ikhlas*. Kota Bharu: Matbaah al-Hikmah, 1954.
As-Shouwy, Ahmad et al. *Mukjizat Al-Qur'ān dan As-Sunnah tentang IPTEK*. 2 volumes. Jakarta: Gema Insani Press, vol. 1, 1997.
Ash-Shiddieqy, M. Hasbi. *Tafsier Al-Qurânul Madjied 'An-Nur'*. Jakarta: Bulan Bintang, 1956.
Barlas, Asma. *Believing Women in Islam: Unreading Patriarchal Interpretations of the Qur'ān*. Texas: The University of Texas Press, 2004.
Barlas, Asma. *Cara Quran membebaskan Perempuan*, trans. Cecep Lukman Yasin. Jakarta: Serambi, 2005.
Bucaille, Maurice. *Bibel, Quran dan Sains Modern*, trans. H.M. Rasjidi. Jakarta: Bulan Bintang, 1992.
Hamidy, Zainuddin and Fachruddin HS. *Tafsir Qurän: Naskah Asli, Terdjemah, Keterangan Lengkap 30 Djuz*. Jakarta: Penerbit Widjaya, 1967.
Hamka. *Tafsir al-Azhar*. Singapore: Pustaka Nasional Pte Ltd, 2001.
Hassan, A. *Al-Furqan: Al-Qur'ān, Terjemah dan Tafsir*. Bandung: Persis Press, 2020.
Hosen, Nadirsyah. *Tafsir Al-Qur'an di Medsos: Mengkaji Makna dan Rahasia Ayat Suci pada Era Media Sosial*. Yogyakarta: Bunyan, 2017.
Ibn Aziz Al-Zindani, Abdul Majid et al. *Mukjizat Al-Qur'an dan As-Sunnah tentang IPTEK*. 2 Volumes. Jakarta: Gema Insani Press, vol. 2, 1997.
Lajnah Pentashihan Mushaf Al-Quran and Balitbangdik Kementerian Agama RI dengan LIPI. *Tafsir Ilmi: Waktu dalam Perspektif al-Qur'an dan Sains*. Jakarta: Lajnah Pentashihahan Mushaf Al-Qur'an, 2013.
Madjid, Nurcholish. *Islam Agama Peradaban: Membangun Makna dan Relevansi Doktrin Islam dalam Sejarah*. Jakarta: Paramadina, 2000.
Madjid, Nurcholish. *Islam, Doktrin dan Peradaban: Sebuah Telaah Kritis tentang Masalah Keimanan, Kemanusiaan dan Kemodernan*. Jakarta: Paramadina, 2000.
Mustaqim, Abdul. "Argumentasi Keniscayaan Tafsir Maqashidi Sebagai Basis Moderasi Islam." Pidato Pengukuhan Guru Besar dalam Bidang Ulumul Qur'an Disampaikan di Hadapan Rapat Senat Terbuka Universitas Islam Negeri Sunan Kalijaga Yogyakarta 16 Desember 2019 ["Argumentation of the Necessity of Maqashidi Tafsir as the Basis of Islamic Moderation." Inaugural Speech of Professor in the Field of Ulumul Qur'an delivered before the Open Senate Meeting of Sunan Kalijaga State Islamic University, Yogyakarta, December 16, 2019].
Purwanto, Agus. *Ayat-ayat Semesta: Sisi-sisi al-Quran yang Terlupakan*. Bandung: Mizan, 2015.
Purwanto, Agus. *Nalar Ayat-ayat Semesta: Menjadikan Al-Quran Sebagai Basis Konstruksi Ilmu Pengetahuan*. Bandung: Mizan, 2015.
Quṭb, Sayyid. *Tafsir fi Zilalil Qur'an: Di Bawah Bayangan al-Qur'an*, trans. Yusoff Zaky Hj. Yaacob. 17 volumes. Kota Bharu: Pustaka Aman Press, 1986–1989.
Sa'id b. Umar Qadi Qadahi, Muhammad. *Tafsir Nur al-Ihsan*. Pulau Pinang: Dar al-Ma'arif, 1971.
Shihab, M. Quraish. *Tafsir al-Mishbah: Pesan, Kesan dan Keserasian al-Qur'an*. Ciputat: Lentera Hati, 2000.
Shihab, M. Quraish. *Wawasan Al-Quran: Tafsir Maudhu'i atas Pelbagai Persoalan Umat*. Bandung: Mizan, 1997.
Tafsir Surah Yasin. Manuscript 848111062. Perpustakaan Negara Malaysia, Kuala Lumpur.
Umar, Nasaruddin. *Argumen Kesetaraan Jender Perspektif Al-Qur'an*. Jakarta: Paramadina, 1999.

Wadud, Amina. *Qur'ān Menurut Perempuan: Meluruskan Bias Gender dalam Tradisi Tafsir*, trans. Abdullah Ali. Jakarta: Serambi, 2001.
Wadud, Amina. *Rereading the Sacred Text from a Woman's Perspective*. New York: Oxford University Press, 1999.
Wijaya, Aksin and Shofiyullah Muzammil, "*Maqāṣidī Tafsīr*: Uncovering and Presenting *Maqāṣid Ilāhī-Qur'anī* into Contemporary Context," *Al-Jami'ah: Journal of Islamic Studies* 59, no. 2 (2021): 449–78.
Yunus, Mahmud. *Tafsir Qurān Karim Bahasa Indonesia*. Jakarta: Hidakarya Agung, 1993.

5.2 Secondary Literature

Abdullah, Mustaffa. *Khazanah Tafsir di Nusantara*. Kuala Lumpur: University of Malaya Press, 2011.
Abdullah, M., S. Arifin, and K. Ahmad. "The Influence of Egyptian Reformists and its Impact on the Development of the Literature of Quranic Exegesis Manuscripts in the Malay Archipelago." *Arts and Social Sciences Journal* (2012): 1–8.
Abu Baka, Sahlawati, Haziyah Hussin, and Wan Nasyrudin Wan Abdulla. "Analisis Perkembangan Penulisan Tafsir Maudu'i di Malaysia." *Al-Irsyad: Journal of Islamic and Contemporary Issues* 4 (2019): 147–49.
Baidan, Nashruddin. *Perkembangan Tafsir Al-Qur'an di Indonesia*. Yogyakarta: Tiga Serangkai, 2003.
Daneshgar, Majid. "God and Magog in Malay-Indonesian Islamic Exegetical Works," in *Gog and Magog: Contributions toward a World History of an Apocalyptic Motif*, edited by Georges Tamer, Andrew Mein and Lutz Greisiger (Berlin: De Gruyter, 2023): 597–615.
Federspiel, Howard M. *Popular Indonesian Literature of the Qur'an*. Ithaca, NY: Cornell Modern Indonesian Project, 1994.
Fikriyati, U. and A. Fawaid, "Pop-Tafsir on Indonesian YouTube Channel: Emergence, Discourses and Contestations," *Proceedings of the 19th-Annual International Conference on Islamic Studies, AICIS 2019, 1–4 October 2019*, Jakarta-Indonesia.
Hussin, Haziyah, Mazlan Ibrahim, Latifah Abdul Majid, Kharuddin Mohd Amin, Sabri Mohamad, Fadlan Mohd Othman, Mohd Arif Nazri, and Faisal Ahmad Shah. "The Trend of Malay Quranic Commentary Writing in Malaysia in the 20th Century." *Journal of Applied Sciences Research* 8 (2012): 4343–49.
Johns, A.H. "Quranic Exegesis in the Malay world: In Search of a Profile." In *Approaches to the History of the Interpretation of the Qur'ān*, edited by Andrew Rippin. New York: Oxford University Press, 1988.
Lukman, Fahdli. "Digital Hermeneutics and A New Face of the Qur'an Commentary: The Qur'an in Indonesian's Facebook. *Al-Jami'ah: Journal of Islamic Studies* 56, no. 1 (2018): 95–120.
Nurtawab, Ervan. "Jalalayn Pedagogical Practice: Styles of Qur'an and Tafsir Learning in Contemporary Indonesia." PhD diss., Monash University, 2018.
Nurtawab, Ervan. "*Tafsīr al-Jalālayn* at the Crossroads: Interpreting the Qur'ān in Modern Indonesia," *Australian Journal of Islamic Studies*, 6, no. 4 (2021): 4–24.
Pink, Johanna, "Eight Shades of Ibn Kathīr: the Afterlives of a Premodern Qur'ānic Commentary in Contemporary Indonesian Translations," in *Malay-Indonesian Islamic Studies: A Festschrift in Honor of Peter G. Riddell*, edited by Majid Daneshgar and Ervan Nurtawab (Leiden: Brill, 2023): 109–33.
Pink, Johanna, "Tradition, Authority and Innovation in Contemporary *Sunnī Tafsīr*: Towards a Typology of Qur'ān Commentaries from the Arab World, Indonesia and Turkey," *Journal of Qur'anic Studies* 12 (2010): 56–82.
Pink, Johanna. *Muslim Qur'ānic Interpretation Today: Media, Genealogies and Interpretive Communities*. Sheffield: Equinox, 2019.
Riddell, Peter G. "The Sources of `Abd al-Ra'ūf's Tarjumān al-Mustafīd." *JMBRAS* 57, no. 2 (1984): 113–18.

Riddell, Peter G. *Islam and the Malay-Indonesian World: Transmission and Responses*. Singapore: Horizon Books, 2001.

Rohman, Izza, "The Pursuit of New Interpretive Approaches to the Qur'ān in Contemporary Indonesia," in *The Qur'ān in the Malay-Indonesian World: Context and Interpretation*, edited by Majid Daneshgar, Peter G. Riddell and Andrew Rippin (London: Routledge, 2016): 83–136.

Ross, Samuel J. *Qur'ān Commentary and the Biblical Turn: A History of Muslim Exegetical Engagement with the Biblical Text*. Berlin: De Gruyter, 2023.

Ar-Rūmī, Fahd b. ʿAbd ar-Raḥmān b. Sulaymān. *Buḥūth fī uṣūl at-tafsīr wa-manāhijihi* (Investigation into the Foundations and Methods of Quranic Exegesis). Riyadh: Maktabat at-Tawba, 1996.

Van Bruinessen, Martin. "Kitab Kuning: Books in Arabic Script used in the Pesantren Milieu: Comments on a new collection in the KITLV Library." *BKI* 146 (1990): 226–69.

Yahaya, Mohammad Zaini et al. *Tafsir Maudhu'iy Ibadah*. Kuala Lumpur: Dewan Bahasa dan Pustaka, 2014.

Yusuff, Mohd Sholeh Sheh, and Mohd Nizam Sahad. "Bacaan Intertekstual Terhadap Tafsīr Nur al-Ihsan: Satu Kajian Menurut Kaedah Parallel." *Labuan e-Journal of Muamalat and Society* 7 (2013): 48–59.

Christine Schirrmacher

Representatives of Progressive Qur'ānic Hermeneutics in the Twenty-First Century

1 Introduction

While pre-modern Qur'ānic exegesis was characterized by verse-by-verse commentary on the entire Qur'ānic text (*tafsīr musalsal*), this approach has changed to a topic-oriented exegesis in modern times,[1] which presents the text as a unit. Whereas pre-modern exegesis derived a universal rule of conduct by giving consideration to individual verses, thus proceeding from the smallest unit, i.e., the meaning of a term, twentieth-century Qur'ānic hermeneutics proceeded by working out a superordinate principle and, with the help of this principle, carries out interpretation of individual Qur'ānic verses via illustrations.[2] Also in the European context, twentieth-century Qur'ānic hermeneutics tended more towards an exegesis that transcends text and verse, placing the Qur'ān under the heading of an overarching topic and no longer commenting on it verse by verse.

1.1 Characteristics of Contemporary Qur'ān Hermeneutics

This tendency had already become apparent in representatives of early "Reform Islam" in the late eighteenth and nineteenth centuries. It intensified over the course of the twentieth century and is also important for present day Qur'ānic commentaries. Further twentieth-century developments included a more apologetic orientation of Qur'ānic exegesis and, due to the emergence of the mass media, a more user-oriented popularization of exegesis that was less limited to scholarly circles.[3] These apologetic aspects came to the fore in Qur'ānic hermeneutics at the beginning of the twentieth century, increasingly emphasizing the compatibility of Islam with the findings of science and rejecting the charge that Islam was irrational.[4] They increasingly developed as responses to the en-

1 Thomas Würtz, "Koranexegese und gesellschaftlicher Wandel," in *Christentum und Islam als politische Religion. Ideenwandel im Spiegel gesellschaftlicher Entwicklungen*, eds. Oliver Hidalgo, Holger Zapf, and Philipp Hildmann (Wiesbaden: VS Springer, 2017), 268.
2 Johanna Pink, *Sunnitischer Tafsīr in der modernen islamischen Welt. Akademische Traditionen, Popularisierung und nationalstaatliche Interessen* (Leiden: E.J. Brill, 2011), 35.
3 Johanna Pink, "Modern and Contemporary Interpretation of the Qur'ān," in *The Wiley Blackwell Campanion to the Qur'ān*, eds. Andrew Rippin and Jawid Majaddedi (Hoboken: Wiley, 2017²), 485.
4 Rotraud Wielandt, "Main Trends of Islamic Theological Thought from the Late Nineteenth Century to Present Times," in *The Oxford Handbook of Islamic Theology*, ed. Sabine Schmidtke (Oxford: Oxford University Press, 2016), 724.

https://doi.org/10.1515/9783111320069-027

counter and confrontation with the West. This resulted in two main strands in the orientation of Qur'ānic hermeneutics: a political Islamist strand and a progressive strand.

Thus, development towards the mentioned new forms of Qur'ān hermeneutics did not only begin in the twenty-first century. Moreover, they did not begin in the context of confrontation with Europe: classical Qur'ānic interpretation was aware of the inclusion of "occasions of revelation" (*asbāb an-nuzūl*), with the help of which, for instance, Qur'ānic calls to fight were mostly related to Muḥammad's war campaigns. Furthermore, the universal validity of these calls to fight was restricted and limited to these historical occasions.

In the progressive interpretation of the Qur'ān in the twentieth and twenty-first centuries, which this handbook contribution deals with by considering four protagonists, the focus is less on the compatibility of the Qur'ān with the findings of scholarship. In the Western context, it is more focused on justifying the compatibility of Islam with democracy, human rights, women's rights, and civil rights and the derivation of these rights from Islamic theology. Furthermore, it emphasizes that Islam is a religion of peace open to dialogue and that Muslims are peaceful members of Western societies. This tendency has become noticeably stronger after the 2001/1422 terrorist attacks on the World Trade Center in the USA and after additional following acts of violence. Also, this has been all the more the case in Western countries where these acts of violence have occurred.

A characteristic of contemporary Qur'ān hermeneutics is – and this also applies to progressive interpretations in the European context – that no longer only theologians with classical Islamic education appear as protagonists. There are also intellectuals from other disciplines who to some extent present far-reaching blueprints. With regard to Western progressive interpretations of the Qur'ān, some common tendencies are recognizable:

– With different justification, progressive Qur'ānic exegetes in the European and Western context hold on to the fact that the Qur'ān is to be regarded as God's word and was sent down to mankind as revelation. Admittedly, other perspectives have given a new orientation to the recognition of the process of revelation; at the same time, there is an emphasis on there being no arbitrariness in hermeneutics, no openness to questioning the Qur'ān and no capriciousness. A progressive interpretation of the Qur'ān would also have to orient itself toward given objective preconditions.

– All representatives of a progressive Qur'ānic hermeneutics in the Western context are opposed to a literalistic interpretation which, from their perspective, never did justice to the text in the past and does not do so today. In general, the assumption that the text is to be transferred literally and ahistorically into today and that man is only a passive recipient of the divine message is rejected. The basic assumption is rather that the reader or listener today is actively involved in the process of revelation, just as Muḥammad was at that time.

– Turning away from a literalistic interpretation of the Qur'ānic text is often associated with progressive representatives calling for Islam to experience an enlightenment and a distancing from fundamentalist political interpretations of the text. This applies in particular to justifying the exercise of violence, the restriction of women's rights, and the degradation of religious minorities. Progressive commentators on the Qur'ān in the Western context have also frequently noted the necessity of an opening up in view of the much-lamented intellectual stagnation of societies characterized by an Islamic majority.

– It includes the consideration that the text of the Qur'ān was revealed and written down in the historical, political, and cultural context of late antiquity. Since today's context is a completely different one, an adapted exegesis of the text is absolutely necessary. This adapted exegesis of the text cannot be a generally valid one for all epochs, just as the conventional interpretation cannot be. This also applies to the hermeneutics of the founders of the four schools of law. Today's exegetes must relate the horizon of understanding of the contemporary recipient to the horizon that emerges from the text of the seventh century AD, so that a new understanding of the text can be gained by taking changed circumstances into account.

– Progressive representatives emphasize that there is not just one single, objective interpretation of the Qur'ānic text that is valid for all times and circumstances. Rather, there are great numbers of interpretations which are justifiable, depending on the time frame, the culture, and the concrete circumstances of a society. Another dependency is the (subjective) position of the commentator, which shapes one's exegesis. An exegesis cannot claim absolute validity either; therefore, even different textual interpretations are equally legitimate.[5] Thus, progressive Qur'ānic hermeneutics represents a counter-position to Islamist-political and Salafist exegesis of the Qur'ān, which seeks to rediscover the one supposedly objectively correct meaning of the Qur'ānic text[6] that was already conveyed in the process of revelation.

– Progressive Qur'ān commentators in the European context emphasize the dialogical structure of the Qur'ānic text and thus emphasize active participation on the part of the individual in the process of revelation: From their perspective, the Qur'ān certainly testifies to a dialogical process between sender (God) and receiver (human). God entered into a dialogue with Muḥammad; likewise, today's reader or listener of the text is no mere passive recipient. If the text enters into dialogue with today's reader, his or her perspective is expanded for interpretations arising from today's time and circumstances. This is essential if the text aims to credibly claim universal validity for all times and circumstances. Education, knowledge, and the human mind

---

5 Victoria S. Harrison, "Hermeneutics, Religious Language and the Qur'an," *Islam and Christian-Muslim Relations* 21, no. 3 (2010): 217.

6 Adis Duderija, "Neo-Traditional Salafi Qur'an-Sunna Hermeneutics and Its Interpretational Implications," *Religion Compass* 5, no. 7 (2011): 320.

are considered by progressive interpreters to be important instruments for the appropriate transmission of the message of revelation within today's circumstances.

– A number of progressive representatives explicitly favor a humanistic interpretation of Islam and the Qur'ānic text. They shift the theological focus of their textual consideration from the proclamation of the majesty of God to man, his needs, perceptions, and limitations. Indeed, the frame of reference remains the process of revelation and thus God's claim on man. However, the definition of the consequences to be drawn from it (such as the view that non-Muslims are unbelievers or that the detailed practice of the five pillars paves the way to paradise), which is traditionally widely accepted in conventional theology, is in many respects extended or reinterpreted.

2 Influential Protagonists

In the following, three protagonists of progressive Qur'ānic hermeneutics from the German-speaking region and a representative from the Pacific context who presently teaches in Australia will be presented. Since all four start from the basic assumption that the Qur'ān is the revealed word of God, they should not be called "liberal" voices. Also, the term "modernist" does not seem to be very appropriate since the participants are not concerned with wanting to be "more modern" than earlier commentators; on the contrary, they often emphasize that they see their hermeneutics as a continuation of approaches from early Islamic times. Their concern is essentially to build a bridge between the revelational context of the seventh century AD and that of an appropriate application of Qur'ānic statements to present circumstances.

The selection of these four protagonists does not imply any evaluation or devaluation of other participants in the field.[7] Rather, as influential Muslim theologians and intellectuals, they are representatives of a generation of German-speaking and English-speaking voices making well-founded demands for a contemporary interpretation of the Qur'ān and a reconciliation of human rights and civil rights with Islamic theology. Of their diverse publications, only those have been considered here which are dedicated to the topic of Qur'ān hermeneutics.

7 For example, the Swiss Islamic scholar Tariq Ramadan (born in 1962) could have been included here. He wishes to distinguish between Islam as a religion and the traditions associated with it and recommends certain adaptations with respect to the application of Sharia law in Europe.

2.1 Abdel-Hakim Ourghi

Abdel-Hakim Ourghi was born in 1968 in Oran, Algeria, and is an Islamic scholar, philosopher, and religious educator. In 2006 he received his doctorate in Islamic Studies from the University of Freiburg with a dissertation on the Ibāḍiyya reform movement. Since 2009 he has been working on his post-doctoral thesis entitled "The Dialogue between God and the Devil. A demythologizing Hermeneutics."[8] He has been head of the Department of Islamic Theology/Religious Education at the Freiburg University of Education since 2011.

Ourghi, who calls himself a "liberal Muslim,"[9] has released several publications, calling for a critical re-evaluation of the sources of Islam and a reorientation of the traditional interpretation of the Qur'ān. He calls for the establishment of a humanist Islam that condemns extremism, discrimination – especially against women and those of other faiths – and the use of violence in the name of religion, and instead establishes democracy and human rights. Its main focus is on the demand for an enlightenment of Islam. Ourghi sees the justification for demanding increased civil rights in the justice of God.

2.1.1 The Call for Enlightenment

His goal is neither the restoration of Islam in its supposedly genuine early form nor an Islamization of modernity. His goal is also not a dissolution of Islam in the fog of critical reason. Abdel-Hakim Ourghi's central concern is the demand for a humanistic enlightenment of Islam in Kant's sense: he strongly advocates the necessity for reform within Islam and the acceptance of humanistic values such as freedom of opinion, freedom of religion as well as equal rights for men and women and Muslims and non-Muslims. For Ourghi, enlightenment within Islam is clearly linked to the adoption of Western values, because Ourghi's goal is "the modernization of Islam and its adaptation to Western culture,"[10] since an Islam that is not critically reflected upon cannot be reconciled with democracy, Germany's Basic Law (Germany's constitutional document), and human rights. In contrast, an enlightenment of Islam, according to Ourghi, leads to freedom of faith (namely to believe or not to believe) and to freedom of critical thinking and action guided by reason. This freedom only ends at the point where it would harm others.

Ourghi is not only concerned with rational consideration of the Qur'ān. Rather, in his view, it is essential that a completely independent Islam develops in Europe, one which critically examines its religious and cultural heritage and follows "the princi-

8 "Der Dialog zwischen Gott und dem Teufel. Eine entmythologisierende Hermeneutik".
9 "[L]iberaler Muslim": Abdel-Hakim Ourghi, *Reform des Islam. 40 Thesen* (Munich: Claudius, 2017), 12.
10 "Die Modernisierung des Islam und seine Anpassung an die westliche Kultur": ibid., 71.

ples of Western modernity." In his view, this may well lead to a "loss of certainties."[11] However, this loss of certainties does not mean Islam will become unfamiliar. Rather, it would make the historical Muḥammad visible again, who for Ourghi was buried under several layers of hagiographic alienation when his character was reinterpreted from that of a fallible human being into a mythological figure of salvation. Also, the traditions attributed to Muḥammad are for Ourghi in essence a "human construct,"[12] which in many cases contradict the ethics of Islam.

For this urgently needed enlightenment "without taboos, prohibitions of thought, and dogmas,"[13] the development of a historical-critical Qur'ān exegesis is, from Ourghi's point of view, an essential precondition that justifies distancing from "statements in Islamic texts legitimizing violence."[14] Historico-critical observation must free these texts from parts people have added to Islam over the course of history, thus leading to a variety of different readings. It will then again be possible to rediscover Islam as spirituality, as a spiritual-religious movement, and as a binding of the individual to God, and not as a state order or a political system as it is understood by conservative voices and representatives of political Islam.

The politico-legal part of the Qur'ān no longer has any significance for Ourghi in modern times; nor do the calls to fight, the (partly pejorative) statements about Jews and Christians, and the orders to have women's rights curtailed.[15] In particular, the "chronology of violence"[16] from the early history of Islam to the present must be worked through and overcome. That the early history of Islam has not yet been worked through, but is rather exemplarily regarded as the Golden Age of Islam, and thus serves as the basis for the Islamic community's fantasies of superiority, is regarded by Ourghi as "nostalgia with no content." He explains its existence with a "dynamic of conscious repression,"[17] which he wants to overcome through a fact-oriented reappraisal of history.

From Ourghi's point of view, blind imitation, obedience to outdated traditions and uncritically assumed scholarly authority must be replaced by individual, rational analysis of religious texts. In doing so, he has far less confidence in texts of tradition than in the Qur'ānic text: it is his view that their writing ultimately served political purposes; many of the texts are considered by Ourghi to be highly dubious. Also, the views of scholars from the early Islamic period, who have been recognized as authoritative, were subject to the historical, political and social conditions of their time. They are therefore to be understood as temporary and no longer compatible with the present

11 "Die Grundsätze der westlichen Moderne"; "Verlust von Gewissheiten": Abdel-Hakim Ourghi, *"Aufklärung" des Islam?* Analysen und Argumente 211 (Berlin: Konrad Adenauer Stiftung, 2016), 9.

12 "Menschliches Konstrukt": ibid., 13.

13 "Ohne Tabus, Denkverbote und Dogmen": ibid., 6.

14 "Gewaltlegitimierenden Aussagen in islamischen Texten": ibid., 1.

15 Ourghi, *Reform*, 46.

16 "Chronologie der Gewalt": Ourghi, *Aufklärung*, 2.

17 "[I]nhaltslose Nostalgie"; "Dynamik des bewussten Verdrängens": ibid.

and altered circumstances. Ourghi's perspective views the attempt of certain actors to claim sovereignty over the interpretation of Islam as altogether a sign of abuse of power, as is the ban on individual critical reasoning and the inviolability of the sources and the person of Muḥammad. From Ourghi's standpoint, claiming an interpretation of the text as the only correct one is strictly rejected. Nor can he acknowledge a justification for a Muslim claim to solely be in possession of the truth with Islam.

2.1.2 The Qur'ān – Word of God or Word of Man?

Ourghi basically understands the Qur'ānic text, which was originally revealed in oral form, to be God's word that came about in the form of a communicative dialogue. However, when the Qur'ān was later set down in written form, the communicative aspect of the Qur'ān, which it had still had in Muḥammad's dialogue with his listeners, ceased to exist. Through the codification of the text and its exegesis in light of the respective historical circumstances, the Qur'ān was increasingly supplemented over the course of time. It thereby emancipated itself from its author, became "autonomous,"[18] and gradually acquired the character of the human word;[19] therefore, it is Ourghi's assessment that the Qur'ān in its present form is only "indirectly God's word."[20] Also, Muslims today are free to perform exegesis within the framework of their culture. However, this path only opens up for the individual if the Qur'ān in its present form and in its entirety is no longer called God's word. For Ourghi, the definition of the Qur'ān as a human word is therefore a prerequisite for an enlightenment and liberatory exegesis of the text.

This enlightened exegesis is also necessary because a literal interpretation of the Qur'ān would not do it justice, indeed making it a "dead word."[21] The Qur'ān only comes to life through a de novo dialogical process with the respective age's exegesis. Ourghi therefore pleads for an interpretation of the Qur'ān that today can establish a humanistic Islam: for the Qur'ān only has eternal meaning as a document of humanistic ethics, not with respect to its politico-legal instructions for action. He regards the politico-legal instructions as no longer valid today. They have to be reinterpreted. If one were to cling to them, although they are no longer feasible today, one would surrender the Qur'ān to insignificance.

Ourghi sees the historico-critical reading of the Qur'ān as having been achieved in an exemplary manner by the Sudanese Sufi scholar and reformer Maḥmūd Muḥammad Ṭaha (1909/1911–1985 / 1327/1328–1405). In the manner of Maḥmūd Muḥammad Ṭaha, Ourghi wants to distinguish between ethical (Meccan) and political (Medinan) components within Islam: it is Ourghi's view that provisions that are invalid

18 "autonom": Ourghi, *Reform*, 75.
19 Ibid.
20 "indirekt Gottes Wort": ibid., 74.
21 "[Zu einem] toten Wort": ibid., 81.

today include, for example, the assurance formulated in Q 3:110 that Muslims are the best community on earth. Also, the headscarf is "un-Islamic" in Ourghi's view and is a sign of "male dominance over Islamic women," a sign of "degradation" and "harassment."[22] These regulations were only addressed to Muslims during Muḥammad's lifetime and were therefore only situationally valid. Ourghi's viewpoint is that the permission to use violence against non-Muslims or to restrict freedom of religion and freedom of opinion was only temporarily allowed. Indeed, Ourghi regards the whole of sharia law as a "man-made construct."[23]

On the other hand, in Ourghi's view, people's ethical behavior called for in the Qur'ān stands for humanity's search for God: it enables peaceful coexistence with other people. In this context, Ourghi considers two messages of the Qur'ān to be central: God's mercy and forgiveness, and secondly, God's love for human beings and human beings' love for God.[24] When man implements the peace and tolerance intended by God, he or she simultaneously approaches his Creator.

For Ourghi, the starting point and prerequisite for the establishment of such a humanistic Islam is the adaptation of exegesis to the circumstances of the present. Starting from these contemporary circumstances, the canonical sources of Islam can be freed from ideological ballast and can now be interpreted under different auspices than at the time of the origin of Islam. This also means, however, that the understanding of the Qur'ānic text may be manifold: indeed, from Ourghi's point of view, the Qur'ān itself already confirms that there may be different sorts of knowledge gained at different times.[25] Similarly, it is Ourghi's assessment that the Qur'ān emphasizes an individual's freedom to set out in search of that individual's (subjective) truth.[26] In this context, it is quite legitimate for Ourghi to reject individual statements of the Qur'ān if the exegete concludes that the statement in question was an order derived from Muḥammad's environment in Medina that was only valid for a limited period of time.[27]

On the basis of an individual's self-determination, the reader will therefore only do justice to the text if he or she draws different conclusions in the present than the reader did in the past. For this reason, the act of understanding by the reader can never come to an end; on the contrary, it moves on continuously. This leads to a mul-

---

22 "Unislamisch"; "[ein Zeichen der] männlichen Dominanz über die islamische Frau"; "[der] Erniedrigung [und der] Schikane": ibid., 199.

23 "von Menschen erdachtes Konstrukt": Abdel-Hakim Ourghi. *Ihr müsst kein Kopftuch tragen. Aufklären statt Verschleiern* (Munich: Claudius, 2018), 12.

24 Ourghi, *Reform*, 147.

25 Abdel-Hakim Ourghi, "Ein postmodern-humanistischer Islam. Die Autonomie des Korantextes und die Freiheit des Lesers," *Aufgang. Jahrbuch für Denken, Dichten, Kunst* 13 (2016): 152.

26 In support of this statement, Ourghi cites Q 17:36, a verse that warns against pursuing things of which people have no knowledge: ibid., 160.

27 Ourghi, *Reform*, 81.

titude of textual interpretations, among which no absolute can be claimed. This paves the way to tolerance and acceptance of other views.[28]

2.1.3 The Search for the Absolute

On the one hand, Ourghi repeatedly weakens the claim that the Qur'ān is the only true revelation and that Muslims are in sole possession of the truth. Indeed, he considers the Torah, the Bible, and the Qur'ān to be equal signposts to God in terms of their truth content,[29] for no religion can, in his view, claim to possess the absolute truth. With this argumentation, Ourghi essentially underpins his calls for tolerance and charity. However, because Muslims have put aside these central messages of the Qur'ān, they now need a "theology of grace" in order to once again place the ethical message of the Qur'ān at the center of their faith.[30]

On the other hand, there are by all means absolute parameters in Ourghi's discourse of enlightenment, even if he demands openness for a pluralistic reading of the Qur'ān and firmly rejects a monopoly on a single, correct interpretation of the text. Thus, he does not want to deny the Qur'ān "its nature as God's book."[31] He is not concerned with a "devaluation" or "dissolution of the collective identity of Islam."[32] Rather, he is concerned with critical reflection, without which the "Islamic faith becomes a dangerous religion."[33] The search for an adequate enlightened interpretation of the ancient texts is for him the attempt "to arrive at the truth"[34] and not only at relative knowledge. As a possible consequence, he himself mentions that this absolute truth might be relativized by the manifold approaches to the Qur'ān that he calls for. However, in principle he welcomes this consequence since no religion is allowed to claim absolute truth.[35] All in all, Ourghi wants to see his efforts to work out a critically reflected Islam understood as a "declaration of love for Muslims."[36]

---

28 Ourghi, *Islam*, 156.

29 "[. . .] die Thora, die Bibel oder der Koran [. . .] sind nur religiöse Schriften, die verschiedene Wege zu ihm weisen. Keiner dieser Schriften ist besser als die andere [. . .]" (English translation: "[. . .] the Torah, the Bible or the Qur'ān [. . .] are just religious writings that point out different ways to get to him. None of these writings is better than the other [. . .]"): Ourghi, *Reform*, 149.

30 "Gnadentheologie": ibid., 151.

31 "seine Natur als Gottes Buch absprechen": Ourghi, *Islam*, 151.

32 "Abwertung"; "Auflösung der kollektiven Identität des Islams": Ourghi, *Aufklärung*, 2.

33 "[der] islamische Glaube [wird] zu einer gefährlichen Religion": ibid., 12.

34 "[der Versuch,] zur Wahrheit zu gelangen": Ourghi, *Islam*, 157.

35 Ibid., 167f.

36 "Liebeserklärung an die Muslime": Ourghi, *Kopftuch*, 17.

2.2 Ahmad Milad Karimi

Ahmad Milad Karimi was born in Kabul in 1979/1399, received his doctorate in philosophy at the University of Freiburg, and is now active as a philosopher of religion, Islamic scholar, and poet. Since 2016/1437 he has been professor of Kalām, Islamic Philosophy and Mysticism, and deputy director of the Center for Islamic Theology at the University of Münster.

Karimi confronts less traditional theological views in a contentious manner rather than expanding them with the aid of philosophical-theological concepts from Christian theologies, European intellectual history, and mysticism, shifting them to self-chosen focal points. One of Karimi's main areas of research is Qur'ān hermeneutics, which he has dealt with in several articles and monographs; he also presented his own translation of the Qur'ān in 2009/1430, while noting: "Translation is constant failure."[37]

He often emphasizes the necessity of a humanistic interpretation of the Qur'ān, taking into account the specific circumstances of history and tradition. He calls for an advent of an understanding of the Qur'ān that correlates with modernity but at the same time takes aesthetic aspects into account; indeed, the actual focus of Karimi's philosophical reflections on Qur'ānic hermeneutics is upon the topic of the Qur'ān's aesthetics. His intrinsic concern is therefore to convey the Qur'ān as an aesthetic work of art. He sees Islamic theology as faced with the task of designing an aesthetic theology, because in his view ethics can only be founded upon aesthetics. Thus, for Karimi, Islam is essentially about an aesthetic education of the individual. This is because aesthetics, which for him is expressed in the beauty of God, in the beautiful role model of Muḥammad, and in individuals acting beautifully, shows the way to "adequate freedom."[38] This prevents a politicization of religion, as it is for instance expressed in Wahhabism and Salafism.

2.2.1 The Qur'ān as an Aesthetic Event

For Karimi, the Qur'ān represents a loving connection between God and man, care or a "declaration of God's love,"[39] to which an individual can respond with faith. The Qur'ān is also something which must take place in the heart of an individual. The loving connection between God and man is expressed in the creation of humankind, in

37 "Übersetzung ist ein ständiges Scheitern": Ahmad Milad Karimi, "Versuch einer ästhetischen Hermeneutik des Qur'ān," in *Herausforderungen an die Islamische Theologie in Europa – Challenges for Islamic Theology in Europe*, eds. Mouhanad Khorchide and Klaus von Stosch (Freiburg: Herder, 2012), 23.

38 "adäquate Freiheit": Ahmad Milad Karimi, *Warum es Gott nicht gibt und er doch ist* (Freiburg: Herder, 2018), 96.

39 "Liebeserklärung Gottes": ibid., 139.

God's mercy, and in the process of revelation, which is a "declaration of God's love for humankind;"[40] indeed, it is on the love of God that the event of revelation is based.

For Karimi, the Qur'ān thereby receives a dialogical, multi-layered character: God's word has for him not become a "book." God was not transformed into the book of the Qur'ān; rather, God "occurs"[41] in the Qur'ān. It is, on the one hand, the word of man, and on the other hand, it is the word of God. In Karimi's view, both terms easily lead to misunderstandings, since with the term "word of man"[42] it is easy to lose sight of the fact that the Qur'ān is a revelation. If, on the other hand, the term "Word of God" is used, the question could arise whether God's Word has become human in the sense of it being transitory through having been sent down and made accessible to the (limited, earth-bound) knowledge of man?[43] The solution for Karimi lies in applying both terms, word of man and word of God, to the Qur'ān.

Karimi considers the idea of aesthetics to be the central element but also the resolution of this apparent contradiction. For him the Qur'ān represents the "glory of God" in theological terms.[44] The Word of God initially became a phonetic experience for Muḥammad and the first listeners and thus an aesthetic experience through which the presence of God was revealed. This means that in the 7th century the Qur'ān only became the Qur'ān when it was recited and listened to, and in the same way it only becomes the Qur'ān for today's listeners through recitation.[45]

For Karimi the presence of God is not reflected in the dead letter but only through the recitation of the Qur'ān: in the aesthetic experience of the recited revelation of the Qur'ān, a work of art, God becomes perceptible as the loving one, the beautiful one, and the one touching individuals, when the aesthetic listening experience overwhelms the recipient and seizes his or her heart for God. Whoever is touched in this way has entered into the presence of God. Thus, the recitation of the Qur'ān is an ever-new process of revelation, and revelation is not an event that is closed once and for all but rather "everlasting care coming from the Lord of the worlds."[46] Revelation, then, constitutes a perpetual dialogue between God and man. At the same time, the recitation of the text also leads the listener to an encounter with Muḥammad, since

40 "Liebeserklärung Gottes an die Menschen": Ahmad Milad Karimi, "Zur Frage der Erlösung des Menschen im religiösen Denken des Islam," in *Streitfall Erlösung*, eds. Klaus von Stosch and Aaron Langenfeld (Paderborn: Schöningh, 2015), 31.

41 "[E]reignet sich": Ahmad Milad Karimi, "Der Koran – eine Liebeserklärung Gottes? Überlegungen zum Islam als eine Offenbarungsreligion," *Freiburger Universitätsblätter* 201 (2013): 79.

42 "Menschenwort": ibid., 77.

43 Ibid., 77.

44 "[D]ie Herrlichkeit Gottes": Karimi, "Versuch," 27.

45 Ahmad Milad Karimi, *Hingabe. Grundlagen der systematisch-islamischen Theologie* (Freiburg: Rombach, 2015), 195.

46 "[E]ine sich immerfort ereignende Zuwendung des Herrn der Welten": Ahmad Milad Karimi, "Die Beziehung von Mensch und Gott aus islamischer Perspektive," in *„Der stets größere Gott". Gottesvorstellungen in Christentum und Islam*, eds. Andreas Renz et al. (Regensburg: Friedrich Pustet, 2012), 235.

the listener now in a way repeats his experience, as it were, and the recitation renews the act of Revelation. Thus, the recitation of the text leads to a union of all participants: God, Muḥammad, and the listener.

For Karimi, on the other hand, the written Qur'ān as a report has "a decidedly secondary character"[47] that covers up the actual character of the Qur'ān, for the proclamation of the Qur'ān was first of all about the "acoustic, atmospheric and consequently aesthetic experience"[48] of listening. Through its written form, the Qur'ān was ultimately cast in human language.

2.2.2 To think of God as Mercy

Karimi does not consider God's mercy only as one of his properties (*ṣifāt al-fiʿl*); rather, God himself is mercy for Karimi and cannot but act mercifully.[49] This mercy is expressed in God's longing for humankind. Since the Qur'ān allows people to experience God's presence, as a revelation it is also to be viewed as an act of mercy.[50]

And yet God remains the one who is withdrawn from the world. Also, the question arises whether man can at all adequately describe God in his transcendence. How is the relationship between immanence and transcendence comprehensible? How can God be thought of on the human level and at the same time maintain his absoluteness without being drawn down to the worldly level? From Karimi's point of view, neither transcendence nor immanence can be dissolved. God remains divorced from the world. In the process of revelation, He must be thought of as immanent and transcendent at the same time. This tension is dissolved by the meeting of the two realms with God's working in Muḥammad's heart: God touches his heart and thereby conquers it.[51] From this it becomes clear that the transcendence of God, who meets Muḥammad in the immanent world, is preserved.

2.2 Not Blindly Bowing down to Reason

In contrast to Abdel-Hakim Ourghi, Milad Karimi is not a fundamental proponent of an enlightenment of Islam. That an Enlightenment of Christianity had taken place

47 "[E]inen entschieden sekundären Charakter"; Karimi, "Versuch," 19.

48 "[A]kustische, atmosphärische und mithin ästhetische Erfahrung": Karimi, "Versuch," 21f.

49 Ahmad Milad Karimi, "Wie Gott als Barmherzigkeit gedacht werden kann," in *Theologie der Barmherzigkeit? Zeitgemäße Fragen und Antworten des Kalām*, eds. Mouhanad Khorchide, Ahmad Milad Karimi, and Klaus von Stosch (Münster: Waxmann, 2014), 107f., 114.

50 Ibid., 115.

51 Ahmad Milad Karimi, "Von der Immanenz Gottes im Islam," in *Gottesvorstellungen im Islam. Zur Dialektik von Transzendenz und Immanenz*, ed. Amir Dziri (Freiburg: Kalām, 2013), 104, 107.

could be explained in this way: "Christianity provokes, as it were, the Enlightenment [. . .] at least the medieval-Christian molded conception of man and the world" had to experience an enlightenment in a comprehensive way; in Islam, however, the enlightenment has already been "internalized,"[52] so that there is no need for a similar development. For Karimi, the Enlightenment criticism of Christianity and church institutions was also owed to the special relationship between the state and religion in the Western context, which is not easily transferrable. The necessity of enlightenment should therefore not be imposed equally on all religions; rather, the special religious and historical context must be taken into account.[53] Karimi thus defends himself against the demand that other religions, such as Islam, also are in need of enlightenment: If criticism of religion is actually directed against an inappropriate understanding of religion that humiliates and enslaves man, but not against religion per se, then not every religion must be confronted with the demand for enlightenment. From Karimi's point of view, reason and criticism have their place, but one should also not "blindly [. . .] kneel before reason."[54]

Therefore, Karimi believes that anyone who demands an enlightenment of Islam is wrong – however, anyone who believes that Islam does not need an enlightenment is also wrong, since such an individual would preserve Islam in its tradition. Karimi emphasizes that discussion and questions are essential, as are doubt and searching, arguing with God, and tentative understanding. Islam must not be understood as a static entity. It is not lifeless and frozen. Rather, it transpires again and again in the relationship between God and man as dialogue and dialectic.[55]

2.2.4 Neither a Mummification of the Text nor Arbitrariness

Karimi describes the Qur'ān as a revelation imbued with ambiguity. Therefore, the interpretation of the Qur'ān is basically open to many interpretations, as well as being open to having new contexts of meaning found. A static interpretation would not do justice to the living character of the text; it would outright "mummify" it.[56]

Karimi equally rejects arbitrariness in hermeneutics in explicit terms. It is not about arbitrariness but about appropriate hermeneutical guidelines. Karimi views these guidelines as the "result of the overall composition of the Qur'ān" as well as resulting "alongside the precisely worked out criteria that can be taken from the spirit

52 "Das Christentum provoziert gleichsam die Aufklärung [. . .] zumindest das mittelalterlich-christlich geprägte Menschen- und Weltbild"; [die Aufklärung ist] "internalisiert": Karimi, *Warum es Gott nicht gibt*, 96.

53 Ibid., 88–90.

54 "[B]lind vor der Vernunft in die Knie [. . .] gehen": ibid., 96.

55 Ibid., 93f.

56 "[M]ummifzieren": ibid., 142, 144.

of this genuine revelation."[57] Since the Qur'ān is a message that is addressed to all people, not only to Muslims, the Qur'ān cannot be reduced to a single idea or orientation. Its character as a communicative text prohibits this. Thus, Karimi affirms the diversity of Qur'ānic interpretations, which for him do not detract from the character of the Qur'ān as absolute truth,[58] but he does so without naming certain criteria that could reveal hermeneutical guidelines.

2.3 Mouhanad Khorchide

Mouhanad Khorchide was born in 1971/1391 in Beirut. He studied sociology in Vienna and Islamic Studies in Beirut. He received his doctorate in Vienna with a thesis on Islamic religious education in public schools and is now professor of Islamic religious education at the Center for Religious Studies (CRS) and head of the Centre for Islamic Theology (CIT) at the University of Münster, Germany.

One of Khorchide's primary areas of research is "Modern Approaches to Qur'ān Hermeneutics." Within the German-speaking world, he is the one individual who, in several book publications, anthologies, and numerous professional articles, has probably most extensively dealt with the topic of progressive Qur'ān hermeneutics in the European context. He also began work on publishing a seventeen-volume commentary on the Qur'ān.[59] The interpretative framework of his Qur'ānic hermeneutics is a "theology of mercy."[60]

2.3.1 The Call for a Humanistic Exegesis of the Qur'ān

Mouhanad Khorchide advocates a historico-critical, humanistic Qur'ānic exegesis, which for him cannot be a literal interpretation of the text. Instead, a historical contextualization of the text is indispensable for Khorchide. If this approach is not used to reconcile the Qur'ānic text with life in today's society, it leads either to persisting in a fidelity to the text that does not meet the demands of modernity or to being completely absorbed into modernity in a way that would be decoupled from the religion. Khorchide rejects both.

57 "[A]us der Gesamtkomposition des Korans"; "entlang der exakt erarbeiteten Kriterien, die aus dem Geiste dieser genuinen Offenbarung heraus zu entnehmen sind": ibid., 142.

58 Karimi, *Hingabe*, 198.

59 Mouhanad Khorchide, *Gottes Offenbarung in Menschenwort. Der Koran im Lichte der Barmherzigkeit* (Freiburg: Herder, 2018).

60 "Theologie der Barmherzigkeit".

For Khorchide, the Qur'ān was revealed as discourse. It is the final result of "dialogue, debate, argumentation, acceptance, and rejection."[61] There can therefore be no single-line view of the text, and that is why in Islamic theology today the dialogical relationship between God and man is also of central importance.[62] Indeed, from Khorchide's point of view, the Qur'ān virtually encourages questioning. However, Khorchide does not consider it helpful to use Islamic theology's widely recognized occasions of revelation (*asbāb an-nuzūl*) as the key to Qur'ānic hermeneutics, since the majority of Qur'ānic verses do not provide any information about them. Moreover only a few Qur'ān specialists are capable of applying this method.[63]

A humanistic exegesis of the Qur'ān is instead oriented towards the question of intentions and goals, the "maxims" (*maqāṣid*). First of all, there is the matter of establishing universally valid "maxims" such as the love and mercy of God, which for Khorchide have a universal claim to validity. Specific maxims comprise a further category, i.e., parts of revelation like ritual law but also eschatological statements. By elaborating these "maxims," the reader is prevented from an arbitrary exegesis and selective awareness of the text. In this context, Khorchide warns against "plays on words" with the help of which "one can ultimately let the Qur'ān say everything and thus nothing."[64] With all warnings against arbitrariness, Khorchide, on the other hand, is not pursuing a literal interpretation of the texts but rather working out the fundamental messages of the Qur'ān. In the center of this are the preservation and fulfillment of peoples' interests: for Khorchide they represent a legitimate orientation for every Qur'ānic exegesis, for it is the human being who is the center of the Qur'ānic message.[65]

For Khorchide, the specific maxims in the Qur'ān, which can be traced back to certain historical circumstances, include the doctrine of God and man, judgment and the hereafter, ethics and acts of worship, as well as the doctrine of the freedom, equality, and dignity of all people. Those who do not orientate themselves towards these maxims run the risk of carrying out a selective Qur'ānic exegesis.[66]

---

61 "Dialog, Debatte, Argumentation, Annahme und Zurückweisung": Mouhanad Khorchide, "Auf dem Weg zu einer humanistischen Qur'ānhermeneutik," in *Moderne Zugänge zum Islam. Plädoyer für eine dialogische Theologie,* eds. Hamideh Mohagheghi and Klaus von Stosch (Paderborn: Ferdinand Schöningh, 2010), 51.

62 Mouhanad Khorchide, "Mit Gott im Dialog – die Notwendigkeit eines Paradigmenwechsels in der Islamischen Systematischen Theologie (Kalām) – am Beispiel der Bestimmung des Menschen als Diener Gottes," in *Herausforderungen an die Islamische Theologie in Europa – Challenges for Islamic Theology in Europe,* eds. Mouhanad Khorchide and Klaus von Stosch (Freiburg: Herder, 2012), 68.

63 Khorchide, "Weg," 35.

64 "Wortspielereien" [mit deren Hilfe] "man den Qur'ān letztlich alles und damit nichts sagen lassen kann": ibid., 47.

65 Ibid., 39, 47.

66 Ibid., 51.

To follow these maxims means to recognize that the core of Muḥammad's message was an ethical one, which can be separated from its temporary, political message in Medina. If the legal and political regulations of Islam dating back to Medinan times were to be regarded as equally timeless (e.g., regulations regarding inheritance law, a man's right to castigate his wife, regulations concerning Jews and Christians, concerning the testimony of women in court, and concerning the age of marriage), then a reconciliation of the message of Islam with modernity is hardly possible.[67]

In any case, for Khorchide many of these verses belong to the set of ambiguous verses in the Qur'ān giving legal instructions depending on the context, while the unambiguous verses contain ethical instructions and are of timeless validity. Therefore, following the legal regulations from the early days of Islam is not only unnecessary for Khorchide. It is virtually contrary to Islam.[68] "Sharia" can therefore only mean the eternally valid ethical guidelines from the early Meccan period because only they are universal guidelines: justice, freedom, equality of all people, inviolability of their dignity, monotheism, and social responsibility.[69]

2.3.2 Theology of Mercy

Khorchide's core concern is to establish a theology of mercy for Islam. For Khorchide, God's mercy is absolute; it extends to this world as well as to the hereafter. In this world, God offers man his love and mercy and out of mercy wants to receive him into "eternal bliss" in the hereafter.[70] The only exception is the individual who refuses this invitation by God. In Khorchide's estimation, to think of God himself as freedom and loving mercy means to know that, in His perfection, He does not need the veneration of men. Rather, man is free to enter into a relationship with God[71] and experience His presence through the recitation of the Qur'ān.

God has chosen humankind to be in communion with Him and invites individuals into a relationship of love with Him through the transmission of his message through prophets and their proclamations.[72] When individuals have entered into a state of complete perfection – not of flawlessness, but with the will to strive for perfection – the relationship of love to God also becomes perfect.

---

67 Ibid., 44–51.

68 Mouhanad Khorchide, *Islam ist Barmherzigkeit. Grundzüge einer modernen Religion* (Freiburg: Herder, 2012), 116.

69 Ibid., 144.

70 "[E]wige Glücksseligkeit": Mouhanad Khorchide, "Anstöße zu einer Theologie der Barmherzigkeit," in *Theologie der Barmherzigkeit? Zeitgemäße Fragen und Antworten des Kalām*, eds. Mouhanad Khorchide, Milad Karimi, and Klaus von Stosch (Münster: Waxmann, 2014), 24.

71 Khorchide, *Gottes Offenbarung*, 109, 127.

72 Ibid., 104.

The mercy of God must be the center of Qur'ānic exegesis, such that any Qur'ānic interpretation that contradicts God's mercy is rejected by Khorchide as unlawful. God's mercy is unconditional and extends to everyone. It even exceeds his love, since love is not an absolute characteristic of God.[73] Even the statement that God is simultaneously merciful and punishes does not hold true for Khorchide, for the Qur'ānic speech about God's punishment can only be understood pedagogically by him in order to incite man to good deeds and perfection.[74]

Khorchide does not consider hell as a location where man experiences punishment and damnation. Rather, it is for him a symbol of the suffering and torment that an individual experiences on the way to perfection, where a person is confronted with God's love as well as with his or her own weaknesses and mistakes.[75] However, this is pared down by Khorchide in that the symbolic understanding of paradise and hell is not the only conceivable one. Rather, it is only one option of interpretation of the relevant passages of the Qur'ān.

With regard to perfecting human beings, Khorchide believes that a dynamic understanding of faith is important, according to which confession alone is not sufficient for belonging to Islam. According to his understanding, God is primarily interested in whether a person does what is good and right, regardless of whether he is Muslim, Jew, Christian, believer or non-believer.[76] Therefore, according to his understanding, Islam is essentially about inner attitude, not about merely performing rites. Accordingly, he views everyone who practices love and mercy as a "Muslim, even if he does not believe in God."[77] Conversely, a person, even if he or she calls himself or herself a Muslim, is in reality an unbeliever if he or she refuses the love and mercy of God and does not practice it – for instance, because he or she is haughty – but prays and fasts.[78] True worship for Khorchide is therefore less the observance of the Five Pillars of Islam than service to one's fellow man[79] to reduce misery and suffering on earth.

Khorchide defines this humanistic Islam as a religion of God's mercy and love and of the freedom of humankind. Individuals possess this freedom since, according to Khorchide's view, it was less the intention of God at the creation of human beings to let himself be worshiped by his creatures as the Almighty; that would produce for Khorchide the image of an "inferior, selfish 'dictator'."[80] Rather, God created man in order to communicate with him and to offer him his love: "God seeks fellow lovers"

73 Khorchide, "Anstöße," 34f.
74 Khorchide, *Islam ist Barmherzigkeit*, 46.
75 Ibid., 50f.
76 Ibid., 58.
77 "Muslim, auch wenn er nicht an Gott glaubt": ibid., 88.
78 Ibid., 90.
79 Khorchide, "Mit Gott im Dialog," 72f., 75.
80 [das Bild eines] "minderwertigen, egoistischen ‚Diktators'": ibid., 69.

because "God is absolute mercy."[81] When an individual in turn expresses this love and mercy towards other people, he or she serves God and his or her neighbor.

For Khorchide, the relationship between God and man is thus not hierarchical; rather, it is a "relationship of friendship, indeed, a relationship of love"[82]: The individual becomes "God's partner."[83] God and individuals "cooperate"[84] when man serves God in all freedom, expresses love and mercy to other people, and thus perfects himself.

2.3.3 The Qur'ān as the Communicative Word of God

For Khorchide it is undisputed that the Qur'ān is God's revelation. He regards it as God's self-revelation, in which he makes himself present: in the Qur'ān God meets people and invites them into his love. Through the process of revelation, God enters the world of human beings, as it were.[85] Thus the Qur'ān possesses a divine-human language ("God-human-speech"), and it is neither a completely divine nor a completely human text.[86] The divine part is made clear, for instance, by the aesthetics of the Qur'ānic language which touch people even if they do not understand Arabic.[87] As the word of God, however, the Qur'ān is not a monologue of God's speech but has been sent down in a communication form. For Khorchide this is evident from numerous verses in which God addresses words of God to humankind ("You believers . . .") or where an individual is allowed to speak to God himself.

At the same time, the Qur'ān is also human for Khorchide in the sense that it enters into contact with people. It first of all entered into contact with the people at the time of Muḥammad's life, speaking into their everyday life. Then, however, the text was cast into writings along with its special instructions for the time of early Islam, such that it congealed. God's communication with people today therefore no longer takes place through the text, since the text is fixed, but rather through the interpretation of the text. In this way God also communicates with people today and inspires them by the Qur'ān. Since God still communicates with people in this way, there cannot be one interpretation of the Qur'ān that is valid for all times – otherwise the Qur'ān would not need to be read again and again.[88] The task of today's reader is to harmonize his or her current life reality with that of the Qur'ān in the text of the 7th

81 "Gott sucht Mitliebende"; "Gott ist die absolute Barmherzigkeit": ibid., 71.
82 "Freundschaftsbeziehung, ja als Liebesbeziehung": Khorchide, *Islam ist Barmherzigkeit*, 29.
83 "Gottes Partner": Khorchide, *Gottes Offenbarung*, 108.
84 "kooperieren": ibid.
85 Ibid., 78.
86 Mouhanad Khorchide, "The Authors of the Qur'an are Still Alive: The Qur'an as an Act of Communication," in *Transformative Readings of Sacred Scriptures: Christians and Muslims in Dialogue*, eds. Simone Sinn, Dina El Omari, and Anne Hege Grung (Genf: The Lutheran World Federation 2017), 97f.
87 Ibid., 93.
88 Ibid., 95.

century revelation. This is done by decoding the text in order to understand God's message for the here and now.

Although Khorchide singles out the dialogical character of the encounter between God and humanity and assigns God some human qualities such as being emotionally touched, experiencing the sense of being aggrieved, and sensing pain due to rejection of his offer of love,[89] God remains transcendent and not really graspable in Khorchide's view. Humanity cannot dispose over God. Also, there is no one who is in possession of the truth, because God alone is the truth and individuals remain seekers.

As Khorchide sums it up, it is for this reason that exegesis is ultimately the result of what oneself is as an individual: if mercy dwells in one's heart, he or she will find God's mercy in the Qur'ān; but if hatred dwells there, one's hatred will be read into the Qur'ān. Thus, the exegesis of the Qur'ān says more about the exegete than about God or the text: "How we interpret the Qur'ān reveals more about us than about God or the Qur'ān itself."[90]

2.4 Abdullah Saeed

Abdullah Saeed was born in 1960/1379 on the island of Feydhoo in the Maldives. Beginning in 1976/1396, he attended an *ahl al-ḥadīth* madrasa in Faisalabad related to the Ḥanafī school of law. He then studied at the Islamic University of Medina. In 1992/1412 he received his doctorate from the University of Melbourne with a thesis on Islamic banking and is now the Sultan of Oman Professor of Arab and Islamic Studies at the University of Melbourne, Australia.

Abdullah Saeed's research focuses on Islamic law (freedom of religion, minority rights, women's rights, the ban on interest), the Islamic history of ideas, Islam in Australia, and the Qur'ān and Qur'ānic hermeneutics. He has published three monographs and several academic articles on the latter two topics. Saeed has dealt with the elaboration of a hermeneutic key to the establishment of a "contextualized" Qur'ān interpretation adapted to modernity, which he describes as dynamic. His work has been more extensive than the work done by what have up to now been European purveyors. This dynamic arises from the relationship between God, the speaker, the text, and the recipient (Muḥammad and the Islamic community, respectively).

89 Khorchide, *Gottes Offenbarung*, 139–41.

90 Khorchide, "Authors," 98.

2.4.1 The Qur'ān as God's Revelation

Saeed precedes all considerations with the observation that it is generally impossible to be unbiased when considering the Qur'ānic text as revelation. This is due to the fact that the very act of reading is an interpretation. From his point of view, therefore, the only question that arises is how the text is exegeted – no interpreter can lay claim to being neutral and objective anyway.[91]

Saeed's conviction is that Muḥammad's role in the process of revelation was not passive. Revelation still occurs today, in the sense that God has sent his word down from the invisible to the visible world such that it has become part of the daily life of Muslims, inspiring people and directing them to do the right thing. As new interpretations are formulated among believers today and are accepted and implemented by the community, original revelation, through its interpretation, experiences enduring continuity.[92]

Again and again, Saeed emphasizes that his thoughts are not the introduction of new hermeneutics. Rather, he sees his approach as a further development of what had already been recognized practice in earlier centuries, indeed even at the time of the Prophet's companions (*ṣaḥāba*). In this context, Saeed repeatedly cites the second caliph ʿUmar b. al-Khaṭṭāb (d. 644/12). To some extent, Saeed already recognizes in ʿUmar b. al-Khaṭṭāb a contextual approach to revelation.[93] Indeed, the fact that Muḥammad himself left no comprehensive, systematic commentary on the Qur'ān allows Saeed to conclude that there cannot be one, absolute meaning of a text.[94]

It is beyond question for Saeed that the Qur'ān is the word of God, "a revelation [. . .] directly from God," written without human intervention.[95] This is the case even if the Qur'ān was written in human language. It is all the more important for Saeed to arrive at an appropriate interpretation of this word of God. Interpretation is always manifold for him. For Saeed, the assumption that there was a uniform interpretation of the Qur'ān's text at any point in the history of Islam is a misconception; rather, from the very beginning, various traditions of interpretation developed, such as linguistically oriented hermeneutics as well as exegesis based on rationality or exegesis that was mystically or traditionally oriented.[96]

---

91 Abdullah Saeed, *Reading the Qur'an in the Twenty-First Century. A Contextualist Approach* (London: Routledge, 2014), 13; Abdullah Saeed, *Interpreting the Qur'an. Towards a Contemporary Approach* (London: Routledge, 2006), 102f.

92 Ibid., 57f.

93 Ibid., 4.

94 Saeed. *Interpreting*, 108.

95 Abdullah Saeed, *The Qur'an. An Introduction* (London: Routledge, 2008), 29.

96 Saeed reaches this distinction in his foundational work: *Reading the Qur'an in the Twenty-First Century*, 15.

At the center of Saeed's thoughts is the question of how the Qur'ān can be freed from its now ossified tradition of interpretation and thus acquire new relevance for modernity. For Saeed, the distinction between actual revelatory contents and later appended interpretative traditions and "fabricated" traditions,[97] which in time received the rank of revelatory content, is indispensable for this purpose. Since not all traditions and interpretations of texts can claim authority, Saeed considers it wrong to follow all the individual instructions of the often-contradictory traditions – this inevitably results in "chaos" for him.[98] To avoid this, it is imperative that a distinction be made in the Qur'ān between content that is timeless and instructions that were temporary. For him, this distinction is the gateway to a Qur'ānic exegesis that enables a reconciliation of seventh century CE texts with modernity. In doing so, Saeed emphatically emphasizes that he does not at all reject the authority of scholars of pre-modernity from the fields of theology and jurisprudence but rather wishes to develop their authority further.[99]

2.4.2 Qur'ānic Exegesis in Context

Saeed opts for a context-based approach to the Qur'ān instead of a literalist exegesis that hides the context. Admittedly, the Qur'ānic text must also be taken into account as a first step in linguistic terms. Afterwards, however, in order to approach the meaning of the text from Saeed's perspective, it is essential to be aware of the historical, political, economic and social conditions of Arabia and its spiritual-intellectual climate at the time of revelation. He calls these circumstances "macro context 1" which means the place and the target audience of revelation, their religious, political and cultural concepts. They comprise, in short, the context in which the revelation took place. Saeed refers to "macro context 2" as the environment of today's interpreter with his or her current cultural, religious, and political norms. The task of hermeneutics for Saeed is to now relate macro context 1 to macro context 2[100] in order to make the text from late antiquity fruitful for today. There is no way to achieve this without knowledge of the circumstances surrounding revelation's origin. Furthermore, parallel texts can also be consulted for additional understanding.

Saeed regards this "contextualist" approach to treating the Qur'ān to be an antithesis to the "textual" approach which is dominant in Islamic theology and aims at a literal interpretation. Saeed explains that contextual hermeneutics also includes the freedom to decide whether a certain text is an unchanging and timelessly valid text or whether it

97 Ibid., 76.
98 Saeed, *Interpreting*, 19.
99 Saeed, *Reading the Qur'an in the Twenty-First Century*, 4.
100 Ibid., 5.

can be adapted to the conditions of modernity by modification once the nature of a particular text has been determined.[101]

In general, understanding the text is not about following the letter of the text but about grasping the "spirit" of the Qur'ānic message[102] and crystallizing guidelines that can also be practiced in the here and now. In Saeed's view, the development of new guidelines for the interpretation of a text is permitted, indeed indispensable, as long as the foundations of Islam are not called into question.[103]

Under changing social and political circumstances, Saeed believes that above all the legal and moral regulations of the Qur'ān ("ethico-legal texts") must be subjected to an examination of their applicability in the present. They are a mirror of Arabian society in the seventh century but are no longer practicable today. They include, for example, the Qur'ānic regulations on marriage and divorce, inheritance and slavery, the position of non-Muslim minorities, and the definition of the rights of men and women. If the wording of these texts were to be applied today, this would ultimately lead to the texts being pushed to one side completely because they are no longer practicable. With that, the Qur'ān would ultimately be damaged.[104] In order to preserve revelation, it is therefore essential to adapt these provisions to the realities of modernity. In any case, the Qur'ān is not primarily a code of law but a text containing universally valid principles that can be applied at all times.[105]

2.4.3 Timeless Validity and Changeability in the Qur'ān

However, it is Saeed's viewpoint that not all contents of the Qur'ān are to be reassessed in the context of the present time. Therefore, he counters the reproach that his views establish a new theology. Accounts by the prophets, theological statements about God, man, creation, or the hereafter have timeless validity for Saeed. The context-based understanding of the text is to be applied only to those passages of the Qur'ānic text that have no universal validity.

In order to make it clear that a literalist understanding of the Qur'ān is misguided and does not do justice to its own claim, Saeed proposes to divide the texts of the Qur'ān into four categories:[106] First, there are the theological texts that deal with hidden things, such as the teachings about God and his throne, paradise, and the angels. These texts describe a reality that is removed from human experience, such that hu-

---

101 Saeed, *Interpreting*, 3.

102 Saeed, *Reading the Qur'an in the Twenty-First Century*, 36.

103 Ibid., 4, 88f.

104 Ibid., 6f.

105 Saeed, Abdullah, "Reading the Quran," in *A Companion to the Muslim World*, ed. Amyn B. Sajoo (London: I.B. Tauris, 2009), 74.

106 Saeed, *Interpreting*, 90ff.

manity can only have an inadequate conception of it. Those who are aware that they are always acting with mental makeshift constructions at this point will not be able to take a position that understands these texts literally.

The Qur'ān, then, contains texts with historico-narrative content, such as verses that report on the battle of Uḥud 625/3, without giving dates, places, or names and can therefore benefit from consulting non-Qur'ānic sources such as the Bible or archaeology. These texts therefore require supplemental wording if their significance is to be brought to bear.

The Qur'ān also contains parables to illustrate Islamic teachings and often uses metaphors for this purpose that are also not to be understood literally. Finally, the fourth category includes texts that convey ethical guidelines and regulations. Here, the context of revelation must in particular be taken into account, in order to avoid arriving at false conclusions. For example, the definition of a half inheritance for women originated in the social circumstances of that time. This cannot be readily carried over to today from its culturally-specific seventh-century A.D. wording;[107] Saeed's view applies similarly to the institution of slavery, polygyny, and relations between Muslims and non-Muslims. From Saeed's point of view, the fact that the Qur'ān's ethical-legal regulations do not have to be applied literally for all times and cultures is obvious because the Qur'ān itself declares texts from an early period to have been abrogated later.[108] All categories of texts thus require, according to Saeed's conclusions, interpretation and partly also adaptation; a literal understanding, on the other hand, does not do them justice. In rare cases, provisions can even be dropped altogether.[109]

Also, with regard to the value of revelatory contents, Saeed has produced a ranking.[110] Saeed considers texts that convey "obligatory values" to include those that deal with basic beliefs such as faith in God, angels, prophets, holy scriptures, judgment, and life after death, the duties of faith (prayer, fasting, and pilgrimage), and observance of what is permitted (*ḥalāl*) and forbidden (*ḥarām*). These texts on "obligatory" values are followed by texts that convey "fundamental values," which are repeatedly confirmed in the Qur'ān as authoritative norms. These include, for example, the protection of life and family and of property. What can be derived from this includes equality of all people before the law, freedom of opinion, protection against torture, arbitrary detention, and exile, as well as the presumption of innocence, human rights for the disadvantaged, and environmental protection.[111]

For Saeed, a third category of "protectional values" includes instructions that explain the "fundamental values" in their practical application, such as regulations that

107 Saeed, *The Qur'an. An Introduction*, 78f.
108 Ibid., 170f.
109 Saeed, *Reading the Qur'an in the Twenty-First Century*, 71.
110 Ibid., 64ff.
111 Saeed, *Interpreting*, 132f.

prohibit theft and thus underpin the fundamental value of protecting property. Saeed views the fourth category as "implementational values" to put "protectional values" into practice, such as the instruction to cut off a thief's hand. Since this fourth category has merely to do with subordinate instructions corresponding to the culture of the time, they do not belong to the fundamental values of revelation and are therefore to be adapted for today.

Lastly, the final category includes special case regulations ("instructional values"). They are context-dependent and often formulate ambiguous admonitions, instructions, or recommendations. Particular caution must be exercised here with regard to a general application through recommending or imitating. Criteria for the significance of these regulations are the frequency they are mentioned in the Qur'ānic text, their significance at the time of Muḥammad, and their significance in the early days of Islam.[112]

3 Methods – Orientation – Scope of the Approaches

In the German-speaking world up to the recent past there have hardly been any Qur'ān interpretations with alternative hermeneutic approaches. The blueprints presented in the meantime are the result of Muslim minorities' decades-long residence in Europe and of an increasing academization of Islamic theology in the Western context, particularly in connection with the establishment of university-affiliated institutes for Islamic theology. This is true, although not all progressive blueprints were presented by theologians.

In view of the different emphases found within the approaches, one can up to now only speak to a limited extent within the European context of the development of an academically well-founded and differentiated methodology of progressive Qur'ān hermeneutics that is consistently applied to the Qur'ānic text. A broad scholarly discussion, further development, and wider acceptance of the basic ideas and basic lines of approach developed so far are still lacking. However, the short time span since the establishment of Islamic theology as an academic discipline at German universities must be taken into account.

The central concern of these progressive approaches is to build a bridge between the social reality of a text, the writing of which dates back some fifteen centuries, and a western modernity of the twenty-first century. There is agreement on the rejection of a literal and political understanding of the text; however, the views on what replaces it are naturally much more complex: New perspectives of hermeneutical textual treatment have been brought into play (such as the focus on individuals' needs as the subject of God's love and mercy), but traditional patterns of argumentation have not been flatly rejected. Thus, progressive Qur'ānic exegetes, in one place, deal critically with the

112 Ibid., 139.

texts of tradition and reject some of them as untrustworthy. However, in another place, – and similar to traditional theological blueprints – they treat texts of tradition as historically reliable proof of their argumentation. A well-founded justification for this different handling of texts is largely missing. Thus, a hermeneutical key that is applicable to all texts has not been found so far among the approaches of European progressive Qur'ān hermeneutics. This could provoke a reproach of arbitrariness.

The lack of a consistent hermeneutic methodology also raises the question of the acceptance and enforceability of these models. Or, more sharply formulated: does the desired result of establishing Islam as a religion of peace and the Qur'ān as compatible with democracy and human rights determine the approach to the texts and the method of interpretation?

Also, very fundamental questions of Islamic history and theology, such as the compilation of the Qur'ān and tradition or the sharia that has come into being through post-Qur'ānic interpretation in the form of numerous individual regulations, have not yet been touched upon. So far, it is primarily large thematic arcs that have been made to establish a humanistic understanding of the Qur'ān, which raise further questions. In questions such as the relationship between religion and politics, the relationship between Arab culture and revelation, women's and human rights, and the relationship to other religions or atheism, it is not enough to reject traditional views influenced by sharia law such as the legal discrimination of women without raising far-reaching consequences resulting from it, for example with regard to the binding nature of sharia law in civil law. The development of consistent hermeneutical keys and comprehensible scholarly arguments will be indispensable for the formulation of a progressive theology if the scope and acceptance of progressive Qur'ān hermeneutics is to be increased.

How academic theology extends into the communities of the faithful also seems important in this context. Up to now, progressive commentators on the Qur'ān have been less concerned with the interpretation of individual verses than with pointing out major lines of interpretation and defining overriding themes such as God's mercy and communication with human beings. Their publications have so far provided only few concrete ethical instructions and recommendations for action on individual issues. However, if progressive interpretation models are to be adaptable beyond the academic framework, the question of ethical orientation will be an essential one for the Muslim community.

Despite the desiderata mentioned above, the influence and scope of the blueprints of progressive Qur'ānic exegetes are increasing. The development of further approaches can be expected, as well as the discussion, substantiation, and rejection of previous blueprints. In any case, the blueprints presented so far bear witness to the unbroken importance of the Qur'ān for the Muslim diaspora in the Western context of the 21st century and to a hitherto unrecognizable attitude of questioning the Qur'ān as God's revelation itself. Despite the pluralization of interpretations in the twenty-first century and the relativization of some of its statements – at least from the perspective of more conservative protagonists – the Qur'ān has also retained its outstanding importance as a text of revelation in the Western context in the twenty-first century.

4 Conclusion

Given all the open questions, perhaps the greatest merit of progressive Qur'ānic exegesis to date is the historicization of the Qur'ānic text under the basic assumption that the Qur'ānic text can only be understood against the background of the particular historical, cultural, and political circumstances of Muḥammad's time and cannot be ahistorically transferred to modernity. This is a great leap for the development of a progressive form of theology; further research results are eagerly awaited.

5 Bibliography

5.1 Sources

5.1.1 Abdel-Hakim Ourghi

Ourghi, Abdel-Hakim. *"Aufklärung" des Islam?* Analysen und Argumente 211. Berlin: Konrad Adenauer Stiftung, 2016.

Ourghi, Abdel-Hakim. "Ein postmodern-humanistischer Islam. Die Autonomie des Korantextes und die Freiheit des Lesers." *Aufgang. Jahrbuch für Denken, Dichten, Kunst* 13 (2016): 145–71.

Ourghi, Abdel-Hakim. *Einführung in die islamische Religionspädagogik*. Ostfildern: Matthias Grünewald, 2017.

Ourghi, Abdel-Hakim. *Reform des Islam. 40 Thesen*. Munich: Claudius, 2017.

Ourghi, Abdel-Hakim. *Ihr müsst kein Kopftuch tragen. Aufklären statt Verschleiern*. Munich: Claudius, 2018.

5.1.2 Ahmad Milad Karimi

Jürgasch, Thomas, and Ahmad Milad Karimi. "'Nicht vernunftgemäß zu handeln, ist dem Wesen Gottes zuwider'. Überlegungen zum Begriff einer vernünftigen Religion." In *Gegenwart der Einheit. Zum Begriff der Religion*, edited by Thomas Jürgasch et al. Freiburg: Rombach, 2008: 167–85.

Karimi, Ahmad Milad. "Moderne Qur'ānhermeneutik – Zeitgemäße Lesarten des Qur'ān." In *Herausforderungen an die Islamische Theologie in Europa – Challenges for Islamic Theology in Europe*, edited by Mouhanad Khorchide and Klaus von Stosch. Freiburg: Herder, 2012: 14–30.

Karimi, Ahmad Milad. "Die Beziehung von Mensch und Gott aus islamischer Perspektive." In *„Der stets größere Gott". Gottesvorstellungen in Christentum und Islam*, edited by Andreas Renz et al. Regensburg: Friedrich Pustet, 2012: 231–40.

Karimi, Ahmad Milad. "Versuch einer ästhetischen Hermeneutik des Qur'ān." In *Herausforderungen an die Islamische Theologie in Europa – Challenges for Islamic Theology in Europe*, edited by Mouhanad Khorchide and Klaus von Stosch. Freiburg: Herder, 2012: 14–30.

Karimi, Ahmad Milad. "Der Koran – eine Liebeserklärung Gottes? Überlegungen zum Islam als eine Offenbarungsreligion." *Freiburger Universitätsblätter* 201 (2013): 73–81.

Karimi, Ahmad Milad. "Von der Immanenz Gottes im Islam." In *Gottesvorstellungen im Islam. Zur Dialektik von Transzendenz und Immanenz*, edited by Amir Dziri. Freiburg: Kalām, 2013: 95–108.

Karimi, Ahmad Milad. *Osama bin Laden schläft bei den Fischen. Warum ich gerne Muslim bin und wieso Marlon Brando viel damit zu tun hat*. Freiburg: Herder, 2013. 2015^2.
Karimi, Ahmad Milad. *Der Koran*, edited by Bernhard Uhde. Freiburg: Herder, 2014.
Karimi, Ahmad Milad. "Wie Gott als Barmherzigkeit gedacht werden kann." In *Theologie der Barmherzigkeit? Zeitgemäße Fragen und Antworten des Kalām*, edited by Mouhanad Khorchide, Ahmad Milad Karimi, and Klaus von Stosch. Münster: Waxmann, 2014: 107–15.
Karimi, Ahmad Milad. *Die Blumen des Koran oder: Gottes Poesie*. Freiburg: Herder, 2015.
Karimi, Ahmad Milad. *Hingabe. Grundlagen der systematisch-islamischen Theologie*. Freiburg: Rombach, 2015.
Karimi, Ahmad Milad. "Vom Verstand zur Offenbarung. Zum antithetischen Charakter der Offenbarung." In *Zwischen Glaube und Wissenschaft. Theologie in Christentum und Islam*, edited by Mohammad Gharaibeh et al. Regensburg: Pustet, 2015: 131–42.
Karimi, Ahmad Milad. "Zur Frage der Erlösung des Menschen im religiösen Denken des Islam." In *Streitfall Erlösung*, edited by Klaus von Stosch and Aaron Langenfeld. Paderborn: Schöningh, 2015: 17–37.
Karimi, Ahmad Milad. *Warum es Gott nicht gibt und er doch ist*. Freiburg: Herder, 2018.

5.1.3 Mouhanad Khorchide

Abdel-Samad, Hamed and Mouhanad Khorchide. *Ist der Islam noch zu retten? Eine Streitschrift in 95 Thesen*. Munich: Droemer, 2017.
Khorchide, Mouhanad. "Auf dem Weg zu einer humanistischen Qur'ānhermeneutik." In *Moderne Zugänge zum Islam. Plädoyer für eine dialogische Theologie*, edited by Hamideh Mohagheghi and Klaus von Stosch. Paderborn: Ferdinand Schöningh, 2010: 31–58.
Khorchide, Mouhanad. *Islam ist Barmherzigkeit. Grundzüge einer modernen Religion*. Freiburg: Herder, 2012.
Khorchide, Mouhanad. "Mit Gott im Dialog – die Notwendigkeit eines Paradigmenwechsels in der Islamischen Systematischen Theologie (Kalām) – am Beispiel der Bestimmung des Menschen als Diener Gottes." In *Herausforderungen an die Islamische Theologie in Europa – Challenges for Islamic Theology in Europe*, edited by Mouhanad Khorchide and Klaus von Stosch. Freiburg: Herder, 2012: 66–76.
Khorchide, Mouhanad. "Die Rede von Gott im Islam – aber von welchem Gott?" In *Gottesvorstellungen im Islam. Zur Dialektik von Transzendenz und Immanenz*, edited by Amir Dziri. Freiburg: Kalām, 2013: 63–79.
Khorchide, Mouhanad. *Scharia – der missverstandene Gott. Der Weg zu einer modernen islamischen Ethik*. Freiburg: Herder, 2013.
Khorchide, Mouhanad. "Anstöße zu einer Theologie der Barmherzigkeit." In *Theologie der Barmherzigkeit? Zeitgemäße Fragen und Antworten des Kalām*, edited by Mouhanad Khorchide, Milad Karimi, and Klaus von Stosch. Münster: Waxmann, 2014: 15–36.
Khorchide, Mouhanad. *Gott glaubt an den Menschen. Mit dem Islam zu einem neuen Humanismus*. Freiburg: Herder, 2015.
Khorchide, Mouhanad. "The Authors of the Qur'an are Still Alive: The Qur'an as an Act of Communication." In *Transformative Readings of Sacred Scriptures: Christians and Muslims in Dialogue*, edited by Simone Sinn, Dina El Omari, and Anne Hege Grung. Genf: The Lutheran World Federation 2017: 91–98.
Khorchide, Mouhanad and Klaus von Stosch. *Der andere Prophet. Jesus im Koran*. Freiburg: Herder, 2018.
Khorchide, Mouhanad. *Gottes Offenbarung in Menschenwort. Der Koran im Lichte der Barmherzigkeit*. Herders Theologischer Koran-Kommentar 1. Freiburg: Herder, 2018.

5.1.4 Abdullah Saeed

Akbar Ali, and Abdullah Saeed. "Interpretation and Mutability: Socio-Legal Texts of the Quran. Three Accounts from Contemporary Iran." *MES* 54, no. 3 (2018): 442–58.

Saeed, Abdullah. "Rethinking 'Revelation' as a Precondition for Reinterpreting the Qur'an. A Qur'anic Perspective." *JQS* 1, no. 1 (1999): 93–114.

Saeed, Abdullah. *Interpreting the Qur'an. Towards a Contemporary Approach*. London: Routledge, 2006.

Saeed, Abdullah. "Some Reflections on the Contextualist Approach to Ethico-Legal Texts of the Quran." *BSOAS* 71, no. 2 (2008): 221–37.

Saeed, Abdullah. *The Qur'an. An Introduction*. London: Routledge, 2008.

Saeed, Abdullah. "Reading the Quran." In *A Companion to the Muslim World*, edited by Amyn B. Sajoo. London: I.B. Tauris, 2009: 55–85.

Saeed, Abdullah. "Authority in Qur'ānic Interpretation and Interpretive Communities." In *Communicating the Word. Revelation, Translation, and Interpretation in Christianity and Islam*, edited by David Marshall. Washington: Georgetown University Press, 2011: 115–23.

Saeed, Abdullah. "How Muslims View the Scriptures of the People of the Book: Towards a Reassessment?." In *Religion and Ethics in a Globalizing World*, edited by Luca Anceschi et al. New York: Palgrave Macmillan 2011: 191–210.

Saeed, Abdullah. "The Islamic Case for Religious Liberty. A Close Reading of the Qur'an and the Prophet Leads to Supporting Religious Tolerance." *First Things* 2017 (November 2011): 33–36.

Saeed, Abdullah. "The Self-Perception and the Originality of the Qur'ān." In *Communicating the Word. Revelation, Translation, and Interpretation in Christianity and Islam*, edited by David Marshall. Washington: Georgetown University Press, 2011: 98–104.

Saeed, Abdullah. *Reading the Qur'an in the Twenty-First Century. A Contextualist Approach*. London: Routledge, 2014.

5.2 Secondary Literature

Duderija, Adis. "Neo-Traditional Salafi Qur'an-Sunna Hermeneutics and Its Interpretational Implications." *Religion Compass* 5, no. 7 (2011): 314–25.

Harrison, Victoria S. "Hermeneutics, Religious Language and the Qur'an." *Islam and Christian-Muslim Relations* 21, no. 3 (2010): 207–20.

Pink, Johanna. *Sunnitischer Tafsīr in der modernen islamischen Welt. Akademische Traditionen, Popularisierung und nationalstaatliche Interessen*. Leiden: E.J. Brill, 2011.

Pink, Johanna. "Striving for a New Exegesis of the Qur'ān." In *The Oxford Handbook of Islamic Theology*, edited by Sabine Schmidtke. Oxford: Oxford University Press, 2016: 765–92.

Pink, Johanna. "Modern and Contemporary Interpretation of the Qur'ān." In *The Wiley Blackwell Campanion to the Qur'ān*, edited by Andrew Rippin, and Jawid Majaddedi. Hoboken: Wiley, 2017[2]: 479–91.

Wielandt, Rotraud. "Main Trends of Islamic Theological Thought from the Late Nineteenth Century to Present Times." In *The Oxford Handbook of Islamic Theology*, edited by Sabine Schmidtke. Oxford: Oxford University Press, 2016: 707–64.

Würtz, Thomas. "Koranexegese und gesellschaftlicher Wandel." In *Christentum und Islam als politische Religion. Ideenwandel im Spiegel gesellschaftlicher Entwicklungen*, edited by Oliver Hidalgo, Holger Zapf, and Philipp Hildmann. Wiesbaden: VS Springer, 2017: 263–86.

List of Contributors

Ramy Abdin
Research Assistant with focus on Islam
Bavarian Research Center for Interreligious Discourses
Friedrich-Alexander-Universität Erlangen-Nürnberg
Germany

Katajun Amirpur
Professor of Islamic Studies
Universität zu Köln
Germany

Moncef Ben Abdeljelil
Professor emeritus, Arabic and Islamic Studies
University of Sousse
Tunisia

Derya Buğur
Research Assistant at the Chair of Islamic-Religious Studies with the focus Text Analysis and Jurisprudence
Department of Islamic-Religious Studies
Friedrich-Alexander-Universität Erlangen-Nürnberg
Germany

Massimo Campanini (d. 09.10.2020)
formerly Associate Professor of Islamic Studies at the University of Naples l'Orientale and University of Trento
Italy

Kathrin Eith
Lecturer for Turkish Language and Literature
Oriental Institute
Martin Luther University Halle-Wittenberg
Germany

Jaʿfar Faḍlallāh
Lecturer in Islamic Jurisprudence at the Islamic Sharia Institute and in Religious Anthropology at the Lebanese University
Beirut, Lebanon

Rahel Fischbach
Professor of Islamic Studies
Department of Religious Studies
College of William & Mary
Williamsburg
USA

Mohammad Gharaibeh
Professor for Islamic Intellectual History
Berlin Institute for Islamic Theology
Humboldt-Universität zu Berlin
Germany

Ursula Günther
Director of Studies for Interreligious Dialogue
Medicine and Ethics at the Catholic Academy
Hamburg

Thomas Hildebrandt
Postdoctoral Researcher at the Chair for Comparative Theology from an Islamic Perspective
Berlin Institute for Islamic Theology
Humboldt-Universität zu Berlin
Germany

Heidi Hirvonen
Free researcher, specializing in interreligious and intercultural dialogue, with a focus on Muslim-Christian and Muslim-Western relations

Munirul Ikhwan
Assistant Professor of Islamic Studies
School of Graduate Studies
Sunan Kalijaga State Islamic University
Yogyakarta
Indonesia

Felix Körner, S.J.
Nicholas of Cusa Chair for Theology of Religions
Institute of Catholic Theology
Humboldt-Universität zu Berlin
Germany

https://doi.org/10.1515/9783111320069-028

Mohammad Mahdi Mojahedi
Assistant Professor
Institute for the Humanities and Cultural Studies
Tehran
Iran

Josep Puig Montada
Professor emeritus
Instituto Universitario de Ciencias de las Religiones
Universidad Complutense de Madrid
Spain

Seyed Hossein Morakabi
Researcher at Imam Moussa Sadr
Cultural & Research Institute
Tehran
Iran

Ervan Nurtawab
Lecturer, Institut Agama Islam
Negeri/IAIN Metro, Indonesia
and
Humboldt Postdoctoral Fellow, FRIAS
Albert-Ludwigs-Universität Freiburg
Germany

Margherita Picchi
Research fellow at the Department
for the Study of Religions
University of Cape Town
South Africa

Abbas Poya
Professor of Reform Discourses in Contemporary Islam
Department of Islamic-Religious Studies
Friedrich-Alexander-Universität Erlangen-Nürnberg
Germany

Peter Riddell
Professorial Research Associate
SOAS University of London
United Kingdom

Shehzad Saleem
Research Fellow
Al-Mawrid Institute of Islamic Sciences
Lahore
Pakistan

Christine Schirrmacher
Professor of Islamic Studies
Rheinische Friedrich-Wilhelms-Universität Bonn
Germany

Irene Schneider
Professor at the Department of Arabic/Islamic Studies
University of Göttingen
Distinguished Fellow at the Max-Weber-Kolleg
University of Erfurt
Germany

Nimet Şeker
Professor of Islamic Textual Studies
(Qur'ān and Hadith)
Berlin Institute for Islamic Theology
Humboldt-Universität zu Berlin
Germany

Georges Tamer
Professor of Oriental Philology and Islamic Studies
Friedrich-Alexander-Universität Erlangen-Nürnberg
Germany

Joas Wagemakers
Associate Professor of Islamic and Arabic Studies
Department of Philosophy and Religious Studies
Utrecht University
Netherlands

Stefan Wild
Professor emeritus of Islamic Studies
Rheinische Friedrich-Wilhelms-Universität Bonn
Germany

Taraneh R. Wilkinson
Postdoctoral fellow at FSCIRE
Bologna
Italy